Cannabinoids and the Brain

Attila Köfalvi
Editor

Cannabinoids and the Brain

Editorial and Chapters 1, 9, 14, 22 were proofed
by Zsófia Gombár

With 44 illustrations and 16 tables

Springer

Attila Köfalvi
Center for Neurosciences of Coimbra
Faculty of Medicine
University of Coimbra
Coimbra, 3000-045 Portugal

ISBN-13: 978-0-387-74348-6 e-ISBN-13: 978-0-387-74349-3

Library of Congress Control Number: 2007933065

Printed on acid-free paper.

9 8 7 6 5 4 3 2 1

springer.com

Editorial

Did you know that if you take aspirin or some other type of painkillers, you simply upregulate your endocannabinoid system against your endovanilloid system? If it happens to be a completely new piece of information to you, then this book is for you! Seriously speaking, the first part of the book you are holding in your hands is an exhaustive source of scientific reviews on the molecular biology, pharmacology, anatomy, and physiology of the endocannabinoid and related lipid mediator systems. The second part of the book, however, covers the involvement of these signaling systems in metabolic, neurological, and psychiatric disorders, and gives an overview on clinical trials and on recent advances in cannabinoid-based medicine. Therefore, the target audience for this book are (a) physicians, especially endocrinologists, neurologists, psychiatrists, and neuroscientists who want to update their knowledge about metabolism, basic brain physiology, molecular biology, and pathology and about novel therapeutic opportunities; (b) graduate and undergraduate students who also wish to broaden their knowledge about endocrinology, neuroscience, neurology, and psychiatry, or may need orientation to determine their future scientific goals; (c) politicians and health care employers who hesitate whether marijuana or cannabinoid-based medications should be legalized; and last but not least, (d) journalists who can help the scientists to convey their message to a larger audience. All the authors of the present volume are world's leading neuroscientists and physicians, who are also regarded to be pioneers in the cannabinoid research area. Here I would like to gratefully thank them for all their altruistic contributions, and for sparing their precious time on this work.

The very first idea of writing this book occurred to me in 2005 when I had an interesting conversation with a neurologist professor from the USA, after his exciting lecture about the impact of adenosine receptors on epilepsy. I asked him whether he would be interested in the role of cannabinoid receptors also besides adenosine receptors. I noticed a faint note of indignation in his answer when he said: "No, I do not treat drug addicts, but epilepsy patients." He was apparently unaware of those facts which are extensively reviewed in this book, especially the CB_1 receptor that is believed to have the highest density among metabotropic receptors in the nervous tissue, and, together with its endogenous agonists, they represent a unique signaling system, which seems to be a goldmine of therapeutic targets against many neuropsychiatric disorders. The reaction of the professor may be excusable, since the body's own cannabinoid system as well as the body's opioid system or the nicotinic receptors were discovered in the quest to find the specific targets for drugs of abuse, such as marijuana, morphine, heroin, and tobacco's nicotine. Importantly, the last 16 years of constant research has discovered a much broader role for endocannabinoids than for the opioid or nicotinic acetylcholine signaling. Nevertheless, this role does not seem to receive sufficient recognition by those who otherwise should find it important in their professional activity. At present, I have the growing belief that the endocannabinoid system and related systems of lipid mediators, such as eicosanoids and endovanilloids, constitute a major modulator/messenger supersystem, which is at least as important as the monoaminergic, purinergic, and cholinergic systems. Furthermore, these modulator systems work hand in hand, and thus they cannot be viewed as solitary therapeutic targets. The borders between classical pharmacological areas are likely to be forgotten. Therefore we, the authors, consider ourselves extremely fortunate to make this book happen and to disseminate challenging up-to-date reviews on the role of cannabinoids in the brain.

Now I would like to take the opportunity of addressing a few challenging ideas to the cannabinoid research area. There are some minor and major problems cannabinoid researchers normally encounter, which could be easily alleviated. For instance, it seems to be ironic and even ridiculous to some extent that permission is required for using certain cannabinoid research tools, such as Δ^9-THC and its potent derivative HU-210. More importantly, their experimental usage is further hindered by other rules in certain places. I will never forget the incident when the police appeared in my lab, inquiring how I had used Δ^9-THC and for what purpose. Absurdly enough, at that point of time, I still had not received the shipment of the compound from the pharmaceutical company due to permission issues. It is no more than pure hypocrisy, knowing that there are several other even more selective, potent, and efficacious cannabinoid ligands available, causing even more expressed effects than Δ^9-THC in animals. It is understandable that Δ^9-THC requires permission, it being the major constituent of marijuana. Nonetheless, the price of Δ^9-THC and HU-210 appears to be so high, especially considering the remarkably little buyable amounts, that selling these products for research purposes without permission would not represent a gross criminal risk.

Normalization of chemical names would also be desirable. For instance, researchers may face a considerable challenge to find all the articles of the popular nonselective potent cannabinoid agonist WIN55212-2 in searchable databases, since the ligand is variously termed WIN-55,212-2, WIN 55212-2, WIN 55,212-2, WIN-2 or R-(+)-WIN55212, R-WIN55212, R-WIN 55212, R-WIN 55,212, etc. with all possible permutations. The same is true for other compounds, such as the popular CB_1 receptor antagonist AM251. It is frequently used as AM 251, and a search for the terms AM and 251 in a database may result in a lot of additional unrelated articles. Thus, combining two or more ligands in one search is definitely a vain idea. The problem could be solved with only a slight common effort to standardize chemical names. It is also unfortunate that several old-fashioned journals still force the authors to use the long cumbersome chemical names of cannabinoid compounds even in the abstract of the article, for example, R(+)-[2,3-Dihydro-5-methyl-3-[(morpholinyl)methyl]pyrrolo[1,2,3-de]-1,4-benzoxazinyl]-(1-naphthalenyl) methanone mesylate or [N-(piperidin-1-yl)-5-(4-chlorophenyl)-1-(2,4-dichlorophenyl)-4-methyl-1H-pyrazole-carboxamidehydrochloride]. Deciphering this long chemical name or similar ones would represent an enormous challenge to almost every researcher in the field. Even a chemist would spend several hours to realize that these terms mean WIN55212-2 and SR141716A (Rimonabant, Acomplia™). Apparently, the reason for these unnecessary complications is again the limited knowledge about the cannabinoid field (including the lack of information about the most common chemical tools used in cannabinoid pharmacology) in the general scientific community.

My other growing concern arises from the rapidly increasing number of publications (in 2006 and 2007, it was ~100 articles per month; see Fig. 2 in Chap. 1). Thus it seems difficult to keep up to date with the physiology, pharmacology, molecular biology, and pathology of cannabinoids. Recently, it has become easier to publish "unorthodox" research findings, as most of them proved to be valid, since they resulted from complex interactions between the endocannabinoid system and other signaling systems, and between new ligands, new receptors, and other targets. Although many laboratories are making an enormous effort to rule out the underlying mechanisms of these unorthodox findings, concomitantly, the same unusual pharmacological or physiological actions are recurrently rediscovered and reported occasionally by new research groups. To be more explicit, I would mention here the pharmacology of cholinergic, purinergic, GABAergic, or glutamatergic signaling, in which commonly accepted ligands, such as methyllicaconitine, nicotine, ATP, PPADS, CGS21680, CNQX, AP5, bicuculline, etc. with well-established maximal selective nanomolar or micromolar concentrations can be found. These concentrations are never to be exceeded because it is common knowledge that it would question the reliability of conclusions about the observations. In contrast, ligands of low nanomolar or picomolar affinity are often used in the micromolar range in the cannabinoid research field. There are research reports in which SR141716A and WIN55212-2 were used even at 10–100 μM in vitro and the authors claimed that the observed effects were CB_1 receptor mediated. Chapter 9 in this book thus tries to establish a bottom line for the pharmacology of cannabinoid research, listing common "side effects" and unorthodox mechanisms that can be easily misinterpreted as actions at novel receptors.

Another chapter also tackles the question of inverse agonism. Several antagonists of the cannabinoid receptors are known as inverse agonists (such as SR141716A and AM251; see Chap. 7). Nonetheless, recent data shed new light on this question by indicating an apparent lack of inverse agonism in the absence of endocannabinoids (which are otherwise generally present in most experimental preparations); in other words, these antagonists would not cause an effect opposite to the agonists. This is topped by reports on novel CB_1 receptor-selective neutral/silent antagonists. Thus, it might be worth solving this problem; otherwise one may eventually conclude that a neutral antagonist inhibits the binding of only the synthetic agonists at the CB_1 receptor, but not that of the endogenous agonists.

As a concluding remark, I would like to express again my gratitude to the contributing authors and to Joseph Burns from Springer-Verlag for recognizing the compelling need for the present volume and for giving me the opportunity to make this work happen. We (the authors) apologize for not discussing many significant publications in the present volume; it is entirely unintentional and completely due to space limitations. Nevertheless, the book the reader may hold right now in his hands has made a serious attempt to give a comprehensive overview of all the essential literature concerning the endocannabinoid and related systems in the nervous tissue.

Coimbra, June 2007

Attila Köfalvi

Contents

Contributors

David Baker
Neuroimmunology Unit, Neuroscience Centre, Institute of Cell and Molecular Science, Barts and the London School of Medicine and Dentistry, Queen Mary University of London, 4 Newark Street, Whitechapel, London E1 2AT, UK, david.baker@qmul.ac.uk

Heather Bradshaw
Psychological and Brain Sciences, The Kinsey Institute for Research in Sex, Gender and Reproduction, Indiana University Bloomington, IN 47405 hbbradsh@indiana.edu

Veronica A. Campbell
Department of Physiology and, Trinity College Institute of Neuroscience, Trinity College, Dublin, Ireland

María L.de Ceballos
Instituto Cajal, CSIC, Doctor Arce, 37, 28002 Madrid, Spain mceballos@cajal.csic.es

Aldemar Degroot
Astellas Pharma Europe B.V., Elisabethhof 1, 2350 AC Leiderdorp, The Netherlands, Aldemar.deGroot@eu.astellas.com

Javier Fernández-Ruiz
Departamento de Bioquímica y Biología Molecular III, Facultad de Medicina, Universidad Complutense, 28040-Madrid, Spain

Christopher J. Fowler
Department of Pharmacology and Clinical Neuroscience, Umeå University, SE901 87 Umeå, Sweden

Markus Fritzsche
Praxis für Innere Medizin, Soodstrasse 13, 8134 Adliswil, Switzerland

Gregory Gerdeman
c/o John Schindler, 6001 E. Pima Street, Apt. 162, Tucson, Arizona 85712

Manuel Guzman
Department of Biochemistry and Molecular Biology I, School of Biology
Complutense University, 28040 Madrid, Spain

Tibor Harkany
Institute of Medical Sciences, University of Aberdeen, Foresterhill
AB25 2ZD, Aberdeen, Scotland, UK

Andrew J. Irving
Neurosciences Institute, Division of Pathology & Neuroscience,
Ninewells Hospital and Medical School, University of Dundee,
Dundee, Scotland, DD1 9SY, UK

Rohini Kuner
Pharmacology Institute, University of Heidelberg, Im Neuenheimer Feld 366,
69120 Heidelberg, Germany

Beat Lutz
Institute of Physiological Chemistry and Pathobiochemistry
Johannes Gutenberg-University Mainz, Duesbergweg 6, 55099 Mainz, Germany
blutz@uni-mainz.de

Giovanni Marsicano
U 862 Centre de Recherche INSERM François Magendie,
Université Bordeaux 2, 146, rue Léo Saignat, 33077 Bordeaux, France
giovanni.marsicano@bordeaux.inserm.fr

Vincenzo Di Marzo
Istituto di Chimica Biomolecolare, Consiglio Nazionale delle Ricerche
Via Campi Flegrei 34, Comprensorio Olivetti, 80078 Pozzuoli (NA), Italy
vdimarzo@icmib.na.cnr.it

Isabel Matias
Istituto di Chimica Biomolecolare, Consiglio Nazionale delle Ricerche
Via Campi Flegrei 34, Comprensorio Olivetti, 80078 Pozzuoli (NA), Italy
vdimarzo@icmib.na.cnr.it

Istvan Nagy
Department of Anaesthetics, Pain Medicine and Intensive Care,
Imperial College London, Chelsea and Westminster Hospital, 369 Fulham Road,
London SW10 9NH, UK

Linda A. Parker
Department of Psychology, University of Guelph, Guelph, ON N1G 2W1,
Canada, parkerl@uoguelph.ca

Roger G. Pertwee
Professor of Neuropharmacology, Institute of Medical Sciences
University of Aberdeen, Aberdeen AB25 2ZD, Scotland, UK
rgp@abdn.ac.uk

Paul L. Prather
Dept of Pharmacology & Toxicology, Mail Slot 611, College of Medicine
University of Arkansas for Medical Sciences, 4301 W. Markham St.
Little Rock, AR 72205, PratherPaulL@uams.edu

Julián Romero
Laboratorio de Apoyo a la Investigación, Fundación Hospital Alcorcón,
C/ Budapest 1. 28922, Alcorcón, Madrid, Spain

Takayuki Sugiura
Faculty of Pharmaceutical Sciences, Teikyo University, Sagamiko, Sagamihara,
Kanagawa 229-0195, Japan

Eleni Tzavara
CR1 INSERM, INSERM U-513 Neurobiologie et Psychiatrie
Faculté de Médecine de Créteil, 8 rue du Général SARRAIL,
F-94010, Créteil, France

Part I
Molecular Biology, Pharmacology, Anatomy, and Physiology of the Endocannabinoid and Related Lipidergic Signaling Systems in the Brain

Chapter 1
An Historical Introduction to the Endocannabinoid and Endovanilloid Systems

Istvan Nagy, John P.M. White, Cleoper C. Paule, and Attila Köfalvi

Abstract Cannabis and chili pepper have been used for medical, gastronomical and recreational purposes for at least 8,000 years. Nevertheless, it was discovered only eight years ago that the cloned neuronal targets of their active principles, delta^9-tetrahydrocannabinol (Δ^9-THC) and capsaicin are related to each other, as they all can be activated by some arachidonic acid-derivative endogenous ligands. Here, we will summarize the history of man's relationship with cannabis and capsaicin, and we will detail the most important scientific keystones in the evolution of cannabinoid and vanilloid research, featuring the list of cannabinoid and capsaicin effects, the discovery of endogenous ligands and the cloning of receptors, namely, the CB_1 and the CB_2 cannabinoid receptor as well as the $TRPV_1$ vanilloid receptor, where the endogenous and the plant-derived substances act upon. This chapter serves, therefore, as an introduction to *Cannabinoids and the Brain*, the book which will extensively describe the neuronal and, to some extent, the peripheral cannabinoid and vanilloid systems in molecular, pharmacological, physiological, pathological and neuropsychiatric viewpoints.

Introduction

The History of Cannabis

The Asiatic plant *cannabis* or *hemp* (*Cannabis sativa/indica* = *useful/Indian Cannabis*) has been used for more than 8,000 years due to its medical and psychotropic effects. It is most likely that the original Sumerian word "kunibu" developed into the forms "kan(n)ab(is)" and "hanaba", then "hennep" and finally, hemp. The plant cannabis belongs to the family *Cannabaceae* and the order *Urticales*. Its leaves and flowering tops are used to produce marijuana and hashish (also known as charas, bhang, ganja, dagga, grass, pot). Seeds of cannabis were found in 8,000 years old Chinese food remains. Interestingly, the first written note about the medical use of cannabis was also discovered in China, which dates back to 2727 B.C. The Atharvaveda, the sacred text of Hinduism, also mentions the use of cannabis for medical purposes in India between 1200 and 800 B.C. The psychotropic properties

A. Köfalvi (ed.), *Cannabinoids and the Brain.*

of cannabis were first described in a Chinese medical book around 100 B.C. It is believed that *Cannabis sativa* was first introduced in Europe by the Scythians, as recorded by Herodotus in 430 B.C. In 100 A.D., Dioskurides inferred that cannabis was a Roman medical plant, whereas Galen highlighted its psychotropic action in 170 A.D. The medieval Europe was first informed about the popularity of cannabis in Asia by Marco Polo. Later cannabis was used mainly as a medicine in England. Even Queen Victoria was prescribed cannabis by her doctor in 1890. Consequently, cannabis was declared harmless and legalized in 1901. However, in 1925, the Geneva Convention included cannabis and hashish in the list of dangerous and illicit drugs. In the USA, cannabis was also used for medical purposes from 1840, but the Mexican Revolution in 1910 changed the general opinion about cannabis. It became a symbol of terrible sins. Until 1931, 29 states prohibited the use of marijuana, and from 1937, the Federal Law proclaimed marijuana as an illicit drug. Still, it regained its popularity when both president Kennedy and president Johnson suggested that cannabis should be legalized. Presumably, due to these propositions, 200–250 million cannabis users were reported by the UN worldwide till 1970. In the last 40 years, the debate on the safety and legal status of cannabis-based medical treatments is becoming increasingly intense at both political and scientific levels (see for example Wall et al., 2001; Hayry, 2004; Comeau, 2006). Although clinical trials with cannabinoid ligands are allowed in many countries, the general use of marijuana to treat the pain and eating problems of cancer patients as well as to reduce intraocular pressure in glaucoma is still not legalized. The main issue is that the concentration of beneficial constituents in the smoke of cannabis cannot be controlled, and furthermore, conservative politics can hardly agree with the medical use of an illicit drug. Nonetheless, we should mention that morphine, codeine, lidocaine and procaine, which are all illicit drug-derivatives, are commonly used in medicine. Moreover, nicotine is regarded as one of the most addictive drugs, yet its use is perfectly legal. All the same, we have to acknowledge certain concerns, as chronic marijuana consumption (ca. ≥50 times) can induce schizophrenia in susceptible persons (see Chap. 22).

The History of Capsaicin

Chili peppers (*Capsicum frutescens var. longum*) are members of the nightshade family (*Solanaceae*), and have been domesticated since about 7500 B.C. in the Americas (Perry et al., 2007). Christopher Columbus was one of the first Europeans who found chili peppers and subsequently transferred a certain amount to the Old World. He accidentally named them "peppers" because of their similarity in taste with the black peppers of the *Piper* genus. This pungent taste is due to capsaicin, a neurotoxin derived from the chili pepper plant. Capsaicin has been extensively used for centuries both as a herbal remedy and a food product prized in the cuisine of many societies. Capsaicin is also responsible for the reduced sensitivity of the mouth to high temperatures and painful mechanical stimuli which results from regular chili pepper consumption. The fact that capsaicin also relieves the spontaneous pain associated with inflammation prompted healers in many different cultures to employ hot peppers to treat painful conditions of varying aetiologies over the centuries. Thus, Native Americans rubbed their gums with hot peppers as a cure for tooth ache, while Europeans used an alcoholic extract prepared from chilies for a similar purpose (Szallasi and Blumberg, 1999). Capsaicin is still commonly used for treating painful conditions as an "over-the-counter" remedy in the form of capsaicin-containing ointments.

The Discovery of the Endocannabinoid and Endovanilloid Systems

The Endocannabinoid System

Although Eastern cultures have been using marijuana as medicine for centuries, Western cultures started to recognize the therapeutic potential of marijuana only recently. For instance, cannabis extract was a licensed medicine and sold under the name of "Tincture of Cannabis" in the UK (Gill et al., 1970). The first observed medicinal benefits encompassed anesthetic, airway opening, antihypertensive, eye pressure reducing (in glaucoma) as well as antiemetic actions, but for decades, the underlying physiological and molecular mechanisms were unknown. The first isolated plant-derived (phyto-) cannabinoid was cannabinol, found in the red oil extract of hemp more than a century ago, and in the 1930s, its chemical structure was elucidated (Pertwee, 2006). Although tetrahydrocannabinols (THCs) and cannabidiols were discovered and isolated from hemp extracts in the following years, the structure and stereochemistry of the naturally occurring (–)-cannabidiol (Mechoulam and Shvo, 1963) and (–)-trans-Δ^9-tetrahydrocannabinol (Δ^9-THC), the main psychoactive constituent of marijuana and hashish (Gaoni and Mechoulam, 1964) were unraveled in the decade when hippies also became interested in cannabis preparations. The major phytocannabinoid structure was identified as a tricyclic ring constituted from a phenol ring, having a 5-carbon alkyl chain meta to the hydroxyl, a central pyran ring, and a mono-unsaturated cyclohexyl ring (Fig. 1; Howlett et al., 2004). Raphael Mechoulam and his laboratory pioneered the discovery and synthesis of numerous novel phytocannabinoids, which enumerate at least 66 distinctive ones hitherto (Mechoulam and Hanus, 2000; Pertwee, 2006). In parallel with their discovery, hemp constituents were tested for psychotropic and motor effects in man and in animal models, mostly in mice, rats, rabbits, and dogs. THCs proved to be the most effective among all phytocannabinoids, whereas among THCs, Δ^9-THC seems to be responsible for the vast majority of effects such as motor disturbances and catalepsy, corneal areflexia (in rabbits), scratching, euphoria and dysphoria, anxiety, drowsiness, altered time and audiovisual perceptions, panic attacks and impaired memory (Haagen-Smit et al., 1940; Loewe, 1946; Paton and Pertwee, 1973; Howlett et al., 2004). The following years then proved that the more psychotropic a cannabinoid substance is

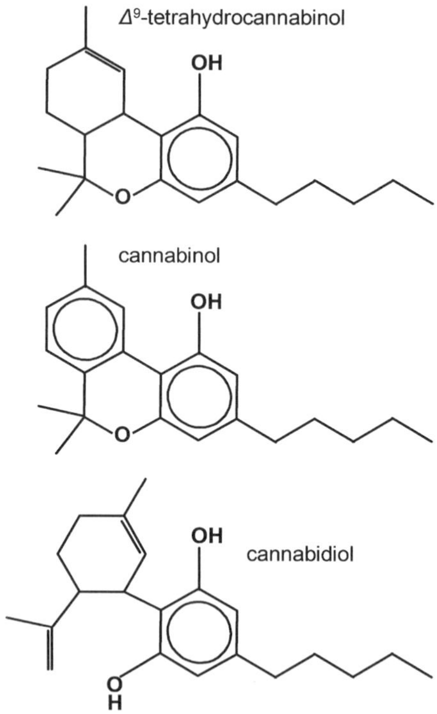

Fig. 1 The most important constituents of *Cannabis sativa* L., namely Δ^9-tetrahydrocannabinol ((–)-(6aR,10aR)-6,6,9-trimethyl-3-pentyl-6a,7,8,10a-tetrahydro-6H-benzo[c]chromen-1-ol; Δ^9-THC), cannabinol (6,6,9-trimethyl-3-pentyl-6H-benzo[c]chromen-1-ol) and cannabidiol (2-((1S,6S)-3-methyl-6-(prop-1-en-2-yl)cyclohex-2-enyl)-5-pentylbenzene-1,3-diol)

the greater motor disturbances it causes. Furthermore, among phytocannabinoids, Δ^9-THC is the most potent and effective psychomotor compound. The underlying mechanisms for these effects were mostly believed to result from "non-specific" interactions between the lipophilic Δ^9-THC and the cell membranes, changing the fluidity and structure of the latter, therefore affecting most cell types (Lawrence and Gill, 1975; Hillard et al., 1985). Nonetheless, a nearly identical molecule, Δ^8-THC, was much less potent and efficacious than Δ^9-THC, and most other phytocannabinoids were devoid of effect, which all weakened the hypothesis of changing membrane fluidity. The next important cornerstone was the discovery that Δ^9-THC inhibits cAMP accumulation (Howlett and Fleming, 1984), and the recognition of specific cannabinoid binding sites in the brain (Devane et al., 1988). These two findings from Allyn Howlett's laboratory predicted that the discovery of atleast one cannabinoid receptor was imminent. And indeed, in 1990, both the rat and the human CB_1 receptors were characterized (Gérard et al., 1990, 1991; Matsuda et al., 1990), and the first study on its distribution found the receptor at an unexpectedly high density in the brain (Herkenham et al., 1991). Right at the moment when we write

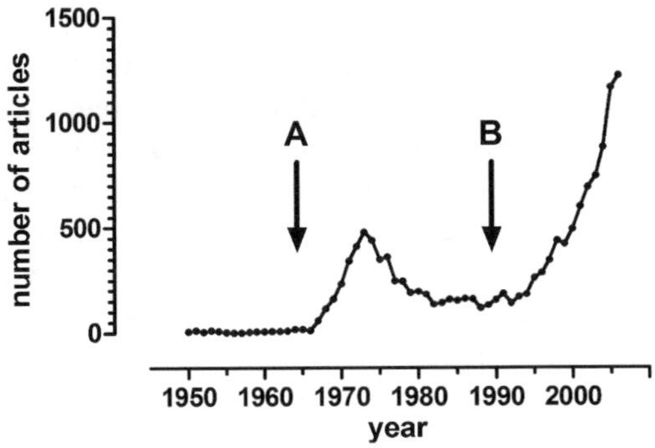

Fig. 2 The yearly number of research and review articles on the cannabinoid field from 1950. The *two arrows* indicate the onset of the two booms: (**a**) Gaoni and Mechoulam (1964) report the structure of Δ^9-THC; (**b**) Gérard and colleagues (1990) and Matsuda and colleagues (1990) report the cloning of the first cannabinoid receptor (the CB_1 receptor) from human and rat (See text for further explanations.)

these lines, there are 14,000 articles published in relation to cannabinoids. The first one – listed by PubMed – is from 1909. As Fig. 2 demonstrates, the first "boom" of cannabinoid research occurred after 1964, the year when Gaoni and Mechoulam reported the structure of Δ^9-THC. The second boom – which is related to the discovery of the CB_1 receptor – resulted in a continuously increasing number of publications in the last 15 years, and in 2006 and in the first five months of 2007, it reached a peak of 100 publications per month. Thus, recognizing the significance of cannabinoid research very early, Di Mahadeen and Rik Musty founded the International Cannabinoid Research Society in 1991 (http://www.cannabinoidsociety.org for further information). Soon, another cannabinoid receptor, the CB_2 receptor was discovered, but its expression was found to be restricted mainly to immune tissues (Munro et al., 1993). Importantly, both cannabinoid receptors are G protein-coupled seven-transmembrane-domain receptors of the rhodopsin type (see Chaps. 5 and 6). In the meantime, the first endogenous cannabinoid ligand, arachidonoylethanolamine or anandamide, was found in porcine brain (Devane et al., 1992), which was followed by 2-arachidonoylglycerol (2-AG) described by two independent laboratories in the same year (Mechoulam et al., 1995; Sugiura et al., 1995; see Chap. 2). As one can notice from their name, both ligands are arachidonic acid derivatives, and interestingly, one of them, namely anandamide, is capable of activating the $TRPV_1$ receptor as well (Zygmunt et al., 1999; and see below). Later in this book, we will mention some new cannabinoid receptors and endocannabinoid candidates (Chaps. 4 and 9). However, hitherto these four molecules received the biggest attention. The last 20 years provided a major boost to the renaissance of the synthesis of novel cannabinoid ligands as well, and further readings can be found in recent reviews (Mechoulam and Hanus, 2000; Howlett et al., 2004; Pertwee, 2006; see Chap. 7). However, we should highlight 1994 when the first selective CB_1 receptor antagonist, SR141716A or Rimonabant, was reported (Rinaldi-Carmona et al., 1994) and, has been marketed in 2006 in Europe under the name Acomplia™ as a promising alternative medicine against cardiovascular and metabolic risk factors (see Chap. 14). Significantly, the wide-spectrum roles of CB_1 and CB_2 receptors were not determined only by the action of antagonists. CB_1 receptor knockout mice strains were engineered by Zimmer and co-workers (1999) and Ledent and co-workers (1999). Most importantly, the vast majority of behavioural and physiological responses to cannabinoid ligands were no longer observed in these mice, compared to findings in the CB_2 receptor knockout mouse (Buckley et al., 2000). These findings underlined that the major cannabinoid receptor of the nervous tissue is the CB_1 receptor. As the conditional knockout technology became widespread, novel, neuron-specific conditional CB_1 receptor knockout animals were also generated (Marsicano et al., 2003).

The Endovanilloid System

Capsaicin was isolated as the active ingredient of chili peppers in the mid-nineteenth century (Thresh, 1846). However, its exact structure was elucidated only some seventy years later (Nelson, 1919). The revealed structure shows that capsaicin possesses a vanilloid moiety, which results in it being assigned to the vanilloid family (see Fig. 2 in Chap. 8). Surprisingly, however, few vanilloids – other than capsaicin itself and the even more potent resiniferatoxin (RTX) – possess the characteristic pungency of capsaicin. Capsaicin in the nanomolar range specifically, and selectively, acts on a large sub-population of nociceptive primary sensory neurons. This remarkable property of capsaicin was first described by two Hungarian scientists, Janos Porszasz and Nicholas Jancso (1959), who reported that, following capsaicin application, sensory fibres fail to produce action potentials on subsequent exposure to capsaicin. Nicholas Jancso's son, Gabor Jancso, subsequently showed that those primary sensory neurons that are sensitive to capsaicin belong to the small diameter sub-population of dorsal root ganglion neurons, which are generally considered to be nociceptive in function (Jancso et al., 1977). His group also showed that capsaicin application in neonates ultimately results in degeneration of the capsaicin-sensitive sensory neurons. The fact that capsaicin appeared to be highly specific in its effects on nociceptive primary sensory neurons, coupled with its "desensitizing" effect, kindled major interest in the mechanisms involved in these phenomena among academics and, also, in the pharmaceutical industry. The search for the molecule, or molecules, which mediate the effects of capsaicin in primary sensory neurons, had an uncertain beginning. Voltage-gated Na^+ channels and K^+ channels were proposed as candidate molecules (Dubois, 1982; Yamanaka et al., 1984; Erdelyi et al., 1987). However, while Porszasz and Jancso (1959) had observed an excitatory effect of capsaicin on mammalian sensory fibres, an inhibitory effect on these voltage-gated ion channels was observed in recordings from frog, snail and crayfish neurons when exposed to capsaicin in micromolar concentrations (see Chap. 9). Notwithstanding this, an attempt was made to argue that this inhibitory effect could be responsible for the loss of responsiveness of sensory neurons found following prolonged, or repeated, exposure to capsaicin. The first satisfactory account of the underlying mechanism for capsaicin-induced pain sensation was provided by Heyman and Rang (1985). These authors reported that capsaicin produces rapid depolarization in a sub-population of rat dorsal root ganglion neurons. They also showed that the capsaicin-sensitive neurons are of the slow-conducting unmyelinated variety of sensory neurons, which confirmed that capsaicin indeed activates nociceptive neurons. The current–voltage relationship of the capsaicin-induced responses in their study clearly suggested that capsaicin increases, rather than decreases, the membrane conductance in capsaicin-sensitive cells. This conductance increase was later analysed and shown to produce inward Na^+ and Ca^{2+} currents and an outward K^+ current (Marsh et al., 1987). Attention then focused on identifying the receptor responsible for mediating these capsaicin effects, because several pieces of evidence suggested the likelihood of the existence of a specific receptor for capsaicin. First, RTX was found to produce responses similar to those produced by capsaicin both in vivo and in vitro (Szallasi and Blumberg, 1989; Winter et al., 1990), indicating the existence of several agonists for a receptor. Second, blockers of the putative capsaicin receptor were found and developed. Thus, the inorganic dye, ruthenium red, was shown to block capsaicin-evoked activation of primary sensory neurons (Bleakman et al., 1990), while a competitive capsaicin antagonist, capsazepine, which blocks the effect of capsaicin both in vitro and in vivo was developed (Dickenson and Dray, 1991; Bevan et al., 1992; Perkins and Campbell, 1992). Third, RTX and capsaicin were shown to compete for a binding site on membrane preparations from primary sensory neurons (Szallasi and Blumberg, 1990). Subsequently, RTX binding was also demonstrated in the dorsal spinal cord where nociceptive fibres terminate and in selective areas of the hypothalamus (Szallasi et al., 1995; Acs et al., 1996). These binding sites were consistent with previous in vivo findings that capsaicin induces pain by activating nociceptive primary sensory neurons and induces hypothermia by activating hypothalamic nuclei involved in thermoregulation (Jancso-Gabor et al., 1970; Jancso et al., 1977). However, the honour of identifying the receptor responsible for the action of capsaicin on primary sensory neurons went to Caterina and colleagues (1997). In a series of elegant experiments, Caterina's group prepared a cDNA library from dorsal root ganglia. Subsequently, pools of this library were created and used to transfect human embryonic kidney 293 cells. The transfected cells were monitored for exhibiting capsaicin-evoked Ca^{2+} influx. The pool producing capsaicin-sensitive cells was then sub-divided until a single clone was found. This capsaicin-responsive molecule was then denominated vanilloid receptor 1 (VR_1) (consult with Fig. 1 in Chap. 8). The predicted structure of VR_1, comprising six transmembrane domains, with both the C-and N-termini being located intracellularly, and with a pore-forming intramembrane loop connecting transmembrane domains 5 and 6, proved to be similar to the structure of known members of the transient receptor ion channel (TRP) superfamily (Montell and Rubin, 1989). Subsequent investigations have revealed five homologues for VR_1 (Nilius et al., 2007). This structural similarity and the existence of these homologues led to the capsaicin receptor being re-named the "transient receptor potential vanilloid type-1 ion channel ($TRPV_1$)". The identification of the $TRPV_1$ ion channel led to frenzied activity as scientists endeavoured to elucidate its expression pattern, function and the mechanisms involved in the regulation of its expression and activation. It has emerged that $TRPV_1$ is a polymodal receptor which responds to various ligands, as well as to heat above ~42 °C, protons and post-translational modifications. Moreover, $TRPV_1$ is capable of integrating the effect of these activators (Caterina et al., 1997; Tominaga et al., 1998; Chuang et al., 2001). In addition to its expression in primary sensory neurons, $TRPV_1$ is also expressed by various neurons in the brain and by non-neuronal cells at the periphery (Nagy et al., 2004). Generation of mice lacking $TRPV_1$ has shown that the capsaicin receptor is indispensable for the development of inflammatory heat hyperalgesia (Caterina et al., 2000; Davis et al., 2000) and bladder hyper-reflexia (Charrua et al., 2007). Finally, recent publications have drawn attention to the role of endovanilloids in the activation of $TRPV_1$ in pathological conditions (Dinis et al., 2004; Singh Tahim et al., 2005). Interestingly, certain of the endogenous $TRPV_1$ ligands, such as anandamide, are also endocannabinoids (Zygmunt et al., 1999). The existence of these shared endogenous ligands has led to the theory that cannabinoid receptors and $TRPV_1$ may be the reverse sides of the same coin, constituting the G protein-coupled receptors and the ligand-gated ion channels of the same sensory system. In Chap. 8, we describe the structure and function of this remarkable ion channel and focus, in particular, on the mechanisms involved in its activation.

Acknowledgement Attila Köfalvi is grateful for the III/BIO/56/2005 grant and for the Fundação para a Ciência e Tecnologia of the Portuguese Government (POCI 2010/SFRH/BPD/18506/2004).

References

Acs G, Palkovits M, Blumberg PM (1996) Specific binding of [^{3}H]resiniferatoxin by human and rat preoptic area, locus ceruleus, medial hypothalamus, reticular formation and ventral thalamus membrane preparations. Life Sci 59:1899–1908.

Bevan S, Hothi S, Hughes G, James IF, Rang HP, Shah K, Walpole CS, Yeats JC (1992) Capsazepine: a competitive antagonist of the sensory neuron excitant capsaicin. Br J Pharmacol 107:544–552.

Bleakman D, Brorson, JR, Miller RJ (1990) The effect of capsaicin on voltage-gated calcium currents and calcium signals in cultured dorsal root ganglion cells. Br J Pharmacol 101:423–431.

Buckley NE, McCoy KL, Mezey E, Bonner T, Zimmer A, Felder CC, Glass M, Zimmer A (2000) Immunomodulation by cannabinoids is absent in mice deficient for the cannabinoid CB_2 receptor. Eur J Pharmacol 396:141–149.

Caterina MJ, Schumacher MA, Tominaga M, Rosen TA, Levine JD, Julius D (1997) The capsaicin receptor: a heat-activated ion channel in the pain pathway. Nature 389:816–824.Caterina MJ, Leffler A, Malmberg AB, Martin WJ, Trafton J, Petersen-Zeitz KR, Koltzenburg M, Basbaum AI, Julius D (2000) Impaired nociception and pain sensation in mice lacking the capsaicin receptor. Science 288:306–313.

Charrua A, Cruz CD, Cruz F, Avelino A (2007) Transient receptor potential vanilloid subfamily 1 is essential for the generation of noxious bladder input and bladder overactivity in cystitis. J Urol 177:1537–1541.

Chuang HH, Prescott ED, Kong H, Shields S, Jordt SE, Basbaum AI, Chao MV, Julius D (2001) Bradykinin and nerve growth factor release the capsaicin receptor from PtdIns(4,5)P2-mediated inhibition. Nature 411:957–962.

Comeau P (2006) Cut to marijuana research sends strong message. CMAJ 175:1507–1508.

Davis JB, Gray J, Gunthorpe MJ, Hatcher JP, Davey PT, Overend P, Harries MH, Latcham J, Clapham C, Atkinson K, Hughes SA, Rance K, Grau E, Harper AJ, Pugh PL, Rogers DC, Bingham S, Randall A, Sheardown SA (2000) Vanilloid receptor-1 is essential for inflammatory thermal hyperalgesia. Nature 405:183–187.

Devane WA, Dysarz FA, Johnson MR, Melvin LS, Howlett AC (1988) Determination and characterization of a cannabinoid receptor in rat brain. Mol Pharmacol 34:605–613.

Devane WA, Hanus L, Breuer A, Pertwee RG, Stevenson LA, Griffin G, Gibson D, Mandelbaum A, Etinger A, Mechoulam R (1992) Isolation and structure of a brain constituent that binds to the cannabinoid receptor. Science 258:1946–1949.

Dickenson AH, Dray A (1991) Selective antagonism of capsaicin by capsazepine: evidence for a spinal receptor site in capsaicin-induced antinociception. Br J Pharmacol 104:1045–1049.

Dinis P, Charrua A, Avelino A, Yaqoob M, Bevan S, Nagy I, Cruz F (2004) Anandamide-evoked activation of vanilloid receptor 1 contributes to the development of bladder hyperreflexia and nociceptive transmission to spinal dorsal horn neurons in cystitis. J Neurosci 24:11253–11263.

Dubois JM (1982) Capsaicin blocks one class of K^+ channels in the frog node of Ranvier. Brain Res 245:372–375.

Erdelyi L, Such G, Jancso G (1987) Intracellular and voltage clamp studies of capsaicin-induced effects on a sensory neuron model. Acta Physiol Hung 69:481–492.

Gaoni Y, Mechoulam R (1964) Isolation, structure and partial synthesis of an active constituent of hashish. J Am Chem Soc 86:1646–1647.

Gerard C, Mollereau C, Vassart G, Parmentier M (1990) Nucleotide sequence of a human cannabinoid receptor cDNA. Nucleic Acids Res 18:7142.

Gerard CM, Mollereau C, Vassart G, Parmentier M (1991) Molecular cloning of a human cannabinoid receptor which is also expressed in testis. Biochem J 279:129–134.

Gill EW, Paton WDM, Pertwee RG (1970) Preliminary experiments on the chemistry and pharmacology of cannabis. Nature 228:134–136.

Haagen-Smit AJ, Wawra CZ, Koepfli JB, Alles GA, Feigen GA, Prater AN (1940) A physiologically active principle from Cannabis sativa (marihuana). Science 91:602–603.

Hayry M (2004) Prescribing cannabis: freedom, autonomy, and values. J Med Ethics 30:333–336.

Herkenham M, Lynn AB, Johnson MR, Melvin LS, De Costa BR, Rice KC (1991) Characterization and localization of cannabinoid receptors in rat brain: a quantitative *in vitro* autoradiographic study. J Neurosci 11:563–583.

Heyman I, Rang HP (1985) Depolarizing responses to capsaicin in a subpopulation of rat dorsal root ganglion cells. Neurosci Lett 56:69–75.

Hillard CJ, Harris RA, Bloom AS (1985) Effects of the cannabinoids on physical properties of brain membranes and phospholipid vesicles: fluorescence studies. J Pharmacol Exp Ther 232:579–588.

Howlett AC, Fleming RM (1984) Cannabinoid inhibition of adenylate-cyclase: pharmacology of the response in neuroblastoma cell membranes. Mol Pharmacol 26:532–538.Howlett AC, Breivogel CS, Childers SR, Deadwyler SA, Hampson RE, Porrino LJ (2004) Cannabinoid physiology and pharmacology: 30 years of progress. Neuropharmacology 47 Suppl 1:345–358.

Jancso G, Kiraly E, Jancso-Gabo A (1977) Pharmacologically induced selective degeneration of chemosensitive primary sensory neurons. Nature 270:741–743.

Jancso-Gabor A, Szolcsanyi J, Jancso N (1970) Stimulation and desensitization of the hypothalamic heat-sensitive structures by capsaicin in rats. J Physiol 208:449–459.

Lawrence DK, Gill EW (1975) The effects of Δ^9-tetrahydrocannabinol and other cannabinoids on spin-labeled liposomes and their relationship to mechanisms of general anesthesia. Mol Pharmacol 11:595–602.

Ledent C, Valverde O, Cossu G, Petitet F, Aubert JF, Beslot F, Bohme GA, Imperato A, Pedrazzini T, Roques BP, Vassart G, Fratta W, Parmentier M (1999) Unresponsiveness to cannabinoids and reduced addictive effects of opiates in CB_1 receptor knockout mice. Science 283:401–404.

Loewe (1946) Studies on the pharmacology and acute toxicity of compounds with marihuana activity. J Pharmacol Exp Ther 88:154–161.

Marsh SJ, Stansfeld CE, Brown DA, Davey R, McCarthy D (1987) The mechanism of action of capsaicin on sensory C-type neurons and their axons *in vitro*. Neuroscience 23:275–289.

Marsicano G, Goodenough S, Monory K, Hermann H, Eder M, Cannich A, Azad SC, Cascio MG, Gutierrez SO, van der Stelt M, Lopez-Rodriguez ML, Casanova E, Schutz G, Zieglgansberger W, Di Marzo V, Behl C, Lutz B (2003) CB_1 cannabinoid receptors and on-demand defense against excitotoxicity. Science 302:84–88.

Matsuda LA, Lolait SJ, Brownstein MJ, Young AC, Bonner TI (1990) Structure of a cannabinoid receptor and functional expression of the cloned cDNA. Nature 346:561–564.

Mechoulam R, Hanus L (2000) A historical overview of chemical research on cannabinoids. Chem Phys Lipids 108:1–13.

Mechoulam R, Shvo Y (1963) The structure of cannabidiol. Tetrahedron 19:2073–2078.

Mechoulam R, Ben-Shabat S, Hanus L, Ligumsky M, Kaminski NE, Schatz AR, Gopher A, Almog S, Martin BR, Compton DR, Pertwee RG, Griffin G, Bayewitch M, Barg J, Vogel Z (1995) Identification of an endogenous 2-monoglyceride, present in canine gut, that binds to cannabinoid receptors. Biochem Pharmacol 50:83–90.

Montell C, Rubin GM (1989) Molecular characterization of the Drosophila trp locus: a putative integral membrane protein required for phototransduction. Neuron 2:1313–1323.

Munro S, Thomas KL, Abu-Shaar M (1993) Molecular characterization of a peripheral receptor for cannabinoids. Nature 365:61–65.

Nagy I, Santha P, Jancso G, Urban L (2004) The role of the vanilloid (capsaicin) receptor ($TRPV_1$) in physiology and pathology. Eur J Pharmacol 500:351–369.

Nelson EK (1919) The constitution of capsaicin—the pungent principle of capsicum. J Am Chem Soc 41:1115–1117.

Nilius B, Mahieu F, Karashima Y, Voets T (2007) Regulation of TRP channels: a voltage-lipid connection. Biochem Soc Trans 35:105–108.

Paton WDM, Pertwee RG (1973). The pharmacology of cannabis in animals. In: Marijuana, Chemistry, Pharmacology, Metabolism and Clinical Effects. Mechoulam R, ed. New York: Academic Press, pp. 191–285.

Perkins MN, Campbell EA (1992) Capsazepine reversal of the antinociceptive action of capsaicin *in vivo*. Br J Pharmacol 107:329–333.

Perry L, Dickau R, Zarrillo S, Holst I, Pearsall DM, Piperno DR, Berman MJ, Cooke RG, Rademaker K, Ranere AJ, Raymond JS, Sandweiss DH, Scaramelli F, Tarble K, Zeidler JA (2007) Starch fossils and the domestication and dispersal of chili peppers (Capsicum spp. L.) in the Americas. Science 315:986–988.

Pertwee RG (2006) Cannabinoid pharmacology: the first 66 years. Br J Pharmacol 147: S163–S171.

Porszasz J, Jancso N (1959) Studies on the action potentials of sensory nerves in animals desensitized with capsaicine. Acta Physiol Acad Sci Hung 16:299–306.

Rinaldi-Carmona M, Barth F, Heaulme M, Shire D, Calandra B, Congy C, Martinez S, Maruani J, Neliat G, Caput D, Ferrara P, Soubrié P, Brelière J-C, Fur GL (1994) SR141716A, a potent and selective antagonist of the brain cannabinoid receptor. FEBS Lett 350:240–244.

Singh Tahim A, Santha P, Nagy I (2005) Inflammatory mediators convert anandamide into a potent activator of the vanilloid type 1 transient receptor potential receptor in nociceptive primary sensory neurons. Neuroscience 136:539–548.

Sugiura T, Kondo S, Sukagawa A, Nakane S, Shinoda A, Itoh K, Yamashita A, Waku K (1995) 2-Arachidonoylglycerol: a possible endogenous cannabinoid receptor ligand in brain. Biochem Biophys Res Commun 215:89–97.

Szallasi A, Blumberg, PM (1989) Resiniferatoxin, a phorbol-related diterpene, acts as an ultrapotent analog of capsaicin, the irritant constituent in red pepper. Neuroscience 30:515–520.

Szallasi A, Blumberg PM (1990) Specific binding of resiniferatoxin, an ultrapotent capsaicin analog, by dorsal root ganglion membranes. Brain Res 524:106–111.

Szallasi A, Blumberg PM (1999) Vanilloid (capsaicin) receptors and mechanisms. Pharmacol Rev 51:159–212.

Szallasi A, Nilsson S, Farkas-Szallasi T, Blumberg PM, Hokfelt T. Lundberg JM (1995) Vanilloid (capsaicin) receptors in the rat: distribution in the brain, regional differences in the spinal cord, axonal transport to the periphery, and depletion by systemic vanilloid treatment. Brain Res 703:175–183.

Thresh LT (1846) Isolation of capsaicin. Pharm J 6:941.

Tominaga M, Caterina MJ, Malmberg AB, Rosen TA, Gilbert H, Skinner K, Raumann BE, Basbaum AI, Julius D (1998) The cloned capsaicin receptor integrates multiple pain-producing stimuli. Neuron 21:531–543.

Wall J, Davis S, Ridgway S (2001) Cannabis: its therapeutic use. Nurs Stand 16:39–44.

Winter J, Dray A, Wood JN, Yeats JC, Bevan S (1990) Cellular mechanism of action of resiniferatoxin: a potent sensory neuron excitotoxin. Brain Res 520:131–140.

Yamanaka K, Kigoshi S, Muramatsu I (1984) Conduction-block induced by capsaicin in crayfish giant axon. Brain Res 300:113–119.

Zimmer A, Zimmer AM, Hohmann AG, Herkenham M, Bonner TI (1999) Increased mortality, hypoactivity, and hypoalgesia in cannabinoid CB_1 receptor knockout mice. Proc Natl Acad Sci USA 96:5780–5785.

Zygmunt PM, Petersson J, Andersson DA, Chuang H, Sorgard M, Di Marzo V, Julius D, Hogestatt ED (1999) Vanilloid receptors on sensory nerves mediate the vasodilator action of anandamide. Nature 400:452–457.

Chapter 2
Biosynthesis of Anandamide and 2-Arachidonoylglycerol

Takayuki Sugiura

Abstract Anandamide (*N*-arachidonoylethanolamine) can be synthesized from free arachidonic acid and ethanolamine by the action of a fatty acid amide hydrolase acting in reverse or from preexisting *N*-arachidonoyl phosphatidylethanolamine by the action of a phosphodiesterase (phospholipase D). Evidence is accumulating that anandamide is synthesized mainly by the latter pathway rather than the former in various mammalian tissues and cells. 2-Arachidonoylglycerol can be synthesized from arachidonic acid-containing diacylglycerol derived from increased inositol phospholipid metabolism by the action of a diacylglycerol lipase. 2-Arachidonoylglycerol can also be formed via other pathways such as the hydrolysis of the diacylglycerol derived from phosphatidylcholine and phosphatidic acid by the action of a diacylglycerol lipase and the hydrolysis of arachidonic acid-containing lysophosphatidic acid by the action of a phosphatase. The relative importance of these pathways may depend on the types of cells and stimuli. In this review, I have summarized the pathways and enzymes involved in the synthesis of anandamide and 2-arachidonoylglycerol.

Introduction

Two types of arachidonic acid-containing lipid molecules, anandamide (*N*-arachidonoylethanolamine, AEA) and 2-AG (*sn*2-arachidonoylglycerol), have been identified as the endogenous ligands (endocannabinoids) for the cannabinoid receptors. The first endocannabinoid to be found was anandamide that was isolated from pig brain by Devane and coworkers (1992). Anandamide binds to both the central and peripheral cannabinoid receptors with high affinity and exhibits a variety of cannabimimetic activities such as the inhibition of mouse twitch response, reduction of spontaneous motor activities, immobility, hypothermia, analgesia, impairment of memory, and inhibition of long-term potentiation (Mechoulam et al., 1998b; Piomelli et al., 1998; Di Marzo et al., 2002). The second endocannabinoid to be found was 2-AG, an arachidonic acid-containing species of 2-monoacylglycerol. Sugiura and colleagues (1995) isolated 2-AG from rat brain, and Mechoulam and colleagues (1995) isolated it from canine gut. 2-AG binds to the cannabinoid

A. Köfalvi (ed.), *Cannabinoids and the Brain.*

receptors (CB_1 and CB_2) with high affinity, although its affinity was somewhat lower than that of anandamide. 2-AG exhibits various pharmacological activities in vitro and in vivo similar to anandamide (see reviews Mechoulam et al., 1998b; Piomelli et al., 1998; Sugiura and Waku, 2000; Di Marzo et al., 2002; Sugiura et al., 2006a). A number of studies have thus far been carried out on anandamide and 2-AG, and it has widely been accepted that these lipid molecules act as important intercellular mediators in various mammalian tissues and cells. In this review, we focused on anandamide and 2-AG and summarized the pathways and enzymes involved in their synthesis.

Biosynthesis of Anandamide

Anandamide Generation

In the late 1970s to early 1980s, Schmid and coworkers (1990) reported that large amounts of *N*-acylethanolamines such as *N*-palmitoylethanolamine and *N*-stearoylethanolamine were produced in degenerating tissues such as infarcted hearts and ischemic brains. However, they did not mention the generation of arachidonic acid-containing species, i.e., anandamide, because these studies were carried out before the discovery of anandamide. The generation of anandamide in stimulated cells was first described by Di Marzo and colleagues (1994). They demonstrated that rat brain neurons generated anandamide when stimulated with ionomycin or with several membrane-depolarizing agents such as kainate, high K^+, and 4-aminopyridine. Di Marzo and coworkers also demonstrated the generation of anandamide in ionomycin-treated J774 macrophages (Di Marzo et al., 1996a), ionomycin-treated RBL-2H3 cells (Bisogno et al., 1997a), and phospholipase D-treated N18TG2 neuroblastoma cells (Di Marzo et al., 1996a). Hansen and coworkers (1995) demonstrated the generation of *N*-acylethanolamine including anandamide in stimulated mouse cortical neurons in culture. The generation of anandamide has also been detected in Δ^9-THC-stimulated N18TG2 cells (Burstein and Hunter, 1995), in ionomycin-stimulated rat macrophages (Wagner et al., 1997), in LPS-, platelet-activating factor-, and Δ^9-THC-stimulated RAW264.7 mouse macrophages (Pestonjamasp and Burstein, 1998), in *N*-arachidonoylglycine-stimulated RAW264.7 cells (Burstein et al., 2002), in mouse peritoneal macrophages in culture supplemented with ethanolamine (Kuwae et al., 1999), in rat testis following the injection of cadmium chloride (Kondo et al., 1998b), in ratridine-, 4-aminopyridine-, and A23187-stimulated SK-N-SH neuroblastoma cells (Basavarajappa and Hungund, 1999), in the periaqueductal gray region of the rat brain following electrical stimulation and the subcutaneous injection of formalin (Walker et al., 1999), in rat brain injected intracerebrally with *N*-methyl-D-aspartate (NMDA) (Hansen et al., 2001a,b), in rat cortical neurons following simultaneous activation of the NMDA receptor and the acetylcholine receptor (Stella and Piomelli, 2001), in capsaicin-stimulated rat sensory neurons (Ahluwalia et al., 2003), in the medial prefrontal cortex of mice reexposed to a tone 24 h after conditioning (Marsicano et al., 2002), in *Clostridium difficile* toxin A-treated rat ileum (McVey et al., 2003), and in endothelin-1-stimulated mouse astrocytes (Walter et al., 2002). On the other hand, Berdyshev and colleagues (2001) reported that treatment of human platelets or P388D1 macrophages with platelet-activating factor did not affect the cellular levels of anandamide. Beaulieu and colleagues (2000) demonstrated that there was no significant difference between the levels of anandamide in the control rat paw skin and inflamed paw skin. Oka and colleagues (2006) also reported that the level of anandamide in mouse ear did not change markedly following acute inflammation induced by TPA or oxazolone.

Anandamide Synthesis

Two enzyme pathways have been reported with respect to the synthesis of anandamide (Sugiura et al., 2006b).

a) The first pathway is the direct *N*-acylation of ethanolamine (the condensation pathway). The second pathway is the synthesis through the combined actions of a transacylase and a phosphodiesterase (Schmid pathway) (Fig. 1). The enzymatic formation of *N*-acylethanoamines from free fatty acids and ethanolamine was first described by Udenfriend and coworkers (Colodzin et al., 1963), although they did not mention the case of arachidonic acid. The enzymatic formation of anandamide (*N*-arachidonoylethanolamine) from free arachidonic acid and ethanolamine was first reported by Deutsch and Chin (1993). Several investigators also demonstrated that anandamide can be enzymatically formed from free arachidonic acid and ethanolamine (Devane and Axelrod, 1994; Kruszka and Gross, 1994; Ueda et al., 1995; Sugiura et al., 1996b). Nevertheless, the physiological significance of this pathway is questioned for the following reasons: (1) The fatty acid profile of the *N*-acyl moiety of *N*-acylethanolamine is quite different from that of the free fatty acids in the same tissues. (2) Large amounts of substrates, especially ethanolamine, are required to form anandamide through this pathway (Devane and Axelrod, 1994; Kruszka and Gross, 1994; Ueda et al., 1995; Sugiura et al., 1996b). In fact, it has been shown that the formation of anandamide through the condensation pathway is catalyzed by a fatty acid amide hydrolase operating in reverse (Ueda et al., 1995). Thus, the formation of anandamide through this pathway may not be physiologically relevant, although the possibility remains that a significant amount of anandamide can be formed via this pathway if high concentrations of arachidonic acid and ethanolamine are colocalized at some sites within the cell.
b) The second pathway for the biosynthesis of anandamide is the formation from *N*-arachidonoyl phosphatidylethanolamine (PE) through the action of a phosphodiesterase (Fig. 1). This enzyme reaction has been assumed to be the

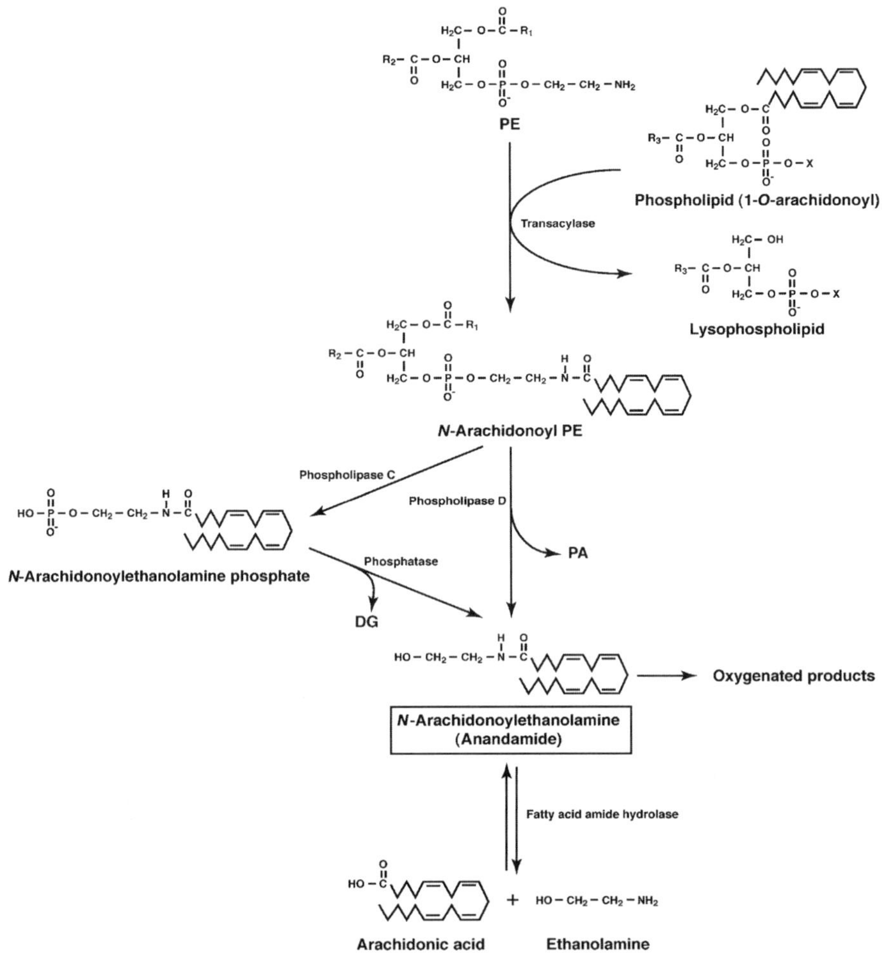

Fig. 1 Metabolic pathways for anandamide

major synthetic route for various *N*-acylethanolamines such as *N*-palmitoyl- and *N*-stearoyl-ethanolamine in mammalian tissues (Schmid et al., 1990). Schmid and coworkers (Epps et al., 1979, 1980) demonstrated the accumulation of various species of *N*-acyl PE (NAPE) in addition to *N*-acylethanolamines in several degenerating tissues. They found that a phosphodiesterase (phospholipase D-type) catalyzes the formation of *N*-acylethanolamine from the corresponding NAPE (Schmid et al., 1983). The addition of Triton X-100 stimulated the enzyme activity, whereas sodium dodecyl sulfate and alkyltrimethylammonium bromide were inhibitory. Ca^{2+} was inhibitory above 1 mM, while up to 0.5 mM it slightly stimulated the enzyme activity. In the absence of detergents, *N*-acyl lysoPE and glycerophospho(*N*-acyl)ethanolamine acted as better substrates than NAPE.

However, they did not examine whether the arachidonic acid-containing species, i.e., anandamide, can be formed via this pathway. Soon after the discovery of anandamide, Di Marzo and coworkers (1994) provided evidence that rat brain neurons contain *N*-arachidonoyl PE and that it can be hydrolyzed by a phosphodiesterase to yield anandamide. This was further confirmed using N18TG2 cells and J774 cells (Di Marzo et al., 1994). Sugiura and colleagues (1996a,b) also provided evidence that rat brain and testis contain substantial amounts of *N*-arachidonoyl PE and a phosphodiesterase activity to form anandamide from *N*-arachidonoyl PE. Importantly, the fatty acid composition of the *N*-acyl moiety of NAPE resembles that of *N*-acylethanolamine present in the same tissue, suggesting that a large portion of *N*-acylethanolamine present in the tissues is derived from the corresponding NAPE through the action of a phosphodiesterase (NAPE-specific phospholipase D). Enzymatic hydrolysis of NAPE was also studied by several investigators. Petersen and Hansen (1999) demonstrated that the NAPE-specific phospholipase D lacks the ability to transphosphatidylate. Moesgaard and colleagues (2000) demonstrated that the NAPE-specific phospholipase D activity substantially increased during the early development of the rat brain. A breakthrough with regard to this enzyme was recently achieved by Ueda and coworkers. Okamoto and colleagues (2004) purified NAPE-specific phospholipase D from rat heart and cloned the gene encoding the protein. This enzyme is composed of 393–396 amino acids and has no homology with the known phospholipase D enzymes but is classified as a member of the zinc metallohydrolase family with the β-lactamase fold. The recombinant enzyme generated anandamide and other *N*-acylethanolamines from their corresponding NAPE at comparable rates. Interestingly, this enzyme did not hydrolyze phosphatidylcholine (PC) and PE. The enzyme activity was stimulated by Ca^{2+} and Mg^{2+} and was inhibited by *p*-chloromercuribenzoic acid and cetyltrimethylammonium chloride. Functional analysis of single mutants of NAPE-phospholipase D revealed that the mutation of Asp-147, His-185, His-187, Asp-189, His-190, His-253, Asp-284, and Cys-224 abolished or caused a remarkable reduction in the catalytic activity (Wang et al., 2006). NAPE-specific phospholipase D is widely distributed in murine organs with higher levels in the brain, kidney, and testis (Okamoto et al., 2004; see Chap. 10 for further information). Interestingly, the strongest activity was detected in the thalamus in the brain (Morishita et al., 2005). The expression of NAPE-specific phospholipase D is well correlated with the enzyme activity and the levels of anandamide in tissues (Guo et al., 2005), and the overexpression of NAPE-specific phospholipase D caused a decrease in the total amount of NAPEs by 50–90% with a 1.5-fold increase in the total amount of *N*-acylethanolamines (Okamoto et al., 2005). These results clearly indicated that NAPE-phospholipase D actually utilizes endogenous NAPE as a substrate to release *N*-acylethanolamines in living cells. In any case, the discovery of NAPE-specific phospholipase D strongly suggests that this enzyme is a physiologically and/or pathophysiologically important one and that saturated or monoenoic species of *N*-acylethanolamines, in addition to anandamide, may play some yet unknown essential roles in mammalian tissues and cells.

c) Very recently, Liu and colleagues (2006) reported another pathway for the synthesis of anandamide. In this pathway, *N*-arachidonoyl PE was first hydrolyzed by phospholipase C to generate *N*-arachidonoylethanolamine phosphate which was then hydrolyzed by a phosphatase to yield anandamide. They demonstrated that this pathway is involved in bacterial endotoxin-induced synthesis of anandamide in macrophages. The relative importance of this pathway and the NAPE-specific phospholipase D pathway in various mammalian tissues and cells remains to be determined. The enzyme activity involved in the synthesis of NAPE was first investigated by Schmid and coworkers in the early 1980s. Natarajan and colleagues (1982, 1983) demonstrated using the dog heart, dog brain, and rat brain that the fatty acids esterified at the *sn*-1 position of the glycerophospholipids are transferred to the amino group of PE through the action of a transacylase to form NAPE. Ca^{2+} is required for this transacylase activity, suggesting that the entry of Ca^{2+} into the cells may trigger the formation of NAPE. Interestingly, the transacylase activity in the rat brain was very high at birth but declined shortly thereafter (Natarajan et al., 1986; Moesgaard et al., 2000), and there are marked species differences in the enzyme activity in heart tissues (Moesgaard et al., 2002). In any case, the transacylation reaction has been assumed to be responsible for the formation of various species of NAPE which accumulate in several degenerating tissues. However, until the mid-1990s, it remained to be determined whether this enzyme reaction was involved in the formation of the arachidonic acid-containing species of NAPE, i.e., *N*-arachidonoyl PE. A decade ago, Sugiura and colleagues (1996a,b) provided evidence that microsomal fractions obtained from the rat brain and testis contain a Ca^{2+}-dependent transacylation activity which catalyzes the formation of *N*-arachidonoyl PE from PE and arachidonic acid esterified at the *sn*-1 position of phospholipids (Fig. 1). Various types of fatty acid esterified at the *sn*-1 position were transferred to PE to form *N*-acyl PE via this pathway. On the contrary, fatty acids esterified at the *sn*-2 position were not transferred. Di Marzo and coworkers (1996b) confirmed that the N18TG2 cell homogenate contains an enzyme activity catalyzing the formation of *N*-arachidonoyl PE from PE and arachidonic acid esterified at the *sn*-1 position of phospholipids. Cadas and colleagues (1997) also detected this enzyme activity in the rat brain particulate fraction. They demonstrated the enhanced formation of *N*-arachidonoyl PE in ionomycin-stimulated neurons and potentiation of the Ca^{2+}-dependent *N*-acyl PE synthesis by agents which augment the level of cyclic AMP (Cadas et al., 1996). The Ca^{2+}-dependent, membrane-associated transacylase responsible for the above reaction has not yet been cloned. Recently, Ueda and coworkers (Jin et al., 2007) demonstrated that lecithin-retinol acyltransferase-like protein (RLP)-1, catalyzed the transfer of a radioactive acyl group from PC to PE, resulting in the formation of radioactive NAPE. In contrast to the Ca^{2+}-dependent transacylase, the RLP-1 activity was detected mainly in the cytosolic rather than the membrane fraction and was little stimulated by Ca^{2+}. Moreover, RLP-1 did not show selectivity with respect to the *sn*-1 and *sn*-2 positions of PC as an acyl donor. These results suggest that RLP-1 may function in the *N*-acylation of PE, catalytically distinguishable from the known Ca^{2+}-dependent transacylase. Further studies are needed to clarify whether this enzyme is involved in the synthesis of NAPE in living tissues. Concerning the Ca^{2+}-dependent transacylase mentioned before, arachidonic acid esterified at the *sn*-1 position of the glycerophospholipids is utilized for the formation of NAPE. However, it is well known that arachidonic acid is usually esterified at the *sn*-2 position of glycerophospholipids in mammalian tissues. For example, 0.3–0.5% of the fatty acyl moiety of the *sn*-1 position of glycerophospholipids in the rat brain is accounted for by arachidonic acid (Sugiura et al., 1996b; Cadas et al., 1997). Thus, it is not possible to generate a large amount of anandamide via this pathway. This is consistent with the observation that the tissue levels of anandamide are generally low except in a few cases (Sugiura et al., 2006b).

Biosynthesis of 2-Arachidonoylglycerol

2-Arachidonoylglycerol Generation

In the early 1980s, Prescott and Majerus (1983) demonstrated the generation of arachidonoylglycerol in thrombin-stimulated platelets. Several investigators also demonstrated the generation of arachidonoylglycerol in stimulated cells such as platelet-derived growth factor-stimulated Swiss 3T3 cells (Hasegawa-Sasaki, 1985) and bradykinin-stimulated rat dorsal ganglion neurons (Gammon et al., 1989), yet these authors did not mention the possible role of this molecule as a cannabinoid receptor ligand. Stimulus-induced generation of 2-AG as an endogenous ligand for the cannabinoid receptors was first described by Bisogno and colleagues (1997a,b) for ionomycin-stimulated N18TG2 cells and by Stella and colleagues (1997) for electrically stimulated rat hippocampal slices and ionomycin-stimulated neurons. Sugiura and coworkers also reported that 2-AG is rapidly produced in rat brain homogenate during incubation in the presence of Ca^{2+} (Kondo et al., 1998a), in thrombin- or A23187-stimulated human umbilical vein endothelial cells (Sugiura et al., 1998), in the picrotoxinin-stimulated rat brain (Sugiura et al., 2000), and in the rat brain after decapitation (Sugiura et al., 2001). The generation of 2-AG was also observed in the carbachol-treated rat aorta (Mechoulam et al., 1998a), in methacholine-stimulated bovine coronary endothelial cells (Gauthier et al., 2005), in ethanol-treated cerebellar granule neurons in culture (Basavarajappa et al., 2000), in the mouse brain following traumatic brain injury (Panikashvili et al., 2001), in NMDA-stimulated rat cortical neurons (Stella and Piomelli, 2001), in the mouse cerebral cortex after sham surgery (Franklin et al., 2003), in the medial prefontal cortex of mice reexposed to a tone which elicits aversive memories (Marsicano et al., 2002), in the rat midbrain periaqueductal gray matter after electric foot shock (Hohmann et al., 2005), in the rat hypothalamic slices after high frequency stimulation (Di et al., 2005a), in the *Clostridium difficile* toxin A-treated rat ileum (McVey et al., 2003), in the cholera toxin-treated mouse small intestine (Izzo et al., 2003), in ATP- or ionomycin-stimulated mouse microglia cells (Walter et al., 2003; Witting et al., 2004), in a macrophage colony stimulating factor-stimulated rat microglia cell line (Carrier et al., 2004), in endothelin 1-stimulated mouse astrocytes (Walter and Stella, 2003), in ATP-stimulated mouse astrocytes (Walter et al., 2004), in pilocarpine-induced temporal lobe epileptiform seizures in the brain (Wallace et al., 2003), in U46619 (a thromboxane A_2-mimetic)-stimulated rat middle artery (Rademacher et al., 2005), in glucocorticoid-treated rat hypothalamic slices (Di et al., 2005b), in 12-*O*-tetradecanoylphorbol 13-acetate (TPA)-induced acute inflammation in mouse ear (Oka et al., 2005), and in oxazolone-induced contact dermatitis in mouse ear (Oka et al., 2006). In contrast, Beaulieu and coworkers (2000) reported that the levels of 2-AG in rat paw skin did not differ markedly between control and formalin-induced inflammation groups. Several types of blood cells or inflammatory cells also generate 2-AG when stimulated; 2-AG has been shown to be produced in lipopolysaccharide (LPS)-stimulated rat platelets (Varga et al., 1998), in LPS-stimulated rat macrophages and LPS- or ionomycin-stimulated J774 macrophage-like cells (Di Marzo et al., 1999), in platelet-activating factor (PAF)-stimulated human platelets (Berdyshev et al., 2001), in PAF-stimulated P388D1 macrophages (Berdyshev et al., 2001), and in PAF-stimulated RAW264.7 cells (Liu et al., 2003).

2-Arachidonoylglycerol Synthesis

As for the pathways involved in the synthesis of 2-AG, Sugiura and colleagues (1995) pointed out that 2-AG can be formed from arachidonic acid-containing membrane phospholipids such as inositol phospholipids through the combined actions of phospholipase C and diacylglycerol lipase or through the combined actions of phospholipase A_1 and phospholipase C (Fig. 2). 2-AG can also be formed via other pathways such as the hydrolysis of arachidonic acid-containing lysophosphatidic acid by the action of a phosphatase (Nakane et al., 2002).

(a) The first pathway, involving the rapid hydrolysis of inositol phospholipids by phospholipase C and subsequent hydrolysis of the resultant diacylglycerol by a diacylglycerol lipase (DG lipase), was described by Prescott and Majerus (1983) as a degradation pathway for arachidonic acid-containing diacylglycerols in platelets. Stella and colleagues (1997) demonstrated that these enzyme activities (phospholipase C and DG lipase) participate in the ionomycin-induced generation of 2-AG in cultured neurons using metabolic inhibitors. Kondo and colleagues (1998a) confirmed that this pathway is important for the Ca^{2+}-induced generation of 2-AG in rat brain homogenate. Phosphatidylinositol (PI) is the most preferred substrate in the generation of 2-AG in brain homogenate (Sugiura et al., 2006a). Interestingly, the addition of GTP S markedly enhanced the generation of 2-AG in brain homogenate in the presence of a low concentration of Ca^{2+} (Sugiura et al., 2006a), suggesting that phospholipase C, regulated by G proteins, is involved in

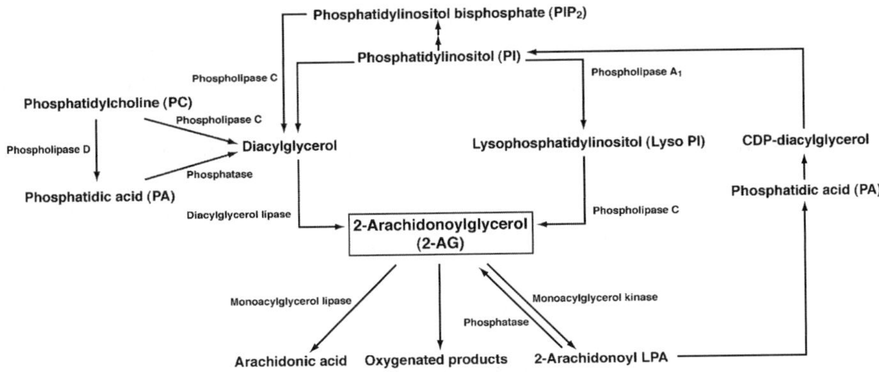

Fig. 2 Metabolic pathways for 2-AG

the generation of 2-AG in the brain. The involvement of phospholipase C in the formation of 2-AG in stimulated brain tissues has also been suggested by a number of investigators employing phospholipase C inhibitors and gene-knockout mice (Melis et al., 2004; Maejima et al., 2005; Jung et al., 2005; Straiker and Mackie, 2005; Hashimotodani et al., 2005; Safo and Regehr, 2005; Edwards et al., 2006). On the other hand, limited information had been available on DG lipase until recently. A breakthrough was recently accomplished by Bisogno and colleagues (2003). They cloned the genes encoding DG lipases. They identified two closely related genes (α and β): both enzymes are mostly expressed in the 10,000×*g* membrane fraction and exhibit optimal activity at pH 7. The activities of these enzymes were stimulated by Ca^{2+} and blocked by *p*-hydroxymercuribenzoate, $HgCl_2$, and RHC-80267, a DG lipase inhibitor. Tetrahydrolipstatin, another DG lipase inhibitor, also suppressed the activities of both enzymes and decreased the ionomycin-induced generation of 2-AG. These results strongly suggest that the enzymes mentioned above actually contribute to the biosynthesis and release of 2-AG. The possible involvement of DG lipase in the formation of 2-AG has also been demonstrated by other investigators employing DG lipase inhibitors (Melis et al., 2004; Jung et al., 2005; Edwards et al., 2006).

(b) The enzyme activities involved in the second pathway, i.e., the hydrolysis of phosphatidylinositol (PI) by phospholipase A_1 and hydrolysis of the resultant lysoPI by a specific phospholipase C, were studied by Okuyama and coworkers in the mid-1990s. Interestingly, lysoPI-specific phospholipase C is distinct from various other types of phospholipase C which act on other inositol phospholipids and is located in the synaptosomes (Tsutsumi et al., 1994). It seems possible, therefore, that this unique enzyme is involved in the metabolism of lysoPI and the generation of lysoPI-derived lipid molecules such as 2-AG in the synapses.

(c) 2-AG can also be formed through the conversion of 2-arachidonoyl lysophosphatidic acid (LPA) to 2-AG (Fig. 2). Nakane and colleagues (2002) detected a substantial amount of arachidonic acid-containing LPA in the rat brain (0.84 nmol/g tissue). About 63% of the arachidonic acid was esterified at the *sn*-2 position. They also detected a phosphatase activity which hydrolyzes 2-arachidonoyl LPA to yield 2-AG in a rat brain homogenate (Nakane et al., 2002). Thus, it is plausible that 2-arachidonoyl LPA acts as a substrate for the synthesis of 2-AG under certain conditions in the brain.

(d) Several types of diacyl glycerophospholipids other than inositol phospholipids have also been shown to serve as precursor molecules in the synthesis of 2-AG. Bisogno and colleagues (1999) demonstrated that 2-AG is formed from phosphatidic acid (PA) in ionomycin-stimulated N18TG2 neuroblastoma cells by employing several metabolic inhibitors. In this case, 2-arachidonoyl PA was converted first to 1-acyl-2-arachidonoylglycerol and then to 2-AG. They described that the breakdown of inositol phospholipids is not involved in the generation of 2-AG. Carrier and coworkers (2004) also demonstrated that 2-AG was formed from PA in a macrophage colony-stimulating factor-stimulated mouse microglia cell line. Some of 2-AG may also be derived from other arachidonic acid-containing phospholipids such as 1-acyl-2-arachidonoyl-*sn*-glycero-3-phosphocholine. Recently, Oka and colleagues (2005) demonstrated that various species of 2-monoacylglycerol were generated in the TPA-treated mouse ear. The rank order of the amounts of 2-monoacylglycerol generated was 2-palmitoylglycerol plus 2-oleoylglycerol plus 2-*cis*-vaccenoylglycerol > 2-linoleoylglycerol > 2-AG. This is quite different from the case for stimulated brain tissues (Sugiura et al., 2000). It seems unlikely that inositol phospholipids act as the sole source of 2-AG in inflamed tissues, because a large proportion of the fatty acyl moiety at the *sn*-2 position of inositol phospholipids in mammalian tissues is usually accounted for by arachidonic acid. Furthermore, U73122, a PI-specific phospholipase C inhibitor, affected the level of 2-AG only modestly, whereas D609, a PC-specific phospholipase C inhibitor, and butanol, an inhibitor of PA generation, exerted more pronounced effects on the level of 2-AG in inflamed ear. It is apparent, therefore, that the biosynthetic pathways for 2-AG differ, depending on the types of tissues and cells and the types of stimuli. Further detailed studies are necessary for a full understanding of the mechanism underlying the biosynthesis of 2-AG in mammalian tissues.

Concluding Remarks

It has been established that anandamide can be formed through several metabolic pathways. Yet, no selective and efficient synthetic pathway for anandamide has hitherto been found. Considering that anandamide acts as a partial agonist toward the cannabinoid receptors (CB_1 and CB_2), it is rather questionable that anandamide acts as an endogenous CB_1 and CB_2 receptor agonist with profound physiological significance. Despite this, however, the discovery of NAPE-specific phospholipase D strongly suggests that this enzyme is a physiologically and/or pathophysiologically important one and that saturated or monoenoic species of *N*-acylethanolamines, in addition to anandamide, may play some yet unknown essential roles in mammalian tissues and cells. A thorough elucidation of the physiological and pathophysiological roles of various *N*-acylethanolamines including anandamide awaits further investigation. In contrast to anandamide, 2-AG acts as a full agonist toward the cannabinoid receptors (CB_1 and CB_2). This strongly suggests that 2-AG rather than anandamide is the true natural ligand for the cannabinoid receptors. 2-AG can be formed via several metabolic pathways in various mammalian tissues; the most important pathway appears to be the formation from arachidonic acid-containing diacylglycerol, derived from increased phospholipid metabolism, by the action of a diacylglycerol lipase. Whatever the precise mechanism of synthesis, it is apparent that 2-AG is a key molecule which links increased phospholipid metabolism upon stimulation with the functions of the cannabinoid receptors (CB_1 and CB_2). Further detailed studies on the metabolism of 2-AG are thus essential to gain insight into the physiological and pathophysiological significance of the endocannabinoid system.

References

Ahluwalia J, Yaqoob M, Urban L, Bevan S, Nagy I (2003) Activation of capsaicin-sensitive primary sensory neurons induces anandamide production and release. J Neurochem 84:585–591.

Basavarajappa BS, Hungund BL (1999) Chronic ethanol increases the cannabinoid receptor agonist anandamide and its precursor *N*-arachidonoylphosphatidylethanolamine in SK-N-SH cells. J Neurochem 72:522–528.

Basavarajappa BS, Saito M, Cooper TB, Hungund BL (2000) Stimulation of cannabinoid receptor agonist 2-arachidonylglycerol by chronic ethanol and its modulation by specific neuromodulators in cerebellar granule neurons. Biochim Biophys Acta 1535:78–86.

Beaulieu P, Bisogno T, Punwar S, Farquhar-Smith WP, Ambrosino G, Di Marzo V, Rice AS (2000) Role of the endogenous cannabinoid system in the formalin test of persistent pain in the rat. Eur J Pharmacol 396:85–92.

Berdyshev EV, Schmid PC, Krebsbach RJ, Schmid HHO (2001) Activation of PAF receptors results in enhanced synthesis of 2-arachidonoylglycerol (2-AG) in immune cells. FASEB J 15:2171–2178.

Bisogno T, Maurelli S, Melck D, De Petrocellis L, Di Marzo V (1997a) Biosynthesis, uptake, and degradation of anandamide and palmitoylethanolamide in leukocytes. J Biol Chem 272: 3315–3323.

Bisogno T, Sepe N, Melck D, Maurelli S, De Petrocellis L, Di Marzo V (1997b) Biosynthesis, release and degradation of the novel endogenous cannabimimetic metabolite 2-arachidonoylglycerol in mouse neuroblastoma cells. Biochem J 322:671–677.

Bisogno T, Melck D, De Petrocellis L, Di Marzo V (1999) Phosphatidic acid as the biosynthetic precursor of the endocannabinoid 2-arachidonoylglycerol in intact mouse neuroblastoma cells stimulated with ionomycin. J Neurochem 72:2113–2119.

Bisogno T, Howell F, Williams G, Minassi A, Cascio MG, Ligresti A, Matias I, Schiano-Moriello A, Paul P, Williams EJ, Gangadharan U, Hobbs C, Di Marzo V, Doherty P (2003) Cloning of the first sn1-DAG lipases points to the spatial and temporal regulation of endocannabinoid signaling in the brain. J Cell Biol 163:463–468.

Burstein SH, Hunter SA (1995) Stimulation of anandamide biosynthesis in *N*-18TG2 neuroblastoma cells by delta 9-tetrahydrocannabinol (THC). Biochem Pharmacol 49:855–858.

Burstein SH, Huang SM, Petros TJ, Rossetti RG, Walker JM, Zurier RB (2002) Regulation of anandamide tissue levels by *N*-arachidonylglycine. Biochem Pharmacol 64:1147–1150.

Cadas H, Gaillet S, Beltramo M, Venance L, Piomelli D (1996) Biosynthesis of an endogenous cannabinoid precursor in neurons and its control by calcium and cAMP. J Neurosci 16:3934–3942.

Cadas H, Di Tomaso E, Piomelli D (1997) Occurrence and biosynthesis of endogenous cannabinoid precursor, *N*-arachidonoyl phosphatidylethanolamine, in rat brain. J Neurosci 17:1226–1242.

Carrier EJ, Kearn CS, Barkmeier AJ, Breese NM, Yang W, Nithipatikom K, Pfister SL, Campbell WB, Hillard CJ (2004) Cultured rat microglial cells synthesize the endocannabinoid 2-arachidonylglycerol, which increases proliferation via a CB_2 receptor-dependent mechanism. Mol Pharmacol 65:999–1007.

Colodzin M, Bachur NR, Weissbach H, Udenfriend S (1963) Enzymatic formation of fatty acid amides of ethanolamine by rat liver microsomes. Biochem Biophys Res Commun 10:165–171.

Deutsch DG, Chin SA (1993) Enzymatic synthesis and degradation of anandamide, a cannabinoid receptor agonist. Biochem Pharmacol 46:791–796.

Devane WA, Axelrod J (1994) Enzymatic synthesis of anandamide, an endogenous ligand for the cannabinoid receptor, by brain membranes. Proc Natl Acad Sci USA 91:6698–6701.

Devane WA, Hanus L, Breuer A, Pertwee RG, Stevenson LA, Griffin G, Gibson D, Mandelbaum A, Etinger A, Mechoulam R (1992) Isolation and structure of a brain constituent that binds to the cannabinoid receptor. Science 258:1946–1949.

Di S, Boudaba C, Popescu IR, Weng FJ, Harris C, Marcheselli VL, Bazan NG, Tasker JG (2005a) Activity-dependent release and actions of endocannabinoids in the rat hypothalamic supraoptic nucleus. J Physiol 569:751–760.

Di S, Malcher-Lopes R, Marcheselli VL, Bazan NG, Tasker JG (2005b) Rapid glucocorticoid-mediated endocannabinoid release and opposing regulation of glutamate and gamma-aminobutyric acid inputs to hypothalamic magnocellular neurons. Endocrinology 146:4292–4301.

Di Marzo V, Fontana A, Cadas H, Schinelli S, Cimino G, Schwartz JC, Piomelli D (1994) Formation and inactivation of endogenous cannabinoid anandamide in central neurons. Nature 372:686–691.

Di Marzo V, De Petrocellis L, Sepe N, Buono A (1996a) Biosynthesis of anandamide and related acylethanolamides in mouse J774 macrophages and N18 neuroblastoma cells. Biochem J 316:977–984.

Di Marzo V, De Petrocellis L, Sugiura T, Waku K (1996b) Potential biosynthetic connections between the two cannabimimetic eicosanoids, anandamide and 2-arachidonoyl-glycerol, in mouse neuroblastoma cells. Biochem Biophys Res Commun 227:281–288.

Di Marzo V, Bisogno T, De Petrocellis L, Melck D, Orlando P, Wagner JA, Kunos G (1999) Biosynthesis and inactivation of the endocannabinoid 2-arachidonoylglycerol in circulating and tumoral macrophages. Eur J Biochem 264:258–267.

Di Marzo V, De Petrocellis L, Bisogno T, Berger A, Mechoulam R (2002) Biology of endocannabinoids. In: Onaivi ES (ed), Biology of Marijuana. Taylor & Francis, London, pp. 125–173.

Edwards DA, Kim J, Alger BE (2006) Multiple mechanisms of endocannabinoid response initiation in hippocampus. J Neurophysiol 95:67–75.

Epps DE, Schmid PC, Natarajan V, Schmid HHO (1979) *N*-Acylethanolamine accumulation in infarcted myocardium. Biochem Biophys Res Commun 90:628–633.

Epps DE, Natarajan V, Schmid PC, Schmid HHO (1980) Accumulation of *N*-acylethanolamine glycerophospholipids in infarcted myocardium. Biochim Biophys Acta 618:420–430.

Franklin A, Parmentier-Batteur S, Walter L, Greenberg DA, Stella N (2003) Palmitoylethanolamide increases after focal cerebral ischemia and potentiates microglial cell motility. J Neurosci 23:7767–7775.

Gammon CM, Allen AC, Morell P (1989) Bradykinin stimulates phosphoinositide hydrolysis and mobilization of arachidonic acid in dorsal root ganglion neurons. J Neurochem 53:95–101.

Gauthier KM, Baewer DV, Hittner S, Hillard CJ, Nithipatikom K, Reddy DS, Falck JR, Campbell WB (2005) Endothelium-derived 2-arachidonylglycerol: an intermediate in vasodilatory eicosanoid release in bovine coronary arteries. Am J Physiol Heart Circ Physiol 288:1344–1351.

Guo Y, Wang H, Okamoto Y, Ueda N, Kingsley PJ, Marnett LJ, Schmid HH, Das SK, Dey SK (2005) *N*-acylphosphatidylethanolamine-hydrolyzing phospholipase D is an important determinant of uterine anandamide levels during implantation. J Biol Chem 280:23429–23432.

Hansen HS, Lauritzen L, Strand AM, Moesgaard B, Frandsen A (1995) Glutamate stimulates the formation of *N*-acylphosphatidylethanolamine and *N*-acylethanolamine in cortical neurons in culture. Biochim Biophys Acta 1258:303–308.

Hansen HH, Ikonomidou C, Bittigau P, Hansen SH, Hansen HS (2001a) Accumulation of the anandamide precursor and other *N*-acylethanolamine phospholipids in infant rat models of *in vivo* necrotic and apoptotic neuronal death. J Neurochem 76:39–46.

Hansen HH, Schmid PC, Bittigau P, Lastres-Becker I, Berrendero F, Manzanares J, Ikonomidou C, Schmid HHO, Fernandez-Ruiz JJ, Hansen HS (2001b) Anandamide, but not 2-arachidonoylglycerol, accumulates during *in vivo* neurodegeneration. J Neurochem 78:1415–1427.

Hasegawa-Sasaki H (1985) Early changes in inositol lipids and their metabolites induced by platelet-derived growth factor in quiescent Swiss mouse 3T3 cells. Biochem J 232:99–109.

Hashimotodani Y, Ohno-Shosaku T, Tsubokawa H, Ogata H, Emoto K, Maejima T, Araishi K, Shin HS, Kano M (2005) Phospholipase Cβ serves as a coincidence detector through its Ca^{2+} dependency for triggering retrograde endocannabinoid signal. Neuron 45:257–268.

Hohmann AG, Suplita RL, Bolton NM, Neely MH, Fegley D, Mangieri R, Krey JF, Walker JM, Holmes PV, Crystal JD, Duranti A, Tontini A, Mor M, Tarzia G, Piomelli D (2005) An endocannabinoid mechanism for stress-induced analgesia. Nature 435:1108–1112.

Izzo AA, Capasso F, Costagliola A, Bisogno T, Marsicano G, Ligresti A, Matias I, Capasso R, Pinto L, Borrelli F, Cecio A, Lutz B, Mascolo N, Di Marzo V (2003) An endogenous cannabinoid tone attenuates cholera toxin-induced fluid accumulation in mice. Gastroenterology 125:765–774.

Jin XH, Okamoto Y, Morishita J, Tsuboi K, Tonai T, Ueda N (2007) Discovery and characterization of a Ca^{2+}-independent phosphatidylethanolamine *N*-acyltransferase generating the anandamide precursor and its congeners. J Biol Chem 282:3614–3623.

Jung KM, Mangieri R, Stapleton C, Kim J, Fegley D, Wallace M, Mackie K, Piomelli D (2005) Stimulation of endocannabinoid formation in brain slice cultures through activation of group I metabotropic glutamate receptors. Mol Pharmacol 68:1196–1202.

Kondo S, Kondo H, Nakane S, Kodaka T, Tokumura A, Waku K, Sugiura T (1998a) 2-Arachidonoylglycerol, an endogenous cannabinoid receptor agonist: identification as one of the major species of monoacylglycerols in various rat tissues, and evidence for its generation through Ca^{2+}-dependent and -independent mechanisms. FEBS Lett 429:152–156.

Kondo S, Sugiura T, Kodaka T, Kudo N, Waku K, Tokumura A (1998b) Accumulation of various *N*-acylethanolamines including *N*-arachidonoylethanolamine (anandamide) in cadmium chloride-administered rat testis. Arch Biochem Biophys 354:303–310.

Kruszka KK, Gross RW (1994) The ATP- and CoA-independent synthesis of arachidonoylethanolamide. A novel mechanism underlying the synthesis of the endogenous ligand of the cannabinoid receptor. J Biol Chem 269:14345–14348.

Kuwae T, Shiota Y, Schmid PC, Krebsbach R, Schmid HHO (1999) Biosynthesis and turnover of anandamide and other *N*-acylethanolamines in peritoneal macrophages. FEBS Lett 459:123–127.

Liu J, Batkai S, Pacher P, Harvey-White J, Wagner JA, Cravatt BF, Gao B, Kunos G (2003) Lipopolysaccharide induces anandamide synthesis in macrophages via CD14/MAPK/phosphoinositide 3-kinase/NF-κB independently of platelet-activating factor. J Biol Chem 278:45034–45039.

Liu J, Wang L, Harvey-White J, Osei-Hyiaman D, Razdan R, Gong Q, Chan AC, Zhou Z, Huang BX, Kim HY, Kunos G (2006) A biosynthetic pathway for anandamide. Proc Natl Acad Sci USA 103:13345–13350.

Maejima T, Oka S, Hashimotodani Y, Ohno-Shosaku T, Aiba A, Wu D, Waku K, Sugiura T, Kano M (2005) Synaptically driven endocannabinoid release requires Ca^{2+}-assisted metabotropic glutamate receptor subtype 1 to phospholipase Cβ4 signaling cascade in the cerebellum. J Neurosci 25:6826–6835.

Marsicano G, Wotjak CT, Azad SC, Bisogno T, Rammes G, Cascio MG, Hermann H, Tang J, Hofmann C, Zieglgansberger W, Di Marzo V, Lutz B (2002) The endogenous cannabinoid system controls extinction of aversive memories. Nature 418:530–534.

McVey DC, Schmid PC, Schmid HHO, Vigna SR (2003) Endocannabinoids induce ileitis in rats via the capsaicin receptor (VR_1). J Pharmacol Exp Ther 304:713–722.

Mechoulam R, Ben-Shabat S, Hanus L, Ligumsky M, Kaminski NE, Schatz AR, Gopher A, Almog S, Martin BR, Compton DR, Pertwee RG, Griffin G, Bayewitch M, Barg J, Vogel Z (1995) Identification of an endogenous 2-monoglyceride, present in canine gut, that binds to cannabinoid receptors. Biochem Pharmacol 50:83–90.

Mechoulam R, Fride E, Ben-Shabat S, Meiri U, Horowitz M (1998a) Carbachol, an acetylcholine receptor agonist, enhances production in rat aorta of 2-arachidonoyl glycerol, a hypotensive endocannabinoid. Eur J Pharmacol 362:1–3.

Mechoulam R, Fride E, Di Marzo V (1998b) Endocannabinoids. Eur J Pharmacol 359:1–18.

Melis M, Perra S, Muntoni AL, Pillolla G, Lutz B, Marsicano G, Di Marzo V, Gessa GL, Pistis M (2004) Prefrontal cortex stimulation induces 2-arachidonoyl-glycerol-mediated suppression of excitation in dopamine neurons. J Neurosci 24:10707–10715.

Moesgaard B, Petersen G, Jaroszewski JW, Hansen HS (2000) Age dependent accumulation of *N*-acyl-ethanolamine phospholipids in ischemic rat brain. A ^{31}P NMR and enzyme activity study. J Lipid Res 41:985–990.

Moesgaard B, Petersen G, Mortensen SA, Hansen HS (2002) Substantial species differences in relation to formation and degradation of *N*-acyl-ethanolamine phospholipids in heart tissue: an enzyme activity study. Comp Biochem Physiol B, Biochem Mol Biol 131:475–482.

Morishita J, Okamoto Y, Tsuboi K, Ueno M, Sakamoto H, Maekawa N, Ueda N (2005) Regional distribution and age-dependent expression of *N*-acylphosphatidylethanolamine-hydrolyzing phospholipase D in rat brain. J Neurochem 94:753–762.

Nakane S, Oka S, Arai S, Waku K, Ishima Y, Tokumura A, Sugiura T (2002) 2-Arachidonoyl-*sn*-glycero-3-phosphate, an arachidonic acid-containing lysophosphatidic acid: occurrence and rapid enzymatic conversion to 2-arachidonoyl-*sn*-glycerol, a cannabinoid receptor ligand, in rat brain. Arch Biochem Biophys 402:51–58.

Natarajan V, Reddy PV, Schmid PC, Schmid HHO (1982) *N*-Acylation of ethanolamine phospholipids in canine myocardium. Biochim Biophys Acta 712:342–355.

Natarajan V, Schmid PC, Reddy PV, Zuzarte-Augustin ML, Schmid HHO (1983) Biosynthesis of *N*-acylethanolamine phospholipids by dog brain preparations. J Neurochem 41:1303–1312.

Natarajan V, Schmid PC, Schmid HHO (1986) *N*-acylethanolamine phospholipid metabolism in normal and ischemic rat brain. Biochim Biophys Acta 878:32–41.

Oka S, Yanagimoto S, Ikeda S, Gokoh M, Kishimoto S, Waku K, Ishima Y, Sugiura T (2005) Evidence for the involvement of the cannabinoid CB_2 receptor and its endogenous ligand 2-arachidonoylglycerol in 12-*O*-tetradecanoylphorbol-13-acetate-induced acute inflammation in mouse ear. J Biol Chem 280:18488–18497.

Oka S, Wakui J, Ikeda S, Yanagimoto S, Kishimoto S, Gokoh M, Nasui M, Sugiura T (2006) Involvement of the cannabinoid CB_2 receptor and its endogenous ligand 2-arachidonoylglycerol in oxazolone-induced contact dermatitis in mice. J Immunol 177:8796–8805.

Okamoto Y, Morishita J, Tsuboi K, Tonai T, Ueda N (2004) Molecular characterization of a phospholipase D generating anandamide and its congeners. J Biol Chem 279:5298–5305.

Okamoto Y, Morishita J, Wang J, Schmid PC, Krebsbach RJ, Schmid HH, Ueda N (2005) Mammalian cells stably overexpressing *N*-acylphosphatidylethanolamine-hydrolysing phospholipase D exhibit significantly decreased levels of *N*-acylphosphatidylethanolamines. Biochem J 389:241–247.

Panikashvili D, Simeonidou C, Ben-Shabat S, Hanus L, Breuer A, Mechoulam R, Shohami E (2001) An endogenous cannabinoid (2-AG) is neuroprotective after brain injury. Nature 413:527–531.

Pestonjamasp VK, Burstein SH (1998) Anandamide synthesis is induced by arachidonate mobilizing agonists in cells of the immune system. Biochim Biophys Acta 1394:249–260.

Petersen G, Hansen HS (1999) *N*-acylphosphatidylethanolamine-hydrolysing phospholipase D lacks the ability to transphosphatidylate. FEBS Lett 455:41–44.

Piomelli D, Beltramo M, Giuffrida A, Stella N (1998) Endogenous cannabinoid signaling. Neurobiol Dis 5:462–473.

Prescott SM, Majerus PW (1983) Characterization of 1,2-diacylglycerol hydrolysis in human platelets. Demonstration of an arachidonoyl-monoacylglycerol intermediate. J Biol Chem 258:764–769.

Rademacher DJ, Patel S, Ho WS, Savoie AM, Rusch NJ, Gauthier KM, Hillard CJ (2005) U-46619 but not serotonin increases endocannabinoid content in middle cerebral artery: evidence for functional relevance. Am J Physiol Heart Circ Physiol 288:H2694–2701.

Safo PK, Regehr WG (2005) Endocannabinoids control the induction of cerebellar LTD. Neuron 48:647–659.

Schmid PC, Reddy PV, Natarajan V, Schmid HHO (1983) Metabolism of *N*-acylethanolamine phospholipids by a mammalian phosphodiesterase of the phospholipase D type. J Biol Chem 258:9302–9306.

Schmid HHO, Schmid PC, Natarajan V (1990) *N*-acylated glycerophospholipids and their derivatives. Prog Lipid Res 29:1–43.

Stella N, Piomelli D (2001) Receptor-dependent formation of endogenous cannabinoids in cortical neurons. Eur J Pharmacol 425:189–196.

Stella N, Schweitzer P, Piomelli D (1997) A second endogenous cannabinoid that modulates long-term potentiation. Nature 388:773–778.

Straiker A, Mackie K (2005) Depolarization-induced suppression of excitation in murine autaptic hippocampal neurones. J Physiol 569:501–517.

Sugiura T, Waku K (2000) 2-Arachidonoylglycerol and the cannabinoid receptors. Chem Phys Lipids 108:89–106.

Sugiura T, Kondo S, Sukagawa A, Nakane S, Shinoda A, Itoh K, Yamashita A, Waku K (1995) 2-Arachidonoylglycerol: a possible endogenous cannabinoid receptor ligand in brain. Biochem Biophys Res Commun 215:89–97.

Sugiura T, Kondo S, Sukagawa A, Tonegawa T, Nakane S, Yamashita A, Waku K (1996a) Enzymatic synthesis of anandamide, an endogenous cannabinoid receptor ligand, through *N*-acylphosphatidylethanolamine pathway in testis: involvement of Ca^{2+}-dependent transacylase and phosphodiesterase activities. Biochem Biophys Res Commun 218:113–117.

Sugiura T, Kondo S, Sukagawa A, Tonegawa T, Nakane S, Yamashita A, Ishima Y, Waku K (1996b) Transacylase-mediated and phosphodiesterase-mediated synthesis of *N*-arachidonoylethanolamine, an endogenous cannabinoid-receptor ligand, in rat brain microsomes. Comparison with synthesis from free arachidonic acid and ethanolamine. Eur J Biochem 240:53–62.

Sugiura T, Kodaka T, Nakane S, Kishimoto S, Kondo S, Waku K (1998) Detection of an endogenous cannabimimetic molecule, 2-arachidonoylglycerol, and cannabinoid CB_1 receptor mRNA in human vascular cells: is 2-arachidonoylglycerol a possible vasomodulator? Biochem Biophys Res Commun 243:838–843.

Sugiura T, Yoshinaga N, Kondo S, Waku K, Ishima Y (2000) Generation of 2-arachidonoylglycerol, an endogenous cannabinoid receptor ligand, in picrotoxinin-administered rat brain. Biochem Biophys Res Commun 271:654–658.

Sugiura T, Yoshinaga N, Waku K (2001) Rapid generation of 2-arachidonoylglycerol, an endogenous cannabinoid receptor ligand, in rat brain after decapitation. Neurosci Lett 297:175–178.

Sugiura T, Kishimoto S, Oka S, Gokoh M (2006a) Biochemistry, pharmacology and physiology of 2-arachidonoylglycerol, an endogenous cannabinoid receptor ligand. Prog Lipid Res 45:405–446.

Sugiura T, Oka S, Ikeda S, Waku K (2006b) Occurrence, biosynthesis, and metabolism of endocannabinoid. In: Onaivi ES, Sugiura T, Di Marzo V (eds), Endocannabinoids: The Brain and Body's Marijuana and Beyond. Taylor & Francis, Boca Raton, pp. 177–214.

Tsutsumi T, Kobayashi T, Ueda H, Yamauchi E, Watanabe S, Okuyama H (1994) Lysophosphoinositide-specific phospholipase C in rat brain synaptic plasma membranes. Neurochem Res 19:399–406.

Ueda N, Kurahashi Y, Yamamoto S, Tokunaga T (1995) Partial purification and characterization of the porcine brain enzyme hydrolyzing and synthesizing anandamide. J Biol Chem 270:23823–23827.

Varga K, Wagner JA, Bridgen DT, Kunos G (1998) Platelet- and macrophage-derived endogenous cannabinoids are involved in endotoxin-induced hypotension. FASEB J 12:1035–1044.

Wagner JA, Varga K, Ellis EF, Rzigalinski BA, Martin BR, Kunos G (1997) Activation of peripheral CB_1 cannabinoid receptors in haemorrhagic shock. Nature 390:518–521.

Walker JM, Huang SM, Strangman NM, Tsou K, Sanudo-Pena MC (1999) Pain modulation by release of the endogenous cannabinoid anandamide. Proc Natl Acad Sci USA 96:12198–12203.

Wallace MJ, Blair RE, Falenski KW, Martin BR, DeLorenzo RJ (2003) The endogenous cannabinoid system regulates seizure frequency and duration in a model of temporal lobe epilepsy. J Pharmacol Exp Ther 307:129–137.

Walter L, Stella N (2003) Endothelin-1 increases 2-arachidonoyl glycerol (2-AG) production in astrocytes. Glia 44:85–90.

Walter L, Franklin A, Witting A, Moller T, Stella N (2002) Astrocytes in culture produce anandamide and other acylethanolamides. J Biol Chem 277:20869–20876.

Walter L, Franklin A, Witting A, Wade C, Xie Y, Kunos G, Mackie K, Stella N (2003) Nonpsychotropic cannabinoid receptors regulate microglial cell migration. J Neurosci 23:1398–1405.

Walter L, Dinh T, Stella N (2004) ATP induces a rapid and pronounced increase in 2-arachidonoylglycerol production by astrocytes, a response limited by monoacylglycerol lipase. J Neurosci 24:8068–8074.

Wang J, Okamoto Y, Morishita J, Tsuboi K, Miyatake A, Ueda N (2006) Functional analysis of the purified anandamide-generating phospholipase D as a member of the metallo-β-lactamase family. J Biol Chem 281:12325–12335.

Witting A, Walter L, Wacker J, Moller T, Stella N (2004) $P2X_7$ receptors control 2-arachidonoylglycerol production by microglial cells. Proc Natl Acad Sci USA 101:3214–3219.

Chapter 3
Removal of Endocannabinoids by the Body: Mechanisms and Therapeutic Possibilities

Christopher J. Fowler and Lina Thors

Abstract The actions of anandamide and 2-arachidonoylglycerol are terminated by cellular uptake followed by metabolism. In the case of anandamide, the uptake process was originally suggested to be achieved by a process of facilitated diffusion, but the mechanism(s) involved are a matter of controversy at present. The main hydrolytic enzyme for anandamide is fatty acid amide hydrolase, and inhibitors of this enzyme have been found to have beneficial effects in animal models of inflammatory pain, inflammation, anxiety and depression. Anandamide is also a substrate for cyclooxygenase-2 and lipoxygenases, and the cyclooxygenase-derived products, the "prostamides" have biological actions of their own. 2-Arachidonoylglycerol can be metabolized by a range of enzymes, including cyclooxygenase-2, monoacylglycerol lipase and fatty acid amide hydrolase. In the brain, monoacylglycerol lipase is probably the most important of these enzymes. However, no selective inhibitors of this enzyme are presently available with which to establish the potential of this enzyme as a target for drug development.

Introduction

Signalling molecules require mechanisms for their removal to produce effective and discrete cellular signalling. The endocannabinoids are no exception in this regard, and considerable effort has been made to delineate the mechanisms by which they are metabolized, to design compounds inhibiting these processes and to ascertain whether or not such compounds have therapeutic promise. Most of this work has been concerned with anandamide (AEA) and this will be reflected in the bias of this section.

Basic Pathways of AEA Metabolism

The main pathways for AEA metabolism are shown schematically in Fig. 1. One year after its discovery, Deutsch and Chin (1993) reported that extracellular AEA was accumulated by cells and hydrolysed to give arachidonic acid. The hydrolytic enzyme

A. Köfalvi (ed.), *Cannabinoids and the Brain.*

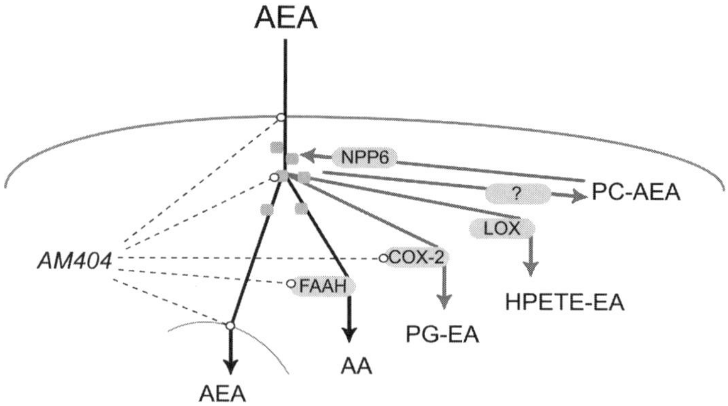

Fig. 1 Representation of the uptake and metabolism of AEA. After uptake, mediated either by passive or alternatively facilitated diffusion across the membrane, the endocannabinoid is primarily hydrolysed to arachidonic acid (AA) by FAAH although it can also be metabolized by COX-2 or by lipoxygenase enzymes to give prostaglandin ethanolamides (PG-EAs) or hydroperoxyeicosatetraenoyl ethanolamides (HPETE-EAs), respectively (Kozak and Marnett, 2002). A pathway for the synthesis of *O*-phosphorylcholine-anandamide (PC-AEA) by an as yet unidentified enzyme (? in the figure) and its dephosphorylation has also been described (Mulder and Cravatt, 2006). Hydrolysis of the AEA by FAAH helps to retain the extracellular–intracellular gradient and thereby to some extent "drives" the uptake. Free intracellular AEA has also been hypothesized to be regulated by intracellular "shuttling" proteins (shown as the small *blobs* in the figure) and by intracellular sequestration (shown at the *bottom* of the figure). AM404 affects the uptake of AEA, partly by acting as an alternative substrate for FAAH (Lang et al., 1999) and theoretically by affecting one or more of the other uptake mechanisms. AM404 is also an inhibitor of COX-2 (Högestätt et al., 2005), can interact with both cannabinoid and $TRPV_1$ receptors as well as with Na^+ channels, affects calcium homeostasis and can produce oxidative stress (Fowler, 2006), which makes for a somewhat complex pharmacology

was termed "anandamide amidase" but was subsequently found to be the same enzyme as the amidohydrolase characterized in the 1980s by Schmid and colleagues using oleoylethanolamide as substrate (Schmid et al., 1985). The enzyme, currently termed fatty acid amide hydrolase, was cloned in 1995 and its structure at the level of 2.8 Å has been elucidated (for a review of the structure and reaction mechanisms of FAAH, see McKinney and Cravatt, 2005). A second FAAH, present in primates but not in rats or mice, has been identified (Wei et al., 2006) although the functional importance of this enzyme has not been established as yet. An *N*-acyl acid amidase (NAAA), with a lower pH optimum than FAAH, has also been characterized (Ueda et al., 2001), but in rat brain membrane fractions it appears not to contribute to any large extent to the total hydrolysis of AEA (Alajakku and Fowler, unpublished data).

The Nature of Cellular AEA Uptake

Ever since 1993 when it was established that anandamide (AEA), an uncharged hydrophobic molecule, was taken up by cells in culture and thereafter metabolized by FAAH (Deutsch and Chin, 1993), a variety of mechanisms have been suggested and debated. The mechanism that has received the most attention is that of a facilitated transporter (Di Marzo et al., 1994; Hillard et al., 1997). However, this suggestion is by no means proven, and there is evidence both in favour and against such a process (Table 1). Two additional mechanisms, passive diffusion gated by FAAH (Deutsch et al., 2001; Kaczocha et al., 2006) and intracellular sequestration (Hillard and Jarrahian, 2000) are discussed in the Table, but at the outset it should be said that the most likely scenario is for AEA to utilize more than one mechanism for its cellular accumulation. The following citation can be instructive in this respect: "Translocation… across the plasma membrane is achieved by a concert of co-existing mechanisms. These lipids can passively diffuse, but transport can also be accelerated by certain membrane proteins as well as lipid rafts" (Ehehalt et al., 2006). The citation in question refers to the cellular uptake of long chain fatty acids, which have been the subject of investigation long before the discovery of AEA, but which have grappled with the same sort of issues as those for AEA uptake. In the absence of a published report concerning the identification and cloning of an AEA transporter

Table 1 The case for/against a facilitated transport mechanism of AEA uptake

Evidence in favour of facilitated transport	Alternative explanations/observations
AEA is concentrated in cells to a greater extent than would be predicted on the basis of a passive diffusion process[a]	The extracellular–intracellular gradient is partly driven in some (but not all) cells by FAAH-catalysed metabolism of AEA[i,j]; intracellular sequestration of AEA[a] would also act to keep the free $[AEA]_i$ low and allow for further passive diffusion
Uptake is time and temperature dependent and shows saturability, but is not coupled to any ion gradient or dependent upon ATP[b,c]	Initial rapid uptake is not saturable, and its temperature dependency reflects effects upon AEA availability rather than the uptake process itself[j,k]
Vesicles prepared from plasma membranes accumulate AEA whereas vesicles prepared from intracellular membranes do not; the FAAH activity of the cells is associated with the latter rather than the former; Cholesterol depletion of some (but not all) cells reduces uptake[d,e]	*N*-Acylethanolamines form complexes with cholesterol[l] which is present in plasma membranes at much greater concentrations than in intracellular membranes[d]
Uptake can be inhibited pharmacologically, even in cells lacking FAAH[f,g]; the inhibitors have pharmacological effects in vivo[h]	FAAH is inhibited by the compounds, and in the absence of this enzyme, the in vitro potencies of the uptake inhibitors are modest (see Fig. 2b); the compounds show little effect on the initial AEA uptake[j]; acyl-derived inhibitors inhibit the retention of AEA by plastic wells at similar concentrations[f,m], suggesting that the specificity of the process is rather limited

Selected references (with apologies to authors not included here): [a]Hillard and Jarrahian (2000); [b]Di Marzo et al. (1994); [c]Hillard et al. (1997); [d]Oddi et al. (2005); [e]Bari et al. (2005); [f]Ortega-Gutiérrez et al. (2004); [g]Fegley et al. (2004); [h]reviewed in Fowler et al. (2005); [i]Deutsch et al. (2001); [j]Kaczocha et al. (2006); [k]Thors and Fowler (2006); [l]Ramakrishnan et al. (2002); [m]Fowler et al. (2004)

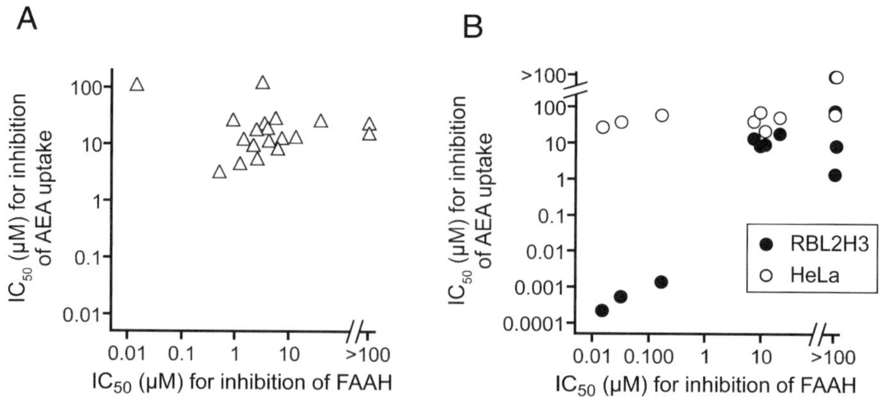

Fig. 2 Comparison of potencies for a series of AEA head group analogues (**a**) and a series of compounds with different primary structures (**b**) with respect to inhibition of FAAH and inhibition of AEA uptake. In (**a**), the data are taken from Jarrahian and colleagues (2000) and the inhibition of rat forebrain FAAH is compared with the inhibition of AEA uptake into rat cerebral granule cell cultures (2-min incubation period). The authors tested 24 compounds for uptake, and 21 of these for inhibition of FAAH. Of these, two were excluded from the analysis, due to inexact measures of IC_{50} values (>10 and >50 μM). IC_{50} values >100 μM, however, are shown in the figure. Despite the wide variation in potency as FAAH inhibitors, there is little variation in potency as uptake inhibitors, suggesting that FAAH is not a major determinant of the rate of uptake in these cells. In (**b**), the data are taken from Dickason-Chesterfield and colleagues (2006) and the inhibition of purified rat ΔTM-FAAH (a truncated version of FAAH) is compared with the uptake of AEA into either RBL2H3 cells and HeLa cells (16–40 h incubation period at room temperature, so the relevance to the rapid uptake of AEA is unclear). For the HeLa cells, the wide variation in FAAH inhibitory potency is not matched by a variation in the potency of the compounds as uptake inhibitors, which is to be expected given that these cells do not express FAAH. For the RBL2H3 cells, which express FAAH, the three most potent FAAH inhibitors (LY2077855, LY2183240 and the "standard" FAAH inhibitor URB597) are also the three most potent uptake inhibitors

protein, any model of AEA uptake should be treated with caution. Thus, for example, interpretation of data supporting an endocytotic delivery of AEA to FAAH (McFarland et al., 2004) and the identification of a membrane binding site for a potent non-acyl uptake inhibitor (Moore et al., 2005) are limited by the lack of selectivity of the compounds used (Alexander and Cravatt, 2006; Thors et al., 2007). Our own current working model is primarily influenced by the finding that AEA can cross plasma membranes very rapidly (Bojesen and Hansen, 2005). This rapidity can be demonstrated in functional studies investigating the ability of AEA to activate $TRPV_1$ (vanilloid) receptors, which is mediated by its binding to a domain of the receptor on the intracellular face of the membrane (Jordt and Julius, 2002). Extracellular AEA produces a $TRPV_1$ receptor-mediated calcium signal within seconds (Jerman et al., 2002), indicating that significant uptake of this endocannabinoid must have occurred within this time frame. Most uptake assays use much longer incubation times (generally 4–10 min), which would suggest that the processes being studied are intracellular redistribution events (shown schematically in Fig. 1) rather than transport across the plasma membrane. The saturability of the AEA uptake seen at 4–10-min incubation times may therefore be due to the limited capacity of the intracellular protein(s). This does not mean that plasma membrane constituents are not of importance: a case can be made whereby such constituents anchor the AEA within the plasma membrane prior to its cellular delivery. The only requirement for such constituents, which can well have other primary functions, is that they bind AEA. In mouse cortical neurons in culture (but not necessarily in other cells), CB_1 receptors may play such a role, since the uptake of AEA is reduced by the CB_1 receptor antagonist/inverse agonist rimonabant (Ortega-Gutiérrez et al., 2004). Initial saturability of AEA uptake would not be expected if the translocation, albeit fast (Bojesen and Hansen, 2005) is the rate-limiting step rather than the dissociation processes from the receptors. Whether or not this working model will stand the test of time remains to be seen; nonetheless, in view of the potential utility of AEA uptake inhibitors for a variety of disorders including multiple sclerosis and neuropathic pain (see e.g., La Rana et al., 2006; Ligresti et al., 2006), it remains a matter of importance to identify once and for all the key mechanisms involved in the cellular accumulation of AEA.

FAAH Inhibitors and Their Therapeutic Potential

The identification of FAAH as the enzyme responsible for AEA hydrolysis was accompanied by the chance discovery that phenylmethylsulphonyl fluoride (PMSF) inhibited this process (Deutsch and Chin, 1993). In vivo, PMSF was found, at a dose of 30 mg/kg, not to produce effects itself on thermal nociception, spontaneous activity and mobility, but to potentiate the effects of AEA in these tests (Compton and Martin, 1997). The ability to potentiate AEA effects without producing direct CB_1 receptor-mediated effects has been seen with more selective inhibitors (Kathuria et al., 2003; Lichtman et al., 2004a) and in animals with a genetic deletion of the enzyme (Cravatt et al., 2001, 2004; Wise et al., 2007). With respect to the latter, brain and liver samples from $FAAH^{-/-}$ mice lose >97% of the capacity to hydrolyse AEA in brain, spinal cord, liver and testis tissues, and this is accompanied by a large increase in the levels of AEA in these tissues (as compared with either wild type or $FAAH^{+/-}$ mice) (Cravatt et al., 2001, 2004). The $FAAH^{-/-}$ mice show reduced sensitivities to carrageenan and formalin, but their thermal hypersensitivity following ligation of the sciatic nerve is unchanged (Lichtman et al., 2004b). The reduced oedema response to intraplantar carrageenan was retained in mice with a genetic deletion in peripheral, but not central FAAH, whereas their reduced sensitivity to thermal pain was lost (Cravatt et al., 2004). $FAAH^{-/-}$ mice have subsequently been utilized to establish the potential importance of this enzyme in a wide variety of physiological and pathological processes, including reproduction (Wang et al., 2006a), colonic inflammation (Massa et al., 2004), and neurodegeneration (Bilsland et al., 2006). The potential of FAAH as a target for drug development has not been overlooked, as adjudged by the patent literature and more recently, by the publication of novel high throughput screening strategies and their implementation. Thus, for example, scientists from Wyeth and Abbott have reported the development of high throughput methods and their implementation in the screening of 457,073 and >650,000 compounds, respectively (Wang et al., 2006b; Kage et al., 2007). Most published work with selective FAAH inhibitors concerns the carbamate compound URB597 (cyclohexylcarbamic acid 3′carbamoylbiphenyl-3-yl ester) and the α-ketoheterocycle OL-135 (1-oxo-1[5-(2-pyridyl)-2-yl]-7-phenylheptane). At the outset, it should be pointed out that the word "selective" is a relative term, and all compounds will interact with other targets if present in sufficient amounts. These compounds are no exception, and can inhibit carboxyesterase enzymes if given at a concentration of 10 μM (Zhang et al., 2007), but this concentration is ≥80-fold higher than the IC_{50} values for the inhibition of FAAH by these compounds (Kathuria et al., 2003; Lichtman et al., 2004a) (see also Lichtman et al., 2004b; Clapper et al., 2006, with respect to the ability of URB597 to interact with, but not inhibit the activity of triacylglycerol hydrolase). In vivo, these two compounds increase the levels of AEA in the brain, and produce potentially useful effects in models of inflammatory pain and inflammation and neurobehavioural disturbances following neglect parenting without adversely affecting working memory (Lichtman et al., 2004a; Holt et al., 2005; Jayamanne et al., 2006; Chang et al., 2006; Varvel et al., 2006; Marco et al., 2007). Their usefulness with respect to anxiety and depression (Kathuria et al., 2003; Gobbi et al., 2005) is less clear (Naidu et al., 2007). FAAH inhibition (or genetic deletion) is not efficacious in models of neuropathic pain (Lichtman et al., 2004b; Jayamanne et al., 2006; Jhaveri et al., 2006; but see Chang et al., 2006) and it has been argued that the contribution of FAAH to endocannabinoid metabolism is altered in a tissue-dependent manner in neuropathic animals (Jhaveri et al., 2006). *N*-Arachidonoylserotonin is also an interesting compound, having both FAAH inhibitory and $TRPV_1$ receptor antagonistic properties, and showing analgesic, antiproliferative and anti-inflammatory effects in vivo (Bifulco et al., 2004; D'Argenio et al., 2006; Maione et al., 2007).

Biochemical Consequences of FAAH Inhibition

When considering the wide range of therapeutic possibilities afforded to FAAH inhibitors, it is important to remember that the enzyme is not restricted to the metabolism of AEA alone. Indeed, the enzyme has wide substrate specificity, and can hydrolyse a number of lipid classes, including other *N*-acylethanolamines (such as palmitoylethanolamide and oleoylethanolamide), *N*-acyltaurines (which activate TRPV ion channels) and fatty acid amides such as the sleep-inducing lipid oleamide (Schmid et al., 1985; Cravatt et al., 1996; Saghatelian et al., 2006). Thus, FAAH inhibition or genetic deletion results in increased levels not only of AEA but of these other lipid classes (Lichtman et al., 2004a; Saghatelian et al., 2006). Such changes can contribute to the net effect of FAAH inhibition: palmitoylethanolamide, for example, has anti-inflammatory and analgesic effects of its own, and recent data has suggested that these are brought about via an activation of the nuclear peroxisome proliferator-activated receptor α (PPARα) (Lo Verme et al., 2005, 2006). Similarly, oleoylethanolamide produces effects upon appetite in a manner involving activation of PPARα (Fu et al., 2003) and GPR119 orphan receptors (Overton et al., 2006) (for further information see Chap. 9). A second issue concerns the potential for diverting AEA (and other *N*-acylethanolamine) metabolism from hydrolysis to production of other biologically active metabolites as a result of FAAH inhibition or genetic deletion. Data is beginning to emerge to support this contention, primarily as a result of the use of $FAAH^{-/-}$ mice. The most recent example is the finding of Mulder and Cravatt (2006) who reported increased levels of phosphocholine derivatives of *N*-acylethanolamines in $FAAH^{-/-}$ mice. AEA is also a substrate for cyclooxygenase-2 (COX-2) and lipoxygenases (Kozak and Marnett, 2002). COX-2-derived prostaglandin ethanolamide (prostamide) levels in the body are normally below the level of detection, but are measurable in $FAAH^{-/-}$ mice treated with AEA (Weber et al., 2004). Prostamides (or analogues with alkyl substituents on the ethanolamide head group) have weak, if any, effects upon CB and prostanoid receptors and do not feedback inhibit AEA (or MAGL) (Berglund et al., 1999; Ross et al., 2002; Matias et al., 2004; Fowler and Tiger, 2005), although they activate a separate "prostamide receptor" in the cat iris (Woodward et al., 2007). Rockwell and Kaminski (2004) provided indirect evidence that in mouse splenocytes, AEA inhibits secretion of the proinflammatory cytokine interleukin-2 via an interaction of prostamides with PPARγ. The reverse pattern of effects can also be considered, namely that inhibition of cyclooxygenase-2 will affect AEA levels. Spinally administered flurbiprofen and indomethacin produce antinociceptive effects in the formalin test of inflammatory pain in a manner blocked by CB_1 receptor antagonists of genetic deletion of the enzyme (Gühring et al., 2002; Ates et al., 2003). The authors suggested that these effects may be related to a build up of arachidonic acid which is then utilized for endocannabinoid synthesis. The ability of acidic non-steroidal anti-inflammatory agents to inhibit FAAH (Fowler et al., 1997, 2003) may further contribute to an increased endocannabinoid tone. In this respect, Guindon and colleagues (2006a,b) have reported that the intraplantar administration of ibuprofen does not potentiate N-acylethanolamine levels in the paw or produce CB_1 receptor-mediated antinociceptive effects in the formalin test, but does potentiate the effects of AEA, the synergy being prevented by a CB_1 receptor antagonist. Furthermore, the combination of AEA and ibuprofen increased *N*-acylethanolamine levels over and above those seen with either compound alone (Guindon et al., 2006b). A final point concerns the products of FAAH action, which in the case of *N*-acylethanolamines are the fatty acids and ethanolamine. In vivo, these are rapidly taken care of by the cell, but this is not necessarily the case in vitro. This has been neatly illustrated in a recent study demonstrating that AEA protects N18TG2 neuroblastoma cells against apoptosis induced by low serum concentrations as a result of its FAAH catalysed conversion to ethanolamine (Matas et al., 2007).

Risk Groups for FAAH Inhibitors

The findings in animal models that FAAH inhibitors produce beneficial effects in models of inflammatory pain, inflammation, anxiety and depression without producing "cannabis-like" actions in the brain makes them an attractive target for drug development. However, given the ubiquity of the cannabinoid system and its involvement in so many different biological processes, it is unavoidable that compounds designed to potentiate endocannabinoid signalling in this way will produce unwanted actions or be inappropriate for certain types of patients. Conditions, for example, whereby FAAH inhibition alters the balance between excitatory and inhibitory signalling in the brain may give rise to unwanted effects (see Clement et al., 2003, for a study of the seizure activity of $FAAH^{-/-}$ mice). Two potential risk groups are women who are both pregnant or planning pregnancy, and individuals with a propensity for drug abuse. With respect to the former, it is now clear that FAAH is an important regulator of reproduction (Maccarrone and Finazzi-Agrò, 2004). AEA plays a key role in the implantation of fertilized eggs (Wang et al., 2003), and mice treated with URB597 show impaired oviductal embryo transport (Wang et al., 2006a). In women, plasma levels of AEA vary during the different stages of labour (Habayeb et al., 2004), leading the authors to suggest that "successful pregnancy implantation and progression requires low levels of AEA". Consistent with this suggestion, low levels of lymphocyte FAAH are associated with spontaneous miscarriages (Maccarrone et al., 2000). With respect to drug abuse, there is evidence that AEA produces a release of dopamine in the nucleus accumbens. URB597 does not produce any release per se, but potentiates the release induced by AEA (Solinas et al., 2006), and potentiates voluntary ethanol consumption (as does genetic ablation of FAAH) (Hansson et al., 2007; Blednov et al., 2007). Conversely, Alko alcohol-preferring rats show a lower expression of FAAH in the prefrontal cortex than Alko non-preferring rats (Hansson et al., 2007). In man, a missense mutation (P129T) of FAAH, which results in a decreased expression of the enzyme, has been associated with street drug abuse and drug/alcohol abuse but not with schizophrenia or methamphetamine dependence (Chiang et al., 2004; Sipe et al., 2002; Morita et al., 2005; Flanagan et al., 2006). An association of this variant with weight issues (Sipe et al., 2005) has not been replicated in a recent study (Jensen et al., 2007).

Metabolism of 2-AG

The processes involved in the metabolism of 2-AG are less well studied than for AEA. With respect to the uptake of 2-AG, the process is saturable, shows an apparent temperature sensitivity, is gated to some extent by its hydrolysis and subsequent esterification into phospholipids, is affected by methyl-β-cyclodextrin treatment and is inhibited by the acyl compounds AM404 and VDM11 (Piomelli et al., 1999; Beltramo and Piomelli, 2000; Bisogno et al., 2001; Hájos et al., 2004; Bari et al., 2006). Thus the uptake of 2-AG shows similar properties (and is thereby subject to the same limitations in terms of data interpretation) as that of AEA (Hermann et al., 2006). 2-AG is metabolized by a variety of enzymes including COX-2, lipoxygenases and monoacylglycerol kinases (see Simpson et al., 1991; Kozak and Marnett, 2002) and the COX-2-derived prostaglandin E_2 glycerol ester has been shown to have biological activity, mobilizing intracellular calcium and modulating hippocampal synaptic transmission (Nirodi et al., 2004; Sang et al., 2006). With respect to hydrolytic enzymes, 2-AG is a substrate for FAAH, monoacylglycerol lipase (MAGL), the esterase domain of neuropathy target esterase and possibly others (Goparaju et al., 1998; Dinh et al., 2002; van Tienhoven et al., 2002; Dinh et al., 2004; Vandevoorde et al., 2007), but most focus has been upon FAAH and MAGL. Treatment with URB597 results in an increased level of 2-AG in the paw, but not in the brain (Kathuria et al., 2003; Jhaveri et al., 2006), suggesting that FAAH is important for 2-AG metabolism in the former, but not in the latter tissue. Indeed, Dinh and colleagues (2002) have argued that MAGL is more important in the brain, and the finding that FAAH and MAGL have different cellular localizations in the hippocampus, cerebellum and amygdala (Gulyás et al., 2004; see Chap. 10) led those authors to suggest that "FAAH may set the resting level of anandamide close to its sites of synthesis, while MAGL may help to inactivate 2-AG close to its sites of action". In contrast to the situation for FAAH, selective inhibitors of MAGL are not yet available. The enzyme is sensitive to the serine hydrolase inhibitor methylarachidonoylfluorophosphonate (MAFP, see Dinh et al., 2002), and this compound has been used to explore the potential role of MAGL in regulating retrograde signalling processes (Hashimotodani et al., 2007). MAFP, however, is by no means selective and can potently inhibit both FAAH (see Ueda et al., 2001) and even 2-AG metabolism by the esterase domain of human neuropathy target esterase (van Tienhoven et al., 2002). In our hands, MAFP inhibits human recombinant MAGL and rat brain FAAH with IC_{50} values of 510 and 39 pM, respectively (Lenman, Alajakku, Jacobsson and Fowler, unpublished results). Nonetheless, physiological actions that are blocked by 2-AG synthesis inhibition [with the caveat that tetrahydrolipstatin, the compound in question, may have an antagonistic effect at CB_1 receptors (Palomäki et al., 2007)], potentiated by MAFP, but not affected by selective FAAH inhibitors (such as in the study of Hashimotodani et al., 2007), can with some confidence be ascribed to 2-AG and its regulation by MAGL. Two compounds, URB602 and URB754, have been proposed as selective MAGL inhibitors, but the selectivity of the former has been questioned, and the activity of the latter was found to be due to contamination with bis(methylthio)mercurane (Hohmann et al., 2005; Makara et al., 2007; Vandevoorde et al., 2007). There is thus a need for the identification of novel MAGL-selective inhibitors for the characterization of this enzyme to determine its physiological importance and its possible role as a target for drug development.

Concluding Remarks

In the short time since the discovery of AEA and the identification of 2-AG as an endocannabinoid, much has been learnt about the metabolism of these compounds and the potential therapeutic utility of agents blocking the metabolism. FAAH inhibitors have now their own momentum, whilst the mechanism(s) governing AEA uptake remain a matter of controversy. We lack selective MAGL inhibitors, and it is not clear as to whether MAGL as described by Dinh and colleagues (2002) represents the only enzyme that hydrolyses 2-AG in its capacity as an endocannabinoid. Our knowledge of the importance of the COX-2 pathway is in its infancy with respect to the endocannabinoid system, but there is evidence to suggest that this enzyme may both regulate the availability of 2-AG in the hippocampus (Kim and Alger, 2004; Hashimotodani et al., 2007) and provide the body with biologically active prostaglandin ethanolamides and glycerol esters. Hopefully, these areas will be the focus of increasing attention in the years to come.

Acknowledgements The authors would like to thank the Swedish Research Council (Grant no. 12158, medicine) and the Research Funds of the Medical Faculty, Umeå University for financial support for their research on endocannabinoid metabolism.

References

Alexander JP, Cravatt BF (2006) The putative endocannabinoid transport blocker LY2183240 is a potent inhibitor of FAAH and several other brain serine hydrolases. J Am Chem Soc 128:9699–9704.

Ates M, Hamza M, Seidel K, Kotalla CE, Ledent C, Gühring H (2003) Intrathecally applied flurbiprofen produces an endocannabinoid-dependent antinociception in the rat formalin test. Eur J Neurosci 17:597–604.

Bari M, Battista N, Fezza F, Finazzi-Agrò A, Maccarrone M (2005) Lipid rafts control signaling of type-1 cannabinoid receptors in neuronal cells. Implications for anandamide-induced apoptosis. J Biol Chem 280:12212–12220.

Bari M, Spagnuolo P, Fezza F, Oddi S, Pasquariello N, Finazzi-Agrò A, Maccarrone M (2006) Effect of lipid rafts on Cb_2 receptor signaling and 2-arachidonoyl-glycerol metabolism in human immune cells. J Immunol 177:4971–4980.

Beltramo M, Piomelli D (2000) Carrier-mediated transport and enzymatic hydrolysis of the endogenous cannabinoid 2-arachidonylglycerol. Neuroreport 11:1231–1235.

Berglund BA, Boring DL, Howlett AC (1999) Investigation of structural analogs of prostaglandin amides for binding to and activation of CB_1 and CB_2 cannabinoid receptors in rat brain and human tonsils. Adv Exp Med Biol 469:527–533.

Bifulco M, Laezza C, Valenti M, Ligresti A, Portella G, Di Marzo V (2004) A new strategy to block tumor growth by inhibiting endocannabinoid inactivation. FASEB J 18:1606–1608.

Bilsland LG, Dick JRT, Pryce G, Petrosino S, Di Marzo V, Baker D, Greensmith L (2006) Increasing cannabinoid levels by pharmacological and genetic manipulation delay disease progression in SOD1 mice. FASEB J 20:1003–1005; full article E180–E190.

Bisogno T, Maccarrone M, De Petrocellis L, Jarrahian A, Finazzi-Agrò A, Hillard C, Di Marzo V (2001) The uptake by cells of 2-arachidonoylglycerol, an endogenous agonist of cannabinoid receptors. Eur J Biochem 268:1982–1989.

Blednov YA, Cravatt BF, Boehm SL, Walker D, Harris RA (2007) Role of endocannabinoids in alcohol consumption and intoxication: studies of mice lacking fatty acid amide hydrolase. Neuropsychopharmacology, doi:10.1038/sj.npp.1301274.

Bojesen IN, Hansen HS (2005) Membrane transport of anandamide through resealed human red blood cell membranes. J Lipid Res 46:1652–1659.

Chang L, Luo L, Palmer JA, Sutton S, Wilson SJ, Barbier AJ, Breitenbucher JG, Chaplan SR, Webb M (2006) Inhibition of fatty acid amide hydrolase produces analgesia by multiple mechanisms. Br J Pharmacol 148:102–113.

Chiang KP, Gerber AL, Sipe JC, Cravatt BF (2004) Reduced cellular expression and activity of the P129T mutant of human fatty acid amide hydrolase: evidence for a link between defects in the endocannabinoid system and problem drug use. Hum Mol Genet 13:2113–2119.

Clapper JR, Duranti A, Tontini A, Mor M, Tarzia G, Piomelli D (2006) The fatty-acid amide hydrolase inhibitor URB597 does not affect triacylglycerol hydrolysis in rat tissues. Pharmacol Res 54:341–344.

Clement AB, Hawkins EG, Lichtman AH, Cravatt BF (2003) Increased seizure susceptibility and proconvulsant activity of anandamide in mice lacking fatty acid amide hydrolase. J Neurosci 23:3916–3923.

Compton DR, Martin BR (1997) The effect of the enzyme inhibitor phenylmethylsulfonyl fluoride on the pharmacological effect of anandamide in the mouse model of cannabimimetic activity. J Pharmacol Exp Ther 283:1138–1143.

Cravatt BF, Giang, DK, Mayfield SP, Boger DL, Lerner RA, Gilula NB (1996) Molecular characterization of an enzyme that degrades neuromodulatory fatty-acid amides. Nature 384:83–87.

Cravatt BF, Demarest K, Patricelli MP, Bracey MH, Giang DK, Martin BR, Lichtman AH (2001) Supersensitivity to anandamide and enhanced endogenous cannabinoid signaling in mice lacking fatty acid amide hydrolase. Proc Natl Acad Sci USA 98:9371–9376.

Cravatt BF, Saghatelian A, Hawkins EG, Clement AB, Bracey MH, Lichtman AH (2004) Functional disassociation of the central and peripheral fatty acid amide signaling systems. Proc Natl Acad Sci USA 101:10821–10826.

D'Argenio G, Valenti M, Scaglione G, Cosenza V, Sorrentini I, Di Marzo V (2006) Up-regulation of anandamide levels as an endogenous mechanism and a pharmacological strategy to limit colon inflammation. FASEB J 20:568–570.

Deutsch DG, Chin SA (1993) Enzymatic synthesis and degradation of anandamide, a cannabinoid receptor agonist. Biochem Pharmacol 46:791–796.

Deutsch DG, Glaser ST, Howell JM, Kunz JS, Puffenbarger RA, Hillard CJ, Abumrad N (2001) The cellular uptake of anandamide is coupled to its breakdown by fatty acid amide hydrolase (FAAH). J Biol Chem 276:6967–6973.

Dickason-Chesterfield AK, Kidd SR, Moore SA, Schaus JM, Liu B, Nomikos GG, Felder CC (2006) Pharmacological characterization of endocannabinoid transport and fatty acid amide hydrolase inhibitors. Cell Mol Neurobiol 26:405–421.

Di Marzo V, Fontana A, Cadas H, Schinelli S, Cimino G, Schwartz J-C, Piomelli D (1994) Formation and inactivation of endogenous cannabinoid anandamide in central neurons. Nature 372:686–691.

Dinh TP, Carpenter D, Leslie FM, Freund TF, Katona I, Sensi SL, Kathuria S, Piomelli D (2002) Brain monoglyceride lipase participating in endocannabinoid inactivation. Proc Natl Acad Sci USA 99:10819–10824.

Dinh TP, Kathuria S, Piomelli D (2004) RNA interference suggests a primary role for monoacylglycerol lipase in the degradation of the endocannabinoid 2-arachidonoylglycerol. Mol Pharmacol 66:1260–1264.

Ehehalt R, Füllekrug J, Pohl J, Ring A, Herrmann T, Stremmel W (2006) Translocation of long chain fatty acids across the plasma membrane – lipid rafts and fatty acid transport proteins. Mol Cell Biochem 284:135–140.

Fegley D, Kathuria S, Mercier R, Li C, Goutopoulos A, Makriyannis A, Piomelli D (2004) Anandamide transport is independent of fatty-acid amide hydrolase activity and is blocked by the hydrolysis-resistant inhibitor AM1172. Proc Natl Acad Sci USA 101:8756–8761.

Flanagan JM, Gerber AL, Cadet JL, Beutler E, Sipe JC (2006) The fatty acid amide hydrolase 385 A/A (P129T) variant: haplotype analysis of an ancient missense mutation and validation of risk for drug addiction. Hum Genet 120:581–588.

Fowler CJ (2006) The cannabinoid system and its pharmacological manipulation – a review, with emphasis upon the uptake and hydrolysis of anandamide. Fundam Clin Pharmacol 20:549–562.

Fowler CJ, Tiger G (2005) Cyclooxygenation of the arachidonoyl side chain of 1-arachidonoylglycerol and related compounds block their ability to prevent anandamide and 2-oleoylglycerol metabolism by rat brain *in vitro*. Biochem Pharmacol 69:1241–1245.

Fowler CJ, Tiger G, Stenström A (1997) Ibuprofen inhibits rat brain deamidation of anandamide at pharmacologically relevant concentrations. Mode of inhibition and structure-activity relationship. J Pharmacol Exp Ther 283:729–734.

Fowler CJ, Holt S, Tiger G (2003) Acidic nonsteroidal anti-inflammatory drugs inhibit rat brain fatty acid amide hydrolase in a pH-dependent manner. J Enz Inhib Med Chem 18:55–58.

Fowler CJ, Tiger G, Ligresti A, López-Rodíguez ML, Di Marzo V (2004) Selective inhibition of anandamide cellular uptake versus enzymatic hydrolysis – a difficult issue to handle. Eur J Pharmacol 492:1–11.

Fowler CJ, Holt S, Nilsson O, Jonsson K-O, Tiger G, Jacobson SOP (2005) The endocannabinoid signaling system: pharmacological and therapeutic aspects. Pharmacol Biochem Behav 81:248–262.

Fu J, Gaetani S, Oveisi F, Lo Verme J, Serrano A, Rodríguez de Fonseca F, Rosengath A, Luecke H, Di Giacomo B, Tarzia G, Piomelli D (2003) Oleylethanolamide regulates feeding and body weight through activation of the nuclear receptor PPAR-α. Nature 425:90–93.

Gobbi G, Bambico FR, Mangieri R, Bortolato M, Campolongo P, Solinas M, Cassano T, Morgese MG, Debonnel G, Duranti A, Tontini A, Tarzia G, Mor M, Trezza V, Goldberg SR, Cuomo V, Piomelli D (2005) Antidepressant-like activity and modulation of brain monoaminergic transmission by blockade of anandamide hydrolysis. Proc Natl Acad Sci USA 102:18620–18625.

Goparaju SK, Ueda N, Yamaguchi H, Yamamoto S (1998) Anandamide amidohydrolase reacting with 2-arachidonoylglycerol, another cannabinoid receptor ligand. FEBS Lett 422:69–73.

Gühring H, Hamza M, Sergejeva M, Ates M, Kotalla CE, Ledent C, Brune K (2002) A role for endocannabinoids in indomethacin-induced spinal antinociception. Eur J Pharmacol 454:153–163.

Guindon J, De Léan A, Beaulieu P (2006a) Local interactions between anandamide, an endocannabinoid, and ibuprofen, a nonsteroidal anti-inflammatory drug, in acute and inflammatory pain. Pain 121:85–93.

Guindon J, LoVerme J, De Léan A, Piomelli D, Beaulieu P (2006b) Synergistic antinociceptive effects of anandamide, an endocannabinoid, and nonsteroidal anti-inflammatory drugs in peripheral tissue: a role for endogenous fatty-acid ethanolamides? Eur J Pharmacol 550:68–77.

Gulyas AI, Cravatt BF, Bracey MH, Dinh TP, Piomelli D, Boscia F, Freund TF (2004) Segregation of two endocannabinoid-hydrolyzing enzymes into pre- and postsynaptic compartments in the rat hippocampus, cerebellum and amygdala. Eur J Neurosci 20:441–458.

Habayeb OMH, Taylor AH, Evans MD, Cooke MS, Taylor DJ, Bell SC, Konje JC (2004) Plasma levels of the endocannabinoid anandamide in women – a potential role in pregnancy maintenance and labor? J Clin Endocrinol Metab 89:5482–5487.

Hájos N, Kathuria S, Dinh T, Piomelli D, Freund TF (2004) Endocannabinoid transport tightly controls 2-arachidonoyl glycerol actions in the hippocampus: effects of low temperature and the transport inhibitor AM404. Eur J Neurosci 19:2991–2996.

Hansson AC, Bermúdez-Silva FJ, Malinen H, Hyytiä P, Sanchez-Vera I, Rimondini R, Rodriguez de Fonseca F, Kunos G, Sommer WH, Heilig M (2007) Genetic impairment of frontocortical endocannabinoid degradation and high alcohol preference. Neuropsychopharmacology 32:117–126.

Hashimotodani Y, Ohno-Shosaku T, Kano M (2007) Presynaptic monoacylglycerol lipase activity determines basal endocannabinoid tone and terminates retrograde endocannabinoid signaling in the hippocampus. J Neurosci 27:1211–1219.

Hermann A, Kaczocha M, Deutsch DG (2006) 2-Arachidonoylglycerol (2-AG) membrane transport: history and outlook. AAPS J 8:E409–E412.

Hillard CJ, Jarrahian A (2000) The movement of *N*-arachidonoylethanolamine (anandamide) across cellular membranes. Chem Phys Lipids 108:123–134.

Hillard CJ, Edgemond WS, Jarrahian A, Campbell WB (1997) Accumulation of *N*-arachidonoylethanolamine (anandamide) into cerebellar granule cells occurs via facilitated diffusion. J Neurochem 69:631–638.

Högestätt ED, Jönsson BAG, Ermund A, Andersson DA, Björk H, Alexander JP, Cravatt BF, Basbaum AI, Zygmunt PM (2005) Conversion of acetaminophen to the bioactive N-acyl phenolamine AM404 via fatty acid amide hydrolase-dependent arachidonic acid conjugation in the nervous system. J Biol Chem 280:31405–31412.

Hohmann AG, Suplita RL, Bolton NM, Neely MH, Fegley D, Mangieri R, Krey JF, Walker JM, Holmes PV, Crystal JD, Duranti A, Tontini A, Mor M, Tarzia G, Piomelli D (2005) An endocannabinoid mechanism for stress-induced analgesia. Nature 435:1108–1112.

Holt S, Comelli F, Costa B, Fowler CJ (2005) Inhibitors of fatty acid amide hydrolase reduce carrageenan-induced hind paw inflammation in pentobarbital-treated mice: comparison with indomethacin and possible involvement of cannabinoid receptors. Br J Pharmacol 146:467–476.

Jarrahian A, Manna S, Edgemond WS, Campbell WB, Hillard CJ (2000) Structure-activity relationships among *N*-arachidonylethanolamine (anandamide) head group analogues for the anandamide transporter. J Neurochem 75:2597–2606.

Jayamanne A, Greenwood R, Mirtchell VA, Aslan S, Piomelli D, Vaughan CW (2006) Actions of the FAAH inhibitor URB597 in neuropathic and inflammatory chronic pain models. Br J Pharmacol 147:281–288.

Jensen DP, Andreasen CH, Andersen MK, Hansen L, Eiberg H, Borch-Johnsen K, Jørgensen T, Hansen T, Pedersen O (2007) The functional Pro129Thr variant of the *FAAH* gene is not associated with various fat accumulation phenotypes in a population-based cohort of 5,801 whites. J Mol Med 85:445–449.

Jerman JC, Gray J, Brough SJ, Ooi L, Owen D, Davis JB, Smart D (2002) Comparison of effects of anandamide at recombinant and endogenous rat vanilloid receptors. Br J Anaesth 89:882–887.

Jhaveri MD, Richardson D, Kendall DA, Barrett DA, Chapman V (2006) Analgesic effects of fatty acid amide hydrolase inhibition in a rat model of neuropathic pain. J Neurosci 26:13318–13327.

Jordt S-E, Julius D (2002) Molecular basis for species-specific sensitivity to "hot" chili peppers. Cell 108:421–430.

Kaczocha M, Hermann A, Glaser ST, Bojesen IN, Deutsch DG (2006) Anandamide uptake is consistent with rate-limited diffusion and is regulated by the degree of its hydrolysis by fatty acid amide hydrolase. J Biol Chem 281:9066–9075.

Kage KL, Richardson PL, Traphagen L, Severin J, Pereda-Lopez A, Lubben T, Davis-Taber R, Vos MH, Bartley D, Walter K, Harlan J, Solomon L, Warrior U, Holzman TF, Faltynek C, Surowy CS, Scott VE (2007) A high throughput fluorescent assay for measuring the activity of fatty acid amide hydrolase. J Neurosci Methods 161:47–54.

Kathuria S, Gaetani S, Fegley D, Valiño F, Duranti A, Tontini A, Mor M, Tarzia G, La Rana G, Calignano A, Giustino A, Tattoli M, Palmery M, Cuomo V, Piomelli D (2003) Modulation of anxiety through blockade of anandamide hydrolysis. Nat Med 9:76–81.

Kim J, Alger BE (2004) Inhibition of cyclooxygenase-2 potentiates retrograde endocannabinoid effects in hippocampus. Nat Neurosci 7:697–698.

Kozak KR, Marnett LJ (2002) Oxidative metabolism of endocannabinoids. Prostaglandins Leukot Essent Fatty Acids 66:211–220.

Lang W, Qin C, Lin S, Khanolkar AD, Goutopoulos A, Fan P, Abouzid K, Meng Z, Biegel D, Makriyannis A. (1999) Substrate specificity and stereoselectivity of rat brain microsomal anandamide amidohydrolase. J Med Chem 42:896–902.

La Rana G, Russo R, Campolongo P, Bortolato M, Mangieri RA, Cuomo V, Iacono A, Raso GM, Meli R, Piomelli D, Calignano A (2006) Modulation of neuropathic and inflammatory pain by the endocannabinoid transport inhibitor AM404 [*N*-(4-hydroxyphenyl)-eicosa-5,8,11,14-tetraenamide]. J Pharmacol Exp Ther 317:1365–1371.

Lichtman AH, Leung D, Shelton CC, Saghatelian A, Hardouin C, Boger DL, Cravatt BF (2004a) Reversible inhibitors of fatty acid amide hydrolase that promote analgesia: evidence for an unprecedented combination of potency and selectivity. J Pharmacol Exp Ther 311:441–448.

Lichtman AH, Shelton CC, Advani T, Cravatt BF (2004b) Mice lacking fatty acid amide hydrolase exhibit a cannabinoid receptor-mediated phenotypic hypoalgesia. Pain 109:319–327.

Ligresti A, Cascio MG, Pryce G, Kulasegram S, Beletskaya I, De Petrocellis L, Saha B, Mahadevan A, Visintin C, Wiley JL, Baker D, Martin BR, Razdan RK, Di Marzo V (2006) New potent and selective inhibitors of anandamide reuptake with antispastic activity in a mouse model of multiple sclerosis. Br J Pharmacol 147:83–91.

Lo Verme J, Fu J, Astarita G, La Rana G, Russo R, Calignano A, Piomelli D (2005) The nuclear receptor PPAR-α mediates the antiinflammatory actions of palmitoylethanolamide. Mol Pharmacol 67:15–19.

Lo Verme J, Russo R, La Rana G, Fu J, Farthing J, Mattace-Raso G, Meli R, Hohmann A, Calignano A, Piomelli D (2006) Rapid broad-spectrum analgesia through activation of peroxisoma proliferator-activated receptor-α. J Pharmacol Exp Ther 319:1051–1061.

Maccarrone M, Finazzi-Agrò A (2004) Anandamide hydrolase: a guardian angel of human reproduction? Trends Pharmacol Sci 25:353–357.

Maccarrone M, Valensise H, Bari M, Lazzarin N, Romanini C, Finazzi-Agrò A (2000) Relation between decreased anandamide hydrolase concentrations in human lymphocytes and miscarriage. Lancet 355:1326–1329.

Maione S, De Petrocellis L, de Novellis V, Schiano Moriello A, Petrosino S, Palazzo E, Sca Rossi F, Woodward DF, Di Marzo V (2007) Analgesic actions of *N*-arachidonoyl-serotonin, a fatty acid amide hydrolase inhibitor with antagonistic activity at vanilloid $TRPV_1$ receptors. Br J Pharmacol 150:766–781.

Makara JK, Mor M, Fegley D, Szabó SI, Kathuria S, Astarita G, Duranti A, Tontini A, Tarzia G, Rivara S, Freund TF, Piomelli D (2007) Corrigendum: selective inhibition of 2-AG enhances endocannabinoid signaling in hippocampus. Nat Neurosci 10:134. Original article published in 2005 in Nat Neurosci 8:1139–1141.

Marco EM, Adriani W, Canese R, Podo F, Viveros MP, Laviola G (2007) Enhancement of endocannabinoid signalling during adolescence: modulation of impulsivity and long-term consequences on metabolic brain parameters in early maternally deprived rats. Pharmacol Biochem Behav 86:334–345.

Massa F, Marsicano G, Hermann H, Cannich A, Monory K, Cravatt BF, Ferri G-L, Sibaev A, Storr M, Lutz B (2004) The endogenous cannabinoid system protects against colonic inflammation. J Clin Invest 113:1202–1209.

Matas D, Juknat A, Pietr M, Klin Y, Vogel Z (2007) Anandamide protects from low serum-induced apoptosis via its degradation to ethanolamine. J Biol Chem 282:7885–7892.

Matias I, Chen J, De Petrocellis L, Bisogno T, Ligresti A, Fezza F, Krauss AH-P, Shi L, Protzman CE, Li C, Liang Y, Nieves AL, Kedzie KM, Burk RM, Di Marzo V, Woodward DF (2004) Prostaglandin ethanolamides (prostamides): *in vitro* pharmacology and metabolism. J Pharmacol Exp Ther 309:745–757.

McFarland MJ, Porter AC, Rakhshan FR, Rawat DS, Gibbs RA, Barker EL (2004) A role for caveolae/lipid rafts in the uptake and recycling of the endogenous cannabinoid anandamide. J Biol Chem 279:41991–41997.

McKinney MK, Cravatt BF (2005) Structure and function of fatty acid amide hydrolase. Annu Rev Biochem 74:411–432.

Moore SA, Nomikos GG, Dickason-Chesterfield AK, Schober DA, Schaus JM, Ying B-P, Xu Y-C, Phebus L, Simmons RMA, Li D, Iyengar S, Felder CC (2005) Identification of a high-affinity binding site involved in the transport of endocannabinoids. Proc Natl Acad Sci USA 102:17852–17857.

Morita Y, Ujike H, Tanaka Y, Uchida N, Nomura A, Ohtani K, Kishimoto M, Morio A, Imamura T, Sakai A, Inada T, Harano M, Komiyama T, Yamada M, Sekine Y, Iwata N, Iyo M, Sora I, Ozaki N, Kuroda S (2005) A nonsynonymous polymorphism in the human fatty acid amide hydrolase gene did not associate with either methamphetamine dependence or schizophrenia. Neurosci Lett 376:182–187.

Mulder AH, Cravatt BF (2006) Endocannabinoid metabolism in the absence of fatty acid amide hydrolase (FAAH): discovery of phosphorylcholine derivatives of N-acyl ethanolamines. Biochemistry 45:11267–11277.

Naidu PS, Varvel SA, Ahn K, Cravatt BF, Martin BR, Lichtman AH (2007) Evaluation of fatty acid amide hydrolase inhibition in murine models of emotionality. Psychopharmacology 192:61–70.

Nirodi CS, Crews BC, Kozak KR, Morrow JD, Marnett LJ (2004) The glyceryl ester of prostaglandin E2 mobilizes calcium and activates signal transduction in RAW264.7 cells. Proc Natl Acad Sci USA 101:1840–1845.

Oddi S, Bari M, Battista N, Barsacchi D, Cozzani I, Maccarrone M (2005) Confocal microscopy and biochemical analysis reveals spatial and functional separation between anandamide uptake and hydrolysis in human keratinocytes. Cell Mol Life Sci 62:386–395.

Ortega-Gutiérrez S, Hawkins EG, Viso A, López-Rodríguez ML, Cravatt BF (2004) Comparison of anandamide transport in FAAH wild-type and knockout neurons: evidence for contributions by both FAAH and the CB_1 receptor to anandamide uptake. Biochemistry 43:8184–8190.

Overton HA, Babbs AJ, Doel SM, Fyfe MCT, Gardner LS, Griffin G, Jackson HC, Procter MJ, Rasamison CM, Tang-Christensen M, Widdowson PS, Williams GM, Reynet C (2006) Deorphanization of a G protein-coupled receptor for oleoylethanolamide and its use in the discovery of small-molecule hypophagic agents. Cell Metab 3:167–175.

Palomäki VA, Lehtonen M, Savinainen JR, Laitinen JT (2007) Visualization of 2-arachidonoylglycerol accumulation and cannabinoid CB_1 receptor activity in rat brain cryosections by functional autoradiography. J Neurochem 101:972–981.

Piomelli D, Beltramo M, Glasnapp S, Lin SY, Goutopoulos A, Xie XQ, Makriyannis A (1999) Structural determinants for recognition and translocation by the anandamide transporter. Proc Natl Acad Sci USA 96:5802–5807.

Ramakrishnan M, Kenoth R, Kamlekar RK, Chandra MS, Radhakrishnan TP, Swamy MJ (2002) *N*-Myristoylethanolamine-cholesterol (1:1) complex: first evidence from differential scanning calorimetry, fast-atom-bombardment mass spectrometry and computational modelling. FEBS Lett 531:343–347.

Rockwell CE, Kaminski NE (2004) A cyclooxygenase metabolite of anandamide causes inhihbition of interleukin-2 secretion in murine splenocytes. J Pharmacol Exp Ther 311:683–690.

Ross RA, Craib SJ, Stevenson LA, Pertwee RG, Henderson A, Toole J, Ellington HC (2002) Pharmacological characterization of the anandamide cyclooxygenase metabolite: prostaglandin E_2 ethanolamide. J Pharmacol Exp Ther 301:900–907.

Saghatelian A, McKinney MK, Bandell M, Patapoutian A, Cravatt BF (2006) A FAAH-regulated class of N-acyl taurines that activates TRP ion channels. Biochemistry 45:9008–9015.

Sang N, Zhang J, Chen C (2006) PGE_2 glycerol ester, a COX-2 oxidative metabolite of 2-arachidonoyl glycerol, modulates inhibitory synaptic transmission in mouse hippocampal neurons. J Physiol 572:735–745.

Schmid PC, Zuzarte-Augustin ML, Schmid HHO (1985) Properties of rat liver *N*-acylethanolamine amidohydrolase. J Biol Chem 260:14145–14149.

Simpson CMF, Itabe H, Reynolds CN, King WC, Glomset JA (1991) Swiss 3T3 cells preferentially incorporate sn-2-arachidonoyl monoacylglycerol into sn-1-stearoyl-2-arachidonoyl phosphatidylinositol. J Biol Chem 266:15902–15909.

Sipe JC, Chiang K, Gerber AL, Beutler Em Cravatt BF (2002) A missense mutation in human fatty acid amide hydrolase associated with problem drug use. Proc Natl Acad Sci USA 99:8394–8399.

Sipe JC, Waalen J, Gerber A, Beutler E (2005) Overweight and obesity associated with a missense polymorphism in fatty acid amide hydrolase (FAAH). Int J Obesity 29:755–759.

Solinas M, Justinova Z, Goldberg SR, Tanda G (2006) Anandamide administration alone and after inhibition of fatty acid amide hydrolase (FAAH) increases dopamine levels in the nucleus accumbens shell in rats. J Neurochem 98:408–419.

Thors L, Fowler CJ (2006) Is there a temperature-dependent uptake of anandamide into cells? Br J Pharmacol 149:173–181.

Thors L, Alajakku K, Fowler CJ (2007) The "specific" tyrosine kinase inhibitor genistein inhibits the enzymic hydrolysis of anandamide. Implications for anandamide uptake. Br J Pharmacol 150:951–960.

Ueda N, Yamanaka K, Yamamoto S (2001) Purification and characterization of an acid amidase selective for N-palmitoylethanolamine, a putative endogenous anti-inflammatory substance. J Biol Chem 276:35552–35557.

Vandevoorde S, Jonsson K-O, Labar G, Persson E, Lambert DM, Fowler CJ (2007) Lack of selectivity of URB602 for 2-oleoylglycerol compared to anandamide hydrolysis *in vitro*. Br J Pharmacol 150:186–191.

van Tienhoven M, Atkins J, Li Y, Glynn P (2002) Human neuropathy target esterase catalyzes hydrolysis of membrane lipids. J Biol Chem 277:20942–20948.

Varvel SA, Cravatt BF, Engram AE, Lichtman AH (2006) Fatty acid amide hydrolase (–/–) mice exhibit an increased sensitivity to the disruptive effects of anandamide or oleamide in a working memory water maze task. J Pharmacol Exp Ther 317:251–257.

Wang H, Matsumoto H, Guo Y, Paria BC, Roberts RL, Dey SK (2003) Differential G protein-coupled cannabinoid receptor signaling by anandamide directs blastocyst activation for implantation. Proc Natl Acad Sci USA 100:14914–14919.

Wang H, Xie H, Guo Y, Zhang H, Takahashi T, Kingsley PJ, Marnett LJ, Das SK, Cravatt BF, Dey SK (2006a) Fatty acid amide hydrolase deficiency limits early pregnancy events. J Clin Invest 116:2122–2131.

Wang Y, Ramirez F, Krishnamurthy G, Gilbert A, Kadakia N, Xu J, Kalgaonkar G, Ramarao MK, Edris W, Rogers KE, Jones PG (2006b) High-throughput screening for the discovery of inhibitors of fatty acid amide hydrolase using a microsome-based fluorescent assay. J Biomol Screen 11:519–527.

Weber A, Ni J, Ling K-HJ, Acheampong A, Tang-Liu DD-S, Burk R, Cravatt BF, Woodward D (2004) Formation of prostamides from anandamide in FAAH knockout mice analyzed by HPLC with tandem mass spectrometry. J Lipid Res 45:757–763.

Wei BQ, Mikkelsen TS, McKinney MK, Lander ES, Cravatt BF (2006) A second fatty acid amide hydrolase with variable distribution among placental mammals. J Biol Chem 281:36569–36578.

Wise LE, Shelton CC, Cravatt BF, Martin BR, Lichtman AH (2007) Assessment of anandamide's pharmacological effects in mice deficient of both fatty acid amide hydrolase and cannabinoid CB_1 receptors. Eur J Pharmacol 557:44–48.

Woodward DF, Krauss AH, Wang JW, Protzman CE, Nieves AL, Liang Y, Donde Y, Burk RM, Landsverk K, Struble C (2007) Identification of an antagonist that selectively blocks the activity of prostamides (prostaglandin-ethanolamides) in the feline iris. Br J Pharmacol 150:342–352.

Zhang D, Saraf A, Kolasa T, Bhatia P, Zheng GZ, Patel M, Lannoye GS, Richardson P, Stewart A, Rogers JC, Brioni JD, Surowy CS (2007) Fatty acid amide hydrolase inhibitors display broad selectivity and inhibit multiple carboxylesterases as off-targets. Neuropharmacology 52:1095–1105.

Chapter 4
Other Cannabimimetic Lipid Signaling Molecules

Heather B. Bradshaw

Abstract The endogenous lipids anandamide and 2-arachidonoylglycerol (2-AG) play predominant signaling roles through G protein-coupled receptors (GPCRs) and at least one transient receptor potential channel (TRP). Additional structurally similar lipid signaling molecules that have cannabinoid-like (cannabimimetic) activity in which they produce similar cellular, physiological, and behavioral phenotypes as anandamide and 2-AG have recently been discovered. Like the endogenous cannabinoids, many of the actions of these structurally similar endogenous lipids are known to occur through both GPCRs and TRPs. The cannabimimetic lipid signaling molecules of *N*-arachidonoyl glycine, *N*-arachidonoyl dopamine, *N*-arachidonoyl serine, the family of *N*-acyl ethanolamines, and 2-acyl glycerols and their roles in cellular signaling and physiology are discussed here.

Introduction

The endogenous lipids anandamide (*N*-arachidonoylethanolamine; AEA; Devane et al., 1992) and 2-arachidonoylglycerol (2-AG; Mechoulam et al., 1995; Sugiura et al., 1995) have been shown to play predominant signaling roles in both the endogenous cannabinoid system through the G protein-coupled (GPCR) cannabinoid receptors 1 and 2 (CB_1, CB_2) and in the endovanilloid activation through transient receptor potential channels (TRPs), and are discussed in depth elsewhere in this book. Our understanding of the families of lipid signaling molecules that activate GPCRs and TRPs is rapidly growing. Lipid signaling molecules that are structurally similar and that have cannabinoid-like activity (cannabimimetic) in which they produce similar cellular, physiological, and behavioral phenotypes as AEA and 2-AG have recently been discovered. Like the endogenous cannabinoids, many of the actions of these structurally similar endogenous lipids are known to occur through both GPCRs and TRPs. A structural commonality between AEA and 2-AG is that they both have a fatty acid (arachidonic acid) conjugated to an additional molecule at the carboxylic acid (ethanolamine and glycerol, respectively). There are other bioactive lipids that, likewise, share this structural similarity and are arachidonic acid conjugates to amino acids and dopamine that have cannabimimetic activity. Additionally, there are

A. Köfalvi (ed.), *Cannabinoids and the Brain.*

other fatty acids such at palmitic acid, stearic acid, oleic acid, linoleic acid, and linolenic acid, which are comprised of varying carbon chain lengths and degrees of saturation that are conjugated to ethanolamine and glycerol to form bioactive lipids. It is hypothesized that the cannabimimetic activity of these lipids has a direct relationship to the fact that they share this structural homology and by extension share the same or similar biosynthetic and metabolic enzymes as well as activate receptors that share similar cellular functions.

N-Arachidonoyl Glycine

N-arachidonoyl glycine (NAGly; Fig. 1b) was initially synthesized by Sheskin and colleagues (1997) as part of a structure–activity relationship study of AEA. NAGly differs from AEA by the oxidation state of carbon beta to the amido

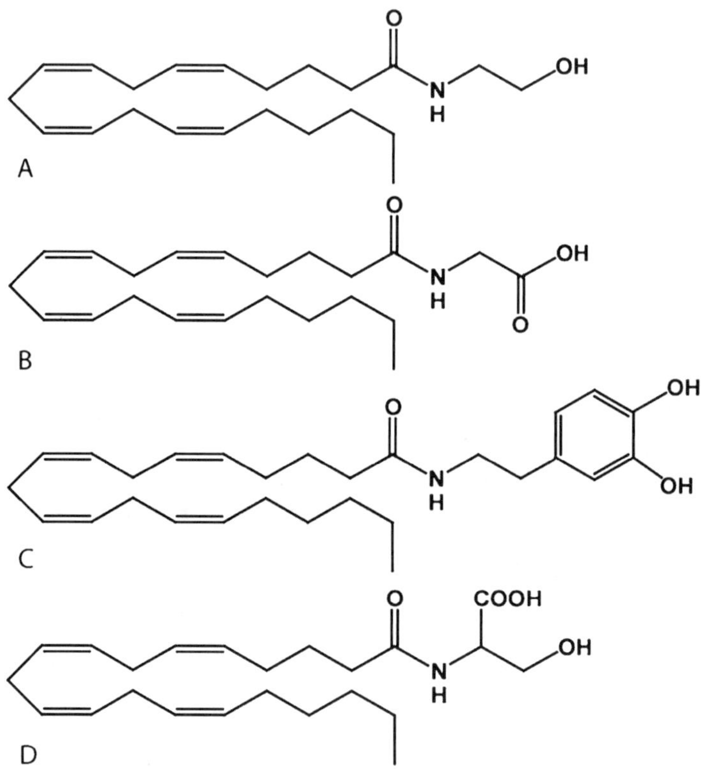

Fig. 1 *N*-Arachidonoyl amides: (**a**) *N*-arachidonoyl ethanolamine; (**b**) *N*-arachidonoyl glycine; (**c**) *N*-arachidonoyl dopamine; (**d**) *N*-arachidonoyl serine

nitrogen (Fig. 1a, b); yet, that modification significantly reduces its activity at both cannabinoid receptors (Sheskin et al., 1997). NAGly was subsequently shown to have antinociceptive and anti-inflammatory effects on mice (Burstein et al., 2000). Huang and colleagues (2001) demonstrated that NAGly is an endogenous compound found throughout the body in amounts equivalent to AEA and confirmed its antinociceptive properties. NAGly was then shown to be a substrate for cyclooxygenase-2 (COX-2), the primary enzyme in prostaglandin synthesis, producing at least three novel NAGly oxygenated metabolites with as-yet unknown biological activity (Prusakiewicz et al., 2002). Pancreatic beta cells were shown to mobilize intracellular calcium in response to application of NAGly in a manner that regulates insulin release (Ikeda et al., 2005). NAGly has also been shown to inhibit the glycine transporter, $GlyT_{2a}$, through direct, though noncompetitive, interactions (Wiles et al., 2006). More recently, NAGly was proposed to be an endogenous ligand for GPR18, an orphan G protein-coupled receptor (Kohno et al., 2006). These data demonstrate that NAGly is an endogenous signaling molecule with multiple biological activities.

N-Arachidonoyl Dopamine

N-arachidonoyl dopamine (NADA; Fig. 1c) was originally synthesized for the study of the endovanilloid system due to its structural similarity to capsaicin in which the dopamine molecule conjugated to arachidonic acid is similar to the vanilloid moiety of capsaicin that is conjugated to an acyl chain (Bisogno et al., 1997). NADA was then identified as an endogenous compound that is primarily localized in the striatum, hippocampus, and cerebellum with a small amount produced in dorsal root ganglion cells (Huang et al., 2002). It was shown to activate CB_1 receptors (K_i 0.5 ± 0.2 μM) and induce analgesia following systemic administration (Bisogno et al., 1997; Huang et al., 2002). In following with its vanilloid structural homology, NADA, like AEA, mobilizes intracellular calcium via activation of transient receptor potential vanilloid type-1 ($TRPV_1$) receptor (Huang et al., 2002; Toth et al., 2003; Gavva et al., 2004). Premkumar and colleagues (2004) hypothesize that it is acting on $TRPV_1$ receptor in a PKC-dependent manner by demonstrating that NADA-induced currents could be blocked by the PKC inhibitor, bisindolymaleimide. They also demonstrated that NADA-induced changes in current were increased ~30-fold by applying NADA intracellularly suggesting that the increased access to the $TRPV_1$ receptor facilitated this change (Premkumar et al., 2004). The distribution of endogenous NADA in various brain areas differs from that of AEA with the highest levels found in the striatum and hippocampus (Huang et al., 2002). Given that it also occurs in the dorsal root ganglion, this suggests that it may serve a role in pain and sensory modulation. Patch-clamp studies of cultured DRG neurons showed that NADA elicited immediate and reversible responses which were blocked by both the CB_1 antagonist, SR141617A and the nonselective $TRPV_1$ antagonist, capsazepine (Sagar et al., 2004). Electrophysiological recordings from the dorsal horn in anesthetized rats showed that neuronal responses to mechanical stimulation were inhibited by 5 μg of NADA (Sagar et al., 2004). When low levels of mechanical pressure were applied, the effect was blocked by SR141716A. Conversely, the $TRPV_1$ receptor antagonist iodoresiniferatoxin (I-RTX) blocked the effects of NADA when higher levels of mechanical pressure were tested (Sagar et al., 2004). In behaving animals, a 5-μg dose of NADA caused thermal hyperalgesia when administered peripherally in rats (Huang et al., 2002) and primates (Butelman et al., 2003). More recently, Huang and Walker (2006) found that when administrated into the receptive fields of the dorsal horn nociceptive neurons on the plantar surface of the ipsilateral hindpaw, NADA caused both increased spontaneous and heat-evoked firing in spinal nociceptive neurons. This NADA-induced neural hypersensitivity was dose dependent (EC_{50}, 1.55 μg) and $TRPV_1$ receptor dependent, but CB_1 receptor independent. Harrison and colleagues (2003) showed that NADA initiates contractions in both pig bronchi and urinary bladder in a manner similar to that of AEA and capsaicin. Additionally, NADA was shown to inhibit T-cell activation, IL-2 and TNFα gene activation, as well as inhibit NF-κB-dependent transcriptional activity (Sancho et al., 2004). Given that NADA is capable of eliciting analgesia upon systemic administration, hyperalgesia upon intradermal injection, inhibition of neuronal responses to mechanical stimulation, inhibition of immune responses, and initiates smooth muscle contraction, it is possible that endogenous NADA may activate $TRPV_1$, CB_1, or an additional as yet unknown receptor depending on location and circumstance.

N-Arachidonoyl Serine

Milman and colleagues (2006) recently discovered the existence of *N*-arachidonoyl serine (ARA-S; Fig. 1d) in bovine brain. They showed that ARA-S does not bind appreciably to CB_1, CB_2, or $TRPV_1$ receptor. It was shown to act as a vasorelaxant of mesenteric arteries, which mimics the pharmacological profile of abnormal cannabidiol (ABN-CBD; see Chap. 9), though it was not blocked by the ABN-CBD antagonist, O-1918. In addition, in a macrophage cell line, ARA-S was shown to inhibit zymosan-induced reactive oxygen species, as well as inhibit LPS-induced NO production and TNFα production.

Additional *N*-Acyl Ethanolamines

Dihomo-γ-Linolenoylethanolamide and Docosatetraenoylethanolamide

In addition to the endocannabinoid, AEA, there are other *N*-acyl ethanolamides that have shown biological activity. Due to their structural similarity, they have been used as analogs to AEA in many structure–activity relationship studies. Although dihomo-γ-linolenoylethanolamide and docosatetraenoylethanolamide bind to CB_1 receptors (Hanus et al., 1993), very little is known about their in vivo activity. The other acyl conjugates with ethanolamine do no show any appreciable binding to either CB_1 or CB_2 receptors; however, they have all been identified in various mammalian and invertebrate tissues and more is known about their roles in physiology and cellular signaling (Di Marzo et al., 1996; Maccarrone et al., 2001b; Schuel et al., 2002; Salzet and Stefano, 2002).

N-Palmitoyl Ethanolamide

N-palmitoyl ethanolamide (PEA; Fig. 2a) is a 16 carbon saturated fatty acid conjugated to ethanolamine and was identified nearly five decades ago as the principle

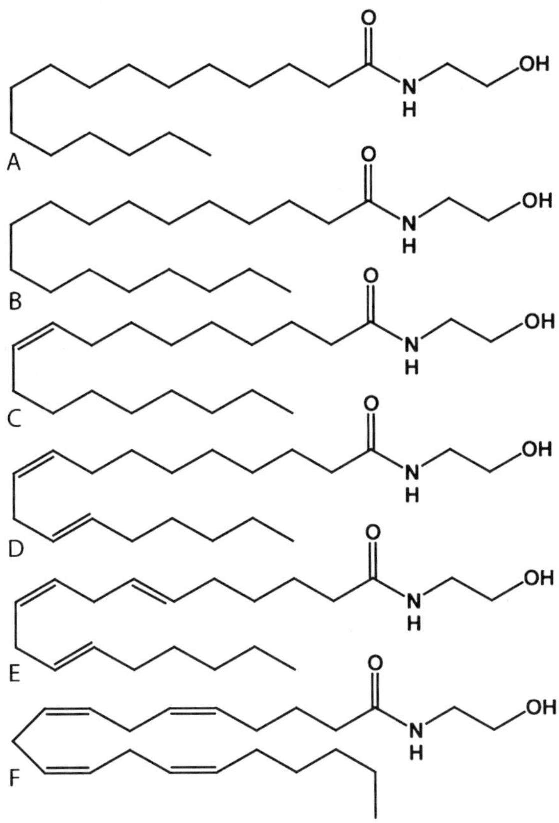

Fig. 2 *N*-Acyl ethanolamines: (**a**) *N*-palmitoyl ethanolamine; (**b**) *N*-stearoyl ethanolamine; (**c**) *N*-oleoyl ethanolamine; (**d**) *N*-linoleoyl ethanolamine; (**e**) *N*-linolenoyl ethanolamine; (**f**) *N*-arachidonoyl ethanolamine

anti-inflammatory agent in lipid extracts of various natural products (Kuehl et al., 1957). There is a considerable amount of evidence to support the role of PEA as an endogenous signaling molecule and many more-inclusive reviews of PEA have recently been published (Lambert et al., 2002; Schmid and Berdyshev, 2002; Darmani et al., 2005). The actions of PEA that are more closely related to the endocannabinoid, AEA, will be discussed here. PEA produces anti-inflammatory and antinociceptive effects when administered exogenously (Facci et al., 1995; Mazzari et al., 1996; Calignano et al., 1998; Jaggar et al., 1998). Synergistic effects in antinociception were observed with coadministration of AEA and PEA, and abolished by either CB_1 or CB_2 receptor antagonists, respectively (Calignano et al., 1998, 2001). PEA produced a twofold decrease in the K_i value for AEA binding at $TRPV_1$ receptor, an effect that was not due to inhibition of AEA hydrolysis (De Petrocellis et al., 2001); nor does it appear that this effect was caused by blocking the putative AEA transporter (Rakhshan et al., 2000). Although PEA exhibits poor affinity for CB_1 and CB_2 receptors (Sheskin et al., 1997; Lambert and Di Marzo 1999), the antinociceptive effects of PEA were blocked by the CB_2 antagonist SR144528 suggesting possible activation of a non-CB_2 receptor of which the molecular nature, location, and signal transduction mechanisms are unknown (Calignano et al., 1998, 2001). A recent study by Darmani and colleagues (2005) investigated the role of PEA in vivo in humans with two chronic pain conditions and in an animal model of diabetic-induced neuropathic pain. They measured the production of PEA after osteopathic manipulation in patients with chronic lower-back pain, in colonic biopsies of patients with ulcerative colitis, and in the skin of mice with a diabetic-induced neuropathic pain. Plasma PEA levels were unchanged in controls after osteopathic manipulations; however, the levels of PEA were significantly elevated after osteopathic manipulations in the chronic pain group. The colonic biopsies from the ulcerative colitis patients showed a significantly higher production of PEA from the controls. Finally, there was also a significant increase in the level of PEA in the paw skin of diabetic-induced mice that demonstrated neuropathic pain responses. This evidence further suggests that PEA is playing a role in the response to chronic inflammation and pain.

N-Oleoyl Ethanolamide

N-oleoyl ethanolamide (OEA; Fig. 2b) is an 18 carbon fatty acid with one point of desaturation (therefore making a double bond) conjugated to ethanolamine. In contrast to PEA, OEA inhibited AEA uptake (Rakhshan et al., 2000) and degradation (Karava et al., 2001). Like AEA, OEA has been implicated in the neural regulation of feeding behaviors by acting on peripheral sensory fibers (Rodriguez et al., 2001). OEA levels significantly decreased during starvation (Rodriquez et al., 2001); however, in contrast, AEA levels increased during starvation (Gomez et al., 2002) suggesting a reciprocal effect of the two compounds within this system. A more recent study localized the decrease of OEA to the duodenum and jejunum with no effect on the ileum accompanied by a rapid increase in production in those two areas 10 min after refeeding (Fu et al., 2007). OEA has negligible affinity for both CB_1 and CB_2 receptors. However, OEA activates the nuclear receptor, peroxisome proliferator-activated receptor α (PPARα; Fu et al., 2003; Guzman et al., 2004), which may explain its effects on feeding (Fu et al., 2003) (see Chap. 14). OEA also activates the $TRPV_1$ receptor in a PKC-dependent manner (Ahern, 2003) and the orphan GPCR, GPR119 (Overton et al., 2006). Even though OEA does not appreciably bind to CB_1 receptors, measurements of the endogenous levels of OEA revealed significant increases in cortical levels in CB_1 receptor knockout mice relative to wild type mice at 2 months of age. At 6 months of age, there were no differences between the wild type and the knockouts (Maccarrone et al., 2001a,b). Conversely, the levels of OEA in the hippocampus of CB_1 receptor knockout mice were significantly lower than the wild type at 2 months with a further reduction at 6 months (Maccarrone et al., 2001a,b). PEA and *N*-stearoyl ethanolamide (SEA; Fig. 2c), an 18 carbon saturated fatty acid conjugated to ethanolamine, showed similar changes in levels in this CB_1 receptor knockout model. These data add to the evidence that OEA, PEA, and SEA may function in concert with endocannabinoids to regulate physiological processes.

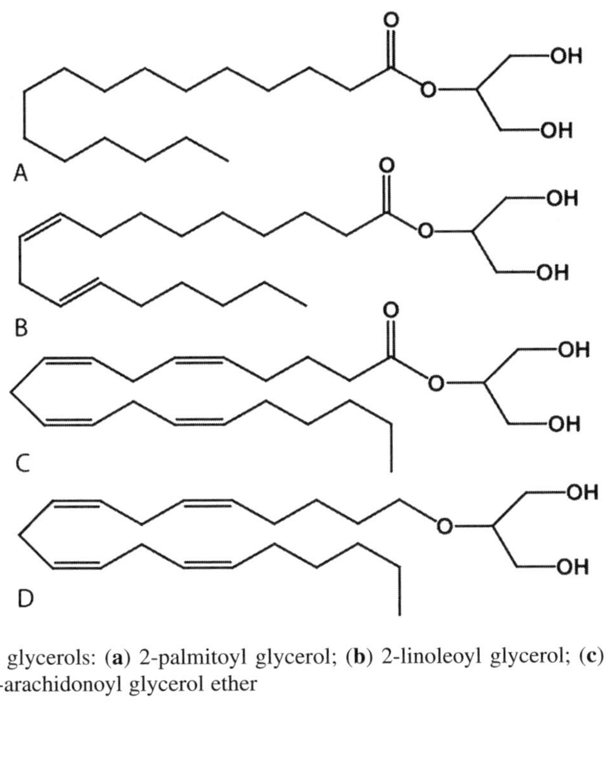

Fig. 3 2-Acyl glycerols: (**a**) 2-palmitoyl glycerol; (**b**) 2-linoleoyl glycerol; (**c**) 2-arachidonolyl glycerol; (**d**) 2-arachidonoyl glycerol ether

N-Linoleoyl Ethanolamide and *N*-Linolenoyl Ethanolamide

N-Linoleoyl ethanolamide that has an 18 carbon fatty acid with two double bonds conjugated to ethanolamine (LinEA; Fig. 2d), *N*-linolenoyl ethanolamide that has an 18 carbon fatty acid with three double bonds conjugated to ethanolamine (LenEA; Fig. 2e), PEA, SEA, and OEA, were isolated from mouse J774 macrophages and N18 neuroblastoma cells (DiMarzo et al., 1996) as well as RBL-2H3 leukocytes (Bisogno et al., 1997). The levels of these compounds were significantly increased by addition of ionomycin in each system (DiMarzo et al., 1996; Bisogno et al., 1997). LinEA inhibits fatty acid amide hydrolase (FAAH; Maurelli et al., 1995; Maccarrone et al., 1998), and was shown to inhibit sea urchin fertilization (Berdyshev, 1999). The production LinEA and LenEA was shown to dramatically increase upon refeeding from a 24-h period of starvation in the duodenum and jejunum of the rat (Fu et al., 2007). Taken together, these data demonstrate that the *N*-acyl ethanolamines have a wide range of bioactivity.

Acyl Glycerols

2-Linoleoyl Glycerol and 2-Palmitoyl Glycerol

2-Linoleoyl glycerol has an 18 carbon-chain fatty acid with two double bonds conjugated to a glycerol at the second carbon (2-LG; Fig. 3a) and 2-palmitoyl glycerol that has a fully saturated 16 carbon-chain fatty acid conjugated to a glycerol at the second carbon (2-PG; Fig. 3b) and they share structural homology with the endogenous cannabinoid 2-ararchidonoyl glycerol (Fig. 3c). 2-LG and 2-PG were isolated in mouse gut (Mechoulam et al., 1995), brain (Sugiura et al., 1995), spleen (Ben-Shabat et al., 1998), and breast milk (Fride et al., 2001). Whereas neither 2-LG nor 2-PG binds appreciably to CB_2 receptors, when combined with 2-AG in the same percentages measured in tissue, these compounds markedly potentiated the binding of 2-AG to CB_2 receptors causing a decrease in the K_i for 2-AG from 1,640 ± 260 nM to 273 ± 22 nM (Ben-Shabat et al., 1998). The same synergistic effects were demonstrated in the aforementioned behavioral tests (Ben-Shabat et al., 1998). Coinjections of 2-LG and 2-PG with the CB_1 antagonist, SR141716A and 2-AG in neonatal pups delayed morality rates induced by injection of SR141716A and 2-AG alone. These data provide additional evidence of the enhancement or synergistic effects of these structurally similar, though CB_1 receptor-inactive biological lipids.

Noladin Ether

2-Arachidonyl glycerol ether (noladin ether; Hanus et al., 2001) is also structurally similar in that it consists of arachidonic acid and glycerol with the exception that the linkage to the glycerol moiety is an ether vs. an ester as is the case for the other compounds in this class (Fig. 3d). Noladin ether was identified by Hanus and coworkers (2001) and confirmed by Fezza and coworkers (2002) in which it was demonstrated to occur in relatively high amounts in dissected thalamus. Oka and colleagues (2003) and Richardson and colleagues (2007), however, failed to measure noladin ether in nervous tissue. Therefore, the endogenous production of noladin ether remains in question; however, it has been shown to possess a range of biological activity and may, therefore, represent an additional avenue for therapy. Hanus and colleagues (2001) showed that the compound produces analgesic effects in the hot plate test following systemic administration in mice (20 mg/kg, *i.p.*), binds to CB_1 but not CB_2 receptors, produces hypothermia, catalepsy, and decreases in locomotor activity. Additionally, noladin ether was more effective and demonstrated a more persistent response in decreasing intraocular pressure than either AEA or 2-AG (Laine et al., 2002).

Concluding Remarks

The discovery of the endogenous cannabinoids led the way for the discovery and characterization of entire families of lipid signaling molecules that have an ever-increasing repertoire of biological activity. Most of the lipid signaling molecules discussed here are found throughout the body and brain and have been shown to activate both GPCRs and TRP channels. Much work is needed to fully elucidate the roles of each of these cannabimimetic lipid signaling molecules.

References

Ahern GP (2003) Activation of $TRPV_1$ by the satiety factor oleoylethanolamide. J Biol Chem 278:30429–30434.

Ben-Shabat S, Fride E, Sheskin T, Tamiri T, Rhee MH, Vogel Z, Bisogno T, De Petrocellis L, Di Marzo V, Mechoulam R (1998) An entourage effect: inactive endogenous fatty acid glycerol esters enhance 2-arachidonoyl-glycerol cannabinoid activity. Eur J Pharmacol 353:23–31.

Berdyshev EV (1999) Inhibition of sea urchin fertilization by fatty acid ethanolamides and cannabinoids. Comp Biochem Physiol C Pharmacol Toxicol Endocrinol 122:327–330.

Bisogno T, Sepe N, Melck D, Maurelli S, De Petrocellis L, Di Marzo V (1997) Biosynthesis, release and degradation of the novel endogenous cannabimimetic metabolite 2-arachidonoyl-glycerol in mouse neuroblastoma cells. Biochem J 322:671–677.

Burstein SH, Rossetti RG, Yagen B, Zurier RB (2000) Oxidative metabolism of anandamide. Prostaglandins Other Lipid Mediat 61:29–41.

Butelman ER, Ball JW, Harris TJ, Kreek MJ (2003) Topical capsaicin-induced allodynia in unanesthetized primates: pharmacological modulation. J Pharmacol Exp Ther 306:1106–1114.

Calignano A, La Rana G, Giuffrida A, Piomelli D (1998) Control of pain initiation by endogenous cannabinoids. Nature 394:277–281.

Calignano A, La Rana G, Piomelli D (2001) Antinociceptive activity of the endogenous fatty acid amide, palmitylethanolamide. Eur J Pharmacol 419:191–198.

Darmani NA, Izzo AA, Degenhardt B, Valenti M, Scaglione G, Capasso R, Sorrentini I, Di Marzo V (2005) Involvement of the cannabimimetic compound, N-palmitoyl-ethanolamine, in inflammatory and neuropathic conditions: review of the available pre-clinical data, and first human studies. Neuropharmacology 48:1154–1163.

De Petrocellis L, Bisogno T, Maccarrone M, Davis JB, Finazzi-Agro A, Di Marzo V (2001) The activity of anandamide at vanilloid VR_1 receptors requires facilitated transport across the cell membrane and is limited by intracellular metabolism. J Biol Chem 276:12856–12863.

Devane WA, Hanus L, Breuer A, Pertwee RG, Stevenson LA, Griffin G, Gibson D, Mandelbaum A, Etinger A, Mechoulam R (1992) Isolation and structure of a brain constituent that binds to the cannabinoid receptor. Science 258:1946–1949.

Di Marzo V, De Petrocellis L, Sepe N, Buono A (1996) Biosynthesis of anandamide and related acylethanolamides in mouse J774 macrophages and N18 neuroblastoma cells. Biochem J 316:977–984.

Facci L, Dal Toso R, Romanello S, Buriani A, Skaper SD, Leon A (1995) Mast cells express a peripheral cannabinoid receptor with differential sensitivity to anandamide and palmitoylethanolamide. Proc Natl Acad Sci USA 92:3376–3380.

Fezza F, Bisogno T, Minassi A, Appendino G, Mechoulam R, Di Marzo V (2002) Noladin ether, a putative novel endocannabinoid: inactivation mechanisms and a sensitive method for its quantification in rat tissues. FEBS Lett 513:294–298.

Fride E, Ginzburg Y, Breuer A, Bisogno T, Di Marzo V, Mechoulam R (2001) Critical role of the endogenous cannabinoid system in mouse pup suckling and growth. Eur J Pharmacol 419:207–214.

Fu J, Gaetani S, Oveisi F, Lo Verme J, Serrano A, Rodriguez De Fonseca F, Rosengarth A, Luecke H, Di Giacomo B, Tarzia G, Piomelli D (2003) Oleylethanolamide regulates feeding and body weight through activation of the nuclear receptor PPAR-alpha. Nature 425:90–93.

Fu J, Astarita G, Gaetani S, Kim J, Cravatt BF, Mackie K, Piomelli D (2007) Food intake regulates oleoylethanolamide formation and degradation in the proximal small intestine. J Biol Chem 282:1518–1528.

Gavva NR, Klionsky L, Qu Y, Shi L, Tamir R, Edenson S, Zhang TJ, Viswanadhan VN, Toth A, Pearce LV, Vanderah TW, Porreca F, Blumberg PM, Lile J, Sun Y, Wild K, Louis JC, Treanor JJ (2004) Molecular determinants of vanilloid sensitivity in $TRPV_1$. J Biol Chem 279:20283–20295.

Gomez R, Navarro M, Ferrer B, Trigo JM, Bilbao A, Del Arco I, Cippitelli A, Nava F, Piomelli D, Rodriguez de Fonseca F (2002) A peripheral mechanism for CB_1 cannabinoid receptor-dependent modulation of feeding. J Neurosci 22:9612–9617.

Guzman M, Lo Verme J, Fu J, Oveisi F, Blazquez C, Piomelli D (2004) Oleoylethanolamide stimulates lipolysis by activating the nuclear receptor peroxisome proliferator-activated receptor alpha (PPAR-α). J Biol Chem 279:27849–27854.

Hanus L, Gopher A, Almog S, Mechoulam R (1993) Two new unsaturated fatty acid ethanolamides in brain that bind to the cannabinoid receptor. J Med Chem 36:3032–3034.

Hanus L, Abu-Lafi S, Fride E, Breuer A, Vogel Z, Shalev DE, Kustanovich I, Mechoulam R (2001) 2-arachidonyl glyceryl ether, an endogenous agonist of the cannabinoid CB_1 receptor. Proc Natl Acad Sci USA 98:3662–3665.

Harrison S, De Petrocellis L, Trevisani M, Benvenuti F, Bifulco M, Geppetti P, Di Marzo V (2003) Capsaicin-like effects of N-arachidonoyl-dopamine in the isolated guinea pig bronchi and urinary bladder. Eur J Pharmacol 475:107–114.

Huang SM, Walker JM (2006) Enhancement of spontaneous and heat-evoked activity in spinal nociceptive neurons by the endovanilloid/endocannabinoid N-arachidonoyldopamine (NADA). J Neurophysiol 95:1207–1212.

Huang SM, Bisogno T, Petros TJ, Chang SY, Zavitsanos PA, Zipkin RE, Sivakumar R, Coop A, Maeda DY, De Petrocellis L, Burstein S, Di Marzo V, Walker JM (2001) Identification of a new class of molecules, the arachidonyl amino acids, and characterization of one member that inhibits pain. J Biol Chem 276:42639–42644.

Huang SM, Bisogno T, Trevisani M, Al-Hayani A, De Petrocellis L, Fezza F, Tognetto M, Petros TJ, Krey JF, Chu CJ, Miller JD, Davies SN, Geppetti P, Walker JM, Di Marzo V (2002) An endogenous capsaicin-like substance with high potency at recombinant and native vanilloid VR1 receptors. Proc Natl Acad Sci USA 99:8400–8405.

Ikeda Y, Iguchi H, Nakata M, Ioka RX, Tanaka T, Iwasaki S, Magoori K, Takayasu S, Yamamoto TT, Kodama T, Yada T, Sakurai T, Yanagisawa M, Sakai J (2005) Identification of N-arachidonylglycine, U18666A, and 4-androstene-3,17-dione as novel insulin Secretagogues. Biochem Biophys Res Commun 333:778–786.

Jaggar SI, Sellaturay S, Rice AS (1998) The endogenous cannabinoid anandamide, but not the CB_2 ligand palmitoylethanolamide, prevents the viscero-visceral hyper-reflexia associated with inflammation of the rat urinary bladder. Neurosci Lett 253:123–126.

Karava V, Fasia L, Siafaka-Kapadai A (2001) Anandamide amidohydrolase activity, released in the medium by Tetrahymena pyriformis. Identification and partial characterization. FEBS Lett 508:327–331.

Kohno M, Hasegawa H, Inoue A, Muraoka M, Miyazaki T, Oka K, Yasukawa M (2006) Identification of N-arachidonylglycine as the endogenous ligand for orphan G-protein-coupled receptor GPR18. Biochem Biophys Res Commun 347:827–832.

Kuehl FA, Jacob TA, Ganley OH, Ormond RE, Meisinger MAP (1957) The identification of N-3(hydroxyethyl)-palmitamide as a naturally occurring anti-inflammatory agent. J Am Chem Soc 79:5577–5578.

Laine K, Jarvinen K, Mechoulam R, Breuer A, Jarvinen T (2002) Comparison of the enzymatic stability and intraocular pressure effects of 2-arachidonylglycerol and noladin ether, a novel putative endocannabinoid. Invest Ophthalmol Vis Sci 43:3216–3222.

Lambert DM, Di Marzo V (1999) The palmitoylethanolamide and oleamide enigmas: are these two fatty acid amides cannabimimetic? Curr Med Chem 6:757–773.

Lambert DM, Vandevoorde S, Jonsson KO, Fowler CJ (2002) The palmitoylethanolamide family: a new class of anti-inflammatory agents? Curr Med Chem 9:663–674.

Maccarrone M, van der Stelt M, Rossi A, Veldink GA, Vliegenthart JF, Agro AF (1998) Anandamide hydrolysis by human cells in culture and brain. J Biol Chem 273:32332–32339.

Maccarrone M, Attina M, Bari M, Cartoni A, Ledent C, Finazzi-Agro A (2001a) Anandamide degradation and N-acylethanolamines level in wild-type and CB_1 cannabinoid receptor knockout mice of different ages. J Neurochem 78:339–348.

Maccarrone M, Bari M, Battista N, Di Rienzo M, Finazzi-Agro A (2001b) Endogenous cannabinoids in neuronal and immune cells: toxic effects, levels and degradation. Funct Neurol 16:53–60.

Maurelli S, Bisogno T, De Petrocellis L, Di Luccia A, Marino G, Di Marzo V (1995) Two novel classes of neuroactive fatty acid amides are substrates for mouse neuroblastoma 'anandamide amidohydrolase'. FEBS Lett 377:82–86.

Mazzari S, Canella R, Petrelli L, Marcolongo G, Leon A (1996) N-(2-hydroxyethyl)hexadecana mide is orally active in reducing edema formation and inflammatory hyperalgesia by down-modulating mast cell activation. Eur J Pharmacol 300:227–236.

Mechoulam R, Ben-Shabat S, Hanus L, Ligumsky M, Kaminski NE, Schatz AR, Gopher A, Almog S, Martin BR, Compton DR, et al. (1995) Identification of an endogenous 2-monoglyceride, present in canine gut, that binds to cannabinoid receptors. Biochem Pharmacol 50:83–90.

Milman G, Maor Y, Abu-Lafi S, Horowitz M, Gallily R, Batkai S, Mo FM, Offertaler L, Pacher P, Kunos G, Mechoulam R (2006) N-arachidonoyl L-serine, an endocannabinoid-like brain constituent with vasodilatory properties. Proc Natl Acad Sci USA 103:2428–2433.

Oka S, Tsuchie A, Tokumura A, Muramatsu M, Suhara Y, Takayama H, Waku K, Sugiura T (2003) Ether-linked analogue of 2-arachidonoylglycerol (noladin ether) was not detected in the brains of various mammalian species. J Neurochem 85:1374–1381.

Overton HA, Babbs AJ, Doel SM, Fyfe MC, Gardner LS, Griffin G, Jackson HC, Procter MJ, Rasamison CM, Tang-Christensen M, Widdowson PS, Williams GM, Reynet C (2006)

Deorphanization of a G protein-coupled receptor for oleoylethanolamide and its use in the discovery of small-molecule hypophagic agents. Cell Metab 3:167–175.
Premkumar LS, Qi ZH, Van Buren J, Raisinghani M (2004) Enhancement of potency and efficacy of NADA by PKC-mediated phosphorylation of vanilloid receptor. J Neurophysiol 91:1442–1449.
Prusakiewicz JJ, Kingsley PJ, Kozak KR, Marnett LJ (2002) Selective oxygenation of N-arachidonylglycine by cyclooxygenase-2. Biochem Biophys Res Commun 296:612–617.
Rakhshan F, Day TA, Blakely RD, Barker EL (2000) Carrier-mediated uptake of the endogenous cannabinoid anandamide in RBL-2H3 cells. J Pharmacol Exp Ther 292:960–967.
Richardson D, Ortori CA, Chapman V, Kendall DA, Barrett DA (2007) Quantitative profiling of endocannabinoids and related compounds in rat brain using liquid chromatography-tandem electrospray ionization mass spectrometry. Anal Biochem 360:216–226.
Rodriguez de Fonseca F, Navarro M, Gomez R, Escuredo L, Nava F, Fu J, Murillo-Rodriguez E, Giuffrida A, LoVerme J, Gaetani S, Kathuria S, Gall C, Piomelli D (2001) An anorexic lipid mediator regulated by feeding. Nature 414:209–212.
Sagar DR, Smith PA, Millns PJ, Smart D, Kendall DA, Chapman V (2004) $TRPV_1$ and CB_1 receptor-mediated effects of the endovanilloid/endocannabinoid N-arachidonoyl-dopamine on primary afferent fibre and spinal cord neuronal responses in the rat. Eur J Neurosci 20:175–184.
Salzet M, Stefano GB (2002) The endocannabinoid system in invertebrates. Prostaglandins Leukot Essent Fatty Acids 66:353–361.
Sancho R, Macho A, de La Vega L, Calzado MA, Fiebich BL, Appendino G, Munoz E (2004) Immunosuppressive activity of endovanilloids: N-arachidonoyl-dopamine inhibits activation of the NF-kappa B, NFAT, and activator protein 1 signaling pathways. J Immunol 172:2341–2351.
Schmid HH, Berdyshev EV (2002) Cannabinoid receptor-inactive N-acylethanolamines and other fatty acid amides: metabolism and function. Prostaglandins Leukot Essent Fatty Acids 66:363–376.
Schuel H, Burkman LJ, Lippes J, Crickard K, Forester E, Piomelli D, Giuffrida A (2002) N-Acylethanolamines in human reproductive fluids. Chem Phys Lipids 121:211–227.
Sheskin T, Hanus L, Slager J, Vogel Z, Mechoulam R (1997) Structural requirements for binding of anandamide-type compounds to the brain cannabinoid receptor. J Med Chem 40:659–667.
Sugiura T, Kondo S, Sukagawa A, Nakane S, Shinoda A, Itoh K, Yamashita A, Waku K (1995) 2-Arachidonoylglycerol: a possible endogenous cannabinoid receptor ligand in brain. Biochem Biophys Res Commun 215:89–97.
Toth A, Kedei N, Wang Y, Blumberg PM (2003) Arachidonyl dopamine as a ligand for the vanilloid receptor VR_1 of the rat. Life Sci 73:487–498.
Wiles AL, Pearlman RJ, Rosvall M, Aubrey KR, Vandenberg RJ (2006) N-Arachidonyl-glycine inhibits the glycine transporter, $GLYT_{2a}$. J Neurochem 99:781–786.

Chapter 5
CB_1 Cannabinoid Receptors: Molecular Biology, Second Messenger Coupling and Polarized Trafficking in Neurons

Andrew J. Irving, Neil A. McDonald, and Tibor Harkany

Abstract The type 1 cannabinoid receptor (CB_1 receptor) is considered to be the most abundant G protein-coupled receptor (GPCR) in the mammalian brain. The presence and highly compartmentalized cellular distribution of CB_1 receptors in neurons localized to corticolimbic areas, basal ganglia, cerebellum, and brainstem accounts for the majority of behavioral actions associated with cannabinoid drugs. The discovery of endocannabinoids led to an avalanche of data showing that signaling at this GPCR is critical for, e.g., neurogenesis, neural development, synaptic plasticity, learning and memory, food intake, and energy metabolism. In contrast, deficient CB_1 receptor expression or coupling to downstream signal transduction cascades contributes to the neuropathogenesis of a broad variety of neurological and metabolic disorders with selective pharmacological modulation of CB_1 receptor availability and activity being a prime target for therapeutic intervention. Here, we summarize contemporary knowledge on the regulation of CB_1 receptor expression in the central nervous system and describe the context-dependent recruitment of second messengers to this receptor. Finally, we present the concept that CB_1 receptor bioavailability together with its momentary signaling activity on neuronal membranes defines the efficacy of endocannabinoid signaling such that a fine-tuned control of synaptic efficacy and plasticity may be achieved.

Introduction

The Physiological Significance of CB_1 Cannabinoid Receptors in the Central Nervous System

The endogenous cannabinoid system plays pivotal roles in regulating diverse and fundamental (patho-)physiological processes including, e.g., the control of food intake, pain sensation, inflammation, and cognition (Lutz, 2004; Di Marzo and Izzo, 2006; Mackie, 2006; Mackie and Stella, 2006; Pertwee, 2006). Cellular actions of both *Cannabis spp.*-derived phytocannabinoids and endocannabinoids,

A. Köfalvi (ed.), *Cannabinoids and the Brain.*

endogenous compounds with functional similarity to Δ^9-tetrahydrocannabinol (Δ^9-THC) from marijuana, are mediated by at least three types of receptors in the body: CB_1 and CB_2 cannabinoid receptors (CB_1/CB_2 receptors) and the orphan G protein-coupled receptor GPR55 (Mackie, 2006; Mackie and Stella, 2006; Petitet et al., 2006; Harkany et al., 2007). According to the broadly accepted concept, the CB_1 receptor is the predominant cannabinoid-sensing receptor subtype expressed on neural cells in the central nervous system (CNS). In particular, CB_1 receptors exhibit highest concentrations on perisynaptic axons segments of both γ-aminobutyric acid (GABA)-containing and glutamatergic neurons (Katona et al., 1999, 2006), although functional evidence also supports their availability also on neuronal somata, albeit at lower levels (Freund et al., 2003). The major physiological effect of CB_1 receptor activation in the CNS is to modulate synaptic communication between neurons, and this occurs primarily via the presynaptic regulation of neurotransmitter release (Freund et al., 2003). CB_1 receptor expression on astrocytes and microglia is yet controversial and is at a considerably lower density than on neurons (Molina-Holgado et al., 2002; Ramirez et al., 2005). In contrast, CB_2 receptor are commonly associated with the regulation of immune function. However, although recent findings indicate the presence of CB_2 receptors on brainstem neurons (Van Sickle et al., 2005) thus providing a more complex view on the cellular regulation of (endo)cannabinoid functions in the CNS. Intriguingly, the CB_1 receptor is one of the most abundantly expressed G protein-coupled receptors (GPCRs) within the brain (Howlett, 1998) with an unprecedented propensity of signaling interactions with other neurotransmitter systems to establish, maintain, or refine synaptic communication between neurons (Howlett, 1995; Irving et al., 2000; Alger, 2002; Berghuis et al., 2007; Harkany et al., 2007). Here, we discuss various aspects of the regulation of CB_1 receptor expression and functions, covering the molecular biology of the CB_1 receptor, its intracellular signaling principles, and its trafficking in neurons, to reveal the backbones of endocannabinoid signaling and the relevance of CB_1 receptor as a therapeutic target to the future treatment of neurological and metabolic diseases.

Molecular Biology of the CB_1 Receptor: Cloning and Initial Characterization

CB_1 Receptor Structure and Sequence Homology

The CB_1 receptor belongs to the GPCR superfamily containing seven α-helical transmembrane domains (Fig. 1a) with 68% amino acid homology within the transmembrane domains, and with a 44% overall homology to the CB_2 receptor (in humans) (Munro et al., 1993). The CB_1 receptor signals through the preferential

A

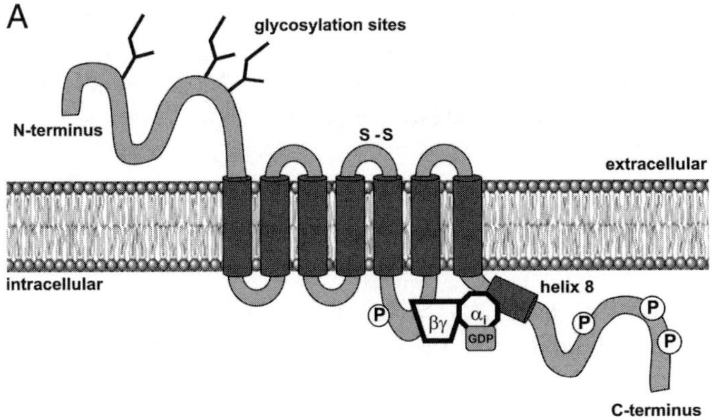

B

```
Rat     MKSILDGLADTTFRTITTDLLYVGSNDIQYEDIKGDMASKLGYFPQKFPLTSFRGSPFQE 60
Human   MKSILDGLADTTFRTITTDLLYVGSNDIQYEDIKGDMASKLGYFPQKFPLTSFRGSPFQE 60
CNR1A                                                             MAL 3
CNR1B   MKSILDGLADTTFRTITTDLL---------------------------------GSPFQE 27
        N-terminal Domain

Rat     KMTAGDNSPLVPAGDTTNITEFYNKSLSSFKENEENIQCGENFMDMECFMILNPSQQLAI 120
Human   KMTAGDNPQLVPA-DQVNITEFYNKSLSSFKENEENIQCGENFMDIECFMVLNPSQQLAI 119
CNR1A   QIPPSAPSPLTSC-TWAQMTFSTRTS---FKENEENIQCGENFMDIECFMVLNPSQQLAI 59
CNR1B   KMTAGDNPQLVPA-DQVNITEFYNKSLSSFKENEENIQCGENFMDIECFMVLNPSQQLAI 86

Rat     AVLSLTLGTFTVLENLLVLCVILHSRSLRCRPSYHFIGSLAVADLLGSVIFVYSFVDFHV 180
Human   AVLSLTLGTFTVLENLLVLCVILHSRSLRCRPSYHFIGSLAVADLLGSVIFVYSFIDFHV 179
CNR1A   AVLSLTLGTFTVLENLLVLCVILHSRSLRCRPSYHFIGSLAVADLLGSVIFVYSFIDFHV 119
CNR1B   AVLSLTLGTFTVLENLLVLCVILHSRSLRCRPSYHFIGSLAVADLLGSVIFVYSFIDFHV 146
        TM1 | ICL1 | TM2

Rat     FHRKDSPNVFLFKLGGVTASFTASVGSLFLTAIDRYISIHRPLAYKRIVTRPKAVVAFCL 240
Human   FHRKDSRNVFLFKLGGVTASFTASVGSLFLTAIDRYISIHRPLAYKRIVTRPKAVVAFCL 239
CNR1A   FHRKDSRNVFLFKLGGVTASFTASVGSLFLTAIDRYISIHRPLAYKRIVTRPKAVVAFCL 179
CNR1B   FHRKDSRNVFLFKLGGVTASFTASVGSLFLTAIDRYISIHRPLAYKRIVTRPKAVVAFCL 206
        ECL1 | TTM3 | ICL2

Rat     MWTIAIVIAVLPLLGWNCKKLQSVCSDIFPLIDETYLMFWIGVTSVLLLFIVYAYMYILW 300
Human   MWTIAIVIAVLPLLGWNCEKLQSVCSDIFPHIDETYLMFWIGVTSVLLLFIVYAYMYILW 299
CNR1A   MWTIAIVIAVLPLLGWNCEKLQSVCSDIFPHIDETYLMFWIGVTSVLLLFIVYAYMYILW 239
CNR1B   MWTIAIVIAVLPLLGWNCEKLQSVCSDIFPHIDETYLMFWIGVTSVLLLFIVYAYMYILW 266
        TMD4 | ECL2 | TMD5

Rat     KAHSHAVRMIQRGTQKSIIIHTSEDGKVQVTRPDQARMDIRLAKTLVLILVVLIICWGPL 360
Human   KAHSHAVRMIQRGTQKSIIIHTSEDGKVQVTRPDQARMDIRLAKTLVLILVVLIICWGPL 359
CNR1A   KAHSHAVRMIQRGTQKSIIIHTSEDGKVQVTRPDQARMDIRLAKTLVLILVVLIICWGPL 299
CNR1B   KAHSHAVRMIQRGTQKSIIIHTSEDGKVQVTRPDQARMDIRLAKTLVLILVVLIICWGPL 326
        ICL3 | TMD6

Rat     LAIMVYDVFGKMNKLIKTVFAFCSMLCLLNSTVNPIIYALRSKDLRHAFRSMFPSCEGTA 420
Human   LAIMVYDVFGKMNKLIKTVFAFCSMLCLLNSTVNPIIYALRSKDLRHAFRSMFPSCEGTA 419
CNR1A   LAIMVYDVFGKMNKLIKTVFAFCSMLCLLNSTVNPIIYALRSKDLRHAFRSMFPSCEGTA 359
CNR1B   LAIMVYDVFGKMNKLIKTVFAFCSMLCLLNSTVNPIIYALRSKDLRHAFRSMFPSCEGTA 386
        | ECL3 | TMD7 |
        helix-8

Rat     QPLDNSMGDSDCLHKHANNTASMHRAAESCIKSTVKIAKVTMSVSTDTSAEAL 473
Human   QPLDNSMGDSDCLHKHANNAASVHRAAESCIKSTVKIAKVTMSVSTDTSAEAL 472
CNR1A   QPLDNSMGDSDCLHKHANNAASVHRAAESCIKSTVKIAKVTMSVSTDTSAEAL 412
CNR1B   QPLDNSMGDSDCLHKHANNAASVHRAAESCIKSTVKIAKVTMSVSTDTSAEAL 439
        C-terminal Domain
```

Fig. 1 (**a**) Structure of the CB_1 receptor indicating its major functional domains and sites of posttranslational modifications. (**b**) Amino acid sequence alignment of the rat and human CB_1 receptor, and its splice variants *CNR1A* and *CNR1B*. *ECL* extracellular loop; *ICL* intracellular loop; *TM* transmembrane domain

recruitment of $G_{i/o\alpha}$ proteins (Mackie, 2006) with a potential switch in G protein coupling to G_s ($G_{q/11}$) proteins as determined by ligand availability (Lauckner et al., 2005) and receptor interactions (Wager-Miller et al., 2002; Kearn et al., 2005; Rios et al., 2006; Harkany et al., 2007) (see Chap. 9).

Receptor Expression and Splice Variants

The CB_1 receptor was originally cloned as an orphan GPCR from a rat cDNA library based on its homology to the bovine substance K receptor (Matsuda et al., 1990; Westlake et al., 1994) with its gene (*CNR1*) located on chromosome 6q14-q15. The functional identity of CB_1 receptor has been revealed by the matching overlap between the distribution of its mRNA throughout the brain and of the specific binding sites for [^{3}H]CP55940, a synthetic cannabinoid (Herkenham et al., 1990) with highest CB_1 receptor density concentrating in the basal ganglia (substantia nigra *pars reticulata* and globus pallidus), hippocampus, and cerebellum. Its human homolog has subsequently been identified (Gerard et al., 1991). A splice variant of the CB_1 receptor mRNA; Ryberg et al., 2005; see chapter 9 has also been identified in human and rat tissues (Rinaldi-Carmona et al., 1996 ($CNR1_A$)) but the confirmed existence of a translated protein product is as yet elusive. More recently, two splice variant of the human CB_1 receptor, ($CNR1_B$), generated by in-frame deletion of amino acids within the N terminus, has been identified and shown to be expressed at very low levels in various tissues (Ryberg et al., 2005) (Fig. 1b; see Chap. 9). Based on the recent association of single nucleotide polymorphisms with obesity-related phenotypes and polysubstance abuse (Zhang et al., 2004; Russo et al., 2007), the concept emerges that genetic variation(s) in the *CNR1* gene can pose increased risk to, e.g., metabolic abnormalities and psychiatric illnesses.

N-Terminal Truncation and Plasma Membrane Expression

CB_1 receptor isoforms of varying molecular sizes have been found in several cellular systems (Wager-Miller et al., 2002; Nordstrom and Andersson, 2006). Notably, the CB_1 receptor contains a 116 amino acid residue-long N-terminal extracellular domain (Fig. 1a), which plays a role in determining the efficiency of receptor biogenesis and plasma membrane expression. Like the majority of GPCRs, the CB_1 receptor does not contain a cleavable N-terminal signal peptide (Andersson et al., 2003) and is thought to rely on transmembrane domains acting as internal signal sequences to direct correct protein translocation into the endoplasmic reticulum (ER) membrane. Partial truncation of the N-terminal tail of the CB_1 receptor has been detected in various cell lineages in vitro, and is due to the fast proteolytic processing of de novo synthesized receptors in the cytoplasm prior to their translocation over the ER via a mechanism independent of the proteasome (Nordstrom and Andersson, 2006). Studies by Andersson et al. (2003) demonstrated that shortening the N-terminus of the CB_1 receptor or the inclusion of a signal peptide greatly increases receptor stability, and results in its increased targeting to the cell surface. In contrast, the large N-terminus of the endogenous CB_1 receptor is thought to inhibit efficient receptor translocation across the ER, leading to high levels of misfolded CB_1 receptors that are rerouted toward proteasomal degradation. Consistent with this, increasing the length of the N-terminus with a green fluorescence protein (GFP) fusion construct inhibits its surface expression, presumably by greatly increasing the bulk of that region. However, cell-surface expression can be rescued by inclusion of an artificial, signal peptide upstream of GFP (McDonald et al., 2007b). Thus, based on the altered ligand-binding properties of hCB_{1A} and hCB_{1B} variants (Ryberg et al., 2005), it may be assumed that N-terminal processing of the CB_1 receptor is a powerful means to regulate ligand specificity and cell-surface receptor availability (truncated isoforms will be more efficiently expressed than the full-length receptor). Our understanding of CB_1 receptor processing, trafficking to the cell surface, and conformational modifications brought upon by posttranslational modifications are of direct therapeutic significance as these changes generate novel receptors with substantially different signaling properties.

CB_1 Receptor Homodimerization Generates Functional Receptor Diversity

Besides the cell-type-specific generation of truncated CB_1 receptors, receptor homo-/heterodimerization provides an attractive alternative for modifying second messenger signaling (Devi, 2000; Wager-Miller et al., 2002; Rios et al., 2006). During the past decade, it has become increasingly apparent that many, if not all, GPCRs exist as multimers, and these may be considered as the functional units of GPCR signaling. GPCR multimerization plays a critical role in enriching the signaling repertoire of these receptors. The existence of a "CB_1 receptor homodimer" was first demonstrated by immunoprecipitation analysis using an antibody directed against a C-terminal CB_1 receptor epitope (Wager-Miller et al., 2002). Subsequent comparative neuroanatomical studies employing this "anti-homodimer" antibody and other N- and C-terminal antibodies recognizing both receptor monomers and dimers revealed indistinguishable labeling patterns between the monomeric and multimeric forms of the receptor (Katona et al., 2001). An appealing interpretation of these data is that CB_1 receptors usually exist as dimers or higher order multimers (Mackie, 2005). Although many structural and biochemical aspects of CB_1 receptor homodimerization remains unresolved, the emerging existence of CB_1 receptor homodimers as potential signaling units may contribute to the many faces of endocannabinoid actions, and unexpected pharmacological behaviors of several ligands under specific cellular conditions (see Chap. 9).

Natural and Synthetic Ligands with High Affinity for the CB_1 Receptor

Identification of the CB_1 receptor was paralleled by the discovery of its endogenous ligands (Piomelli, 2003; Pertwee, 2006). Endocannabinoids, including arachidonoylethanolamide (anandamide, AEA) and 2-arachidonoylglycerol (2-AG), are all eicosanoids and are synthesized *on demand* in multistep enzymatic pathways (see Chap. 2). Besides eicosanoids, the best-known examples of CB_1 receptor agonists include (1) the "classical" cannabinoids, Δ^9-THC and (–)-11-hydroxy- Δ^8-THC-dimethylheptyl (HU-210), (2) the "nonclassical" cannabinoid, CP55940, and (3) the aminoalkylindole cannabinoid, R-(+)-WIN55212 (Pertwee, 2006; see Chap. 7). In turn, *O*-arachidonoylethanolamine (virodhamine) appears to act as an endogenous antagonist at this receptor (Porter et al., 2002). Nowadays, a key focus is directed toward the design of highly selective and potent agonists, inverse agonists, and neutral antagonists that can be of therapeutic significance in the treatment of a variety of diseases (ALS, multiple sclerosis, Alzheimer's (AD) and Parkinson's diseases, cancer) and pathophysiological conditions (obesity, metabolic disorders) (Guzman, 2003; Di Marzo and Izzo, 2006; Di Marzo and Petrocellis, 2006; Mackie, 2006; Galve-Roperh et al., 2007; and Part II of this book).

CB_1 Receptor Signal Transduction: Adenylyl Cyclase Inhibition, Effector Kinases, and Coupling to Ion Channels

Classically the CB_1 receptor is linked to $G_{i/o}$ mediated inhibition of adenylyl cyclase activity and a concordant decrease in cytosolic cAMP levels. Stimulation of effector kinase cascades, closure of Ca^{2+} channels, and opening of K^+ channels have also been documented (Piomelli, 2003; Szabo and Schlicker, 2005). The recruitment of particular signaling mechanisms translating CB_1 receptor activity into biological output appears to be dictated by the cellular context at which signaling occurs.

Inhibition of Adenylyl Cyclase Activity and Receptor Convergence

Activation of CB_1 receptors and the subsequent liberation of $G_{i/o\alpha}$ proteins couples to the inhibition of adenylate cyclases (Childers et al., 1993). The subsequent depletion of intracellular cAMP levels leads to the inactivation of the protein kinase A (PKA) phosphorylation pathway. The complexity of agonist-induced CB_1 receptor activation with downstream adenylyl cyclase inhibition in the brain is exemplified by the phenomenon termed receptor convergence when CB_1 receptors and other $G_{i/o}$-linked receptors (e.g., $GABA_B$ receptors) are coexpressed in particular neurons where they share common effector systems (adenylyl cyclase catalytic units) but not common G proteins. This sharing of effector mechanisms underpins that agonist stimulation of distinct receptor types can produce the same biological response.

Kinase Signaling

CB_1 receptor activation recruits complex networks of intracellular protein kinases that are physiologically critical in, e.g., producing lasting changes in synaptic strength. Cannabinoid agonists are particularly potent in stimulating the extracellular signal-regulated kinase (ERK) and focal adhesion kinase (FAK) cascades both in vitro and in vivo (Derkinderen et al., 1996; Derkinderen et al., 2003). This activation is mimicked by inhibitors of cAMP-dependent kinase and is abolished by cell-permeant cAMP analogues thus implying that ERK and FAK become activated upon a decrease in intracellular cAMP concentrations. The fundamental roles ERK and FAK kinases play in synaptic plasticity suggests that their cannabinoid-induced activation is a pivotal determinant of synaptic functions with long-term modifications to synaptic structure and efficacy brought about by the selective regulation of several synaptic plasticity-related genes. In addition, Δ^9-THC and endocannabinoids were shown to activate c-Jun N-terminal kinases 1/2 and the p38 mitogen-activated protein kinase (Rueda et al., 2000) with long-term activation of these kinase pathways being involved in Δ^9-THC-induced cell death. However, more complex protein phosphorylation cascades involving the release of G protein $\beta\gamma$ subunits and activation of phosphoinositide-3-kinase/Akt/glycogen synthase kinase 3β (Ozaita et al., 2007) and protein kinase B cascades (Galve-Roperh et al., 2002) are also triggered by CB_1 receptors and underscore persistent neuronal adaptations that accompany cannabinoid administration. Notably, stimulation, rather than inhibition, of adenylyl cyclases via G_s proteins has also been described. Decisions on which of these pathways will be modulated by CB_1 receptor activation is critically dependent on the cellular context, interacting proteins, temporal coincidence of active second messenger pathways, and the particular ligands activating the CB_1 receptor.

Ca^{2+} Channels

Agonist stimulation of CB_1 receptors commonly inhibits N- and P/Q-type voltage-activated Ca^{2+} channels in neuronal cell lines (Caulfield and Brown, 1992; Mackie and Hille, 1992; Mackie et al., 1995) and cultured neurons (Twitchell et al., 1997) through the direct interaction of $G_{i/o}$ protein $\beta\gamma$ subunits with these channels (Wilson and Nicoll, 2002). The requirement of G proteins in this inhibition is corroborated by the pertussis toxin sensitivity of this mechanism. This action appears physiologically critical when CB_1 receptors depress the release of the inhibitory neurotransmitter GABA at hippocampal and neocortical synapses (Hoffman and Lupica, 2000). Intriguingly, AEA exerts ligand-specific inhibition of T-type channels, an effect that is though independent of CB_1 receptor activation (Chemin et al., 2001). In contrast, cannabinoid ligands may enhance L-type Ca^{2+} currents as shown in immortalized neuronal cells (Rubovitch et al., 2002).

Voltage-Gated K^+ Channels

Agonist stimulation of the CB_1 receptor can also couple to multiple K^+ channels: stimulation of inwardly rectifying K^+ channels I_{Kir} has been commonly observed (Mackie et al., 1995; McAllister et al., 1999) together with the enhancement of potassium A currents (Deadwyler et al., 1995). In contrast, cannabinoid ligands inhibit both I_M (Schweitzer, 2000) and I_K currents in hippocampal neurons in vitro (Hampson et al., 2000; and consult with chapter 9). Physiologically, cannabinoid regulation of voltage-gated K^+ channels has been implicated in presynaptic inhibition at both GABAergic and glutamatergic synapses (Robbe et al., 2001; Kreitzer et al., 2002). These responses are also pertussis toxin sensitive implying that they are mediated by $G_{i/o}$ proteins, but it remains unclear whether signal transduction is mediated directly (βγ subunit) or indirectly (second messengers).

Agonist-Induced CB_1 Receptor Desensitization

In common with many GPCRS, CB_1 receptors undergo agonist-induced desensitization and this cellular process is thought to underlie the distinct pattern of tolerance that develops to cannabinoids (Martin et al., 2004). Short-term desensitization of CB_1 receptors involves receptor internalization and G protein uncoupling (Jin et al., 1999) and is dependent on G protein-coupled receptor kinase 3 and β-arrestin 2 (Jin et al., 1999). In various cell lines, agonist exposure results in rapid, clathrin-dependent endocytosis of CB_1 receptors (Rinaldi-Carmona et al., 1998; Hsieh et al., 1999; Jin et al., 1999; Keren and Sarne, 2003). However, studies in neurons suggest that agonist-induced CB_1 receptor internalization is much slower, with maximal effects achieved after 12–48 h (Coutts et al., 2001; Leterrier et al., 2006; Tappe-Theodor et al., 2007). Four amino acids (460–463) in the CB_1 receptor C-terminal tail are required for agonist-induced endocytosis, notably a region distinct from that required for functional desensitization (Hsieh et al., 1999; Jin et al., 1999). Receptor recycling following acute agonist exposure (20 min) requires both endosomal acidification and phosphatase activity (Hsieh et al., 1999) and does not involve new protein synthesis. However, protein synthesis is required for recovery of expression following longer incubation times (>90 min; Hsieh et al., 1999). Recent studies suggest that sustained CB_1 receptor activation leads to lysosomal targeting and receptor degradation, via a mechanism involving GPCR–Associated Sorting Protein 1 (GASP1; Martini et al., 2007). Chronic administration of cannabinoids in vivo is also associated with CB_1 receptor down-regulation, although the extent of this varies between CNS regions (Sim-Selley et al., 2006). Interestingly, in the spinal cord, an interaction between GASP1 and CB_1 receptors is thought to underlie receptor down-regulation and the development of analgesic tolerance to cannabinoids (Tappe-Theodor et al., 2007) thus underpinning a key therapeutic window for CB_1 receptor-selective ligands.

Domain-Specific CB_1 Receptor Endocytosis and Axonal Targeting

The correct trafficking of CB_1 receptors to the axonal surface (Fig. 2) is clearly critical for their physiological role in regulating synaptic transmission. It has been demonstrated that this process reflects domain-specific endocytosis (Leterrier et al., 2006; McDonald et al., 2007a) with somatodendritic receptors being internalized more rapidly than those on the axonal plasma membrane leading to a net receptor accumulation in the axon. Studies in HEK293 cells (Leterrier et al., 2004; Ellis et al., 2006), and neurons (Leterrier et al., 2006; McDonald et al., 2007a), suggest that CB_1 receptors undergo constitutive endocytosis and recycling, leading to a pronounced intracellular pool of receptors at steady state. Blockade of endocytosis pathways, using dominant-negative mutants of dynamin-1, dynamin-2, eps15, or rab5 results in a dramatic change in the distribution of cell surface CB_1 receptors from the axon to a nonpolarized state with pronounced somatodendritic plasma membrane expression (Leterrier et al., 2006; McDonald et al., 2007a). The effect of dominant-negative dynamin has been demonstrated for both recombinant CB_1

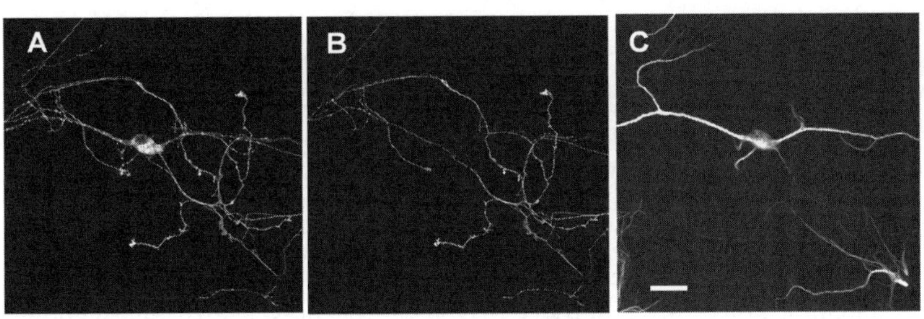

Fig. 2 Cell surface CB_1 receptors are highly polarized toward the axon. Representative z projection images of a cultured hippocampal neuron (9 days in vitro) expressing recombinant, N-terminally tagged GFP-CB_1 (**a**) and probed for surface expression of GFP (anti-GFP) (**b**), and intracellular MAP2 to label somatodendritic regions (anti-MAP2) (**c**). Note surface expression of GFP-CB_1 is restricted to the MAP2-negative axon. Scale bar = 20 μm

receptor expression and for native CB_1 receptors expressed in interneurons (McDonald et al., 2007a). Indeed, domain-specific endocytosis is recognized as a key mechanism for limiting the surface expression of a range of axonal proteins (Sampo et al., 2003; Wisco et al., 2003). It is also likely that transcytotic delivery of CB_1 receptors to the axonal plasma membrane from the somatodendritic cell surface contributes to the generation of CB_1 receptor cell-surface polarity and this may act as a salvage pathway for somatodendritc receptors. The precise mechanisms that underlie the preferential endocytosis of the CB_1 receptor within the somatodendritic compartment are unclear at present. Specific anchoring proteins present in axons may bind CB_1 receptors and stabilize them within the plasma membrane; however, these remain to be identified. Differences in the internalization machinery may also play a role; for example, the expression of dynamin subtypes varies between axonal and somatodendritic compartments (Gray et al., 2003) and adaptor complexes directing intracellular trafficking may be selectively targeted to distinct subcellular domains (Seong et al., 2005).

Constitutive CB_1 Receptor Activity is Not Required for Axonal Targeting

Constitutive receptor activity is the condition when ligand availability does not per se limit the recruitment of G proteins and downstream signaling. The extent to which the CB_1 receptor displays constitutive activity and its effects on CB_1 receptor trafficking are, however, controversial. For example, HEK293 cells expressing C-terminally-tagged fluorescent-CB_1 receptor chimeras display high levels of constitutive endocytosis, leading to a marked intracellular localization at steady state (D'Antona et al., 2006; Ellis et al., 2006; Leterrier et al., 2006). However, in AT20 cells expressing wild-type CB_1 receptor, the vast majority of receptor-derived fluorescence is membrane associated (Jin et al., 1999). In some studies, exposure to CB_1 receptor antagonists can lead to an up-regulation of cell surface CB_1 expression, for example in HEK293 cells (D'Antona et al., 2006; Ellis et al., 2006; Leterrier et al., 2006), CHO cells (Rinaldi-Carmona et al., 1998) and hippocampal neurons (Leterrier et al., 2006), which is thought to reflect inhibition of constitutive endocytosis. However, other work suggests that prolonged exposure of hippocampal neurons to CB_1 receptor antagonists does not lead to an up-regulation of wild-type cell surface CB_1 receptors expressed at the axonal plasma membrane (Coutts et al., 2001; McDonald et al., 2007a,b). An opportunity to rationalize these conflicting data is emerging from studies investigating apparent constitutive activity driven by the presence of endogenous ligands either produced by the cells themselves (Turu et al., 2007) or present in the serum component of cell culture media (Stoddart et al., 2007). It is suggested that cell-derived endocannabinoids may underlie CB_1 receptor basal activity in neuronal and nonneuronal cells, which can in turn stimulate receptor endocytosis (Turu et al., 2007). Evidence for a mechanism of CB_1 receptor axonal targeting independent of constitutive activity comes from studies with mutant GFP-CB_1 receptor chimeras that prevent agonist-induced endocytosis (Hsieh et al., 1999; Roche et al., 1999) and constitutive activation (D164N; Roche et al., 1999), which do not affect CB_1 receptor cell-surface polarity (McDonald et al., 2007a). Thus, constitutive endocytosis in the somatodendritic compartment is suggested to be a distinct process and likely to involve different motifs/conformational states within the CB_1 receptor than those utilized by agonist-induced internalization. Distinct pathways for clathrin-mediated GPCR enodcytosis have been reported in other GPCRs (Diviani et al., 2003; Mundell et al., 2006), and may involve a direct, β-arrestin-independent interaction with the AP2 complex (Diviani et al., 2003). Differences in structural/conformational requirements for constitutive and agonist-promoted endocytosis have also been identified in other GPCR systems (Waldhoer et al., 2003; Whistler et al., 2002). Importantly, further studies aimed at identifying the regions of the CB_1 receptor that are involved in constitutive endocytosis and to identify the protein(s) that interact with these sites will be required to conclusively define the contribution of constitutive CB_1 receptor activity to the many known functions of this receptor class.

Concluding Remarks

Cannabinoids produce the majority of their psychoactive effects through interaction with CB_1 receptors. CB_1 receptors are expressed predominantly in the CNS and transduce intracellular signals through coupling to $G_{i/o}$ proteins with downstream modulation of a broad array of signaling mechanisms. Interestingly, different CB_1 agonists can distinctly regulate multiple effectors. The many aspects and probable outcomes of context-dependent signaling through CB_1 receptors in the CNS suggest that bolstering our understanding of the regulatory mechanisms controlling the biosynthesis, bioavailability, and recycling of this receptor may provide new vistas for therapeutic interventions.

Acknowledgments This work was supported by the Swedish Medical Research Council (T.H.), Stiftelsen Ragnhild och Einar Lundströms Minne (T.H.), the Alzheimer's Association (T.H.), the Anonymous Trust (A.I.) and TENOVUS Scotland (A.I.).

References

Alger BE (2002) Retrograde signaling in the regulation of synaptic transmission: focus on endocannabinoids. Prog Neurobiol 68:247–286.

Andersson H, D'Antona AM, Kendall DA, Von Heijne G, Chin CN (2003) Membrane assembly of the cannabinoid receptor 1: impact of a long N-terminal tail. Mol Pharmacol 64:570–577.

Berghuis P, Rajnicek AM, Morozov YM, Ross RA, Mulder J, Urban GM, Monory K, Marsicano G, Matteoli M, Canty A, Irving AJ, Katona I, Yanagawa Y, Rakic P, Lutz B, Mackie K, Harkany T (2007) Hardwiring the brain: endocannabinoids shape neuronal connectivity. Science 316:1212–1216.

Caulfield MP, Brown DA (1992) Cannabinoid receptor agonists inhibit Ca^{2+} current in NG108–15 neuroblastoma cells via a pertussis toxin-sensitive mechanism. Br J Pharmacol 106:231–232.

Chemin J, Monteil A, Perez-Reyes E, Nargeot J, Lory P (2001) Direct inhibition of T-type calcium channels by the endogenous cannabinoid anandamide. EMBO J 20:7033–7040.

Childers SR, Pacheco MA, Bennett BA, Edwards TA, Hampson RE, Mu J, Deadwyler SA (1993) Cannabinoid receptors: G-protein-mediated signal transduction mechanisms. Biochem Soc Symp 59:27–50.

Coutts AA, Anavi-Goffer S, Ross RA, MacEwan DJ, Mackie K, Pertwee RG, Irving AJ (2001) Agonist-induced internalization and trafficking of cannabinoid CB_1 receptors in hippocampal neurons. J Neurosci 21:2425–2433.

D'Antona AM, Ahn KH, Kendall DA (2006) Mutations of CB_1 T210 produce active and inactive receptor forms: correlations with ligand affinity, receptor stability, and cellular localization. Biochemistry 45:5606–5617.

Deadwyler SA, Hampson RE, Mu J, Whyte A, Childers S (1995) Cannabinoids modulate voltage sensitive potassium A-current in hippocampal neurons via a cAMP-dependent process. J Pharmacol Exp Ther 273:734–743.

Derkinderen P, Toutant M, Burgaya F, Le Bert M, Siciliano JC, de Franciscis V, Gelman M, Girault JA (1996) Regulation of a neuronal form of focal adhesion kinase by anandamide. Science 273:1719–1722.

Derkinderen P, Valjent E, Toutant M, Corvol JC, Enslen H, Ledent C, Trzaskos J, Caboche J, Girault JA (2003) Regulation of extracellular signal-regulated kinase by cannabinoids in hippocampus. J Neurosci 23:2371–2382.

Devi LA (2000) G-protein-coupled receptor dimers in the lime light. Trends Pharmacol Sci 21:324–326.

Di Marzo V, Izzo AA (2006) Endocannabinoid overactivity and intestinal inflammation. Gut 55:1373–1376.

Di Marzo V, Petrocellis LD (2006) Plant, synthetic, and endogenous cannabinoids in medicine. Annu Rev Med 57:553–574.

Diviani D, Lattion AL, Abuin L, Staub O, Cotecchia S (2003) The adaptor complex 2 directly interacts with the alpha 1b-adrenergic receptor and plays a role in receptor endocytosis. J Biol Chem 278:19331–19340.

Ellis J, Pediani JD, Canals M, Milasta S, Milligan G (2006) Orexin-1 receptor-cannabinoid CB_1 receptor heterodimerization results in both ligand-dependent and -independent coordinated alterations of receptor localization and function. J Biol Chem 281:38812–38824.

Freund TF, Katona I, Piomelli D (2003) Role of endogenous cannabinoids in synaptic signaling. Physiol Rev 83:1017–1066.

Galve-Roperh I, Rueda D, Gomez dP, Velasco G, Guzman M (2002) Mechanism of extracellular signal-regulated kinase activation by the CB_1 cannabinoid receptor. Mol Pharmacol 62:1385–1392.

Galve-Roperh I, Aguado T, Palazuelos J, Guzman M (2007) The endocannabinoid system and neurogenesis in health and disease. Neuroscientist 13:109–114.

Gerard CM, Mollereau C, Vassart G, Parmentier M (1991) Molecular cloning of a human cannabinoid receptor which is also expressed in testis. Biochem J 279:129–134.

Gray NW, Fourgeaud L, Huang B, Chen J, Cao H, Oswald BJ, Hemar A, McNiven MA (2003) Dynamin 3 is a component of the postsynapse, where it interacts with $mGluR_5$ and Homer. Curr Biol 13:510–515.

Guzman M (2003) Cannabinoids: potential anticancer agents. Nat Rev Cancer 3:745–755.

Hampson RE, Mu J, Deadwyler SA (2000) Cannabinoid and kappa opioid receptors reduce potassium K current via activation of G_s proteins in cultured hippocampal neurons. J Neurophysiol 84:2356–2364.

Harkany T, Guzman M, Galve-Roperh I, Berghuis P, Devi LA, Mackie K (2007) The emerging functions of endocannabinoid signaling during CNS development. Trends Pharmacol Sci 28:83–92.

Herkenham M, Lynn AB, Little MD, Johnson MR, Melvin LS, de Costa BR, Rice KC (1990) Cannabinoid receptor localization in brain. Proc Natl Acad Sci USA 87:1932–1936.

Hoffman AF, Lupica CR (2000) Mechanisms of cannabinoid inhibition of $GABA_A$ synaptic transmission in the hippocampus. J Neurosci 20:2470–2479.
Howlett AC (1995) Pharmacology of cannabinoid receptors. Annu Rev Pharmacol Toxicol 35:607–634.
Howlett AC (1998) The CB_1 cannabinoid receptor in the brain. Neurobiol Dis 5:405–416.
Hsieh C, Brown S, Derleth C, Mackie K (1999) Internalization and recycling of the CB_1 cannabinoid receptor. J Neurochem 73:493–501.
Irving AJ, Coutts AA, Harvey J, Rae MG, Mackie K, Bewick GS, Pertwee RG (2000) Functional expression of cell surface cannabinoid CB_1 receptors on presynaptic inhibitory terminals in cultured rat hippocampal neurons. Neuroscience 98:253–262.
Jin W, Brown S, Roche JP, Hsieh C, Celver JP, Kovoor A, Chavkin C, Mackie K (1999) Distinct domains of the CB_1 cannabinoid receptor mediate desensitization and internalization. J Neurosci 19:3773–3780.
Katona I, Sperlagh B, Sik A, Köfalvi A, Vizi ES, Mackie K, Freund TF (1999) Presynaptically located CB_1 cannabinoid receptors regulate GABA release from axon terminals of specific hippocampal interneurons. J Neurosci 19:4544–4558.
Katona I, Rancz EA, Acsady L, Ledent C, Mackie K, Hajos N, Freund TF (2001) Distribution of CB_1 cannabinoid receptors in the amygdala and their role in the control of GABAergic transmission. J Neurosci 21:9506–9518.
Katona I, Urban GM, Wallace M, Ledent C, Jung KM, Piomelli D, Mackie K, Freund TF (2006) Molecular composition of the endocannabinoid system at glutamatergic synapses. J Neurosci 26:5628–5637.
Kearn CS, Blake-Palmer K, Daniel E, Mackie K, Glass M (2005) Concurrent stimulation of cannabinoid CB_1 and dopamine D_2 receptors enhances heterodimer formation: a mechanism for receptor cross-talk? Mol Pharmacol 67:1697–1704.
Keren O, Sarne Y (2003) Multiple mechanisms of CB_1 cannabinoid receptors regulation. Brain Res 980:197–205.
Kreitzer AC, Carter AG, Regehr WG (2002) Inhibition of interneuron firing extends the spread of endocannabinoid signaling in the cerebellum. Neuron 34:787–796.
Lauckner JE, Hille B, Mackie K (2005) The cannabinoid agonist WIN55,212-2 increases intracellular calcium via CB_1 receptor coupling to $G_{q/11}$ G proteins. Proc Natl Acad Sci USA 102:19144–19149.
Leterrier C, Bonnard D, Carrel D, Rossier J, Lenkei Z (2004) Constitutive endocytic cycle of the CB_1 cannabinoid receptor. J Biol Chem 279:36013–36021.
Leterrier C, Laine J, Darmon M, Boudin H, Rossier J, Lenkei Z (2006) Constitutive activation drives compartment-selective endocytosis and axonal targeting of type 1 cannabinoid receptors. J Neurosci 26:3141–3153.
Lutz B (2004) On-demand activation of the endocannabinoid system in the control of neuronal excitability and epileptiform seizures. Biochem Pharmacol 68:1691–1698.
Mackie K (2005) Cannabinoid receptor homo- and heterodimerization. Life Sci 77:1667–1673.
Mackie K (2006) Cannabinoid receptors as therapeutic targets. Annu Rev Pharmacol Toxicol 46:101–122.
Mackie K, Hille B (1992) Cannabinoids inhibit N-type calcium channels in neuroblastoma-glioma cells. Proc Natl Acad Sci USA 89:3825–3829.
Mackie K, Stella N (2006) Cannabinoid receptors and endocannabinoids: evidence for new players. AAPS J 8:E298–E306.
Mackie K, Lai Y, Westenbroek R, Mitchell R (1995) Cannabinoids activate an inwardly rectifying potassium conductance and inhibit Q-type calcium currents in AtT20 cells transfected with rat brain cannabinoid receptor. J Neurosci 15:6552–6561.
Martin BR, Sim-Selley LJ, Selley DE (2004) Signaling pathways involved in the development of cannabinoid tolerance. Trends Pharmacol Sci 25:325–330.
Martini L, Waldhoer M, Pusch M, Kharazia V, Fong J, Lee JH, Freissmuth C, Whistler JL (2007) Ligand-induced down-regulation of the cannabinoid 1 receptor is mediated by the G-protein-coupled receptor-associated sorting protein GASP1. FASEB J 21:802–811.

Matsuda LA, Lolait SJ, Brownstein MJ, Young AC, Bonner TI (1990) Structure of a cannabinoid receptor and functional expression of the cloned cDNA. Nature 346:561–564.
McAllister SD, Griffin G, Satin LS, Abood ME (1999) Cannabinoid receptors can activate and inhibit G protein-coupled inwardly rectifying potassium channels in a xenopus oocyte expression system. J Pharmacol Exp Ther 291:618–626.
McDonald NA, Henstridge CM, Connolly CN, Irving AJ (2007a) An essential role for constitutive endocytosis, but not activity, in the axonal targeting of the CB_1 cannabinoid receptor. Mol Pharmacol 71:976–984.
McDonald NA, Henstridge CM, Connolly CN, Irving AJ (2007b) Generation and functional characterization of fluorescent, N-terminally tagged CB_1 receptor chimeras for live-cell imaging. Mol Cell Neurosci 35:237–248.
Molina-Holgado E, Vela JM, Arevalo-Martin A, Almazan G, Molina-Holgado F, Borrell J, Guaza C (2002) Cannabinoids promote oligodendrocyte progenitor survival: involvement of cannabinoid receptors and phosphatidylinositol-3 kinase/Akt signaling. J Neurosci 22:9742–9753.
Mundell SJ, Luo J, Benovic JL, Conley PB, Poole AW (2006) Distinct clathrin-coated pits sort different G protein-coupled receptor cargo. Traffic 7:1420–1431.
Munro S, Thomas KL, Abu-Shaar M (1993) Molecular characterization of a peripheral receptor for cannabinoids. Nature 365:61–65.
Nordstrom R, Andersson H (2006) Amino-terminal processing of the human cannabinoid receptor 1. J Recept Signal Transduct Res 26:259–267.
Ozaita A, Puighermanal E, Maldonado R (2007) Regulation of PI_3K/Akt/GSK-3 pathway by cannabinoids in the brain. J Neurochem 102:1105–1114.
Pertwee RG (2006) The pharmacology of cannabinoid receptors and their ligands: an overview. Int J Obes (Lond) 30(Suppl. 1):S13–S18.
Petitet F, Donlan M, Michel A (2006) GPR55 as a new cannabinoid receptor: still a long way to prove it. Chem Biol Drug Des 67:252–253.
Piomelli D (2003) The molecular logic of endocannabinoid signalling. Nat Rev Neurosci 4:873–884.
Porter AC, Sauer JM, Knierman MD, Becker GW, Berna MJ, Bao J, Nomikos GG, Carter P, Bymaster FP, Leese AB, Felder CC (2002) Characterization of a novel endocannabinoid, virodhamine, with antagonist activity at the CB_1 receptor. J Pharmacol Exp Ther 301:1020–1024.
Ramirez BG, Blazquez C, del Pulgar TG, Guzman N, de Ceballos MAL (2005) Prevention of Alzheimer's disease pathology by cannabinoids: neuroprotection mediated by blockade of microglial activation. J Neurosci 25:1904–1913.
Rinaldi-Carmona M, Calandra B, Shire D, Bouaboula M, Oustric D, Barth F, Casellas P, Ferrara P, Le Fur G (1996) Characterization of two cloned human CB_1 cannabinoid receptor isoforms. J Pharmacol Exp Ther 278:871–878.
Rinaldi-Carmona M, Le Duigou A, Oustric D, Barth F, Bouaboula M, Carayon P, Casellas P, Le Fur G (1998) Modulation of CB_1 cannabinoid receptor functions after a long-term exposure to agonist or inverse agonist in the Chinese hamster ovary cell expression system. J Pharmacol Exp Ther 287:1038–1047.
Rios C, Gomes I, Devi LA (2006) Mu opioid and CB_1 cannabinoid receptor interactions: reciprocal inhibition of receptor signaling and neuritogenesis. Br J Pharmacol 148:387–395.
Robbe D, Alonso G, Duchamp F, Bockaert J, Manzoni OJ (2001) Localization and mechanisms of action of cannabinoid receptors at the glutamatergic synapses of the mouse nucleus accumbens. J Neurosci 21:109–116.
Roche JP, Bounds S, Brown S, Mackie K (1999) A mutation in the second transmembrane region of the CB_1 receptor selectively disrupts G protein signaling and prevents receptor internalization. Mol Pharmacol 56:611–618.
Rubovitch V, Gafni M, Sarne Y (2002) The cannabinoid agonist DALN positively modulates L-type voltage-dependent calcium-channels in N18TG2 neuroblastoma cells. Brain Res Mol Brain Res 101:93–102.
Rueda D, Galve-Roperh I, Haro A, Guzman M (2000) The CB_1 cannabinoid receptor is coupled to the activation of c-Jun N-terminal kinase. Mol Pharmacol 58:814–820.

Russo P, Strazzullo P, Cappuccio FP, Tregouet DA, Lauria F, Loguercio M, Barba G, Versiero M, Siani A (2007) Genetic variations at the endocannabinoid type 1 receptor gene (CNR1) are associated with obesity phenotypes in men. J Clin Endocrinol Metab 92:2382–2386.

Ryberg E, Vu HK, Larsson N, Groblewski T, Hjorth S, Elebring T, Sjogren S, Greasley PJ (2005) Identification and characterisation of a novel splice variant of the human CB_1 receptor. FEBS Lett 579:259–264.

Sampo B, Kaech S, Kunz S, Banker G (2003) Two distinct mechanisms target membrane proteins to the axonal surface. Neuron 37:611–624.

Schweitzer P (2000) Cannabinoids decrease the K^+ M-current in hippocampal CA1 neurons. J Neurosci 20:51–58.

Seong E, Wainer BH, Hughes ED, Saunders TL, Burmeister M, Faundez V (2005) Genetic analysis of the neuronal and ubiquitous AP-3 adaptor complexes reveals divergent functions in brain. Mol Biol Cell 16:128–140.

Sim-Selley LJ, Schechter NS, Rorrer WK, Dalton GD, Hernandez J, Martin BR, Selley DE (2006) Prolonged recovery rate of CB_1 receptor adaptation after cessation of long-term cannabinoid administration. Mol Pharmacol 70:986–996.

Stoddart LA, Brown AJ, Milligan G (2007) Uncovering the pharmacology of the G protein-coupled receptor GPR40: high apparent constitutive activity in guanosine 5 -*O*-(3-[^{35}S]thio)triphosphate binding studies reflects binding of an endogenous agonist. Mol Pharmacol 71:994–1005.

Szabo B, Schlicker E (2005) Effects of cannabinoids on neurotransmission. Handb Exp Pharmacol 327–365.

Tappe-Theodor A, Agarwal N, Katona I, Rubino T, Martini L, Swiercz J, Mackie K, Monyer H, Parolaro D, Whistler J, Kuner T, Kuner R (2007) A molecular basis of analgesic tolerance to cannabinoids. J Neurosci 27:4165–4177.

Turu G, Simon A, Gyombolai P, Szidonya L, Bagdy G, Lenkei Z, Hunyady L (2007) The role of diacylglycerol lipase in constitutive and angiotensin AT_1 receptor-stimulated cannabinoid CB_1 receptor activity. J Biol Chem 282:7753–7757.

Twitcheli W, Brown S, Mackie K (1997) Cannabinoids inhibit N- and P/Q-type calcium channels in cultured rat hippocampal neurons. J Neurophysiol 78:43–50.

Van Sickle MD, Duncan M, Kingsley PJ, Mouihate A, Urbani P, Mackie K, Stella N, Makriyannis A, Piomelli D, Davison JS, Marnett LJ, Di Marzo V, Pittman QJ, Patel KD, Sharkey KA (2005) Identification and functional characterization of brainstem cannabinoid CB_2 receptors. Science 310:329–332.

Wager-Miller J, Westenbroek R, Mackie K (2002) Dimerization of G protein-coupled receptors: CB_1 cannabinoid receptors as an example. Chem Phys Lipids 121:83–89.

Waldhoer M, Casarosa P, Rosenkilde MM, Smit MJ, Leurs R, Whistler JL, Schwartz TW (2003) The carboxyl terminus of human cytomegalovirus-encoded 7 transmembrane receptor US28 camouflages agonism by mediating constitutive endocytosis. J Biol Chem 278:19473–19482.

Westlake TM, Howlett AC, Bonner TI, Matsuda LA, Herkenham M (1994) Cannabinoid receptor binding and messenger RNA expression in human brain: an in vitro receptor autoradiography and in situ hybridization histochemistry study of normal aged and Alzheimer's brains. Neuroscience 63:637–652.

Whistler JL, Gerber BO, Meng EC, Baranski TJ, von Zastrow M, Bourne HR (2002) Constitutive activation and endocytosis of the complement factor 5a receptor: evidence for multiple activated conformations of a G protein-coupled receptor. Traffic 3:866–877.

Wilson RI, Nicoll RA (2002) Endocannabinoid signaling in the brain. Science 296:678–682.

Wisco D, Anderson ED, Chang MC, Norden C, Boiko T, Folsch H, Winckler B (2003) Uncovering multiple axonal targeting pathways in hippocampal neurons. J Cell Biol 162:1317–1328.

Zhang PW, Ishiguro H, Ohtsuki T, Hess J, Carillo F, Walther D, Onaivi ES, Arinami T, Uhl GR (2004) Human cannabinoid receptor 1: 5' exons, candidate regulatory regions, polymorphisms, haplotypes and association with polysubstance abuse. Mol Psychiatry 9:916–931.

Chapter 6
CB_2 Cannabinoid Receptors: Molecular, Signaling, and Trafficking Properties

Paul L. Prather

Abstract Two G protein-coupled receptors, CB_1 and CB_2, have thus far been identified and are responsible for most of the effects produced by cannabinoids. Cannabinoids, such as Δ^9-THC, produce psychoactive effects through activation of neuronal CB_1 receptors, while CB_2 receptors mediate the immune properties of this class of drugs. The molecular, signaling, and trafficking properties of CB_2 receptors will be the focus of this review. The cloning of CB_2 receptors will be described, along with evidence that individual cannabinoid ligands, differing subtly in structure, might bind to CB_2 receptors in distinct fashions. In addition, potential mechanisms underlying the dramatic upregulation of CB_2 receptors in response to inflammatory stimuli will be discussed. Next, the currently known signal transduction pathways associated with CB_2 receptor activation will be detailed, from G protein coupling to regulation of intracellular effectors. Evidence for the ability of different CB_2 receptor agonists to distinctly regulate multiple effectors, known as agonist-directed trafficking of response (ADTR), will also be presented. Furthermore, a potential relationship between CB_2 receptor ADTR and immune cell function will be discussed. Lastly, two distinct aspects of CB_2 receptor signaling will be described that may help to explain the well-documented interactions of cannabinoids with other receptor systems. It is hoped that this brief review will provide a basic understanding of CB_2 receptor signaling necessary to appreciate the exciting future approaching for the development of potentially therapeutic selective CB_2 receptor ligands.

Overview of Cannabinoid Receptors

Cannabis sativa has been used both therapeutically and recreationally for centuries (see Chap. 1). Δ^9-Tetrahydrocannabinol (Δ^9-THC) has been acknowledged to be the main psychoactive ingredient in marijuana and mediates its effects primarily through activation of two G protein-coupled receptors, CB_1 and CB_2 (Howlett, 1995). Identified in 1990 (Matsuda et al., 1990), the human CB_1 receptor was found to be primarily localized in central and peripheral nervous tissue (Herkenham et al., 1990; Ishac et al., 1996). The CB_1 receptor has been identified as a therapeutic

A. Köfalvi (ed.), *Cannabinoids and the Brain.*

target in a variety of disease states, such as obesity (Ravinet et al., 2002), alcohol dependence (Racz et al., 2003), Parkinson's disease (Brotchie, 2003), and pain (Iversen and Chapman, 2002) (for further details, consult with Chaps. 14, 21, 22). The second G protein-coupled cannabinoid receptor, CB_2, was cloned two years later (Munro et al., 1993). These receptors are prevalently found in immune tissues, most abundantly in the spleen and leukocytes (Galiegue et al., 1995). As the localization of the CB_2 receptors might indicate, selective CB_2 receptor ligands have potential therapeutic use as immune modulators for tumor suppression (Klein et al., 2003) and inflammation (Conti et al., 2002). Recently, CB_2 receptor agonists have also been shown to produce potent and efficacious analgesia of neuropathic pain (Ibrahim et al., 2003; Scott et al., 2004). This finding is of particular benefit due to the localization of CB_2 receptors outside of the CNS; therefore, agonists that selectively activate the CB_2 receptor may produce effective analgesia without the unwanted psychoactive CNS effects associated with CB_1 receptor agonists (Cravatt and Lichtman, 2004).

Molecular Biology of CB_2 Receptors

Cloning of the CB_2 Receptor

The cDNA for the human CB_2 (hCB_2) receptor was initially cloned from HL-60 cells, a human promyelocytic leukaemic cell line (Munro et al., 1993). The hCB_2 receptor is a protein consisting of 360 amino acids forming a structure that is predicted to span the plasma membrane seven times, characteristic of G protein-coupled receptors. Employing splenocyte cDNA libraries, CB_2 receptors have also subsequently been cloned from the mouse (Shire et al., 1996) and the rat (Brown et al., 2002). In contrast to the CB_1 receptor which has been cloned from a diverse array of vertebrates such as mammals (Gerard et al., 1991), birds (Soderstrom and Johnson, 2000), fish (Yamaguchi et al., 1996), and amphibians (Soderstrom et al., 2000), the CB_2 receptor has only been cloned in mammals. The hCB_2 receptor is also quite different from the human CB_1 (hCB_1) receptor on a structural basis, sharing only 44% overall homology, increasing to 68% identical amino acid identity when only the seven transmembrane domains are considered. Furthermore, the amino terminal domain of the hCB_2 receptor is much shorter than, and has no significant conservation with, the hCB_1 receptor. When comparing the CB_2 receptor across species, a high degree of homology exists when hCB_2, mCB_2, and rCB_2 receptors are aligned except in the carboxyl terminus. In this region, the rCB_2 receptor is 50 and 63 amino acid residues longer than the hCB_2 and mCB_2 receptors, respectively. The genes encoding for the mCB_2 and hCB_2 receptors have been mapped to distal locations on their respective chromosomes [mouse #4, human #1P36, (Valk et al., 1997)] and are encoded by a single exon. However, a subsequent study reported that the rCB_2 receptor was the first known example of an expressed cannabinoid receptor encoded by multiple exons (Brown et al., 2002).

Binding Characteristics of the CB_2 Receptor

Mutagenesis of cannabinoid receptors has revealed insight into the basis for CB_2/CB_1 receptor selectivity. As predicted by molecular modeling, mutation of a phenylalanine in transmembrane domain 5 (F5.46) of the hCB_2 receptor decreased the affinity of the CB_2 receptor-preferring ligand WIN55212–2 for the hCB_2 receptor by 14-fold. In contrast, mutation of a valine in the analogous position (V5.46) of the hCB_1 receptor to phenylalanine increased the affinity of WIN55212–2 for the hCB_1 receptor by 12-fold (Song et al., 1999). Furthermore, a comparison of CB_2 and CB_1 receptor binding sites by docking studies of WIN55212–2 complexed with both hCB_1 and hCB_2 receptors suggests that CB_2/CB_1 receptor selectivity is determined primarily by interaction with serine (S3.31) and F5.46 residues in the hCB_2 receptor (Tao et al., 1999; Tuccinardi et al., 2006). Specifically, it is proposed that selectivity for the CB_2 receptor may be enhanced by developing ligands with a lipophilic group able to interact with F5.46 of hCB_2 and a group able to form a H bond with S3.31. In addition to selectivity, there is an increasing amount of evidence that individual ligands, differing subtly in structure, might bind the CB_2 receptor in distinct fashions. For example, while substitution of F5.46 with valine in transmembrane domain 5 of the hCB_2 receptor decreases the affinity of the aminoakylindole cannabinoid WIN55212–2 for the hCB_2 receptor, the affinities for the classical cannabinoid HU-210, the nonclassical cannabinoid CP55940, or the eicosanoid cannabinoid anandamide are unchanged (Song et al., 1999). In studies examining the selectivity of the cannabinoid antagonist SR144528 for the CB_2 receptor, mutation of amino acids adjacent to transmembrane domain 4 (serine 161, serine 165, or cysteine 175), eliminates CB_2 receptor binding by SR144528, but has minimal effect on the affinity of CP55490 or WIN55212–2 (Gouldson et al., 2000). If cannabinoid ligands derived from diverse structural classes bind to CB_1 and CB_2 receptors in distinct manners, it is likely that individual agonists might selectively activate signal transduction pathways (i.e., agonist-directed trafficking of response, ADTR). If so, agonists might be developed that at optimal concentrations preferentially activate signal transduction pathways responsible for the therapeutic effects of cannabinoids (i.e., antinociception), while avoiding activation of other pathways potentially mediating undesirable actions (i.e., disruption of short-term memory).

Regulation of CB_2 Receptor Expression

Although recent studies have suggested the presence of low levels of functional CB_2 receptors in the CNS (van Sickle et al., 2005; Gong et al., 2006; Onaivi et al., 2006; see Chap. 10), CB_2 receptors are predominantly expressed in immune cells (Herkenham et al., 1990; Ishac et al., 1996). However, during chronic inflammation associated with several diseases affecting the CNS, CB_2 receptor levels are dramatically upregulated in inflamed neural tissues (Benito et al., 2003; Ramirez et al., 2005; Shoemaker et al., 2007). The increase in the density of CB_2 receptors appears to occur primarily in activated microglia, the resident immune cells of the CNS. Few studies have attempted to investigate the mechanisms underlying CB_2 receptor upregulation in response to inflammation. There is evidence, however, indicating a role for specific cytokines (Maresz et al., 2005) and the cyclic AMP-protein kinase A signaling pathway (Mukhopadhyay et al., 2006). For example, Maresz and colleagues (2005) demonstrated that microglial cells cultured with combinations of gamma-interferon and granulocyte macrophage-colony stimulating factor, which both promote microglial cell activation and are expressed in the CNS during many neuroinflammatory diseases, produce a synergistic eightfold to tenfold increase in the levels of CB_2 receptors within these cells. In another recent study, CB_2 receptors in cultured RAW 264.7 macrophages increase following exposure to the bacterial cell wall component lipopolysaccharide (Mukhopadhyay et al., 2006). CB_2 receptor upregulation was partially blocked by cyclohexamide or the protein kinase A and C inhibitors H8 and bis-indolylmaleimide. Furthermore, application of dibutyryl cyclic AMP or activation of adenylyl cyclase by forskolin increased CB_2 receptor levels. This data suggest that the regulation of CB_2 receptor expression in macrophages following exposure to inflammatory stimuli, such as lipopolysaccharide, involves the cyclic AMP-protein kinase A-cyclic AMP response element pathway.

CB_2 Receptor Signal Transduction

G Protein Coupling

Both CB_1 and CB_2 receptors are G protein-coupled receptors that traverse the plasma membrane seven times and regulate the activity of intracellular effectors through activation of intracellular G proteins. Heterotrimeric G proteins are composed of three distinct subunits, α (39–50 kDa), β (35–36 kDa), and γ (6–10 kDa) and their activation by G protein-coupled receptors produces an exchange of GTP for GDP on the subunits. This results in the dissociation of the G protein from the receptor and the separation of the α GTP from the $\beta\gamma$ subunits. Both the free αGTP and $\beta\gamma$ subunits then proceed to regulate various downstream effectors (Gudermann et al., 1997). Pertussis toxin (PTX)-sensitive G proteins (i.e., $G_{i\alpha}$ and $G_{o\alpha}$-subtypes) appear to mediate the physiological effects of cannabinoids acting on CB_1 and CB_2 receptors (Howlett, 1995). However, other studies also suggest that CB_1 receptors may regulate intracellular signaling via

PTX-insensitive $G_{s\alpha}$ as well (Glass and Felder, 1997; Maneuf and Brotchie, 1997; Felder et al., 1998; see Chap. 9).

Effector Regulation

CB_1 and CB_2 receptors couple to multiple intracellular effectors. Both CB_1 and CB_2 receptors regulate the activity of adenylyl cyclase (Howlett, 1985) and the extracellular regulated kinase subgroup of the mitogen-activated protein kinases (ERK-MAPK) (Bouaboula et al., 1995). Activation of CB_1 (Sugiura et al., 1997) and CB_2 (Sugiura et al., 2000) receptors also evokes a rapid, transient increase in intracellular free Ca^{2+} in neuronal and immune cells. Chronic CB_1 and CB_2 receptor stimulation results in elevation of intracellular levels of ceramide, associated with decreased proliferation and apoptosis in glioma cells (Guzman et al., 2001). More recently, it has been shown that cannabinoids, acting at both CB_1 and CB_2 receptors, also promote survival of cortical neurons and oligodendrocyte progenitors through stimulation of the phosphoinositide 3-kinase/protein kinase B (PI_3K/Akt) signaling pathway (Molina-Holgado et al., 2002; Molina-Holgado et al., 2005). Interestingly, only CB_1, but not CB_2 (Felder et al., 1995; McAllister et al., 1999), additionally couples to certain ion channels, producing inhibition of voltage-gated Ca^{2+} channels (Mackie and Hille, 1992) and activation of inwardly rectifying K^+ channels (Mackie et al., 1995). The specific regulation of each of these intracellular effectors by the CB_2 receptor will be briefly discussed below.

Adenylyl Cyclase

Initial studies demonstrated that cannabinoids produce concentration-dependent inhibition of adenylyl cyclase activity in CHO (Bayewitch et al., 1995) or COS (Slipetz et al., 1995) cells, transfected with the CB_2 receptor. Cannabinoids also reduce intracellular cAMP levels in human lymphocytes and mouse spleen cells expressing endogenous CB_2 receptors (Howlett and Mukhopadhyay, 2000). In all studies, CB_2 receptor-dependent adenylyl cyclase inhibition is PTX-sensitive, indicating the requirement for $G_{i/o\alpha}$ subtypes of G proteins in the signaling cascade. It has been suggested that the regulation of immune function by the CB_2 receptor is mediated, in part, by a reduction in adenylyl cyclase activity (Kaminski et al., 1994). Surprisingly, in cells pretreated with PTX to eliminate $G_{i/o\alpha}$-coupling, CB_1 (but not CB_2) receptor agonists are still able to couple to $G_{s\alpha}$ to produce stimulation of adenylyl cyclase activity (Glass and Felder, 1997; Maneuf and Brotchie, 1997; Felder et al., 1998). These data demonstrate that, in addition to being unable to regulate ion channels (Felder et al., 1995; McAllister et al., 1999), CB_2 receptors also cannot couple to $G_{s\alpha}$. Collectively, these studies importantly indicate that CB_1 and CB_2 receptors transduce intracellular signals in significantly different manners.

ERK-MAPK

Activation of CB_2 receptors by cannabinoids also stimulates the activity of p42/p44 ERK-MAPK in HL-60 cells endogenously expressing CB_2 receptors (Kobayashi et al., 2001) and in CB_2 receptor-transfected CHO cells (Bouaboula et al., 1996). In both studies, the cannabinoid-mediated effect on ERK-MAPK was time- and concentration-dependent and blocked by pretreatment with either PTX or the selective CB_2 receptor antagonist SR144528. In PC-3 cells, a human prostate epithelial cell line, the activation of ERK-MAPK by cannabinoids appears to be mediated via a PI_3K/Akt pathway that produces membrane translocation of Raf-1 with subsequent phosphorylation of p42/p44 ERK-MAPK (Sanchez et al., 2003). This response was blocked by pretreatment of cells with SR144528, indicating the involvement of CB_2 receptors. CB_2 receptor-mediated activation of ERK-MAPK by endogenous cannabinoids in immune cells appears to be associated with their migration. For example, in HL-60 cells differentiated into a macrophage-like state, the endogenous cannabinoid 2-arachidonoylglycerol (2-AG) produces marked migration through a CB_2 receptor- and ERK-MAPK-dependent pathway (Kishimoto et al., 2003). 2-AG also results in pronounced ERK-MAPK-dependent migration of myeloid precursor cells via overexpressed CB_2 receptors (Jorda et al., 2002). Microglial cell migration, a neuroinflammatory response to dying neurons, is initiated in response to CB_2 receptor activation by 2-AG and is dependent on ERK-MAPK activation (Walter et al., 2003). Lastly, cannabinoids can also inhibit ERK-MAPK in stimulated mouse splenocytes, presumably via CB_2 receptor (although not directly demonstrated) (Kaplan et al., 2003). By use of the mitogen-activated kinase (MEK) inhibitor PD098059, the authors suggest that cannabinoid-mediated reduction in ERK-MAPK may inhibit IL-2 production in these cells, contributing to the mechanism for immunosuppression commonly observed with cannabinoids.

Ca^{2+} Transients

Stimulation of CB_2 receptors produces transient increases in intracellular free Ca^{2+} concentration via a phospholipase-Cβ (PLCβ-mediated mechanism in HL-60 cells expressing endogenous CB_2 cannabinoid receptors (Sugiura et al., 2000) and in CHO cells stably transfected with CB_2 (Shoemaker et al., 2005b). In both studies, the Ca^{2+} transients produced were concentration-dependent and blocked by pretreatment with either PTX or selective CB_2 antagonists. In CHO-CB_2 cells, the cannabinoid-elicited rise in intracellular free Ca^{2+} concentration was blocked by preincubation with the active (U73122), but not the inactive (U73343), inhibitor of PLCβ. This provides rather strong evidence that activation of PLCβ is involved in the observed CB_2 receptor-mediated production of Ca^{2+}transients in transfected CHO cells. Interestingly, a previous study reported that activation of transfected CB_2 receptors in CHO cells is unable to elevate intracellular free Ca^{2+}

concentration (Felder et al., 1995). While the exact reasons for the differences between these studies are not known, one potential explanation might be due to the choice of agonists employed. While Felder and colleagues (1995) observed no effect on intracellular calcium concentrations in response to WIN55212–2, anandamide, and HU-210, the agonist evaluated by Shoemaker and colleagues (2005a,b) (2-AG) was not examined.

Ceramide Synthesis

Several early reports demonstrated that ceramide accumulation participates in cannabinoid-induced apoptosis of glioma cells (Galve-Roperh et al., 2000; Gomez del Pulgar et al., 2002), a mechanism that appears to rely on the activation of the CB_2 receptor (Sanchez et al., 2001). Recent studies employing Jurkat cells, a human leukemia cell line expressing endogenous CB_2 receptors, further showed that CB_2 receptor activation signals apoptosis via a ceramide-dependent stimulation of the mitochondrial intrinsic pathway (Herrera et al., 2006). Specifically, cannabinoid treatment resulted in a CB_2 receptor-dependent stimulation of ceramide biosynthesis, and inhibition of this pathway prevented cannabinoid-induced mitochondrial hypopolarization and cytochrome-*c* release. These results indicate that ceramide acts at a premitochondrial level. Ceramide synthesis inhibition in this study also prevented caspase activation and apoptosis. Collectively, these reports demonstrate that CB_2 receptor signaling plays a major role in the proapoptotic effect of cannabinoids and suggest that selective CB_2 cannabinoids might be developed as useful agents to slow tumor growth in various forms of cancer.

PI_3K/Akt Pathway

Survival signaling of many cell types, including neurons, has been clearly demonstrated to be associated with the PI_3K/Akt pathway (Brunet et al., 2001). Cannabinoids, acting at both CB_1 and CB_2 receptors, also promote survival of cortical neurons and oligodendrocyte progenitors through stimulation of the PI_3K/Akt signaling pathway (Molina-Holgado et al., 2002, 2005). Specifically, the nonselective cannabinoid agonist HU-210 inhibits the death of rat primary cortical neurons induced by the neurotoxin (S)-alpha-amino-3-hydroxy-5-methyl-4-isoxazole propionic acid (S-AMPA). The neuroprotective effect of HU-210 is reversed by antagonists selective for either CB_1 or CB_2 receptors. HU-210 triggers activation of Akt, but not activation of the ERK-MAPK, JNK-MAPK, or p38-MAPK signaling pathways. Furthermore, the PI_3K inhibitors LY294002 and wortmannin prevent phosphorylation of Akt in response to HU-210, and reversed the neuroprotective effect of HU-210. As such, the authors suggest that the PI_3K/Akt signaling pathway mediates the neuroprotective effect of the exogenous cannabinoid

HU-210 acting at CB_1 and CB_2 receptors in primary cultured CNS neurons (Molina-Holgado et al., 2005).

CB_2 Receptor Internalization and Trafficking in Response to Acute and Chronic Ligand Exposure

Acute and chronic ligand exposure significantly regulates both CB_1 and CB_2 receptor signaling. A great deal of research has been conducted on this topic concerning CB_1 receptor signaling and has recently been reviewed extensively by Sim-Selley (2003) (see Chap. 5). In contrast, much less has been reported about the effect of acute and chronic ligand exposure on signaling by CB_2 receptors. As observed with other G protein-coupled receptors, studies with CB_2 transfected CHO cells demonstrate that upon initial exposure to the full agonist CP55940, serine 353 is extensively phosphorylated and phosphorylation is maintained for up to 8 h (Bouaboula et al., 1999b). CB_2 receptor phosphorylation by CP55940 can be reversed by preincubation with the CB_2 receptor-selective antagonist/inverse agonist SR144528. Furthermore, CB_2 receptors desensitize in a time-and concentration-dependent manner following prolonged agonist exposure, such that cellular responses are abolished in response to challenge with CB_2 receptor agonists following chronic exposure to either CP55940 (Bouaboula et al., 1999b) or to the putative endogenous cannabinoid noladin ether (Shoemaker et al., 2005a). If exposure to CP55940 is extended to 24 hours, CB_2 receptors are also down-regulated, reflected by over a 90% loss of receptors as measured by receptor binding (Shoemaker et al., 2005a). Interestingly, similar chronic exposure to noladin ether produces significantly less CB_2 receptor down-regulation, resulting in only approximately a 50% loss of receptor binding (Shoemaker et al., 2005a). Several recent studies have revealed some very interesting findings concerning CB_2 receptor localization in immune cells and the effect of acute cannabinoid exposure on CB_2 receptor trafficking within these cells (Walter et al., 2003; Carrier et al., 2004; Rayman et al., 2004). Microglial cell lines and primary cultures of microglia exist in an activated state when maintained in culture (Becher and Antel, 1996). In cultured (activated) mouse microglial BV-2 cells, rat microglial RTMGL1 cells, and mouse microglial primary cultured cells, CB_1 receptors appear to be localized in the intracellular compartment (Walter et al., 2003; Carrier et al., 2004). In marked contrast, CB_2 receptors are expressed heterogeneously throughout the activated microglial cells, both at the cell surface and internally. Even more interesting is the observation that CB_2 receptors are expressed in relatively high density at the leading edge of the lamellipodia of activated microglial cells (Walter et al., 2003). This critical positioning suggests that CB_2 receptors might participate in the migration of microglial cells occurring in response to inflammatory stimuli. Indeed, microglial cell migration is initiated following exposure to the endogenous cannabinoid agonist 2-AG, an effect mediated by both CB_2 and abnormal cannabidiol-sensitive receptors (Walter et al.,

2003). Furthermore, exposure of microglial cells to 2-AG significantly increases CB_2 receptor internalization, but not degradation (Carrier et al., 2004). In lymphoid tissues, CB_2 receptors are also expressed in distinct patterns, depending on receptor activation status (Rayman et al., 2004). For example, active CB_2 receptors are present mainly in the germinal centers, while inactive CB_2 receptors are confined to the mantle and marginal zones of the secondary follicles where resting cells reside. Collectively, these studies suggest that activated CB_2 receptors are selectively trafficked within immune cells to specific regions, critically posed to participate in important immune cell functions such as proliferation and migration.

CB_2 Receptor ADTR

Definition and Observation of ADTR at CB_2 Receptors

Evidence suggests that G protein-coupled receptors exist in multiple active receptor conformations (Kenakin, 2002). It has been predicted that binding of a particular agonist to a GPCR results in enrichment of a unique set of receptor conformations based on the microaffinity of the agonist for each conformation. Because distinct conformations could presumably couple receptors differently to specific G proteins and intracellular effectors, individual agonists could ultimately produce distinct effects. Numerous studies provide support that individual agonists acting at several different classes of G protein-coupled receptors (Figini et al., 1997; Berg et al., 1998; Wiens et al., 1998), including CB_1 receptors (Bonhaus et al., 1998), are able to traffic intracellular responses in a ligand-dependent manner. Furthermore, utilizing plasmon waveguide resonance spectroscopy, Alves and colleagues have recently provided direct evidence for the existence of distinct topographical configurations of human delta opioid receptors with discrete affinities between individual G protein subclasses and different ligand-induced states (Alves et al., 2003). Very recently, evidence for ADTR by endocannabinoids acting at CB_2 receptors has been provided (Shoemaker et al., 2005b). Specifically, in CHO-CB_2 cells it was shown that 2-AG, acting through CB_2 receptors, most potently activates ERK-MAPK, requiring greater concentrations to inhibit adenylyl cyclase, and even higher amounts to stimulate Ca^{2+}transients. In contrast, two other cannabinoids tested (noladin ether and CP55940) most potently inhibit adenylyl cyclase, necessitating higher concentrations to stimulate ERK-MAPK and Ca^{2+}transients.

Potential Relationship of CB_2 ADTR to Function

If ADTR occurs at CB_2 receptors, the preferential activation of the ERK-MAPK pathway by 2-AG, relative to noladin ether and CP55940 demonstrated by Shoemaker and coworkers (2005a,b), might provide insight into the cellular basis for well-documented agonist selective actions reported for cannabinoids in immune cells. For example, in HL-60 cells differentiated into a macrophage-like state, 2-AG produces marked migration through a CB_2 receptor- and ERK-MAPK-dependent pathway (Kishimoto et al., 2003). In contrast, noladin ether only weakly stimulates migration, while anandamide, CP55940, WIN55212–2, and several other cannabinoids have no effect. 2-AG also results in pronounced ERK-MAPK-dependent migration of myeloid precursor cells via overexpressed CB_2 receptors, whereas anandamide produces near negligible effects and other cannabinoids are devoid of activity (Jorda et al., 2002). Microglial cell migration, a neuroinflammatory response to dying neurons, is initiated in response to CB_2 receptor activation by 2-AG, but not by two other putative endocannabinoids and is dependent on ERK-MAPK activation (Walter et al., 2003). Although involvement of ERK-MAPK was not tested, activation of CB_2 receptors by 2-AG induces the migration of EoL-1 human eosinophilic leukemia cells, noladin ether is only weakly effective, and anandamide does not induce migration (Oka et al., 2004). In all the cited studies, 2-AG induces pronounced migration of cells while other endogenously occurring or synthetically derived cannabinoids produce only modest or no effects at all. In addition, migration induced by 2-AG was shown to occur through activation of CB_2 receptors and ERK-MAPK. As such, it is tempting to speculate that this rather selective, robust ability of 2-AG to induce migration of variety of cell types might be due to the ability of 2-AG to preferentially regulate ERK-MAPK via CB_2 receptors relative to other cannabinoids.

CB_2 Receptor Interactions

Inactivation of Other G_i/G_o-Coupled Receptor Signaling by CB_2 Receptors

Many G protein-coupled receptors exhibit constitutive activity, producing spontaneous regulation of effectors in the absence of activation by agonists (Kenakin, 2001). Ligands that can reduce or abolish this spontaneous, agonist-independent activity are termed inverse agonists (Strange, 2002; Prather, 2004). CB_2 receptors are constitutively active (Bouaboula et al., 1999b). The CB_2 inverse agonist JTE-907 demonstrates anti-inflammatory actions in several animal models (Maekawa et al., 2006; Ueda et al., 2007). Furthermore, a novel CB_2 inverse agonist has recently been shown to inhibit leukocyte recruitment induced by several different chemokines (Lunn et al., 2006). While the mechanism for the blockade of leukocyte recruitment was not examined, constitutively active CB_2 and CB_1 receptors appear to be able to sequester $G_{i/o}$ type G proteins away from other G protein-coupled receptors, interfering with their function (Bouaboula et al., 1999a; Vasquez and Lewis, 1999). Since chemokine receptors produce immune cell migration via activation of $G_{i/o}$ type G proteins, it is possible that CB_2 inverse agonists (such as JTE-907) might reduce inflammation by interfering with this critical step in the immune response mediated by chemokines. This indicates that CB_2 inverse agonists might be potentially developed as drugs to treat a variety of inflammatory disorders.

Transcriptional Regulation of Other Receptors by CB_2 Receptors

Very recently, CB_2 receptor activation has been shown to be regulating the expression of CB_1, μ-, and δ-opioid receptors in the $CD4^+$ T cell line Jurkat (Borner et al., 2006, 2007). Specifically, the upregulation of all three receptors involves activation of CB_2 receptors followed by phosphorylation of signal transducer and activator of transcription 5 (STAT5) with subsequent transactivation of the gene encoding for interleukin-4 (IL-4). Transactivation of CB_1, μ-, and δ-opioid receptor genes in response to IL-4 is then mediated by phosphorylation of the signal transducer and activator of transcription 6 (STAT6). Increasing the levels of CB_1 receptors in T lymphocytes, and possibly other immune cells, in response to CB_2 receptor stimulation would be expected to enhance the immunomodulatory effects mediated by cannabinoids in these cells. Furthermore, if CB_2-mediated upregulation of μ- or δ-opioid receptors also occurs in neurons, it might help explain the well-documented synergistic analgesic effects between cannabinoids and opioids (Cichewicz, 2004).

Concluding Remarks

Cannabinoids produce the majority of their effects through interaction with CB_1 and CB_2 receptors. CB_2 receptors (the subject of this review) are expressed predominantly in immune tissues and transduce intracellular signals through coupling to the G_i/G_o subtype of G proteins. Upon receptor activation by agonists, CB_2 receptors regulate the activity of multiple intracellular effectors, including adenylyl cyclase, ERK-MAPK, Ca^{2+} transients, ceramide synthesis, and PI_3K/Akt. Interestingly, different CB_2 agonists bind uniquely to CB_2 receptors and distinctly regulate multiple effectors. This type of intracellular signaling has been described as agonist-directed trafficking of response (ADTR). The ability of CB_2 ligands to selectively traffic intracellular responses, coupled with their selective expression profile in inflamed tissues, and pronounced anti-inflammatory and neuroprotective properties, suggest an exciting future is approaching for the development this novel class of drugs for the treatment of a variety of inflammatory disorders.

Acknowledgments This work was supported in part by National Institute on Drug Abuse (grant RO1-DA13660), Amyotrophic Lateral Sclerosis Association (ALSA) (grant 1311), and University of Arkansas for Medical Sciences (Tobacco Award).

References

Alves ID, Salamon Z, Varga E, Yamamura HI, Tollin G, Hruby VJ (2003) Direct observation of G protein binding to the human delta-opioid receptor using plasmon-waveguide resonance spectroscopy. J Biol Chem 278:48890–48897.

Bayewitch M, Avidor-Reiss T, Levy R, Barg J, Mechoulam R, Vogel Z (1995) The peripheral cannabinoid receptor: adenylate cyclase inhibition and G protein coupling. FEBS Lett 375:143–147.

Becher B, Antel JP (1996) Comparison of phenotypic and functional properties of immediately ex vivo and cultured human adult microglia. Glia 18:1–10.

Benito C, Nunez E, Tolon RM, Carrier EJ, Rabano A, Hillard CJ, Romero J (2003) Cannabinoid CB_2 receptors and fatty acid amide hydrolase are selectively overexpressed in neuritic plaque-associated glia in Alzheimer's disease brains. J Neurosci 23:11136–11141.

Berg KA, Maayani S, Goldfarb J, Scaramellini C, Leff P, Clarke WP (1998) Effector pathway-dependent relative efficacy at serotonin type 2A and 2C receptors: evidence for agonist-directed trafficking of receptor stimulus. Mol Pharm 54:94–104.

Bonhaus DW, Chang LK, Kwan J, Martin GR (1998) Dual activation and inhibition of adenylyl cyclase by cannabinoid receptor agonists: evidence for agonist-specific trafficking of intracellular responses. J Pharmacol Exp Ther 287:884–888.

Borner C, Hollt V, Kraus J (2006) Cannabinoid receptor type 2 agonists induce transcription of the mu-opioid receptor gene in Jurkat T cells. Mol Pharm 69:1486–1491.

Borner C, Hollt V, Sebald W, Kraus J (2007) Transcriptional regulation of the cannabinoid receptor type 1 gene in T cells by cannabinoids. J Leukoc Biol 81:336–343.

Bouaboula M, Poinot-Chazel C, Bourrie B, Canat X, Calandra B, Rinaldi-Carmona M, Le Fur G, Casellas P (1995) Activation of mitogen-activated protein kinases by stimulation of the central cannabinoid receptor CB_1. Biochem J 312:637–641.

Bouaboula M, Poinot-Chazel C, Marchand J, Canat X, Bourrie B, Rinaldi-Carmona M, Calandra B, Le Fur G, Casellas P (1996) Signaling pathway associated with stimulation of CB_2 peripheral cannabinoid receptor. Involvement of both mitogen-activated protein kinase and induction of Krox-24 expression. Eur J Biochem 237:704–711.

Bouaboula M, Desnoyer N, Carayon P, Combes T, Casellas P (1999a) G protein modulation induced by a selective inverse agonist for the peripheral cannabinoid receptor CB_2: implication for intracellular signalization cross-regulation. Mol Pharm 55:473–480.

Bouaboula M, Dussossoy D, Casellas P (1999b) Regulation of peripheral cannabinoid receptor CB2 phosphorylation by the inverse agonist SR 144528. Implications for receptor biological responses. J Biol Chem 274:20397–20405.

Brotchie JM (2003) CB_1 cannabinoid receptor signalling in Parkinson's disease. Curr Opin Pharmacol 3:54–61.

Brown SM, Wager-Miller J, Mackie K (2002) Cloning and molecular characterization of the rat CB_2 cannabinoid receptor. Biochim Biophys Acta 1576:255–264.

Brunet A, Datta SR, Greenberg ME (2001) Transcription-dependent and -independent control of neuronal survival by the PI_3K-Akt signaling pathway. Curr Opin Neurobiol 11:297–305.

Carrier EJ, Kearn CS, Barkmeier AJ, Breese NM, Yang W, Nithipatikom K, Pfister SL, Campbell WB, Hillard CJ (2004) Cultured rat microglial cells synthesize the endocannabinoid 2-arachidonylglycerol, which increases proliferation via a CB2 receptor-dependent mechanism. Mol Pharm 65:999–1007.

Cichewicz DL (2004) Synergistic interactions between cannabinoid and opioid analgesics. Life Sci 74:1317–1324.
Conti S, Costa B, Colleoni M, Parolaro D, Giagnoni G (2002) Antiinflammatory action of endocannabinoid palmitoylethanolamide and the synthetic cannabinoid nabilone in a model of acute inflammation in the rat. Br J Pharmacol 135:181–187.
Cravatt B, Lichtman A (2004) The endogenous cannabinoid system and its role in nociceptive behavior. J Neurobiol 61:149–160.
Felder CC, Joyce KE, Briley EM, Mansouri J, Mackie K, Blond O, Lai Y, Ma AL, Mitchell RL (1995) Comparison of the pharmacology and signal transduction of the human cannabinoid CB1 and CB2 receptors. Mol Pharm 48:443–450.
Felder CC, Joyce KE, Briley EM, Glass M, Mackie KP, Fahey KJ, Cullinan GJ, Hunden DC, Johnson DW, Chaney MO, Koppel GA, Brownstein M (1998) LY320135, a novel cannabinoid CB1 receptor antagonist, unmasks coupling of the CB1 receptor to stimulation of cAMP accumulation. J Pharmacol Exp Ther 284:291–297.
Figini M, Emanueli C, Bertrand C, Sicuteri R, Regoli D, Geppetti P (1997) Differential activation of the epithelial and smooth muscle NK1 receptors by synthetic tachykinin agonists in guinea-pig trachea. Br J Pharmacol 121:773–781.
Galiegue S, Mary S, Marchand J, Dussossoy D, Carriere D, Carayon P, Bouaboula M, Shire D, Le Fur G, Casellas P (1995) Expression of central and peripheral cannabinoid receptors in human immune tissues and leukocyte subpopulations. Eur J Biochem 232:54–61.
Galve-Roperh I, Sanchez C, Cortes ML, del Pulgar TG, Izquierdo M, Guzman M (2000) Antitumoral action of cannabinoids: involvement of sustained ceramide accumulation and extracellular signal-regulated kinase activation. Nat Med 6:313–319.
Gerard CM, Mollereau C, Vassart G, Parmentier M (1991) Molecular cloning of a human cannabinoid receptor which is also expressed in testis. Biochem J 279:129–134.
Glass M, Felder CC (1997) Concurrent stimulation of cannabinoid CB_1 and dopamine D_2 receptors augments cAMP accumulation in striatal neurons: evidence for a G_s linkage to the CB_1 receptor. J Neurosci 17:5327–5333.
Gomez del Pulgar T, Velasco G, Sanchez C, Haro A, Guzman M (2002) De novo-synthesized ceramide is involved in cannabinoid-induced apoptosis. Biochem J 363:183–188.
Gong JP, Onaivi ES, Ishiguro H, Liu QR, Tagliaferro PA, Brusco A, Uhl GR (2006) Cannabinoid CB_2 receptors: immunohistochemical localization in rat brain. Brain Res 1071:10–23.
Gouldson P, Calandra B, Legoux P, Kerneis A, Rinaldi-Carmona M, Barth F, Le Fur G, Ferrara P, Shire D (2000) Mutational analysis and molecular modelling of the antagonist SR 144528 binding site on the human cannabinoid CB_2 receptor. Eur J Pharmacol 401:17–25.
Gudermann T, Schoneberg T, Schultz G (1997) Functional and structural complexity of signal transduction via G protein-coupled receptors. Annu Rev Neurosci 20:399–427.
Guzman M, Galve-Roperh I, Sanchez C (2001) Ceramide: a new second messenger of cannabinoid action. Trends Pharmacol Sci 22:19–22.
Herkenham M, Lynn AB, Little MD, Johnson MR, Melvin LS, de Costa BR, Rice KC (1990) Cannabinoid receptor localization in brain. Proc Natl Acad Sci USA 87:1932–1936.
Herrera B, Carracedo A, Diez-Zaera M, Gomez del Pulgar T, Guzman M, Velasco G (2006) The CB_2 cannabinoid receptor signals apoptosis via ceramide-dependent activation of the mitochondrial intrinsic pathway. Exp Cell Res 312:2121–2131.
Howlett AC (1985) Cannabinoid inhibition of adenylate cyclase. Biochemistry of the response in neuroblastoma cell membranes. Mol Pharm 27:429–436.
Howlett AC (1995) Pharmacology of cannabinoid receptors. Annu Rev Pharmacol Toxicol 35:607–634.
Howlett AC, Mukhopadhyay S (2000) Cellular signal transduction by anandamide and 2-arachidonoylglycerol. Chem Phys Lipids 108:53–70.
Ibrahim MM, Deng H, Zvonok A, Cockayne DA, Kwan J, Mata HP, Vanderah TW, Lai J, Porreca F, Makriyannis A, Malan Jr TP (2003) Activation of CB2 cannabinoid receptors by AM1241 inhibits experimental neuropathic pain: pain inhibition by receptors not present in the CNS. Proc Natl Acad Sci USA 100:10529–10533.

Ishac E, Jiang L, Lake K, Varga K, Abood M, Kunos G (1996) Inhibition of exocytotic noradrenaline release by presynaptic CB_1 receptors on peripheral sympathetic nerves. Br J Pharmacol 118:2023–2028.

Iversen L, Chapman V (2002) Cannabinoids: a real prospect for pain relief. Curr Opin Pharmacol 2:50–55.

Jorda MA, Verbakel SE, Valk PJ, Vankan-Berkhoudt YV, Maccarrone M, Finazzi-Agro A, Lowenberg B, Delwel R (2002) Hematopoietic cells expressing the peripheral cannabinoid receptor migrate in response to the endocannabinoid 2-arachidonoylglycerol. Blood 99:2786–2793.

Kaminski NE, Koh WS, Yang KH, Lee M, Kessler FK (1994) Suppression of the humoral immune response by cannabinoids is partially mediated through inhibition of adenylate cyclase by a pertussis toxin-sensitive G protein-coupled mechanism. Biochem Pharmacol 48:1899–1908.

Kaplan BL, Rockwell CE, Kaminski NE (2003) Evidence for cannabinoid receptor-dependent and -independent mechanisms of action in leukocytes. J Pharmacol Exp Ther 306:1077–1085.

Kenakin T (2001) Inverse, protean, and ligand-selective agonism: matters of receptor conformation. FASEB J 15:598–611.

Kenakin T (2002) Drug efficacy at G protein-coupled receptors. Annu Rev Pharmacol Toxicol 42:349–379.

Kishimoto S, Gokoh M, Oka S, Muramatsu M, Kajiwara T, Waku K, Sugiura T (2003) 2-arachidonoylglycerol induces the migration of HL-60 cells differentiated into macrophage-like cells and human peripheral blood monocytes through the cannabinoid CB_2 receptor-dependent mechanism. J Biol Chem 278:24469–24475.

Klein T, Newton C, Larsen K, Lu L, Perkins I, Liang N, Friedman H (2003) The cannabinoid system and immune modulation. J Leukoc Biol 74:486–496.

Kobayashi Y, Arai S, Waku K, Sugiura T (2001) Activation by 2-arachidonoylglycerol, an endogenous cannabinoid receptor ligand, of p42/44 mitogen-activated protein kinase in HL-60 cells. J Biochem 129:665–669.

Lunn CA, Fine JS, Rojas-Triana A, Jackson JV, Fan X, Kung TT, Gonsiorek W, Schwarz MA, Lavey B, Kozlowski JA, Narula SK, Lundell DJ, Hipkin RW, Bober LA (2006) A novel cannabinoid peripheral cannabinoid receptor-selective inverse agonist blocks leukocyte recruitment *in vivo*. J Pharmacol Exp Ther 316:780–788.

Mackie K, Hille B (1992) Cannabinoids inhibit N-type calcium channels in neuroblastoma-glioma cells. Proc Natl Acad Sci USA 89:3825–3829.

Mackie K, Lai Y, Westenbroek R, Mitchell R (1995) Cannabinoids activate an inwardly rectifying potassium conductance and inhibit Q-type calcium currents in AtT20 cells transfected with rat brain cannabinoid receptor. J Neurosci 15:6552–6561.

Maekawa T, Nojima H, Kuraishi Y, Aisaka K (2006) The cannabinoid CB2 receptor inverse agonist JTE-907 suppresses spontaneous itch-associated responses of NC mice, a model of atopic dermatitis. Eur J Pharmacol 542:179–183.

Maneuf YP, Brotchie JM (1997) Paradoxical action of the cannabinoid WIN 55,212-2 in stimulated and basal cyclic AMP accumulation in rat globus pallidus slices. Br J Pharmacol 120:1397–1398.

Maresz K, Carrier EJ, Ponomarev ED, Hillard CJ, Dittel BN (2005) Modulation of the cannabinoid CB2 receptor in microglial cells in response to inflammatory stimuli. J Neurochem 95:437–445.

Matsuda L, Lolait S, Brownstein M, Young A, Bonner T (1990) Structure of a cannabinoid receptor and functional expression of the cloned cDNA. Nature 346:561–564.

McAllister SD, Griffin G, Satin LS, Abood ME (1999) Cannabinoid receptors can activate and inhibit G protein-coupled inwardly rectifying potassium channels in a xenopus oocyte expression system. J Pharmacol Exp Ther 291:618–626.

Molina-Holgado E, Vela JM, Arevalo-Martin A, Almazan G, Molina-Holgado F, Borrell J, Guaza C (2002) Cannabinoids promote oligodendrocyte progenitor survival: involvement of cannabinoid receptors and phosphatidylinositol-3 kinase/Akt signaling. J Neurosci 22:9742–9753.

Molina-Holgado F, Pinteaux E, Heenan L, Moore JD, Rothwell NJ, Gibson RM (2005) Neuroprotective effects of the synthetic cannabinoid HU-210 in primary cortical neurons are mediated by phosphatidylinositol 3-kinase/Akt signaling. Mol Cell Neurosci 28:189–194.

Mukhopadhyay S, Das S, Williams EA, Moore D, Jones JD, Zahm DS, Ndengele MM, Lechner AJ, Howlett AC (2006) Lipopolysaccharide and cyclic AMP regulation of CB_2 cannabinoid receptor levels in rat brain and mouse RAW 264.7 macrophages. J Neuroimmunol 181:82–92.

Munro S, Thomas K, Abu-Shaar M (1993) Molecular characterization of a peripheral receptor for cannabinoids. Nature 365:61–65.

Oka S, Ikeda S, Kishimoto S, Gokoh M, Yanagimoto S, Waku K, Sugiura T (2004) 2-arachidonoylglycerol, an endogenous cannabinoid receptor ligand, induces the migration of EoL-1 human eosinophilic leukemia cells and human peripheral blood eosinophils. J Leukoc Biol 76:1002–1009.

Onaivi ES, Ishiguro H, Gong JP, Patel S, Perchuk A, Meozzi PA, Myers L, Mora Z, Tagliaferro P, Gardner E, Brusco A, Akinshola BE, Liu QR, Hope B, Iwasaki S, Arinami T, Teasenfitz L, Uhl GR (2006) Discovery of the presence and functional expression of cannabinoid CB_2 receptors in brain. Ann N Y Acad Sci 1074:514–536.

Prather PL (2004) Inverse agonists: tools to reveal ligand-specific conformations of G protein-coupled receptors. Sci STKE 2004:1.

Racz I, Bilkei-Gorzo A, Toth Z, Michel K, Palkovits M, Zimmer A (2003) A critical role for the cannabinoid CB_1 receptors in alcohol dependence and stress-stimulated ethanol drinking. J Neurosci 23:2453–2458.

Ramirez BG, Blazquez C, Gomez del Pulgar T, Guzman M, de Ceballos ML (2005) Prevention of Alzheimer's disease pathology by cannabinoids: neuroprotection mediated by blockade of microglial activation. J Neurosci 25:1904–1913.

Ravinet T, Arnone M, Delgorge C, Gonalons N, Keane P, Maffrand J, Soubrie P (2002) Anti-obesity effect of SR141716, a CB_1 receptor antagonist, in diet-induced obese mice. Am J Physiol Regul Integr Comp Physiol 284:345–353.

Rayman N, Lam KH, Laman JD, Simons PJ, Lowenberg B, Sonneveld P, Delwel R (2004) Distinct expression profiles of the peripheral cannabinoid receptor in lymphoid tissues depending on receptor activation status. J Immunol 172:2111–2117.

Sanchez C, de Ceballos ML, del Pulgar TG, Rueda D, Corbacho C, Velasco G, Galve-Roperh I, Huffman JW, Ramon y Cajal S, Guzman M (2001) Inhibition of glioma growth *in vivo* by selective activation of the CB_2 cannabinoid receptor. Cancer Res 61:5784–5789.

Sanchez MG, Ruiz-Llorente L, Sanchez AM, Diaz-Laviada I (2003) Activation of phosphoinositide 3-kinase/PKB pathway by CB_1 and CB_2 cannabinoid receptors expressed in prostate PC-3 cells. Involvement in Raf-1 stimulation and NGF induction. Cell Signal 15:851–859.

Scott D, Wright C, Angus J (2004) Evidence that CB_1 and CB_2 cannabinoid receptors mediate antinoception in neuropathic pain in the rat. Pain 109:124–131.

Shire D, Calandra B, Rinaldi-Carmona M, Oustric D, Pessegue B, Bonnin-Cabanne O, Le Fur G, Caput D, Ferrara P (1996) Molecular cloning, expression and function of the murine CB_2 peripheral cannabinoid receptor. Biochim Biophys Acta 1307:132–136.

Shoemaker JL, Joseph BK, Ruckle MB, Mayeux PR, Prather PL (2005a) The endocannabinoid noladin ether acts as a full agonist at human CB_2 cannabinoid receptors. J Pharmacol Exp Ther 314:868–875.

Shoemaker JL, Ruckle MB, Mayeux PR, Prather PL (2005b) Agonist-directed trafficking of response by endocannabinoids acting at CB_2 receptors. J Pharmacol Exp Ther 315:828–838.

Shoemaker JL, Seely KA, Reed RL, Crow JP, Prather PL (2007) The CB_2 cannabinoid agonist AM-1241 prolongs survival in a transgenic mouse model of amyotrophic lateral sclerosis when initiated at symptom onset. J Neurochem 101:87–98.

Sim-Selley LJ (2003) Regulation of cannabinoid CB_1 receptors in the central nervous system by chronic cannabinoids. Crit Rev Neurobiol 15:91–119.

Slipetz DM, O'Neill GP, Favreau L, Dufresne C, Gallant M, Gareau Y, Guay D, Labelle M, Metters KM (1995) Activation of the human peripheral cannabinoid receptor results in inhibition of adenylyl cyclase. Mol Pharm 48:352–361.

Soderstrom K, Johnson F (2000) CB_1 cannabinoid receptor expression in brain regions associated with zebra finch song control. Brain Res 857:151–157.
Soderstrom K, Leid M, Moore FL, Murray TF (2000) Behavioral, pharmacological, and molecular characterization of an amphibian cannabinoid receptor. J Neurochem 75:413–423.
Song ZH, Slowey CA, Hurst DP, Reggio PH (1999) The difference between the CB_1 and CB_2 cannabinoid receptors at position 5.46 is crucial for the selectivity of WIN55212-2 for CB_2. Mol Pharm 56:834–840.
Strange PG (2002) Mechanisms of inverse agonism at G protein-coupled receptors. Trends Pharmacol Sci 23:89–95.
Sugiura T, Kodaka T, Kondo S, Nakane S, Kondo H, Waku K, Ishima Y, Watanabe K, Yamamoto I (1997) Is the cannabinoid CB_1 receptor a 2-arachidonoylglycerol receptor? Structural requirements for triggering a Ca^{2+} transient in NG108-15 cells. J Biochem 122:890–895.
Sugiura T, Kondo S, Kishimoto S, Miyashita T, Nakane S, Kodaka T, Suhara Y, Takayama H, Waku K (2000) Evidence that 2-arachidonoylglycerol but not N-palmitoylethanolamine or anandamide is the physiological ligand for the cannabinoid CB_2 receptor. Comparison of the agonistic activities of various cannabinoid receptor ligands in HL-60 cells. J Biol Chem 275:605–612.
Tao Q, McAllister SD, Andreassi J, Nowell KW, Cabral GA, Hurst DP, Bachtel K, Ekman MC, Reggio PH, Abood ME (1999) Role of a conserved lysine residue in the peripheral cannabinoid receptor (CB_2): evidence for subtype specificity. Mol Pharm 55:605–613.
Tuccinardi T, Ferrarini PL, Manera C, Ortore G, Saccomanni G, Martinelli A (2006) Cannabinoid CB_2/CB_1 selectivity. Receptor modeling and automated docking analysis. J Med Chem 49:984–994.
Ueda Y, Miyagawa N, Wakitani K (2007) Involvement of cannabinoid CB_2 receptors in the IgE-mediated triphasic cutaneous reaction in mice. Life Sci 80:414–419.
Valk PJ, Hol S, Vankan Y, Ihle JN, Askew D, Jenkins NA, Gilbert DJ, Copeland NG, de Both NJ, Lowenberg B, Delwel R (1997) The genes encoding the peripheral cannabinoid receptor and alpha-L-fucosidase are located near a newly identified common virus integration site, Evi11. J Virol 71:6796–6804.
Van Sickle MD, Duncan M, Kingsley PJ, Mouihate A, Urbani P, Mackie K, Stella N, Makriyannis A, Piomelli D, Davison JS, Marnett LJ, Di Marzo V, Pittman QJ, Patel KD, Sharkey KA (2005) Identification and functional characterization of brainstem cannabinoid CB_2 receptors. Science 310:329–332.
Vasquez C, Lewis DL (1999) The CB_1 cannabinoid receptor can sequester G proteins, making them unavailable to couple to other receptors. J Neurosci 19:9271–9280.
Walter L, Franklin A, Witting A, Wade C, Xie Y, Kunos G, Mackie K, Stella N (2003) Nonpsychotropic cannabinoid receptors regulate microglial cell migration. J Neurosci 23:1398–1405.
Wiens BL, Nelson CS, Neve KA (1998) Contribution of serine residues to constitutive and agonist-induced signaling via the D2S dopamine receptor: evidence for multiple, agonist-specific active conformations. Mol Pharm 54:435–444.
Yamaguchi F, Macrae AD, Brenner S (1996) Molecular cloning of two cannabinoid type 1-like receptor genes from the puffer fish Fugu rubripes. Genomics 35:603–605.

Chapter 7
CB_1 and CB_2 Receptor Pharmacology

Roger G. Pertwee

Abstract This review describes compounds that are currently most widely used in preclinical research to activate or block cannabinoid CB_1 and CB_2 receptors. Some of these compounds are ligands that display significant selectivity as CB_1 or CB_2 receptor agonists or antagonists, the remainder consisting of agonists each of which exhibits more or less equal potency at CB_1 and CB_2 receptors. The cannabinoid receptor antagonists most often used as pharmacological tools behave as inverse agonists in at least some assay systems and possible explanations for this inverse agonism are briefly discussed. Also considered in this review are actual and potential therapeutic applications for CB_1 and CB_2 receptor ligands.

Introduction

So far, two G protein-coupled cannabinoid receptors, namely, the CB_1 and the CB_2 receptors, were cloned from several vertebrata species including humans (see Chaps. 1, 5, 6). A large number of exogenous/synthetic agonists and antagonists of these receptors were identified and/or designed. The first ligands were engineered principally after the structure of the active ingredients of marijuana (see Chap. 1); however, several novel molecules were also created which are strikingly different from (–)-Δ^9-tetrahydrocannabinol (Δ^9-THC, the main psychoactive constituent of marijuana). The present review summarizes the pharmacological properties of these ligands, and gives an insight in the receptor–ligand interactions as well as the therapeutical potential of these substances.

CB_1 and CB_2 Receptor Agonists

Several of the compounds most often used in the laboratory as CB_1 or CB_2 receptor agonists activate each of these receptor types with approximately equal potency. As detailed elsewhere (Howlett et al., 2002; Pertwee, 1999, 2005b), these compounds include (1) the classical cannabinoids Δ^9-THC, (–)-Δ^8-THC (Δ^8-THC, another active constituent of marijuana) and (–)-11-hydroxy Δ^8-THC-dimethylheptyl (HU-210), (2) the non-classical cannabinoid CP55940, (3) the aminoalkylindole R-(+)-WIN55212 (WIN55212-2) and (4) the endogenous agonists eicosanoids, *N*-arachidonoylethanolamine (anandamide) and 2-arachidonoylglycerol (2-AG). The CB_1 and CB_2 receptor binding affinities of these compounds are shown in Table 1. It is also worth noting that

(a) Δ^9-THC, the main psychoactive constituent of cannabis, is a partial agonist for both CB_1 and CB_2 receptors that exhibits even less efficacy at CB_2 than at CB_1 receptors to the extent that it has been found to behave as a CB_2 receptor antagonist in one bioassay system (Bayewitch et al., 1996).

Table 1 K_i values of CB_1 and CB_2 receptor-selective ligands for the in vitro displacement of [^{3}H]CP55940 or [^{3}H]HU-243 from CB_1 and CB_2 receptor-specific binding sites

Ligand	CB_1 K_i value (nM)	CB_2 K_i value (nM)
CB_1 receptor-selective antagonists/inverse agonists		
SR141716A	1.8 to 12.3	514 to 13,200
AM281	12	4,200
AM251	7.49	2,290
LY320135	141	14,900
CB1 receptor-selective agonists		
ACEA	1.4 or 5.29	195 or >2,000
O-1812	3.4	3,870
ACPA	2.2	715
(–)-3-(1-Adamantyl)-Δ^8-THC (AM411)[a]	6.8	52
2-Arachidonyl glyceryl ether (noladin ether)	21.2	>3,000
R-(+)-methanandamide	17.9 to 28.3	815 or 868
Oleamide[b]	1,140	>100,000
Agonists without significant CB1 or CB2 receptor-selectivity		
HU-210	0.06 to 0.73	0.17 to 0.52
CP55940	0.5 to 5.0	0.69 to 2.8
WIN55212-2	1.89 to 123	0.28 to 16.2
(–)-Δ^9-THC	5.05 to 80.3	3.13 to 75.3
(–)-Δ^8-THC	44 or 47.6	39.3 or 44
Anandamide	61 to 543	279 to 1,940
2-Arachidonoyl glycerol	58.3 or 472	145 or 1,400
CB2-selective agonists		
AM1241	280	3.4
JWH-133	677	3.4
GW405833[c]	273 or 4,772	3.6 or 3.92
JWH-015	383	13.8
HU-308	>10,000	22.7
CB2 receptor-selective antagonists/inverse agonists		
SR144528	50.3 to >10,000	0.28 to 5.6
AM630	5152	31.2

[a]Lu et al. (2005); [b]Leggett et al. (2004); [c]Valenzano et al. (2005); for other references and further details see Pertwee (2005b)

(b) Δ^8-THC resembles Δ^9-THC both in its affinities for CB_1 and CB_2 receptors and in its CB_1 receptor efficacy.
(c) HU-210 has CB_1 and CB_2 receptor affinity and efficacy values that greatly exceed those of Δ^9-THC, its efficacies at CB_1 and CB_2 receptors matching those of CP55940 and WIN55212-2 (see below).
(d) CP55940 and WIN55212-2 each has CB_1 and CB_2 receptor affinities in the low nanomolar range and exhibits relatively high HU-210-like efficacy at both these receptor types.
(e) Anandamide binds a little more readily to CB_1 than to CB_2 receptors and resembles Δ^9-THC in its CB_1 affinity, in behaving as a partial agonist at CB_1 and CB_2 receptors and in exhibiting lower CB_2 than CB_1 efficacy.
(f) 2-AG has been found in several investigations to display higher efficacy than anandamide at CB_1 and CB_2 receptors but to possess anandamide-like affinity for each of these receptor types.

Structure–Activity Relationships of CB_1 and CB_2 Receptor Agonists

Non-Selective Agonists

The structures of CP55940 and other non-classical cannabinoids are quite similar to those of classical cannabinoids such as HU-210, an important distinguishing feature of the CP55940 molecule being that it lacks a pyran ring and hence is bicyclic rather than tricyclic. In contrast, the structure of WIN55212-2 is markedly different from that of both classical and non-classical cannabinoids and, line with this structural difference, there is evidence that it also binds differently to the CB_1 receptor than both HU-210 and CP55940 (Howlett et al., 2002; Pertwee, 1997). However, despite this difference, mutual displacement between WIN55212-2 and non-aminoalkylindole cannabinoids does still occur at CB_1 receptor binding sites. Another difference between WIN55212-2 and these other two cannabinoids is that it exhibits a slightly greater affinity for CB_2 than for CB_1 receptors (Table 1).

CB_1 Receptor-Selective Agonists

Anandamide exhibits marginal CB_1 selectivity and, as indicated in Table 1, it has proved possible to enhance this selectivity by modifying the structure of this ligand to form compounds such as R-(+)-methanandamide (R-methanandamide), arachidonyl-2 -chloroethylamide (ACEA), arachidonylcyclopropylamide (ACPA) and O-1812 (Howlett et al., 2002; Pertwee, 1999, 2005b). In contrast to

ACEA and ACPA, R-methanandamide and O-1812 are also more resistant to enzymic hydrolysis than anandamide (Di Marzo et al., 2001; Howlett et al., 2002; Pertwee, 2005b). Two other notable CB_1 receptor-selective agonists (Table 1) are (–)-3-(1-adamantyl)-Δ^8-THC (Lu et al., 2005) and the putative endocannabinoid, 2-arachidonylglyceryl ether (noladin ether), which when compared to CP55940, exhibits similar CB_1 efficacy but less CB_1 potency (Savinainen et al., 2001, 2003). Another putative endocannabinoid, oleamide, has also been reported to behave as a CB_1 receptor-selective agonist (Leggett et al., 2004). However, its affinity for the CB_1 receptor is markedly less than that of noladin ether (Table 1).

CB_2 Receptor-Selective Agonists

As to CB_2 receptor-selective agonists (Table 1), those most commonly used for research purposes have been JWH133, which is a classical cannabinoid, and the less selective JWH015, which is an aminoalkylindole (Howlett et al., 2002; Pertwee, 2000, 2005b). HU-308, AM1241, and the Merck Frosst compounds, L-759633 and L-759656, are also notable CB_2 receptor-selective agonists (Howlett et al., 2002; Pertwee, 2005b), as is the Glaxo Smith Kline compound, GW405833, a potent CB_2 receptor partial agonist (Valenzano et al., 2005). Although (racemic) AM1241 also behaves as a CB_2 receptor partial agonist in some in vitro assay systems, in others it behaves either as a CB_2 receptor antagonist or as a CB_2 receptor inverse agonist, prompting the hypothesis that it is a CB_2 receptor "protean agonist" (Yao et al., 2006).

The Role of Chiral Centres in Cannabinoid Receptor Agonist Activity

A number of cannabinoids contain chiral centres that affect their potencies as CB_1 and/or CB_2 receptor agonists. For classical and non-classical cannabinoids, it is the (–)-*trans* (6a*R*, 10a*R*) enantiomers that generally exhibit the greatest agonist activity at CB_1 or CB_2 receptors (Howlett et al., 2002; Pertwee, 1999, 2005b). Thus, Δ^9-THC, HU-210 and CP55940 are all (–)-*trans* (6a*R*, 10a*R*) ligands and exhibit significantly greater potency as cannabinoid receptor agonists than their (+)-*cis* (6a*S*, 10a*S*) enantiomers. WIN55212-2 also exhibits stereoselectivity, its *S*-(–)-enantiomer WIN55212-3 behaves in vitro at concentrations in the low micromolar range as a CB_1 receptor partial/inverse agonist and as a CB_2 receptor neutral antagonist (Savinainen et al., 2005). Similarly, R-methanandamide has significantly greater affinity for CB_1 receptors than its *S*-(–)-isomer (Abadji et al., 1994). There are no chiral centres in anandamide.

CB_1 and CB_2 Receptor Antagonists

Turning now to cannabinoid receptor antagonists, the first of these to be developed was the CB_1 receptor-selective SR141716A (rimonabant; Acomplia™) (Rinaldi-Carmona et al., 1994; Howlett et al., 2002; Pertwee, 1999, 2005b). Other CB_1 receptor-selective antagonists include AM251 and AM281, which are both structural analogues of SR141716A and are particularly widely used as research tools, and the less potent LY320135 (Howlett et al., 2002; Pertwee, 1999, 2005b). The best-known CB_2 receptor-selective antagonists are SR144528, (Rinaldi-Carmona et al., 1998) and 6-iodopravadoline (AM630) (Ross et al., 1999a). It is worth noting that although the antagonists just mentioned exhibit marked selectivity as CB_1 or CB_2 receptor antagonists, none of them is completely CB_1 or CB_2 receptor-specific (Table 1). As a result, although these ligands will exhibit selectivity when administered at doses or concentrations that lie within their CB_1 or CB_2 receptor "selectivity window", there will be higher doses or concentrations at which they are capable of blocking both these receptor types equally well. Similarly, cannabinoid receptor agonists that can selectively target CB_1 or CB_2 receptors will only display such selectivity when administered at doses or concentrations that fall within their CB_1 or CB_2 receptor "selectivity window".

The Question of Inverse Agonism

In some experiments performed in vivo or with CB_1 receptor-containing tissues, SR141716A, AM251, AM281 and LY320135 have been found to elicit responses that are opposite in direction from those elicited by CB_1 receptor agonists. Sometimes, this may have resulted from a direct antagonism of responses evoked at CB_1 receptors by released endocannabinoids or, as proposed by Savinainen and colleagues (2003), from antagonism of adenosine when this is being released onto adenosine A_1 receptors. However, the production of such effects in some instances at least most probably reflects an ability of these compounds to induce inverse agonism by reducing spontaneous coupling of CB_1 receptors to their effector mechanisms in the absence of exogenously added or endogenously released CB_1 agonists (Pertwee, 2005a,b). There is evidence that CB_2 receptors can also exist in such a "constitutively active" state and that SR144528 and AM630 are both CB_2 receptor inverse agonists (Howlett et al., 2002; Pertwee, 1999, 2005b; Ross et al., 1999a,b). The likelihood that this "first generation" of CB_1 and CB_2 receptor antagonists are all inverse agonists has prompted a search for a ligand possessing high affinity and selectivity for CB_1 or CB_2 receptors but lacking significant efficacy as either a CB_1 or CB_2 agonist or a CB_1 or CB_2 inverse agonist. The availability of such a "neutral antagonist" would be of interest not least because it would then become easier to distinguish between tonic cannabimimetic activity arising from ongoing endocannabinoid release onto CB_1 or CB_2 receptors, which a neutral antagonist would be expected to oppose, and tonic activity arising from the presence of constitutively active CB_1 or CB_2 receptors, which it would not be expected to alter. Compounds that have been reported to behave as neutral CB_1 receptor antagonists include a sulphonamide analogue of Δ^8-THC with an acetylenic side chain (O-2050), 6″-azidohex-2″-yne-cannabidiol (O-2654) and two structural analogues of SR141716A (VCHR and NESS 0327) (Ruiu et al., 2003; Pertwee, 2005a,b). There are no reports as yet of the development of a neutral CB_2 receptor antagonist. Interestingly, Leterrier and co-workers (2006) have obtained evidence that endogenously induced CB_1 receptor signalling may explain why the somatodendritic surface of neurons is normally so much less populated with CB_1 receptors than the axonal surface. Their data suggest that such signalling causes CB_1 receptors to undergo endocytosis and that this process is restricted mainly to the somatodendritic region of neurons. They also found that the CB_1 receptor-selective antagonist/inverse agonist, AM281, reduces this endocytosis thereby causing a selective upregulation of CB_1 receptors on somatodendritic plasma membranes. Results obtained by Turu and colleagues (2007) suggest that CB_1 receptor endocytosis in the somatodendritic region of neurons may result from increased CB_1 receptor signalling induced by endogenously produced 2-AG rather than by CB_1 receptor constitutive activity.

Concluding Remarks

The following chapters of this book will thoroughly review the role of the endocannabinoid system in neuropsychiatric and metabolic disorders. Described in this chapter are actual or potential therapeutic applications for cannabinoid receptor agonists and antagonists (Pertwee and Thomas, 2008). Cannabinoid receptor ligands already used as medicines are SR141716A, Δ^9-THC and nabilone, which is a structural analogue of Δ^9-THC. SR141716A (rimonabant; Acomplia™, Sanofi-Aventis) is prescribed to treat obesity and related metabolic risk factors. Δ^9-THC is prescribed as Marinol™ (Unimed Pharmaceuticals) for the suppression of nausea and vomiting induced by cancer chemotherapy, as is Nabilone (under the name of Cesamet™ in the US, UK and Canada). Marinol™ is also used to stimulate appetite, particularly in AIDS patients who are experiencing excessive loss of body weight. Another medicine that contains Δ^9-THC is Sativex™ (GW Pharmaceuticals) which is prescribed for the symptomatic relief of neuropathic pain in adults with multiple sclerosis. Sativex™ has the non-psychoactive plant cannabinoid, cannabidiol, as a second major constituent. Cannabinoid receptor ligands also have other potential uses. For CB_1/CB_2 receptor agonists, these include the management of various kinds of pain, some types of cancer, inflammation, glaucoma, cough and cholestatic pruritis, and the amelioration of certain symptoms of multiple sclerosis and spinal cord injury, of Alzheimer's disease, of amyotrophic lateral sclerosis, of tardive dyskinesia induced in psychiatric patients by neuroleptic drugs, of Tourette's syndrome, of anxiety disorders, of attention deficit hyperactivity disorder, of some gastrointestinal disorders and of atherosclerosis and certain other cardiovascular disorders (Pertwee and Thomas, 2008). For CB_1 receptor antagonists, potential clinical applications include the management of nicotine dependence, of impaired fertility in some women, of stroke, of the hypotension of endotoxaemic shock triggered by advanced liver cirrhosis and of intestinal hypomotility in paralytic ileus (Izzo and Coutts, 2005; Le Foll and Goldberg, 2005; Pertwee, 2005c). As to CB_2 receptor inverse agonists, these exhibit therapeutic potential as anti-inflammatory agents (Lunn et al., 2006). Other possible future uses for CB_1/CB_2 receptor agonists or antagonists include the clinical management of motor impairment and tremor in Parkinson's disease, of dyskinesia induced by L-DOPA in patients with this disease and of osteoporosis (Fernández-Ruiz and González, 2005; Idris et al., 2005; Pertwee, 2005c; Robson, 2005; Ofek et al., 2006). For some of these disorders it is unclear at present whether an agonist or an antagonist should be used as the medicine. There is currently considerable interest in strategies that would improve the selectivity of cannabinoid receptor agonists as therapeutic agents (Pertwee and Thomas, 2008). For the production of analgesia, one possibility would be to target CB_2 receptors that mediate relief from inflammatory and neuropathic pain by using a CB_2 receptor-selective agonist as a medicine. A second possible strategy would be to exploit the ability of cannabinoid receptors in the spinal cord and skin, to mediate pain relief by administering a CB_1 and/or CB_2 receptor agonist intrathecally or topically. A third possibility would be to target peripheral CB_1 and CB_2 receptors that mediate relief from inflammatory and neuropathic pain by treating patients with a CB_1 and/or CB_2 receptor agonist that does not readily cross the blood brain barrier. It may also be possible to achieve greater selectivity by exploiting the ability of a low dose of a cannabinoid receptor agonist to interact synergistically with a non-cannabinoid to produce a sought-after effect. Thus, there is evidence that such synergism takes place between Δ^9-THC and opioid receptor agonists such as morphine or codeine for the production of analgesia and between Δ^9-THC and the 5-HT_3 receptor antagonist, ondansetron, for the suppression of vomiting and retching (Cichewicz, 2004; Kwiatkowska et al., 2004). Finally, there is evidence that a number of disorders or unwanted symptoms are associated with a selective increase in the expression levels and/or coupling efficiencies of particular populations of cannabinoid CB_1 and/or CB_2 receptors, the activation of which leads to an amelioration of symptoms (Pertwee, 2005c). These disorders/symptoms include neuropathic pain, intestinal inflammation, colitis, diarrhoea, prostate cancer, hypertension, atherosclerosis and Parkinson's disease. Upregulation of this kind would be expected to improve the benefit-to-risk ratio of a cannabinoid receptor full or partial agonist by increasing the potency with which it produces its sought-after effect(s) without affecting the potency with which it produces unwanted effects. For a partial agonist such as Δ^9-THC, but not for a full agonist, such upregulation is also expected to produce a selective augmentation of the maximal degree of symptom relief that can be produced, thereby further enhancing the selectivity of a partial agonist and so favouring its use as a medicine over that of a full agonist.

Acknowledgements The writing of this chapter was supported by funding from GW Pharmaceuticals, the BBSRC and NIDA.

References

Abadji V, Lin S, Taha G, Griffin G, Stevenson LA, Pertwee RG, Makriyannis A (1994) (*R*)-methanandamide: a chiral novel anandamide possessing higher potency and metabolic stability. J Med Chem 37:1889–1893.

Bayewitch M, Rhee M-H, Avidor-Reiss T, Breuer A, Mechoulam R, Vogel Z (1996) (–)-Δ^9-tetrahydrocannabinol antagonizes the peripheral cannabinoid receptor-mediated inhibition of adenylyl cyclase. J Biol Chem 271:9902–9905.

Cichewicz DL (2004) Synergistic interactions between cannabinoid and opioid analgesics. Life Sci 74:1317–1324.

Di Marzo V, Bisogno T, De Petrocellis L, Brandi I, Jefferson RG, Winckler RL, Davis JB, Dasse O, Mahadevan A, Razdan RK, Martin BR (2001) Highly selective CB_1 cannabinoid receptor ligands and novel CB_1/VR_1 vanilloid receptor "hybrid" ligands. Biochem Biophys Res Commun 281:444–451.

Fernández-Ruiz J, González S (2005) Cannabinoid control of motor function at the basal ganglia. In: Pertwee RG, ed. *Cannabinoids. Handbook of Experimental Pharmacology*. Heidelberg: Springer-Verlag, Vol. 168, pp. 479–507.

Howlett AC, Barth F, Bonner TI, Cabral G, Casellas P, Devane WA, Felder CC, Herkenham M, Mackie K, Martin BR, Mechoulam R, Pertwee RG (2002) International Union of Pharmacology. XXVII. Classification of cannabinoid receptors. Pharmacol Rev 54:161–202.

Idris AI, Van't Hof RJ, Greig IR, Ridge SA, Baker D, Ross RA, Ralston SH (2005) Regulation of bone mass, bone loss and osteoclast activity by cannabinoid receptors. Nat Med 11:774–779.

Izzo AA, Coutts AA (2005) Cannabinoids and the digestive tract. In: Pertwee RG, ed. *Cannabinoids. Handbook of Experimental Pharmacology*. Heidelberg: Springer-Verlag, Vol. 168, pp. 573–598.

Kwiatkowska M, Parker LA, Burton P, Mechoulam R (2004) A comparative analysis of the potential of cannabinoids and ondansetron to suppress cisplatin-induced emesis in the *Suncus murinus* (house musk shrew). Psychopharmacology 174:254–259.

Le Foll B, Goldberg SR (2005) Cannabinoid CB_1 receptor antagonists as promising new medications for drug dependence. J Pharmacol Exp Ther 312:875–883.

Leggett JD, Aspley S, Beckett SRG, D'Antona AM, Kendall DA, Kendall DA (2004) Oleamide is a selective endogenous agonist of rat and human CB_1 cannabinoid receptors. Br J Pharmacol 141:253–262.

Leterrier C, Laine J, Darmon M, Boudin H, Rossier J, Lenkei Z (2006) Constitutive activation drives compartment-selective endocytosis and axonal targeting of type 1 cannabinoid receptors. J Neurosci 26:3141–3153.

Lu D, Meng Z, Thakur GA, Fan P, Steed J, Tartal CL, Hurst DP, Reggio PH, Deschamps JR, Parrish DA, George C, Järbe TUC, Lamb RJ, Makriyannis A (2005) Adamantyl cannabinoids: a novel class of cannabinergic ligands. J Med Chem 48:4576–4585.

Lunn CA, Fine JS, Rojas-Triana A, Jackson JV, Fan X, Kung TT, Gonsiorek W, Schwarz MA, Lavey B, Kozlowski JA, Narula SK, Lundell DJ, Hipkin RW, Bober LA (2006) A novel cannabinoid peripheral cannabinoid receptor-selective inverse agonist blocks leukocyte recruitment *in vivo*. J Pharmacol Exp Ther 316:780–788.

Ofek O, Karsak M, Leclerc N, Fogel M, Frenkel B, Wright K, Tam J, Attar-Namdar M, Kram V, Shohami E, Mechoulam R, Zimmer A, Bab I (2006) Peripheral cannabinoid receptor, CB_2, regulates bone mass. Proc Natl Acad Sci USA 103:696–701.

Pertwee RG (1997) Pharmacology of cannabinoid CB_1 and CB_2 receptors. Pharmacol Ther 74:129–180.

Pertwee RG (1999) Pharmacology of cannabinoid receptor ligands. Curr Med Chem 6:635–664.

Pertwee RG (2000) Cannabinoid receptor ligands: clinical and neuropharmacological considerations relevant to future drug discovery and development. Exp Opin Investig Drugs 9:1553–1571.

Pertwee RG (2005a) Inverse agonism and neutral antagonism at cannabinoid CB_1 receptors. Life Sci 76:1307–1324.

Pertwee RG (2005b) Pharmacological actions of cannabinoids. In: Pertwee RG, ed. *Cannabinoids. Handbook of Experimental Pharmacology*. Heidelberg: Springer-Verlag, Vol. 168, pp. 1–51.

Pertwee RG (2005c) The therapeutic potential of drugs that target cannabinoid receptors or modulate the tissue levels or actions of endocannabinoids. AAPS J 7:E625–E654.

Pertwee RG, Thomas A (2008) Therapeutic applications for agents that act at CB_1 and CB_2 receptors. In: Reggio P, ed. *The Cannabinoid Receptors* (in press).

Rinaldi-Carmona M, Barth F, Héaulme M, Shire D, Calandra B, Congy C, Martinez S, Maruani J, Néliat G, Caput D, Ferrara P, Soubrié P, Brelière JC, Le Fur G (1994) SR141716A, a potent and selective antagonist of the brain cannabinoid receptor. FEBS Lett 350:240–244.

Rinaldi-Carmona M, Barth F, Millan J, Derocq J-M, Casellas P, Congy C, Oustric D, Sarran M, Bouaboula M, Calandra B, Portier M, Shire D, Brelière J-C, Le Fur G (1998) SR 144528, the first potent and selective antagonist of the CB_2 cannabinoid receptor. J Pharmacol Exp Ther 284:644–650.

Robson P (2005) Human studies of cannabinoids and medicinal cannabis. In: Pertwee RG, ed. *Cannabinoids. Handbook of Experimental Pharmacology*. Heidelberg: Springer-Verlag, Vol. 168, pp. 719–756.

Ross RA, Brockie HC, Stevenson LA, Murphy VL, Templeton F, Makriyannis A, Pertwee RG (1999a) Agonist-inverse agonist characterization at CB_1 and CB_2 cannabinoid receptors of L759633, L759656 and AM630. Br J Pharmacol 126:665–672.

Ross RA, Gibson TM, Stevenson LA, Saha B, Crocker P, Razdan RK, Pertwee RG (1999b) Structural determinants of the partial agonist-inverse agonist properties of 6′-azidohex-2′-yne-Δ^8-tetrahydrocannabinol at cannabinoid receptors. Br J Pharmacol 128:735–743.

Ruiu S, Pinna GA, Marchese G, Mussinu J-M, Saba P, Tambaro S, Casti P, Vargiu R, Pani L (2003) Synthesis and characterization of NESS 0327: a novel putative antagonist of the CB_1 cannabinoid receptor. J Pharmacol Exp Ther 306:363–370.

Savinainen JR, Järvinen T, Laine K, Laitinen JT (2001) Despite substantial degradation, 2-arachidonoylglycerol is a potent full efficacy agonist mediating CB_1 receptor-dependent G-protein activation in rat cerebellar membranes. Br J Pharmacol 134:664–672.

Savinainen JR, Saario SM, Niemi R, Järvinen T, Laitinen JT (2003) An optimized approach to study endocannabinoid signaling: evidence against constitutive activity of rat brain adenosine A_1 and cannabinoid CB_1 receptors. Br J Pharmacol 140:1451–1459.

Savinainen JR, Kokkola T, Salo OMH, Poso A, Järvinen T, Laitinen JT (2005) Identification of WIN55212-3 as a competitive neutral antagonist of the human cannabinoid CB_2 receptor. Br J Pharmacol 145:636–645.

Turu G, Simon A, Gyombolai P, Szidonya L, Bagdy G, Lenkei Z, Hunyady L (2007) The role of diacylglycerol lipase in constitutive and angiotensin AT_1 receptor-stimulated cannabinoid CB_1 receptor activity. J Biol Chem 282:7753–7757.

Valenzano KJ, Tafesse L, Lee G, Harrison JE, Boulet JM, Gottshall SL, Mark L, Pearson MS, Miller W, Shan S, Rabadi L, Rotshteyn Y, Chaffer SM, Turchin PI, Elsemore DA, Toth M, Koetzner L, Whiteside GT (2005) Pharmacological and pharmacokinetic characterization of the cannabinoid receptor 2 agonist, GW405833, utilizing rodent models of acute and chronic pain, anxiety, ataxia and catalepsy. Neuropharmacology 48:658–672.

Yao BB, Mukherjee S, Fan Y, Garrison TR, Daza AV, Grayson GK, Hooker BA, Dart MJ, Sullivan JP, Meyer MD (2006) In vitro pharmacological characterization of AM1241: a protean agonist at the cannabinoid CB_2 receptor? Br J Pharmacol 149:145–154.

Chapter 8
Functional Molecular Biology of the $TRPV_1$ Ion Channel

Istvan Nagy, John P.M. White, Cleoper C. Paule, Mervyn Maze, and Laszlo Urban

Abstract This chapter offers an introduction to the structure and function of $TRPV_1$ receptor. In terms of activators and sites of interaction, we elaborate at least five mechanisms by which $TRPV_1$ receptor may be activated, including ligand binding, protonation, post-translational changes, thermal energy, and electrical energy. The sub-cellular expression of $TRPV_1$ receptor and its expression in primary sensory neurons in the central nervous system and by non-neuronal cells are also examined, as is co-expression of $TRPV_1$ receptor with the cannabinoid 1 receptor. The cellular responses to $TRPV_1$ receptor activation are discussed, including its role in generating ionic influx into primary sensory neurons and desensitisation of $TRPV_1$ receptor. Finally, consideration is afforded to the role of $TRPV_1$ receptor in physiological and pathological conditions.

Introduction

The transient receptor potential vanilloid type 1 ion channel ($TRPV_1$ receptor), which is found in a sub-population of primary sensory neurons, is now well recognised as being a transducer for noxious heat (Caterina et al., 1997). $TRPV_1$ receptor is a ligand-gated ion channel. Such channels are also often referred to as "ionotropic receptors". Receptors of this type control the fastest changes in membrane conductance in the nervous system by transiently increasing the permeability of the neuronal membrane to particular ions, such as Na^+, K^+, Ca^{2+}, or Cl^-. The $TRPV_1$ receptor ion channel, when activated, is permeable to cations only. Such cation-selective ligand-gated ion channels produce, on activation, a net inward current which depolarises the membrane and increases the probability of action potential generation. It is believed that fast ionotropic receptors on primary sensory neurons are ideally designed for facilitating the rapid transduction of mechanical, thermal, or chemical stimuli into electrical signals within a time frame which may enable the organism to react to such stimuli. Although the $TRPV_1$ ion channel is selectively permeable to cations, it exhibits no preference for any of those found in the extracellular fluid, thus, it is permeable to Na^+, K^+, and Ca^{2+}. $TRPV_1$ receptor exhibits

A. Köfalvi (ed.), *Cannabinoids and the Brain.*

all of the general characteristics of ionotropic channels, but it also possesses certain special features. It is considered that the most important of these special features is that in addition to $TRPV_1$ receptor agonists, such as capsaicin and anandamide, which activate the receptor through a direct binding mechanism, it responds to a multiplicity of other stimuli. These "non-agonist" activators of $TRPV_1$ receptor may mediate their effect through one (or a combination) of several mechanisms, including "protonation", "electrical energy-mediated gating", "thermal energy-mediated gating", and "post-translational changes mediated gating", respectively. Protonation refers to the activation in acidic (low pH) conditions of the ion channel by hydrogen ions. Electrical energy-mediated gating refers to activation of the channel by depolarisation of the membrane potential. Thermal energy-mediated gating refers to the effect of heat in activating the channel directly, or in facilitating its activation by other activators. Finally, post-translational changes-mediated gating refers to activation of the channel as a result of the ligand-binding of receptors which are co-expressed with $TRPV_1$ receptor. Further consideration is afforded to each of these gating mechanisms later in this chapter. Remarkably, $TRPV_1$ receptor is capable of integrating these stimuli and translating those into membrane currents. In primary sensory neurons, the membrane currents generated by neuronal $TRPV_1$ receptor ion channels affect neuronal firing patterns to constitute and initiate the electrical transfer to the brain of an integrated picture of pain-inducing stimuli: *E pluribus unum*. The detail of $TRPV_1$ receptor activation and function has yet to be elucidated, but our present knowledge is sufficient to demonstrate that the subject is one of tremendous complexity. In this chapter, we will trace the outline of the components of this complex process in a manner which will serve to inform the uninitiated reader of their essential features.

Structure of the $TRPV_1$ Ion Channel

The term "$TRPV_1$ receptor" is generally applied to describe the ion channel of that name which is comprised (usually) of four $TRPV_1$ receptor molecules. These $TRPV_1$ receptor molecules represent the so-called $TRPV_1$ receptor *channel subunits*. It is these molecules which are coded for in the genome; but, when made, these molecules possess the features which are necessary to cause them to combine to constitute the $TRPV_1$ receptor ion channel. The $TRPV_1$ receptor ion channel thus constituted is, therefore, a multi-molecular structure. The core functional component of this molecular complex, which constitutes the $TRPV_1$ receptor ion channel, is an aqueous pore which is opened and closed (gated) to allow the cation influx necessary to generate changes in membrane potential which affect neuronal excitability. $TRPV_1$ receptor ion channels are located in the plasma membrane or in other internal membranes where they can function in relation to the generation of cation currents. The functioning of these ion channels is intimately affected by the activity of other so-called *auxiliary molecules* which are co-localised with the $TRPV_1$ receptor molecular complex. Finally, small variations in the execution of the genetic code for individual $TRPV_1$ molecules (i.e., channel sub-units) result in the production of molecules, and ultimately, of ion channels, which, although substantially similar structurally, nevertheless may differ in important respects as regards their pharmacological profile. These "non-standard" $TRPV_1$ molecules are referred to as "splice variants". The subsequent discussion sequentially addresses: the structure of standard $TRPV_1$ channel sub-units, the related issue of their "splice variants", and the structure of the $TRPV_1$ receptor ion channel.

The $TRPV_1$ Receptor "Subunit"

The $TRPV_1$ receptor cDNA was first isolated by Caterina and colleagues (1997) from rat dorsal root ganglia and it contains an open reading frame of 2,514 nucleotides that encodes a protein of 838 amino acids with a predicted relative molecular mass of 95,000. Each of the channel subunits has six transmembrane segments (S1–S6), with the pore region between the fifth and sixth segments, and cytoplasmic N- and C-termini (see Fig.1). $TRPV_1$ receptor channels contain three ankyrin domains in the N-terminus, which are thought to interact with cytosolic proteins (Clapham, 2003). The N-terminus of the $TRPV_1$ molecule does not exhibit a strong homophilic interaction and does not associate with other full-length $TRPV_1$ subunits. However, the C-terminus does exhibit a homophilic interaction and a segment comprising E684 and R721 constitutes an association domain of the protein which acts as a molecular determinant of the assembly into functional channels (Garcia-Sanz et al., 2004).

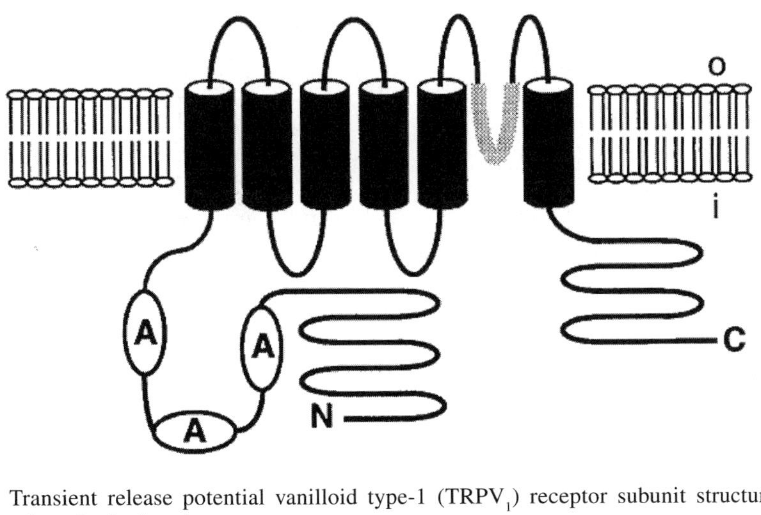

Fig. 1 Transient release potential vanilloid type-1 ($TRPV_1$) receptor subunit structure. "A" indicates the ankyrin domains

Splice Variants

Splice variants of a standard molecule are functionally important because, although substantially similar in structure to the standard molecule, they may exhibit different pharmacological responses. Since the identification of the $TRPV_1$ molecule, several splice variants of the standard molecule have been found. Their function, however, remains unknown. The $TRPV_1$ receptor 5′ splice variant (VR_1 5′ splice variant) is found in many areas of the nervous system. However, the VR_1 5′ splice variant is not responsive to any of the "classical" $TRPV_1$ activators, such as vanilloids, protons, or heat (Schumacher et al., 2000; Sanchez et al., 2001). The so-called *stretch-inhibitable $TRPV_1$ channel* and the *$TRPV_1$(VAR) channel* are splice variants which are both expressed in the kidney (Suziki et al., 1999; Tian et al., 2006). The 563 amino acid protein stretch-inhibitable channel shares the six transmembrane domains with the standard $TRPV_1$ molecule. Again, it is not sensitive to any of the classical activators. Moreover, as its name implies, its activation is blocked by mechanical stimuli (Suzuki et al., 1999). The 248 amino acid N-terminus protein, $TRPV_1$(VAR), when co-expressed with the standard $TRPV_1$ molecule in human embryonic kidney cells, potentiates $TRPV_1$ ion channel responses to resiniferatoxin. However, in another cell line (COS-7), this splice variant partially blocks resiniferatoxin-evoked responses (Tian et al., 2006). The reason for this difference in the effect of $TRPV_1$(VAR) on $TRPV_1$-responsiveness is not known. The most recently identified $TRPV_1$ splice variants are the $TPRV_{1b}$ and $TRPV_{1\beta}$ molecules (Wang et al., 2004; Lu et al., 2005). $TRPV_{1b}$ is expressed in human, while $TRPV_{1\beta}$ is the murine homologue. $TRPV_{1b}$ lacks the whole of exon 7, which encodes a 60 amino acid segment of the $TRPV_1$ molecule between the third ankyrin and the first transmembrane domains. $TRPV_{1\beta}$, on the other hand, appears to lack only a 10 amino acid sequence encoded by exon 7. Nevertheless, the human and murine homologues appear to possess almost identical properties. While neither splice variant responds to vanilloids or protons, $TRPV_{1b}$, when expressed in *Xenopus laevis* oocytes, can be activated by noxious heat. However, the activation threshold is at 47 °C, which is significantly higher than that of $TRPV_1$ (~42 °C) (Wang et al., 2004). Vos and colleagues (2006) examined the expression profile and relative abundance of the standard $TRPV_1$ ion channel and the $TRPV_{1b}$-constituted ion channel in 35 different human tissues using quantitative RT-PCR. $TRPV_{1b}$ was most abundant in foetal brain, adult cerebellum, and dorsal root ganglia. Recombinant $TRPV_{1b}$ forms multimeric complexes with $TRPV_1$, and is found in the plasma membrane of cells, which shows that the lack of channel function is not due to defects in complex formation or cell surface expression. When $TRPV_{1b}$ is co-expressed with $TRPV_1$, it inhibits $TRPV_1$ channel function in response to capsaicin, protons, noxious heat, and endogenous vanilloids. This inhibitory effect depends on the ratio of $TRPV_{1b}$ and $TRPV_1$ receptors. Charrua and colleagues (2005) have recently studied the changes of $TRPV_{1b}$ expression in inflammatory conditions. In such conditions, the level of $TRPV_1$ ion channel activation is increased, and so also is $TRPV_1$ expression. However, $TRPV_{1b}$ mRNA is down-regulated in cyclophosphamide-induced cystitis, suggesting that $TRPV_{1b}$ is a naturally existing inhibitory modulator of $TRPV_1$ in non-inflammatory conditions (Charrua et al., 2005).

The Tetrameric Ion Channel

The functional $TPRV_1$ channel is a multimer both in its native and recombinant forms, with a tetramer as the predominant form. In the tetramer, the $TRPV_1$ receptor monomers appear to assemble with fourfold symmetry around a central aqueous pore (Kedei et al., 2001; Kuzhikandathil et al., 2001). $TRPV_1$ receptor channel subunits do not combine arbitrarily. On the contrary, they appear to predominantly assemble through an interaction of protein moieties located between transmembrane segments 1–6. Both cytosolic termini and transmembrane segments synergistically contribute to the overall affinity between $TRPV_1$ channel subunits and control the selectivity of homo- and heteromeric assembly of the pore-forming subunits. Thus, inter-subunit interaction between $TRPV_1$ subunits also involves the transmembrane portion of the protein. This is consistent with the fact that the hexahelical channel subunits that are flanking the pore probably come into close contact with their transmembrane segments 5 and 6 and also their pore loops to stabilise the closed pore conformation of the inactive channel complex or to maintain the selectivity filter upon gating (Hellwig et al., 2005). As a heteromer, $TRPV_1$ receptor, in addition to splice variants, could be assembled with other TRP molecules. For example, $TRPV_1$ receptor can form heteromer with the transient receptor potential vanilloid type 3 ($TRPV_3$ receptor) that does not respond to capsaicin, but does respond to heat with a threshold of about 39 °C. $TRPV_3$ receptor is co-expressed in dorsal root ganglion neurones with $TRPV_1$ receptor. The association of $TRPV_1$ with $TRPV_3$ receptor may modulate the responses of the former, because of the different sensitivity of the respective molecules to various activators (Smith et al., 2002). Other members of the TRPV family (Clapham, 2003) may also form heteromers with $TRPV_1$ receptor. Indeed Rutter and colleagues (2005) have reported that the high threshold heat-sensitive TRPV channel, $TRPV_2$ receptor, may form heteromers in a small sub-population of primary sensory neurons. Cheng and colleagues (2007) have shown that $TRPV_1$, $TRPV_2$, $TRPV_3$, and $TRPV_4$ receptors form heteromers when co-transfected and that the responses of the heteromeric channels had properties "inherited" from the sub-units. However, others found that TRPV channels prefer to form homomers in transfection systems (Hellwig et al., 2005). Nevertheless, the probability that $TRPV_1$ receptor forms heteromers with sub-units which modify its responses may contribute to functional diversity (Nagy and Rang, 1999).

Auxiliary Molecules

Many ion channels associate with auxiliary proteins that regulate the trafficking, or biophysical and pharmacological properties, of the pore-forming subunits. Many of the TRP channels also associate with auxiliary proteins and form *transducisomes* or *signalplexes* (Korschen et al., 1999). The existence of such auxilliary proteins in association with native $TRPV_1$ receptor may explain in part, at least the differences in pharmacological responses to vanilloids or protons of native and recombinant $TRPV_1$ ion channels, respectively, when the latter characteristically function without such association. One such regulatory auxiliary protein is Fas-associated factor 1 (FAF1), which is co-expressed with $TRPV_1$ receptor, and forms an integral component of the $TRPV_1$ receptor complex. FAF1 is an adapter protein which is associated with the Fas receptor and is known to induce apoptosis or to augment Fas-induced apoptosis. FAF1 interacts physically with $TRPV_1$ receptor and controls its activity constitutively, reducing its sensitivity to capsaicin, acid, and heat. The mechanism, by which FAF1 controls the sensitivity of $TRPV_1$ receptor to capsaicin, acid, and heat, is unknown, although it has been suggested that it may do so by stabilizing $TRPV_1$ receptor from ligand activation (Kim et al., 2006). The vesicular proteins, snapin and synaptotagmin IX, associate and co-localise with $TRPV_1$ receptor, and strongly interact with the $TRPV_1$ receptor domain. However, they do not affect $TRPV_1$ ion channel function. Instead, these proteins augment $TRPV_1$ receptor expression in the plasma membrane in a protein kinase C (PKC)-dependent manner (Morenilla et al., 2004). Tubulin is a cytoplasmic protein and the principal constituent of microtubules. The C-terminus of $TRPV_1$ receptor interacts with and stabilises microtubules in vitro. This interaction is Ca^{2+}-sensitive and affects microtubule properties, such as microtubule sensitivity to low temperatures (Goswami et al., 2004). Activation of $TRPV_1$ receptor results in rapid disassembly of dynamic microtubules, but not of the actin or neurofilament cytoskeletons. The C-terminal fragment of $TRPV_1$ receptor exerts a stabilizing effect on microtubules when over-expressed in F11 cells (Goswami et al., 2006). The C-terminus of $TRPV_1$ receptor neither interacts with soluble actin nor with soluble neurofilaments, but specifically interacts with the components of the microtubule cytoskeleton, preferring to interact with β-tubulin rather than with α-tubulin (Goswami et al., 2007a). $TRPV_1$ receptor is physically and functionally present at dynamic neuronal extensions, and their growth cones. Activation of $TRPV_1$ receptor resulting in disassembly of microtubules occasions growth cone retraction and collapse, and formation of varicosities along axons (Goswami et al., 2007b). Human eferin is a protein of unknown function. The mouse eferin that is highly homologous to human eferin, interacts with $TRPV_1$ receptor. When co-transfected into HEK cells, $TPRV_1$ and eferin largely co-localise. $TRPV_1$ receptor and eferin are also co-localised in rat dorsal root ganglion cells. Eferin, however, exhibits no significant effect on $TRPV_1$ receptor channel activation in response to capsaicin (Lee, 2005). Receptor-tyrosine kinase A (TrkA) is the high affinity receptor for NGF which is released during inflammation or injury and causes hyperalgesia. Immunoprecipitation studies revealed that TrkA could be a part of the $TRPV_1$ receptor signalling complex (Chuang et al., 2001). In addition to TrkA, components of its signalling pathways, such as phosphoinositide-3-kinase (PI_3K) and phospholipase C (PLC)-gamma, could also be of the $TRPV_1$ receptor signalling complex (Chuang et al., 2001; Stein et al., 2006).

$TRPV_1$ Receptor Activators and Sites of Interaction

The $TRPV_1$ ion channel can be activated by any of at least five mechanisms, namely, direct ligand binding, protonation, post-translational changes, thermal energy, and electrical energy.

Exogenous Vanilloids: Activation and Binding Sites

Consideration of the role of external substances in activating $TRPV_1$ ion channels has long focused upon the dramatic effect of the application of capsaicin to those channels. In the search for physiological structures involved in mediating pain, the fact that a known pain-inducing substance, capsaicin, produces an effect on the ion channels, which are now denominated $TRPV_1$ receptor, led to the denomination of those channels by reference to this exogenous activator. Thus, these channels were initially denominated as "capsaicin receptors" and, subsequently (and equally illogically) as "vanilloid receptors", on the basis that capsaicin is a member of the family of vanilloids that activate these channels (Szallasi and Blumberg, 1990). The characterisation of these ion channels by reference to one set of exogenous activators is highly illogical and confusing, since it suggests that nature developed the $TRPV_1$ ion channel for the purpose of saving that part of humanity which indulges in chilli peppers from the consequences of over-indulgence. Obviously, the fact that $TRPV_1$ receptor is activated by capsaicin, and other plant-derived vanilloids, is a coincidence, and $TRPV_1$ receptor was not intended to be redundant in the preponderance of human kind which is not consumer of chilli peppers. On the contrary, $TRPV_1$ receptor was in fact developed to serve as a receptor for an endogenous ligand, or ligands, and also to mediate the pain of inflammation so as to elicit a self-protective response to injury on the part of the organism. The vanilloids have acquired a special place as research tools in the investigation of the characteristics of $TRPV_1$ receptor and are discussed first for that reason. The archetypical vanilloid, capsaicin, and its ultrapotent counterpart, resiniferatoxin, have been serving as the main research tools for studying $TRPV_1$ receptor, and, obviously, in the search for that channel's vanilloid-binding site (Fig. 2; Caterina et al., 1997). The intracellular/intramembranous residues, Y511 and T550, which are in, and adjacent to the

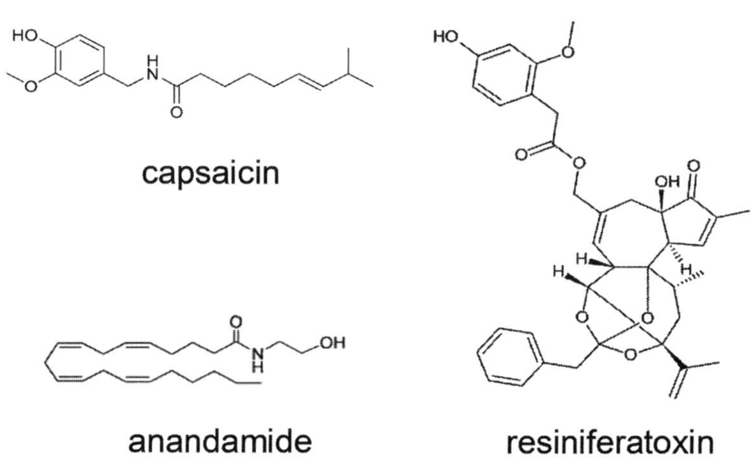

Fig. 2 Chemical structure of commonly used potent activators of the $TRPV_1$ receptor

third and fourth transmembrane domains, are critical for capsaicin binding (Jordt and Julius, 2002; Gavva et al., 2004). However, binding of resiniferatoxin depends on residues T550 and M547 (Gavva et al., 2004). The capsaicin-binding site is also the locus for binding competitive antagonists, such as capsazepine (Bevan et al., 1992; Jordt and Julius, 2002). Extracellular sites may also be involved in vanilloid binding to $TRPV_1$ receptor. Thus, Rami and co-workers (2004) reported that a $TRPV_1$ receptor antagonist containing a quarternary amine was effective only when applied to the external surface of membrane in patch clamp studies. Moreover, Vyklicky and colleagues (2003) found that intracellular application of vanilloids is insufficient for activating $TRPV_1$ receptor channels when they are expressed heterologously in HEK cells. Data from single channel recordings indeed suggest that $TRPV_1$ receptor may bind capsaicin at multiple sites (Hui et al., 2003), in contrast to the single vanilloid-binding site suggested by other studies (Jordt and Julius, 2002). In agreement with the putative existence of multiple vanilloid binding sites, mutations on both the C- and N-termini seem to modify capsaicin sensitivity and binding (Jung et al., 2002; Vlachova et al., 2003). The two amino acids, which may be essential for the hydrophilic interactions of $TRPV_1$ receptor with vanilloids, are R114 and E761 on the C- and N-termini, respectively (Jung et al., 2002). These sites are also in the intracellular side of the molecule. Mutation of residues in the sixth transmembrane domain also disrupts the ability of capsaicin and resiniferatoxin to activate the channel (Kuzhikandathil et al., 2001). Single channel recordings suggest that partial as well as full binding by capsaicin may open the channel. Capsaicin association occurs preferentially to the closed channel. However, when the channel is activated, multiple open states are accessible irrespective of the level of binding (Hui et al., 2003). Capsaicin binding induces conformational changes in the $TRPV_1$ receptor ion channel, which differ from the conformational changes that are induced by proton binding or by thermal activation. The fact that this difference exists is consistent with there being independent activating pathways of $TRPV_1$ receptor for capsaicin, protons, and heat. The structural re-arrangements induced by capsaicin binding include changes in the putative pore domain and reveal the location of an intracellular domain that contributes to the positive co-operativity observed on activation by capsaicin (Welch et al., 2000).

Endogenous Activators

A number of endogenous agents have been proposed as direct activators of the $TRPV_1$ receptor ion channel. These endogenous agonists include anandamide (Fig. 2; Zygmunt et al., 1999), N-arachidonoyl-dopamine (NADA) (Huang et al., 2002), N-oleoyldopamine (OLDA) (Chu et al., 2003), lipoxygenase products, such as 12-or 15-(S)-HPETE (Hwang et al., 2000), and unsaturated C18 N-acylethanolamines (Movahed et al., 2005). All of these agonists have been shown to compete for the capsaicin-binding site, leading to their characterisation as "endovanilloids". Of major interest is the finding that anandamide and NADA, in addition to being agonists at the $TRPV_1$ ion channel, also activate the cannabinoid 1 (CB_1) receptor (Zygmunt et al., 1999; O'Sullivan et al., 2004). Among the endogenous $TRPV_1$ receptor agonists, anandamide is the best characterised. Anandamide is a member of the group of bioactive lipids known as "long chain C18 N-acylethanolamines (NAEs)". Anandamide is generated following the hydrolysis of membrane N-acylphosphatidylethanolamine (NAPE) in a reaction catalysed by phospholipase D-like enzymes (see Chap. 2). Several endogenous NAEs, many of which are more abundant than anandamide in rat tissues, are capable of activating $TRPV_1$ receptor and may therefore play a role as endogenous $TRPV_1$ receptor modulators (Movahed et al., 2005). Anandamide has been found not to induce desensitisation of $TRPV_1$ receptor in certain conditions (Dinis et al., 2004). This is in contrast to the desensitisation effect by all the known exogenous ligands of $TRPV_1$ receptor, and the majority of the known endogenous agonists. The concept of desensitisation is afforded consideration later in this chapter. The capacity of anandamide to activate $TRPV_1$ receptor in normal physiological conditions is very limited. This limitation could be necessary to prevent unnecessary activity of $TRPV_1$ receptor, thereby signalling pain, in the absence of a relevant pain-inducing stimulus. However, when $TRPV_1$ receptor is activated by other stimuli, such as inflammatory mediators, anandamide becomes a powerful activator of $TRPV_1$ receptor (Olah et al., 2001b; Singh Tahim et al., 2005). Anandamide and other endogenous activators of $TRPV_1$ receptor have therefore been described as "conditional activators" of this ion channel.

Activation by Protons

Protons are able to activate $TRPV_1$ receptor at pH below 6.5. Capsaicin binding, the temperature threshold, and channel gating are all affected by pH. Lowering pH enhances the apparent binding affinity of capsaicin, and lowers the heat threshold for activation of the channel. It also promotes the occurrence of long openings and short closures, and stabilises at least one of the open conformations of the channel. Moreover, capsaicin binding and protonation of the channel interact allosterically, where the effect of one can be offset by the effect of the other (Ryu et al., 2003). Jordt and colleagues (2000) thought that protons modulate $TRPV_1$ receptor activity by interacting with specific amino acid residues on the extracellular surface of the channel protein. The response of $TRPV_1$ receptor to protons involves at least two different mechanisms, namely, activation of the channel and potentiation of the currents generated by an already activated channel (Ryu et al., 2003). These mechanisms appear to be distinct and separate although originating at the site where protonation is initiated. Mutations at the extracellular E648 residue selectively abrogate proton-evoked channel activation without diminishing the channel's responses to other stimuli (Jordt et al., 2000). However, while mutation at this site blocks proton-evoked activation, it does not affect proton-evoked potentiation. Another extracellular residue, E600, in the region linking the fifth transmembrane domain with the putative pore-forming region of the channel, constitutes the key regulatory site involved in the dynamic potentiation of the $TRPV_1$ receptor response to capsaicin or heat. It has been suggested that this site could set the sensitivity to other noxious stimuli in response to changes in extracellular proton concentration. In addition to protons, excess positive charges carried by various ions seem also to be able to activate $TRPV_1$ receptor. Ahern and colleagues (2005) showed that extracellular Na^+, Mg^{2+}, and Ca^{2+} can open the $TRPV_1$ ion channel. Moreover, these extracellular cations can sensitise the ion channel to other activators. Interestingly, the activating and sensitizing effect of these cations seem to occur via electrostatic interactions with the residues identified as sites for proton interaction (E600 and E648). At low concentration, the rare earth metal, gadolinium, also activates $TRPV_1$ receptor (Tousova et al., 2005). While the sites for this latter interaction are the same as those for protonation (E600 and E648), the site of action of other cations has not been identified.

The Temperature Sensor

Sensitivity to noxious heat is one of the most distinctive features of $TRPV_1$ receptor. Heat may contribute to $TRPV_1$ receptor activation in two distinct ways. First, heat reduces the threshold for the activation of $TRPV_1$ receptor by other ligands. Second, heat above ~43 °C independently activates $TRPV_1$ receptor (Tominaga et al., 1998). At less than 43 °C, openings of $TRPV_1$ receptor channels are few and brief. However, raising the ambient temperature rapidly increases the frequency of openings. Despite the large temperature coefficient of the apparent activity, the unitary current, the open dwell-times, and the intra-burst closures show only weak temperature dependence. Instead, heat exercises a localised effect on the reduction of long closures between bursts and the elongation of burst durations. Both membrane lipids and the ionic strength of the bath solution affect the threshold of the activation, but neither diminishes the response (Liu et al., 2003). There is also some evidence from human embryonic kidney cell studies that the presence of either reducing, or oxidising agents, results in an increased response to heat by $TRPV_1$ receptor channels (Susankova et al., 2006). A series of other thermo-sensitive ion channels have been identified more recently (Clapham, 2003). As expected, the main structure of these ion channels is very similar and they seem to exhibit a "modular" C-terminus (Brauchi et al., 2006). Brauchi and colleagues (2006) prepared chimera from $TRPV_1$ receptor and the cold-sensing TRP channel, $TRPM_8$, by altering their C-termini. They found that the $TRPV_1$ receptor which had the C terminus of $TRPM_8$ became cold sensitive, while the $TRPM_8$ with the C terminus of $TRPV_1$ receptor became sensitive to heat. These findings suggest that the thermo sensor resides in the C-terminus. The amino acids which govern the sensitivity of the channel to heat and which constitute the coupling machinery that converts thermal energy into mechanical work (pore opening) remain unknown. It has been suggested that the ultimate TRPV1 activator is heat, since protons, capsaicin, and indirect activators, such as bradykinin, all reduce the temperature threshold of the channel (Jordt et al., 2000; Liang et al., 2001; Babes et al., 2002; Sugiura et al., 2002).

The Voltage Sensor

Voets and co-workers (2004) reported that $TRPV_1$ receptor, in common with other TRP channels, can be activated by depolarizing the cell membrane. The voltage–activity curve of $TRPV_1$ receptor is much wider than that of other voltage-activated channels, such as voltage-gated Na^+, K^+, or Ca^{2+} channels. Thus, while voltage-activated Na^+, K^+, or Ca^{2+} channels reach their maximal opening probability from zero within a narrow voltage range (~50 mV), full $TRPV_1$ receptor activation occurs within a much wider voltage range (~150 mV). This difference has been attributed to the paucity of positive charges found in the fourth transmembrane domain (S4; Voets et al., 2004). However, the voltage sensor of $TRPV_1$ receptor has yet to be identified. Voets and colleagues' (2004) recent findings suggest that all activators may act through changing the voltage-evoked activation properties of $TRPV_1$ receptor. $TRPV_1$ receptor is not governed by a single characteristic thermal threshold; instead the temperature sensitivity is modulated by the transmembrane voltage, and changes in ambient temperature result in graded shifts of the voltage dependence of channel activation. Furthermore, the voltage-evoked activation properties of $TRPV_1$ receptor are dependent upon the presence, or absence, of other activators. At lower temperatures, the activation curve of the channel shifts towards more positive membrane potentials, while at higher temperatures, or in the presence of capsaicin, it shifts towards negative membrane potentials. Eventually, the shift exceeds the resting membrane potential. Brauchi and colleagues (2004) proposed another model in which the voltage-dependent and temperature-dependent activation of the channel occurs through separate structures, which are able to sense specific stimuli and act through allosteric mechanisms. They hypothesise that, by doing so, the structures transfer the electrical or thermal energy to the gate.

Activation by Indirect Activators

Several agents activate $TRPV_1$ receptor indirectly through a process called "sensitisation". Sensitisation increases the responsiveness of the $TRPV_1$ receptor ion channel through post-translational modification of the ion channel. The overwhelming majority of the $TRPV_1$ receptor-sensitizing agents are so-called *inflammatory mediators*, which are produced and released during tissue inflammation. Among the best known of these agents which sensitise $TRPV_1$ receptor, are bradykinin, prostaglandins, and nerve growth factor (NGF). Prostaglandin E2 (PGE_2) and prostaglandin I2 (PGI_2) are the products of arachidonic acid metabolism through the cyclooxygenase pathway. Prostaglandins act upon a family of pharmacologically distinct prostanoid receptors, including EP_1, EP_2, EP_3, EP_4, and IP that activate several different G protein-coupled signalling pathways (Narumiya et al., 1999). PGE_2 and PGI_2 each increase $TRPV_1$ receptor responses through EP_1 or IP receptors, respectively, predominantly in a protein kinase C (PKC)-dependent manner in both human embryonic kidney cells expressing $TRPV_1$ receptor and in mouse dorsal root ganglion neurones. The temperature threshold for $TRPV_1$ receptor activation is reduced below 35 °C in the presence of either PGE_2 or PGI_2 so that $TRPV_1$ receptor may be activated at normal body temperature possibly leading to spontaneous pain sensation (Moriyama et al., 2005). The enhanced thermal sensitivity found in wild type mice as a result of PGE_2 injection is reduced in mice lacking the neuronal-specific isoform of the type I regulatory subunit of protein kinase A (PKA) (Malmberg et al., 1997). This suggests that PKA may also be involved in PGE_2-induced $TRPV_1$ receptor sensitisation. Bradykinin is a nonapeptide, which acts on two main sub-types of bradykinin receptors: B_1 and B_2. Most of the bradykinin-evoked effects are mediated through the B_2 receptors (Dray and Perkins, 1993). Bradykinin activates $TRPV_1$ receptor through at least two mechanisms. First, bradykinin induces the production of 12-lipoxygenase metabolites of arachidonic acid, which, in turn, act as agonists at $TRPV_1$ receptor (Shin et al., 2002). Second, through post-translational changes, bradykinin lowers the temperature threshold of the channel for heat activation (Liang et al., 2001; Sugiura et al., 2002). This sensitizing effect of bradykinin was characterised by Cesare and colleagues (1999) who found that bradykinin, by activating protein kinase C (PKC), induces phosphorylation of $TRPV_1$ receptor; and that the ε isoform of PKC is responsible for bradykinin-induced $TRPV_1$ receptor sensitisation. PKC-mediated sensitisation is partly due to the recruitment of a pool of vesicular receptors to the plasma membrane (Morenilla-Palao et al., 2004). In addition, bradykinin may initiate the first step in the process of $TRPV_1$ receptor activation; it induces the removal of the auxiliary molecule PIP_2 from $TRPV_1$ receptor, which has been shown to inhibit $TRPV_1$ receptor (Chuang et al., 2001; but see below). There are conflicting views as to the molecular mechanisms by which NGF sensitises $TRPV_1$ receptor. PKA has been reported to be a member of the signalling pathway activated by the binding of NGF to the NGF receptor (TrkA), resulting in $TRPV_1$ receptor sensitisation (Shu and Mendell, 2001), also seen in PI_3K (Bonnington and McNaughton, 2003). A competing hypothesis is that activation of PLCγ by TrkA leads to hydrolysis of PIP_2 and release of $TRPV_1$ receptor from inhibition by endogenous PIP_2 (Chuang et al., 2001). Zhang and colleagues (2005) have proposed that NGF, acting on the TrkA receptor, activates a signalling pathway in which PI_3K plays a crucial early role, with Src kinase as the downstream element which binds to, and phosphorylates, $TRPV_1$ receptor. Phosphorylation of $TRPV_1$ receptor at a single tyrosine residue, Y200, followed by insertion of $TRPV_1$ receptor channels into the surface membrane, is claimed to account for most of the rapid sensitising action of NGF (Zhang et al., 2005). More recently, Stein and co-workers (2006) have proposed a model for NGF-mediated sensitisation in which physical coupling of $TRPV_1$ receptor and PI_3K in a signal transduction complex facilitates trafficking of $TRPV_1$ receptor to the plasma membrane. In contrast to the findings by Chuang and colleagues (2001) that PIP_2 inhibits the activity of $TRPV_1$ receptor, Stein and co-workers propose a model in which PIP_2 binding to $TRPV_1$ receptor sensitises the activity of the channel, while PI_3K activity is required for NGF-mediated sensitisation, with that sensitisation consisting of an increase in the number of channels in the plasma membrane (Stein et al., 2006). Phosphorylation by PKA occurs at S116, T144, T370, S502, S774, and S800 (Bhave et al., 2002; Mohapatra and Nau, 2005; Jeske et al., 2006). PKC phosphorylates $TRPV_1$ receptor at S502, T704, D744, S800, and S820 (Numazaki et al., 2002; Bhave et al., 2003). Control of $TRPV_1$ receptor trafficking to the plasma membrane depends on phosphorylation of Y199 in rat (Y200 in humans) by the tyrosine kinase Src (Zhang et al., 2005). PIP_2-binding domains of ion channels are loosely characterised by clusters of basic residues interspersed with hydrophobic amino acids, an arrangement that may facilitate interactions with the negatively charged head groups of the phospholipid. The segment of the C-terminal cytoplasmic domain of $TRPV_1$ receptor comprising amino acids 777 to 820 has been identified as the PIP2-binding site (Prescott and Julius, 2003).

Transcriptional Regulation

$TRPV_1$ receptor expression is not static. Various pathological events, including inflammation or injury of peripheral nerves, result in changes in $TRPV_1$ receptor expression and are accompanied by changes in transcription. For example, axotomy results in the down-regulation of $TRPV_1$ mRNA expression in dorsal root ganglion cells (Michael and Priestley, 1999). Protein expression is also differentially affected in injured dorsal root ganglion neurones after sciatic nerve injury with the altered level of expression of $TRPV_1$ receptor being dependent on the nature of the injury. At the same time, $TRPV_1$ expression is up-regulated in uninjured neurones after partial nerve injury (Hudson et al., 2001). It has been suggested that Ras, that is, a small GTPase involved in intracellular signalling, plays a crucial role in the regulation of $TRPV_1$ receptor expression in DRG neurones. It has been hypothesised that a certain level of Ras activation is required to keep the transcriptional machinery active to produce $TRPV_1$ receptor. Removal of essential Ras-activating stimuli, like neurotrophic factors, leads to a shutdown of this transcription programme. Overstimulation of Ras resulting from, for example, increased levels of NGF at sites of inflammation, may lead to over-production of $TRPV_1$ receptor. The effects of Ras are, in part mediated via ERK and possibly also via PI_3K, but activation of ERK, either alone or together with PI_3K, is not sufficient (Bron et al., 2003).

$TRPV_1$ Receptor Expression and Distribution

Subcellular Expression

$TRPV_1$ receptor protein expression is found in at least three cellular compartments, namely, in the plasma membrane, in the membrane of cytoplasmic vesicles, and in the membrane of the endoplasmic reticulum. In fact, most $TRPV_1$ receptor ion channels appear to be located at internal membranes (Olah et al., 2002). The expression of $TRPV_1$ receptor in membranes surrounded by an aqueous environment is consistent with the predicted structure of the channel as comprising hydrophilic termini, with a hydrophobic area between those termini. $TRPV_1$ receptor ion channels in the plasma membrane of neurones allow inward currents on activation resulting in depolarisation, increased probability of action potential generation and transmitter release. $TRPV_1$ receptors located in cytoplasmic vesicles are thought to serve as a reserve, which can be quickly translocated to the plasma membrane following, for example, PKC activation. The activity of $TRPV_1$ receptor in the endoplasmic reticulum results in the release of calcium from endoplasmic stores and also facilitates Ca^{2+} entry from outside the cell (Eun et al., 2001). The complex sub-cellular localisation of $TRPV_1$ receptor, coupled with barriers to agonist access from outside the cell, means that different $TRPV_1$ receptor agonists may exhibit a substantially different time-course of action, as some agonists penetrate the cell more slowly than others (Lazar et al., 2006). Thus, natural and synthetic capsaicin analogues – known as capsaicinoids – may not produce the same calcium response as capsaicin. For example, highly lipophylic compounds may cause only a slight Ca^{2+} influx, via $TRPV_1$ receptor channels localised in the plasma membrane, and may not be able to activate $TRPV_1$ channels found in the endoplasmic reticulum (Morita et al., 2006).

Expression by Primary Sensory Neurons

$TRPV_1$ receptors find their most prominent expression in a sub-population of nociceptive primary sensory neurons (for details on $TRPV_1$ receptor expression in the nervous tissue see Chap. 10). About 40% of the total neuronal population of primary sensory neurons express $TRPV_1$ receptor. These C-fibre nociceptive neurones can be divided into two groups based on growth factor dependency and isolectin B4 (IB4) binding. The first group comprises isolectin B4 (IB4)-binding non-peptidergic neurones which are dependent on glial cell-derived neurotrophic factor (GDNF) for survival during post-natal development. The second group consists of IB4-non-binding peptidergic neurones which are dependent on NGF for survival during the same period (Nagy, 2004). These two populations of nociceptive neurones also innervate different peripheral tissues and terminate in distinct regions of the superficial spinal cord (Guo et al., 1999; Avelino et al., 2002). Moreover, IB4-positive neurones have smaller noxious heat-activated currents than IB4-negative neurons (Stucky and Lewin, 1999). $TRPV_1$ receptor function and expression are selectively increased in IB4-positive neurons during inflammation (Breese et al., 2005). The reason for the differences between these neuronal groups is unclear; but the composition of the $TRPV_1$ receptor signalling complex, comprising the channel itself together with its auxiliary proteins, may differ between the two groups of neurones. $TRPV_1$ receptor is expressed in the perikarya as well as in both the central and peripheral termini of primary sensory neurons. In peripheral tissues, $TRPV_1$ receptor-expressing sensory fibres can be found in the dermis, along the epidermal/dermal junction, epidermis, and also in Meissner's corpuscles (Guo et al., 1999; Pare et al., 2001). In the viscera, $TRPV_1$ receptor immunopositive fibres are found in the mucous membrane, submucous, and muscular layer (Avelino et al., 2002; Ward et al., 2003). $TRPV_1$ receptor-expressing fibres also innervate the Langerhans islands in the pancreas (Gram et al., 2007; Razavi et al., 2006; see Chap. 14). They are also found in the synovial membrane of certain joints (Sato et al., 2005). The central terminals of TRPV1-expressing primary sensory neurons terminate primarily in the superficial dorsal horn of the spinal cord (Guo et al., 1999). However, some TRPV1-expressing fibres can also be found in the deep dorsal horn and around the central canal (Tominaga et al., 1998).

Expression in the Central Nervous System

Many neurons in the central nervous system also express $TRPV_1$ receptor. The olfactory nuclei, cerebral and cerebellar cortex, thalamus, hypothalamus, lateral and dorsal septal nuclei, periaqueductal grey, locus coeruleus, substantia nigra, inferior olive, dentate gyrus, and hippocampus express $TRPV_1$ receptor (Mezey et al., 2000; Roberts et al., 2004; Cristino et al., 2006). $TRPV_1$ receptor ion channels expressed in neurons of the thermoregulatory nucleus apparently respond to capsaicin as evidenced by the failure of thermoregulation in animals injected systemically with capsaicin, or directly into the medial preoptic hypothalamic nucleus, at neonatal age (Jancso-Gabor et al., 1970). $TRPV_1$ receptor is also expressed on GABA-ergic terminals in the medial preoptic nucleus which is evidenced by the excitatory action of capsaicin in those structures (Karlsson et al., 2005).

Expression by Non-Neuronal Cells

Some non-neuronal cells have also been shown to express $TRPV_1$ receptor, but the function of these ion channels generally remains unknown. $TRPV_1$ receptor-expressing cells are found in the inner ear where they include inner and outer hair cells, inner and outer pillar cells, Hensen's cells, spiral ganglion neurons, Scarpa's ganglionic neurons, and satellite cells (Balaban et al., 2003; Zheng et al., 2003). Both capsaicin and resiniferatoxin increase the threshold for auditory nerve compound action potential generation and reduce the magnitude of cochlear microphonic and electrically evoked oto-acoustic emissions suggesting that capsaicin receptors are functional in the inner ear (Zheng et al., 2003). A sub-population of keratinocytes also express $TRPV_1$ receptor ion channels which appear to be functional (Ioue et al., 2002; Southall et al., 2003). A sub-population, at least, of cultured rat gastric epithelial cells express $TRPV_1$ receptor, but these ion channels are peculiar in that they are not desensitised or damaged by exposure to capsaicin (Kato et al., 2003). This peculiarity is shared by $TRPV_1$ receptor ion channels found in the basal and superficial layers of the urothelium. These cells are functional as they respond to capsaicin and resiniferatoxin application with a $TRPV_1$ receptor-mediated increase in intracellular calcium concentration (Birder et al., 2001). $TRPV_1$ receptor ion channels are expressed, and functionally active, in human prostate cancer cells (Sanchez et al., 2005). Cardiomyocytes express $TRPV_1$ receptor ion channels during their development, but whether these are functional is unknown (Dvorakova and Kummer, 2001). $TRPV_1$ receptor ion channels are also found in cervical cancer cells (Contassot et al., 2004).

Co-Expression of $TRPV_1$ Receptor with the Cannabinoid Receptors

The endogenous $TRPV_1$ receptor ligands, anandamide and NADA, are remarkable in that they are also endogenous ligands of the inhibitory CB_1 receptor. However, sharing ligands is not the only connection between the vanilloid and cannabinoid systems, because $TRPV_1$ receptor and the CB_1 receptor are co-expressed in groups of neurons in both the peripheral and central nervous systems. At the periphery, virtually all $TRPV_1$ receptor-expressing neurons express the CB_1 receptor (Ahluwalia et al., 2000). In agreement with the co-expression, anandamide is capable of mediating dual effects on capsaicin-sensitive primary sensory neurones, namely: that anandamide exerts an inhibitory effect on these neurones by its action at CB_1 receptors, while it exerts an excitatory effect by its action at $TRPV_1$ receptor (Ahluwalia et al., 2003a). The extensive co-expression of $TRPV_1$ and the CB_1 receptor in primary sensory neurones is, however, disputed (Bridges et al., 2003; Price et al., 2004). Co-expression of CB_1 and $TRPV_1$ receptors in the brain is now known to be extensive. Neurons which co-express the CB_1 and $TRPV_1$ receptor have been found in the hippocampus, basal ganglia, thalamus, hypothalamus, cerebral peduncle, pontine nuclei, periaqueductal grey matter, cerebellar cortex, dentate cerebellar nucleus, the globus pallidus, and substantia nigra (Cristino et al., 2006). Collectively, these findings suggest that anandamide may regulate the activity of groups of neurons in both the peripheral and central nervous systems. In fact, in rats, elevation of endocannabinoid levels in the ventrolateral periaqueductal grey affects descending nociceptive pathways via both CB_1 receptors and $TRPV_1$ ion channels (Maione et al., 2006).

Cellular Responses to $TRPV_1$ Activation

Ionic Influx in Primary Sensory Neurones

As mentioned above, $TRPV_1$ receptor is a non-selective cationic channel and is, therefore, permeable to Na^+, Ca^{2+}, and K^+. Thus, in physiological conditions, $TRPV_1$ receptor ion channels, when activated, induce inward Na^+ and Ca^{2+} currents and an outward K^+ current in the primary sensory neurones in which they are expressed. These currents depolarise the $TRPV_1$ receptor-expressing neurones, resulting in the activation of voltage-gated ion channels, which leads to an increase in the probability of action potential generation and Ca^{2+} influx- and Ca^{2+}-dependent transmitter release. Ca^{2+} is an important intracellular messenger. There are at least four mechanisms which contribute to increasing free intracellular Ca^{2+} when $TRPV_1$ receptor ion channels are activated, namely: direct increase from the opening of $TRPV_1$ receptor channels at the plasma membrane and endoplasmic reticulum, store-operated Ca^{2+} entry, and calcium-induced calcium release (Eun et al., 2001; Marshall et al., 2003; Karai et al., 2004). The $TRPV_1$ receptor-mediated Ca^{2+} release from intracellular stores occurs from the ryanodine-sensitive store exclusively, and not from IP_3-sensitive stores (Eun et al., 2001). The amount of calcium within the $TRPV_1$ receptor-gated compartment of the endoplasmic reticulum is finite, can be depleted and replenished, displays store-operated features, and overlaps with a thapsigargin-sensitive pool of Ca^{2+}. The depletion by either agent appears to leave some residual calcium (7–15%) in the endoplasmic reticulum, which can be released by further treatment with either agent. One of the most interesting consequences of $TRPV_1$ receptor activation-evoked Ca^{2+} influx in primary sensory neurons is that it results in the production of the endocannabinoid/endovanilloid, anandamide (Ahluwalia et al., 2003b; van der Stelt et al., 2005). Anandamide binds to $TRPV_1$ receptor at the capsaicin-binding site on the intracellular side of the channel (Jordt and Julius, 2002) to, again, activate the channel. Thus, $TRPV_1$ receptor activation-evoked anandamide production appears to constitute a mechanism that potentiates the activation of $TRPV_1$ receptor. Another significant, and probably one of the most-studied consequences of $TRPV_1$ receptor activation-evoked Ca^{2+} entry is the Ca^{2+}-evoked cytotoxic effect. This effect has been extensively used both as an experimental (Jancso et al., 1977) and therapeutic tool (Cruz, 2004). At least three vital organelles are immediately damaged following excessive $TRPV_1$ receptor activation (Olah et al., 2001a). These are the nucleus, mitochondria, and endoplasmic reticulum, with the latter reacting with abrupt fragmentation. Since these are vital cell organelles, disruption of their function results in the elimination of $TRPV_1$ receptor-expressing cells (Olah et al., 2001a). Activation of $TRPV_1$ receptor by capsaicin results in a significant increase in the cytoplasmic Ca^{2+} concentration, but the application of capsaicin also produces a profound and sustained suppression of high voltage-activated Ca^{2+} channels in primary sensory neurons. This effect is abolished by iodoresiniferatoxin, a highly specific $TRPV_1$ receptor antagonist, demonstrating that capsaicin inhibits high voltage-activated channels through the action of $TRPV_1$ receptor. This inhibitory effect is mediated by both Ca^{2+} influx and release from the intracellular stores, and can be inhibited by blocking calcineurin, a Ca^{2+}-dependent phosphatase (Wu et al., 2005).

Desensitisation of $TRPV_1$ Channels

At high doses, or with prolonged exposure, capsaicin induces $TRPV_1$ receptor desensitisation before inducing cytotoxicity (Xu et al., 2005). Indeed, $TRPV_1$ receptor desensitisation may be regarded as a protective mechanism which guards against potential excitotoxicity. The desensitisation of $TRPV_1$ receptor ion channels is dependent upon several factors, namely: the presence of extracellular calcium, the concentration of the stimulating ligand, and the duration of stimulation of the receptor by that ligand. Where persistent stimulation of the receptor has not resulted in degeneration of the cell beyond recovery, there is an obligatory period of delay after termination of the stimulus before recovery from desensitisation occurs. Liu and colleagues (2005) found that prolonged application of capsaicin leads to nearly complete desensitisation of the channel and its functional recovery from desensitisation requires a high concentration of intracellular ATP. Neither inhibition nor activation of protein kinases prevents recovery of the channel from desensitisation. However, blockade of lipid kinases, in particular, phosphatidlinositol-4-kinase, abolishes recovery, as does activation of membrane receptors that stimulate hydrolysis of PIP_2. Depletion of PIP_2 occurs concomitantly with activation of $TRPV_1$ receptor and its replenishment in the membrane determines the recovery of the channel from desensitisation (Liu et al., 2005). There is evidence that activation of both PKC-ε and PKA decreases desensitisation of $TRPV_1$ receptor by directly phosphorylating the channel (Mandadi et al., 2006; Mohapatra and Nau, 2005). Inhibiting calcineurin also significantly decreases $TRPV_1$ receptor desensitisation evoked by the application of capsaicin or protons.

The Role of $TRPV_1$ Receptor in Physiological and Pathological Conditions

Pain

The $TRPV_1$ receptor ion channel is well established as the principal mediator of the pain sensation which results from inflammation. This was demonstrated in behavioural experiments by Davis and colleagues (2000) and Caterina and colleagues (2000) in which mice lacking the $TRPV_1$ receptor gene were exposed to behavioural testing of their pain experience when exposed to various types of painful stimuli. These “knockout mice” appear normal in a wide range of behavioural tests, including responses to acute noxious thermal stimuli, but their ability to develop thermal hyperalgesia after inflammation is absent. Thus, these authors concluded that the $TRPV_1$ receptor ion channel is required for inflammatory sensitisation to noxious thermal stimuli but normal sensation of noxious heat does not depend on these ion channels (Davis et al., 2000; Caterina et al., 2000). The role of $TRPV_1$ receptor in relation to neuropathic pain remains uncertain at the moment.

$TRPV_1$ Receptor and Inflammatory Pain

Tissue injury is normally associated with inflammation and inflammatory pain. Inflammatory pain is induced by inflammatory mediators released in the injured tissue, such as PGE_2, NGF, and bradykinin, acting on nociceptors in peripheral nerve terminals. Another prominent mediator generated in injured tissue is protons which result in tissue acidosis. Inflammatory pain is characterised by hyperalgesia (an increased response to noxious stimulation) and allodynia (noxious response to previously innocuous stimulation). The development of hyperalgesia following inflammation involves an increased level of $TRPV_1$ receptor expression as well as the sensitisation of existing $TRPV_1$ receptor channels. Levels of both NGF and GDNF increase following inflammation and contribute to inflammatory hyperalgesia via an increase in $TRPV_1$ receptor expression. The increase in the level of NGF and GDNF follows different time courses and they act on distinct populations of DRG neurones (Amaya et al., 2004). Inflammatory mediators, such as ATP, bradykinin, and NGF, increase the temperature and proton sensitivity of $TRPV_1$ receptor and contribute to enhanced $TRPV_1$ receptor ion channel activity. Chuang and colleagues (2001) injected wild-type and $TRPV_1$ receptor-deficient mice with bradykinin or NGF and measured paw withdrawal latencies from a radiant heat source before, and after, treatment. Each agent produced substantial sensitisation in wild-type animals but not in the knockout mice, demonstrating that $TRPV_1$ receptor is essential for the development of bradykinin-induced or NGF-induced thermal hypersensitivity in vivo. Increased activation of $TRPV_1$ receptor channels expressed with the bradykinin B_2 receptors in HEK cells also results from bradykinin, while NGF produces increased responses in proton-evoked currents in oocytes expressing both $TRPV_1$ receptor and the NGF receptor, TrkA. Both bradykinin and NGF therefore mediate their effect by activation of $TRPV_1$ and their own receptors (Chuang et al., 2001). Increased expression of $TRPV_1$ in dorsal root ganglion neurons is also found in conditions of inflammation, which may contribute to sustained hyperalgesia (Amaya et al., 2003). Moreover, the contribution of $TRPV_1$ to other important elements of inflammatory pain, such as mechanical hyperalgesia and allodynia, remains to be addressed.

$TRPV_1$ Receptor and Neuropathic Pain

Although the role of $TRPV_1$ receptor in mediating inflammatory pain conditions is well established, the extent of the involvement of $TRPV_1$ receptor in neuropathic pain conditions remains unknown. There is, however, evidence that in relation to neuropathic pain, the contribution made by $TRPV_1$ receptor depends on the context of origin of the neuropathic pain condition and, perhaps, even on the species of animal. Caterina and colleagues (2000) found that in a model of partial spinal nerve ligation, there is no difference between the level of mechanical and thermal nociceptive responses of $TRPV_1$ receptor null mice as opposed to wild-type mice. On the other hand, after nerve injury, a distinct difference in the regulation of $TRPV_1$ receptor is observed (Hudson et al., 2001; Fukuoka et al., 2002; Kanai et al., 2005). After sciatic nerve ligation, intrathecal application of capsazepine, at $TRPV_1$ receptor antagonist, blocks A-delta fibre-evoked responses in the dorsal horn neurones of rats (Kelly and Chapman, 2002). There is also evidence that the contribution made by $TRPV_1$ receptor ion channels to neuropathic pain conditions may even be species dependent. Capsazepine reverses mechanical hyperalgesia in guinea pig model of partial sciatic nerve ligation, but has no effect in rat or mouse models of neuropathic pain (Walker et al., 2003). Recently, Christoph and colleagues reported that in an in vivo rat model of spinal nerve ligation, both intravenous application of the $TRPV_1$ receptor antagonist thioxo-BCTC and intrathecal administration of the antisense oligonucleotide against $TRPV_1$ receptor reduce mechanical hypersensitivity in a similar manner, evidencing the involvement of $TRPV_1$ receptor in such neuropathic pain conditions in rat (Christoph et al., 2006, 2007). Again, intrathecal injection of a siRNA against $TRPV_1$ receptor reduces cold allodynia of mononeuropathic rats by more than 50% over a period of approximately five days (Christoph et al., 2006, 2007).

$TRPV_1$ Receptor and Itch

The sensation of itch is, like pain, a noxious sensory experience. Itch results in debilitating illness with a severe impact on the sufferer's sense of well-being and quality of life. The noxious stimulus which provokes the sensation of itch is the presence or application of histamine. Histamine induces this sensation by exciting primary sensory neurones. This it achieves by activating $TRPV_1$ receptor ion channels through stimulation of these channels by phospholipase A_2 and products of lipoxogenases. Mice lacking $TRPV_1$ receptor show markedly reduced histamine-induced scratching compared with wild-type mice (Shim et al., 2007).

Visceral Hyper-Reflexia and Pain

Capsaicin-sensitive sensory fibres have long been known to have a role in the micturition reflex, because capsaicin instillation first induces contraction, then desensitisation of the urinary bladder. Moreover, capsaicin instillation also evokes the expression of the early gene, *c-Fos* in the dorsal spinal cord, which indicates nociceptive input to second order sensory neurons. Recently, Charrua and colleagues (2007) demonstrated that $TRPV_1$ receptor is responsible for both the pain sensation associated with overfilled bladder and the hyper-reflexia associated with inflammation, because both bladder distension-evoked spinal *c-Fos* expression and inflammation-evoked hyper-reflexia failed to occur in mice lacking $TRPV_1$ receptor. The degree of contribution from $TRPV_1$ receptor expressed by bladder afferents and by $TRPV_1$ receptor expressed by urothelial cells to the development of pain and hyper-reflexia, however, remains to be established.

Diabetes

Perhaps, one of the most unexpected putative roles for $TRPV_1$ receptor has been suggested recently by Razavi and co-workers (2006). They have reported that $TRPV_1$ receptor-expressing nerve fibres innervate the Langerhans islets in the pancreas. $TRPV_1$ receptor in the non-obese diabetic mouse model of type 1 diabetes shows polymorphism (P322A and D734E), which results in reduced sensitivity of $TRPV_1$ receptor to capsaicin. These authors argue that there is a negative feedback between $TRPV_1$ receptor-expressing primary sensory terminals and β-cells. Insulin, by increasing the activity of $TRPV_1$ receptor (Sathianathan et al., 2003; Santha and Nagy, 2005; Baiou et al., 2007), induces the release of neuropeptides, such as substance P and CGRP, which, in turn, inhibit insulin secretion. When this regulatory mechanism becomes unbalanced due to reduced $TRPV_1$ receptor responsiveness, the reduced local levels of neuropeptides produce insulin resistance, β cell stress, and a local proinflammatory milieu. The inflammatory milieu sustains Schwann cell and islet-specific T cells infiltration, while degenerating β cells present auto-antigens. Gram and colleagues (2007) have reported recently that systemic injection of capsaicin prevents the development of hyperglycaemia and reduced insulin secretion in Zucker Diabetic Fatty rats, which are regarded in certain aspects as a model of human type 2 diabetes mellitus. The preventive effects of systemic capsaicin injection were accompanied by complete loss of CGRP- and $TRPV_1$ receptor-coexpressing fibers in the islets. The authors hypothesise that enhanced release of CGRP from sensitised $TRPV_1$ receptor-expressing fibres reduces insulin secretion from β cells (Kozlova and Jansson, 2005). The sensitisation may result from increased release of inflammatory cytokines and anandamide from the enhanced amount of adipose tissue (Sopasakis et al., 2005; see Chap. 14). In addition, hyperglycaemia-induced NGF release from the β cells also contributes to $TRPV_1$ receptor sensitisation. Whether $TRPV_1$ receptor itself has any role in the putative enhanced activity of the islet-innervating fibres awaits further elucidation (see Chap. 14 as well).

Obesity

Zhang and colleagues (2007) have reported recently that $TRPV_1$ receptor is expressed by preadipocytes and adipose tissues. $TRPV_1$ receptor expressed in these cells and receptor respond to capsaicin with Ca^{2+} influx and, more importantly, by inhibiting adipogenesis (see Fig. 2 in Chap. 14). Overweight individuals as well as obese laboratory animals have lower levels of $TRPV_1$ receptor expression than their lean counterparts. Furthermore, capsaicin prevents the development of obesity in wild-type, but not in $TRPV_1$ receptor KO, mice.

Other Roles for $TRPV_1$ Receptor

$TRPV_1$ receptor plays a role in thermoregulation (Jancso-Gabor et al., 1970). $TRPV_1$ receptor antagonists representing various chemotypes cause an increase in body temperature (hyperthermia) (Gavva et al., 2007). Fever evoked by lipopolysaccharide is reduced in $TRPV_1$ receptor knock-out mice (Iida et al., 2005), but the absence of any indication of central neuronal activation in fever suggests that such involvement of $TRPV_1$ receptor may exist in the periphery, rather than in the brain. Neurones of the organum vasculosum lamina terminalis require the expression of $TRPV_1$ receptor to maintain their intrinsic osmosensitivity. $TRPV_1$ receptor may be involved in the development of thirst since hypertonic solution-evoked cellular and behavioural responses are absent in $TRPV_1$ receptor knock-out mice (Ciura and Bourque, 2006). Finally, there is now emerging evidence which implicates $TRPV_1$ receptor in the regulation of anxiety-related behaviours, conditioned fear, and long-term potentiation of excitatory postsynaptic potentials in the hippocampus (Marsch et al., 2007).

Concluding Remarks

There is little doubt that the incidence of $TRPV_1$ receptor in the human body will be found to be considerably greater than that which has been disclosed by recent studies. The widespread expression of $TRPV_1$ receptor in man suggests that this receptor subserves an array of vitally important functions. Our appreciation of this fact offers wonderful opportunities for therapeutic interventions. But, at the same time, this fact introduces a tremendous complication in developing a drug to serve in one context which may have profound implications for normal $TRPV_1$ receptor functioning in other non-pathological contexts.

References

Ahern GP, Brooks IM, Miyares RL, Wang X (2005) Extracellular cations sensitise and gate capsaicin receptor $TRPV_1$ modulating pain signalling. J Neurosci 25:5109–5116.

Ahluwalia J, Urban L, Capogna M, Bevan S, Nagy I (2000) Cannabinoid 1 receptors are expressed in nociceptive primary sensory neurons. Neuroscience 100:685–688.

Ahluwalia J, Urban L, Bevan S, Nagy I (2003a) Anandamide regulates neuropeptide release from capsaicin-sensitive primary sensory neurons by activating both the cannabinoid 1 receptor and the vanilloid receptor 1 *in vitro*. Eur J Neurosci 17:2611–2618.

Ahluwalia J, Yaqoob M, Urban L, Bevan S, Nagy I (2003b) Activation of capsaicin-sensitive primary sensory neurones induces anandamide production and release. J Neurochem 84:585–591.

Amaya F, Oh-hashi K, Naruse Y, IIjima N, Ueda M, Shimosato G, Tominaga M, Tanaka Y, Tanaka M (2003) Local inflammation increases vanilloid receptor 1 expression within distinct subgroups of DRG neurons. Brain Res 963:190–196.

Amaya F, Shimosato G, Nagano M, Ueda M, Hashimoto S, Tanaka Y, Suzuki, Tanaka M (2004) NGF and GDNF differentially regulate $TRPV_1$ expression that contributes to development of inflammatory thermal hyperalgesia. Eur J Neurosci 20:2303–2310.

Avelino A, Cruz C, Nagy I, Cruz F (2002) Vanilloid receptor 1 expression in the rat urinary tract. Neuroscience 109:787–798.

Babes A, Amuzescu B, Krause U, Scholz A, Flonta ML, Reid G (2002) Cooling inhibits capsaicin-induced currents in cultured rat dorsal root ganglion neurones. Neurosci Lett 317:131–134.

Baiou D, Santha P, Avelino A, Charrua A, Bacskai T, Matesz K, Cruz F, Nagy I (2007) Neurochemical characterization of insulin receptor-expressing primary sensory neurons in wild-type and vanilloid type 1 transient receptor potential receptor knockout mice. J Comp Neurol 503:334–347.

Balaban CD, Zhou J, Li HS (2003) Type 1 vanilloid receptor expression by mammalian inner ear ganglion cells. Hear Res 175:165–170.

Bevan S, Hothi S, Hughes G, James IF, Rang HP, Shah K, Walpole CS, Yeats JC (1992) Capsazepine: a competitive antagonist of the sensory neurone excitant capsaicin. Br J Pharmacol 107:544–552.

Bhave G, Zhu W, Wang H, Brasier DJ, Gereau RW (2002) cAMP-dependent protein kinase regulates desensitization of the capsaicin receptor (VR_1) by direct phosphorylation. Neuron 35:721–731.

Bhave G, Hu HJ, Glauner KS, Zhu W, Wang H, Brasier DJ (2003) Protein kinase C phosphorylation sensitizes but does not activate the capsaicin receptor transient receptor potential vanilloid 1 ($TRPV_1$). Proc Natl Acad Sci USA 100:12480–12485.

Birder LA, Kanai AJ, De Groat WC, Kiss S, Nealen ML, Burke NE, Kineley KE, Watkins S, Reynolds IJ, Caterina MJ (2001) Vanilloid receptor expression suggests a sensory role for urinary bladder epithelial cells. Proc Natl Acad Sci USA 98:13396–13401.

Bonnington JK, McNaughton PA (2003) Signalling pathways involved in the sensitisation of mouse nocicpetive neurones by nerve growth factor. J Physiol 551:433–446.

Brauchi S, Orio P, Latorre R (2004) Clues to understanding cold sensation: thermodynamics and electrophysiological analysis of the cold receptor $TRPM_8$. Proc Natl Acad Sci USA 101:15494–15499.

Brauchi S, Orta G, Salazar M, Rosenmann E, Latorre R (2006) A hot-sensing cold receptor: C-terminal domain determines thermosensation in transient receptor potential channels. J Neurosci 26:4835–4840.

Breese NM, George AC, Pauers LE, Stucky CL (2005) Peripheral inflammation selectively increases $TRPV_1$ function in IB4-positive sensory neurons from adult mouse. Pain 115:37–49.

Bridges D, Rice AS, Egertova M, Elphick MR, Winter J, Michael GJ (2003) Localisation of cannabinoid receptor 1 in rat dorsal root ganglion using in situ hybridisation and immunohistochemistry. Neuroscience 119:803–812.

Bron R, Klesse LJ, Shah K, Parada LF, Winter J (2003) Activation of Ras is necessary and sufficient for upregulation of vanilloid receptor type 1 in sensory neurons by neurotrophic factors. Mol Cell Neurosci 22:118–132.

Caterina MJ, Schumacher MA, Tominaga M, Rosen TA, Levine JD, Julius D (1997) The capsaicin receptor: a heat-activated ion channel in the pain pathway. Nature 389:816–824.

Caterina MJ, Leffler A, Malmberg AB, Martin WJ, Trafteon J, Persen-Zeitz KR, Koltzenburg M, Basbaum AI, Julius D (2000) Impaired nociception and pain sensation in mice lacking the capsaicin receptor. Science 288:306–313.

Cesare P, Dekker LV, Sardini A, Parker PJ, McNaughton PA (1999) Specific involvement of PKC-e in sensitization of the neuronal response to painful heat. Neuron 23:617–624.

Charrua A, Reguenga C, Nagy I, Cruz F, Avelino A (2005) Expression of trpv1 and trpv1b in dorsal root ganglia innervating the inflamed rat urinary bladder. Program No 170.8 Abstract Viewer/Itinerary Planner, Washington, DC: Society for Neuroscience, Online.

Charrua A, Cruz CD, Cruz F, Avelino A (2007) Transient receptor potential vanilloid subfamily 1 is essential for the generation of noxious bladder input and bladder overactivity in cystitis. J Urol 177:1537–1541.

Cheng W, Yang F, Takanishi CL, Zheng J (2007) Thermosensitive TRPV channel subunits coassemble into heteromeric channels with intermediate conductance and gating properties. J Gen Physiol 129:191–207.

Christoph T, Grunweller A, Mika J, Schafter MK-H, Wade EJ, Weihe E, Erdmann VA, Frank R, Gillen C, Kurreck J (2006) Silencing of vanilloid receptor $TRPV_1$ by RNAi reduces neuropathic and visceral pain *in vivo*. Biochem Biophys Res Commun 350:238–243.

Christoph T, Gillen C, Mika J, Grunweller A, Schater MKH, Schiene K, Frank R, Jostock R, Bahrenberg G, Weihe E, Erdmann VA, Kurreck J (2007) Antinociceptive effect of antisense oligonucleotides against the vanilloid receptor VR_1/$TRPV_1$. Neurochem Int 50:281–290.

Chu CJ, Huang SM, De Petrocellis L, Bisogno T, Ewing SA, Miller JD, Zipkin RE, Daddario N, Appendino G, Di Marzo V, Walker JM (2003) N-oleoyldopamine, a novel endogenous capsaicin-like lipid that produces hyperalgesia. J Biol Chem 278:13633–13639.

Chuang HH, Prescottt ED, Kong H, Shields S, Jordt SE, Basbaum AI, Caho MV, Julius D (2001) Bradykinin and nerve growth factor release the capsaicin receptor from $PtdIns_{4,5}P2$-mediated inhibition. Nature 411:957–962.

Ciura S, Bourque CW (2006) Transient receptor potential vanilloid 1 is required for intrinsic osmoreception in organum vasculosum lamina terminalis neurons and for normal thirst responses to systemic hyperosmolality. J Neurosci 26:9069–9075.

Clapham DE (2003) TRP channels as cellular sensors. Nature 426:517–524.

Contassot E, Tenan M, Schnuriger V, Pelte MF, Dietrick PY (2004) Arachidoyyl ethanolamide induces apoptosis of uterine cervix cancer cells via aberrantly expressed vanilloid receptor-1. Gynecol Oncol 93:182–188.

Cristino L, De Petrocellis L, Pryce G, Baker D, Guglielmotti V, Ci Marzo V (2006) Immunohistochemical localisazation of cannabinoid type 1 and vanilloid transient receptor potential vanilloid type 1 receptors in the mouse brain. Neuroscience 139:1405–1415.

Cruz F (2004) Mechanisms involved in new therapies for overactive bladder. Urology 63:65–73.

Davis JB, Gray J, Gunthorpe MJ, Hatcher JP, Davey PT, Overend P, Harries MH, Latcham J, Clapham C, Atkinson K, Huges SA, Rance K, Grau E, Harper AJ, Pugh PL, Rogers DC, Bingham S, Randall A, Sheardown SA (2000) Vanilloid receptor-1 is essential for inflammatory thermal hyperalgesia. Nature 405:183–187.

Dinis P, Charrua A, Avelino A, Yaqoob, Bevan S, Nagy I, Cruz F (2004) Anandamide-evoked activation of vanilloid receptor 1 contributes to the development of bladder hyperreflexia and nociceptive transmission to spinal dorsal horn neurons in cystitis. J Neurosci 24:11253–11263.

Dray A, Perkins M (1993) Bradykinin and inflammatory pain. Trends Neurosci 16:99–104.

Dvorakova M, Kummer W (2001) Transient expression of vanilloid receptor subtype 1 in rat cardiomyocytes during development. Histochem Cell Biol 116:223–225.

Eun SY, Jung SJ, Park YK, Kwak J, Kim SJ, Kim J (2001) Effects of capsaicin on Ca^{2+} release from the intracellular Ca^{2+} stores in the dorsal root ganglion cells of adult rats. Biochem Biophys Res Commun 285:1114–11120.

Fukuoka T, Tokunaga A, Tachibana T, Dai Y, Yamanaka H, Noguchi K (2002) VR_1, but not P_2X_3 increases in the spared L4 DRG in rats with L5 spinal nerve ligation. Pain 99:111–120.

Garcia-Sanz N, Fernandez-Carvajal A, Morenilla-Palao C, Planells-Cases R, Fajardo-Sanchez E, Fernandez-Ballester G, Ferrer-Montiel A (2004) Identification of a tetramerization domain in the C terminus of the vanilloid receptor. J Neurosci 24:5307–5314.

Gavva NR, Klinonsky L, Qu Y, Shi L, Tamir R, Edenson S, Zhang TJ, Viswanadhan VN, Toth A, Pearce LHV, Vanderah TW, Porreca F, Blumberg PM, Lile J, Sun Y, Wild K, Louis JC, Treanor JJ (2004) Molecular determinants of vanilloid sensitivity in TRPV1. J Biol Chem 279:20283–20295.

Gavva NR, Bannon AW, Surapaneni S, Hovland DN, Lehto SG, Gore A, Juan T, Deng H, Han B, Klionsky L, Kuang R, Le A, Tamir R, Wang J, Youngblood B, Zhu D, Norman MH, Magal E,

Treanor JJ, Louis JC (2007) The vanilloid receptor $TRPV_1$ is tonically activated *in vivo* and involved in body temperature regulation. J Neurosci 27:3366–3374.
Goswami C, Dreger M, Jahnel R, Bogen O, Gillen C, Hucho F (2004) Identification and characterization of a Ca^{2+}-sensitive interaction of the vanilloid receptor $TRPV_1$ with tubulin. J Neurochem 91:1092–1103.
Goswami C, Dreger M, Otto H, Schwappach, Hucho F (2006) Rapid disassembly of dynamic microtubules upon activation of the capsaicin receptor $TRPV_1$. J Neurochem 96:254–266.
Goswami C, Hucho TB, Hucho F (2007a) Identification and characterisation of novel tubulin-binding motifs located within the C-terminus of $TRPV_1$. J Neurochem 101:250–262.
Goswami C, Schmidt H, Hucho F (2007b) $TRPV_1$ at nerve endings regulates growth cone morphology and movement through cytoskeleton reorganization. FEBS J 274:760–772.
Gram DX, Ahren B, Nagy I, Olsen UB, Brand CL, Sundler F, Tabanera R, Svendsen O, Carr RD, Santha P, Wierup N, Hansen AJ (2007) Capsaicin-sensitive sensory fibers in the islets of Langerhans contribute to defective insulin secretion in Zucker diabetic rat, an animal model for some aspects of human type 2 diabetes. Eur J Neurosci 25:213–223.
Guo A, Vulchanova L, Wang J, Li X, Elde R (1999) Immunocytochemical localisation of the vanilloid receptor (VR_1): relationship to neuropeptides, the $P2X_3$ purinoceptor and IB4 binding sites. Eur J Neurosci 11:946–958.
Hellwig N, Albrecht N, Harteneck C, Schultz G, Schaefer M (2005) Homo- and heteromeric assembly of $TRPV_1$ channel subunits. J Cell Sci 118:917–928.
Huang SM, Bisogno T, Trevisani M, Al-Hayani A, De Petrocellis L, Fezza F, Tognetto M, Petros TJ, Krey JF, Chu CJ, Miller JD, Davies SN, Geppetti P, Walker JM, Di Marzo V (2002) An endogenous capsaicin-like substance with high potency at recombinant and native vanilloid VR_1 receptors. Proc Natl Acad Sci USA 99:8400–8405.
Hudson LJ, Bevan S, Wotherspoon G, Gentry C, Fox A, Winter J (2001) VR_1 protein expression increases in undamaged DRG neurons after partial nerve injury. Eur J Neurosci 13:2105–2114.
Hui K, Liu B, Qin F (2003) Capsaicin activation of the pain receptor, VR_1: multiple open states from both partial and full binding. Biophys J 84:2957–2968.
Hwang SW, Cho H, Kwak J, Lee SY, Kang CJ, Jang J, Cho S, Min KH, Suh YG, Kim D, Oh U (2000) Direct activation of capsaicin receptors by products of lipoxygenases: endogenous capsaicin-like substances. Proc Natl Acad Sci USA 97:6155–6160.
Iida T, Shimizu I, Nealen ML, Campbell A, Caterina M (2005) Attenuated fever response in mice lacking $TRPV_1$. Neurosci Lett 378:28–33.
Ioue K, Koizumi S, Fuziwara S, Denda S, Inoue K, Denda M (2002) Functional vanilloid receptors in cultured normal human epidermal keratinocytes. Biochem Biophys Res Commun 291:124–129.
Jancso G, Kiraly E, Jancso-Gabor A (1977) Pharmacologically induced selective degeneration of chemosensitive primary sensory neurones. Nature 270:741–743.
Jancso-Gabor A, Szolecsanyi J, Jancso N (1970) Stimulation and desensitization of the hypothalamic heat-sensitive structure by capsaicin in rats. J Physiol 208:444.
Jeske NA, Patwardhan AM, Gamper N, Price TJ, Alopian An, Hargreaves KM (2006) Cannabinoid WIN 55,212-2 regulates $TRPV_1$ phosphorylation in sensory neurons. J Biol Chem 281:32879–32890.
Jordt SE, Julius D (2002) Molecular basis for species-specific sensitivity to "hot" chili peppers. Cell 2108:421–430.
Jordt SE, Tominaga M, Julius D (2000) Acid potentiation of the capsaicin receptor determined by a key extracellular site. Proc Natl Acad Sci USA 97:8134–8139.
Jung J, Lee SY, Hwang SW, Cho H, Shin J, Kang YS, Kim S, Oh U (2002) Agonist recognition sites in the cytosolic tails of vanilloid receptor 1. J Biol Chem 277:44448–444454.
Kanai, Nakazato E, Jujiuchi A, Hara T, Imai A (2005) Involvement of an increased spinal $TRPV_1$ sensitization through its up-regulation in mechanical allodynia of CCI rats. Neuropharmacology 49:977–984.

Karai LJ, Russell JT, Iadarola MJ, Olah Z (2004) Vanilloid receptor 1 regulates multiple calcium compartments and contributes to Ca^{2+}-induced Ca^{2+} release in sensory neurons. J Biol Chem 279:16377–16387.

Karlsson U, Sundgren-Andersson AK, Johansson S, Krupp JJ (2005) Capsaicin augments synaptic transmission in the rat medial preoptic nucleus. Brain Res 1043:1–11.

Kato S, Aihara E, Nakamura A, Xin H, Matsui H, Kohama K, Takeuchi K (2003) Expression of vanilloid receptors in rat gastric epithelial cells: role in cellular protection. Biochem Pharmacol 66:1115–1121.

Kedei N, Szabo T, Lile JD, Treanor JJ, Olah Z, Iadarola MJ, Blumberg PM (2001) Analysis of the native quaternary structure of vanilloid receptor 1. J Biol Chem 276:28613–28619.

Kelly S, Chapman V (2002) Spinal administration of capsazepine inhibits noxious evoked responses of dorsal horn neurons in non-inflamed and carrageenan inflamed rats. Brain Res 935:103–108.

Kim S, Kang C, Shin CY, Hwang SW, Yang YD, Shim WS, Park M-Y, Kim E, Kim M, Kim B-M, Cho H, Shin Y, Oh U (2006) $TRPV_1$ recapitulates native capsaicin receptor in sensory neurons in association with Fas-Associated Factor 1. J Neurosci 26:2403–2412.

Korschen HG, Beyermann M, Muller F, Heck M, Vantler M, Koch KW, Kellner R, Wolfrum U, Bode C, Hofmann KP, Kaupp UB (1999) Interaction of glutamic-acid-rich proteins with the cGMP signalling pathway in rod photoreceptors. Nature 400:761–766.

Kozlova EN, Jansson L (2005) *In vitro* interactions between insulin-producing beta cells and embryonic dorsal root ganglia. Pancreas 31:380–384.

Kuzhikandathil EV, Wang H, Szabo T, Morozova N, Blumberg PM, Oxford GS (2001) Functional analysis of capsaicin receptor (vanilloid receptor subtype 1) multimerization and agonist respsonsiveness using a dominant negative mutation. J Neurosci 21:8697–8706.

Lazar J, Braun DC, Toth A, Wang Y, Pearce LV, Pavlyukovets VA, Blumberg PM, Garfield SH, Wincovitch S, Choi, HK, Lee J (2006) Kinetics of penetration influence the apparent potency of vanilloids on $TRPV_1$. Mol Pharmacol 69:1166–1173.

Lee SY (2005) Identification of a protein that interacts with the vanilloid receptor. Biochem Biophys Res Commun 331:1445–1451.

Liang YF, Haake B, Reeh PW (2001) Sustained sensitization and recruitment of rat cutaneous nociceptors by bradykinin and a novel theory of its excitatory action. J Physiol 523:229–239.

Liu B, Hui K, Qin F (2003) Thermodynamics of heat activation of single capsaicin ion channels VR_1. Biophys J 85:2988–3006.

Liu B, Zhang C, Qin F (2005) Functional recovery from desensitization of vanilloid receptor $TRPV_1$ requires resynthesis of phosphatidylinositol 4,5-bisphospahte. J Neurosci 25:4835–4843.

Lu G, Henderson D, Liu L, Reinhart PH, Simon SA (2005) $TRPV_{1b}$, a functional human vanilloid receptor splice variant. Mol Pharmacol 67:1119–1127.

Maione S, Bisogno T, de Novellis V, Palazzo E, Cristino L, Valenti M, Petrosino S, Guglielmotti V, Rossi F, Di Marzo V (2006) Elevation of endocannabinoid levels in the ventrolateral periaqueductal grey through inhibition of fatty acid amide hydrolase affects descending nociceptive pathways via both cannabinoid receptor type 1 and transient receptor potential vanilloid type 1 receptors. J Pharmacol Exp Ther 316:969–982.

Malmberg AB, Brandon EP, Idzerda RL, Liu H, McKnight GS, Basbaum AI (1997) Diminished inflammation and nociceptive pain with preservation of neuropathic pain in mice with a targeted mutation of the type 1 regulatory subunit of cAMP-dependent protein kinase. J Neurosci 19:7462–7470.

Mandadi S, Tominaga T, Numazaki M, Murayama N, Saito N, Armati PJ, Roufogalis BD, Tominaga M (2006) Increased sensitivity of desensitized $TRPV_1$ by PMA occurs through PKCε-mediated phosphorylation at S800. Pain 123:106–116.

Marsch R, Foeller E, Rammes G, Bunck M, Kossl M, Holsboer F, Zieglgansberger W, Landgraf R, Lutz B, Wotjak CT (2007) Reduced anxiety, conditioned fear, and hippocampal long-term potentiation in transient receptor potential vanilloid type 1 receptor-deficient mice. J Neurosci 27:832–839.

Marshall IC, Owen DE, Cripps TV, Davis JB, McNulty S, Smart D (2003) Activation of vanilloid receptor 1 by resiniferatoxin mobilizes calcium from inositol 1,4,5-trisphospahte-sensitive stores. Br J Pharmacol 138:172–176.

Mezey E, Toth ZE, Cortright DN, Arzubi MK, Krause JE, Elde R, Guo A, Blumberg PM, Szallasi A (2000) Distribution of mRNA for vanilloid receptor subtype 1 (VR_1), and VR_1-like immunoreactivity, in the central nervous system of the rat and human. Proc Natl Acad Sci USA 97:3655–3660.

Michael GJ, Priestley JV (1999) Differential expression of the mRNA for the vanilloid receptor subtype 1 in cells of the adult rat dorsal root and nodose ganglia and its downregulation by axotomy. J Neurosci 19:1844–1854.

Mohapatra DP, Nau C (2005) Regulation of Ca^{2+}-dependent desensitization in the vanilloid receptor $TRPV_1$ by calcineurin and cAMP-dependent protein kinase. J Biol Chem 280:13424–13432.

Morenilla-Palao C, Planells-Cases R, Garcia-Sanz N, Ferrer-Montiel A (2004) Regulated exocytosis contributes to protein kinase C potentiation of vanilloid receptor activity. J Biol Chem 279:25665–25672.

Morita A, Iwasaki Y, Kobata K, Iida T, Higashi T, Oda K, Suzuki A, Narukawa M, Sasakuma S, Yokogoshi H, Yazawa S, Tominaga M, Watanabe T (2006) Lipophilitic of capsaicinoids and capsinoids influences the multiple activation process of rat $TRPV_1$. Life Sci 79:2303–2310.

Moriyama T, Higashi T, Togashi K, Iida T, Segi E, Sugimoto Y, Tominaga T, Narumiya S, Tominaga M (2005) Sensitization of $TRPV_1$ by EP_1 and IP reveals peripheral nociceptive mechanism of prostaglandins. Mol Pain 1:3.

Movahed P, Jonsson BAG, Birnir B, Wingstrand JA, Jorgensen TK, Ermund A, Sterner O, Zygmunt PM, Hogestatt ED (2005) Endogenous unsaturated C18 N-acylethanolamines are vanilloid receptor ($TRPV_1$) agonists. J Biol Chem 280:38496–38504.

Nagy I (2004) Sensory processing: primary afferent neurons/DRG. In: Anesthetic Pharmacology: Physiologic Principles and Clinical Practice, eds: Evers and Maze. pp. 187–197, Chirchill Livingstone, Philadelphia.

Nagy I, Rang HP (1999) Similarities and differences between the responses of rat sensory neurons to noxious heat and capsaicin. J Neurosci 19:10647–10655.

Narumiya S, Sugimoto Y, Ushikubi F (1999) Prostanoid receptors: structures, properties, and functions. Physiol Rev 79:1193–1226.

Numazaki M, Tominaga T, Toyooka H, Tominaga M (2002) Direct phosphorylation of capsaicin receptor VR_1 by protein kinase Ce and identification of two target serine residues. J Biol Chem 277:13375–13378.

Olah A, Szabo T, Karai L, Hough C, Fields RD, Caudle RM, Blumberg PM, Iadarola MJ (2001a) Ligand-induced dynamic membrane changes and cell deletion conferred by vanilloid receptor 1. J Biol Chem 276:11021–11030.

Olah Z, Karai L, Iadarola MJ (2001b) Anandamide activates vanilloid receptor 1 (VR_1) at acidic pH in dorsal root ganglia neurons and cells ectopically expressing VR_1. J Biol Chem 276:31163–31170.

Olah Z, Karai L, Iadarola MJ (2002) Protein kinase C (alpha) is required for vanilloid receptor 1 activation. Evidence for multiple signalling pathways. J Biol Chem 277: 35752–35759.

O'Sullivan SE, Kendall DA, Randall MD (2004) Characterisation of the vasorelaxant properties of the novel endocannabinoid N-arachidonoyl-dopamine (NADA). Br J Pharmacol 141:803–812.

Pare M, Elde R, Mazurkiewicz JE, Smith AM, Rice FL (2001) The Meissner corpuscle revised: a multiafferented mechanoreceptor with nociceptor immunochemical properties. J Neurosci 21:7236–7246.

Prescott ED, Julius D (2003) A modular PIP_2 binding site as a determinant of capsaicin receptor sensitivity. Science 300:1284–1288.

Price TJ, Patwardhan A, Akopian AN, Hargreaves KM, Flores CM (2004) Modulation of trigeminal sensory neuron activity by the dual cannabinoid-vanilloid agonists anandamide, N-arachidonoyl-dopamine and arachidonyl-2-chloroethylamide. Br J Pharmacol 141:1118–1130.

Rami HK, Thompson M, Whyman P, Jerman JC, Egerton J, Brough S, Stevens AJ, Randall AD, Smart D, Gunthorpe MJ, Davis JB (2004) Discovery of small molecule antagonists of TRPV1. Bioorg Med Chem Lett 14:3631–3634.

Razavi R, Chan Y, Afifiyan FN, Liu XJ, Wan X, Yantha J, Tsui H, Tang L, Tsai S, Santamaria P, Driver JP, Serreze D, Salter MW, Dosch HM (2006) $TRPV_1$ + sensory neurons control beta cell stress and islet inflammation in autoimmune diabetes. Cell 127:1123–1135.

Roberts JC, Davis JB, Benham CK (2004) [^{3}H]Resiniferatoxin autoradiography in the CNS of wild-type and TRPV1 null mice defines $TRPV_1$ (VR-1) protein distribution. Brain Res 995:176–183.

Rutter AR, Ma QP, Leveridge M, Bonnert TP (2005) Heteromerization and colocalization of $TrpV_1$ and $TrpV_2$ in mammalian cell lines and rat dorsal root ganglia. Neuroreport 16:1735–1739.

Ryu S, Liu B, Qin F (2003) Low pH potentiates both capsaicin binding and channel gating of VR_1 receptors. J Gen Physiol 122:45–61.

Sanchez JF, Krause JE, Cortright DN (2001) The distribution and regulation of vanilloid receptor VR_1 and VR_1 5′ splice variant RNA expression in rat. Neuroscience 107:373–381.

Sanchez MG, Sangchez AM, Collado B, Malagarie-Cazenave S, Olea N, Carmena MJ, Prieto JC, Diaz-Laviada I (2005) Expression of the transient receptor potential vanilloid 1 ($TRPV_1$) in LNCaP and PC-3 prostate cancer cells and in human prostate tissue. Eur J Pharmacol 515:20–27.

Santha P, Nagy I (2005) Insulin-induced membrane currents in capsaicin-sensitive primary sensory neurones. J Physiol 565P: C117.

Sathianathan V, Avelino A, Charrua A, Santha P, Matesz K, Cruz F, Nagy I (2003) Insulin induces cobalt uptake in a subpopulation of rat cultured primary sensory neurons. Eur J Neurosci 18:2477–2486.

Sato J, Segami N, Yoshitake Y, Kaneyama K, Abe A, Yoshimura H, Fujimura K (2005) Expression of capsaicin receptor TRPV-1 in synovial tissues of patients with symptomatic internal derangement of the temporomandibular joint and joint pain. Oral Surg Oral Med Oral Pathol Oral Radiol Endod 100:674–681.

Schumacher MA, Jong BE, Frey SL, Sudanagunta SP, Capra NF, Levine JD (2000) The stretch-inactivated channel, a vanilloid receptor variant, is expressed in small-diameter sensory neurons in the rat. Neurosci Lett 287:215–218.

Shim WS, Tak MH, Lee MH, Kim M, Kim M, Koo JY, Lee CH, Kim M, Oh U (2007) TRPV1 mediates histamine-induced itching via the activation of phospholipase A_2 and 12-lipoxygenase. J Neurosci 27:2331–2337.

Shin J, Cho H, Hwang SW, Jung J, Shin Cu, Lee S-Y, Kim SH, Lee MG, Choi YH, Kim J, Haber NA, Reichling DB, Khasar S, Levine JD, Oh U (2002) Bradykninin-12-lipoxygenase-VR1 signalling pathway for inflammatory hyperalegesia. Proc Natl Acad Sci USA 99:10150–10155.

Shu X, Mendell LM (2001) Acute sensitization by NGF of the response of small-diameter sensory neurones to capsaicin. J Neurophysiol 86:2931–2938.

Singh Tahim A, Santha P, Nagy I (2005) Inflammatory mediators convert anandamide into a potent activator of the vanilloid type 1 transient receptor potential receptor in nociceptive primary sensory neurons. Neuroscience 136:539–548.

Smith GD, Gunthorpe MJ, Kelsell RE, Hayes PD, Reilly P, Facer P, Wright JE, Jerman JC, Walhin JP, Ooi L, Egerton J, Charles KJ, Smart D, Randall AD, Anand P, Davis JB (2002) $TRPV_3$ is a temperature-sensitive vanilloid receptor-like protein. Nature 418:186–190.

Sopasakis VR, Nagaev I, Smith U (2005) Cytokine release from adipose tissue of non-obese individuals. Int J Obes (Lond) 29:1144–1147.

Southall MD, Li T, Gharibova Ls, Pei Y, Nicol GD, Travers JB (2003) Activation of epidermal vanilloid receptor-1 induces release of proinflammatory mediators in human keratinocytes. J Pharmacol Exp Ther 304:217–222.

Stein AT, Ufret-Vincenty CA, Hua L, Santana Lf, Gordon SE (2006) Phosphoinositide 3-kinase binds to TRPV1 and mediates NGF-stimulated $TRPV_1$ trafficking to the plasma membrane. J Gen Physiol 28:509–522.

Stucky CL, Lewin GR (1999) Isolectin B4-positive and -negative nociceptors are functionally distinct. J Neurosci 19:6497–6505.

Sugiura T, Tominaga M, Katsuya H, Mizumura K (2002) Bradykinin lowers the threshold temperature for heat activation of vanilloid receptor 1. J Neurophysiol 88:544–548.

Susankova K, Tousova K, Vyklicky L, Teisinger J, Vlachova V (2006) Reducing and oxidizing agents sensitize heat-activated vanilloid receptor ($TRPV_1$) current. Mol Pharmacol 70:383–394.

Suzuki R, Chapman V, Dickenson AH (1999) The effectiveness of spinal and systemic morphine on rat dorsal horn neuronal responses in the spinal nerve ligation model of neuropathic pain. Pain 80:215–228.

Szallasi A, Blumberg P (1990) Resiniferatoxin and its analogues provide novel insights into the pharmacology of the vanilloid (capsaicin) receptor. Life Sci 47:1399–1408.

Tian W, Fu Y, Wang DH, Cohen DM (2006) Regulation of TRPV1 by a novel renally expressed rat $TRPV_1$ splice variant. Am J Physiol Renal Physiol 290:F117–F126.

Tominaga M, Caterina MJ, Malmberg AB, Rosen TA, Gilbert H, Skinner K, Raumann BE, Basbaum AI, Julius D (1998) The cloned capsaicin receptor integrates multiple pain-producing stimuli. Neuron 21:531–543.

Tousova K, Vyklicky L, Susankova K, Benedikt J, Vlachova V (2005) Gadolinium activates and sensitizes the vanilloid receptor $TRPV_1$ through the external protonation sites. Mol Cell Neurosci 30:207–217.

Van der Stelt M, Trevisani M, Vellani V, De Petrocellis L, Schiano Moriello A, Campi B, McNaughton P, Geppetti P, Di Marzo V (2005) Anandamide acts as an intracellular messenger amplifying Ca^{2+} influx via $TRPV_1$ channels. EMBO J 24:3026–3037.

Vlachova V, Teisinger J, Susankova K, Lyfenko A, Ettrich R, Vyklicy L (2003) Functional role of C-terminal cytoplasmic tail of rat vanilloid receptor 1. J Neurosci 23:1340–1350.

Voets T, Droogmans G, Wissenbach U, Janssens A, Flockerzi V, Nilius B (2004) The principle of temperature-dependent gating in cold- and heat-sensitive TRP channels. Nature 430:748–754.

Vos MH, Neelands TR, McDonald HA, Choi W, Kroeger PE, Puttfarcken PS, Faltynek CR, Moreland RB, Han P (2006) $TRPV_{1b}$ overexpression negatively regulates $TRPV_1$ responsiveness to capsaicin, heat and low pH in HEK293 cells. J Neurochem 99:1088–1102.

Vyklicky L, Lyfenko A, Kuffler DP, Vlachova V (2003) Vanilloid receptor $TRPV_1$ is not activated by vanilloids applied intracellularly. Neuroreport 14:1061–1065.

Walker K, Urban L, Medhurst SJ, Patel S, Panesar M, Fox AJ, McIntyre P (2003) The VR_1 antagonist, capsazepine, reverses hyperalgesia in models of inflammatory and neuropathic pain. J Pharmacol Exp Ther 304:56–62.

Wang C, Hu HZ, Colton CK, Wood JD, Zhu MX (2004) An alternative splicing product of the murine $trpv_1$ gene dominant negatively modulates the activity of $TRPV_1$ channels. J Biol Chem 279:37423–37430.

Ward SM, Bayguinov J, Won KJ, Grundy D, Berthoud HR (2003) Distribution of the vanilloid receptor (VR_1) in the gastrointestinal tract. J Comp Neurol 465:121–135.

Welch JM, Simon SA, Reinhart PH (2000) The activation mechanism of rat vanilloid receptor 1 by capsaicin involves the pore domain and differs from the activation by either acid or heat. Proc Natl Acad Sci USA 97:13889–13894.

Wu ZZ, Chen SR, Pan HL (2005) Transient receptor potential vanilloid type 1 activation downregulates voltage-gated calcium channels through calcium-dependent calcineurin in sensory neurons. J Biol Chem 280:18142–18151.

Xu H, Blair NT, Clapham DE (2005) Camphor activates and strongly desensitises the transient receptor potential vanilloid subtype 1 channel in a vanilloid-independent mechanism. J Neurosci 25:8924–8937.

Zhang X, Huang J, McNaughton PA (2005) NGF rapidly increases membrane expression of TRPV1 heat-gated ion channels. EMBO J 24:4211–4223.

Zhang LL, Yan Liu D, Ma LQ, Luo ZD, Cao TB, Zhong J, Yan ZC, Wang LJ, Zhao ZG, Zhu SJ, Schrader M, Thilo F, Zhu ZM, Tepel M (2007) Activation of transient receptor potential vanilloid type-1 channel prevents adipogenesis and obesity. Circ Res 100:1063–1070.

Zheng J, Dai C, Steyger PS, Kim Y, Vass Z, Ren T, Nuttall AL (2003) Vanilloid receptors in hearing: altered cochlear sensitivity by vanilloids and expression of TRPV1 in the organ of corti. J Neurophysiol 90:444–455.

Zygmunt PM, Peterson J, Anderson DA, Chuang H, Sorgard M, Di Marzo V (1999) Vanilloid receptors on sensory nerves mediate the vasodilator action of anandamide. Nature 400:452–457.

Chapter 9
Alternative Interacting Sites and Novel Receptors for Cannabinoid Ligands

Attila Köfalvi

Abstract The previous chapters provided us with detailed reviews on the molecular biology and pharmacology of major endocannabinoid and endovanilloid ligands (namely, anandamide and 2-arachidonoylglycerol) and their receptors (CB_1, CB_2 and $TRPV_1$ receptors), which altogether can be termed as "canonical knowledge". Still, experimental findings often display mismatches with this canonical knowledge: in the last decade, a rapidly increasing number of studies have reported "non-canonical", "unusual" pharmacological profiles for certain cannabinoid ligands and receptors. Furthermore, from time to time results are explained by suggesting the involvement of a "new receptor". The present chapter attempts to give a helpful guideline about how to evaluate "non-canonical" results in the cannabinoid field. All the major topics of "non-canonical" cannabinoid pharmacology, namely interactions of endogenous and exogenous cannabinoid and vanilloid ligands with (1) CB_1 receptor splice variants, (2) CB_1 receptor heterodimers and other non-ionotropic receptors, (3) ligand- and voltage-gated ion channels and finally, (4) neurotransmitter uptake systems are thoroughly reviewed. For sake of simplicity, studies reporting unknown cannabinoid receptors without sufficient investigation of other already defined targets are not discussed here. Finally, this review highlights that the "unorthodox sites of action" may be an unavoidable consequence of evolution, providing novel ideas and pharmaceutical targets to modulate signaling systems in neuropsychiatric disorders.

Introduction

Phylogenetically, the endocannabinoid and the endovanilloid systems have been present for a very long period. Several invertebrates possess the same neuroactive hybrid endocannabinoid/endovanilloid (mostly arachidonic acid-derivative) ligands like mammalians. In addition, these substances activate both ionotropic and metabotropic receptors, being basically similar to the mammalian counterparts (Salzet and Stefano, 2002; McPartland and Glass, 2003; Anday and Mercier, 2005). It seems entirely plausible for cells and organisms to detect and respond to the changes of the external milieu with the changes in a physicochemically sensitive

A. Köfalvi (ed.), *Cannabinoids and the Brain.*

relay system, which can be found between the external and the internal space, i.e. the plasma membrane. Since plasma membranes are rich in arachidonic acid-derivative lipids, it seems logical that deliberation of these substances upon changes in membranes' physicochemical properties could be responsible for signaling. Then the signal has to be transduced by relay proteins (ancient metabotropic receptors) or translated into ion entry, which either changed the electrical charge of the membrane or interacted with intracellular cation- (Ca^{2+}?) sensing proteins to evoke responses. For instance, ancient simple multicellular aquatic animals needed to avoid water zones of disadvantageous pH and temperature. A prototypic vanilloid receptor may have been this kind of thermo- and pH-sensor, since the $TRPV_1$ vanilloid receptor is the main proton- and heat-gated ion channel in vertebrates, responding to arachidonic acid-derivative substances such as anandamide (Nagy et al., 2004; see Chap. 8). In line with this hypothesis, it is assumed that prototypic vanilloid receptors preceded cannabinoid receptors in evolution (McPartland and Glass, 2003). With the appearance of more complex and bigger animal organisms in the evolution process, changes had to be detected also in the internal extracellular milieu. Given the complexity of the desired responses, several novel ion channels evolved, which gave relatively selective access to Na^+ and/or Ca^{2+} or K^+ or Cl^-/ HCo_3^- ions. The phylogeny tree of ion channels indicates that one ion channel was the likely ancestor of several present-day plasma membrane ion channels together with the vanilloid receptor (Moran et al., 2004). Taking into consideration that ancient endocannabinoid/endovanilloid-sensing ion channels had to exist early in evolution and that many present time ligand- and voltage-gated ion channels interact with endocannabinoid/endovanilloid substances and other cannabinoid ligands (see later), it is assumed that the ancestor of a number of present-time ion channels was originally a vanilloid receptor prototype, which evolved into several new types, whose main function became different from the function of recognition of lipid substances. Nonetheless, they still preserve something from the original receptor–ligand interacting site. Although this assumption is based on certain genetic and functional evidence, recent novel approaches have revealed alternative evolutionary trajectories for the development of homologue, ortologue and paralogue receptors and enzymes for the endocannabinoid/endovanilloid systems (Elphick and Egertová, 2005; McPartland et al., 2006). In the following, the non-conventional ("unorthodox") interactions between ligands and target proteins will be summarized, and they all are also listed in Table 1.

CB_1 Receptor-Mediated Non-Canonical Effects

Alternative Signaling at CB_1 Receptor Homodimers

Before jumping to the conclusion of the involvement of a hypothetical "CB_3" receptor, one should first consider whether results are not due to variations in the pharmacological

Table 1 Interaction of frequent endogenous and exogenous cannabinoid, vanilloid and related substances with rodent and human intra- and extracellular targets

	2-AG	AEA, mAEA	WIN-2	WIN-3	Δ^9-THC	HU-210	CP55940	ABN-CBD	Cannabidiol	Cannabinol	NADA
CB_1R	+	+	+	ø	+	+	+	ø	–	+	+
$CB_{1A}R$	+	ø	+		+	+	+				
$CB_{1B}R$	+	ø	+		+	+	+				
CB_2R	+	+	+	–	+	+	+	ø		+	+
$TRPV_1R$	ø	+	ø	ø	ø	ø	ø	ø	+		+
ABN-CBDR	ø	+	ø		ø			+	–		
GPR55	+	+	ø		+		+				
Imidazoline R		+	–				+				
A_1R											
$M_{1/4}R$		–	ø								
PPARα											
PPARγ	+				+						
$5\text{-}HT_{3A}R$		–	–		–		–		–		
α7 nAChR	–	–	ø		ø		ø				
GlyR	–	–+[a]	–ø		+						
Ca_v1	–	–	ø		ø		ø	–			
Ca_v2		–	–	–	–		–				–
Ca_v3		–	ø		ø	–	ø				
VGSCs		–	–								
$K_v1.2$		–			–						
DR K_v		–	–								

(continued)

Table 1 (continued)

	2-AG	AEA, mAEA	WIN-2	WIN-3	Δ^9-THC	HU-210	CP55940	ABN-CBD	Cannabidiol	Cannabinol	NADA
BK		+[b]									
TASK-1	ø	–	–		ø	ø	–				
TASK-3		–									–
HCN1											
$TRPV_4R$	+	+									
$TRPA_1R$			+		+					+	
$TRPC_1R$	ø	ø			+	+	ø			+	
NMDAR		+			ø						
ENaC											
DAT		–	–	–							
SerT		–	–								
$GluT_{1,2}$			–	–							
$GlyT_{1A}$	+	+									

	PEA	OEA	Virodhamine	Noladin ether	SR141716A	AM251	JWH015	AM404	Capsaicin	Ccapsazepine	I-RTX
CB_1R	+ø	+ø	+–[c]	+	–	–	ø	ø	ø	ø	ø
$CB_{1A}R$			ø	ø	–						
$CB_{1B}R$			ø	ø	–						
CB_2R	–		+	+ø	ø	–ø[d]	+	ø	ø	ø	ø
$TRPV_1R$		+			ø	ø	ø	+	+	–	–
ABN-CBDR			+	+	–	ø					
GPR55	+		+		–	–	+				
Imidazoline R					–						

A_1R			–	–					
$M_{1/4}R$			ø						
PPARα	+	+							
PPARγ									
$5\text{-}HT_{3A}R$					–				
α7 nAChR								–	
GlyR									
Ca_v1									
Ca_v2			–	–			–	–	–
Ca_v3			–						
VGSCs	–			–		–	–	–	
$K_v1.2$									
DR K_v									
BK_{Ca}			–						
TASK-1									
TASK-3									
HCN1								–	
$TRPV_4R$									
$TRPA_1R$							ø		
$TRPC_1R$			–		–				
NMDAR									
ENaC								–	
DAT				–					

(continued)

Table 1 (continued)

	PEA	OEA	Virodhamine	Noladin ether	SR141716A	AM251	JWH015	AM404	Capsaicin	Ccapsazepine	I-RTX
SerT											
$GluT_{1,2}$						–					
$GlyT_{1A}$											

For further information, see text. For efficacy and potency values at CB_1 and CB_2 receptors consult the previous chapter. Only those interactions are listed that develop up to 10μM concentration of the ligand. *Symbols:* +, facilitation, activation, or (partial) agonism; –, inhibition, direct blockade, or antagonism; ø, reported lack of effect; + ø, reports exist on weak agonism and lack of effect as well; empty cells, data not available. *Abbreviations for ligands:* 2-AG, 2-arachidonoyl glycerol; AEA, anandamide; mAEA, *R*-methanandamide; WIN-2, WIN55212-2; WIN-3, WIN55212-3, the CB_1 receptor receptor-inactive entantiomer; Δ^9-THC, Δ^9-tetrahydrocannabinol, the main psychoactive constituent of marijuana; ABN-CBD, abnormal-cannabidiol; NADA, *N*-arachidonoyl dopamine; PEA, palmitoylethanolamide; OEA, N-oleoylethanolamide; I-RTX, iodoresiniferatoxin. *Abbreviations for targets:* CB_1R, CB_1 receptor; $CB_{1A}R$ and $CB_{1B}R$, human CB_1 receptor splice variants A and B; CB_2R, CB_2 receptor; $TRPV_1R$, transient release potential family "Vanilloid-type 1"; ABN-CBDR, abnormal-cannabidiol receptor; GPR55, G protein-coupled orphan receptor Nr 55; A_1R, adenosine A_1 receptor; $M_{1/4}R$, muscarinic M_1 and M_4 receptors; PPARα and PPARγ, peroxisome proliferator-activated receptors α and γ; $5\text{-}HT_{3A}R$, serotonin $5\text{-}HT_3$ receptor; α7 nAChR, α7 nicotinic acetylcholine receptor; GlyR, glycine receptor; Ca_v1, L-type Ca^{2+} channels; Ca_v2, N-, P/Q- and R-type channels; Ca_v3, T-type Ca^{2+} channels; VGSCs, voltage-gated Na^+ channels, $K_v1.2$, Shaker-type voltage-sensitive potassium channels; DR K_v, delayed rectifier voltage-sensitive K^+ channels; BK_{Ca}, Ca^{2+}-activated large-conductance K^+ channel type BK; TASK-1 and TASK-3, two-pore-domain acid sensitive background K^+ channel types 1 and 3; HCN1, hyperpolarization-activated cyclic nucleotide-gated channel type 1; $TRPV_4R$, transient release potential family "Vanilloid-type 4" receptor; $TRPA_1R$, transient release potential family "Ankyrin-type 1" receptor; $TRPC_1R$, transient release potential family "Canonical-type 1" receptor; NMDAR, *N*-methyl-D-aspartate NR1/NR2A receptor; ENaC, Amiloride-sensitive epithelial Na^+ channel; DAT, dopamine transporter; Sert, serotonin transporter; $GluT_{1,2}$, glutamate transporters 1 and 2; $GlyT_{1A}$, glycine transporter 1A. AEA and mAEA are taken together because apparently there is no major difference in their effects on the targets listed here

[a]Depending on the concentration of glycine and the site of the receptor (see text)

[b]Mediated by a presumable cytosolic factor

[c]Weak endogenous agonist/inverse agonist

[d]Inverse agonism was reported

profile of CB_1 receptors. The pharmacological profile of G protein-coupled receptors is usually dependent on splice variants, heterodimerization, alternative coupling, the brain area and the cell types where the receptor is situated. As a short note, I draw the reader's attention to the fact that functional CB_2 receptors have been reported in brain neurons; therefore, one might assume that certain non-CB_1 receptor-mediated effects in neurons are due to CB_2 receptor activation. Currently, the neuronal presence of functional CB_2 receptors is a subject of hot debates (see Chap. 10). In contrast, using different antibodies and comparing Western-blotting results from wild-type and CB_2 receptor knockout mouse spleen, liver and brain, we concluded that CB_2 receptors are undetectable in whole brain membrane preparation of mice (Köfalvi et al., 2006b). Furthermore, the inhibitory effect of the mixed CB_1/CB_2 receptor agonist WIN55212-2 (100 nM–1 µM) on K^+-evoked Ca^{2+} entry and transmitter releases in rat hippocampi is abolished by the CB_1 receptor-selective antagonist AM251 (500 nM), whereas the CB_2 receptor-selective agonist JWH133 and antagonist AM630 are devoid of effects in these assays (Köfalvi et al., 2007). Altogether, further studies are required to reveal the impact of neuronal CB_2 receptors in the brain. Hereinafter, I provide the reader with a brief outline of the pharmacology and signaling properties of the CB_1 receptor homodimers.

a. Agonists display different efficacy and potency at homomeric CB_1 receptors. This topic is thoroughly reviewed in Chap. 7. Depending on the assay, the rank order of potency for commonly used agonists ("averaged" as it is referred to in several studies) is usually: HU-210 > CP55940 ≈ 2-arachidonoylglycerol (2-AG) > Δ^9-THC ≥ levonantradol ≈ WIN55212-2 > anandamide; whereas the rank order of efficacy is WIN55212-2 > levonantradol > HU-210 ≈ CP55940 ≈ 2-AG ≥ Δ^9-THC ≥ anandamide. Hence anandamide is questioned to be a significant CB_1 receptor agonist (Sugiura et al., 1999). Most notably, in its effective concentration range, anandamide has numerous other targets (see below), which complicates the evaluation of its CB_1 receptor-mediated effects. Interestingly, synthetic agonists can antagonize the effect of the endogenous agonists at the CB_1 receptor via desensitization (Sugiura et al., 1999). Therefore, it should not be surprising if the effects of CB_1 receptor antagonists and exogenous agonist appear to be similar.
b. Homomeric CB_1 receptors can couple to different effector systems, depending on the length of drug exposure and brain area. It was shown that the number of G proteins and the $G_{i/o\alpha}$ subtypes, coupled to the CB_1 receptor, can vary between brain areas. Furthermore, the EC_{50} of WIN55212-2 to activate different subtypes may vary in a 30-fold range in the same brain area (Prather et al., 2000). In other words, if one investigates a particular change in the biological system that is weakly coupled to the CB_1 receptor (i.e., to observe any change requires high concentrations of the agonist), then it is easy to antagonize it with a potent antagonist. Here again, if there is another effect measured, which couples to the CB_1 receptor with high efficacy, it is perhaps not blocked by the same concentration of the antagonist. Therefore, the latter may appear as a CB_1 receptor-independent effect. Although CB_1 receptors are generally viewed as to inhibit adenylyl cyclase activity with the consequent decrease in cAMP levels, it does not prove to be a general rule. When acutely activated, CB_1 receptors inhibit adenylyl cyclase

types I, V, VI and VIII, but activate types II, IV and VII, respectively (Rhee et al., 1998). In contrast, chronic CB_1 receptor stimulation superactivates adenylyl cyclase types I, III, V, VI and VIII. Activation of these adenylyl cyclase types might be due to alternative CB_1 receptor coupling to $G_{s\alpha}$ (Rhee et al., 2000). Last but not least, different agonists (e.g., WIN55212-2, anandamide, HU-210 or Δ^9-THC) can induce different conformational changes in the CB_1 receptor, which in turn will recognize different G proteins (Glass and Northup, 1999). This can produce different predominant effects for each ligand in the same system, falsely suggesting the involvement of other receptors. For instance, it was demonstrated recently that WIN55212-2 (but not R-methanandamide, 2-AG, HU-210, Δ^9-THC, cannabidiol or CP55940) can stimulate CB_1 receptor coupling to $G_{q/11}$ G proteins and, consequently, induce Ca^{2+} efflux from intracellular stores in hippocampal culture and in transfected HEK293 cells (Lauckner et al., 2005). In other words, CB_1 receptor-mediated effects can greatly differ, depending on the type of agonists used, the length of the stimulation and the cell type or tissue, respectively.

c. To date, two alternative splice variants of the human CB_1 receptor have been reported. The CB_{1A} receptor is 61 amino acid-shorter than the CB_1 receptor, and it also differs in its first 28 amino acids, which is more hydrophobic for the CB_{1A} receptor (Shire et al., 1995). CB_{1A} receptor mRNA is widely present in all tissues investigated and in the brain. Its level is 4–200 times lower than that of the CB_1 receptor, depending on age and brain area. It is important to note here that we have recently observed lower mRNA levels for the CB_1 receptor with an increase in the maximum binding sites and in the membrane-bound CB_1 receptor protein density in type-1 diabetic hippocampus (see Chap. 14). These findings indicate that a low mRNA level may be a consequence of an accelerated translation; therefore, a lower mRNA level does not mean that the protein density will be also low (Duarte et al., 2007). Recently, a novel splice variant, termed as hCB_{1B} receptor, has been described (Ryberg et al., 2005). This novel splice variant is between the two other forms, as it lacks only 33 amino acids from the hCB_1 receptor. The binding constant and the potency and efficacy of HU-210, WIN55212-2, CP55940 and Δ^9-THC are only slightly different for the three splice variants, expressed in HEK-239 cells. However, these values are doubled for 2-AG at the CB_{1A} receptor, while they are tripled at the CB_{1B} receptor. Nevertheless, the real dramatic change is observed in case of anandamide, virodhamine and noladin ether, which all lose their ability to bind to or activate the CB_{1A} and CB_{1B} receptors (Ryberg et al., 2005). From these results, it seems obvious that the tissue distribution and density of these CB_1 receptor splice variants should be determined with novel antibodies which are able to distinguish the two alternative forms from the full-length CB_1 receptor.

Alternative Signaling at CB_1 Receptor Heterodimers

It is a widely accepted fact that metabotropic receptors can form homo- and heterodimers via physical interaction. When forming heterodimers, the new receptor dimer entity often displays a different pharmacological profile and is coupled to alternative signaling cascades, compared with those of the single component receptors (Mackie, 2005). Dimerization may occur intracellularly and then the dimer moves to the plasma membrane; but most often, this phenomenon takes place in lipid raft microdomains of the plasma membrane upon concurrent stimulation by submaximal concentrations of agonists of the component receptors. Therefore, forming a heterodimer is quite likely a part of a dynamically changing state of receptors in the plasma membrane, besides being active alone or in a homomer or being desensitized.

a. The most-studied heterodimer of CB_1 receptors is the one with the D_2 dopamine receptor (Glass and Felder, 1997; Jarrahian et al., 2004; Kearn et al., 2004). Acute activation of this dimer results in a $G_{s\alpha}$-mediated increase in cAMP level and MAPK activation, but chronic stimulation (18 h) of the dimer may switch back to $G_{i/o\alpha}$-mediated signaling – opposed to what happens when CB_1 receptor is chronically stimulated alone (see earlier). This D_2/CB_1 heterodimer may be an important regulator of basal ganglia function and may serve as an attractive therapeutic target in neuropsychiatric disorders in which dopamine and/or cannabinoid signaling is impaired.
b. Very recently, CB_1 receptors have been recognized as true heterodimer partners with the A_{2A} receptor in co-transfected HEK-293 cells and in the rat striatum. Activation of A_{2A} receptors induces cAMP stimulation, which can be counteracted by CB_1 receptor blockade. Furthermore, A_{2A} receptor blockade counteracts motor depressant effects of the intrastriatally administered CB_1 receptor agonist WIN55212-2 (Carriba et al., 2007). This demonstrates that at least in the rat striatum, CB_1 receptor function is highly dependent on A_{2A} receptors.
c. Large body of evidence supports the interaction between the opioid and the endocannabinoid system at several levels (Vigano et al., 2005). In pharmacological assays carried out in rat cortical membranes, Δ^9-THC and cannabidiol accelerated the dissociation of [^{3}H]DAMGO (a μ-opioid receptor ligand) and [^{3}H]naltrindole (a μ-opioid receptor ligand). In addition, Δ^9-THC, cannabidiol and SR141716A all displaced [^{3}H]DAMGO in the pseudo-competition assay, and all of them altered the equilibrium binding of the μ-opioid agonist as well (Kathmann et al., 2006). In the nucleus accumbens, activation of μ-opioid and CB_1 receptors both inhibit the evoked release of GABA and glutamate in a naloxone- and SR141716A-sensitive manner. Apparently, the inhibitory action of the two receptors was synergistic on the release of GABA, but was non-additive on the release of glutamate. Moreover, antagonism by SR141716A was prevented by naloxone, and vice versa, the antagonism of naloxone was prevented by SR141716A (Schoffelmeer et al., 2006). Consequently, it is not surprising that μ-opioid receptors have been reported to form true heterodimers with the CB_1 receptors both in expression systems and in endogenous tissue. Activation of one of the two receptors reciprocally diminishes signaling at the other receptor, data obtained from MAPK, GTPγS binding and Src-Stat3 assays (Rios et al., 2006). Altogether, heterodimers of CB_1 receptors with opioid receptors provide new and

excellent therapeutic targets in pain, addiction, and eventually, in the complications associated with the development of the CNS (Harkány et al., 2007).

d. So far, only pharmacological assays have demonstrated the existence of a possible 5-HT$_2$/CB$_1$ receptor heterodimer. In rat cerebral cortex membranes, oleamide (which displaced [^{3}H]CP55940-binding) and HU-210 both increased the affinity of serotonin for 5-HT$_2$ receptor recognition sites, and potentiated back muscle contraction, induced by the 5-HT$_2$ receptor agonist DOI, in vivo (Cheer et al., 1999). Contrariwise, in rat cerebellar membranes, serotonin has been observed to increase the binding affinity of WIN55212-2, whereas significantly reduced the proportion of high-affinity binding of WIN55212-2 and HU-210. These reported effects of serotonin were prevented by the 5-HT$_2$ antagonist Ritanserin (Devlin and Christopoulos, 2002). Altogether, these data demonstrate that CB$_1$ receptors may be an alternative therapeutic target in neuropsychiatric disorders involving 5-HT$_2$ receptors.

e. CB$_1$ receptors have been shown to form heterocomplexes with some types of receptor tyrosine-kinases (Harkány et al., 2007). CB$_1$ receptor activation triggers the migration of progenitor neurons, whereas attenuates neurotrophin-induced neuronal differentiation and neurite outgrowth (see Chap. 12). The underlying mechanism is thought to be a transactivation of either the brain-derived neurotrophic factor (BDNF) TrkB receptor (Berghuis et al., 2005) or of the fibroblast growth factor (FGF) receptor in the growth cone of developing axons (Williams et al., 2003). In contrast, CB$_1$ receptor activation promotes the migration, transformation and proliferation of cancer cells via transactivation of the epidermal growth factor (EGF) receptor (Hart et al., 2004; Zhao et al., 2005). Apart from this, our laboratory has recently reported that BDNF acutely potentiates the release of glutamate via activation of presynaptic TrkB receptors in rat hippocampus (Pereira et al., 2006). Therefore, CB$_1$ receptor heterodimers with receptor tyrosine kinases may also have an acute neuromodulatory impact with pharmacological profiles different from that of the CB$_1$ receptor homomer.

Other Non-Ionotropic Receptor-Mediated Actions

Abnormal-Cannabidiol (ABN-CBD Receptor)

The endothelium of the rat mesenteric artery is endowed with a novel G protein-coupled (pertussis toxin-sensitive) cannabinoid receptor of unknown molecular identity. Activation of ABN-CBD receptors potentiates microglial cell migration and activates BK$_{Ca}$ currents (see later) enhancing vasorelaxation, respectively (Pertwee, 2005). Furthermore, this receptor may be similar to the one which mediates lypopolysaccharide-induced hypotension and increase in cardiac contractility (Bátkai et al., 2004). The ABN-CBD receptor is the only known target activated by the compound abnormal cannabidiol (ABN-CBD, EC$_{50}$, ~1 μM). ABN-CBD receptors can also be activated by endogenous agonists, anandamide, virodhamine, and presumably by noladin ether; and by the synthetic *R*-methanandamide and the selective synthetic O-1602, but not by WIN55212-2, Δ^9-THC or 2-AG. SR141716A, cannabidiol and the selective O-1918, but not SR144528 or AM630 (CB_2 receptor antagonists) or AM251, are antagonists for the ABN-CBD receptor (Begg et al., 2003, 2005; Ho and Hiley, 2003, 2004; Offertaler et al., 2003; Pertwee, 2004, 2005). A recent study has proposed that a novel hypothetic receptor might partly contribute to effects that were previously discussed as ABN-CBD receptor-mediated (Hoi and Hiley, 2006), and another investigation has revealed a putative cross-talk between ABN-CBD and the CB_1 receptors (Su and Vo, 2007).

GPR55

This orphan receptor has recently been recognized as a novel metabotropic cannabinoid receptor, discovered by in silico patent research. It displays higher sequence similarity to the platelet activating factor receptor and to $P2Y_9$ and $P2Y_5$ receptors than to CB_1 and CB_2 receptors. Among several endogenous substances, palmitoylethanolamide stimulates GTPγS incorporation with the lowest EC_{50} value, but anandamide, 2-AG, virodhamine, Δ^9-THC and CP55940 all display EC_{50} values less than 20 nM. In addition, several CB_2 receptor-selective ligands, but not WIN55212-2, were identified as agonist at the GPR55. Activation of the GPR55 has been observed to induce a slowly developing intracellular Ca^{2+} rise, but likely independently from G_i and G_s proteins. AM251 and SR141716A are presumable antagonists at the receptor (Brown and Wise, 2001; Baker et al., 2006; Mackie and Stella, 2006). Although various studies with the use of antibodies and PCR techniques have reported the presence of the receptor in the brain, data are often controversial. Until comparison of data in the GPR55 knockout mouse becomes available – which will hopefully appear from 2007 – every conclusion about the role of GPR55 in the brain is premature (see also Chap. 10).

Presynaptic Imidazoline Receptors

A certain unique class of imidazoline receptors, differing from the imidazoline 1 and 2 receptors, inhibits noradrenaline release from cardiovascular sympathetic nerve endings. The imidazoline BDF6143- and aganodine-mediated inhibition of noradrenaline release was counteracted by high concentration of rauwolscine and the CB_1 receptor antagonists SR141716A and LY320135. CP55940 and anandamide also inhibited the release of noradrenalin in a rauwolscine- and SR141716A-sensitive fashion. Additionally, these cannabinoid and imidazoline ligands displaced the radiolabeled guanidine derivative [^{3}H]DTG (Gothert et al., 1999; Molderings et al., 1999). In PC12 cell line, the inhibition of veratridine-evoked noradrenaline release by cirazoline, clonidine, aganodine, agmatine and BDF6143 was antagonized by WIN55212-2. In additional experiments, the inhibitory action of clonidine was prevented by rauwolscine and SR141716A as well. Further experiments have suggested that the underlying receptor might be an edg-like lysophospholipid receptor (Molderings et al., 2002). It is also of interest that the hypothermic effect of WIN55212-2 is synergistically augmented by agmatine in rats, whereas agmatine itself was devoid of hypothermic effects (Rawls et al., 2006). Noteworthy, CB_1 receptors show evolutional relationship with edg receptors; therefore, further studies are invited to reveal direct interactions between cannabinoid ligands and edg receptors. Alternatively, it may be possible that a CB_1 receptor/edg-like receptor heterodimer is responsible for the underlying mechanisms. It is also of note that this imidazoline-like receptor does not strikingly differ from the ABN-CBD receptor in its pharmacological profile.

Adenosine and its A_1 Receptor

One of the major presynaptic inhibitory neuromodulator receptors is the A_1 receptor (Cunha, 2001). It has been shown recently that SR141716A and AM251 may block A_1 receptors in the micromolar range (Savinainen et al., 2003). Since several electrophysiological studies have applied these CB_1 receptor antagonists in the micromolar range reporting CB_1 receptor-independent inhibitory actions to synaptic transmission and G protein activation, further studies are required to determine whether those findings were A_1 receptor-mediated or not. Moreover, novel data suggest that adenosine and its analogues (agonists and antagonists) interact with $TRPV_1$ receptors. The capsaicin-evoked Ca^{2+} entry in HEK293/$TRPV_1$ cells and $TRPV_1$ receptor-mediated currents in the dorsal root ganglion are prevented by CGS21680, ZM241385, adenosine and R-phenylisopropyladenosine (R-PIA) with low nanomolar potency. Furthermore, CGS21680, ZM241385, R-PIA, and DPCPX are all able to displace [^{3}H]resiniferatoxin binding in HEK-293/$TRPV_1$ cells, whereas [^{3}H]CGS21680 labels $TRPV_1$ receptor-expressing oocytes. Finally, R-PIA was shown to prevent capsaicin-induced cell death (Puntambekar et al., 2004).

Muscarinic M_1 and M_4 Receptors

It is of particular interest that anandamide and *R*-methanandamide at low micromolar concentrations, but not WIN55212-2 or SR141716A, have been shown to reduce radioligand binding to human M_1 and M_4 receptors, as well as the apparent affinity and the maximal density of binding sites (Lagalwar et al., 1999; Christopoulos and Wilson, 2001). Retrograde endocannabinoid transmission can be elicited by postsynaptic muscarinic receptor activation (Ohno-Shosaku et al., 2003). Consequently, anandamide may exert a feed-back inhibition on its postsynaptic release. Further investigations are needed to prove this hypothesis.

Peroxisome Proliferator-Activated Receptors Alpha and Gamma

Peroxisome proliferator-activated receptors (PPARs) belong to the nuclear receptor family, and have a broad role in energy homeostasis and metabolism. Expression of the subtypes PPARα, PPARδ and PPARγ is tissue specific, and when activated by one of the several natural ligands, they form heterodimer with the retinoic X receptor, promoting transcription of genes (Burstein, 2005). As PPAR-mediated effects involve a critical step, namely, de novo protein expression, measurable changes in a biological system are expected in a minute/hour scale.

a. PPARα: N-oleoylethanolamide has been shown to activate PPARα whereby regulating feeding and body weight (Fu et al., 2003). 2-AG is metabolized by 15-LOX into 15-hydroxyeicosatetraenoic acid glyceryl ester, which in turn can activate the PPARα receptor (Kozak et al., 2002). Palmitoylethanolamide, a weak CB_1 receptor agonist, has been shown to exert its anti-inflammatory effect at least partly via PPARα receptor activation (Lo Verme et al., 2005).
b. PPARγ: It is also called the "ajulemic acid receptor", because ajulemic acid, an analogue of a Δ^9-THC metabolite, can also activate it whereby inducing anti-inflammatory responses (Liu et al., 2003; Ambrosio et al., 2007). PPARγ seems to be activated by 2-AG as well, promoting the differentiation of murine fibroblasts into adipocytes and suppressing the release of the T cell growth factor interleukin 2 (Rockwell et al., 2006). Another study has reported findings that may be consistent with the agonist action of the novel endocannabinoid *N*-arachidonoyl glycine at the PPARγ (Burstein, 2005). Finally, Δ^9-THC at 10 μM has been shown to relax isolated rat arteries via PPARγ activation, followed by protein synthesis, NO and H_2O_2 deliberation (O'Sullivan et al., 2005). In conclusion, activation of PPARs by endogenous and exogenous ligands may provoke long-term changes in the investigated biological system contributing to the diverse effects of cannabinoids. Cannabinoid activation of PPARs has been suggested to be a novel therapeutic target against several cardiometabolic risk factors (Burstein, 2005).

Cannabinoid and Vanilloid Ligand-Sensing Ion Channels I: Channel Inhibition and Blockade

It is now generally accepted that cannabinoid molecules have a new role, namely, ion channel inhibition and blockade (van der Stelt and Di Marzo, 2005; Oz, 2006). The heterologous channel blocker property of cannabinoid ligands has already been employed by the antiemetic medicine Nabilone™ (Canada, US, UK: Cesamet™) (see later). Here I also try to give further indications as to how ligands can serve as pharmaceutical targets.

Serotonin 5-HT$_{3A}$ Receptors

The antiemetic medicine Nabilone™ exerts its beneficial effects mainly via the blockade of vagal 5-HT$_{3A}$ receptors, apart from acting at central antiemetic areas (consult with Chap. 13 as well). Nabilone™ is a Δ^9-THC analogue, which was developed as a molecule having less psychotropic side effects, still keeping the antiemetic properties of Δ^9-THC. Other cannabinoid molecules, which have been tested and which stereoselectively reduced 5-HT$_{3A}$ receptor-mediated currents in the nanomolar range, are Δ^9-THC, WIN55212-2, anandamide, JWH015 and CP55940 (CB$_1$ and CB$_2$ receptor agonists), CP56667 (non-psychoactive enantiomer) and LY320135 (CB$_1$ receptor antagonist) (Fan, 1995; Barann et al., 2002; Oz et al., 2002, 2004b; Godlewski et al., 2003). Given the high extent of co-localization of 5-HT$_3$ and CB$_1$ receptors in hippocampal and dentate gyrus interneurons, it may be feasible that the two receptors act to some extent as a molecular relay in the presence of anandamide (Morales and Backman, 2002), in other words, nanomolar concentrations of anandamide inhibit 5-HT$_3$ and activate CB$_1$ receptors in the same neurons at the same instant.

α7 Nicotinic Acetylcholine Receptors

Nicotinic and 5-HT$_3$ receptors are phylogenetically closely related, since both are members of the Cys-Cys loop ligand-gated ion channel superfamily (Maricq et al., 1991). Furthermore, a lot of behavioural effects of nicotine are mediated through an interaction with the brain serotonergic system (Seth et al., 2002), and serotonin can also reduce ACh-induced currents in α9 nACh receptor (Rothlin et al., 1999). Therefore, it is not surprising that not only 5-HT$_3$, but also nACh receptors can be non-competitively inhibited by cannabinoid ligands: anandamide and its metabolically stable analogue *R*-methanandamide as well as 2-AG, but not WIN55212-2, CP55940 or Δ^9-THC, inhibited currents evoked at the α7 nicotinic acetylcholine (α7 nACh) and at an α7 nACh/5-HT$_3$ chimera receptor in the nanomolar/low micromolar range (Oz et al., 2003, 2004b). Presumably, similar findings could be observed with other nACh receptors, it is simply a question of trial. For instance, Liu and Simon (1997) have reported that low micromolar capsazepine strongly inhibited nicotine- (100 μM) evoked currents in rat trigeminal culture, whereas hexametonium failed to affect capsaicin-evoked currents. Another similarity between the α7 nACh receptor and the TRPV$_1$ receptor is that the effect of anandamide on them is potentiated by ethanol (Trevisani et al., 2002; Oz et al., 2005). This may further suggest that a prototypic vanilloid receptor was the ancestor of several ligand-gated ion channels. All in all, the non-competitive blocking effect of endocannabinoids at physiological concentrations on nicotinic and serotonin receptors may gain importance in certain pathomechanisms of depression and schizophrenia (see Chaps. 22 and 23).

Glycine Receptors

The glycine receptor is also a Cys-Cys loop ligand-gated ion channel, and has fragments in its amino acid sequence that display high level of homology with the binding site of CB_1 and CB_2 receptors (Lozovaya et al., 2005). The authors found that in isolated hippocampal pyramidal and Purkinje cerebellar neurons, anandamide and 2-AG applied at physiological concentrations inhibited the glycine-activated current's amplitude and altered the kinetics of rise time, desensitization and deactivation. Glycine was used at 100 μM, which was close to its measured EC_{50} value (91 μM) in this system. WIN55212-2 only accelerated the rise and the desensitization of the current at 1 μM; and at 5 μM, slightly inhibited the glycine-activated current (Lozovaya et al., 2005). In contrast, another study of the same year reported the virtually opposite action for cannabinoids. In acutely isolated neurons from rat ventral tegmental area and in Xenopus oocytes expressing human homomeric (α1) and heteromeric (α1β1) subunits of glycine receptors, Δ^9-THC and anandamide potentiated glycine- (less than 30 μM) activated currents. Currents activated by glycine at 30 μM were already unaffected by cannabinoids (Hejazi et al., 2005). In conclusion, cannabinoid agonists inhibit presynaptic and potentiate post- and extrasynaptic glycine receptors, depending on the concentration of glycine around the receptor.

Calcium Channels

Plasma membrane Ca^{2+} channels play a basic role in the physiology and pathology of neurons and glia. CB_1 receptors can negatively couple to the major types of high voltage-gated Ca^{2+} channels (VGCCs) whereby modulating neurotransmission (Mackie et al., 1995; Shen and Thayer, 1998; Brown et al., 2004). Nevertheless, endocannabinoids (even at physiological concentration) and synthetic cannabinoid ligands (usually above 1 μM) are capable to inhibit Ca^{2+} influx into cells via direct channel blockade. Three major classes of VGCCs are distinguished, namely the high-voltage-activated L-type (Ca_v1) channels, the N-, P/Q- and R-type channels (Ca_v2) and the low-voltage-activated T-type (Ca_v3) channels (Ertel et al., 2000). T-type Ca^{2+} channels contribute to pacemaker activity and the pathomechanism of epilepsy. They are inhibited by nanomolar concentrations of anandamide, meth-anandamide and SR141716A, and by micromolars of HU-210 (Chemin et al., 2001). At least for anandamide, the binding site must be intracellular (similarly to the vanilloid receptor–anandamide interaction), since blockade of the anandamide transporter prevents anandamide blockade of T-type Ca^{2+} channels. In contrast, they are insensitive to WIN55212-2, CP55940 or Δ^9-THC (Chemin et al., 2001). L-type Ca^{2+} channels have been shown to be directly inhibited by anandamide and 2-AG above 1 μM (Johnson et al., 1993; Oz et al., 2000, 2004a). The latter study has also revealed that these L-type Ca^{2+} channels are not sensitive to CP55940, WIN55212-2 and Δ^9-THC. In the rat mesenteric artery, ABN-CBD is also capable to inhibit

L-type Ca^{2+} channels above 3 μM (Ho and Hiley, 2003). In cultured hippocampal neurons, nanomolar WIN55212-2, but not WIN55212-3, the CB_1 receptor-inactive enantiomer, inhibited the N- and P/Q-type Ca^{2+} channels via CB_1 receptor activation. Above 1 μM, however, both WIN55212-2 and WIN55212-3 directly inhibited the N- and P/Q-type Ca^{2+} channels (Shen and Thayer, 1998). Our extended neurochemical investigations have revealed that the majority of cannabinoid ligands are capable to inhibit Ca^{2+} entry and Ca^{2+}-dependent transmitter release in nerve terminals of the hippocampus and striatum of rats. When using low-strength K^+ stimulation (20 mM for 30 s), which allows detecting G protein-coupled receptor-mediated fine modulation of Ca^{2+} entry and transmitter release, WIN55212-2 (EC_{50}, ~60 nM; E_{max}, ~30%) inhibits the release of GABA and glutamate in the hippocampus via activation of presynaptic CB_1 receptors. Above 1 μM for GABA and 3 μM for glutamate, WIN55212-2 produces another phase of inhibition (E_{max}, ~60%) via direct Ca^{2+} channel blockade. Notably, the CB_1 receptor antagonist AM251 (1 μM or greater) also causes similar inhibition on low-strength K^+ stimulation (Köfalvi et al., 2007). When using high-strength K^+-stimulation (e.g. 25–30 mM K^+ for 2–3 min), the potency of CB_1 receptor agonists to inhibit Ca^{2+}-dependent transmitter release shifts to the right (to the micromolar range), where already their direct Ca^{2+} channel blocker effect dominates. Nonetheless, CB_1 receptors still function, but inhibition by CB_1 receptor agonists cannot be prevented by CB_1 receptor antagonists if Ca^{2+} channels are already directly blocked. Therefore, data can be easily misinterpreted as non-CB_1 receptor-mediated, "putative CB_3 receptor-mediated" inhibition. For example, we found that both in the hippocampus and the striatum, CP55940, WIN55212-2, Δ^9-THC, AM251 and SR141716A all inhibited Ca^{2+} entry and Ca^{2+}-dependent release of glutamate with EC_{50} values of 1–4 μM, and efficacies ranging from 40 to 70% (Köfalvi et al., 2003, 2005, 2006a,b). These data are in agreement with the findings of White and Hiley (1998) that low micromolar SR141716A robustly inhibits VGCCs in the mesenteric artery. In our studies, we also found that the $TRPV_1$ receptor agonist, capsaicin, and antagonists, capsazepine and iodoresiniferatoxin, inhibited Ca^{2+} entry and Ca^{2+}-dependent release of GABA and glutamate. Furthermore, the two structurally and pharmacologically different ligands, namely AM251 and iodoresiniferatoxin competitively antagonized each other's inhibition on the stimulated Ca^{2+} entry without the maximal efficacy being affected (Köfalvi et al., 2006a,b). This further suggests that these two ligands acted on the same site, i.e. on N- and P/Q-type Ca^{2+} channels. In other experimental protocols, micromolar capsaicin has been shown to presynapticaly diminish GABAergic IPSCs in the hippocampus (Drebot et al., 2006), whereas micromolar capsazepine inhibited VGCCs in sensory neurons (Docherty et al., 1997). The former study further supports the notion that at least in the hippocampus, it is most unlikely that functional presynaptic $TRPV_1$ receptors control transmitter release (Köfalvi et al., 2006a,b, 2007). The direct inhibitory action of $TRPV_1$ receptor ligands on VGCCs is not unexpected, since ruthenium red, another antagonist of the $TRPV_1$ receptor, is generally known as a non-selective pore-blocker, i.e., VGCCs antagonist as well (Tapia and Velasco, 1997). All things considered, the non-specific inhibitory action of cannabinoid and vanilloid ligands deserves more attention. The threshold concentration for these substances should be set as 1 μM, because if these low nanomolar affinity ligands are unable to elicit the desired effect up to 1 μM then the receptor in question does not function there. Still, some Ca^{2+} blocker effects develop at nanomolar concentrations of these ligands, thus it is wise to carefully interpret neuroprotective and excitability-depressing effects of cannabinoid and vanilloid substances if they turn to be CB_1/CB_2 receptor independent. Additionally, cannabinoid and vanilloid ligands may serve templates for novel selective Ca^{2+} channel inhibitor medicines.

Na^+ Channels

Opening of voltage-gated sodium channels is the underlying mechanism for axonal depolarization and neuronal firing. Therefore, Na^+ channel blockade (e.g. with tetrodotoxin) abolishes action potential-induced presynaptic Ca^{2+} entry and transmitter release. Interestingly, all cannabinoid and vanilloid ligands tested so far – namely anandamide, palmitoylethanolamide, WIN55212-2, AM251, the anandamide uptake inhibitor VDM11, and the hybrid anandamide uptake inhibitor/$TRPV_1$ receptor agonist (and COX-1/COX-2 inhibitor) AM404 – profoundly block tetrodotoxin-sensitive Na^+ channels in the low micromolar range. These ligands inhibit (1) veratridine-evoked release of GABA and glutamate, (2) binding of [^{3}H]batrachotoxinin-A-20-a-benzoate to site 2 on Na^+ channels, (3) tetrodotoxin-sensitive Na^+ currents in dorsal root ganglion neurons, (4) the network-driven, glutamate- (but not K^+-) induced intracellular Ca^{2+} rise and (5) tetrodotoxin-sensitive sustained repetitive firing in cortical neurones without altering primary spikes, consistent with a state-dependent mechanism (Nicholson et al., 2003; Kelley and Thayer, 2004; Liao et al., 2004; Kim et al., 2005). Another study demonstrated that low micromolar capsaicin and capsazepine decrease membrane bilayer stiffness, which inhibits currents through voltage-gated Na^+ channels (Lundbaek et al., 2005). These studies should prompt careful evaluation of electrophysiology data, because cannabinoid ligands are often used in the concentration range of 1–10 μM to facilitate their wash-in into the slices. Choosing the right concentration can help avoiding the implication of a non-CB_1 receptor-mediated response, when for instance, WIN55212-2 and AM251 (both at 10 μM) block action potential generation and their effects are additive (Matyas et al., 2006). Since acetaminophen (paracetamol) breaks down into AM404 in the nervous system (Högestätt et al., 2005) its analgesic activity might be partly related to Na^+ channel blockade. Cannabinoid and vanilloid ligands, therefore, may serve templates for novel anesthetic/analgesic medicines.

K^+ Channels

K^+ channels play a major role in the plasma membrane excitability. Inwardly rectifying and voltage-sensitive rapidly inactivating A-type (K_{ir} and K_v)

channels are indirectly activated by intracellular messengers upon CB_1 receptor activation (Deadwyler et al., 1995; Mackie et al., 1995). Consequently, the activation of K_v channels largely contributes to the depression of synaptic transmission in the CNS. Although it was believed that K^+ channel activation occurs via the activation of the cAMP-PKA pathway it must be noted that recent findings have shown that activation of tetraethylammonium- and 4-aminopyridine-sensitive K^+ channels upon CB_1 receptor stimulation perhaps does not involve cAMP signaling (del Carmen Godino et al., 2005). In contrast to all these, direct K^+ channel blockade is expected to depolarize the membranes. Anandamide and Δ^9-THC have been shown to directly inhibit the Shaker family $K_v1.2$ channels with IC_{50} values of 2–3 μM (Poling et al., 1996). Another astonishing study has demonstrated that arachidonic acid and anandamide are capable to render non-inactivating delayed rectifier K_v channels rapidly inactivating A-type K_v channels via immobilizing the inactivation domains (Oliver et al., 2004). Accordingly, another study has found that anandamide (IC_{50}, 600 nM) and *R*-methanandamide as well as WIN55212-2 in the low micromolar range block delayed rectifier K_v channels in rat aorta myocytes (Van den Bossche and Vanheel, 2000). The BK_{Ca} subtype of voltage- and Ca^{2+}-sensitive Ca^{2+}-activated K^+ channels are widely expressed in the smooth muscle and the nervous tissue, and are implicated for instance in neuroprotection against excessive depolarization and Ca^{2+} levels (Lawson, 2000). BK_{Ca} channels also account in part for the repolarizatory phase of the action potential. SR141716A at 10 μM has been shown to directly block BK_{Ca} channels in the rat mesenteric artery (White and Hiley, 1998). TASK channels are members of the two-pore domain K^+ channel subfamily. They are sensitive to protons, hypoxia and volatile anesthetics, but are voltage-insensitive, and are responsible for setting the resting membrane potential and input-resistance (Lesage and Lazdunski, 2000). High nanomolar/low micromolar anandamide (IC_{50}, 700 nM), *R*-methanandamide, WIN55212-2 and CP55940 (in this rank order of efficacy) but not 2-AG, Δ^9-THC or HU-210 block the TASK-1 channel, whereas blockade of the TASK-3 develops only by anandamide and from 10 μM (Maingret et al., 2001). This is paralleled by the observation that analgesic, sedative and hypothermic effects of WIN55212-2 are reduced in the TASK-1 knockout mice (Linden et al., 2006). In contrast, Aller and colleagues (2005) reported an unpublished observation that anandamide and WIN55212-2 had not distinguished between TASK-1 and TASK-3. Recently, we observed that both anandamide and *N*-arachidonoyl dopamine, but not WIN55212-2 at low micromolar levels, are capable to depolarize hippocampal nerve terminals and, consequently, release GABA and glutamate in a fashion similar to the action of Ruthenium Red, Zn^{2+} and protons. Since the latter three are selective inhibitors of TASK-3 over TASK-1, we concluded that this effect of NADA and anandamide are possibly mediated by blockade of TASK-3 (Köfalvi et al., 2007). All in all, cannabinoid ligands can directly interfere with depolarization and repolarization of neurons, and can concomitantly influence synaptic transmission and exocytotoxicity.

Hyperpolarization-Activated Cyclic Nucleotide-Gated Channel Type 1

Hyperpolarization-activated cyclic nucleotide-gated channel type 1 (HCN1) is a widely expressed cation channel in several tissues. In neurons, HCN1 controls membrane excitability, pacemaker activity and signaling, and is presumably involved in neuropathic pain (Robinson and Siegelbaum, 2003). HCN1 shares structural similarity with the $TRPV_1$ receptor; and capsazepine has been shown to block the human HCN1 with the IC_{50} of 8 μM, in a reversible voltage- and use-independent fashion (Gill et al., 2004).

Other Assumed Targets

In this paragraph, I mention that two other channel types have been demonstrated to be inhibited by cannabinoids. In contrast to the general belief, these seem to be indirect blockades.

a. Venance and colleagues (1995) reported that anandamide blocks gap junction communication in astrocytes, virtually independently from activation of known cell surface and intracellular targets. Although it is now generally accepted that cannabinoids inhibit gap junctions in various tissues, recent investigations have demonstrated that the inhibitory effect of cannabinoids can be prevented by inhibition of ERK1/2 activation (Brandes et al., 2002; Upham et al., 2003). Therefore, further studies are welcome to explore if every cannabinoid ligand-mediated inhibition of gap junctions occurs indirectly.
b. It is also believed that anandamide directly inhibits kainate-induced currents at the different homo- and heteromers of the $GluR_{1,2,3}$ subunits, based on the report of Akinshola and colleagues (1999). However, in this study, anandamide inhibited currents only in a very high (100–200 μM) concentration range, and its effect was dependent on cAMP. Still, direct blockade at other subunit compositions or by different cannabinoid ligands does not seem to be impossible, regarding the fact that cannabinoids interact with the NMDA channel as well (see later).

Cannabinoid and Vanilloid Ligand-Sensing Ion Channels II: Channel Activation and Potentiation

Anandamide and *N*-arachidonoyl dopamine (but not WIN55212-2, HU-210, CP55940, 2-AG or Δ^9-THC) are potent endogenous activators of the Na^+/Ca^{2+} channel transient release potential family "Vanilloid-type 1" ($TRPV_1$) receptor. Furthermore, as discussed earlier, Δ^9-THC and anandamide can potentiate glycine-activated currents at the glycine receptor. Here I conclude that cannabinoid and vanilloid ligands can activate other currents as well.

Transient Release Potential Family "Vanilloid-Type 4" ($TRPV_4$) Receptor

The $TRPV_4$ receptor shares 45% sequence homology with its most studied relative $TRPV_1$ ("capsaicin") receptor, and is activated by moderate heat (>24 °C), hypotonic cell swelling, mechanical stress, certain endogenous substances such as α-phorbol esters and endogenous substances, for instance the P450-epoxygenase products epoxyeicosatrienoic acids, and finally by the FAAH substrates *N*-acyl taurines and anandamide, as well as by the other endocannabinoid 2-AG (Nilius et al., 2004; Pedersen et al., 2005; Saghatelian et al., 2006). Further studies are required to determine the physiological and pathological roles of endocannabinoid activation of the $TRPV_4$ receptor.

Transient Release Potential Family "Ankyrin-Type 1" Receptor

The transient release potential family "ankyrin-type 1" ($TRPA_1$) (formerly $ANKTM_1$) receptor is a noxious cold-sensitive Ca^{2+}/Na^+ channel, a distant relative of the TRP superfamily, and is widely expressed in $TRPV_1$ receptor-positive sensory nerves and in the mechanosensory epithelia of inner ear (Pedersen et al., 2005; Garcia-Anoveros and Nagata, 2007). It can be activated by noxious cold, isothiocyanates (mustard oil, wasabi, horse radish), garlic (allicin), cinnamon and bradykinin, but not with capsaicin or menthol. Cannabinoid ligands in the low micromolar range activate the receptor in the following rank order of efficacy: WIN55212-2, Δ^9-THC, cannabinol (Jordt et al., 2004; Jeske et al., 2006). Importantly, these cannabinoid ligands can desensitize the $TRPV_1$ receptor in sensory nerves via Ca^{2+} influx triggered at the $TRPA_1$ receptor (Jeske et al., 2006). Although the $TRPA_1$ receptor is called one of the new ionotropic cannabinoid receptors, this term should be used with caution until endogenous cannabinoids are shown to activate it.

Transient Release Potential Family "Canonical-Type 1" Receptor

The transient release potential family "canonical-type 1" ($TRPC_1$) receptor shows the greatest homology to the *Drosophila* trp channels, and is widely expressed throughout the body. It is a non-selective ligand-gated Ca^{2+} channel activated by diacyl-glycerol, and usually forms heterotetramers with other TRPC receptors ($TRPC_{2-7}$). In the brain, it controls several physiological functions (Pedersen et al., 2005). Robust inward Ca^{2+} currents are activated at the homomer $TRPC_1$ receptor of immune cells by low micromolar concentrations of the tricyclic cannabinoids HU-210, cannabinol and Δ^9-THC, which can be antagonized by the CB_1 receptor antagonist SR141716A and by CB_2 receptor antagonists (Rao and Kaminski,

2006a,b). It is of note that CP55940, anandamide or 2-AG, and the CB_1 and CB_2 receptor antagonists fail to elicit Ca^{2+} entry. Further studies are warranted to delineate the clinical impact of these findings, ranging from immunology to neuropsychiatric disorders.

NMDA Receptors

Anandamide and its metabolically stable analogue *R*-methanandamide, but not Δ^9-THC, were shown to potentiate NMDA-evoked currents at the NMDA receptor, both in the hippocampus, cortex and cerebellum, and in oocytes expressing the NR1/NR2A receptor (Hampson et al., 1998). The maximal potentiation (~50%) has been observed at 10 μM anandamide. Apart from being a possible new neuromodulator role for anandamide, this mechanism may be implicated in neuropsychiatric disorders. It is of interest that both glycine and anandamide can interact with glycine and NMDA receptors.

Amiloride-Sensitive Epithelial Na^+ Channel

The degenerin/epithelial Na^+ channel member epithelial Na^+ channel (ENaC) controls Na^+ transport into the cells and through the epithelia. Gating of the ENaC is modulated by a large variety of factors: syntaxin 1A, the copper transporter Murr1, low pH, benzamil, amiloride, cAMP and mechanical stress (Schild, 2004). This channel has been implicated in nociception, peptid-gating and mechanotransduction as well, but its main physiological function is blood pressure regulation by controlling blood Na^+ levels in the kidney. Therefore, it is of high importance that the $TRPV_1$ receptor antagonist, capsazepine, turned to be the first ENaCδ subunit activator chemical agent with the EC_{50} of 8 μM (Yamamura et al., 2004). Eventually, this finding is less surprising when considering that the ENaC is functionally homologous with the $TRPV_1$ receptor to some extent.

Ca^{2+}-Activated Large-Conductance K^+ Channels (BK_{Ca} Channels)

Low micromolar anandamide and *R*-methanandamide have been shown to activate BK_{Ca} currents depending on the presence of $BK_{Ca}\alpha$ subunits, but not on cannabinoid receptors and common intracellular messengers, and this phenomenon was pertussis toxin insensitive. Notwithstanding, it is most likely that this potentiation was mediated by an unknown cytosolic factor which in turn activated BK_{Ca} channels (Sade et al., 2006); therefore, I cannot discuss it as a direct channel opening as in the case of $TRPV_1$ receptors. It is noteworthy that Begg and colleagues (2003,

2005) reported that anandamide and ABN-CBD augment BK_{Ca} currents via activation of the endothelial anandamide/ABN-CBD receptor (see earlier), but this was sensitive to pertussis toxin and SR141716A. All in all, activation/potentiation of BK_{Ca} channels is an attractive therapeutic target against certain neurological disorders.

Direct Interaction of Cannabinoids with Plasma Membrane Transporters

Dopamine and Serotonin Transporters

Several studies have observed inhibitory action of endocannabinoids, their endogenous non-CB_1 receptor-active relatives and exogenous CB_1 receptor ligands on the uptake of dopamine, and in one case on serotonin uptake. Chen and co-workers (2003) have reported that arachidonic acid and its endogenous derivatives inhibit DA uptake in HEK293 cells expressing the human dopamine transporter (DAT). Among them, anandamide possessed the unique feature of greatly inhibiting the V_{max} and slightly the K_m values of dopamine uptake. All effects were independent of CB_1 receptor activation. Price and colleagues (2007) have reported recently that WIN55212-2 and its CB_1 receptor-inactive enantiomer WIN55212-3, as well as AM251, all decreased dopamine uptake independently of CB_1 receptors in striatal synaptosomes with IC_{50} values around 2–4 μM. WIN55212-2, WIN55212-3, *R*-methanandamide and AM251 all displaced the binding of the cocaine analogue [^{3}H]WIN35428, respectively. Finally, WIN55212-2, WIN55212-3 and AM251 all inhibited the clearance of striatally-injected dopamine. As for the underlying mechanism, Steffens and Feuerstein (2004) have proposed that WIN55212-2 and anandamide inhibit the uptake of serotonin and dopamine into cortical synaptosomes partly via impairing the activity of the uptake energy source Na^+/K^+-ATPase. In conclusion, cannabinoid ligands can increase dopamine and serotonin levels in several brain areas via inhibiting the respective transporters. This may highlight the role of cannabinoids in neuropsychiatric disorders of impaired serotonergic and dopaminergic signaling.

Glutamate Transporters

We have recently found that in rat striatal nerve terminals, WIN55212-2 and WIN55212-3 as well as AM251 all inhibited the uptake of glutamate (Köfalvi et al., 2005). A strikingly similar observation to that of Price and colleagues (2007, see earlier) reveals that these three ligands acted in exactly the same concentration range with similar IC_{50} values. This indicates a common mechanism whereby cannabinoid ligands can interfere with the uptake of different transmitters. Furthermore, it means that glutamatergic transmission can be elevated not only by activation of presynaptic CB_1 receptors in GABAergic terminals.

Glycine Transporters

The glycine transporter 1A ($GlyT_{1A}$) is widely expressed in glial cells surrounding the synapse. Anandamide, *R*-methanandamide and 2-AG in the low micromolar range have been shown to facilitate the transport of glycine through the $GlyT_{1A}$ (Pearlman et al., 2003). In other words, high endocannabinoid levels may facilitate glycine clearance, which may eventually impair NMDA receptor-mediated signaling, contributing to the pathomechanisms of schizophrenia (see Chap. 22)

Concluding Remarks

In conclusion, I wish to call the reader's attention to the fact that "unorthodox" cannabinoid actions (all summarized in Table 1) are not equal to unwanted side effects to be concerned. On the contrary, their existence is necessary, since they may provide us with new ideas and new targets in order for us to interfere with (patho)physiological processes.

Acknowledgement Attila Köfalvi is grateful for the III/BIO/56/2005 grant and for the Fundação para a Ciência e Tecnologia of the Portuguese Government (POCI2010/SFRH/BPD/18506/2004).

References

Akinshola BE, Taylor RE, Ogunseitan AB, Onaivi ES (1999) Anandamide inhibition of recombinant AMPA receptor subunits in Xenopus oocytes is increased by forskolin and 8-bromo-cyclic AMP. Naunyn Schmiedebergs Arch Pharmacol 360:242–248.

Aller MI, Veale EL, Linden AM, Sandu C, Schwaninger M, Evans LJ, Korpi ER, Mathie A, Wisden W, Brickley SG (2005) Modifying the subunit composition of TASK channels alters the modulation of a leak conductance in cerebellar granule neurons. J Neurosci 25:11455–11467.

Ambrosio AL, Dias SM, Polikarpov I, Zurier RB, Burstein SH, Garratt RC (2007) Ajulemic acid, a synthetic nonpsychoactive cannabinoid acid, bound to the ligand binding domain of the human peroxisome proliferatoractivatedreceptor gamma. J Biol Chem doi/10.1074/jbc.M702538200.

Baker D, Pryce G, Davies WL, Hiley CR (2006) In silico patent searching reveals a new cannabinoid receptor. Trends Pharmacol Sci 27:1–4.

Barann M, Molderings G, Bruss M, Bonisch H, Urban BW, Gothert M (2002) Direct inhibition by cannabinoids of human $5\text{-}HT_{3A}$ receptors: probable involvement of an allosteric modulatory site. Br J Pharmacol 137:589–596.

Bátkai S, Pacher P, Járai Z, Wagner JA, Kunos G (2004) Cannabinoid antagonist SR-141716 inhibits endotoxic hypotension by a cardiac mechanism not involving CB_1 or CB_2 receptors. Am J Physiol Heart Circ Physiol 287:H595–H600.

Begg M, Mo FM, Offertaler L, Batkai S, Pacher P, Razdan RK, Lovinger DM, Kunos G (2003) G protein-coupled endothelial receptor for atypical cannabinoid ligands modulates a Ca^{2+}-dependent K^+current. J Biol Chem 278:46188–46194.

Begg M, Pacher P, Batkai S, Osei-Hyiaman D, Offertaler L, Mo FM, Liu J, Kunos G (2005) Evidence for novel cannabinoid receptors. Pharmacol Ther 106:133–145.

Berghuis P, Dobszay MB, Wang X, Spano S, Ledda F, Sousa KM, Schulte G, Ernfors P, Mackie K, Paratcha G, Hurd YL, Harkany T (2005) Endocannabinoids regulate interneuron migration and morphogenesis by transactivating the TrkB receptor. Proc Natl Acad Sci USA 102:19115–191120.

Brandes RP, Popp R, Ott G, Bredenkotter D, Wallner C, Busse R, Fleming I (2002) The extracellular regulated kinases (ERK) 1/2 mediate cannabinoid-induced inhibition of gap junctional communication in endothelial cells. Br J Pharmacol 136:709–716.

Brown AJ, Wise A (2001) Identification of modulators of GPR55 activity. Patent Number WO0186305.

Brown SP, Safo PK, Regehr WG (2004) Endocannabinoids inhibit transmission at granule cell to Purkinje cell synapses by modulating three types of presynaptic calcium channels. J Neurosci 24:5623–5631.

Burstein S (2005) PPAR-γ: A nuclear receptor with affinity for cannabinoids. Life Sci 77:1674–1684.

Carriba P, Ortiz O, Patkar K, Justinova Z, Stroik J, Themann A, Müller C, Woods AS, Hope BT, Ciruela F, Casadó V, Canela EI, Lluis C, Goldberg SR, Moratalla R, Franco R, Ferré S (2007) Striatal adenosine A_{2A} and cannabinoid CB_1 receptors form functional heteromeric complexes that mediate the motor effects of cannabinoids. Neuropsychopharmacol doi:10.1038/sj.npp.1301375

Cheer JF, Cadogan AK, Marsden CA, Fone KC, Kendall DA (1999) Modification of 5-HT_2 receptor mediated behaviour in the rat by oleamide and the role of cannabinoid receptors. Neuropharmacology 38:533–541.

Chemin J, Monteil A, Perez-Reyes E, Nargeot J, Lory P (2001) Direct inhibition of T-type calcium channels by the endogenous cannabinoid anandamide. EMBO J 20:7033–7040.

Chen N, Appell M, Berfield JL, Reith ME (2003) Inhibition by arachidonic acid and other fatty acids of dopamine uptake at the human dopamine transporter. Eur J Pharmacol 478:89–95.

Christopoulos A, Wilson K (2001) Interaction of anandamide with the M_1 and M_4 muscarinic acetylcholine receptors. Brain Res 915:70–78.

Cunha RA (2001) Adenosine as a neuromodulator and as a homeostatic regulator in the nervous system: different roles, different sources and different receptors. Neurochem Int 38:107–125.

Deadwyler SA, Hampson RE, Mu J, Whyte A, Childers S (1995) Cannabinoids modulate voltage sensitive potassium A-current in hippocampal neurons via a cAMP-dependent process. J Pharmacol Exp Ther 273:734–743.

del Carmen Godino M, Torres M, Sanchez-Prieto J (2005) The modulation of Ca^{2+} and K^+ channels, but not changes in cAMP, signaling contribute to the inhibition of glutamate release by cannabinoid receptors in cerebrocortical nerve terminals. Neuropharmacology 48:547–557.

Devlin MG, Christopoulos A (2002) Modulation of cannabinoid agonist binding by 5-HT in the rat cerebellum. J Neurochem 80:1095–1102.

Docherty RJ, Yeats JC, Piper AS (1997) Capsazepine block of voltage-activated calcium channels in adult rat dorsal root ganglion neurones in culture. Br J Pharmacol 121:1461–1467.

Drebot II, Storozhuk MV, Kostyuk PG (2006) An unexpected effect of capsaicin on spontaneous GABAergic IPSCs in hippocampal cell cultures. Neurophysiology 38:364–367.

Duarte JM, Nogueira C, Mackie K, Oliveira CR, Cunha RA, Köfalvi A. Increase of cannabinoid CB_1 receptor density in the hippocampus of streptozotocin-induced diabetic rats. Exp Neurol 204:479–484.

Elphick MR, Egertová M (2005) The phylogenetic distribution and evolutionary origins of endocannabinoid signalling. Handbook Exp Pharmacol 168:283–297.

Ertel EA, Campbell KP, Harpold MM, Hofmann F, Mori Y, Perez-Reyes E, Schwartz A, Snutch TP, Tanabe T, Birnbaumer L, Tsien RW, Catterall WA (2000) Nomenclature of voltage-gated calcium channels. Neuron 25:533–555.

Fan P (1995) Cannabinoid agonists inhibit the activation of 5-HT_3 receptors in rat nodose ganglion neurons. J Neurophysiol 73:907–910.

Fu J, Gaetani S, Oveisi F, Lo Verme J, Serrano A, Rodriguez De Fonseca F, Rosengarth A, Luecke H, Di Giacomo B, Tarzia G, Piomelli D (2003) Oleylethanolamide regulates feeding and body weight through activation of the nuclear receptor PPAR-alpha. Nature 425:90–93.

Garcia-Anoveros J, Nagata K (2007) $TRPA_1$. Handbook Exp Pharmacol 179:347–362.

Gill CH, Randall A, Bates SA, Hill K, Owen D, Larkman PM, Cairns W, Yusaf SP, Murdock PR, Strijbos PJ, Powell AJ, Benham CD, Davies CH (2004) Characterization of the human HCN1 channel and its inhibition by capsazepine. Br J Pharmacol 143:411–421.

Glass M, Felder CC (1997) Concurrent stimulation of cannabinoid CB_1 and dopamine D_2 receptors augments cAMP accumulation in striatal neurons: evidence for a Gs linkage to the CB_1 receptor. J Neurosci 17:5327–5333.

Glass M, Northup JK (1999) Agonist selective regulation of G proteins by cannabinoid CB_1 and CB_2 receptors. Mol Pharmacol 56:1362–1369.

Godlewski G, Gothert M, Malinowska B (2003) Cannabinoid receptor-independent inhibition by cannabinoid agonists of the peripheral 5-HT_3 receptor-mediated von Bezold-Jarisch reflex. Br J Pharmacol 138:767–774.

Gothert M, Bruss M, Bonisch H, Molderings GJ (1999) Presynaptic imidazoline receptors. New developments in characterization and classification. Ann N Y Acad Sci 881:171–184.

Hampson AJ, Bornheim LM, Scanziani M, Yost CS, Gray AT, Hansen BM, Leonoudakis DJ, Bickler PE (1998) Dual effects of anandamide on NMDA receptor-mediated responses and neurotransmission. J Neurochem 70:671–676.

Harkány T, Guzmán M, Galve-Roperh I, Berghuis P, Devi LA, Mackie K (2007) The emerging functions of endocannabinoid signaling during CNS development. Trends Pharmacol Sci 28:83–92.

Hart S, Fischer OM, Ullrich A (2004) Cannabinoids induce cancer cell proliferation via tumor necrosis factor alpha-converting enzyme (TACE/ADAM17)-mediated transactivation of the epidermal growth factor receptor. Cancer Res 64:1943–1950.

Hejazi N, Zhou C, Oz M, Sun H, Ye JH, Zhang L (2005) Delta^9-tetrahydrocannabinol and endogenous cannabinoid anandamide directly potentiate the function of glycine receptors. Mol Pharmacol 69:991–997.

Ho WS, Hiley CR (2003) Vasodilator actions of abnormal-cannabidiol in rat isolated small mesenteric artery. Br J Pharmacol 138:1320–1332.

Ho WS, Hiley CR (2004) Vasorelaxant activities of the putative endocannabinoid virodhamine in rat isolated small mesenteric artery. J Pharm Pharmacol 56:869–875.

Hogestatt ED, Jonsson BA, Ermund A, Andersson DA, Bjork H, Alexander JP, Cravatt BF, Basbaum AI, Zygmunt PM (2005) Conversion of acetaminophen to the bioactive N-acylphenolamine AM404 via fatty acid amide hydrolase-dependent arachidonic acid conjugation in the nervous system. J Biol Chem 280:31405–31412.

Hoi PM, Hiley CR (2006) Vasorelaxant effects of oleamide in rat small mesenteric artery indicate action at a novel cannabinoid receptor. Br J Pharmacol 147:560–568.

Jarrahian A, Watts VJ, Barker EL (2004) D_2 dopamine receptors modulate Galpha-subunit coupling of the CB_1 cannabinoid receptor. J Pharmacol Exp Ther 308:880–886.

Jeske NA, Patwardhan AM, Gamper N, Price TJ, Akopian AN, Hargreaves KM (2006) Cannabinoid WIN 55,212-2 regulates $TRPV_1$ phosphorylation in sensory neurons. J Biol Chem 281:32879–32890.

Johnson DE, Heald SL, Dally RD, Janis RA (1993) Isolation, identification and synthesis of an endogenous arachidonic amide that inhibits calcium channel antagonist 1,4-dihydropyridine binding. Prostaglandins, Leukotrienes and Essential Fatty Acids 48:429–437.

Jordt SE, Bautista DM, Chuang HH, McKemy DD, Zygmunt PM, Hogestatt ED, Meng ID, Julius D (2004) Mustard oils and cannabinoids excite sensory nerve fibres through the TRP channel $ANKTM_1$. Nature 427:260–265.

Kathmann M, Flau K, Redmer A, Trankle C, Schlicker E (2006) Cannabidiol is an allosteric modulator at mu- and delta-opioid receptors. N Schmied Arch Pharmacol 372:354–361.

Kearn CS, Blake-Palmer K, Daniel E, Mackie K, Glass M (2004) Concurrent stimulation of cannabinoid CB_1 and dopamine D_2 receptors enhances heterodimer formation: a mechanism for receptor cross-talk? Mol Pharmacol 67:1697–1704.

Kelley BG, Thayer SA (2004) Anandamide transport inhibitor AM404 and structurally related compounds inhibit synaptic transmission between rat hippocampal neurons in culture independent of cannabinoid CB_1 receptors. Eur J Pharmacol 496:33–39.

Kim HI, Kim TH, Shin YK, Lee CS, Park M, Song JH (2005) Anandamide suppression of Na+ currents in rat dorsal root ganglion neurons. Brain Res 1062:39–47.

Köfalvi A, Vizi ES, Ledent C, Sperlágh B (2003) Cannabinoids inhibit the release of $[^3H]$glutamate from rodent hippocampal synaptosomes via a novel CB_1 receptor-independent action. Eur J Neurosci 18:1973–1978.

Köfalvi A, Rodrigues RJ, Ledent C, Mackie K, Vizi ES, Cunha RA, Sperlágh B (2005) Involvement of cannabinoid receptors in the regulation of neurotransmitter release in the rodent striatum: a combined immunochemical and pharmacological analysis. J Neurosci 25:2874–2884.

Köfalvi A, Oliveira CR, Cunha RA (2006a) Lack of evidence for functional $TRPV_1$ vanilloid receptors in rat hippocampal nerve terminals. Neurosci Lett 403:151–156.

Köfalvi A, Rebola N, Rodrigues RJ, Pereira MF, Cunha RA (2006b) Evidence for CB_1Rs, but lack of evidence for presynaptic functional CB_2Rs and $TRPV_1$Rs in the hippocampus. Annual International Cannabinoid Research Society Meeting, Tihany, Hungary.

Köfalvi A, Pereira MF, Rebola N, Rodrigues RJ, Oliveira CR, Cunha RA (2007) Anandamide and NADA bi-directionally modulate presynaptic Ca^{2+} levels and transmitter release in the hippocampus. Br J Pharmacol doi:10.1038/sj.bjp.0707252.

Kozak KR, Gupta RA, Moody JS, Ji C, Boeglin WE, DuBois RN, Brash AR, Marnett LJ (2002) 15-Lipoxygenase metabolism of 2-arachidonylglycerol. Generation of a peroxisome proliferator-activated receptor alpha agonist. J Biol Chem 277:23278–23286.

Lagalwar S, Bordayo EZ, Hoffmann KL, Fawcett JR, Frey WH II (1999) Anandamides inhibit binding to the muscarinic acetylcholine receptor. J Mol Neurosci 13:55–61.

Lauckner JE, Hille B, Mackie K (2005) The cannabinoid agonist WIN55,212-2 increases intracellular calcium via CB_1 receptor coupling to $G_{q/11}$ G proteins. Proc Natl Acad Sci USA 102:19144–19149.

Lawson K (2000) Is there a role for potassium channel openers in neuronal ion channel disorders? Expert Opin Investig Drugs 9:2269–2280.

Lesage F, Lazdunski M (2000) Molecular and functional properties of two-pore-domain potassium channels. Am J Physiol Renal Physiol 279:F793–801.

Liao C, Zheng J, David LS, Nicholson RA (2004) Inhibition of voltage-sensitive sodium channels by the cannabinoid 1 receptor antagonist AM 251 in mammalian brain. Basic Clin Pharmacol Toxicol 94:73–78.

Linden AM, Aller MI, Leppa E, Vekovischeva O, Aitta-Aho T, Veale EL, Mathie A, Rosenberg P, Wisden W, Korpi ER (2006) The *in vivo* contributions of TASK-1-containing channels to the actions of inhalation anesthetics, the $alpha_2$ adrenergic sedative dexmedetomidine, and cannabinoid agonists. J Pharmacol Exp Ther 317:615–626.

Liu J, Li H, Burstein SH, Zurier RB, Chen JD (2003) Activation and binding of peroxisome proliferator-activated receptor gamma by synthetic cannabinoid ajulemic acid. Mol Pharmacol 63:983–992.

Liu L, Simon SA (1997) Capsazepine, a vanilloid receptor antagonist, inhibits nicotinic acetylcholine receptors in rat trigeminal ganglia. Neurosci Lett 228:29–32

Lo Verme J, Fu J, Astarita G, La Rana G, Russo R, Calignano A, Piomelli D (2005) The nuclear receptor peroxisome proliferator-activated receptor-alpha mediates the anti-inflammatory actions of palmitoylethanolamide. Mol Pharmacol 67:15–19.

Lozovaya N, Yatsenko N, Beketov A, Tsintsadze T, Burnashev N (2005) Glycine receptors in CNS neurons as a target for nonretrograde action of cannabinoids. J Neurosci 25:7499–7506.
Lundbaek JA, Birn P, Tape SE, Toombes GE, Sogaard R, Koeppe RE 2nd, Gruner SM, Hansen AJ, Andersen OS (2005) Capsaicin regulates voltage-dependent sodium channels by altering lipid bilayer elasticity. Mol Pharmacol 68:680–689.
Mackie K (2005) Cannabinoid receptor homo- and heterodimerization. Life Sci 77:1667–1673.
Mackie K, Stella N (2006) Cannabinoid receptors and endocannabinoids: evidence for new players. AAPS J 8:E298–306.
Mackie K, Lai Y, Wenstenbroek R, Mitchell R (1995) Cannabinoids activate an inwardly rectifying potassium conductance and inhibit Q-type calcium currents in At20 cells transfected with rat brain cannabinoid receptor. J Neurosci 15:6552–6561.
Maingret F, Patel AJ, Lazdunski M, Honore E (2001) The endocannabinoid anandamide is a direct and selective blocker of the background K^+ channel TASK-1. EMBO J 20:47–54.
Maricq AV, Peterson AS, Brake AJ, Myers RM, Julius D (1991) Primary structure and functional expression of the $5HT_3$ receptor, a serotonin-gated ion channel. Science 254:432–437.
Matyas F, Yanovsky Y, Mackie K, Kelsch W, Misgeld U, Freund TF (2006) Subcellular localization of type 1 cannabinoid receptors in the rat basal ganglia. Neuroscience 137:337–361.
McPartland JM, Glass M (2003) Functional mapping of cannabinoid receptor homologs in mammals, other vertebrates, and invertebrates. Gene 312:297–303.
McPartland JM, Matias I, Di Marzo V, Glass M (2006) Evolutionary origins of the endocannabinoid system. Gene 370:64–74.
Molderings GJ, Likungu J, Gothert M (1999) Presynaptic cannabinoid and imidazoline receptors in the human heart and their potential relationship. N Schmied Arch Pharmacol 360:157–164.
Molderings GJ, Bonisch H, Hammermann R, Gothert M, Bruss M (2002) Noradrenaline release-inhibiting receptors on PC12 cells devoid of $alpha_2$- and CB_1 receptors: similarities to presynaptic imidazoline and edg receptors. Neurochem Int 40:157–167.
Morales M, Backman C (2002) Coexistence of serotonin 3 (5-HT_3) and CB_1 cannabinoid receptors in interneurons of hippocampus and dentate gyrus. Hippocampus 12:756–764.
Moran MM, Xu H, Clapham DE (2004) TRP ion channels in the nervous system. Curr Opin Neurobiol 14:362–369.
Nicholson RA, Liao C, Zheng J, David LS, Coyne L, Errington AC, Singh G, Lees G (2003) Sodium channel inhibition by anandamide and synthetic cannabimimetics in brain. Brain Res 978:194–204.
Nilius B, Vriens J, Prenen J, Droogmans G, Voets T (2004) $TRPV_4$ calcium entry channel: a paradigm for gating diversity. Am J Physiol Cell Physiol 286:C195–205.
Offertaler L, Mo FM, Batkai S, Liu J, Begg M, Razdan RK, Martin BR, Bukoski RD, Kunos G (2003) Selective ligands and cellular effectors of a G protein-coupled endothelial cannabinoid receptor. Mol Pharmacol 63:699–705.
Ohno-Shosaku T, Matsui M, Fukudome Y, Shosaku J, Tsubokawa H, Taketo MM, Manabe T, Kano M (2003) Postsynaptic M_1 and M_3 receptors are responsible for the muscarinic enhancement of retrograde endocannabinoid signalling in the hippocampus. Eur J Neurosci 18:109–116.
Oliver D, Lien CC, Soom M, Baukrowitz T, Jonas P, Fakler B (2004) Functional conversion between A-type and delayed rectifier K^+ channels by membrane lipids. Science 304:265–270.
O'Sullivan SE, Tarling EJ, Bennett AJ, Kendall DA, Randall MD (2005) Novel time-dependent vascular actions of Δ^9-tetrahydrocannabinol mediated by peroxisome proliferator-activated receptor gamma. Biochem Biophys Res Commun 337:824–831.
Oz M (2006) Receptor-independent effects of endocannabinoids on ion channels. Curr Pharm Des 12:227–329.
Oz M, Tchugunova Y, Dunn SMJ (2000) Endogenous cannabinoid anandamide directly inhibits voltage- dependent calcium fluxes in rabbit T-tubule membrane preparations. Eur J Pharmacol 404:13–20.

Oz M, Zhang L, Morales M (2002) Endogenous cannabinoid, anandamide, acts as a noncompetitive inhibitor on 5-HT_3 receptor-mediated responses in Xenopus oocytes. Synapse 46:150–156.
Oz M, Ravindran R, Zhang L, Morales M (2003) Endogenous cannabinoid, anandamide inhibits neuronal nicotinic acethylcholine receptor-mediated responses in Xenopus oocytes. J Pharmacol Exp Ther 306:1003–1010.
Oz M, Tchugunova Y, Dinc M (2004a) Differential effects of endogenous and synthetic cannabinoids on voltage-dependent calcium fluxes in rabbit T-tubule membranes: comparison with fatty acids. Eur J Pharmacol 502:47–58.
Oz M, Zhang L, Ravindran A, Morales M, Lupica CR (2004b) Differential effects of endogenous and synthetic cannabinoids on alpha7-nicotinic acetylcholine receptor-mediated responses in Xenopus Oocytes. J Pharmacol Exp Ther 310:1152–1160.
Oz M, Jackson SN, Woods AS, Morales M, Zhang L (2005) Additive effects of endogenous cannabinoid anandamide and ethanol on alpha7-nicotinic acetylcholine receptor-mediated responses in Xenopus Oocytes. J Pharmacol Exp Ther 313:1272–1280.
Pearlman RJ, Aubrey KR, Vandenberg RJ (2003) Arachidonic acid and anandamide have opposite modulatory actions at the glycine transporter, $GLYT_{1a}$. J Neurochem 84:592–601.
Pedersen SF, Owsianik G, Nilius B (2005) TRP channels: an overview. Cell Calcium 38:233–252.
Pereira DB, Rebola N, Rodrigues RJ, Cunha RA, Carvalho AP, Duarte CB (2006) Trkb receptors modulation of glutamate release is limited to a subset of nerve terminals in the adult rat hippocampus. J Neurosci Res 83:832–844.
Pertwee RG (2004) Novel pharmacological targets for cannabinoids. Curr Neuropharmacol 2:9–29.
Pertwee RG (2005) Pharmacological actions of cannabinoids. Handb Exp Pharmacol 168:1–51.
Poling JS, Rogawski MA, Salem N Jr, Vicini S (1996) Anandamide, an endogenous cannabinoid, inhibits Shaker-related voltage-gated K^+ channels. Neuropharmacology 35:983–991.
Prather PL, Martin NA, Breivogel CS, Childers SR (2000) Activation of cannabinoid receptors in rat brain by WIN 55212-2 produces coupling to multiple G protein alpha-subunits with different potencies. Mol Pharmacol 57:1000–1010.
Price DA, Owens WA, Gould GG, Frazer A, Roberts JL, Daws LC, Giuffrida A (2007) CB_1-independent inhibition of dopamine transporter activity by cannabinoids in mouse dorsal striatum. J Neurochem doi:10.1111/j.1471–4159.2006.04383.x
Puntambekar P, Van Buren J, Raisinghani M, Premkumar LS, Ramkumar V (2004) Direct interaction of adenosine with the $TRPV_1$ channel protein. J Neurosci 24:3663–3671.
Rao GK, Kaminski NE (2006a) Cannabinoid-mediated elevation of intracellular calcium: a structure-activity relationship. J Pharmacol Exp Ther 317:820–829.
Rao GK, Kaminski NE (2006b) Induction of intracellular calcium elevation by $Delta^9$-tetrahydrocannabinol in T cells involves $TRPC_1$ channels. J Leukoc Biol 79:202–213.
Rawls SM, Tallarida RJ, Zisk J (2006) Agmatine and a cannabinoid agonist, WIN 55212-2, interact to produce a hypothermic synergy. Eur J Pharmacol 553:89–98.
Rhee M-H, Bayewitch ML, Avidor-Reiss T, Levy R and Vogel Z (1998) Cannabinoid receptor activation differentially regulates the various adenylyl cyclase isozymes. J Neurochem 71:1525–1534.
Rhee MH, Nevo I, Avidor-Reiss T, Levy R, Vogel Z (2000) Differential superactivation of adenylyl cyclase isozymes after chronic activation of the CB_1 cannabinoid receptor. Mol Pharmacol 57:746–552.
Rios C, Gomes I, Devi LA (2006) mu opioid and CB_1 cannabinoid receptor interactions: reciprocal inhibition of receptor signaling and neuritogenesis. Br J Pharmacol 148:387–395.
Robinson RB, Siegelbaum SA (2003) Hyperpolarization-activated cation currents: from molecules to physiological function. Annu Rev Physiol 65:453–480.
Rockwell CE, Snider NT, Thompson JT, Vanden Heuvel JP, Kaminski NE (2006) Interleukin-2 suppression by 2-arachidonyl glycerol is mediated through peroxisome proliferator-activated receptor gamma independently of cannabinoid receptors 1 and 2. Mol Pharmacol 70:101–111.

Rothlin CV, Katz E, Verbitsky M, Elgoyhen AB (1999) The alpha9 nicotinic acetylcholine receptor shares pharmacological properties with type A gamma-aminobutyric acid, glycine, and type 3 serotonin receptors. Mol Pharmacol 55:248–254.

Ryberg E, Vu HK, Larsson N, Groblewski T, Hjorth S, Elebring T, Sjogren S, Greasley PJ (2005) Identification and characterisation of a novel splice variant of the human CB_1 receptor. FEBS Lett 579:259–264.

Sade H, Muraki K, Ohya S, Hatano N, Imaizumi Y (2006) Activation of large-conductance, Ca^{2+}-activated K^+ channels by cannabinoids. Am J Physiol Cell Physiol 290:C77–C86.

Saghatelian A, McKinney MK, Bandell M, Patapoutian A, Cravatt BF (2006) A FAAH-regulated class of *N*-acyl taurines that activates TRP ion channels. Biochemistry 45:9007–9015.

Salzet M, Stefano GB (2002) The endocannabinoid system in invertebrates. Prostaglandins, Leukotrienes and Essential Fatty Acids 66:353–361.

Savinainen JR, Saario SM, Niemi R, Jarvinen T, Laitinen JT (2003) An optimized approach to study endocannabinoid signaling: evidence against constitutive activity of rat brain adenosine A_1 and cannabinoid CB_1 receptors. Br J Pharmacol 140:1451–1459.

Schild L (2004) The epithelial sodium channel: from molecule to disease. Rev Physiol Biochem Pharmacol 151:93–107.

Schoffelmeer AN, Hogenboom F, Wardeh G, De Vries TJ (2006) Interactions between CB_1 cannabinoid and mu opioid receptors mediating inhibition of neurotransmitter release in rat nucleus accumbens core. Neuropharmacology 51:773–781.

Seth P, Cheeta S, Tucci S, File SE (2002) Nicotinic-serotonergic interactions in brain and behaviour. Pharmacol Biochem Behav 71:795–805.

Shen M, Thayer SA (1998) The cannabinoid agonist Win55,212-2 inhibits calcium channels by receptor-mediated and direct pathways in cultured rat hippocampal neurons. Brain Res 783:77–84.

Shire D, Carillon C, Kaghad M, Calandra B, Rinaldi-Carmona M, Le Fur G, Caput D, Ferrara P (1995) An amino-terminal variant of the central cannabinoid receptor resulting from alternative splicing. J Biol Chem 270:3726–3731.

Steffens M, Feuerstein TJ (2004) Receptor-independent depression of DA and 5-HT uptake by cannabinoids in rat neocortex - involvement of Na^+/K^+-ATPase. Neurochem Int 44:529–538.

Su JY, Vo AC (2007) 2-Arachidonylglyceryl ether and abnormal cannabidiol-induced vascular smooth muscle relaxation in rabbit pulmonary arteries via receptor-pertussis toxin sensitive G proteins-ERK1/2 signaling. Eur J Pharmacol 559:189–195.

Sugiura T, Kodaka T, Nakane S, Miyashita T, Kondo S, Suhara Y, Takayama H, Waku K, Seki C, Baba N, Ishima Y (1999) Evidence that the cannabinoid CB_1 receptor is a 2-arachidonoylglycerol receptor. Structure-activity relationship of 2-arachidonoylglycerol, ether-linked analogues, and related compounds. J Biol Chem 274:2794–2801.

Tapia R, Velasco I (1997) Ruthenium red as a tool to study calcium channels, neuronal death and the function of neural pathways. Neurochem Int 30:137–147.

Trevisani M, Smart D, Gunthorpe MJ, Tognetto M, Barbieri M, Campi B, Amadesi S, Gray J, Jerman JC, Brough SJ, Owen D, Smith GD, Randall AD, Harrison S, Bianchi A, Davis JB, Geppetti P (2002) Ethanol elicits and potentiates nociceptor responses via the vanilloid receptor-1. Nat Neurosci 5:546–551.

Upham BL, Rummel AM, Carbone JM, Trosko JE, Ouyang Y, Crawford RB, Kaminski NE (2003) Cannabinoids inhibit gap junctional intercellular communication and activate ERK in a rat liver epithelial cell line. Int J Cancer 104:12–18.

Van den Bossche I, Vanheel B (2000) Influence of cannabinoids on the delayed rectifier in freshly dissociated smooth muscle cells of the rat aorta. Br J Pharmacol 131:85–93.

van der Stelt M, Di Marzo V (2005) Anandamide as an intracellular messenger regulating ion channel activity. Prostaglandins and other Lipid Mediators 77:111–122.

Venance L, Piomelli D, Glowinski J, Giaume C (1995) Inhibition by anandamide of gap junctions and intercellular calcium signalling in striatal astrocytes. Nature 376:590–594.

Vigano D, Rubino T, Parolaro D (2005) Molecular and cellular basis of cannabinoid and opioid interactions. Pharmacol Biochem Behav 81:360–368.

White R, Hiley CR (1998) The actions of the cannabinoid receptor antagonist, SR 141716A, in the rat isolated mesenteric artery. Br J Pharmacol 125:689–696.
Williams EJ, Walsh FS, Doherty P (2003) The FGF receptor uses the endocannabinoid signaling system to couple to an axonal growth response. J Cell Biol 160:481–486.
Yamamura H, Ugawa S, Ueda T, Nagao M, Shimada S (2004) Capsazepine is a novel activator of the delta subunit of the human epithelial Na^+channel. J Biol Chem 279:44483–44489.
Zhao Q, He Z, Chen N, Cho YY, Zhu F, Lu F, Ma WY, Bode AM, Dong Z (2005) 2-Arachidonoylglycerol stimulates activator protein-1-dependent transcriptional activity and enhances epidermal growth factor-induced cell transformation in JB6 P+ cells. J Biol Chem 280:26735–26742.

Chapter 10
Anatomical Distribution of Receptors, Ligands and Enzymes in the Brain and in the Spinal Cord: Circuitries and Neurochemistry

Giovanni Marsicano and Rohini Kuner

Abstract The endocannabinoid system has emerged during the last two decades as a very important regulator of neuronal and cellular activity in many different body tissues and particularly in the central and peripheral nervous systems. The endocannabinoid system constitutes of lipid signaling molecules (the endocannabinoids), the enzymatic machineries for their synthesis and degradation, and their cellular targets, the cannabinoid receptors. "Bona fide" targets of endocannabinoids are the G protein-coupled cannabinoid receptors type 1 and type 2 (CB_1 and CB_2 receptors, respectively). However, recent evidence indicates that endocannabinoids also have other targets besides "classical" cannabinoid receptors. Furthermore, the steadily growing list of newly discovered elements of endocannabinoid signaling further expands the definition of the endocannabinoid system on an almost daily base. In this chapter, we will describe the anatomical distribution of the various elements of the endocannabinoid system in the brain, the spinal cord and the peripheral nervous system. In particular, we will address the distribution of receptors (CB_1 and CB_2 receptors and other targets of endocannabinoids), of enzymes involved in the synthesis and degradation of endocannabinoids, and of the two major endocannabinoids described so far, anandamide and 2-arachidonoyl-glycerol. Particular emphasis will be given to new findings indicating a larger distribution of the endocannabinoid system in nervous tissues than previously believed. The need for improvement of unbiased techniques for the detection of various elements of the endocannabinoid system will be also underlined, which will allow a more precise identification of the sites where endocannabinoid signalling exerts important physiological and pathophysiological functions.

Introduction

In this chapter, we will address the anatomical distribution of various elements of the endocannabinoid system in the nervous system of adult mammals. Since the first evidence of the existence of cannabinoid receptors (Howlett and Fleming, 1984) and their discovery [CB_1, (Matsuda et al., 1990) and CB_2 (Munro et al.,

A. Köfalvi (ed.), *Cannabinoids and the Brain.*

1993)], followed by the steady-state accumulation of new findings over the course of the last two decades, the question of the anatomical loci where this system exerts its plethora of functions has been addressed by different means (Mackie, 2005b). Given the complexity of the endocannabinoid system and the intrinsic limits of detection systems to identify the precise location of its various elements, this field is constantly evolving, with the addition of new information almost on a daily base. Therefore, in this chapter, we will present the actual state of knowledge concerning the localization of receptors, ligands and related enzymes forming the endocannabinoid system, but the reader should keep always in mind the "golden rule" that, in science, "lack of evidence is not evidence of a lack" and be aware that discoveries that challenge the current view represented here might indeed appear in the literature in the very next future (even during the publication process of this book). This chapter will be divided in subchapters analyzing the known distribution of the generally accepted elements constituting the endocannabinoid system. Of course, as always in biology, the endocannabinoid system interacts at different levels with other systems, resulting in a complex pattern of physiological and pathophysiological activities, which, in turn, can alter the functions of the endocannabinoid system itself. To limit our targets, we will describe the distribution of molecules (receptors, ligands and enzymes) that are generally considered as "*bonafide*" members of the endocannabinoid system. In other words, only enzymatic pathways and ligands known to interfere directly with the activity of cannabinoid receptors in the nervous system will be taken in consideration and, conversely, only established receptor targets of known endocannabinoids will be described. For instance, it is known that dopamine, glutamate, acetylcholine receptors and many others can influence the synthesis of endocannabinoids (Alger, 2002; Doherty and Dingledine, 2003; Piomelli, 2003; van der Stelt and Di Marzo, 2003; Chevaleyre et al., 2006) and can interact with cannabinoid receptors (in some cases, even physically) (Mackie, 2005a). However, these biological elements will not be directly taken in consideration in the following as "parts" of the endocannabinoid system, but only when co-expression data can help identify the anatomical patterns of distribution of the "proper" elements of the endocannabinoid system. Another important general issue concerning the anatomical features of the endocannabinoid system is that the levels of expression of the various constitutive elements do not necessarily reflect the functional significance of the system itself, i.e., a direct proportionality between expression and function is not always given. In some cases, for instance, relatively very low levels of cannabinoid receptors or of endocannabinoids might underlie very important functions of the endocannabinoid system in certain regions or cell types. The reasons of this apparent discrepancy are presently unknown, but, in our opinion, could be related to the typical "on demand" nature of activity of the endocannabinoid system, which is described in greater detail in other chapters of the present book. To summarize this point in the context of this chapter, endocannabinoids are believed to be synthesized, released and degraded in a very specific fashion and their spreading is very likely to be highly limited by their lipid nature and via efficient degradation systems (Piomelli, 2003; Lutz, 2004; Di Marzo et al., 2005; Marsicano and

Lutz, 2006; see Chaps. 2, 3, 11). Therefore, it is quite likely that their concentrations can rapidly reach high levels in small areas over short periods of time. Consequently, despite low general levels, high densities of receptor expression over narrow, specific anatomical domains might be sufficient to exert important biological functions. Again, given the intrinsic limits of detection of endocannabinoids and proteins belonging to the endocannabinoid system, it is possible that such highly sophisticated mechanisms of action might escape anatomical observations. Indeed, sensitive, high resolution tools for detection of these elements are being continuously developed and their future use will certainly refine our current understanding of the anatomy of the endocannabinoid system. Several techniques have been used to detect the localization of elements of the endocannabinoid system in the nervous system. Ligand binding, functional activation of G proteins, immunohistochemistry (IHC) and in situ hybridization (ISH) analysis were used to identify cannabinoid receptors (Mackie, 2005b). Conversely, IHC, ISH and enzymatic activity assays from tissue extracts were used to identify enzymes involved in the synthesis and degradation of endocannabinoids (Freund et al., 2003). Given the lipid nature of endocannabinoids, the only means of their detection are direct biochemical measurements on tissue extracts, which, although very powerful and sensitive, intrinsically lack the spatial resolution inherent to histochemical techniques, such as IHC and ISH. Therefore, the precise identification of the exact loci where endocannabinoids are actually synthesized and exert their functions can only be extrapolated from the expression data on the receptors and the enzymes involved in the synthesis and degradation of endocannabinoids (which, however, are not yet fully identified, see later and Chaps. 2, 3, 11). Thus, the direct identification of the actual presence of the signalling molecules must necessarily rely on detection systems lacking spatial resolution. In the second part of this chapter, we will begin our anatomical description of the endocannabinoid system starting with CB_1, the first cannabinoid receptor identified (Matsuda et al., 1990) and hitherto, the best known and also the most widely expressed cannabinoid receptor in neurons (Herkenham et al., 1991; Mackie, 2005b). In contrast, CB_2 cannabinoid receptors were believed to be predominantly expressed in peripheral cells belonging to the immune system (Munro et al., 1993). However, recent evidence indicates that this receptor subtype is also present in the central and peripheral nervous system, either in neurons, glial or microglial cells. These aspects of the expression of CB_2 receptors will be described in the third part of this chapter. Moreover, endocannabinoids are able to bind and activate other targets than CB_1 and CB_2 (Mackie and Stella, 2006; Pacher et al., 2006; see Chap. 9 for the full list of targets). This exciting but largely unknown aspect of the endocannabinoid system will be briefly taken into consideration with a concise description of the expression patterns of some of these additional putative "endocannabinoid receptors". The presence of endocannabinoids and the enzymes responsible for their synthesis and degradation will be described in the fourth part of this chapter. In addition, in this part we will also touch upon "non-canonical" enzymes that have been described to participate in the synthesis and/or degradation of endocannabinoids.

Distribution of CB_1 Cannabinoid Receptors in the Nervous System

The main cannabinoid receptor expressed in neurons is CB_1, although recent data indicate that the CB_2 receptor is also present in certain neuronal populations (van Sickle et al., 2005), which will be described later. CB_1 receptor is very abundantly expressed in the adult nervous system and represents the seven transmembrane G protein-coupled receptor (GPCR) with highest expression in the brain (Herkenham et al., 1991; Howlett et al., 2002). CB_1 transcript and protein are detectable in brain regions implicated in several vital functions of the CNS, including learning and memory, pain perception, neuroendocrine control, reward and many others. In general, the expression patterns of CB_1 receptor in the brain, spinal cord and peripheral nerves correspond quite well to the known effects of exogenously administered CB_1 receptor agonists and to the reported endogenous functions of CB_1 receptor (Breivogel and Childers, 1998). However, there are some exceptions. For instance, the ratio between estimated amount of CB_1 receptor calculated via direct ligand binding and the G protein activation estimated by functional agonist-induced GTPγ binding assays is not always constant, thus, indicating regional differences in receptor coupling efficiencies (Breivogel and Childers, 1998; see Chap. 9). This is important to consider, because sometimes the endocannabinoid system appears to be functionally very important in regions or cell types where the density of CB_1 receptor is relatively low [e.g. control of pain perception in the brainstem (Walker et al., 1999; Hohmann et al., 2005) or control of epileptiform seizures in hippocampal glutamatergic neurons (Marsicano et al., 2003; Monory et al., 2006). Therefore, the activity of cannabinoids at CB_1 receptors cannot be predicted solely based on the relative receptor density, but other factors, such as efficiency of receptor coupling and local synthesis of endocannabinoids, need to be taken into account. Another general aspect to be considered is the subcellular localization of CB_1 receptor in neurons. CB_1 protein is predominantly, but not exclusively (Freund et al., 2003; Bacci et al., 2004), found in axon terminals of neurons. IHC, ligand binding or functional agonist-induced GTPγ binding techniques allow the detection of the CB_1 protein. However, as described above, the intrinsic detection limits of these approaches might preclude, in some cases, to detect low levels in particular brain regions. On the other hand, detection of CB_1 mRNA expression by ISH or by single-cell reverse transcriptase PCR (single-cell RT-PCR) do not allow to identify the location of the protein, but possess higher sensitivity, making it possible to identify neurons containing relatively low levels of the transcript. As mRNA is normally present in cell bodies, techniques to detect transcript also allow identifying the actual location of the soma of the cells expressing CB_1 receptor. Axonal terminals can indeed be located very distantly from cell bodies. As a consequence, the CB_1 mRNA can have an anatomical distribution quite different from the protein that it is encoded in the same cell. In the case of CB_1 receptors, this is the case in many brain regions. For instance, the substantia nigra in the midbrain or the nucleus accumbens in the ventral forebrain contain very low amounts of CB_1 mRNA but relatively high levels of CB_1 protein on account of incoming axonal projections from other brain regions. In this chapter, therefore, whenever possible, we will take care to refer to the detection system employed when discussing the expression pattern of CB_1 receptors.

Distribution of CB_1 Receptors in Cortical Regions of the Forebrain

a. Olfactory bulb and cortical olfactory areas: By IHC experiments, CB_1 receptors were identified in different olfactory regions of the brain. In the olfactory bulb, they are at highest levels in the inner granule cell layer and lower amounts are expressed in the inner plexiform layer (Herkenham et al., 1990). The external plexiform layer, the mitral cell (glomerular) layer and the accessory olfactory bulb show a low density of CB_1 receptor expression (Herkenham et al., 1991; Tsou et al., 1998a; Egertová and Elphick, 2000). In cortical olfactory areas, the anterior olfactory nucleus contains high levels of CB_1 receptor. In this region, most neurons express CB_1 receptors (Herkenham et al., 1991; Mailleux and Vanderhaeghen, 1992; Matsuda et al., 1993; Tsou et al., 1998a; Marsicano and Lutz, 1999; Egertová and Elphick, 2000). Moreover, a great majority of neurons belonging to the piriform cortex contain CB_1 mRNA (Marsicano and Lutz, 1999; Hermann et al., 2002).
b. Cortex: This brain region contains high levels of CB_1 receptors in all of its subfields, including the prefrontal cortex, the neocortex, the entorhinal and the perirhinal cortex (Herkenham, 1991; Herkenham et al., 1991; Mailleux et al., 1992; Mailleux and Vanderhaeghen, 1992; Matsuda et al., 1993; Glass et al., 1997; Tsou et al., 1998a). The variation in CB_1 expression at protein level across cortical regions has been examined most extensively in human brain using receptor autoradiography. Here, there is variation between regions, with higher levels found in the cingulate gyrus, frontal cortex as well as the secondary somatosensory and motor cortices. Lower levels are found in the primary somatosensory and motor cortices (Glass et al., 1997). The laminar nature of CB_1 expression within the neocortex is striking. The relative levels of expression between regions vary (Glass et al., 1997). For example, in rat somatosensory cortex, CB_1 receptor levels are relatively higher in layers II, upper III, IV and VI and relatively lower in deeper layer III and layer V (Freund et al., 2003). Layer I appears to be almost devoid of CB_1 receptors. Ultrastructural studies have revealed that in the cortex, CB_1 receptor-expressing terminals synapse onto pyramidal cell bodies, apical dendrites and their smaller caliber branches (Freund et al., 2003). The expression pattern of CB_1 receptors in different neuronal populations within the cortical subregions is a good example of how improved detection systems can change pre-existing anatomical concepts. The original IHC and ISH studies concurred in indicating that almost all neurons

expressing CB_1 at high or moderate levels constitute a subpopulation of GABAergic interneurons, mostly belonging to the cholecystokinin (CCK)-positive subgroup (Tsou et al., 1998a; Marsicano and Lutz, 1999; Freund et al., 2003), whereas principal glutamatergic neurons appeared to be depleted of CB_1 receptor. However, CB_1 receptor-mediated effects on glutamatergic transmission have been reported in the cortex (Sjöström et al., 2003, 2004). These apparent discrepancies were recently solved via the use of more sensitive means to detect CB_1 mRNA. First, an improved ISH technique using enhanced detection methods revealed that the great majority of glutamatergic neurons in cortical regions (including neocortex) contain CB_1 mRNA at low but detectable levels as indicated by co-expression with vesicular glutamate transporter 1 ($VGluT_1$), a marker of glutamatergic neurons (Monory et al., 2006) (Fig. 1). Second, single-cell RT-PCR recently revealed that at least 50% of neocortical pyramidal neurons do contain CB_1 mRNA (Hill et al., 2007). Furthermore, these

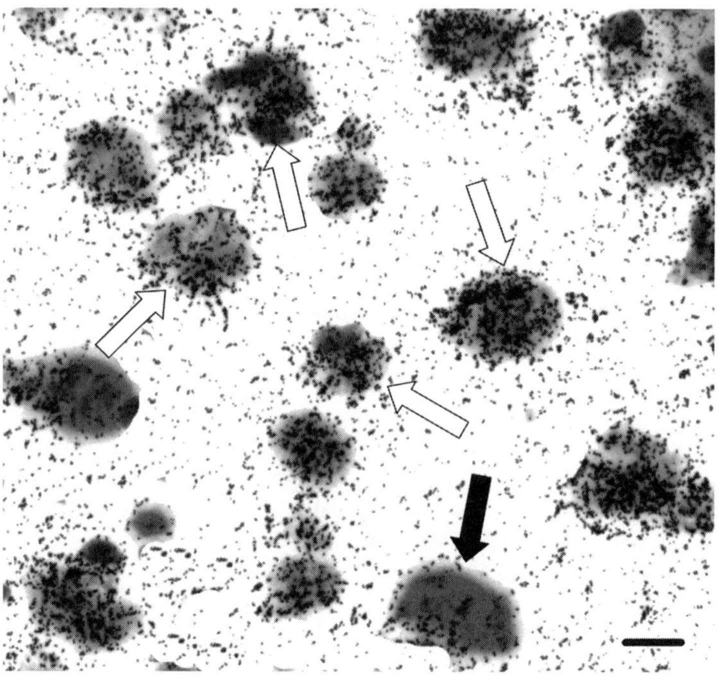

Fig. 1 CB_1 mRNA is expressed in glutamatergic neurons of the cortex. The images show double ISH analysis of CB_1 (uniform grey staining) and $VGluT_1$ mRNA (vesicular glutamate transporter, a marker of cortical glutamatergic neurons, silver grains) in the mouse cortex. Note that neurons expressing relatively low levels of CB_1 mRNA (as indicated by light grey staining, *unfilled arrows*) do contain $VGluT_1$ mRNA, whereas neurons containing high levels of the receptor do not express the transporter due to their GABAergic nature (Marsicano and Lutz, 1999; Monory et al., 2006). Bar: 10 μm

experiments revealed that subgroups of neocortical GABAergic interneurons, which are distinct from the CCK-positive sub-population (i.e. expressing somatostatin or vasoactive intestinal peptide mRNAs), also contain CB_1 mRNA. These new data suggest that the expression pattern of CB_1 in the cortex is quite likely to be much broader than believed previously and that improvement of detection systems is warranted for identifying additional loci where CB_1 receptor might be present at low, but possibly functionally important, levels. In this regard, it is interesting to mention some aspects regarding subcellular localization of CB_1 receptors in neurons. Present anatomical data in the literature indicate that the predominant localization of CB_1 is on axonal terminals in basically all regions examined (Tsou et al., 1998a; Egertová and Elphick, 2000; Freund et al., 2003; Mackie, 2005b). However, recent functional data strongly suggest that endocannabinoids, acting through CB_1 receptors, mediate a form of self-inhibition, which is exerted at the somatodendritic level of neocortical GABAergic interneurons, (Bacci et al., 2004). These data further strengthen the concept that the current picture of the anatomical distribution of CB_1 receptor is far from being definitive and that new studies and improved techniques are needed to complete it.

c. Hippocampal formation: The hippocampus contains high levels of CB_1 receptor, both at protein and mRNA level, as shown in early studies (Herkenham et al., 1991; Mailleux et al., 1992; Mailleux and Vanderhaeghen, 1992; Matsuda et al., 1993). Similarly to the cortex, the most striking expression of CB_1 receptor is detected in GABAergic interneurons, belonging mostly to the CCK-positive subpopulation of basket cells (Tsou et al., 1998a; Marsicano and Lutz, 1999; Egertová and Elphick, 2000; Freund et al., 2003; Mackie, 2005b), though lower levels of expression were also found in other subtypes of interneurons (Marsicano and Lutz, 1999). It is well-established that the intense staining surrounding principal neurons of the hippocampus in IHC experiments using various different types of CB_1 receptor antisera is due to the presence of CB_1 receptor in terminals of GABAergic basket cells (Freund et al., 2003; Mackie, 2005b). This particular type of interneuron can be subclassified into two non-overlapping populations, which are identified by their mutally exclusive expression of CCK or parvalbumin. The fact that CB_1 is present almost exclusively in the former subgroup might have interesting functional consequences (Chen et al., 2003; Freund et al., 2003; Klausberger et al., 2005). CB_1 receptors appear to be present in GABAergic neurons containing the serotonin receptor 5-HT_3 (Hermann et al., 2002; Morales and Backman, 2002), with possible interesting implications concerning the interaction between the serotonergic system and endocannabinoid system (see Chap. 22). The presence of CB_1 receptor in hippocampal glutamatergic neurons has been intensely debated in the recent years. Early studies revealed that CB_1 mRNA is indeed present in CA1 and CA3 pyramidal neurons (Herkenham et al., 1990; Marsicano and Lutz, 1999; Matsuda et al., 1993), whereas IHC studies failed to detect CB_1 protein in this kind of neurons (Tsou et al., 1998a; Katona et al., 1999; Freund et al., 2003). These discrepancies, again, arise likely due to intrinsic limitations in detection systems and the immense variations in expression levels of

CB_1 in different neuronal populations. In fact, CB_1 receptors are expressed at such high levels in cortical GABAergic interneurons that it makes it very difficult to distinguish the extremely low amounts of protein present in glutamatergic neurons from background staining. In the last years, the use of sophisticated genetic approaches that enable specific deletion of CB_1 receptor in certain neuronal populations (i.e., only in GABAergic vs. glutamatergic or vice versa) paved the way to the anatomical identification of CB_1 protein in glutamatergic hippocampal neurons (Marsicano et al., 2003; Lutz, 2004; Lutz et al., 2004). More recently, several independent groups have identified CB_1 protein in glutamatergic hippocampal neurons (Degroot et al., 2006; Katona et al., 2006; Kawamura et al., 2006; Monory et al., 2006) and characterized some of their functions in vitro (Domenici et al., 2006; Monory et al., 2006; Takahashi and Castillo, 2006; Köfalvi et al., 2007) and in vivo (Monory et al., 2006). Amongst the glutamatergic neurons in the hippocampal formation, the only cell-type where CB_1 receptor does not seem to be expressed are the granule cells of the dentate gyrus, whereas CA1 and CA3 pyramidal neurons contain CB_1 mRNA as well as protein. Interestingly, another class of glutamatergic hippocampal neurons, the mossy cells, which reside in the hilus of the dentate gyrus and receive and send glutamatergic projections to granule dentate neurons (Johnston and Amaral, 2004) seem to contain the highest level of CB_1 amongst excitatory hippocampal neurons (Kawamura et al., 2006; Monory et al., 2006). In fact, a dense band of staining with specific CB_1 receptor antisera can be observed in the so-called "inner third" of the molecular layer of the dentate gyrus, where these cells form synapses onto dendrites of the granule cells (Johnston and Amaral, 2004; Kawamura et al., 2006; Monory et al., 2006) (Fig. 2) and where the endocannabinoid system might exert important CB_1 receptor-dependent functions. Interestingly, many CB_1 receptor-positive neurons in the hilus of the dentate gyrus (non-GABAergic and therefore presumably belonging to the mossy cells population) contain dopamine D_2 receptors, suggesting that this brain region might be one site where the

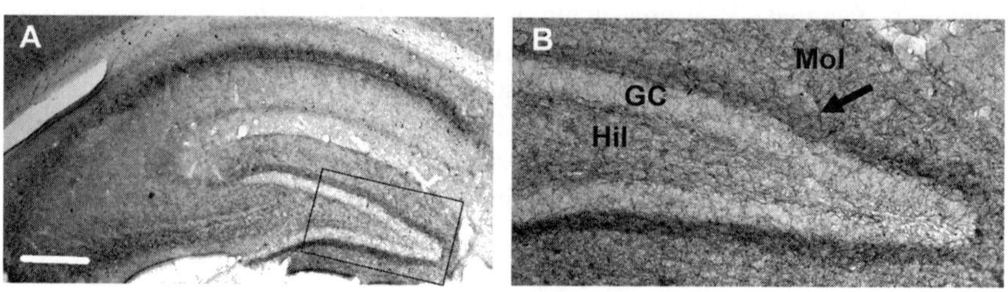

Fig. 2 Immunohistochemical detection of CB_1 in the mouse hippocampal formation. (**a**) A general view of the whole formation. (**b**) Detailed view of the dentate gyrus (corresponding to the square in *panel A*). The intense staining in the inner third of the molecular layer (*arrow*) likely results from the staining of mossy cells terminals (Kawamura et al., 2006; Monory et al., 2006). Hil, hilus of dentate gyrus; GC, granule cell layer; Mol, molecular layer. Bar, 150 μm (**a**), 40 μm (**b**)

interactions between the dopaminergic system and the endocannabinoid system take place (Hermann et al., 2002; Degroot et al., 2006). Biochemical and electrophysiological evidence suggest that other types of cannabinoid receptors might exist in glutamatergic neurons of the hippocampal formation and cortical areas (Freund et al., 2003; Mackie and Stella, 2006; see Chap. 9). Although the recent data summarized earlier suggest that the low levels of CB_1 expression in these neuronal types might explain most of the discrepant results in the literature, the presence of (an)other cannabinoid receptor(s) (i.e., non-CB_1, non-CB2) cannot be excluded at the moment (see later and Chap. 9).

d. Adult neuronal stem cells in the dentate gyrus: During the recent years, the discovery that ongoing neuronal generation occurs in the mammalian central nervous system has attracted a great deal of attention to molecular mechanisms regulating the proliferation, survival and differentiation of adult neuronal progenitor cells (Gross, 2000; Lledo et al., 2006; see Chap. 12). The endocannabinoid system appears to actively participate in these processes (Jiang et al., 2005; Aguado et al., 2005, 2006; Galve-Roperh et al., 2007). A subset of neural progenitor cells in the subgranular zone of the dentate gyrus contain low but detectable levels of CB_1 receptors (Aguado et al., 2005; Galve-Roperh et al., 2007), which participate in cell fate determination (Aguado et al., 2006; Galve-Roperh et al., 2007).
e. Amygaloid nuclei: The amygdala is a complex anatomical component of the forebrain, which plays an important role in the processing of emotional responses (LeDoux, 2000). Several subnuclei form this anatomical entity and CB_1 receptors are differentially expressed in different parts. The amygdala can be grossly differentiated in a "cortical" component, including, among others, the basolateral, lateral and basomedial nuclei, and a "striatal" component, including, among others, the central and the medial nuclei (Swanson and Petrovich, 1998). This arbitrary subdivision is reflected by the different structural organization and neurochemical properties of the different subnuclei. For instance, whereas principal neurons of the "cortical" amygdala use glutamate as main neurotransmitter, the great majority of "striatal amygdala" principal neurons are GABAergic. The expression pattern of CB_1 receptors in the "cortical" part of the amygdala is similar to the one described earlier for other cortical areas: both at protein and mRNA level, CB_1 receptors are abundantly present in GABAergic interneurons mainly belonging to the CCK-positive basket cell population (Marsicano and Lutz, 1999; Katona et al., 2001; McDonald and Mascagni, 2001). Similarly to cortex and hippocampal formation, the expression of CB_1 receptor in glutamatergic neurons of the amygdala has been debated (Marsicano and Lutz, 1999; Katona et al., 2001; Azad et al., 2003; Freund et al., 2003; Domenici et al., 2006). However, recent evidence clearly shows that also in this region, glutamatergic neurons do contain low but significant amounts of CB_1 receptor (Monory et al., 2006). The "striatal" component of amygdala (e.g., central and medial nuclei) contains much lower levels of CB_1 receptor, which are barely detectable by IHC experiments (Katona et al., 2001), but are visible by ISH techniques (Marsicano and Lutz, 1999) and are absent in preparation derived from CB_1 receptor KO mice (Fig. 3). The presence of CB_1 receptor in these brain

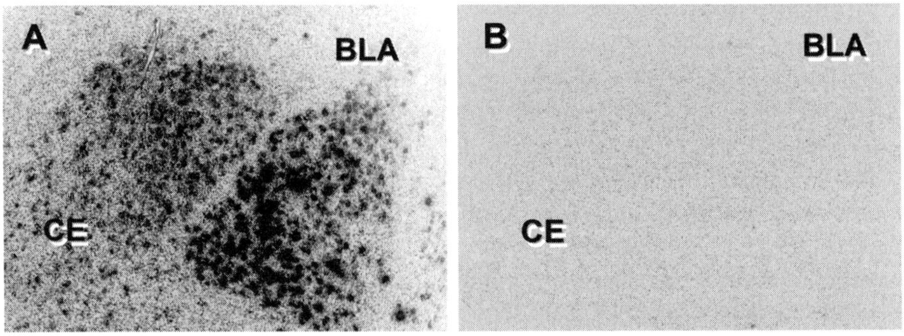

Fig. 3 CB_1 mRNA is detectable in the basolateral nucleus (BLA) and in the central nucleus (CE) of the amygdala. ISH hybridization shows staining of CB_1 mRNA (silver grains) in the amygdaloid region of wild-type (**a**) and CB_1 receptor knock-out mice (**b**). Note that in the BLA, neurons expressing high as well as low levels of the transcript can be seen, whereas in the CE only neurons expressing low levels are observed

regions might be very important for the functions of cannabinoid drugs and of the endocannabinoid system in the processing of emotional responses (Marsicano et al., 2002; Wotjak, 2005).

Distribution of CB_1 Receptors in Subcortical Regions of the Forebrain

a. Basal forebrain: CB_1 receptor is present in many neurons of the basal forebrain, including the medial and lateral septum and the nucleus of the diagonal band, both at protein and mRNA levels (Herkenham et al., 1991; Mailleux et al., 1992; Matsuda et al., 1993; Marsicano and Lutz, 1999). Immunohistochemical studies have revealed that in the tenia tecta, ventral pallidum and substantia innominata, intensely-stained CB_1 receptor-positive fibres are present, whereas in the medial septum and nucleus basalis, the expression of CB_1 receptor appears to be weaker (Harkany et al., 2003; Mackie, 2005b). This work suggested a complete lack of CB_1 protein in cell bodies of basal forebrain cholinergic cells, which appear to express the anandamide-degrading enzyme fatty acid amide hydrolase (FAAH). However, considering that CB_1 receptor is mostly localized on axon terminals, it is possible that expression of CB_1 in cholinergic neurons was missed because cell bodies were analyzed. Indeed, ISH studies show that a great majority of cells in the septum contain CB_1 mRNA (Marsicano and Lutz, 1999). Indeed, a more recent paper showed that in cholchicine-treated rats, where axonal transport of proteins is blocked, CB_1 protein expression is indeed detected in at least one-third of cholinergic neurons, which were identified via immunoreactivity for choline acetyltransferase (ChAT), where co-expression with $GABA_B$ receptors was also observed (Nyiri et al., 2005). Accordingly, another recent study of

Degroot and colleagues (2006) identified that the majority of cholinergic hippocampal nerve terminals are CB_1 receptor-positive, and presynaptic CB_1 receptors control the release of acetylcholine in vivo.

b. Basal ganglia: The expression of CB_1 mRNA in striatal neurons presents a typical gradient, with a great majority of medium spiny neurons (more than 95%) of the dorsolateral part of the caudate expressing moderate levels of the transcript and lower levels of expression in the ventral striatum (Matsuda et al., 1993; Marsicano and Lutz, 1999). These neurons are GABAergic, belong to the subpopulations expressing either D_1 or D_2 dopamine receptors (Hermann et al., 2002) and contain mRNAs for different peptides belonging to the direct as well as the indirect striatal output pathways (Hohmann and Herkenham, 2000). Interestingly, local GABAergic interneurons also seem to contain CB_1 mRNA, whereas large aspiny cholinergic or somatostatin-positive striatal interneurons are devoid of the receptor (Hohmann and Herkenham, 2000). The gradient of expression observed at mRNA level is also observed in IHC and ligand binding experiments, with the dorsolateral part of caudate putamen expressing higher levels of the receptor both in the matrix and patch structures (Tsou et al., 1998a; Egertová and Elphick, 2000). Early functional studies showed that corticostriatal projection neurons express CB_1 receptors (Gerdeman and Lovinger, 2001; Huang et al., 2001a; Gerdeman et al., 2002), which was recently confirmed by IHC studies on tissue sections (Uchigashima et al., 2007) and in nerve terminals (Köfalvi et al., 2005). GABAergic striatal projections to the substantia nigra likely contain CB_1 receptor protein (Matyas et al., 2006). CB_1 mRNA, however, is present also in subthalamic neurons (Matsuda et al., 1993). It is, therefore, possible that glutamatergic subthalamic projections to the substantia nigra also contain CB_1 receptor protein, thereby contributing to the cannabinoid-mediated control of locomotor activity (Sanudo-Pena et al., 2000; Romero et al., 2002; van der Stelt and Di Marzo, 2003).

c. Nucleus accumbens: In the nucleus accumbens, the expression levels of CB_1 mRNA are much lower when compared with the dorsolateral striatum. However, a certain number of neurons in this region do contain detectable levels of the transcript (Matsuda et al., 1993; Monory et al., 2006). Protein expression in this region seems to be associated with glutamatergic transmission derived from prefrontal cortex projections (Robbe et al., 2001). However, the presence of CB_1 receptor on dopaminergic and GABAergic terminals in this region cannot be excluded at the moment (see later).

d. Thalamus: In this brain region, the levels of CB_1 receptor are relatively low (Matsuda et al., 1993; Tsou et al., 1998a; Marsicano and Lutz, 1999; Egertová and Elphick, 2000; Mackie, 2005b). However, some nuclei contain CB_1 mRNA and protein, such as the lateral habenula, the reticular nucleus, the paraventricular thalamic nucleus and the intermediodorsal thalamic nucleus (Matsuda et al., 1993; Tsou et al., 1998a; Marsicano and Lutz, 1999; Mackie, 2005b). Particularly interesting for the functions of the endocannabinoid system might be the relative abundance of CB_1 mRNA in the lateral habenula, which projects to many different brain regions where the receptor might have important functions (Herkenham and Nauta, 1977).

e. Hypothalamus: The endocannabinoid system plays a central role in many functions regulated by different hypothalamic nuclei (Pagotto et al., 2006). It is, therefore, not surprising that CB_1 receptor is present at moderate levels in several hypothalamic areas. Importantly, the hypothalamus is one of the best examples of those regions where the relative amount of CB_1 mRNA and protein does not correlate with the intensity of the CB_1 receptor-mediated signaling events. Indeed, ligand-induced GTPγ assays revealed that in this region, the relatively low levels of expressed CB_1 receptor are more efficiently coupled with G protein activation than in many other regions (Breivogel and Childers, 1998). Relatively sparse information is present in the literature concerning the different hypothalamic nuclei where CB_1 receptors are expressed. CB_1 mRNA and protein is present in several subnuclei, including the paraventricular nucleus, and the dorsomedial, ventromedial and lateral hypothalamic nuclei. In the paraventricular nucleus, CB_1 mRNA is present in neurons synthesizing corticotrophin releasing factor (CRH) and cocaine-amphetamine-regulated transcript (CART) (Cota et al., 2003). In the ventromedial hypothalamus, CB_1 mRNA does not appear to be present in GABAergic neurons (Marsicano and Lutz, 1999). In the lateral hypothalamus, CB_1 mRNA is expressed in a small fraction of neurons expressing pre-pro-orexin and melanin-concentrating hormone (MCH) (Cota et al., 2003). Functional experiments suggest that CB_1 expressed in glutamatergic terminals projecting onto the preoptic anterior hypothalamic nucleus likely mediate hypothermic effects induced by CB_1 agonists (Rawls et al., 2002a,b).

Distribution of CB_1 Receptors in the Midbrain

a. Substantia nigra: CB_1 protein is highly expressed in the substantia nigra (Tsou et al., 1998a; Egertova et al., 2000). This expression likely derives from projecting neurons from other brain regions, such as nuclei of basal ganglia (striatum for GABAergic projections and subthalamic nucleus for glutamatergic ones, see above). These projections likely account for the functional control of CB_1 receptor on nigral activity, e.g., in the control of locomotion (Sanudo-Pena and Walker, 1998; Szabo et al., 2000). Sparse intrinsic nigral neurons might contain very low levels of CB_1 receptor (Matsuda et al., 1993). Nevertheless, this low amount of CB_1 receptor in nigral (and tegmental, see later) neurons might underlie a direct control of CB_1 receptor on dopaminergic transmission, although this has not been shown yet (see Chap. 22 further details). Indeed, CB_1 receptor protein was recently identified in a very low but significant proportion of striatal nerve terminals containing tyrosine hydroxylase (TH, a marker of monoaminergic neurons) (Köfalvi et al., 2005).
b. Ventral tegmental area: CB_1 receptor is centrally involved in the regulation of the activity of dopaminergic neurons in the ventral tegmental area (Szabo et al., 2002a; Melis et al., 2004a,b; Riegel and Lupica, 2004). These actions might partially explain the role of the endocannabinoid system in rewarding processes.

In this region, CB_1 protein is present on glutamatergic as well as GABAergic terminals. Early studies reported no evidence of CB_1 expression in dopaminergic neurons of the VTA either at protein (Herkenham et al., 1991) or mRNA level (Mailleux et al., 1992; Matsuda et al., 1993). However, some data have been published reporting co-expression of CB_1 receptor and TH in this brain region, pointing to the possible direct activity of the endocannabinoid system on dopaminergic neurons (Wenger et al., 2003). Further detailed studies are needed to elucidate this issue.

c. Periaqueductal grey: Low to moderate levels of CB_1 receptor are found in the periaqueductal grey (PAG), where the endocannabinoid system might play a central role in the control of pain sensations (Walker et al., 1999; Hohmann et al., 2005). In contrast to opiate receptors, CB_1 receptor is preferentially (but not exclusively) expressed in the dorsal part of the PAG (Tsou et al., 1998a; Azad et al., 2001).

Distribution of CB_1 Receptors in the Hindbrain

a. Brainstem: CB_1 receptor is expressed in the brainstem region at relatively low levels. In contrast to opiate receptors, it is not found in the respiratory control centres of the medulla (Herkenham et al., 1991; Glass et al., 1997), likely explaining the low mortality caused by cannabinoid intoxication in humans and animals. The dorsal motor nucleus of the vagus and the nucleus of the solitary tract involved in central control of the gastrointestinal activity contain relatively high levels of CB_1 receptors (van Sickle et al., 2001; Mackie, 2005b). CB_1 mRNA is also present in neurons belonging to or surrounding the raphe nucleus, which is the main neuronal source of serotonin in the brain. Here, a low but significant proportion of neurons containing tryptophane hydroxylase 2 mRNA (the rate-limiting enzyme for the synthesis of serotonin) appear to co-express CB_1 mRNA (Haring et al., 2007). Moreover, expression of CB_1 protein was identified on serotonergic terminals in the hippocampus and amygdala (Haring et al., 2007). Therefore, the endocannabinoid system might control serotonergic transmission both by regulating the activity of afferents onto serotonin-producing neurons (Haj-Dahmane and Shen, 2005) and by directly modulating the functions of a subset of serotonergic neurons (Haring et al., 2007).

b. Cerebellum: The expression of CB_1 receptor in the cerebellum is extremely high. Autoradiographic and IHC imaging shows very strong presence of the protein in the molecular layer (Herkenham et al., 1991; Tsou et al., 1998a; Egertová and Elphick, 2000; Mackie, 2005b), whereas CB_1 mRNA is mainly expressed in the granule cell layer (though scattered cells in the molecular layer also express the CB_1 receptor transcript) (Matsuda et al., 1993). This complementary expression of CB_1 protein and mRNA suggests that CB_1 receptors are expressed on climbing fibres and parallel fibres and in basket cells. A certain number of Purkinje cells might express low amounts of CB_1 mRNA (Matsuda et al., 1993). These data indicate that the main glutamatergic and GABAergic inputs onto Purkinje cells are under the control of the endocannabinoid system, as exhaustively shown in several electrophysiological studies (Kreitzer and Regehr, 2001a,b, 2002; Maejima et al., 2001; Diana et al., 2002). The functions of the endocannabinoid system in the regulation of synaptic transmission in this brain region are well established, whereas functional information regarding a role of the endocannabinoid system in the control of cerebellum-dependent functions still remains sparse.

Distribution of CB_1 Receptors in the Spinal Cord

As with the brain, the endocannabinoid system in the spinal cord has attracted a lot of attention and spurred numerous studies. This is not surprising, given its immense therapeutic significance, not only in pain and analgesia, but also in mechanisms governing spinal cord diseases such as multiple sclerosis, traumatic spinal injury and so on. However, the anatomical data concerning the endocannabinoid system in the spinal cord and somatosensory ganglia, such as the dorsal root ganglia (DRG) and trigeminal ganglia, have been somewhat controversial. For instance, although the bulk of current literature favours the view that CB_1 receptors are targeted exclusively to axons and axonal terminals, i.e. a presynaptic localization, the spinal cord represents one of the few regions where postsynaptic localization of CB_1 receptor has been reported (Farquhar-Smith et al., 2000; Salio et al., 2002b). IHC analyses have revealed anti-CB_1 receptor immunoreactivity in the dorsolateral funiculus, laminae I and II inner/III transition and lamina X. In addition to neuronal expression, CB_1 receptor immunoreactivity has also been reported in spinal astrocytes (Salio et al., 2002a). The existence of presynaptic receptors is suggested by the staining of the DRG neurons (see later) as well as the axons of the Lissauer's tract (Salio et al., 2002b). Because the superficial spinal laminae, which show dense immunoreactivity for CB_1 receptor, constitute major termination sites of primary afferent terminals of nociceptive DRG neurons, it is important to clarify what proportion of CB_1 receptor in these pain-processing laminae derives from the peripheral vs. the central nervous system. CB_1 receptor immunoreactivity in these spinal regions has been observed to show little co-localization at the axonal level with primary afferent nociceptive markers. Furthermore, interruption of primary afferent input via dorsal root rhizotomy does not significantly lower CB_1 receptor immunoreactivity in these regions (Farquhar-Smith et al., 2000). This is supported by ligand-binding studies which show that neonatal capsaicin treatment, which leads to an early destruction of nociceptive afferent fibres, reduces ligand binding to CB_1/CB_2 receptors in the superficial dorsal horn by a minor extent only (Hohmann and Herkenham, 1998). Consistent with the above, a nociceptor-specific deletion of CB_1 receptor using conditional gene targeting approaches was observed to reduce spinal CB_1 receptor-specific ligand binding by approximately 20% only and not lead to a major change in the CB_1 receptor immunoreactivity in spinal laminae I and

II (Agarwal et al., 2007). Taken together, these findings suggest that a significant proportion of CB_1 receptor protein in the superficial spinal laminae is derived from spinal, not peripheral, neurons. This is also supported by ISH studies, which have revealed labelling of cells throughout the spinal cord (Fig. 4), which is entirely lost in global CB_1 receptor knock-out mice (Agarwal et al., 2007). Postsynaptic CB_1 receptors in the spinal cord are evidenced by the staining of numerous neurons in lamina II outer as well as lamina X, many of which also express GABA, the neuronal nitric oxide synthase (nNOS) or the protein kinase C subunit gamma, based upon which they have been suggested to be spinal interneurons (Farquhar-Smith et al., 2000; Salio et al., 2002b). Indeed, the postsynaptic localization of CB_1 receptor immunoreactivity in somatic as well as dendritic compartments has been confirmed by electron microscopy (Salio et al., 2002b). However, as discussed in the

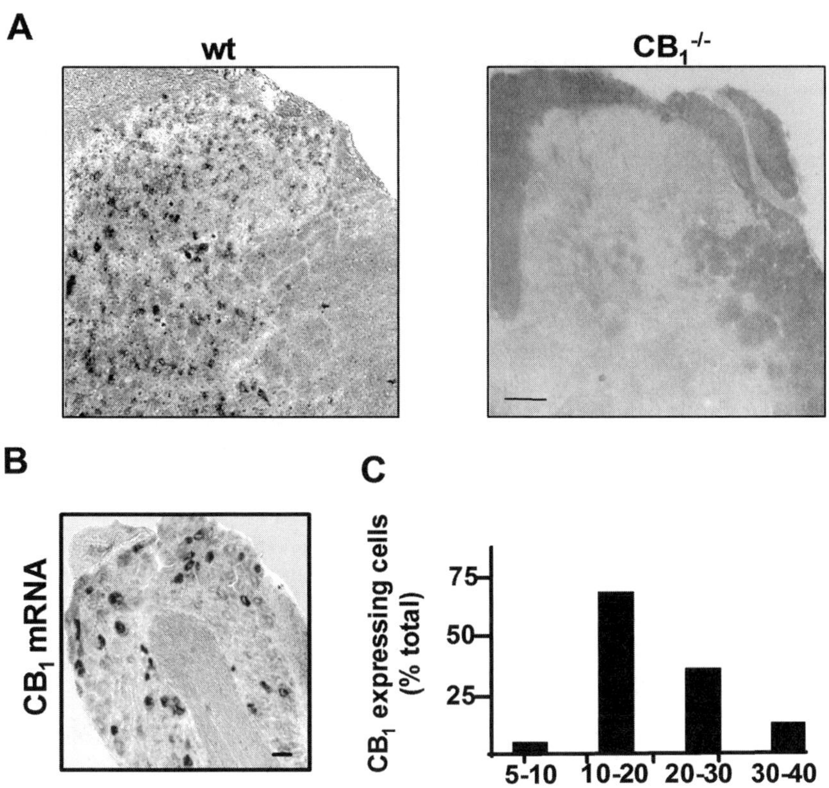

Fig. 4 In situ hybridization analysis of adult mouse spinal cord and DRG using a riboprobe recognizing CB_1 mRNA. (**a**) Widespread labelling is seen in all laminae of the spinal cord of adult wild-type mice, but not of global CB_1 receptor knock-out mice. (**b**) A large number of DRG cells express CB_1 mRNA. (**c**) The corresponding quantitative size analysis of neurons shows that a large fraction of DRG neurons expressing CB_1 mRNA are small-diameter neurons. Bars, 150 µm in (**a**) and 40 µm in (**b**)

introduction, it is important to consider that differences in the origin and specificity of antibodies used in diverse studies and the frequent lack of appropriate controls, e.g., loss of staining in CB_1 receptor knock-out mice, do hinder an unequivocal interpretation of several of these earlier studies. In a recent study, which implemented global CB_1 receptor knock-out mice as a control, spinal CB_1 receptor immunoreactivity was observed not only in the neuropil, but also in small-sized somata of cells in the dorsal horn (Agarwal et al., 2007). However, it remains unclear whether these somata represent neurons or astrocytes. Thus, further studies are required to clarify whether the spinal cord presents an exception to the general rule of an exclusive presynaptic (axonal) localisation of CB_1 receptor.

Distribution of CB_1 Receptors in Dorsal Root Ganglia and Peripheral Nerves

Although numerous past and recent studies have clearly demonstrated the presence of CB_1 receptor in neuronal somata in the DRG and their axons (primary afferents or peripheral nerves), the exact nature of the distribution of CB_1 mRNA and protein in different peripheral neuronal subtypes is very controversial – different studies have reported a varying abundance of CB_1 mRNA or protein in different DRG cell types identified via marker proteins (Table 1). For example, some earlier studies reported that CB_1 receptor is predominantly expressed in non-nociceptive DRG neurons (medium- and large-diameter cells, distinguished by expression of neurofilament 200) and only very poorly in a small proportion of nociceptive DRG neurons (small-diameter cells, distinguished by expression of markers such as substance P, $Na_v1.8$, $TRPV_1$ and by binding to labelled Isolectin B_4). In contrast, more recent studies have suggested that the distribution of CB_1 receptor in nociceptive neurons of the DRG and trigeminal ganglia is much broader than previously thought and support a role for peripheral CB_1 receptor in pain and analgesia. The most likely reasons for these

Table 1 Percentage of marker-positive cells with CB_1 expression in adult DRG

Marker	(Hohmann and Herkenham, 1999) (CB_1 mRNA) (%)	(Bridges et al., 2003) (CB_1 protein) (%)	(Binzen et al., 2006) (CB_1 protein) (%)	(Mitrirattanakul et al., 2006) (CB_1 protein) (%)	(Agarwal et al., 2007) (CB_1 protein) (%)
CGRP or substanceP	13	12 ± 4	Not reported	76–82	40 ± 2
Isolectin B4	Not reported	22 ± 6	Not reported	not reported	38 ± 2
$TRPV_1$	Not reported	7 ± 2	69 ± 7	76–82	90 ± 2
NF200	Not reported	49 ± 1	Not reported	81	94 ± 1

discrepancies lie in species differences (rats vs. mice), the sensitivity of the detection methods applied (e.g., in situ riboprobes of various lengths), diverse labelling methods, various sensitivities and stringencies of protocols or antibodies recognizing different epitopes on CB_1 receptor. In this regard, it is important to note that CB_1 receptor can exist as multiple N-terminal splice variants, some of which are not recognised by antibodies targeted against the N-terminus of CB_1 receptor (Shire et al., 1995; see Chap. 9 for detailed molecular biology and pharmacology of the splice variants). In general, judging from past and recent studies, it is quite likely that CB_1 receptor is expressed at high levels in large-diameter neurons and in a small population of small-diameter neurons (and therefore detected readily in all studies) and is, in addition, expressed at lower levels in other small-diameter neurons in the rat (therefore, detected in some, but not all studies depending on sensitivity of assay used). In this regard, the scenario of CB_1 receptor expression in DRG is not different from the controversies that existed until very recently with respect to CB_1 receptor expression in the brain, which is discussed extensively earlier. Indeed, as with the brain, conditional knockout mice in which CB_1 receptor was deleted specifically in specific populations of DRG neurons (see later) have revealed that CB_1 receptor expressed in peripheral nociceptive neurons, which had been actually missed in initial expression analyses studies, is of very high physiological and therapeutic relevance (Agarwal et al., 2007). These studies have again brought home the message that the abundance of CB_1 receptor expression is not necessarily equivalent to its functional significance. In contrast to the outcome of a previous ISH study (Hohmann and Herkenham, 1999), a riboprobe, which detects CB_1 mRNA with a high degree of sensitivity (Marsicano and Lutz, 1999; Marsicano et al., 2003) as well as specificity (complete lack of signals in global CB_1 receptor knockout mice – shown in Fig. 4) revealed that in mouse DRG, more than 50% of cells expressing CB_1 mRNA are small-diameter neurons (diameter of less than 20 µm) (Agarwal et al., 2007). Consistent with the above, goat-derived C-terminal CB_1 receptor antibody, which recognizes all CB_1 receptor splice variants in a very specific manner [complete lack of staining in global CB_1 receptor knockout mice, see Fig. 5 and (Coutts et al., 2002)], revealed that more than 40% of CB_1 receptor-expressing DRG neurons are nociceptors (Agarwal et al., 2007). In this study conducted on mouse DRGs, CB_1 protein was detected in more than 80% $TRPV_1$ receptor-positive neurons, which is consistent with two very recent studies on rat DRG (Binzen et al., 2006; Mitrirattanakul et al., 2006). Furthermore, 40–45% of $Na_{v1.8}$-expressing neurons of the peptidergic population and of the non-peptidergic Isolectin-B4-binding population of nociceptors were found to express CB_1 protein, in line with previous reports on rat DRG (Binzen et al., 2006; Mitrirattanakul et al., 2006). Thus, the pattern of expression of CB_1 mRNA as well as protein in the DRG is considerably broader than the pattern described in older studies. A recent study has shown, moreover, expression as well as peripheral transport of CB_1 receptor in nociceptors that is further increased in states of peripheral inflammation (Amaya et al., 2006). Taken together, these recent studies suggest that the former view that CB_1 receptor is not expressed to any substantial extent in nociceptors is incorrect. This is also supported by novel functional data, which show a striking loss of a major proportion of cannabinoid analgesia in mice conditionally lacking CB_1 receptor on nociceptors specifically (Agarwal et al., 2007).

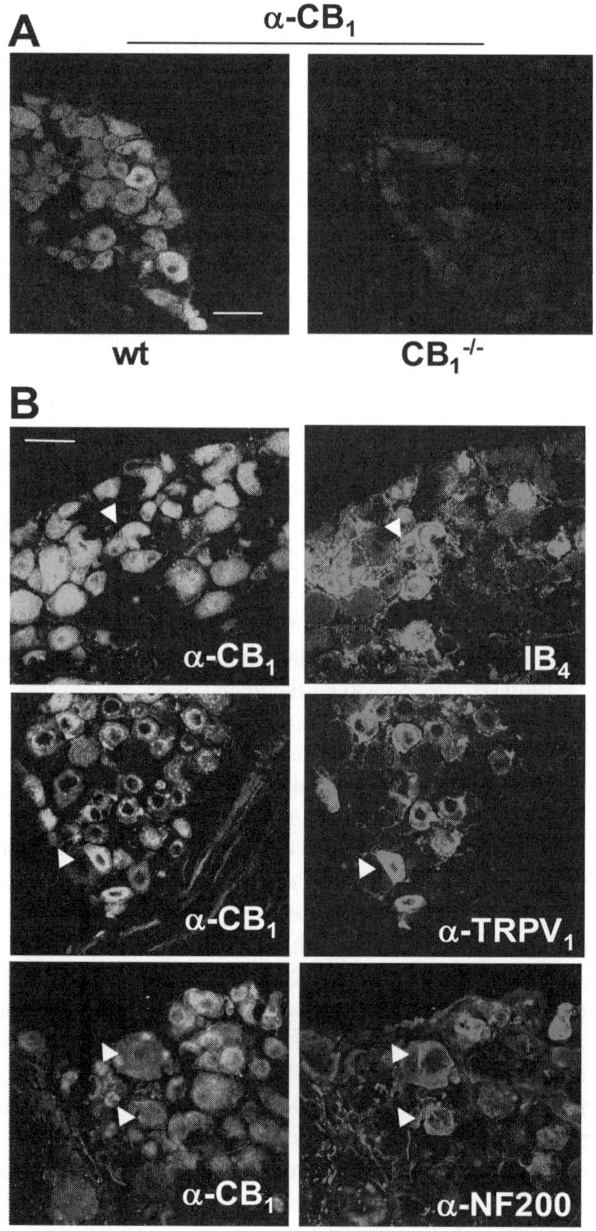

Fig. 5 Immunohistochemical analysis of CB_1 protein expression in adult mouse DRG. (**a**) A goat anti-CB_1 antibody yields specific labelling of DRG neurons, which is entirely lost in DRG of global CB_1 knock-out mice. (**b**) Typical examples of anti-CB_1 receptor immunoreactivity in subpopulations of DRG neurons labelled using IB4, anti-$TRPV_1$ and anti-neurofilament 200 (NF200) antibodies in wild-type mice. Note the large overlap between all labelled populations and CB1 expression. Bars, 40 μm

Distribution of CB_2 Cannabinoid Receptors and Other Endocannabinoid-Target Receptors in the Nervous System

At the current state of knowledge, CB_1 receptors are the most important effectors of the functions of the endocannabinoid system in the CNS as well as of the pharmacological effects of cannabinoid drugs in the brain and spinal cord. The presence of CB_2 receptors has been mainly described in peripheral immune cells. However, recent data suggest that CB_2 receptors might be present in certain neuronal populations (see further for details). Moreover, (endo)cannabinoids have been also shown to act in neurons through the activation of different targets than the "conventional" cannabinoid receptors, CB_1 and CB_2. A full discussion of all these putative "cannabinoid receptors" is present in several exhaustive reviews (Di Marzo et al., 2002; Mackie and Stella, 2006; Baker et al., 2006; Begg et al., 2005; see Chap. 9) and is beyond the scope of the present chapter. Here, we will focus on the expression pattern of some of these potential targets of cannabinoids in the CNS. Among this ever-growing list, we have chosen to elaborate on three interesting candidates: the transient receptor potential vanilloid type 1 receptor ($TRPV_1$ receptor, previously known as vanilloid receptor 1 or VR1), the acid sensitive K^+ channel TASK-1 and the "orphan" G protein-coupled receptor GPR55 (see Chap. 9).

a. CB_2 receptors: Originally, CB_1 receptors were described as the "neuron type" cannabinoid receptors, whereas CB_2 receptors were identified as the "immune type" members of the family (Matsuda et al., 1990; Munro et al., 1993). This differential distribution has been profoundly challenged in the last years. On one hand, the presence of CB_1 receptors in peripheral tissues, including some types of immune cells, has been clearly demonstrated (Pacher et al., 2006; Pagotto et al., 2006). On the other hand, CB_2 receptor expression is now known to be not restricted to peripheral immune cells. Indeed, CB_2 receptor was recently described to be functionally present in neurons of the brainstem, where it can modulate emesis (van Sickle et al., 2005). Some studies even propose a wide distribution of CB_2 receptor in the brain (Gong et al., 2006; Onaivi et al., 2006). However, on account of a lack of stringent controls, such as lack of staining in corresponding tissues from CB_2 receptor knock-out mice under the same staining conditions, it is difficult, in our opinion, to unequivocally pinpoint the expression of CB_2 receptors in the brain. One important noteworthy feature of CB_2 receptors is that their expression appears to be inducible (Wotherspoon et al., 2005). In several studies, CB_2 receptor was not detected in neurons in physiological states, but the expression was induced in pathological states, e.g. in neurons of the dorsal root ganglia following nerve injury (Wotherspoon et al., 2005). Because CB_2 receptors are also expressed by neurons under in vitro conditions (Ross et al., 2001), it is possible that CB_2 receptor expression is induced in "stressed" neurons. Therefore, it may be important in future studies addressing CB_2 receptor expression in neurons to carefully control the conditions at the moment of killing test animals for expression analysis.

b. $TRPV_1$ receptors: The $TRPV_1$ receptor is the first member of a large family of transient potential ion channel receptors (TRPs; see Chap. 8). It was originally discovered as the endogenous target of the "hot" component of chilli peppers, capsaicin (Caterina et al., 1997; Tominaga et al., 1998). $TRPV_1$ receptors are also activated by low pH, heat and, more recently, were shown to be activated by the endocannabinoid anandamide and by a novel putative endocannabinoid, *N*-arachidonoyl-dopamine (Zygmunt et al., 1999; Di Marzo et al., 2002; van der Stelt and Di Marzo, 2004; see Chap. 2, 4, 8). The activity of endocannabinoids at $TRPV_1$ receptors might have very interesting functional consequences. The presence of $TRPV_1$ in the spinal cord and peripheral neurons is well established (see later). In the brain, $TRPV_1$ is present at much lower levels than in peripheral neurons (Sanchez et al., 2001), but the exact distribution of these receptors in the brain is only starting to be elucidated in detail. Original evidence for the presence of $TRPV_1$ in the brain came from ISH and IHC experiments in rats (Mezey et al., 2000), which suggested a widespread distribution of the receptor in different brain regions. Since then, ISH, IHC and ligand-binding studies using the specific radioligand [H^3]RTX have partially confirmed previous observations (Cortright et al., 2001; Sanchez et al., 2001; Szabo et al., 2002b; Roberts et al., 2004; Toth et al., 2005; Cristino et al., 2006). Briefly, $TRPV_1$ is localized in the cortex, hippocampus, dentate gyrus, central amygdala, striatum, hypothalamus, thalamus, substantia nigra, cerebellum, locus coeruleus and other smaller brain nuclei. This wide expression of $TRPV_1$ receptor in the brain supports the notion that this receptor plays broader roles in animal behaviour than the control of pain sensations alone. For instance, it was recently shown that $TRPV_1$ knock-out mice display pain-independent lower anxiety in a range of behavioural tests, a phenotype that is essentially opposite to that seen in CB_1 receptor knock-out mice (Marsch et al., 2007), suggesting that CB_1 and $TRPV_1$ receptors may potentially play opposite functions in the regulation of certain behaviours. Interestingly, CB_1 and $TRPV_1$ receptors can influence each other's function when the two proteins are co-expressed in the same cells in vitro (Hermann et al., 2003; Sagar et al., 2004), and they have been described to be co-expressed in some neurons of the brain (Cristino et al., 2006). Noteworthy though that at least in the hippocampus, the expression pattern is more likely to be post-synaptic (Cristino et al., 2006), and a recent studies also failed to demonstrate functional presynaptic hippocampal $TRPV_1$ receptors in the rat hippocampus (Köfalvi et al., 2006, 2007). Concerning the spinal cord and peripheral neurons, initial studies suggested a selective expression of $TRPV_1$ receptor in heat-response small-diameter neurons of the DRG. However, it is now clear that $TRPV_1$ receptor is also expressed in the spinal cord, in addition to the brain as described earlier. $TRPV_1$ receptor immunoreactivity has been observed in somata in lamina II of the spinal dorsal horn and was also prominently seen in dendrites that are contacted by primary afferent endings in electron microscopy studies (Valtschanoff et al., 2001). In double IHC experiments, about 14% of $TRPV_1$ receptor-expressing dorsal horn cell bodies were also found to express the NK1 receptor, suggesting that endogenous vanilloids may directly modulate the activity of spinal projection neurons (Dolly et al., 2004). Furthermore, co-expression of $TRPV_1$ and CB_1 receptors has also been suggested in the spinal cord (Cristino et al., 2006).

c. TASK-1 K^+ channels: The endocannabinoid anandamide was shown to directly inhibit TASK-1, an acid- and anaesthetic-sensitive K^+ channel, which sets the resting membrane potential of some types of central and peripheral neurons (Di Marzo et al., 2002; Maingret et al., 2001; see Chap. 9). Interestingly, this effect results in a depolarization of cerebellar granule neurons, possibly explaining some "paradoxical" excitatory effects of anandamide (Maingret et al., 2001). TASK-1 belongs to the family of the four transmembrane (4TM) channels, which likely form homodimers with four pore domains (Bayliss et al., 2001; Mathie et al., 2003). The distribution of TASK-1 in the brain has been described (Karschin et al., 2001; Kindler et al., 2000; Talley et al., 2001). In Northern blot analysis, TASK-1 RNA was detected in all subregions of CNS, with highest levels in the cerebellum, medulla and subthalamic nucleus (Kindler et al., 2000). ISH studies have elucidated the expression pattern of TASK-1 in more detail. In the brain, TASK-1 mRNA is present in many different regions, with highest levels in the granule cell layer of the cerebellum, arcuate nucleus of the hypothalamus, the granule cells of olfactory bulb and the locus coeruleus (Karschin et al., 2001; Talley et al., 2001). High levels were also described in islands of Calleja, in the septum and motor nuclei of cranial nerves, all regions containing cholinergic neurons, which appear to be particularly associated with this channel (Karschin et al., 2001). In cortical regions, low to moderate levels of TASK-1 are present in the piriform cortex, neocortex (where few scattered cells expressing high levels are observed) and in glutamatergic neurons of the hippocampal formation and amygdala (Karschin et al., 2001; Talley et al., 2001). Also many subnuclei of the hypothalamus, thalamus, midbrain and brainstem contain detectable levels of TASK-1 mRNA (Karschin et al., 2001; Talley et al., 2001). In ISH studies, the predominant expression of TASK-1 appears to be in neurons, but IHC experiments came to different conclusions (Kindler et al., 2000). In fact, these authors reported a strong association of TASK-1 with glial cells in the hippocampal formation, in the cerebellar cortex and granular layer (here also neuronal staining was observed) and in the white matter of the spinal cord. Neuronal staining was reported in the neocortex, basal ganglia, amygdala, thalamus, midbrain and in Purkinje and granular cells of cerebellum (Kindler et al., 2000). The reasons underlying these discrepancies are not clear yet. Neuronal expression of TASK-1 in serotonergic neurons has also been described (Washburn et al., 2002). Notably though that in the hippocampus, it was recently suggested that the predominant TASK channel, modulated by hybrid endocannabinoid/endovanilloid substances, anandamide and *N*-arachidonoyl dopamine, is the TASK-3 (Köfalvi et al., 2007; see Chap. 9). To date, very few studies have addressed the distribution of TASK channels in the spinal cord and DRG. In peripheral sensory afferents, TASK channels have been implicated in the pain sensory transduction pathway and could constitute a target for anaesthetics and analgesics. TASK-1 has been detected in small- to medium-diameter DRG cell types and it has been proposed that these sensory afferents might contain functional heterodimeric channels (Rau et al., 2006). In the spinal cord, strong TASK-1 immunoreactivity has been mainly reported in ependymal cells lining the central canal and in white matter (Kindler et al., 2000).

d. GPR55: In the last few years, two patents proposed that a cloned orphan G protein-coupled receptor might be a *bona fide* cannabinoid receptor (Baker et al., 2006; see Chap. 9). Despite the fact that a definitive confirmation of this idea has not yet been provided (Petitet et al., 2006), there is accumulating evidence that GPR55 could really be the target of several endogenous and exogenous cannabinoids (Reyes et al., 2006). There is not much information concerning the expression of GPR55 in the brain. The original publication reporting the cloning of the orphan receptor reported the Northern blot analysis of GPR55 expression in human brain (Sawzdargo et al., 1999). The transcript seems to be expressed in human caudate and putamen, but no expression was detected in cortex, hippocampus, thalamus, pons and cerebellum (Sawzdargo et al., 1999). Initial ISH experiments, however, revealed that GPR55 might be present in rat hippocampus, thalamic nuclei and parts of the midbrain (Sawzdargo et al., 1999), as also reported in the Allen Brain Atlas studies (http://www.brainatlas.org/aba/). Recently, immunohistochemical data have also been reported showing that the expression of GPR55 is developmentally regulated in the rat brain. In the embryonic brain, GPR55 expression seems to be strongly associated with various axonal tracts, including the thalamocortical pathway innervating layer IV of the "barrel" cortex. Two weeks after birth, this axonal expression decreases and post-synaptic neuronal staining of the hippocampus, thalamus, striatum and cortex become more evident (Chen et al., 2006). Further studies are mandatory, especially in the soon available GPR55 KO mouse, because of the important implications of this novel putative cannabinoid receptor in the endocannabinoid system.

Distribution of Endocannabinoids and Related Enzymes in the Nervous System

Endocannabinoids: As mentioned earlier and described in details in other chapters of this book, endocannabinoids are lipids in nature, polyunsaturated fatty acid derivatives, which share the ability to modulate the activity of cannabinoid receptors. Several endogenous compounds have been shown to interfere with the activity of cannabinoid receptors, at least under in vitro conditions. They are arachidonoyl ethanolamine (AEA, also known as anandamide) (Devane et al., 1992; see Chap. 2), 2-arachidonoylglycerol (2-AG) (Mechoulam et al., 1995; Sugiura et al., 1995; see Chap. 2), noladin ether (2-AGE) (Hanus et al., 2001; see Chap. 4), virodhamine (Porter et al., 2002) and *N*-arachidonoyldopamine (NADA) (Huang et al., 2001b, 2002; see Chap. 4). Of these compounds, the best studied and the ones clearly shown to act, at least in part, through the activation of cannabinoid receptors are

AEA and 2-AG and this subchapter will deal mainly with their distribution in the CNS. Given the lipophilic nature of endocannabinoids, it is very difficult to identify their presence in biological samples with spatial resolution and the measurement of their levels must necessarily rely on biochemical analysis of tissue extracts. Endocannabinoids are believed to be produced from precursors residing in cell membranes by the activation of more-or-less specific enzymatic machineries, the molecular composition of which is still in the process of elucidation (Piomelli, 2003; De Petrocellis et al., 2004; Di Marzo et al., 2004). The distribution of these agents in the CNS will be the subject of this part of the present chapter. Because they are derived from cell membranes, it is very likely that any cell type potentially possesses the ability to produce endocannabinoids. Indeed, in the CNS, neurons, glial cells and endothelia have been shown to produce endocannabinoids and their levels can be modulated in physiological and pathophysiological conditions (see Chap. 2). Here, we will describe the general distribution of endocannabinoids in the CNS, whereas the pathophysiological regulation of their levels is described in other chapters of this book and in specialized reviews (Di Marzo et al., 2000a, 2004; Piomelli et al., 2000, 2003). Another interesting issue concerning endocannabinoids is their localization and their ability to be released in the extracellular space. It is currently very well-accepted that endocannabinoids act as retrograde neurotransmitters at the synaptic level (Alger, 2002; Kreitzer and Regehr, 2002; Chevaleyre et al., 2006; Marsicano and Lutz, 2006; see Chaps. 11 and 21). However, given the lipophilic nature of these compounds, it is not yet fully understood how they can leave the membrane and travel through the watery extracellular space. Again, it is very difficult to address this issue because a direct observation of endocannabinoids in different tissue compartments is not currently feasible. Nevertheless, the combination of microdialysis and analytical techniques have enabled the determination of endocannabinoids in extracellular dialysates in the striatum (Giuffrida et al., 1999), in the periaqueductal grey (Walker et al., 1999) and more recently, in the nucleus accumbens of rats self-administering cocaine chronically (Caille et al., 2007). Interestingly, in the first two studies, only AEA was detected in the microdialysates, whereas in the latter both AEA and 2-AG were identified. Therefore, there is indirect evidence that, at least under certain conditions, endocannabinoids could be present in extracellular fluids within the CNS. In the brain, endocannabinoids can be found in all brain regions. Early studies revealed that the amount of 2-AG is about two orders of magnitude higher than AEA (Stella et al., 1997). In rat brains, AEA levels range from 10–15 pmols g^{-1} of tissue in the diencephalon, cortex and cerebellum, to 30–60 pmols g^{-1} of tissue in the medulla, the limbic forebrain, the hippocampus, the striatum and the brainstem, the last three regions showing the highest amounts (Bisogno et al., 1999a; Di Marzo et al., 2000b). Conversely, 2-AG is present in amounts ranging between 2 and 14 nmols g^{-1} of tissue, with the highest content (similar to AEA) described in the hippocampus, striatum and brainstem (Bisogno et al., 1999a). Both AEA and 2-AG are also present in the hypothalamus at about 8–9 pmols g^{-1} of tissue and 8 nmols g^{-1} tissue, respectively (Di Marzo et al., 2001). Importantly, in this region, the levels of endocannabinoids are regulated by feeding and by leptin administration, thereby providing one of the first pieces of evidence that the endocannabinoid system is important in the control of food intake (Di Marzo et al., 2001; see Chap. 14). In general, the levels of endocannabinoids in the brain do not correlate well with the levels of expression of CB_1 receptors in the same brain regions (see earlier). This apparent discrepancy might be explained by the fact that endocannabinoids can additionally bind to yet uncharacterized receptors (Mackie and Stella, 2006). However, it is also possible that similar to the coupling efficiency of CB_1 in different brain regions (Breivogel and Childers, 1998), the expression levels of the receptor do not correlate with the activity of the endocannabinoid system in a particular region and further that, the actual local concentrations of receptors and ligands may play a more decisive role in certain functions of the endocannabinoid system. As with the brain, several endocannabinoids, such as 2-AG, anandamide as well as the anandamide cogener, palmitoylethanolamide (PEA) have been detected in the spinal cord (Suplita et al., 2005; Agarwal et al., 2007; Petrosino et al., 2007). Interestingly, spinal endocannabinoid levels were found to be elevated following nerve injury (Petrosino et al., 2007) or continuous foot-shock (Suplita et al., 2005), but not following peripheral paw inflammation (Agarwal et al., 2007).

Enzymes for the Synthesis of Endocannabinoids

a. Synthesis of anandamide: Despite intense efforts in the last 15 years, the enzymatic pathways involved in the synthesis of AEA and other polyunsaturated *N*-acyl-phosphatidylethanolamines are far from being completely understood in their molecular identities. Since Chap. 2 extensively reviews the question of endocannabinoid synthesis, here we only briefly do so: AEA is believed to be synthesized in a two-step pathway (Piomelli, 2003; Basavarajappa, 2007). First, *N*-arachidonoyl-phosphatidylethanolamine (NAPE) is synthesized from phosphatidylethanolamine and arachidonic acid by an *N*-acyltransferase enzymatic activity. The molecular identity of the specific enzymes involved in this step is not known so far, although, recently, a cytosolic enzyme capable of catalyzing the synthesis of NAPE (i.e. with *N*-acyltransferase activity) in a Ca^{2+}-independent manner was identified as RLP-1 (Jin et al., 2007). As a second step, a NAPE-specific phospholipase D (NAPE-PLD) catalyzes the cleavage of NAPE to yield AEA (Piomelli, 2003). Recently, a form of PLD able to catalyze the synthesis of AEA in vitro was cloned and proposed as the NAPE-PLD able to mediate the synthesis of AEA in the body (Ueda et al., 2005). However, the recent generation of NAPE-PLD null mutant mice revealed the existence of alternative pathways leading to the synthesis of AEA (Leung et al., 2006). In particular, the phosphatase PTPN22 (Liu et al., 2006) and the α/β-hydrolase 4 (Abh4) (Simon and Cravatt, 2006) have been recently added to the list of enzymes possibly involved in the synthesis of AEA in vivo. The expression patterns of these enzymes in the CNS are not known in great detail. Western blotting and RT-PCR were used to identify NAPE-PLD expression in different brain regions, showing that the enzyme is present in all nine brain structures examined, with higher levels in the thalamus. Interestingly, an age-dependent increase in NAPE-PLD expression was also observed between 14-day-old and 56-day-old rats in most brain regions (Morishita et al., 2005). The phosphatase PTPN22, capable of synthesizing AEA by dephosphorylation of an alternative precursor, P-AEA, which is derived from the phospholipase C-mediated cleavage of NAPE, is mainly expressed in lymphoid and hematopoietic tissues and at much lower levels in the CNS (Liu et al., 2006). However, Western and Northern blotting experiments have identified the enzyme in brain extracts and initial IHC experiments have started revealing its specific expression in neurons of the hippocampus (Liu et al., 2006) (see also the Allen Brain Atlas, http://www.brainatlas.org/aba/). Abh4 has not been studied in detail concerning its expression in the CNS. However, RT-PCR studies have suggested that the enzyme is present in the brain and spinal cord (Simon and Cravatt, 2006). Further studies are needed, first to clarify the complete molecular identity of the enzymes involved in the synthesis of AEA and second, to elucidate their expression pattern in the CNS in details.

b. Synthesis of 2-AG: The synthetic pathways of this endocannabinoid, abundantly present in the CNS, are also very complex and their molecular components are just starting to be elucidated (Piomelli, 2003; Basavarajappa, 2007; see Chap. 2). Nevertheless, studies in the last few years allowed identification of some of the enzymes involved in the synthesis of 2-AG and their expression patterns in the brain have been described partially. Briefly, 2-AG can be produced mainly via two distinct pathways. In one route, a phospholipase C (PLC) induces the formation of 1,2-diacylglycerol (DAG), which is, in turn, cleaved by a DAG lipase to finally produce 2-AG (Piomelli, 2003). Alternatively, 2-AG can be synthesized via the phospholipase A_1 (PLA_1)-mediated production of a 2-arachidonoyl-lysophospholipid, which, in turn, might be hydrolyzed to 2-AG by lyso-PLC activity (Bisogno et al., 1999b; Piomelli, 2003; Basavarajappa, 2007). The former pathway is better known in terms of the precise molecular identification of the enzymes involved. Indeed, one isoform of PLC involved in the synthesis of 2-AG has been recently identified as the $PLC\beta_1$ subtype (Hashimotodani et al., 2005), and two isoforms of DAG lipase (DAGL α and β) have been recently identified as the ones responsible for the cleavage of DAG (Bisogno et al., 2003). Data are present in the literature concerning the distribution of these enzymes in the CNS. $PLC\beta_1$ is one of the four different isoforms of PLCβ. PLCβ is activated by $G_{q/11}$ proteins, as, for instance, after stimulation of $mGuR_1$ and $mGluR_5$ metabotropic glutamate receptors and different subclasses of muscarinic acetycholine receptors (Gutkind et al., 1991; Abe et al., 1992; Watanabe et al., 1998; see Chapter 11), thereby explaining their influence on endocannabinoid and, in particular, 2-AG synthesis (Hashimotodani et al., 2005). $PLC\beta_1$ appears to be the main isoform involved in endocannabinoid synthesis (Hashimotodani et al., 2005). The mRNA of PLCβ1 has been shown by ISH experiments to be expressed in the whole rat brain, particularly in the olfactory bulb, the cortex, caudate putamen, piriform cortex, lateral septum, hippocampal formation, with weaker expression in the midbrain, cerebellum (including Purkinje cells), medulla and brainstem (Watanabe et al., 1998). Already at the first glance, this expression pattern is strikingly similar to that of CB_1 receptors in the brain. In the hippocampal formation, $PLC\beta_1$ mRNA is present mainly in CA1 and CA3 pyramidal neurons, where it likely co-localizes with CB_1 receptors and in granule cells of the dentate gyrus. GABAergic interneurons seem to be devoid of the enzyme, though they may express very low amounts of the mRNA (Watanabe et al., 1998). Interestingly, mossy cells in the hilus of the dentate gyrus, expressing CB_1 receptor (Monory et al., 2006), also seem to express mRNA of the enzyme (Watanabe et al., 1998). Therefore, $PLC\beta_1$ is expressed in locations that could potentially mediate synthesis of 2-AG both in cells neighbouring CB_1 receptor-positive terminals and in neurons containing CB_1 receptors themselves. DAGLα and β are both present in neurons (Bisogno et al., 2003). Interestingly, the subcellular localization of these proteins appears to undergo a switch during development, with a prominent axonal distribution in early embryonic stages, and a prominent somatodendritic expression in adults (Bisogno et al., 2003). ISH and IHC experiments recently addressed the distribution of these enzymes in the adult rat brain (Yoshida et al., 2006; Uchigashima et al., 2007). DAGLα and β mRNA are expressed in the brain in similar patterns, with the β isoform being expressed at lower levels. DAGLα mRNA is mainly expressed in the olfactory bulb, cortex, caudate putamen, thalamus and Purkinje cells of the cerebellum. Weaker expression is present in other regions, such as midbrain areas and brainstem (Yoshida et al., 2006). Similar patterns were observed for DAGLβ, with the difference that thalamus, caudate putamen, midbrain and brainstem appeared to express very low (if any) amounts of the transcript. The subcellular localization of DAGLα protein has been described in detail in immunofluorescence and electron microscopy studies on the hippocampal formation, the cerebellum (Katona et al., 2006; Yoshida et al., 2006) and the striatum (Uchigashima et al., 2007). In the hippocampus, DAGLα is present in the head and neck of dendritic spines of pyramidal neurons, with the somatodendritic membrane expressing low levels of the protein (Yoshida et al., 2006). In the cerebellum, DAGLα protein is predominantly present in Purkinje cells, both on dendritic and somatic surface (Bisogno et al., 2003; Yoshida et al., 2006). The expression pattern of DAGLα in spines of Purkinje cells is different from the one in hippocampal pyramidal neurons: the enzyme appears to be present on the dendritic membrane, occasionally on the somatic surface and on the membrane forming the neck of dendritic spines, but neither on the main body nor on the head of the spines (Yoshida et al., 2006). In the striatum, DAGLα is also localized at the somatodendritic surface of medium spiny neurons, where, importantly, it co-localizes with $mGluR_5$ and the M_1-type muscarinic acetylcholine receptor (Uchigashima et al., 2007). However, $mGluR_5$ and DAGLα were detected in dendritic spines of these neurons, whereas M_1 receptor was almost excluded from this subcellular compartment, suggesting a differential influence of these two receptors on the activity of the 2-AG synthesizing enzyme (Uchigashima et al., 2007). So far, very little is known about the expression of enzymes involved in the synthesis of endocannabinoids in the spinal cord and DRG. During early embryonic development, DAGLα and β can be detected in axons crossing the floor plate of the spinal cord (Bisogno et al., 2003). RT-PCR analysis suggests that at least DAGLα is expressed at moderate to high levels in the adult mouse spinal cord, but the spatial distribution pattern has not been described (Bisogno et al., 2003). Interestingly, cultured spinal neurons have been reported to contain activities for diacylglycerol and monoacylglycerol lipases, which are further enhanced upon stimulation with glutamate and bradykinin, suggesting that these enzymes may play an important role in spinal glutamatergic processes, such as pain (Farooqui et al., 1993). Recent ISH studies have shown that all four members of the PLCβ family are expressed in adult mouse DRG. $PLC\beta_4$ and $PLC\beta_3$ are present at higher levels, whereas $PLC\beta_1$ and $PLC\beta_2$ are only weakly detectable (Han et al., 2006). $PLC\beta_1$ and $PLC\beta_4$ mRNAs also appear to be expressed in the adult spinal cord, but the identity of cells expressing them has not been clarified (Han et al., 2006).

Enzymes for Degradation of Endocannabinoids

The enzymatic pathways mediating the degradation of endocannabinoids are much better known with respect to molecular identity than endocannabinoid synthetising enzymes (Piomelli, 2003; De Petrocellis et al., 2004; Basavarajappa, 2007; see Chap. 3). An important element of endocannabinoid degradation is the facilitated uptake of the ligands by cell membranes. Despite the accumulating biochemical evidence for the existence of endocannabinoid transporter (ECT) protein(s) (Piomelli, 2003; De Petrocellis et al., 2004; McFarland and Barker, 2004; Basavarajappa, 2007), its (their) molecular identity has not been clarified yet. It is, therefore, yet unfeasible to provide information concerning the neuroanatomical distribution of the ECT(s). However, it can be generally said that all cell types analyzed in vitro so far show a certain degree of ECT activity. Therefore, it could be argued that the distribution of ECT might be ubiquitous, though much more work is necessary, first for its molecular characterization and second, for elucidating the precise localization in different tissues. Interestingly, it was recently suggested that ECT activity might be associated with lipid rafts in specialized membrane compartments (McFarland and Barker, 2004; McFarland et al., 2004).

a. Degradation of anandamide: There is a general consensus that the main enzyme involved in the degradation of AEA is the FAAH, which, however, is also involved in the degradation of other polyunsaturated fatty acid amides (Deutsch and Chin, 1993; Piomelli, 2003; Cravatt and Lichtman, 2002; De Petrocellis et al., 2004; Basavarajappa, 2007; see Chap. 3). The distribution of FAAH in the brain has been described through analysis of its enzymatic activity (Thomas et al., 1997) as well as via direct detection of mRNA (Thomas et al., 1997) and protein (Egertova et al., 2003; Gulyas et al., 2004). The enzymatic activity of

FAAH is present throughout the brain, with higher levels in the hippocampus and cortex, middle levels in cerebellum, thalamus, olfactory bulb and striatum, and lower levels in the hypothalamus and brainstem (Thomas et al., 1997). FAAH mRNA distribution correlates well with the enzymatic activity, with highest levels in the cortex, hippocampus and cerebellum and lower levels in the other brain regions. Interestingly, the subthalamic and pontine nuclei seem to express high levels of the transcript (Thomas et al., 1997). In the hippocampal formation, FAAH mRNA is exclusively detected in CA1/CA3 pyramidal neurons and in granule cells of the dentate gyrus (Thomas et al., 1997). In different brain regions, FAAH protein has been described to be expressed post-synaptically, often juxtaposed to CB_1-containing fibres, supporting the concept of retrograde signalling of endocannabinoids (Egertova et al., 1998, 2003; Tsou et al., 1998b; Gulyas et al., 2004). In the olfactory bulb, FAAH is present in somata and dendrites of mitral cells (Tsou et al., 1998b; Egertova et al., 2003). In cortical regions, including cortex, amygdala and hippocampus, FAAH mRNA and protein seem to be present exclusively in glutamatergic neurons (Thomas et al., 1997; Egertova et al., 1998, 2003; Tsou et al., 1998b; Gulyas et al., 2004). At subcellular level, FAAH protein is present in the perinuclear cytoplasm, dendrites and dendritic spines of principal cells, whereas axon terminals and glial processes do not appear to contain the enzyme (Gulyas et al., 2004). Several subcortical forebrain regions contain FAAH, including lateral septum, caudate putamen, most of thalamic subnuclei and subthalamic nucleus. Interestingly, besides a weak expression of the protein in the anterior hypothalamic area, the hypothalamus appears to show undetectable or very low levels of FAAH protein, suggesting that the functions of the endocannabinoid system in this brain region might be mediated by other ligands than AEA (Egertova et al., 2003). FAAH is present in the midbrain, in particular in the inferior and superior colliculus, in the rhabdoid nucleus and in mesencephalic raphe and trigeminal nuclei (Egertova et al., 2003). In the cerebellum, FAAH protein is associated with cell bodies and dendrites of Purkinje cells (Tsou et al., 1998b; Egertova et al., 2003). In general, there is good agreement between the expression of FAAH mRNA and protein, indicating that the enzyme is predominantly expressed in cell bodies and is, therefore, unlikely to be localized at axonal terminals (Thomas et al., 1997, 1998b; Egertova et al., 2003). In addition to neurons, FAAH is also expressed by oligodendrocytes, ventricular ependymal cells and the choroid plexus in the rodent brain (Egertova et al., 2000, 2003). FAAH immunoreactivity has been reported in spinal motor neurons (Tsou et al., 1998b). However, most of the insights on the significance of FAAH in modulation of the spinal endocannabinoid system in the context of pain and analgesia have come from biochemical and functional studies. For example, paw inflammation, which leads to enhanced pain via peripheral and central mechanisms, has been reported to be associated with reduced FAAH activity in the affected paw as well as in the spinal cord (Holt et al., 2005). Furthermore, two recent studies have demonstrated a therapeutic potential for FAAH inhibitors in alleviating neuropathic pain (Jhaveri et al., 2006). Therefore, it will be interesting to see in future anatomical and functional studies where and how exactly FAAH modulates endocannabinoid system under physiological and pathological conditions in the spinal cord.

b. Degradation of 2-AG: Monoacylglycerol lipase (MAGL) activity is responsible for the degradation of 2-AG in biological tissues (Piomelli, 2003; De Petrocellis et al., 2004; Basavarajappa, 2007; see Chap. 3). One form of this enzyme has been recently identified (Dinh et al., 2002) and its distribution in the CNS has been described partially (Dinh et al., 2002; Gulyas et al., 2004). It is important to note, however, that the cloned enzyme, MAGL, does not seem to be the only isoform present in biological samples possessing the ability to degrade 2-AG. Indeed, RNA interference experiments have revealed that the cloned isoform is responsible for approximately only 50% of the enzymatic activity in the rat brain (Dinh et al., 2004). More recently, a novel 2-AG hydrolyzing enzyme was biochemically characterized in microglial cells, though its molecular identity is still not known (Muccioli et al., 2007). It is interesting to note that this novel enzymatic activity was identified in nuclear and mitochondrial cellular fractions, possibly revealing novel intracellular locations where endocannabinoid metabolism might be functionally important (Muccioli et al., 2007). The brain distribution of the cloned isoform of MAGL was described via Northern blot analysis, ISH (Dinh et al., 2002) as well as via light and electron IHC experiments in the hippocampus, amygdala and cerebellum (Dinh et al., 2002; Gulyas et al., 2004). MAGL mRNA is present in all brain regions analyzed, including the thalamus, cerebellum, cortex, brainstem, striatum, hippocampus and the hypothalamus. In particular, ISH analysis has revealed that the transcript is particularly abundant in different thalamic nuclei (especially anterodorsal nucleus), in layers IV, V and VI of neocortex, in the hippocampal formation and cerebellum. Moderate signals were also detected in the nucleus accumbens shell, islands of Calleja and the pontine nucleus (Dinh et al., 2002). In the hippocampal formation, MAGL mRNA is associated with pyramidal neurons of CA region (but not with granule cells of dentate gyrus) as well as with sparse neurons within other layers (Dinh et al., 2002). Double ISH-neurochemical data are not available so far concerning the identity of these sparse cells, but electron microscopy and IHC studies have shown that MAGL protein is partially associated with axon terminals of GABAergic interneurons (mainly basket cells containing CCK), suggesting that this neuronal type might actively participate in the enzymatic degradation of 2-AG (Gulyas et al., 2004). The strongest association of MAGL in the hippocampal formation is, however, with presynaptic glutamatergic terminals, whereas cell bodies and dendrites do not appear to express the enzyme (Gulyas et al., 2004). In the amygdaloid nuclei, MAGL is expressed mainly in the basolateral amygdala, whereas the central nucleus appears to be devoid of the protein (Gulyas et al., 2004). In the cerebellum, MAGL does not seem to be present in Purkinje cells, whereas their dendrites in the stratum molecularis are surrounded by intensely stained puncta. Weaker punctate staining was observed in the stratum granulosum (Gulyas et al., 2004). Altogether, the distribution patterns of FAAH and MAGL suggest that there is a strikingly exclusive distribution of the two enzymes, with FAAH mainly expressed postsynaptically and MAGL mainly present on presynaptic terminals of different neuronal populations (Gulyas et al., 2004).

c. Distribution of cyclooxygenase-2 (COX-2) in the CNS. In the early 1990s, an inducible isoform of the cyclooxygenase enzyme was cloned and characterized in the brain (Yamagata et al., 1993; Kaufmann et al., 1996). Recently, this enzyme has been implicated in the oxidative metabolism of endocannabinoids, both by direct biochemical evidence (Yu et al., 1997; Kozak et al., 2000,2003; see Chap. 3) and by indirect observations from functional experiments (Kim and Alger, 2004; Slanina and Schweitzer, 2005; see Chap. 3). COX-2 appears to oxygenate 2-AG more efficiently and selectively than anandamide, but the differential role of these two activities is not yet fully elucidated in vivo (Kozak et al., 2004). The COX-2-mediated oxidation of endocannabinoids gives rise to a plethora of bioactive compounds belonging to the family of prostaglandins (Kozak et al., 2004). This phenomenon further expands and complicates the biological activity of endocannabinoids, but a discussion of these implications is beyond the scope of the present chapter. Here, we will dwell on the anatomical distribution of COX-2 in the CNS, because it might present additional ways of mediating the termination of endocannabinoid signalling. COX-2 is encoded by an inducible gene, whose expression can be stimulated by several conditions, including increased neuronal activity, and inhibited by other conditions, such as glucocorticoid activation (Yamagata et al., 1993; Chen et al., 2002). COX-2 mRNA is expressed in the brain, spinal cord as well as sensory ganglia. In particular, high levels were detected in pyramidal and granule cells of the hippocampal formation, in the piriform cortex, in the superficial layers of neocortex and in the amygdala. Lower levels were observed in the caudate putamen, the thalamus and hypothalamus (Yamagata et al., 1993). These observations have been confirmed in IHC experiments, which revealed that COX-2 immunoreactivity is mainly limited to forebrain areas, where the protein localizes to cell bodies, distal dendrites and the spines of excitatory neurons (Kaufmann et al., 1996). Further studies have revealed that COX-2 immunoreactivity is also present in the paraventricular nucleus of the hypothalamus and in the brainstem (dorsal raphe, nucleus of the brachium and subcoeruleus) (Breder et al., 1995). Amongst enzymes currently thought to be involved in the degradation of endocannabinoids, much is known about the expression of COX-2 in the spinal cord and DRG, whereas the precise distribution pattern of FAAH and MAGL at these avenues remains to be elucidated. COX-2 is known to be constitutively expressed in the spinal cord and is strongly upregulated in an interleukin-1β-dependent manner following peripheral inflammation, concurrent to the development of pain hypersensitivity (Samad et al., 2001). However, there is no clear evidence for expression and significance of COX-2 in the DRG (Dou et al., 2004).

Concluding Remarks

The endocannabinoid system is becoming one of the most studied modulatory systems in the body, because not only of its intrinsic scientific interest but also of the plethora of potential and actual therapeutic implications of its functions. The distribution of various elements of the endocannabinoid system is a very important issue in clarifying the mechanisms, through which endocannabinoids exert many different functions in the body and, in particular, in the central and peripheral nervous system. In this chapter, we have tried to present an updated overview of what it is known concerning the tissue and cellular distribution of various molecules constituting the endocannabinoid system. However, this field is in continuous expansion, with an exponential growth of scientific publications during the last years. The daily development of new methods for the detection of endocannabinoid system elements, the discovery of novel molecules possibly participating in the activity of the system, and the steadily growing accumulation of new functional data implying the presence of these elements in previously "unsuspected" locations, make the endocannabinoid field a very exciting and challenging area of research. During the creation of this chapter, we constantly feared missing some new and important piece of information. That this has happened is very likely the case, and we apologize in advance to the readers and the authors of these potentially overlooked studies for these omissions. It could not be helped simply due to the impressive amount of data that are present in the literature. What is known to date concerning the distribution of the endocannabinoid system in the nervous systems accounts for many, but not all the different functions assigned to endocannabinoids. We are anxiously awaiting novel and exciting discoveries in the immediate future to complete and implement our understanding of the fascinating world of endocannabinoid signaling.

Acknowledgements The authors thank Krisztina Monory for reading part of the manuscript and for their useful comments. Giovanni Marsicano is supported by an AVENIR grant of INSERM, in partnership with the Fondation Bettencourt-Schueller.

References

Abe T, Sugihara H, Nawa H, Shigemoto R, Mizuno N, Nakanishi S (1992) Molecular characterization of a novel metabotropic glutamate receptor mGluR5 coupled to inositol phosphate/Ca^{2+} signal transduction. J Biol Chem 267:13361–13368.

Agarwal N, Pacher P, Tegeder I, Amaya F, Constantin C, Brenner GJ, Rubino T, Michalski CW, Marsicano G, Monory K, Mackie K, Marian C, Batkai S, Parolaro D, Fischer MJ, Reeh P, Kunos G, Kress M, Lutz B, Woolf CJ, Kuner R (2007) Cannabinoids mediate analgesia largely via peripheral type 1 cannabinoid receptors in nociceptors. Nat Neurosci doi: 10.1038/nn1916.

Aguado T, Monory K, Palazuelos J, Stella N, Cravatt B, Lutz B, Marsicano G, Kokaia Z, Guzman M, Galve-Roperh I (2005) The endocannabinoid system drives neural progenitor proliferation. FASEB J 19:1704–1706.

Aguado T, Palazuelos J, Monory K, Stella N, Cravatt B, Lutz B, Marsicano G, Kokaia Z, Guzman M, Galve-Roperh I (2006) The endocannabinoid system promotes astroglial differentiation by acting on neural progenitor cells. J Neurosci 26:1551–1561.

Alger E (2002) Retrograde signaling in the regulation of synaptic transmission: focus on endocannabinoids. Prog Neurobiol 68:247–286.

Amaya F, Shimosato G, Kawasaki Y, Hashimoto S, Tanaka Y, Ji RR, Tanaka M (2006) Induction of CB_1 cannabinoid receptor by inflammation in primary afferent neurons facilitates anti-hyperalgesic effect of peripheral CB_1 agonist. Pain 124:175–183.

Azad SC, Marsicano G, Eberlein I, Putzke J, Zieglgansberger W, Spanagel R, Lutz B (2001) Differential role of the nitric oxide pathway on delta^9-THC-induced central nervous system effects in the mouse. Eur J Neurosci 13:561–568.
Azad SC, Eder M, Marsicano G, Lutz B, Zieglgansberger W, Rammes G (2003) Activation of the cannabinoid receptor type 1 decreases glutamatergic and GABAergic synaptic transmission in the lateral amygdala of the mouse. Learn Mem 10:116–128.
Bacci A, Huguenard JR, Prince DA (2004) Long-lasting self-inhibition of neocortical interneurons mediated by endocannabinoids. Nature 431:312–316.
Baker D, Pryce G, Davies WL, Hiley CR (2006) In silico patent searching reveals a new cannabinoid receptor. Trends Pharmacol Sci 27:1–4.
Basavarajappa BS (2007) Critical enzymes involved in endocannabinoid metabolism. Protein Pept Lett 14:237–246.
Bayliss DA, Talley EM, Sirois JE, Lei Q (2001) TASK-1 is a highly modulated pH-sensitive 'leak' K^+ channel expressed in brainstem respiratory neurons. Respir Physiol 129:159–174.
Begg M, Pacher P, Batkai S, Osei-Hyiaman D, Offertaler L, Mo FM, Liu J, Kunos G (2005) Evidence for novel cannabinoid receptors. Pharmacol Ther 106:133–145.
Binzen U, Greffrath W, Hennessy S, Bausen M, Saaler-Reinhardt S, Treede RD (2006) Co-expression of the voltage-gated potassium channel $K_v1.4$ with transient receptor potential channels ($TRPV_1$ and $TRPV_2$) and the cannabinoid receptor CB_1 in rat dorsal root ganglion neurons. Neuroscience 142:527–539.
Bisogno T, Berrendero F, Ambrosino G, Cebeira M, Ramos JA, Fernandez-Ruiz JJ, Di Marzo V (1999a) Brain regional distribution of endocannabinoids: implications for their biosynthesis and biological function. Biochem Bioph Res Co 256:377–380.
Bisogno T, Melck D, De Petrocellis L, Di Marzo V (1999b) Phosphatidic acid as the biosynthetic precursor of the endocannabinoid 2-arachidonoylglycerol in intact mouse neuroblastoma cells stimulated with ionomycin. J Neurochem 72:2113–2119.
Bisogno T, Howell F, Williams G, Minassi A, Cascio MG, Ligresti A, Matias I, Schiano-Moriello A, Paul P, Williams EJ, Gangadharan U, Hobbs C, Di Marzo V, Doherty P (2003) Cloning of the first *sn*1-DAG lipases points to the spatial and temporal regulation of endocannabinoid signaling in the brain. J Cell Biol 163:463–468.
Breder CD, Dewitt D, Kraig RP (1995) Characterization of inducible cyclooxygenase in rat brain. J Comp Neurol 355:296–315.
Breivogel CS, Childers SR (1998) The functional neuroanatomy of brain cannabinoid receptors. Neurobiol Dis 5:417–431.
Bridges D, Rice AS, Egertova M, Elphick MR, Winter J, Michael GJ (2003) Localisation of cannabinoid receptor 1 in rat dorsal root ganglion using in situ hybridisation and immunohistochemistry. Neuroscience 119:803–812.
Caille S, varez-Jaimes L, Polis I, Stouffer DG, Parsons LH (2007) Specific alterations of extracellular endocannabinoid levels in the nucleus accumbens by ethanol, heroin, and cocaine self-administration. J Neurosci 27:3695–3702.
Caterina MJ, Schumacher MA, Tominaga M, Rosen TA, Levine JD, Julius D (1997) The capsaicin receptor: a heat-activated ion channel in the pain pathway. Nature 398:816–824.
Chen C, Magee JC, Bazan NG (2002) Cyclooxygenase-2 regulates prostaglandin E_2 signaling in hippocampal long-term synaptic plasticity. J Neurophysiol 87:2851–2857.
Chen H, Wang Y, Mackie K, Lu H (2006) The expression of GPR55, a putative cannabinoid receptor, is dynamically regulated during pre- and postnatal brain development.
Chen K, Ratzliff A, Hilgenberg L, Gulyas A, Freund TF, Smith M, Dinh TP, Piomelli D, Mackie K, Soltesz I (2003) Long-term plasticity of endocannabinoid signaling induced by developmental febrile seizures. Neuron 39:599–611.
Chevaleyre V, Takahashi KA, Castillo PE (2006) Endocannabinoid-mediated synaptic plasticity in the CNS. Annu Rev Neurosci 29:37–76.
Cortright DN, Crandall M, Sanchez JF, Zou T, Krause JE, White G (2001) The tissue distribution and functional characterization of human VR_1. Biochem Biophys Res Commun 281:1183–1189.

Cota D, Marsicano G, Tschop M, Grubler Y, Flachskamm C, Schubert M, Auer D, Yassouridis A, Thone-Reineke C, Ortmann S, Tomassoni F, Cervino C, Nisoli E, Linthorst AC, Pasquali R, Lutz B, Stalla GK, Pagotto U (2003) The endogenouscannabinoid system affects energy balance via central orexigenic drive and peripheral lipogenesis. J Clin Invest 112:423–431.

Coutts AA, Irving AJ, Mackie K, Pertwee RG, Anavi-Goffer S (2002) Localisation of cannabinoid CB_1 receptor immunoreactivity in the guinea pig and rat myenteric plexus. J Comp Neurol 448:410–422.

Cravatt BF, Lichtman AH (2002) The enzymatic inactivation of the fatty acid amide class of signaling lipids. Chem Phys Lipids 121:135–148.

Cristino L, De PL, Pryce G, Baker D, Guglielmotti V, Di Marzo V (2006) Immunohistochemical localization of cannabinoid type 1 and vanilloid transient receptor potential vanilloid type 1 receptors in the mouse brain. Neuroscience 139:1405–1415.

De Petrocellis L., Cascio MG, Di Marzo V (2004) The endocannabinoid system: a general view and latest additions. Br J Pharmacol 141:765–774.

Degroot A, Köfalvi A, Wade MR, Davis RJ, Rodrigues RJ, Rebola N, Cunha RA, Nomikos GG (2006) CB_1 receptor antagonism increases hippocampal acetylcholine release: site and mechanism of action. Mol Pharmacol 70:1236–1245.

Deutsch DG, Chin SA (1993) Enzymatic synthesis and degradation of anandamide, a cannabinoid receptor agonist. Biochem Pharmacol 46:791–796.

Devane WA, Hanus L, Breuer A, Pertwee RG, Stevenson LA, Griffin G, Gibson D, Mandelbaum A, Etinger A, Mechoulam R (1992) Isolation and structure of a brain constituent that binds to the cannabinoid receptor. Science 258:1946–1949.

Di Marzo V, Bisogno T, De Petrocellis L (2000a) Endocannabinoids: new targets for drug development. Curr Pharm Des 6:1361–1380.

Di Marzo V, Breivogel CS, Tao Q, Bridgen DT, Razdan RK, Zimmer AM, Zimmer A, Martin BR (2000b) Levels, metabolism, and pharmacological activity of anandamide in CB_1 cannabinoid receptor knockout mice: evidence for non-CB_1, non-CB_2 receptor-mediated actions of anandamide in mouse brain. J Neurochem 75:2434–2444.

Di Marzo V, Goparaju SK, Wang L, Liu J, Batkai S, Jarai Z, Fezza F, Miura GI, Palmiter RD, Sugiura T, Kunos G (2001) Leptin-regulated endocannabinoids are involved in maintaining food intake. Nature 410:822–825.

Di Marzo V, De Petrocellis L, Fezza F, Ligresti A, Bisogno T (2002) Anandamide receptors. Prostaglandins Leukot Essent Fatty Acids 66:377–391.

Di Marzo V, Bifulco M, De Petrocellis L (2004) The endocannabinoid system and its therapeutic exploitation. Nat Rev Drug Discov 3:771–784.

Di Marzo V, De PL, Bisogno T (2005) The biosynthesis, fate and pharmacological properties of endocannabinoids. Handb Exp Pharmacol 147–185.

Diana MA, Levenes C, Mackie K, Marty A (2002) Short-term retrograde inhibition of GABAergic synaptic currents in rat purkinje cells is mediated by endogenous cannabinoids. J Neurosci 22:200–208.

Dinh TP, Carpenter D, Leslie FM, Freund TF, Katona I, Sensi SL, Kathuria S, Piomelli D (2002) Brain monoglyceride lipase participating in endocannabinoid inactivation. Proc Natl Acad Sci USA 99:10819–10824.

Dinh TP, Kathuria S, Piomelli D (2004) RNA interference suggests a primary role for monoacylglycerol lipase in the degradation of the endocannabinoid 2-arachidonoylglycerol. Mol Pharmacol 66:1260–1264.

Doherty J, Dingledine R (2003) Functional interactions between cannabinoid and metabotropic glutamate receptors in the central nervous system. Curr Opin Pharmacol 3:46–53.

Doly S, Fischer J, Conrath M (2004) The vanilloid receptor-1 ($TRPV_1$) is expressed in some rat dorsal horn NK1 cells. Brain Res 1004:203–207.

Domenici MR, Azad SC, Marsicano G, Schierloh A, Wotjak CT, Dodt HU, Zieglgansberger W, Lutz B, Rammes G (2006) Cannabinoid receptor type 1 located on presynaptic terminals of principal neurons in the forebrain controls glutamatergic synaptic transmission. J Neurosci 26:5794–5799.

Dou W, Jiao Y, Goorha S, Raghow R, Ballou LR (2004) Nociception and the differential expression of cyclooxygenase-1 (COX-1), the COX-1 variant retaining intron-1 (COX-1v), and COX-2 in mouse dorsal root ganglia (DRG). Prostaglandins Other Lipid Mediat 74:29–43.
Egertová M, Elphick MR (2000) Localisation of cannabinoid receptors in the rat brain using antibodies to the intracellular C-terminal tail of CB_1. J Comp Neurol 422:159–171.
Egertova M, Giang DK, Cravatt BF, Elphick MR (1998) A new perspective on cannabinoid signalling: complementary localization of fatty acid amide hydrolase and the CB_1 receptor in rat brain. Proc Biol Sci 265:2081–2085.
Egertova M, Cravatt BF, Elphick MR (2000) Fatty acid amide hydrolase expression in rat choroid plexus: possible role in regulation of the sleep-inducing action of oleamide. Neurosci Lett 282:13–16.
Egertova M, Cravatt BF, Elphick MR (2003) Comparative analysis of fatty acid amide hydrolase and cb_1 cannabinoid receptor expression in the mouse brain: evidence of a widespread role for fatty acid amide hydrolase in regulation of endocannabinoid signaling. Neuroscience 119:481–496.
Farooqui AA, Anderson DK, Horrocks LA (1993) Effect of glutamate and its analogs on diacylglycerol and monoacylglycerol lipase activities of neuron-enriched cultures. Brain Res 604:180–184.
Farquhar-Smith WP, Egertova M, Bradbury EJ, McMahon SB, Rice AC, Elphick MR (2000) Cannabinoid CB1 receptor expression in rat spinal cord. Mol Cell Neurosci 15:510–521.
Freund TF, Katona I, Piomelli D (2003) Role of endogenous cannabinoids in synaptic signaling. Physiol Rev 83:1017–1066.
Galve-Roperh I, Aguado T, Palazuelos J, Guzman M (2007) The endocannabinoid system and neurogenesis in health and disease. Neuroscientist 13:109–114.
Gerdeman G, Lovinger DM (2001) CB_1 cannabinoid receptor inhibits synaptic release of glutamate in rat dorsolateral striatum. J Neurophysiol 85:468–471.
Gerdeman GL, Ronesi J, Lovinger DM (2002) Postsynaptic endocannabinoid release is critical to long-term depression in the striatum. Nat Neurosci 5:446–451.
Giuffrida A, Parsons LH, Kerr TM, Rodriguez de Fonseca F, Navarro M, Piomelli D (1999) Dopamine activation of endogenous cannabinoid signaling in dorsal striatum. Nat Neurosci 2:358–363.
Glass M, Dragunow M, Faull RL (1997) Cannabinoid receptors in the human brain: a detailed anatomical and quantitative autoradiographic study in the fetal, neonatal and adult human brain. Neuroscience 77:299–318.
Gong JP, Onaivi ES, Ishiguro H, Liu QR, Tagliaferro PA, Brusco A, Uhl GR (2006) Cannabinoid CB_2 receptors: immunohistochemical localization in rat brain. Brain Res 1071:10–23.
Gross CG (2000) Neurogenesis in the adult brain: death of a dogma. Nat Rev Neurosci 1:67–73.
Gulyas AI, Cravatt BF, Bracey MH, Dinh TP, Piomelli D, Boscia F, Freund TF (2004) Segregation of two endocannabinoid-hydrolyzing enzymes into pre- and postsynaptic compartments in the rat hippocampus, cerebellum and amygdala. Eur J Neurosci 20:441–458.
Gutkind JS, Novotny EA, Brann MR, Robbins KC (1991) Muscarinic acetylcholine receptor subtypes as agonist-dependent oncogenes. Proc Natl Acad Sci USA 88:4703–4707.
Haj-Dahmane S, Shen RY (2005) The wake-promoting peptide orexin-B inhibits glutamatergic transmission to dorsal raphe nucleus serotonin neurons through retrograde endocannabinoid signaling. J Neurosci 25:896–905.
Han SK, Mancino V, Simon MI (2006) Phospholipase Cβ3 mediates the scratching response activated by the histamine H_1 receptor on C-fiber nociceptive neurons. Neuron 52:691–703.
Hanus L, Abu-Lafi S, Fride E, Breuer A, Vogel Z, Shalev DE, Kustanovich I, Mechoulam R (2001) 2-arachidonyl glyceryl ether, an endogenous agonist of the cannabinoid CB_1 receptor. Proc Natl Acad Sci USA 98:3662–3665.
Haring M, Marsicano G, Lutz B, Monory K (2007) Identification of the cannabinoid receptor type 1 in serotonergic cells of raphe nuclei in mice. Neuroscience 146(3):1212–1219.
Harkany T, Hartig W, Berghuis P, Dobszay MB, Zilberter Y, Edwards RH, Mackie K, Ernfors P (2003) Complementary distribution of type 1 cannabinoid receptors and vesicular glutamate

transporter 3 in basal forebrain suggests input-specific retrograde signalling by cholinergic neurons. Eur J Neurosci 18:1979–1992.

Hashimotodani Y, Ohno-Shosaku T, Tsubokawa H, Ogata H, Emoto K, Maejima T, Araishi K, Shin HS, Kano M (2005) Phospholipase Cbeta serves as a coincidence detector through its Ca^{2+} dependency for triggering retrograde endocannabinoid signal. Neuron 45:257–268.

Herkenham M (1991) Characterization and localization of cannabinoid receptors in brain: an *in vitro* technique using slide-mounted tissue sections. NIDA Res Monogr 112:129–145.

Herkenham M, Nauta WJ (1977) Afferent connections of the habenular nuclei in the rat. A horse-radish peroxidase study, with a note on the fiber-of-passage problem. J Comp Neurol 173:123–146.

Herkenham M, Lynn AB, Little MD, Johnson MR, Melvin LS, De Costa BR, Rice KC (1990) Cannabinoid receptor localization in brain. Proc Natl Acad Sci USA 87:1932–1936.

Herkenham M, Lynn AB, Johnson MR, Melvin LS, De Costa BR, Rice KC (1991) Characterization and localization of cannabinoid receptors in rat brain: a quantitative *in vitro* autoradiographic study. J Neurosci 11:563–583.

Hermann H, Marsicano G, Lutz B (2002) Coexpression of the cannabinoid receptor type 1 with dopamine and serotonin receptors in distinct neuronal subpopulations of the adult mouse forebrain. Neuroscience 109:451–460.

Hermann H, De Petrocellis L, Bisogno T, Schiano MA, Lutz B, Di Marzo V (2003) Dual effect of cannabinoid CB1 receptor stimulation on a vanilloid VR_1 receptor-mediated response. Cell Mol Life Sci 60:607–616.

Hill EL, Gallopin T, Ferezou I, Cauli B, Rossier J, Schweitzer P, Lambolez B (2007) Functional CB_1 receptors are broadly expressed in neocortical GABAergic and glutamatergic neurons. J Neurophysiol 97:2580–2589.

Hohmann AG, Herkenham M (1998) Regulation of cannabinoid and mu opioid receptors in rat lumbar spinal cord following neonatal capsaicin treatment. Neurosci Lett 252:13–16.

Hohmann AG, Herkenham M (1999) Localization of central cannabinoid CB_1 receptor messenger RNA in neuronal subpopulations of rat dorsal root ganglia: a double-label in situ hybridization study. Neuroscience 90:923–931.

Hohmann AG, Herkenham M (2000) Localization of cannabinoid CB_1 receptor mRNA in neuronal subpopulations of rat striatum: a double-label in situ hybridization study. Synapse 37:71–80.

Hohmann AG, Suplita RL, Bolton NM, Neely MH, Fegley D, Mangieri R, Krey JF, Walker JM, Holmes PV, Crystal JD, Duranti A, Tontini A, Mor M, Tarzia G, Piomelli D (2005) An endocannabinoid mechanism for stress-induced analgesia. Nature 435:1108–1112.

Holt S, Comelli F, Costa B, Fowler CJ (2005) Inhibitors of fatty acid amide hydrolase reduce carrageenan-induced hind paw inflammation in pentobarbital-treated mice: comparison with indomethacin and possible involvement of cannabinoid receptors. Br J Pharmacol 146:467–476.

Howlett AC, Fleming RM (1984) Cannabinoid inhibition of adenylate cyclase. Pharmacology of the response in neuroblastoma cell membranes. Mol Pharmacol 26:532–538.

Howlett AC, Barth F, Bonner TI, Cabral G, Casellas P, Devane WA, Felder CC, Herkenham M, Mackie K, Martin BR, Mechoulam R, Pertwee RG (2002) International Union of Pharmacology. XXVII. Classification of cannabinoid receptors. Pharmacol Rev 54:161–202.

Huang CC, Lo SW, Hsu KS (2001a) Presynaptic mechanisms underlying cannabinoid inhibition of excitatory synaptic transmission in rat striatal neurons. J Physiol 532:731–748.

Huang SM, Bisogno T, Petros TJ, Chang SY, Zavitsanos PA, Zipkin RE, Sivakumar R, Coop A, Maeda DY, De Petrocellis L, Burstein S, Di M, V, Walker JM (2001b) Identification of a new class of molecules, the arachidonyl amino acids, and characterization of one member that inhibits pain. J Biol Chem 276:42639–42644.

Huang SM, Bisogno T, Trevisani M, Al Hayani A, De Petrocellis L, Fezza F, Tognetto M, Petros TJ, Krey JF, Chu CJ, Miller JD, Davies SN, Geppetti P, Walker JM, Di M, V (2002) An endogenous capsaicin-like substance with high potency at recombinant and native vanilloid VR_1 receptors. Proc Natl Acad Sci USA 99:8400–8405.

Jhaveri MD, Richardson D, Kendall DA, Barrett DA, Chapman V (2006) Analgesic effects of fatty acid amide hydrolase inhibition in a rat model of neuropathic pain. J Neurosci 26:13318–13327.

Jiang W, Zhang Y, Xiao L, Van CJ, Ji SP, Bai G, Zhang X (2005) Cannabinoids promote embryonic and adult hippocampus neurogenesis and produce anxiolytic- and antidepressant-like effects. J Clin Invest 115:3104–3116.

Jin XH, Okamoto Y, Morishita J, Tsuboi K, Tonai T, Ueda N (2007) Discovery and characterization of a Ca^{2+}-independent phosphatidylethanolamine *N*-acyltransferase generating the anandamide precursor and its congeners. J Biol Chem 282:3614–3623.

Johnston D, Amaral DG (2004) Hippocampus. In: The synaptic organization of the brain (Shepherd GM, ed.), pp 455–498. Oxford: Oxford University Press.

Karschin C, Wischmeyer E, Preisig-Muller R, Rajan S, Derst C, Grzeschik KH, Daut J, Karschin A (2001) Expression pattern in brain of TASK-1, TASK-3, and a tandem pore domain K^+ channel subunit, TASK-5, associated with the central auditory nervous system. Mol Cell Neurosci 18:632–648.

Katona I, Rancz EA, Acsady L, Ledent C, Mackie K, Hajos N, Freund TF (2001) Distribution of CB_1 cannabinoid receptors in the amygdala and their role in the control of GABAergic transmission. J Neurosci 21:9506–9518.

Katona I, Sperlagh B, Sik A, Köfalvi A, Vizi ES, Mackie K, Freund TF (1999) Presynaptically located CB_1 cannabinoid receptors regulate GABA release from axon terminals of specific hippocampal interneurons. J Neurosci 19:4544–4558.

Katona I, Urban GM, Wallace M, Ledent C, Jung KM, Piomelli D, Mackie K, Freund TF (2006) Molecular composition of the endocannabinoid system at glutamatergic synapses. J Neurosci 26:5628–5637.

Kaufmann WE, Worley PF, Pegg J, Bremer M, Isakson P (1996) COX-2, a synaptically induced enzyme, is expressed by excitatory neurons at postsynaptic sites in rat cerebral cortex. Proc Natl Acad Sci USA 93:2317–2321.

Kawamura Y, Fukaya M, Maejima T, Yoshida T, Miura E, Watanabe M, Ohno-Shosaku T, Kano M (2006) The CB_1 cannabinoid receptor is the major cannabinoid receptor at excitatory presynaptic sites in the hippocampus and cerebellum. J Neurosci 26:2991–3001.

Kim J, Alger BE (2004) Inhibition of cyclooxygenase-2 potentiates retrograde endocannabinoid effects in hippocampus. Nat Neurosci 7:697–698.

Kindler CH, Pietruck C, Yost CS, Sampson ER, Gray AT (2000) Localization of the tandem pore domain K^+ channel TASK-1 in the rat central nervous system. Brain Res Mol Brain Res 80:99–108.

Klausberger T, Marton LF, O'Neill J, Huck JH, Dalezios Y, Fuentealba P, Suen WY, Papp E, Kaneko T, Watanabe M, Csicsvari J, Somogyi P (2005) Complementary roles of cholecystokinin- and parvalbumin-expressing GABAergic neurons in hippocampal network oscillations. J Neurosci 25:9782–9793.

Köfalvi A, Rodrigues RJ, Ledent C, Mackie K, Vizi ES, Cunha RA, Sperlagh B (2005) Involvement of cannabinoid receptors in the regulation of neurotransmitter release in the rodent striatum: a combined immunochemical and pharmacological analysis. J Neurosci 25:2874–2884.

Köfalvi A, Oliveira CR, Cunha RA (2006) Lack of evidence for functional $TRPV_1$ vanilloid receptors in rat hippocampal nerve terminals. Neurosci Lett 403:151–156.

Köfalvi A, Pereira MF, Rebola N, Rodrigues RJ, Oliveira CR, Cunha RA (2007) Anandamide and NADA bi-directionally modulate presynaptic Ca^{2+} levels and transmitter release in the hippocampus. Br J Pharmacol doi:10.1038/sj.bjp.0707252.

Kozak KR, Rowlinson SW, Marnett LJ (2000) Oxygenation of the endocannabinoid, 2-arachidonylglycerol, to glyceryl prostaglandins by cyclooxygenase-2. J Biol Chem 275:33744–33749.

Kozak KR, Prusakiewicz JJ, Rowlinson SW, Prudhomme DR, Marnett LJ (2003) Amino acid determinants in cyclooxygenase-2 oxygenation of the endocannabinoid anandamide. Biochemistry 42:9041–9049.

Kozak KR, Prusakiewicz JJ, Marnett LJ (2004) Oxidative metabolism of endocannabinoids by COX-2. Curr Pharm Des 10:659–667.
Kreitzer AC, Regehr WG (2001a) Cerebellar depolarization-induced suppression of inhibition is mediated by endogenous cannabinoids. J Neurosci 21:RC174.
Kreitzer AC, Regehr WG (2001b) Retrograde inhibition of presynaptic calcium influx by endogenous cannabinoids at excitatory synapses onto Purkinje cells. Neuron 29:717–727.
Kreitzer AC, Regehr WG (2002) Retrograde signaling by endocannabinoids. Curr Opin Neurobiol 12:324–330.
LeDoux JE (2000) Emotion circuits in the brain. Annu Rev Neurosci 23:155–184.
Leung D, Saghatelian A, Simon GM, Cravatt BF (2006) Inactivation of *N*-acyl phosphatidylethanolamine phospholipase D reveals multiple mechanisms for the biosynthesis of endocannabinoids. Biochemistry 45:4720–4726.
Liu J, Wang L, Harvey-White J, Osei-Hyiaman D, Razdan R, Gong Q, Chan AC, Zhou Z, Huang BX, Kim HY, Kunos G (2006) A biosynthetic pathway for anandamide. Proc Natl Acad Sci USA 103:13345–13350.
Lledo PM, Alonso M, Grubb MS (2006) Adult neurogenesis and functional plasticity in neuronal circuits. Nat Rev Neurosci 7:179–193.
Lutz B (2004) On-demand activation of the endocannabinoid system in the control of neuronal excitability and epileptiform seizures. Biochem Pharmacol 68:1691–1698.
Lutz B, Hermann H, Woelfel B, Monory K, Ekker M, Rubenstein JL, Marsicano G (2004) Anatomical and functional localization of CB_1 cannabinoid receptors in principal glutamatergic neurons.
Mackie K (2005a) Cannabinoid receptor homo- and heterodimerization. Life Sci 77:1667–1673.
Mackie K (2005b) Distribution of cannabinoid receptors in the central and peripheral nervous system. Handb Exp Pharmaco l299–325.
Mackie K, Stella N (2006) Cannabinoid receptors and endocannabinoids: evidence for new players. AAPS J 8:E298–E306.
Maejima T, Ohno-Shosaku T, Kano M (2001) Endogenous cannabinoid as a retrograde messenger from depolarized postsynaptic neurons to presynaptic terminals. Neurosci Res 40:205–210.
Mailleux P, Vanderhaeghen JJ (1992) Distribution of neuronal cannabinoid receptor in the adult rat brain: a comparative receptor binding radioautography and in situ hybridization histochemistry. Neuroscience 48:655–668.
Mailleux P, Parmentier M, Vanderhaeghen JJ (1992) Distribution of cannabinoid receptor messenger RNA in the human brain: an in situ hybridization histochemistry with oligonucleotides. Neurosci Lett 143:200–204.
Maingret F, Patel AJ, Lazdunski M, Honore E (2001) The endocannabinoid anandamide is a direct and selective blocker of the background K^+ channel TASK-1. EMBO J 20:47–54.
Marsch R, Foeller E, Rammes G, Bunck M, Kossl M, Holsboer F, Zieglgansberger W, Landgraf R, Lutz B, Wotjak CT (2007) Reduced anxiety, conditioned fear, and hippocampal long-term potentiation in transient receptor potential vanilloid type 1 receptor-deficient mice. J Neurosci 27:832–839.
Marsicano G, Lutz B (1999) Expression of the cannabinoid receptor CB1 in distinct neuronal subpopulations in the adult mouse forebrain. Eur J Neurosci 11:4213–4225.
Marsicano G, Lutz B (2006) Neuromodulatory functions of the endocannabinoid system. J Endocrinol Invest 29:27–46.
Marsicano G, Wotjak CT, Azad SC, Bisogno T, Rammes G, Cascio MG, Hermann H, Tang J, Hofmann C, Zieglgänsberger W, Di Marzo V, Lutz B (2002) The endogenous cannabinoid system controls extinction of aversive memories. Nature 418:530–534.
Marsicano G, Goodenough S, Monory K, Hermann H, Eder M, Cannich A, Azad SC, Cascio MG, Gutierrez SO, Van D, Lopez R, Casanova E, Schutz G, Zieglgänsberger W, Di Marzo V, Behl C, Lutz B (2003) CB1 cannabinoid receptors and on-demand defense against excitotoxicity. Science 302:84–88.
Mathie A, Clarke CE, Ranatunga KM, Veale EL (2003) What are the roles of the many different types of potassium channel expressed in cerebellar granule cells? Cerebellum 2:11–25.

Matsuda LA, Lolait SJ, Brownstein MJ, Young AC, Bonner TI (1990) Structure of a cannabinoid receptor and functional expression of the cloned cDNA. Nature 346:561–564.

Matsuda LA, Bonner TI, Lolait SJ (1993) Localization of cannabinoid receptor mRNA in rat brain. J Comp Neurol 327:535–550.

Matyas F, Yanovsky Y, Mackie K, Kelsch W, Misgeld U, Freund TF (2006) Subcellular localization of type 1 cannabinoid receptors in the rat basal ganglia. Neuroscience 137:337–361.

McDonald AJ, Mascagni F (2001) Localization of the CB_1 type cannabinoid receptor in the rat basolateral amygdala: high concentrations in a subpopulation of cholecystokinin-containing interneurons. Neuroscience 107:641–652.

McFarland MJ, Barker EL (2004) Anandamide transport. Pharmacol Ther 104:117–135.

McFarland MJ, Porter AC, Rakhshan FR, Rawat DS, Gibbs RA, Barker EL (2004) A role for caveolae/lipid rafts in the uptake and recycling of the endogenous cannabinoid anandamide. J Biol Chem 279:41991–41997.

Mechoulam R, Ben Shabat S, Hanus L, Ligumsky M, Kaminski NE, Schatz AR, Gopher A, Almog S, Martin BR, Compton DR (1995) Identification of an endogenous 2-monoglyceride, present in canine gut, that binds to cannabinoid receptors. Biochem Pharmacol 50:83–90.

Melis M, Perra S, Muntoni AL, Pillolla G, Lutz B, Marsicano G, Di M, V, Gessa GL, Pistis M (2004a) Prefrontal cortex stimulation induces 2-arachidonoyl-glycerol-mediated suppression of excitation in dopamine neurons. J Neurosci 24:10707–10715.

Melis M, Pistis M, Perra S, Muntoni AL, Pillolla G, Gessa GL (2004b) Endocannabinoids mediate presynaptic inhibition of glutamatergic transmission in rat ventral tegmental area dopamine neurons through activation of CB_1 receptors. J Neurosci 24:53–62.

Mezey E, Toth ZE, Cortright DN, Arzubi MK, Krause JE, Elde R, Guo A, Blumberg PM, Szallasi A (2000) Distribution of mRNA for vanilloid receptor subtype 1 (VR_1), and VR_1-like immunoreactivity, in the central nervous system of the rat and human. Proc Natl Acad Sci USA 97:3655–3660.

Mitrirattanakul S, Ramakul N, Guerrero AV, Matsuka Y, Ono T, Iwase H, Mackie K, Faull KF, Spigelman I (2006) Site-specific increases in peripheral cannabinoid receptors and their endogenous ligands in a model of neuropathic pain. Pain 126:102–114.

Monory K, Massa F, Egertova M, Eder M, Blaudzun H, Westenbroek R, Kelsch W, Jacob W, Marsch R, Ekker M, Long J, Rubenstein JL, Goebbels S, Nave KA, During M, Klugmann M, Wolfel B, Dodt HU, Zieglgansberger W, Wotjak CT, Mackie K, Elphick MR, Marsicano G, Lutz B (2006) The endocannabinoid system controls key epileptogenic circuits in the hippocampus. Neuron 51:455–466.

Morales M, Backman C (2002) Coexistence of serotonin 3 ($5\text{-}HT_3$) and CB_1 cannabinoid receptors in interneurons of hippocampus and dentate gyrus. Hippocampus 12:756–764.

Morishita J, Okamoto Y, Tsuboi K, Ueno M, Sakamoto H, Maekawa N, Ueda N (2005) Regional distribution and age-dependent expression of *N*-acylphosphatidylethanolamine-hydrolyzing phospholipase D in rat brain. J Neurochem 94:753–762.

Muccioli GG, Xu C, Odah E, Cudaback E, Cisneros JA, Lambert DM, Lopez Rodriguez ML, Bajjalieh S, Stella N (2007) Identification of a novel endocannabinoid-hydrolyzing enzyme expressed by microglial cells. J Neurosci 27:2883–2889.

Munro S, Thomas KL, Abu-Shaar M (1993) Molecular characterization of a peripheral receptor for cannabinoids. Nature 365:61–65.

Nyiri G, Cserep C, Szabadits E, Mackie K, Freund TF (2005) CB_1 cannabinoid receptors are enriched in the perisynaptic annulus and on preterminal segments of hippocampal GABAergic axons. Neuroscience 136:811–822.

Onaivi ES, Ishiguro H, Gong JP, Patel S, Perchuk A, Meozzi PA, Myers L, Mora Z, Tagliaferro P, Gardner E, Brusco A, Akinshola BE, Liu QR, Hope B, Iwasaki S, Arinami T, Teasenfitz L, Uhl GR (2006) Discovery of the presence and functional expression of cannabinoid CB_2 receptors in brain. Ann N Y Acad Sci 1074:514–536.

Pacher P, Batkai S, Kunos G (2006) The endocannabinoid system as an emerging target of pharmacotherapy. Pharmacol Rev 58:389–462.

Pagotto U, Marsicano G, Cota D, Lutz B, Pasquali R (2006) The emerging role of the endocannabinoid system in endocrine regulation and energy balance. Endocr Rev 27:73–100.

Petitet F, Donlan M, Michel A (2006) GPR55 as a new cannabinoid receptor: still a long way to prove it. Chem Biol Drug Des 67:252–253.

Petrosino S, Palazzo E, de N, V, Bisogno T, Rossi F, Maione S, Di Marzo V (2007) Changes in spinal and supraspinal endocannabinoid levels in neuropathic rats. Neuropharmacology 52:415–422.

Piomelli D (2003) The molecular logic of endocannabinoid signalling. Nat Rev Neurosci 4:873–884.

Piomelli D, Giuffrida A, Calignano A, de Fonseca FR (2000) The endocannabinoid system as a target for therapeutic drugs. Trends Pharmacol Sci 21:218–224.

Porter AC, Sauer JM, Knierman MD, Becker GW, Berna MJ, Bao J, Nomikos GG, Carter P, Bymaster FP, Leese AB, Felder CC (2002) Characterization of a novel endocannabinoid, virodhamine, with antagonist activity at the CB_1 receptor. J Pharmacol Exp Ther 301:1020–1024.

Rau KK, Cooper BY, Johnson RD (2006) Expression of TWIK-related acid sensitive K^+ channels in capsaicin sensitive and insensitive cells of rat dorsal root ganglia. Neuroscience 141:955–963.

Rawls SM, Cabassa J, Geller EB, Adler MW (2002a) CB(1) receptors in the preoptic anterior hypothalamus regulate WIN 55212-2 [(4,5-dihydro-2-methyl-4(4-morpholinylmethyl)-1-(1-naphthalenyl-carbonyl)-6H-pyrrolo[3,2,1ij]quinolin-6-one]-induced hypothermia. J Pharmacol Exp Ther 301:963–968.

Rawls SM, Cowan A, Tallarida RJ, Geller EB, Adler MW (2002b) *N*-methyl-D-aspartate antagonists and WIN 55212-2 [4,5-dihydro-2-methyl-4(4-morpholinylmethyl)-1-(1-naphthalenyl-carbonyl)-6 H-pyrrolo[3,2,1-i,j]quinolin-6-one], a cannabinoid agonist, interact to produce synergistic hypothermia. J Pharmacol Exp Ther 303:395–402.

Reyes FR, Cudaback E, Hague C, Moeller T, Mackie K, Stella N (2006) Initial characterization of GPR55, a novel cannabinoid receptor. Program No. 791.1 2006 Abstract.

Riegel AC, Lupica CR (2004) Independent presynaptic and postsynaptic mechanisms regulate endocannabinoid signaling at multiple synapses in the ventral tegmental area. J Neurosci 24:11070–11078.

Robbe D, Alonso G, Duchamp F, Bockaert J, Manzoni OJ (2001) Localization and mechanisms of action of cannabinoid receptors at the glutamatergic synapses of the mouse nucleus accumbens. J Neurosci 21:109–116.

Roberts JC, Davis JB, Benham CD (2004) [^{3}H]Resiniferatoxin autoradiography in the CNS of wild-type and $TRPV_1$ null mice defines $TRPV_1$ (VR-1) protein distribution. Brain Res 995:176–183.

Romero J, Lastres-Becker I, de Miguel R, Berrendero F, Ramos J, Fernandez-Ruiz J (2002) The endogenous cannabinoid system and the basal ganglia. biochemical, pharmacological, and therapeutic aspects. Pharmacol Ther 95:137.

Ross RA, Gibson TM, Brockie HC, Leslie M, Pashmi G, Craib SJ, Di Marzo V, Pertwee RG (2001) Structure-activity relationship for the endogenous cannabinoid, anandamide, and certain of its analogues at vanilloid receptors in transfected cells and vas deferens. Br J Pharmacol 132:631–640.

Sagar DR, Smith PA, Millns PJ, Smart D, Kendall DA, Chapman V (2004) TRPV1 and CB(1) receptor-mediated effects of the endovanilloid/endocannabinoid *N*-arachidonoyl-dopamine on primary afferent fibre and spinal cord neuronal responses in the rat. Eur J Neurosci 20:175–184.

Salio C, Doly S, Fischer J, Franzoni MF, Conrath M (2002a) Neuronal and astrocytic localization of the cannabinoid receptor-1 in the dorsal horn of the rat spinal cord. Neurosci Lett 329:13–16.

Salio C, Fischer J, Franzoni MF, Conrath M (2002b) Pre- and postsynaptic localizations of the CB_1 cannabinoid receptor in the dorsal horn of the rat spinal cord. Neuroscience 110:755–764.

Samad TA, Moore KA, Sapirstein A, Billet S, Allchorne A, Poole S, Bonventre JV, Woolf CJ (2001) Interleukin-1beta-mediated induction of Cox-2 in the CNS contributes to inflammatory pain hypersensitivity. Nature 410:471–475.

Sanchez JF, Krause JE, Cortright DN (2001) The distribution and regulation of vanilloid receptor VR_1 and VR1 5 splice variant RNA expression in rat. Neuroscience 107:373–381.
Sanudo-Pena MC, Walker JM (1998) Effects of intrapallidal cannabinoids on rotational behavior in rats: interactions with the dopaminergic system. Synapse 28:27–32.
Sanudo-Pena MC, Romero J, Seale GE, Fernandez-Ruiz JJ, Walker JM (2000) Activational role of cannabinoids on movement. Eur J Pharmacol 391:269–274.
Sawzdargo M, Nguyen T, Lee DK, Lynch KR, Cheng R, Heng HH, George SR, O'Dowd BF (1999) Identification and cloning of three novel human G protein-coupled receptor genes *GPR52, PsiGPR53* and *GPR55: GPR55* is extensively expressed in human brain. Brain Res Mol Brain Res 64:193–198.
Shire D, Carillon C, Kaghad M, Calandra B, Rinaldi-Carmona M, Le Fur G, Caput D, Ferrara P (1995) An amino-terminal variant of the central cannabinoid receptor resulting from alternative splicing. J Biol Chem 270:3726–2731.
Simon GM, Cravatt BF (2006) Endocannabinoid biosynthesis proceeding through glycerophospho-*N*-acyl ethanolamine and a role for alpha/beta-hydrolase 4 in this pathway. J Biol Chem 281:26465–26472.
Sjöström PJ, Turrigiano GG, Nelson SB (2003) Neocortical LTD via coincident activation of presynaptic NMDA and cannabinoid receptors. Neuron 39:641–654.
Sjöström PJ, Turrigiano GG, Nelson SB (2004) Endocannabinoid-dependent neocortical layer-5 LTD in the absence of postsynaptic spiking. J Neurophysiol 92:3338–3343.
Slanina KA, Schweitzer P (2005) Inhibition of cyclooxygenase-2 elicits a CB_1-mediated decrease of excitatory transmission in rat CA1 hippocampus. Neuropharmacology 49:653–659.
Stella N, Schweitzer P, Piomelli D (1997) A second endogenous cannabinoid that modulates long-term potentiation. Nature 388:773–778.
Sugiura T, Kondo S, Sukagawa A, Nakane S, Shinoda A, Itoh K, Yamashita A, Waku K (1995) 2-Arachidonoylglycerol: a possible endogenous cannabinoid receptor ligand in brain. Biochem Biophys Res Commun 215:89–97.
Suplita RL, Gutierrez T, Fegley D, Piomelli D, Hohmann AG (2005) Endocannabinoids at the spinal level regulate, but do not mediate, nonopioid stress-induced analgesia. Neuropharmacology 50(3):372–379.
Swanson LW, Petrovich GD (1998) What is amygdala? TINS 21:323–331.
Szabo B, Wallmichrath I, Mathonia P, Pfreundtner C (2000) Cannabinoids inhibit excitatory neurotransmission in the substantia nigra pars reticulata. Neuroscience 97:89–97.
Szabo B, Siemes S, Wallmichrath I (2002a) Inhibition of GABAergic neurotransmission in the ventral tegmental area by cannabinoids. Eur J Neurosci 15:2057–2061.
Szabo T, Biro T, Gonzalez AF, Palkovits M, Blumberg PM (2002b) Pharmacological characterization of vanilloid receptor located in the brain. Brain Res Mol Brain Res 98:51–57.
Takahashi KA, Castillo PE (2006) The CB_1 cannabinoid receptor mediates glutamatergic synaptic suppression in the hippocampus. Neuroscience 139:795–802.
Talley EM, Solorzano G, Lei Q, Kim D, Bayliss DA (2001) Cns distribution of members of the two-pore-domain (KCNK) potassium channel family. J Neurosci 21:7491–7505.
Thomas EA, Cravatt BF, Danielson PE, Gilula NB, Sutcliffe JG (1997) Fatty acid amide hydrolase, the degradative enzyme for anandamide and oleamide, has selective distribution in neurons within the rat central nervous system. J Neurosci Res 50:1047–1052.
Tominaga M, Caterina MJ, Malmberg AB, Rosen TA, Gilbert H, Skinner K, Raumann BE, Basbaum AI, Julius D (1998) The cloned capsaicin receptor integrates multiple pain-producing stimuli. Neuron 21:531–543.
Toth A, Boczan J, Kedei N, Lizanecz E, Bagi Z, Papp Z, Edes I, Csiba L, Blumberg PM (2005) Expression and distribution of vanilloid receptor 1 ($TRPV_1$) in the adult rat brain. Brain Res Mol Brain Res 135:162–168.
Tsou K, Brown S, Sanudo-Pena MC, Mackie K, Walker JM (1998a) Immunohistochemical distribution of cannabinoid CB_1 receptors in the rat central nervous system. Neuroscience 83:393–411.

Tsou K, Nogueron MI, Muthian S, Sanudo-Pena MC, Hillard CJ, Deutsch DG, Walker JM (1998b) Fatty acid amide hydrolase is located preferentially in large neurons in the rat central nervous system as revealed by immunohistochemistry. Neurosci Lett 254:137–140.

Uchigashima M, Narushima M, Fukaya M, Katona I, Kano M, Watanabe M (2007) Subcellular arrangement of molecules for 2-arachidonoyl-glycerol-mediated retrograde signaling and its physiological contribution to synaptic modulation in the striatum. J Neurosci 27:3663–3676.

Ueda N, Okamoto Y, Morishita J (2005) *N*-acylphosphatidylethanolamine-hydrolyzing phospholipase D: a novel enzyme of the beta-lactamase fold family releasing anandamide and other *N*-acylethanolamines. Life Sci 77:1750–1758.

Valtschanoff JG, Rustioni A, Guo A, Hwang SJ (2001) Vanilloid receptor VR_1 is both presynaptic and postsynaptic in the superficial laminae of the rat dorsal horn. J Comp Neurol 436:225–235.

van der Stelt M, Di Marzo V (2003) The endocannabinoid system in the basal ganglia and in the mesolimbic reward system: implications for neurological and psychiatric disorders. Eur J Pharmacol 480:133–150.

van der Stelt M, Di Marzo V (2004) Endovanilloids. Putative endogenous ligands of transient receptor potential vanilloid 1 channels. Eur J Biochem 271:1827–1834.

Van Sickle MD, Oland LD, Ho W, Hillard CJ, Mackie K, Davison JS, Sharkey KA (2001) Cannabinoids inhibit emesis through CB_1 receptors in the brainstem of the ferret. Gastroenterology 121:767–774.

Van Sickle MD, Duncan M, Kingsley PJ, Mouihate A, Urbani P, Mackie K, Stella N, Makriyannis A, Piomelli D, Davison JS, Marnett LJ, Di M, V, Pittman QJ, Patel KD, Sharkey KA (2005) Identification and functional characterization of brainstem cannabinoid CB_2 receptors. Science 310:329–332.

Walker JM, Huang SM, Strangman NM, Tsou K, Sanudo-Pena MC (1999) Pain modulation by release of the endogenous cannabinoid anandamide. Proc Natl Acad Sci USA 96:12198–12203.

Washburn CP, Sirois JE, Talley EM, Guyenet PG, Bayliss DA (2002) Serotonergic raphe neurons express TASK channel transcripts and a TASK-like pH- and halothane-sensitive K^+ conductance. J Neurosci 22:1256–1265.

Watanabe M, Nakamura M, Sato K, Kano M, Simon MI, Inoue Y (1998) Patterns of expression for the mRNA corresponding to the four isoforms of phospholipase Cbeta in mouse brain. Eur J Neurosci 10:2016–2025.

Wenger T, Moldrich G, Furst S (2003) Neuromorphological background of cannabis addiction. Brain Res Bull 61:125–128.

Wotherspoon G, Fox A, McIntyre P, Colley S, Bevan S, Winter J (2005) Peripheral nerve injury induces cannabinoid receptor 2 protein expression in rat sensory neurons. Neuroscience 135:235–245.

Wotjak CT (2005) Role of endogenous cannabinoids in cognition and emotionality. Mini Rev Med Chem 5:659–670.

Yamagata K, Andreasson KI, Kaufmann WE, Barnes CA, Worley PF (1993) Expression of a mitogen-inducible cyclooxygenase in brain neurons: regulation by synaptic activity and glucocorticoids. Neuron 11:371–386.

Yoshida T, Fukaya M, Uchigashima M, Miura E, Kamiya H, Kano M, Watanabe M (2006) Localization of diacylglycerol lipase-alpha around postsynaptic spine suggests close proximity between production site of an endocannabinoid, 2-arachidonoyl-glycerol, and presynaptic cannabinoid CB1 receptor. J Neurosci 26:4740–4751.

Yu M, Ives D, Ramesha CS (1997) Synthesis of prostaglandin E_2 ethanolamide from anandamide by cyclooxygenase-2. J Biol Chem 272:21181–21186.

Zygmunt PM, Petersson J, Andersson DA, Chuang H, Sorgard M, Di M, V, Julius D, Hogestatt ED (1999) Vanilloid receptors on sensory nerves mediate the vasodilator action of anandamide. Nature 400:452–457.

Chapter 11
Endocannabinoids at the Synapse: Retrograde Signaling and Presynaptic Plasticity in the Brain

Gregory L. Gerdeman

Abstract In the study of synaptic transmission, the field of endocannabinoid research is flowering. A wealth of recent findings has revealed critical molecular underpinnings of endocannabinoid generation and signal transduction, and the subcellular localization of these processes within neurons has been mapped with increasing detail. Sophisticated techniques of neuronal imaging and electrophysiological recording have combined to yield new insights into the timing and regulation of endocannabinoid signaling at synapses throughout the brain, and these discoveries are beginning to influence models of synaptic computation and plasticity on a profound level. Triggered by membrane depolarization, intracellular Ca^{2+} elevation, and/or activation of $G_{q/11}$-coupled metabotropic receptors, postsynaptically released endocannabinoids act at presynaptic CB_1 receptors to mediate retrograde synaptic inhibition at both excitatory and inhibitory synapses, and on timescales that are either transient (on a scale of seconds) or long lasting. By dynamically modulating synapse reliability, synaptic suppression mediated by endocannabinoids provides a means for postsynaptic neurons to "tune" the sensitivity of their synaptic inputs to afferent patterns of stimulation. This may in turn help to regulate burst firing, or to generate or maintain synchronous membrane oscillations in interconnected neuronal populations. Endocannabinoid-dependent long-term synaptic depression (LTD) has also been recently demonstrated to underlie multiple forms of spike-timing-dependent plasticity (STDP) in the cerebral cortex, long thought to regulate the neuronal representation of sensory maps. In this chapter, I briefly survey updated concepts and mechanisms of endocannabinoid-mediated synaptic plasticity, and discuss the possible functional relevance of these processes to perception and behavior.

Introduction

For decades, the study of cannabinoid effects in the brain was directed by interest in the psychoactive properties of phytocannabinoids derived from *Cannabis ssp.*, especially Δ^9-tetrahydrocannabinol (Δ^9-THC). As elaborated throughout this book, cellular mechanisms discovered in the search for cannabinoid effects have revealed a

A. Köfalvi (ed.), *Cannabinoids and the Brain.*

system of membrane-derived, bioactive lipids – the endocannabinoids – with physiological roles far more extensive than originally expected. Endocannabinoids, represented primarily by anandamide (AEA) and 2-arachidonoylglycerol (2-AG) at present, are now known to act as a widespread system of *retrograde signaling* at central synapses, whereby the stimulus-dependent synthesis of endocannabinoids in postsynaptic neurons leads to the activation of presynaptic CB_1 receptors, and a subsequent inhibition of neurotransmitter release via multiple presynaptic mechanisms. First demonstrated only six years ago (Wilson and Nicoll, 2001), mechanisms of retrograde signaling and endocannabinoid-mediated synaptic plasticity have been studied and elaborated at a rapid pace, and recent findings are altering models of synaptic function and plasticity at many of the numerous brain areas where CB_1 receptors are expressed. Accordingly, this subject has been covered in several excellent recent reviews, to which I direct the reader for comprehensive mechanistic and historical perspectives (Alger, 2003; Freund et al., 2003; Gerdeman and Lovinger, 2003; Piomelli, 2003; Diana and Marty, 2004; Chevaleyre et al., 2006; Hashimotodani et al., 2007b). As the metabolic and anatomical characteristics of the brain endocannabinoid system are reviewed elsewhere in this book (see Chaps. 2, 3, and 10), it is my intention here to provide a point of access into the fast-growing literature on endocannabinoids as retrograde mediators of synaptic suppression, and overview some recent lines of study wherein the "basics" of endocannabinoid physiology are being integrated into models of neuronal function at the level of simple circuits and networks, with considerable implications for mental health and disease.

Endocannabinoids as a Synaptic System

Multiple lines of evidence have led to general consensus that in many brain areas, the molecular synthetic pathways for AEA and 2-AG (as reviewed in Chap. 2) are functionally located postsynaptic to both excitatory and inhibitory synapses (Matyas et al., 2006; Katona et al., 2006; Yoshida et al., 2006; see Chap. 10). A broad schematic depiction of these mechanisms, including details that are known to vary among individual endocannabinoid-releasing synapses, is shown in Fig. 1. The release of AEA from depolarized neurons in primary culture was first described in 1994 (Di Marzo et al., 1994), and was subsequently observed in the striatum of rats in vivo (Giuffrida et al., 1999). Moreover, neuronal activity and elevations of internal Ca^{2+} were demonstrated to increase the enzymatic synthesis of AEA, enhancing *N*-acyltransferase (NAT) activity and the production of *N*-arachidonoyl-phosphatidylethanolamine (NAPE) (Cadas et al., 1996, 1997). Thus, the generation of endocannabinoids was proposed to occur "on demand," consistent with what is now seen as a system of molecules that convey rapid synaptic feedback from somatodendritic neuronal membranes (Piomelli, 2003). The localization of CB_1 receptors on presynaptic terminals (Herkenham et al., 1991; Egertova et al., 1998; Katona et al., 1999), preferential coupling to $G_{i/o}$ proteins (Howlett et al., 2002), and inhibition of voltage-gated Ca^{2+} channels (VCCs)

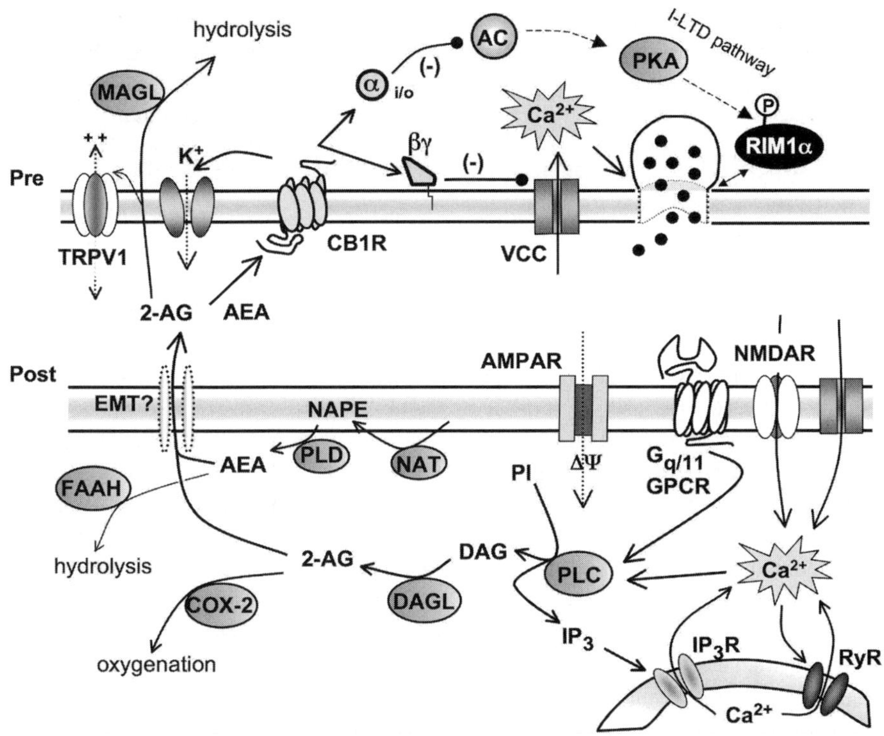

Fig. 1 Molecular players that can mediate or influence endocannabinoid signaling at central synapses. This overview highlights potential factors controlling the physiology of 2-AG or anandamide (AEA) at postsynaptic sites of generation and release, as well as presynaptic signaling by CB_1 receptors (CB_1R). In particular, 2-AG synthesis is stimulated by activation of phospholipase Cβ (PLC), which can serve as a coincidence detector of convergent postsynaptic signaling. Sources of Ca^{2+} vary in synapse-specific ways, as discussed in the text. AEA synthesis is also activated by internal Ca^{2+}, membrane depolarization (ΔΨ), and certain metabotropic receptors (not shown), but in general 2-AG seems to be more prevalent as a fast retrograde synaptic messenger. Presynaptic inhibition of neurotransmitter release via CB_1 receptors can involve multiple effectors, especially including voltage-gated Ca^{2+} channels (VCC), and also K^+ channels. In addition, long-term depression of GABAergic synapses (I-LTD) is mediated by decreased phosphorylation of $RIM_{1\alpha}$ via a signaling cascade involving adenylyl cyclase (AC) and protein kinase A (PKA). Additional details are described in the text. *EMT* putative endocannabinoid membrane transporter; *COX-2* cyclooxygenase-2; *DAG* diacylglycerol; IP_3 inositol 1,4,5-trisphosphate; *NAPE N*-acyl-phosphatid ylethanolamine; *NAT N*-acyltransferase; *PI* phosphoinositol; *RyR* ryanodine receptor

(Twitchell et al., 1997; Shen and Thayer, 1998) all further indicated that endocannabinoids would cause suppression of vesicular neurotransmitter release (see Fig. 1). This has indeed been supported by numerous studies showing that CB_1 receptor activation can inhibit the evoked release of neurotransmitters in a variety of in vitro tissue preparations (Szabo and Schlicker, 2005). Thus, while it is important here to note that there are numerous central neurons that express CB_1 receptors at somatic or postsynaptic sites, the function of presynaptic inhibition has emerged as a dominant theme in the physiological significance of cannabinoids in the brain.

Cannabinoids, Coincidence Detection, and Long-Term Synaptic Plasticity

Several studies in the late 1990s demonstrated that CB_1 receptor activation could have distinct inhibitory effects on both LTD and long-term potentiation (LTP) of excitatory inputs onto CA1 pyramidal neurons of the hippocampus (Sullivan, 2000) and LTD in the cerebellum (Levenes et al., 1998). As a brief background, LTP and LTD have become rather general terms that describe the ability of neurons to adjust the strength of their synaptic connections in a stable and persistent manner. Throughout the brain, numerous distinct types of LTP and LTD can be experimentally elicited by electrical stimulation of afferent synaptic pathways (Malenka and Bear, 2004). Although the various mechanisms of LTD and LTP are evoked by an equally diverse collection of experimental protocols – not always tightly reflective of neuronal firing patterns known to occur in vivo – it is believed that these mechanisms reflect the molecular physiology underlying various forms of learning and memory (Martin et al., 2000; Ito, 2001; Malenka and Bear, 2004), and are integral to the proper organization of synaptic connections during brain development and homeostasis (Feldman and Brecht, 2005; Dan and Poo, 2006; Turrigiano, 2007). With that said, it is valuable for the present discussion to note some common features of LTP and LTD, which are also relevant to short-term endocannabinoid-mediated presynaptic plasticity. Theories of synaptic plasticity underlying learning have evolved exquisite detail since the landmark proposals of Donald Hebb (Hebb, 1949), yet all address the means by which coincident neuronal firing or associative synaptic activation can generate specific signals that lead to persistent changes in the strength of connection. Induction of long-term synaptic plasticity typically involves patterns of stimulation that cause an elevation of internal Ca^{2+} in the postsynaptic cell, activating various downstream effectors. Rises in Ca^{2+} have proven to be an integral component of neuronal *coincidence detection*, and have thus become pivotal to models of LTP and LTD (Bear, 1996; Sjostrom and Nelson, 2002; Malenka and Bear, 2004; Chevaleyre et al., 2006). Sources of Ca^{2+} influx into the cytoplasm – such as through NMDA-sensitive ionotropic glutamate receptors (NMDARs), VCCs, or from intracellular stores of the endoplasmic reticulum – generally require, or are enhanced by, pronounced levels of membrane depolarization typical of coincident or sustained sources of excitation (Bear et al., 1987). Thus, returning to the present topic, the ability of CB_1 receptor activation to acutely block the induction of LTP or LTD might be explained entirely by inhibition of presynaptic glutamate release, which will prevent sufficient postsynaptic depolarization required to elevate Ca^{2+} through voltage-dependent sources like NMDARs and VCCs (Sullivan, 2000; Gerdeman and Lovinger, 2003). This may or may not be a significant physiological function of endocannabinoid-mediated retrograde signaling (Stella et al., 1997; Slanina et al., 2005), to which we will now turn our attention.

General Modes and Mechanisms of Endocannabinoid-Mediated Synaptic Plasticity

The first demonstration of a retrograde synaptic function of endocannabinoids came with the elegant solution to a standing puzzle in neurophysiology (Wilson and Nicoll, 2001). It was known that postsynaptic depolarization of principal neurons in the hippocampus or cerebellum, using whole-cell electrophysiology, can lead to a transient but reproducible decrease in the release of GABA from presynaptic terminals of inhibitory interneurons (Llano et al., 1991; Pitler and Alger, 1992). This phenomenon, known as depolarization-induced suppression of inhibition (DSI, Fig. 2a), can last for several seconds, is dependent upon postsynaptic Ca^{2+} signaling, and requires

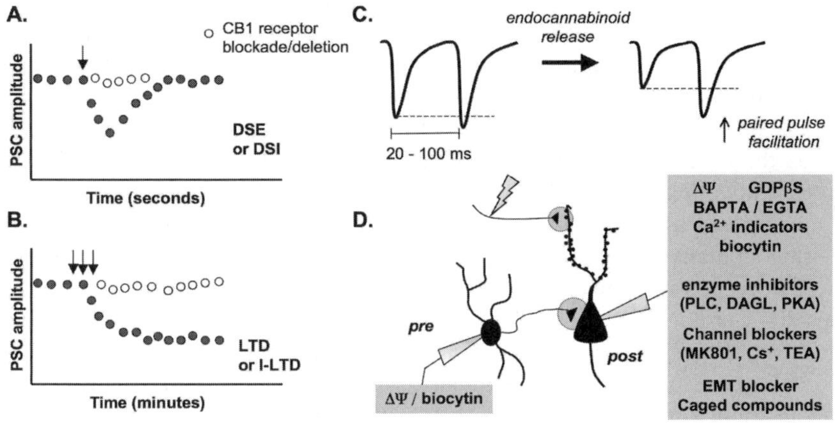

Fig. 2 Basics of synaptic suppression by endocannabinoids, as described by electrophysiology. (**a**) Fast and transient synaptic inhibition by endocannabinoids occurs at many synapses in brain slice or cultured neuronal preparations. In response to particular protocols of postsynaptic stimulation (*arrow*), postsynaptic current (PSC) amplitude is reduced for several seconds, in a manner dependent on CB_1 receptor activation. (**b**) Endocannabinoid-dependent LTD and I-LTD are evoked by different specific protocols of synaptic stimulation and somatic depolarization, and persist stably for many minutes to hours, generally longer than the experiment can be maintained. (**c**) Measurement of paired pulse facilitation (PPF), a form of short-term plasticity that is indicative of changes in neurotransmitter release probability, is a very common technique used to infer a presynaptic locus for endocannabinoid-dependent synaptic depression of evoked PSC amplitudes. Enhanced PPF occurs when PSC inhibition is associated with an increased ratio between the amplitudes of two closely timed evoked responses (PSC #2/PSC #1 = PPF ratio). Thus, if CB_1 receptor activation by postsynaptically generated endocannabinoids is correlated to increased PPF ratio, a presynaptic or retrograde mechanism is indicated. (**d**) Experimental manipulations used to characterize endocannabinoid-dependent synaptic plasticity using whole-cell patch clamp techniques. Mechanisms underlying endocannabinoid generation have been identified and localized in identified neuronal preparations by the introduction of various molecules directly into recorded neurons via the patch pipette, as indicated (described further in text). ΔΨ represents current injection; *Lighting bolt* represents electrical stimulation of afferent axons in the brain slice; *Shaded circles* represent local postsynaptic endocannabinoid release

a retrograde signal to the presynaptic terminal. There is now good evidence that this retrograde messenger is 2-AG, acting at presynaptic CB_1 receptors to inhibit VCCs and thereby decrease vesicular release of GABA (Wilson and Nicoll, 2002; Alger, 2003; Diana and Marty, 2004; Szabo et al., 2006). Such a mechanism was also found to exist at excitatory synapses, and by analogy to DSI was termed DSE (Kreitzer and Regehr, 2001a; Maejima et al., 2001). Endocannabinoid-mediated DSI and DSE have since been demonstrated at many other synapses throughout the brain, including hippocampus (Isokawa and Alger, 2005; Hofmann et al., 2006), neocortex (Trettel and Levine, 2002; Bodor et al., 2005; Fortin and Levine, 2007), basal ganglia (Chap. 21), amygdala (Zhu and Lovinger, 2005), hypothalamus (Hentges et al., 2005), midbrain (Melis et al., 2004), and brainstem (Kushmerick et al., 2004). Endocannabinoid synthesis and retrograde feedback suppression also can be induced by relatively brief, physiologically relevant patterns of afferent synaptic stimulation (Brown et al., 2003; Fortin et al., 2004). Endocannabinoid signaling at synapses is therefore a highly prevalent mechanism for modulating the efficacy of neurotransmitter release in the brain, and has quickly come to represent the most prominent example of retrograde signaling in the nervous system. In addition to transient modes of inhibition, endocannabinoids have also been found to be critically necessary for multiple forms of LTD (Fig. 2b) (Chevaleyre et al., 2006). As was the case for DSI, the first evidence for endocannabinoid-mediated LTD marked a novel explanation for a highly studied form of plasticity. Corticostriatal LTD, expressed as a lasting presynaptic decrease in glutamate release – and thus requiring a putative retrograde messenger (Choi and Lovinger, 1997) – was shown to be CB_1 receptor dependent and mimicked by postsynaptic introduction of AEA (Fig. 2d) (Gerdeman et al., 2002). Subsequent studies have provided further evidence that AEA is a retrograde messenger in striatal LTD (Ronesi et al., 2004; Ade and Lovinger, 2007). Moreover, similar forms of LTD have been discovered using different induction protocols, such that in the striatum and nucleus accumbens, CB_1 receptor-dependent LTD can be generated by several different patterns of afferent stimulation (Robbe et al., 2002; Hoffman et al., 2003; Kreitzer and Malenka, 2005; Ronesi and Lovinger, 2005). The CB_1 receptor is highly expressed at axon terminals of certain populations of GABA-releasing neurons (Katona et al., 1999; Freund et al., 2003; Matyas et al., 2006; and see Chap. 10), where postsynaptically released 2-AG mediates DSI, as referenced above. It is also now well established that presynaptic CB_1 receptors can trigger LTD at many inhibitory synapses, a process most commonly referred to as I-LTD (Chevaleyre et al., 2006). First shown in the basolateral amygdala (BLA), where I-LTD was correlated to the behavioral extinction of conditioned fear responses (but termed "LTDi," Marsicano et al., 2002; Azad et al., 2004), Chevaleyre and Castillo (2003) also demonstrated I-LTD to occur in the hippocampus. More recently, this group has discovered that the presynaptic mechanisms mediating lasting inhibition of vesicular GABA release are similar in both brain areas (Chevaleyre et al., 2007). This novel finding marks an important step in understanding how the CB_1 receptor appears strategically expressed to directly transduce both short- and long-lasting mechanisms of presynaptic suppression of neurotransmitter release, depending on the pattern and duration of postsynaptic activation. In the proposed model by Chevaleyre and coworkers (2007), the presynaptic DSI pathway involves a classic, membrane-delimited and voltage-dependent inhibition of presynaptic VCCs (Wilson et al., 2001; Brown et al., 2004; Foldy et al., 2006), signaled via the G protein β/γ subunit (Ikeda, 1996). The I-LTD pathway, however, is recruited only in situations of relatively prolonged (>5 min) CB_1 receptor activation (Chevaleyre and Castillo, 2003) and involves the $G_{i/o\alpha}$ subunit, decreased activation of protein kinase A (PKA), and likely reduced phosphorylation of $RIM_{1\alpha}$ (Chevaleyre et al., 2007), which is a presynaptic protein intimately associated with the vesicular release machinery (see Fig. 1; Kaeser and Sudhof, 2005). Importantly, $RIM_{1\alpha}$ is already implicated in a presynaptic form of LTP at a glutamatergic synapse (Castillo et al., 2002). It is therefore compelling to consider that regulation of $RIM_{1\alpha}$ may likewise mediate CB_1 receptor-dependent LTD at excitatory synapses, which also is thought to require prolonged CB_1 receptor signaling (e.g., Ronesi et al., 2004). In the striatum, LTD was recently shown to occur only at presynaptic terminals specifically activated by the plasticity-inducing afferent stimulus (Singla et al., 2007), which might relate to a priming of the $G_{i/o}$ signaling cascade leading to decreased activity of $RIM_{1\alpha}$, if the model applies here. In any case, synapse specificity is an important aspect for any synaptic mechanism to fit within accepted learning theories, and the finding by Singla and coworkers (2007) confirms that CB_1 receptor-dependent LTD evoked by high frequency synaptic activation meets this requirement.

Electrophysiological Techniques as Windows into Endocannabinoid Synaptic Function

Since the initial descriptions of endocannabinoid mechanisms in DSI/E and LTD, an impressive array of modern neuroscience methodologies has been applied to questions of endocannabinoid-mediated synaptic plasticity in numerous brain areas (Fig. 2d). For example, populations of endocannabinoid-releasing neurons have been exquisitely identified, such as by filling postsynaptic neurons with biocytin for post hoc microscopic analysis, or by using sophisticated transgenic techniques to select recorded neurons on the basis of gene expression (Wang et al., 2006; Kreitzer and Malenka, 2007). The whole-cell electrophysiology technique has furthermore proven tremendously valuable for identifying cell-signaling pathways necessary for endocannabinoid production in cells. Thus, the involvement of G protein-coupled receptors (GPCRs) and intracellular Ca^{2+} are commonly tested by inclusion of GDPβS or Ca^{2+} chelators (BAPTA or EGTA) into the postsynaptic recording pipette, respectively (e.g., Brown et al., 2003; Galante and Diana, 2004). Elevation of intracellular Ca^{2+} is often necessary for endocannabinoid formation (Piomelli, 2003), although Ca^{2+}-independent mechanisms also occur (Varma et al., 2001; Kim et al., 2002). Moreover, the subcellular localization of Ca^{2+} microdomains required to induce endocannabinoid release have been visualized and quantified by the postsynaptic infusion of fluorescent Ca^{2+} indicators, a method that has added invaluable perspective to the dendritic induction of retrograde signaling (Brenowitz and Regehr, 2003; Brown et al., 2003; Brenowitz et al., 2006; Nevian and Sakmann, 2006; Rancz and Hausser, 2006). At synapses where phospholipase C (PLC) and diacylglycerol lipase (DAGL) have been identified as important for presynaptic CB_1 receptor activation (thus indirectly identifying 2-AG as a released endocannabinoid), this has largely been demonstrated by introducing specific inhibitors of these enzymes directly into a recording pipette (Hashimotodani et al., 2007c). These techniques have the powerful combined ability to both identify enzymatic players involved in synaptic plasticity, and by virtue of being applied only to the single neuron under investigation, also demonstrate convincingly that the target cell supplies the endocannabinoid mediating synaptic plasticity, thereby decisively supporting the retrograde messenger hypothesis. Another important question is the source of Ca^{2+} often required for endocannabinoid generation, and whole-cell techniques have been essential in elaborating these mechanisms as well. For example, the intracellular infusion of MK-801, a pore-blocking inhibitor of NMDARs, demonstrated that postsynaptic NMDARs are necessary for DSE in interneurons of the cerebellum (Beierlein and Regehr, 2006), yet excluded this possibility in layer 5 neurons of the cerebral cortex (implicating presynaptic NMDARs instead: Sjostrom et al., 2003; see below). Other ion channel inhibitors routinely present in recording pipettes include cesium (Cs^+) and tetraethylammonium (TEA), both K^+ channel blockers, which by blocking numerous leak conductances can enhance the ability of depolarizing stimuli to propagate to distal dendrites. While such techniques can be necessary to voltage-clamp distal synapses, it is pertinent to apply careful perspective to experimental results, in the sense that DSE/I protocols in the presence of Cs^+ or TEA may elicit greater endocannabinoid release at synapses than would occur in unperturbed neurons with numerous dendritic K^+ channels available to shunt active currents. This general approach has also been used to test the involvement of the putative endocannabinoid membrane transporter (EMT) in the release of endocannabinoids from postsynaptic membranes. Although no protein has been cloned representing an EMT, specific inhibitors suggest that one exists, functioning by ATP-independent mechanisms of facilitated diffusion (Hillard and Jarrahian, 2000; Piomelli, 2003; see also Chap. 3). Therefore, my colleagues and I reasoned that postsynaptic intracellular application of a competitive blocker for the EMT might prevent the timely release of endocannabinoids needed for retrograde synaptic signaling. Indeed, the inclusion of either AM404 or VDM11 into patch pipettes prevented striatal LTD (Ronesi et al., 2004), a finding that has been also observed in somatosensory cortex (Bender et al., 2006b). This suggests that the rapid and spatially limited release of endocannabinoids (Heinbockel et al., 2005) may be facilitated by a specific EMT activity in postsynaptic membranes, which could be a significant target for therapeutic intervention (Piomelli, 2003). A great deal of information on endocannabinoid signaling has recently been gained by the increasing feasibility of paired whole cell recordings in brain slices (e.g., Sjostrom et al., 2003, 2004, 2007; Freiman et al., 2006). In such experiments, presynaptic inputs of single neurons can be induced by direct current injection to elicit individual action potentials or spike trains. The presynaptic, endocannabinoid-sensitive cell can also be biocytin-filled and co-labeled for other neurochemical markers such as cholecystokinin (CCK), allowing a fine degree of identification, and elaborating the function of CB_1 receptor expressing neurons within local circuits (e.g., Galarreta et al., 2004; Klausberger et al., 2005; Foldy et al., 2006; Glickfeld and Scanziani, 2006).

Presynaptic Inhibition and Synaptic "Tuning"

At this point, it is valuable to emphasize the presynaptic nature of endocannabinoid-mediated plasticity. An important and distinctive feature of presynaptic inhibition is that it alters the sensitivity of synaptic transmission to the *frequency* of input stimuli (Abbott and Regehr, 2004). This is reflected by changes in the short-term plasticity expressed at the synapse in response to subsequent afferent impulses, often measured in terms of paired-pulse facilitation (PPF, Fig. 2c). PPF is related to a nonlinear Ca^{2+} dependence of vesicular release, and the ability of residual presynaptic Ca^{2+} to markedly augment the effectiveness of successively timed stimuli (Zucker and Regehr, 2002). This index is generally greater when the initial probability of release (P_r) in response to an action potential is low, so an increase in the PPF ratio is a useful diagnostic to test whether a reduction in the amplitude of an *evoked* postsynaptic current, such as following an LTD- or DSI-inducing protocol, is actually mediated by *presynaptic* inhibition (a decrease in P_r, see Gerdeman and Lovinger, 2003). By changing the short-term plasticity characteristics of synapses, presynaptic inhibition can have significant implications for neuronal computational functions such as gain control (Abbot and Regehr, 2004). Specifically, presynaptic CB_1 receptor activation – either transient or long-lasting – can "tune" the filtering properties of a given synapse such that the *reliability* of an incoming action potential to evoke a postsynaptic response is decreased for single or low frequency stimuli, but increased in the context of higher frequency afferent bursts (e.g., Oliet et al., 2007). Note that the aforementioned voltage-dependence of VGCC inhibition by $G_{\beta\gamma}$ subunits (Ikeda, 1996) means that a DSI/E mechanism of presynaptic inhibition might be relieved by sustained high frequency inputs, an observation that has been suggested to underlie the context-dependence of some neuronal and behavioral effects of exogenous cannabinoid agonists (Foldy et al., 2006).

$G_{q/11}$-Coupled Metabotropic Receptors and PLCβ

Metabotropic receptors play decisive roles to enhance the postsynaptic generation of endocannabinoids and thus their action as retrograde messengers (see Fig. 1). Of particular importance are GPCRs coupled preferentially to $G_{q/11}$, especially group I mGluRs (mGluR subtypes 1 and 5) (Maejima et al., 2001; Varma et al., 2001) and muscarinic acetylcholine receptors (mAChRs) of the M_1 and M_3 subtypes (Kim et al., 2002; Ohno-Shosaku et al., 2003; Fukudome et al., 2004; Narushima et al.,

2007). Numerous investigations have now addressed the extent to which metabotropic receptor activation activates endocannabinoid signaling, with or without postsynaptic depolarization, and it is clear that mechanisms can vary between synaptic pathways (Hashimotodani et al., 2007c). Even within a single class of neuron, the CA1 hippocampal pyramidal cell, generation of endocannabinoids by DSI, mGluRs, or mAChRs, has been shown to proceed by multiple mechanisms (Edwards et al., 2006). An important observation, however, is that coupling postsynaptic depolarization with simultaneous agonist-induced activation of $G_{q/11}$-coupled receptors leads to greatly enhanced endocannabinoid generation compared to either approach alone (Varma et al., 2001; Kim et al., 2002; Ohno-Shosaku et al., 2002, 2003). In elegant studies, Kano and colleagues demonstrated that the cooperative enhancement of 2-AG release by depolarization and $G_{q/11}$-coupled receptors is a result of PLCβ activity (Hashimotodani et al., 2005; Maejima et al., 2005). In particular, PLCβ is activated both by $G_{q/11}$ and Ca^{2+}, in a manner that acts as a postsynaptic coincidence detector of these two signal transduction pathways, a process which these authors have dubbed Ca^{2+}-assisted receptor-driven endocannabinoid release (Ca-RER) (Maejima et al., 2005). Thus, the sharp Ca^{2+} dependency of $PLCβ_1$ in hippocampal pyramidal neurons (Hashimotodani et al., 2005), and $PLCβ_4$ in Purkinje cells of the cerebellum (Maejima et al., 2005), allow for physiologically relevant Ca^{2+} levels and metabotropic receptor activation to mutually augment their effects on 2-AG generation. Such a mechanism can describe forms of associative synaptic plasticity in which one synaptic input facilitates a decrease in the efficacy (CB_1 receptor-mediated decrease in P_r) of a second, coincidentally active input (Fig. 3a) (Brenowitz and Regehr, 2005). PLCβ activity leads to both the generation of DAG (then 2-AG, via the activity of DAGL) and IP_3, the latter of which can directly increase the concentration of cytoplasmic Ca^{2+} through actions on IP_3 receptors on the endoplasmic reticular membrane (see Fig. 1). Likewise, Ca^{2+}-gated ryanodine receptors, which can modulate 2-AG synthesis in some situations (Isokawa and Alger, 2006), can also further stimulate Ca^{2+} release from intracellular stores. On these grounds, it has been proposed that certain plasticity-evoking stimuli might activate PLCβ in a sustained manner through a regenerative positive feedback loop (Hashimotodani et al., 2007c). This would enhance the ability of synaptic inputs to reliably recruit endocannabinoid signaling. It can also be emphasized that such signaling loops would contain numerous points of interaction with other cellular processes. Although this discussion is perhaps more relevant to 2-AG, which appears to be more frequently utilized by neurons as a retrograde messenger, it is also the case that AEA can be stimulated by either $G_{q/11}$ (Azad et al., 2004; Wettschureck et al., 2006) or G_s-coupled GPCRs (Cadas et al., 1996). In the striatum, AEA is also stimulated by D_2 dopamine receptors (Giuffrida et al., 1999), which preferentially couple to $G_{i/o}$. In the context of LTD induction, this appears to be a heterosynaptic effect of D_2 receptors located on cholinergic interneurons, activation of which leads to decreased Ca^{2+} signaling and AEA synthesis in medium spiny neurons through indirect mechanisms (Wang et al., 2006; see Fig. 2 in Chap. 21). It is, however, also very compelling to note a recent report of functional D_2/CB_1 receptor heterodimers, which paradoxically couple to G_s (Kearn et al., 2005; see Chaps. 9, 23), and may

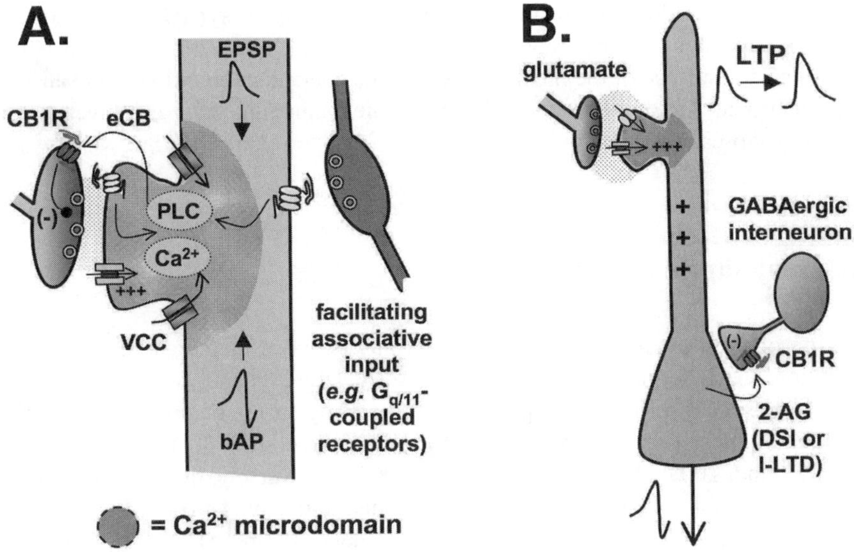

Fig. 3 Endocannabinoids mediate associative synaptic plasticity and heterosynaptic metaplasticity. (**a**) Local and synapse specific endocannabinoid generation can be influenced by numerous signals of coincident cellular activity. The timing, intensity, and/or cellular source of elevated dendritic Ca^{2+} (e.g., from active VCC conductances or mGluR-induced Ca^{2+} release from intracellular stores) is particularly critical. Endocannabinoid-mediated synaptic plasticity can therefore be an associative process that might be important for various functions of neuronal computation. (**b**) Metaplasticity between GABAergic and glutamatergic synapses. Disinhibition of GABAergic interneurons via either DSI or I-LTD facilitates dendritic excitability and the induction of LTP at distal synapses of the CA3-CA1 Schaffer collateral pathway. *bAP* back propagating action potential; *eCB* endocannabinoid; *EPSP* excitatory postsynaptic potential

therefore represent a novel way in which postsynaptic D_2 and CB_1 receptors could stimulate AEA release. In addition to mGluRs and mAChRs, any $G_{q/11}$-coupled receptor could potentially activate the PLCβ-DAGL pathway to generate 2-AG (Piomelli, 2003; Wettschureck et al., 2006). Indeed, 2-AG can be stimulated by $G_{q/11}$-coupled orexin receptors in the dorsal raphe nucleus (Haj-Dahmane and Shen, 2005), oxytocin autoreceptors in magnocellular neurosecretory cells of the hypothalamus (Hirasawa et al., 2004; Oliet et al., 2007), and by serotonin acting at 5-HT_{2C} receptors in cultured fibroblasts (Parrish and Nichols, 2006). Taken together, the above-described mechanisms reveal that endocannabinoid pathways have properties ideally suited to act as signals of postsynaptic coincidence detection, and that the brain expresses a rich diversity of means to utilize this system for synaptic regulation. Either precisely timed release (Brenowitz and Regehr, 2005) or basal release (Narushima et al., 2007) of modulatory neurotransmitters, or "spillover" activation of extrasynaptic mGluRs (Marcaggi and Attwell, 2005, 2007; Wadiche and Jahr, 2005), may facilitate endocannabinoid release in activated regions of the somatodendritic membrane (see Fig. 3a), thereby dampening CB_1 receptor-expressing inputs (e.g., Neu et al., 2007) and/or tuning cellular responses to synaptic excitation in a stimulus-dependent manner (e.g., Oliet et al., 2007). Quite likely, the field is only scratching the surface of the means by which the endocannabinoid system can serve as a functionally relevant readout of the complex signaling state of a given neuron and its synapses at a given point in time.

Heterosynaptic Metaplasticity

The life of most neurons is a constant interplay between membrane excitation and inhibition. GABAergic interneurons often form large synaptic contacts on somatodendritic sites of principal output neurons, exerting strong control over the ability of excitatory postsynaptic potentials (EPSPs) to drive cell firing (McBain and Fisahn, 2001; Freund, 2003; Hestrin and Galarreta, 2005). Conversely, these local inhibitory inputs also shunt somatically generated currents, dampening the invasion of back-propagating actions potentials (bAPs) into the dendritic tree (Hausser et al., 2000; Sjostrom and Nelson, 2002). It therefore was hypothesized that a reduced P_r of GABA from CB_1 receptor-expressing inhibitory interneurons would heterosynaptically influence properties of glutamate-mediated excitatory responses in multiple ways (Fig. 3b). In the hippocampus, LTP has been associated with an increased likelihood to observe a postsynaptic spike in response to an evoked EPSP (a property termed E–S coupling), and Chevaleyre and Castillo (2003) found evidence that this is due to simultaneously induced I-LTD at interneuron-pyramidal cell synapses. In other words, endocannabinoid-mediated I-LTD elicits a lasting increase in the efficiency of EPSPs to drive somatic depolarization to the threshold for firing an action potential. Inversely, the suppression of GABAergic inhibition by 2-AG can *facilitate* the induction of LTP at glutamatergic synapses. Specifically, when postsynaptically released endocannabinoids cause a retrograde suppression of inhibition, whether transiently (DSI, Carlson et al., 2002) or long term (I-LTD, Chevaleyre and Castillo, 2004; Zhu and Lovinger, 2007), hippocampal LTP can be induced by stimuli that are otherwise insufficient. Therefore a heterosynaptic effect of endocannabinoid-mediated disinhibition can lead to LTP-enhancing *metaplasticity* in a circuit critical to many forms of explicit learning (Martin et al., 2000; Malenka and Bear, 2004). This is in contrast to the ability of exogenous cannabinoids to inhibit LTP in other brain slice experiments, as discussed above, or of 2-AG to likewise inhibit LTP induction when the endocannabinoid is generated local to the excitatory inputs (Stella et al., 1997; Slanina et al., 2005). In short, endocannabinoid signaling can play multiple roles within a given neural circuit, with sometimes opposite effects on the valence of synaptic plasticity. The extent to which these physiological roles are mimicked by exogenous cannabinoids depends upon the spatial localization of endocannabinoid release (which varies according to the nature of stimulation; see Brown et al., 2003; Brenowitz et al., 2006), the limits of endocannabinoid diffusion (Hajos et al., 2004; Chevaleyre et al., 2006; Hofmann et al., 2006), and the nature of the pharmacological agent (Whalley et al., 2004; Lauckner et al., 2005; Straiker and Mackie, 2005).

Endocannabinoid Synapses of the Cerebellum

The diverse utility of endocannabinoid retrograde signaling has been most extensively explored in the cerebellum. In addition to being one of the brain areas richest in CB_1 receptor expression (see Chap. 10), the cerebellar cortex is amenable to precise investigation of synaptic pathways due to the well-defined architecture of its axonal projections (Fig. 4) (Palay and Chan-Palay, 1974). The principal output

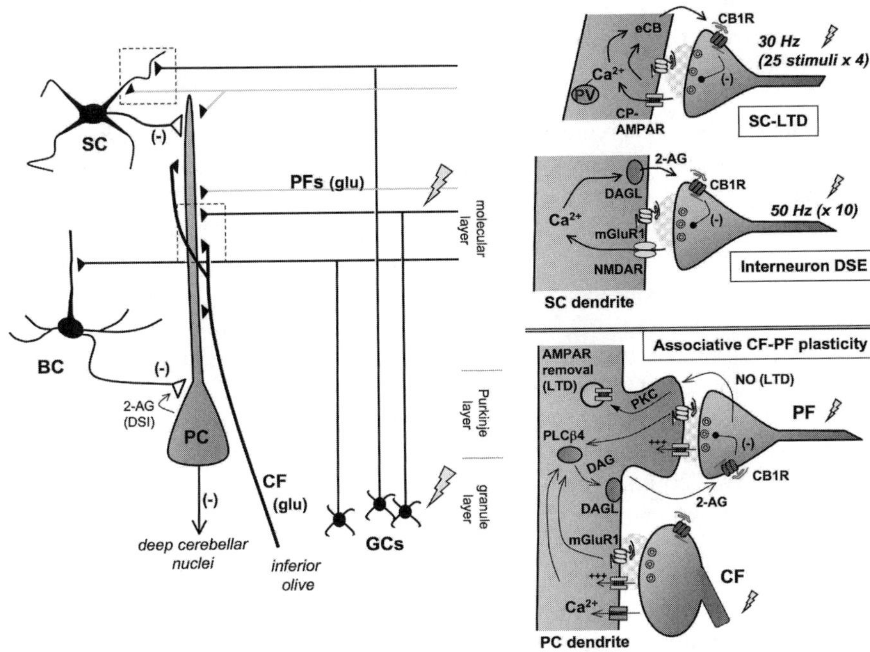

Fig. 4 Diversity and ubiquity of endocannabinoid-mediated synaptic depression in the cerebellar cortex. (*Left*) The neural circuitry of the cerebellar cortex includes excitatory inputs from parallel fibers [PFs, arising from glutamatergic (glu) granule cells GCs)] and climbing fibers (CF), synapsing onto the extensive dendritic tree of Purkinje cell (PC) output neurons. PCs also receive inhibitory input from local interneurons, the stellate cells (SC) and basket cells (BC). Every synapse depicted expresses CB_1 receptors and functional retrograde endocannabinoid signaling. *Dashed boxes* indicate areas illustrated in the cartoons at right, and *lightning bolts* represent different sites of PF stimulation as discussed in the text. (*Top right*) Mechanisms of SC-LTD and DSE in interneurons. In SC-LTD (induced by four bursts of 25 stimuli at 30 Hz), synapse specificity is achieved by the tightly limited diffusion of Ca^{2+} in thin dendrites expressing the Ca^{2+} binding protein parvalbumin (PV). The Ca^{2+} necessary for SC-LTD is gated through Ca^{2+}-permeable AMPA receptors (CP-AMPAR), whereas NMDA receptors provide Ca^{2+} influx mediating DSE (induced by a short, 10-stimuli burst at 50 Hz; this DSE also occurs at PF-BC synapses). *Lower right:* Paired stimulation of CF and PF inputs (or PF stimulation briefly preceding CF) leads to supralinear Ca^{2+} summation and activation of PLCβ4, leading to DSE or LTD of activated PF synapses. Unlike other forms of eCB-mediated LTD, cerebellar LTD is expressed postsynaptically via a PKC-mediated internalization of AMPA receptors, which is triggered by nitric oxide (NO). The source of NO is not entirely certain, and other explanations have been offered for the role of presynaptic CB_1 receptors in facilitating cerebellar LTD (see text)

neurons of the cerebellar cortex are the tonically active Purkinje cells (PCs), which have elaborate, planar-oriented dendritic arbors (ideal for Ca^{2+} visualization) and use GABA as a neurotransmitter. Individual PCs receive many (~100,000) weak glutamatergic inputs from cerebellar granule cells (GCs) via stereotypically arranged parallel fibers (PFs), as well as strong excitatory connections from a single climbing fiber (CF) arising from the inferior olive (Ito, 2001). The PFs also form excitatory synapses onto two primary types of inhibitory interneuron, the basket cells (BCs) and stellate cells (SCs), which in turn form GABAergic, feed-forward synapses onto PCs (Mittmann et al., 2005; Beierlein and Regehr, 2006). Experiments showing endocannabinoid-mediated DSI and DSE were pioneered in PCs (Kreitzer and Regehr, 2001a,b; Diana et al., 2002), yet these early studies utilized depolarizing stimuli that are not likely to occur in vivo. Subsequent studies by Regehr's group and others have greatly refined models of endocannabinoid retrograde signaling in this circuitry. Brown and colleagues (2003) first demonstrated that brief bursts (3–5 impulses at 50 Hz) of stimuli delivered to PF axons in the molecular layer, can induce feedback suppression of PF inputs by endocannabinoids acting at CB_1 receptors. This retrograde inhibition was found to be synapse specific: suppression did not spread to other PFs activated by a separate stimulating electrode, consistent with the observation (using intracellular Ca^{2+} indicators) that dendritic Ca^{2+} influx was only locally enhanced (within $<20\,\mu m$ of the activated PF inputs). It was also confirmed that retrograde inhibition was mediated solely by the recorded neuron (as it was blocked by intracellular BAPTA and GDPβS) (Brown et al., 2003). This is an important distinction from DSI evoked in PCs: DSI induced by strong somatic depolarization (Maejima et al., 2001; Brenowitz and Regehr, 2003; Szabo et al., 2006), or prolonged but physiological Ca^{2+} signaling (Brenowitz et al., 2006), appears to induce 2-AG release throughout the dendritic arbor, inhibiting multiple interneuron inputs. 2-AG release activated by the $mGluR_1$ agonist DHPG was also found to act beyond tight spatial restrictions (Galante and Diana, 2004). Moreover, whereas retrograde suppression of PF to PC inputs is mediated by CB_1 receptor inhibition of presynaptic VGCCs (Brown et al., 2004), DSI appears mediated, at least in part, by an activation of GIRK-like K^+ conductances (Kreitzer et al., 2002). This latter mechanism, by inducing a strong hyperpolarization of the CB_1 receptor-expressing GABAergic interneuron, can spread the effect of DSI far beyond the spatial limits of endocannabinoid diffusion, potentially disinhibiting every PC targeted by that interneuron (Kreitzer et al., 2002). A mechanistically distinct form of interneuron firing suppression mediated by endocannabinoids has also been demonstrated in neocortex (Bacci et al., 2004).

Associative CF-PF Plasticity and LTD

The cooperative interactions observed between dendritic Ca^{2+} and $mGluR_1$ activation to stimulate 2-AG formation led to the proposal that such a mechanism might occur with the proper timing and spatial fidelity to allow for associative mechanisms of plasticity between CF and PF synapses (Brenowitz and Regehr, 2005). Associative processes – whereby the postsynaptic neuron integrates convergent inputs to adapt its synaptic weights and/or circuit function – are key to neuronal learning theories (Schultz and Dickinson, 2000). In cerebellar PCs, LTD of PF inputs can be induced by associative pairing of CF activation with PF stimulation (Ito, 2001). Regehr and colleagues have recently shown that CF–PF paired stimulation induces the $mGluR_1$- and Ca^{2+}-dependent generation of 2-AG, thereby identifying the associative process of suppressing PF glutamate release (see Fig. 4). In response to brief paired stimuli, this associative plasticity is expressed as a *transient* retrograde inhibition (Brenowitz and Regehr, 2005), which might be a means for the circuit to achieve rapid correction of errors during fine motor behaviors (Schultz and Dickinson, 2000). Although mechanisms for rapid associative plasticity have been postulated previously, the endocannabinoid system represents the first molecular pathway anywhere in the brain to definitively mediate such synaptic integration and feedback plasticity (Brenowitz and Regehr, 2005). A similar mechanism was also demonstrated to explain cerebellar LTD (Safo and Regehr, 2005). Thus, PF stimulation leads to a modest level of $mGluR_1$ activation and Ca^{2+} influx, which is enhanced dramatically by a closely following CF-evoked EPSP, leading to supralinear Ca^{2+} summation and the activation of $PLC\beta_4$ and DAGL (Brenowitz and Regehr, 2005; Maejima et al., 2005; Safo and Regehr, 2005). In the case of LTD, which requires repetitive paired stimuli, nitric oxide (NO) signaling is also recruited, leading to PKC-mediated internalization of postsynaptic AMPARs (Ito, 2001). Thus, the final expression of cerebellar LTD is mechanistically distinct from presynaptic forms of endocannabinoid-mediated LTD (Chevaleyre et al., 2006). Cerebellar LTD has been studied for many years, and is believed to be necessary for forms of learning requiring the cerebellum, such as trace eye-blink conditioning (Ito, 2001). This notion has now been further verified by the finding that trace eye-blink conditioning is specifically impaired when CB_1 receptors are inactivated by genetic deletion or with the selective antagonist SR141716A (Kishimoto and Kano, 2006). Some important caveats to these findings have emerged, however, which have further demonstrated both the complex functionality and descriptive power of retrograde endocannabinoid signaling at synapses. For instance, a series of studies by Marcaggi and Attwell (2005, 2007) investigated the stimulation patterns required to elicit retrograde endocannabinoid inhibition of activated PFs. To activate PFs experimentally, most investigators place their stimulating electrode in the *molecular layer* of the cerebellar cortex (see the top *lightning bolt* in the circuit diagram of Fig. 4). As pointed out by Marcaggi and Attwell, however, this approach will activate numerous closely spaced PFs, and result in a synchronous convergence of glutamate release and subsequent activation of extrasynaptic mGluRs. They found that retrograde inhibition of PF synapses (e.g., Brown et al., 2003) did not happen when similar stimulus trains were applied in the *granule layer*, which is more likely to activate spatially separated PF synapses onto patch-clamped PCs (see Fig. 4). Thus, PC endocannabinoid release was proposed to act as a homeostatic mechanism to restore synaptic independence when the "crosstalk" of adjacent PF inputs leads to spillover of synaptic glutamate onto extrasynaptic mGluRs (Marcaggi and Attwell, 2005). A similar finding was more recently demonstrated for CB_1 receptor-dependent associative plasticity and cerebellar LTD (Marcaggi and Attwell, 2007). Consistent with the need for glutamate spillover and extrasynaptic mGluR activation, retrograde 2-AG signaling apparently does not happen at PC dendritic domains where synaptic glutamate is rapidly and efficiently removed by a high expression pattern of the glutamate transporter EAAT4 (Wadiche and Jahr, 2005). Retrograde signaling by endocannabinoids at PF synapses might therefore be somewhat less common than experiments stimulating the molecular layer have suggested, and a mechanism recruited specifically by adjacent, synchronously active PF inputs (Rancz and Hausser, 2006). That this is proposed as homeostatic (Marcaggi and Attwell, 2005) does not negate an important computational role (Turrigiano, 2007), and one which is likely to occur in many areas of the brain where axon pathways are less favorably arranged for experimental isolation. Evidence has also been recently presented for an entirely alternative explanation for the role of CB_1 receptors in LTD (van Beugen et al., 2006). Based on their findings, van Beugen and colleagues (2006) propose that retrograde endocannabinoids act not to directly *induce* LTD, but rather to *unmask* LTD by inhibiting concurrent presynaptic mechanisms of LTP. This scheme, although complicated at first glance, is arguably more parsimonious than the alternative, since there is scant evidence to connect CB_1 receptors with synaptic NO synthesis (see Duguid and Sjostrom, 2006 for commentary). The conclusions of van Beugen and coworkers (2006) are also reminiscent of models emerging to describe findings in other brain areas. That is, LTD and LTP are increasingly seen as closely intertwined, perhaps even competing mechanisms, with LTD being favored over LTP when endocannabinoid-generating pathways are activated (Nevian and Sakmann, 2006; Sjostrom and Hausser, 2006; Ade and Lovinger, 2007; Sjostrom et al., 2007; and see below). Again, the discovery of synaptic endocannabinoids is helping to stir up many novel concepts of neuronal plasticity, highlighting the importance of this system to adaptive brain function.

Interneurons Also Release Endocannabinoids

The cerebellum is exemplary as a circuit "on endocannabinoids," since every synaptic connection diagrammed in Fig. 4 has been shown to functionally express presynaptic CB_1 receptors. Very recent findings have demonstrated that both BC and SC interneurons can release 2-AG in response to either depolarization (DSE) or brief trains (10 pulses at50 Hz) of PF inputs (Beierlein and Regehr, 2006; see Fig. 4). In addition, a more prolonged pattern of PF stimulation (25 pulses at 30 Hz, repeated 4 times) induces LTD of PF inputs onto SCs (SC-LTD) (Soler-Llavina and Sabatini, 2006). Important and novel mechanistic differences distinguish the two, especially that the gating of requisite postsynaptic Ca^{2+} for DSE involves postsynaptic NMDARs (Beierlein and Regehr, 2006), whereas SC-LTD depends not on NMDARs but on Ca^{2+}-permeable AMPARs (CP-AMPARs) (Soler-Llavina and Sabatini, 2006). Moreover, SC-LTD was shown to be specific for activated synapses, a property previously thought unlikely in neurons lacking dendritic spines. Through techniques of Ca^{2+} imaging in the postsynaptic neuron, Soler-Llovina and Sabatini (2006) also shed new light on mechanisms defining the spatial domains of endocannabinoid synthesis (the endocannabinoid was not identified). They found that the diffusion of Ca^{2+} necessary for endocannabinoid formation was limited by the expression of parvalbumin, a Ca^{2+}-binding protein present in these aspiny cells. Both the fast kinetics of CP-AMPARs and the narrow dendritic structure of SCs also contributed to the spatial specificity of endocannabinoid-mediated SC-LTD (Soler-Llavina and Sabatini, 2006). DSE is likely to show similar spatial constraints, at least in SCs, and 2-AG release selectively inhibited CB_1 receptor-expressing PF inputs, without influencing CB_1 receptor-sensitive somatic conductances within the interneurons themselves (Beierlein and Regehr, 2006). This is consistent also with the selective targeting of CB_1 receptors to axon terminals (Leterrier et al., 2006), which would be distant from PF-interneuron inputs. Both BC and SC interneurons play important roles of feed-forward inhibition in the cerebellar cortex, and thereby directly influence PC spike output to the deep cerebellar nuclei (see Fig. 4; Mittmann et al., 2005). DSE and SC-LTD in interneurons might therefore be mechanisms to refine the integrative properties of PC output and thus motor learning or behavior (Ito, 2001; Patel and Hillard, 2001). In summary, brain slices from the CB_1 receptor-rich cerebellar cortex have been a tremendously instructive setting for elucidating synaptic endocannabinoid function, with discoveries ranging from molecular mechanisms, to novel insights regarding the regulation of circuits and behavior.

Bidirectional Synapses and Spike-Timing-Dependent Plasticity

In proposing a synaptic basis of learning and memory, Hebb postulated that changes in synaptic strength would occur in response to persistent simultaneous activation between neurons (Hebb, 1949). For a given neuron or synapse type, the spatiotemporal parameters describing how patterns of activity will induce synaptic plasticity – and in what direction the change will be expressed (LTP vs. LTD) – is known as a *learning rule* (Bear, 1996; Yao and Dan, 2005). Hebbian learning rules include frequency-based rules of homosynaptic plasticity – occurring for example in neurons which exhibit LTD in response to low frequency repetitive stimuli/LTP in response to high frequency stimulation (Bear et al., 1987). Other synaptic learning rules are based on coincidence of inputs, or on the relative timing of pre- and postsynaptic spikes in a pair of synaptically connected neurons (Feldman and Brecht, 2005; Yao and Dan, 2005). The latter case is referred to as spike-timing-dependent plasticity (STDP) (Sjostrom and Nelson, 2002; Dan and Poo, 2004), or famously: "cells that fire together, wire together" (Hebb, 1949). The *bidirectionality* of neuronal plasticity is necessary for information storage in the brain (Bear, 1996; Turrigiano, 2007), and the various synaptic learning rules are believed to describe how patterns of activity

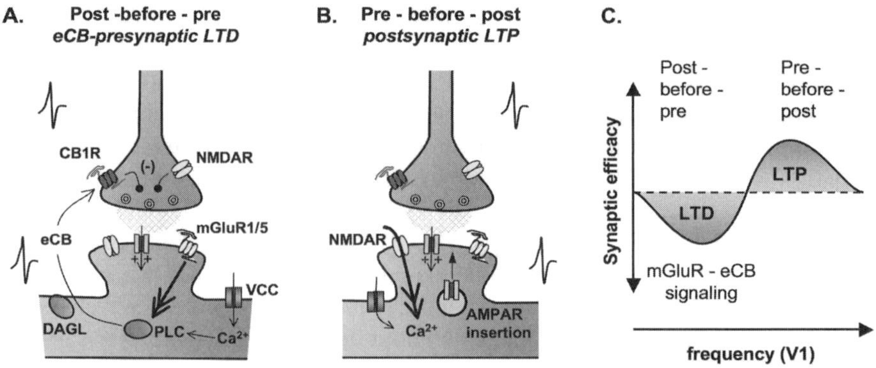

Fig. 5 A model of endocannabinoid-dependent LTD as a coincidence detector in spike-timing-dependent plasticity (STDP) at neocortical glutamatergic synapses. (**a**) When a postsynaptic spike or subthreshold depolarization precedes a presynaptic spike (post before pre), PLC activity is primed by Ca^{2+} influx and enhanced by subsequent $mGluR_{1/5}$ activation, leading to endocannabinoid release and presynaptic LTD. Also, coincident activity of postsynaptic and presynaptic elements is detected by presynaptic CB_1 receptors and NMDARs, respectively, which are both necessary for LTD using STDP protocols in layer 5 neocortex. (**b**) When a presynaptic action potential precedes postsynaptic spiking (pre before post) within ~25 ms, glutamate-bound NMDARs lose their voltage-dependent Mg^{2+} block, and the subsequent influx of Ca^{2+} leads to postsynaptic LTP through classic mechanisms involving protein kinase A (not shown) and insertion of new AMPARs into the dendritic plasma membrane. (**c**) An idealized "learning rule" for bidirectional synapses. The relative timing of coincident neuronal firing leads to either weakening (LTD) or strengthening (LTP) of synaptic efficacy between the neurons. As depicted in (**a**) and (**b**), this model revises influential theories of bidirectional Hebbian plasticity by involving two distinct mechanisms of postsynaptic coincidence detection: for LTD, the coincidence detector is represented by mGluR-eCB-CB_1R signaling. At layer 2/3 synapses of mouse visual cortex (V1), an analogous learning rule describes bidirectional *frequency*-dependent plasticity, raising the strong possibility that endocannabinoid-dependent LTD mediates a form of synaptic depression induced by visual deprivation in vivo

evoked by sensory experience encode functional modifications in neuronal circuits (Feldman and Brecht, 2005). In an exciting development, recent investigations have discovered that the LTD component of numerous neuronal learning rules – including multiple expressions of STDP – are mediated by endocannabinoid retrograde signaling (Fig. 5).

Endocannabinoids as an LTD Coincidence Detector in STDP

Sjostrom and colleagues (2003) demonstrated the first example of a CB_1 receptor-dependent learning rule in STDP. In layer 5 (L5) neurons of visual cortex, as in most known forms of STDP (Dan and Poo, 2004), when presynaptic firing occurs closely before a postsynaptic spike, LTP results from postsynaptic NMDAR-dependent mechanisms (Sjostrom et al., 2003) (see Fig. 5b). Post-before-presynaptic spiking leads to LTD, and these authors showed that in this case it was prevented by a CB_1 receptor antagonist. In addition, the synaptic learning rule was greatly modified to favor LTD in the presence of an EMT blocker or FAAH inhibitor, clearly demonstrating an endocannabinoid mediator (Sjostrom et al., 2003). Endocannabinoid-dependent LTD also occurred when postsynaptic spiking was replaced by subthreshold depolarization (Sjostrom et al., 2004). Further studies by this group have found that when L5 neuron pairs are stimulated to fire coincident, natural spike trains (by a simple current injection), *multiple* expressions of synaptic plasticity are simultaneously induced (Sjostrom et al., 2007). Yet here again, the LTD component of this mix was definitively presynaptic and mediated by CB_1 receptors, blockade of which biased the system toward LTP. L5 cortical neurons extend long apical dendrites across several cortical layers. In addition to L5–L5 synapses just discussed, L5 neurons also integrate synaptic inputs within these superficial cortical layers, especially inputs in L2/3 (Sjostrom and Hausser, 2006). In a clever study, Sjostrom and Hausser (2006) found that when a timing-dependent pairing protocol was used to induce plasticity, the sign of change was dependent on the dendritic location of synapses, and controlled by the success (or failures) of bAPs. Endocannabinoid-dependent LTD was normally favored at more distal L2/3 inputs, whereas at synapses proximal to L5, where bAPs were more robust, the same protocol reliably induced LTP. If, however, the propagation of a bAP further along the apical dendrite was augmented by a distally generated Ca^{2+} spike, the resulting plasticity switched to LTP, a L2/3 synapses as well (Sjostrom and Hausser, 2006). This scheme is similar to that depicted in Fig. 3a – imagine, however, that the convergence of a bAP and EPSP actually "switches" the LTD-expressing synapse to instead express NMDAR-dependent LTP. What do we make of these dynamic interrelationships between mechanisms of plasticity? It is primarily because both synaptic inputs and bAPs represent active Ca^{2+} conductances that they influence synaptic plasticity. The finding just described could therefore be based on a long-standing "Ca^{2+} hypothesis" of bidirectional synaptic plasticity, where the LTD/LTP learning rule is primarily a result of varying levels of internal Ca^{2+} (Bear et al., 1987; Bear, 1996; Sjostrom and Nelson, 2002). In this influential, single coincidence detector model, Ca^{2+} induces either LTD or LTP based largely on its different affinities for functionally opposing enzymatic processes in the postsynaptic neuron (Bear et al., 1987). An alternative model argues, however, that at least for STDP, a second postsynaptic coincidence detector is required to achieve the learning rules observed in many neuronal circuits (Karmarkar and Buonomano, 2002). Recent findings in somatosensory cortex (S1) (Bender et al., 2006b; Nevian and Sakmann, 2006), as well as in the auditory sensory brainstem (Tzounopoulos et al., 2007), strongly support the latter model, and that the second coincidence detector responsible for signaling LTD is the mGluR-dependent generation of retrograde endocannabinoid inhibition. In L2/3 pyramidal neurons, even when spike pairing achieved comparable concentrations of internal Ca^{2+}, the system was biased toward LTD if endocannabinoids were activated (Nevian and Sakmann, 2006). If postsynaptic VGCC activation was followed closely by $mGluR_{1/5}$ and PLC signaling (as with a post-before-pre STDP protocol), the resulting surge in endocannabinoid release and presynaptic CB_1 receptor activation deterministically resulted in LTD (see Fig. 5a). If in the same situation CB_1 receptors were blocked, LTP was the outcome – thus in this case, the STDP learning rule was set mostly by the synaptic endocannabinoid system – a switch to induce LTD in a cortical circuit (Nevian and Sakmann, 2006) (Fig. 5c). Likewise at L4 to L2/3 synapses in S1 cortex, the STDP learning rule is determined by CB_1 receptor-dependent LTD, such that application of the CB_1 receptor antagonist AM251 results in LTP following STDP protocols that normally favor LTD (Bender et al., 2006b). In these studies, the post-before-pre induction of LTD was also blocked by (1) a postsynaptic DAGL inhibitor or (2) postsynaptic VDM11, a selective EMT blocker, evidencing the role of 2-AG efflux through an EMT (Bender et al., 2006b).

Implications for Sensory Map Plasticity

Processes such as STDP in S1 cortex are believed to underlie important processes in the formation and activity-dependent plasticity of cortical somatosensory maps (Feldman and Brecht, 2005). An important corollary of this hypothesis is that processes of in vivo sensory map plasticity ought to reflect mechanisms of STDP seen in reduced experimental preparations. Recently, Bender and coworkers (2006a) indeed demonstrated that sensory map plasticity observed in rat S1 barrel fields after whisker trimming (a well-studied methodology of precise sensory deprivation) exhibited in vivo electrophysiological changes indicative of presynaptic LTD, such as increases in the PPF ratio. While a direct test of endocannabinoid dependence was not reported, the known correlations between whisker deprivation, barrel field plasticity, and STDP seen in S1 slices (Feldman and Brecht, 2005) argue that this form of LTD induced by behavioral experience is a CB_1 receptor-dependent phenomenon (Bender et al., 2006a). Further indirect evidence comes from the observation that CB_1 receptor null mice exhibit altered S1 barrel field morphology (Deshmukh et al., 2007). Synaptic depression induced by *visual* deprivation is one of the earliest and most studied forms of behaviorally induced neuronal plasticity (Bear, 1996). Monocular deprivation occurring early in life results in lost visual responsiveness and plasticity of ocular dominance cortical maps, and it was in fact a model of this phenomenon that led to the original postulation of LTD as an experimentally tractable paradigm (Bear et al., 1987). LTD induced by low frequency afferent stimulation has subsequently become a model for deprivation-induced synaptic depression, and shares overlapping characteristics (Crozier et al., 2007). Explicitly motivated by the growing involvement of endocannabinoid-mediated LTD in STDP of visual cortex (V1) (Sjostrom et al., 2003), Crozier and colleagues (2007) reevaluated mechanisms of low frequency stimulated LTD in L2/3 pyramidal neurons of V1, finding that the phenomenon is CB_1 receptor dependent. Furthermore, this form of endocannabinoid-mediated LTD is mimicked and occluded by prior visual deprivation, indicating redundant mechanisms

(Crozier et al., 2007). Sensory map plasticity – changes in the cortical representation of the sensory environment – is a wide-ranging set of phenomena that is likely to be as complex as the cortex itself (Feldman and Brecht, 2005). It appears, however, that mechanisms of endocannabinoid-mediated LTD are commonly involved with map plasticity in multiple areas of primary sensory cortex, as well as in earlier sensory centers (Tzounopoulos et al., 2007).

Endocannabinoids and the Control of Neuronal Oscillations and Synchrony

One of the ultimate questions driving neuroscience research is how mechanisms of synaptic plasticity in discrete circuits – and the timing of neuronal firing within and between circuits – actually relate to the encoding of lasting memories, or the binding of sensory experiences into a cognitive perceptual reality. A leading mode of thought is that synchronous firing in neuronal ensembles – measurable as rhythmic oscillations of various frequencies – is critical to such higher order functions of neuronal networks (Varela, 2001). Oscillations are windows of active temporal coherence within cooperating neuronal assemblies, and might therefore reflect the distributed representation, storage or retrieval of information within those assemblies (Csicsvari et al., 2003). There is now growing evidence that the brain endocannabinoid system has particular relevance to the generation or maintenance of hippocampal oscillations (Freund, 2003; Klausberger et al., 2005; Robbe et al., 2006), and this may relate to the pronounced effects of *Cannabis* on both memory and perception (Iverson, 2000). In brief, the generation of synchronous oscillations in neuronal networks is highly dependent upon patterns of *inhibitory* modulation. In cortical structures (including hippocampus), the rhythmic firing of GABAergic interneurons can represent a large fraction of field oscillations, and is believed to play critical roles in maintaining network synchrony (McBain and Fisahn, 2001; Freund, 2003; Hestrin and Galarreta, 2005). The exceptional pattern of dense CB_1 receptor expression in subpopulations of GABAergic interneurons in the hippocampus thus led to the first published consideration that these receptors may be vital to the fine-tuning of synchronous rhythms there (Katona et al., 2000). The same group showed in contemporaneous work that cannabinoid agonists could disrupt kainic acid-induced gamma oscillations in hippocampal slices (Hajos et al., 2000). Electrical coupling among functional groups of interneurons is thought to contribute significantly to their role in rhythm generation (Hestrin and Galarreta, 2005), and coupled CB_1 receptor-expressing interneurons might indeed work cooperatively in this regard (Galarreta et al., 2004). Other recently discovered properties of CB_1 receptor expressing interneurons seem to place them in a physiological role of generating oscillations. Compared with the fast-firing, high-fidelity neurons expressing parvalbumin in the hippocampus, the CB_1 receptor- and CCK-expressing interneurons receive weak afferent excitation, which they integrate relatively slowly, requiring the summation of consecutive EPSPs to trigger a spike (Glickfeld and Scanziani, 2006). These neurons therefore require stronger or more global, convergent input to fire. This especially includes feedback excitation from activated pyramidal neurons (Glickfeld and Scanziani, 2006) and from extrinsic inputs related to emotional arousal (Freund, 2003). Both of these observations suggest that endocannabinoid sensitive interneuron networks may be particularly tuned to behaviorally relevant contexts that push them to fire synchronously. In this perspective, the stimuli related to these contexts might transmit emotionally salient inputs – mediated for example, by serotonergic or cholinergic inputs that specifically target this class of endocannabinoid-sensitive interneuron (Freund, 2003). Strong inputs pushing these interneurons to fire may also be related to active exploration, which evokes complex, theta-rhythm burst firing of pyramidal "place cells" occurring when the animal enters a particular area in space (Klausberger et al., 2005). In support of this idea, CB_1 receptor-expressing interneurons have specifically been found to fire action potentials early in the phase procession of a population theta rhythm, a characteristic that phase-locks their firing within place fields (Klausberger et al., 2005). In other words, endocannabinoid-sensitive cells contribute to the behaviorally relevant theta rhythm, and might be fundamentally important to encoding information conveyed by the oscillation, such as a representation of physical space. Stimulus-dependent retrograde inhibition by endocannabinoids, driven perhaps by experience-driven glutamatergic activity or LTP (Chevaleyre and Castillo, 2003; Zhu and Lovinger, 2007), may disinhibit pyramidal cells, dropping them out of population synchrony set up the interneuron network (Freund, 2003; Freund et al., 2003; Hestrin and Galarreta, 2005). Such a finding was reported for mAChR-stimulated theta rhythms in CA1 slices (Reich et al., 2005). If this "dropping out" were driven by patterned inputs relevant to perception, such an effect might be an integral part of encoding the natural world. If activated nonspecifically, however, a widespread loss of synchrony may lead to poor encoding of spatial information – consistent with some of the memory impairments elicited by *Cannabis*, for example. Indeed, there is compelling recent evidence that CB_1 receptor activation by Δ^9-THC, or the more potent CP55940, can markedly reduce the power (synchrony) of hippocampal oscillations in theta (4–12 Hz), gamma (30–80 Hz), and fast ripple (100–200 Hz) bands in awake, behaving rats (Robbe et al., 2006). The cannabinoid-induced reduction in theta power was specifically correlated to impaired performance of a hippocampus-dependent, delayed spatial alternation task (Robbe et al., 2006). Yet there was no effect of CB_1 receptor activation on the overall firing rates of the neurons – only their synchrony within the ensemble. Perhaps the influence of DSI or other intrinsic forms of endocannabinoid synaptic inhibition is likewise to preferentially regulate *synchrony*, rather than overall firing frequencies. It seems a worthy pursuit that is already providing new insight into brain function. This story is considerably more complicated, but my intention is simply to (1) provide a basic discussion of how population rhythms are believed to be generated and (2) emphasize that models increasingly find the CB_1 receptor expressed at synaptic points that are prominent in rhythm generation. Clearly, the two fields of study are productively informing one another. With regard to the complexity of these descriptions, it can be pointed out that endocannabinoids have also been found as critical regulators of an important central pattern generator in a simpler motor system (Kettunen et al., 2005). These authors point to an intrinsic utility of endocannabinoids to mediate rhythmicity, as retrograde molecules that allow postsynaptic cells to become modulators of their own activities.

Future Directions

It has been an exciting decade to be engaged in studying the synaptic physiology of cannabinoids. The growing wealth of evidence, discussed at length in this book, suggests that the endocannabinoid system is fundamental to neurobiology, at least as it has evolved in mammals. This notion is supported by the evolutionary ancientness of this system (McPartland et al., 2006), which has been implicated in presynaptic function and adaptive behavior in diverse species, including very primitive organisms (De Petrocellis et al., 1999; Egertova and Elphick, 2007). Moreover, the excitable membranes of multiple primary sensory systems have been found to utilize endocannabinoids for functional regulation (Yazulla et al., 2000; Czesnik et al., 2007), suggesting mechanisms that predated the evolution of highly developed brains. It is not surprising in this context that the brain itself utilizes endocannabinoid retrograde signaling in myriad ways. The recent pace of discovery has been remarkable, yet many pivotal mysteries remain. In this last section, I will attempt to relate some of the many insightful questions being considered by labs in the field.

Mechanisms Governing the Expression of CB_1 Receptor-Dependent Plasticity

When does CB_1 receptor expression tip the scale between competing mechanisms of LTP and LTD within the same or interconnected synapses (Sjostrom and Hausser, 2006; Ade and Lovinger, 2007; Tzounopoulos et al., 2007)? What is the importance of the tremendous density of presynaptic CB_1 receptors in some axons? Is the apparent reserve of these receptors indicative of dynamic regulation of synaptic reliability by the constant recycling of CB_1 receptors to the presynaptic membrane (Leterrier et al., 2006)? Perhaps the neurophysiology of CB_1 receptors, as the prototype receptor for retrograde signaling and plasticity, can be informed by careful analogy to mechanisms of regulating AMPAR surface expression at *postsynaptic* membranes, now considered a fundamental mechanism of postsynaptic changes in synaptic strength. In other words, insertion of new CB_1 receptors could *silence* synapses (Losonczy et al., 2004; Neu et al., 2007), or re-adjust their filtering properties (Oliet et al., 2007) where there is a tonic endocannabinoid release, or where GABA release already fails to saturate postsynaptic receptors (Biro et al., 2006). Such mechanisms could coexist with those of LTD, which do not require sustained

CB_1 receptor signaling (Chevaleyre and Castillo, 2003; Ronesi et al., 2004). How do synaptic processes adapt to chronic cannabinoid exposure? Recent studies demonstrate that chronic Δ^9-THC can lead to changes in the induction of LTD and LTP in multiple brain areas (Hoffman et al., 2003, 2007; Tonini et al., 2006). Given the diverse role of endocannabinoids both in directly mediating LTD and indirectly regulating heterosynaptic efficacy (metaplasticity), an equally wide range of mechanisms is sure to underlie physiological adaptations to repetitive CB_1 receptor activation by exogenous ligands (Mato et al., 2005). Given the enduring popularity of *Cannabis* as a mind-altering substance in many cultures, as well as the promising and not-distant future of more cannabinoid-based medicines (both antagonists and agonists), this is clearly not a trivial question.

Plasticity and Localization of Endocannabinoid Generation, Release, and Degradation

Enzymes involved in endocannabinoid signaling have been localized to high resolution with electron microscopy, and this has greatly informed models of CB_1 receptor synaptic function and retrograde signaling. The differential expression of DAGLα – for example, at the head vs. the neck of a dendritic spine, or in relative degrees of proximity to $G_{q/11}$-coupled receptors – might play a defining role in the Ca^{2+} dynamics required to elicit stimulus-dependent release of 2-AG (Katona et al., 2006; Yoshida et al., 2006; Uchigashima et al., 2007). Changing the local expression of Ca^{2+} binding proteins could similarly influence the interplay between Ca^{2+} influx and 2-AG release (Rancz and Hausser, 2006; Soler-Llavina and Sabatini, 2006). The nature of lipid microdomains – where NAPE precursors for AEA could hypothetically be enriched – is largely unknown and few tools exist to investigate them within physiological contexts. How plastic are the enzymatic activities responsible for generating and terminating endocannabinoid signaling? In cerebellar PCs, the sustained elevation of dendritic Ca^{2+} for several seconds greatly enhances 2-AG release, a process likened to posttetanic potentiation of presynaptic neurotransmitter release (Brenowitz et al., 2006). It has also recently been shown that endocannabinoid release in the hippocampus can be enhanced in a lasting way following tetanic patterns of synaptic input (Chen et al., 2007; Zhu and Lovinger, 2007). A similar potentiation of DSI has been shown to be an enduring consequence of hyperthermia-induced febrile seizures, with possible clinical relevance to epileptic conditions (Chen et al., 2007). Changing functions of endocannabinoid-mediated synaptic plasticity during normal development are a related issue of significant importance for future investigation (Bernard et al., 2005; Henneberger et al., 2007; Ade and Lovinger, 2007; Crepel, 2007). The ability of synapses to prolong the window of endocannabinoid signaling can clearly enhance short-term plasticity (Hashimotodani et al., 2007a), can bias synapses toward the induction of LTD (Gerdeman et al., 2002; Chevaleyre and Castillo, 2003; Sjostrom et al., 2003), and could influence the stability of network oscillations or the participation of specific cells within a stable cortical rhythm (Freund et al., 2003; Klausberger et al., 2005; Glickfeld and Scanziani, 2006; Robbe et al., 2006). As this chapter is intended to emphasize, the dynamic alteration of a neuron's endocannabinoid signaling repertoire – essentially the functional *lipidomics* of endocannabinoids – is likely to be definitive of its contribution within a neuronal circuit.

Other Targets of Endocannabinoids and Related Lipids

I have deliberately avoided in-depth discussion of non-CB_1 receptor targets for endocannabinoids, as covered by earlier chapters in this book (Chaps. 8–10). The study of such effector systems are nonetheless likely to have significant impact on understanding the role of endocannabinoids and their relationships to other signaling pathways within and between neurons. In particular, $TRPV_1$ receptors, now sometimes referred to as the ionotropic companion to metabotropic cannabinoid receptors, remain relatively unknown in terms of their functional significance within the brain. Early studies were equivocal regarding their expression in the brain (see Chaps. 1 and 8), and only a few studies have directly analyzed their function to regulate synaptic transmission. Can endocannabinoids, acting either at presynaptic $TRPV_1$ receptors (Marinelli et al., 2003) or background K^+ channels (Köfalvi et al., 2007), lead to sufficient axonal depolarization to elicit an antidromic spike? Such a finding would extend the retrograde function of endocannabinoids to a level that would seriously redefine models of information flow in the nervous system. Despite a notable controversy over the last few years, it is now clearly established that CB_1 receptors are functionally expressed on excitatory, glutamatergic terminals (Domenici et al., 2006; Katona et al., 2006; Kawamura et al., 2006; Takahashi and Castillo, 2006; Yoshida et al., 2006; Köfalvi et al., 2007; see Chap. 10), and previously controversial results with the WIN55212-2 agonist are likely to reflect direct effects of this compound on VCCs (Shen and Thayer, 1998; Köfalvi et al., 2007; see Chap. 9). The novel GPR55 receptor, however, is still mostly an unknown player (Mackie and Stella, 2006; see Chaps. 9, 10). Activation of GPR55 by palmitoylethanolamide, a NAE generated in "entourage" with AEA, suggests that this receptor might mediate synaptic functions in close coordination with AEA actions at the CB_1 receptor. Lastly, the existence of *N*-arachidonoyl dopamine (NADA) and other relatively novel endocannabinoid family lipids (Mackie and Stella, 2006) are sure to grow in prominence (Chaps. 4, 8), but are too new for inclusion in this broad review of synaptic mechanisms.

Concluding Remarks

In many ways, the activity and functional potential of a neural circuit is defined by the moment-to-moment state of its many, many synapses (Hebb, 1949; Bear, 1996; Abbott and Regehr, 2004). Recent years have seen the endocannabinoids emerge as a system of modulating synaptic efficacy within a great many brain areas, acting as postsynaptically released retrograde messengers to presynaptic CB_1 receptors. Short- and long-term synaptic suppression by CB_1 receptors is a widespread mechanism to fine-tune synaptic filtering properties, and to facilitate associative modes of plasticity as a coincidence detector. In addition, endocannabinoid-mediated LTD has now redefined many known forms of synaptic plasticity, clarifying these processes on a new level of mechanistic detail. The endocannabinoid system is thus critical for a great many cellular mechanisms of stable neuronal plasticity, a reality that has wide-ranging implications for understanding brain function, as well as the etiology and treatment of neurological diseases.

Acknowledgments I am deeply grateful for the support of Dr. Ed French, Dr. Jason Schechter, and Shinai A. Schindler, and for funding by the National Institute on Drug Abuse (DA14263-04). I also thank Drs. Attila Köfalvi, Pablo Castillo, Alberto Pereda, Laurent Venance, Istvan Katona, John McPartland, and Bela Szabo for helpful correspondence, and to Pablo Castillo for sharing unpublished data. Lastly, the form of this chapter was shaped by numerous invaluable discussions at the 2007 Mind & Life Summer Research Institute, and I am very thankful to each of the participants at the discussion.

References

Abbott LF, Regehr WG (2004) Synaptic computation. Nature 431:796–803.

Ade KK, Lovinger DM (2007) Anandamide regulates postnatal development of long-term synaptic plasticity in the rat dorsolateral striatum. J Neurosci 27:2403–2409.

Alger B (2003) Retrograde signaling in the regulation of synaptic transmission: focus on endocannabinoids. Prog Neurobiol 68:247–286.

Azad SC, Monory K, Marsicano G, Cravatt BF, Lutz B, Zieglgansberger W, Rammes G (2004) Circuitry for associative plasticity in the amygdala involves endocannabinoid signaling. J Neurosci 24:9953–9961.

Bacci A, Huguenard JR, Prince DA (2004) Long-lasting self-inhibition of neocortical interneurons mediated by endocannabinoids. Nature 431:312–316.

Bear MF (1996) A synaptic basis for memory storage in the cerebral cortex. Proc Natl Acad Sci USA 93:13453–13459.

Bear MF, Cooper LN, Ebner FF (1987) A physiological basis for a theory of synapse modification. Science 237:42–48.

Beierlein M, Regehr WG (2006) Local interneurons regulate synaptic strength by retrograde release of endocannabinoids. J Neurosci 26:9935–9943.

Bender KJ, Allen CB, Bender VA, Feldman DE (2006a) Synaptic basis for whisker deprivation-induced synaptic depression in rat somatosensory cortex. J Neurosci 26:4155–4165.

Bender VA, Bender KJ, Brasier DJ, Feldman DE (2006b) Two coincidence detectors for spike timing-dependent plasticity in somatosensory cortex. J Neurosci 26:4166–4177.

Bernard C, Milh M, Morozov YM, Ben-Ari Y, Freund TF, Gozlan H (2005) Altering cannabinoid signaling during development disrupts neuronal activity. Proc Natl Acad Sci USA 102:9388–9393.

Biro AA, Holderith NB, Nusser Z (2006) Release probability-dependent scaling of the postsynaptic responses at single hippocampal GABAergic synapses. J Neurosci 26:12487–12496.

Bodor AL, Katona I, Nyiri G, Mackie K, Ledent C, Hajos N, Freund TF (2005) Endocannabinoid signaling in rat somatosensory cortex: laminar differences and involvement of specific interneuron types. J Neurosci 25:6845–6856.

Brenowitz SD, Regehr WG (2003) Calcium dependence of retrograde inhibition by endocannabinoids at synapses onto Purkinje cells. J Neurosci 23:6373–6384.
Brenowitz SD, Regehr WG (2005) Associative short-term synaptic plasticity mediated by endocannabinoids. Neuron 45:419–431.
Brenowitz SD, Best AR, Regehr WG (2006) Sustained elevation of dendritic calcium evokes widespread endocannabinoid release and suppression of synapses onto cerebellar Purkinje cells. J Neurosci 26:6841–6850.
Brown SP, Brenowitz SD, Regehr WG (2003) Brief presynaptic bursts evoke synapse-specific retrograde inhibition mediated by endogenous cannabinoids. Nat Neurosci 6:1048–1057.
Brown SP, Safo PK, Regehr WG (2004) Endocannabinoids inhibit transmission at granule cell to Purkinje cell synapses by modulating three types of presynaptic calcium channels. J Neurosci 24:5623–5631.
Cadas H, Gaillet S, Beltramo M, Venance L, Piomelli D (1996) Biosynthesis of an endogenous cannabinoid precursor in neurons and its control by calcium and cAMP. J Neurosci 16:3934–3942.
Cadas H, di Tomaso E, Piomelli D (1997) Occurrence and biosynthesis of endogenous cannabinoid precursor, N-arachidonoyl phosphatidylethanolamine, in rat brain. J Neurosci 17:1226–1242.
Carlson G, Wang Y, Alger BE (2002) Endocannabinoids facilitate the induction of LTP in the hippocampus. Nat Neurosci 5:723–724.
Castillo PE, Schoch S, Schmitz F, Sudhof TC, Malenka RC (2002) RIM1alpha is required for presynaptic long-term potentiation. Nature 415:327–330.
Chen K, Neu A, Howard AL, Foldy C, Echegoyen J, Hilgenberg L, Smith M, Mackie K, Soltesz I (2007) Prevention of plasticity of endocannabinoid signaling inhibits persistent limbic hyperexcitability caused by developmental seizures. J Neurosci 27:46–58.
Chevaleyre V, Castillo PE (2003) Heterosynaptic LTD of hippocampal GABAergic synapses: a novel role of endocannabinoids in regulating excitability. Neuron 38:461–472.
Chevaleyre V, Castillo PE (2004) Endocannabinoid-mediated metaplasticity in the hippocampus. Neuron 43:871–881.
Chevaleyre V, Takahashi KA, Castillo PE (2006) Endocannabinoid-mediated synaptic plasticity in the CNS. Annu Rev Neurosci 29:37–76.
Chevaleyre V, Heifets BD, Kaeser PS, Sudhof TC, Purpura DP, Castillo PE (2007) Endocannabinoid-mediated long-term plasticity requires cAMP/PKA signaling and RIM1alpha. Neuron 54:801–812.
Choi S, Lovinger DM (1997) Decreased probability of neurotransmitter release underlies striatal long-term depression and postnatal development of corticostriatal synapses. Proc Natl Acad Sci USA 94:2665–2670.
Crepel F (2007) Developmental changes in retrograde messengers involved in depolarization-induced suppression of excitation at parallel fiber-Purkinje cell synapses in rodents. J Neurophysiol 97:824–836.
Crozier RA, Wang Y, Liu CH, Bear MF (2007) Deprivation-induced synaptic depression by distinct mechanisms in different layers of mouse visual cortex. Proc Natl Acad Sci USA 104:1383–1388.
Csicsvari J, Jamieson B, Wise KD, Buzsaki G (2003) Mechanisms of gamma oscillations in the hippocampus of the behaving rat. Neuron 37:311–322.
Czesnik D, Schild D, Kuduz J, Manzini I (2007) Cannabinoid action in the olfactory epithelium. Proc Natl Acad Sci USA 104:2967–2972.
Dan Y, Poo MM (2004) Spike timing-dependent plasticity of neural circuits. Neuron 44:23–30.
Dan Y, Poo MM (2006) Spike timing-dependent plasticity: from synapse to perception. Physiol Rev 86:1033–1048.
De Petrocellis L, Melck D, Bisogno T, Milone A, Di Marzo V (1999) Finding of the endocannabinoid signalling system in Hydra, a very primitive organism: possible role in the feeding response. Neuroscience 92:377–387.
Deshmukh S, Onozuka K, Bender KJ, Bender VA, Lutz B, Mackie K, Feldman DE (2007) Postnatal development of cannabinoid receptor type 1 expression in rodent somatosensory cortex. Neuroscience 145:279–287.

Diana MA, Marty A (2004) Endocannabinoid-mediated short-term synaptic plasticity: depolarization-induced suppression of inhibition (DSI) and depolarization-induced suppression of excitation (DSE). Br J Pharmacol 142:9–19.

Diana MA, Levenes C, Mackie K, Marty A (2002) Short-term retrograde inhibition of GABAergic synaptic currents in rat Purkinje cells is mediated by endogenous cannabinoids. J Neurosci 22:200–208.

Di Marzo V, Fontana A, Cadas H, Schinelli S, Cimino G, Schwartz JC, Piomelli D (1994) Formation and inactivation of endogenous cannabinoid anandamide in central neurons. Nature 372:686–691.

Domenici MR, Azad SC, Marsicano G, Schierloh A, Wotjak CT, Dodt HU, Zieglgänsberger W, Lutz B, Rammes G (2006) Cannabinoid receptor type 1 located on presynaptic terminals of principal neurons in the forebrain controls glutamatergic synaptic transmission. J Neurosci 26:5794–5799.

Duguid I, Sjostrom PJ (2006) Novel presynaptic mechanisms for coincidence detection in synaptic plasticity. Curr Opin Neurobiol 16:312–322.

Edwards DA, Kim J, Alger BE (2006) Multiple mechanisms of endocannabinoid response initiation in hippocampus. J Neurophysiol 95:67–75.

Egertova M, Elphick MR (2007) Localization of CiCBR in the invertebrate chordate Ciona intestinalis: evidence of an ancient role for cannabinoid receptors as axonal regulators of neuronal signalling. J Comp Neurol 502:660–672.

Egertova M, Giang DK, Cravatt BF, Elphick MR (1998) A new perspective on cannabinoid signalling: complementary localization of fatty acid amide hydrolase and the CB_1 receptor in rat brain. Proc R Soc Lond B Biol Sci 265:2081–2085.

Feldman DE, Brecht M (2005) Map plasticity in somatosensory cortex. Science 310:810–815.

Foldy C, Neu A, Jones MV, Soltesz I (2006) Presynaptic, activity-dependent modulation of cannabinoid type 1 receptor-mediated inhibition of GABA release. J Neurosci 26:1465–1469.

Fortin DA, Levine ES (2007) Differential effects of endocannabinoids on glutamatergic and GABAergic inputs to layer 5 pyramidal neurons. Cereb Cortex 17:163–174.

Fortin DA, Trettel J, Levine ES (2004) Brief trains of action potentials enhance pyramidal neuron excitability via endocannabinoid-mediated suppression of inhibition. J Neurophysiol 92:2105–2112.

Freiman I, Anton A, Monyer H, Urbanski MJ, Szabo B (2006) Analysis of the effects of cannabinoids on identified synaptic connections in the caudate-putamen by paired recordings in transgenic mice. J Physiol 575:789–806.

Freund TF (2003) Interneuron Diversity series: rhythm and mood in perisomatic inhibition. Trends Neurosci 26:489–495.

Freund TF, Katona I, Piomelli D (2003) Role of endogenous cannabinoids in synaptic signaling. Physiol Rev 83:1017–1066.

Fukudome Y, Ohno-Shosaku T, Matsui M, Omori Y, Fukaya M, Tsubokawa H, Taketo MM, Watanabe M, Manabe T, Kano M (2004) Two distinct classes of muscarinic action on hippocampal inhibitory synapses: M_2-mediated direct suppression and M_1/M_3-mediated indirect suppression through endocannabinoid signalling. Eur J Neurosci 19:2682–2692.

Galante M, Diana MA (2004) Group I metabotropic glutamate receptors inhibit GABA release at interneuron-Purkinje cell synapses through endocannabinoid production. J Neurosci 24:4865–4874.

Galarreta M, Erdelyi F, Szabo G, Hestrin S (2004) Electrical coupling among irregular-spiking GABAergic interneurons expressing cannabinoid receptors. J Neurosci 24:9770–9778.

Gerdeman GL, Lovinger DM (2003) Emerging roles for endocannabinoids in long-term synaptic plasticity. Br J Pharmacol 140:781–789.

Gerdeman GL, Ronesi J, Lovinger DM (2002) Post-synaptic endocannabinoid release is critical to long-term depression in the striatum. Nat Neurosci 5:446–451.

Giuffrida A, Parsons LH, Kerr TM, Rodriguez de Fonseca F, Navarro M, Piomelli D (1999) Dopamine activation of endogenous cannabinoid signaling in dorsal striatum. Nat Neurosci 2:358–363.

Glickfeld LL, Scanziani M (2006) Distinct timing in the activity of cannabinoid-sensitive and cannabinoid-insensitive basket cells. Nat Neurosci 9:807–815.

Haj-Dahmane S, Shen RY (2005) The wake-promoting peptide orexin-B inhibits glutamatergic transmission to dorsal raphe nucleus serotonin neurons through retrograde endocannabinoid signaling. J Neurosci 25:896–905.

Hajos N, Katona I, Naiem SS, Mackie K, Ledent C, Mody I, Freund TF (2000) Cannabinoids inhibit hippocampal GABAergic transmission and network oscillations. Eur J Neurosci 12:3239–3249.

Hajos N, Kathuria S, Dinh T, Piomelli D, Freund TF (2004) Endocannabinoid transport tightly controls 2-arachidonoyl glycerol actions in the hippocampus: effects of low temperature and the transport inhibitor AM404. Eur J Neurosci 19:2991–2996.

Hashimotodani Y, Ohno-Shosaku T, Tsubokawa H, Ogata H, Emoto K, Maejima T, Araishi K, Shin HS, Kano M (2005) Phospholipase Cβ serves as a coincidence detector through its Ca^{2+} dependency for triggering retrograde endocannabinoid signal. Neuron 45:257–268.

Hashimotodani Y, Ohno-Shosaku T, Kano M (2007a) Presynaptic monoacylglycerol lipase activity determines basal endocannabinoid tone and terminates retrograde endocannabinoid signaling in the hippocampus. J Neurosci 27:1211–1219.

Hashimotodani Y, Ohno-Shosaku T, Kano M (2007b) Endocannabinoids and synaptic function in the CNS. Neuroscientist 13:127–137.

Hashimotodani Y, Ohno-Shosaku T, Kano M (2007c) Ca^{2+}-assisted receptor-driven endocannabinoid release: mechanisms that associate presynaptic and postsynaptic activities. Curr Opin Neurobiol 17:360–365.

Hausser M, Spruston N, Stuart GJ (2000) Diversity and dynamics of dendritic signaling. Science 290:739–744.

Hebb DO (1949) Organization of Behavior. New York: Wiley.

Heinbockel T, Brager DH, Reich CG, Zhao J, Muralidharan S, Alger BE, Kao JP (2005) Endocannabinoid signaling dynamics probed with optical tools. J Neurosci 25:9449–9459.

Henneberger C, Redman SJ, Grantyn R (2007) Cortical Efferent Control of Subcortical Sensory Neurons by Synaptic Disinhibition. Cereb Cortex 17:2039–2049

Hentges ST, Low MJ, Williams JT (2005) Differential regulation of synaptic inputs by constitutively released endocannabinoids and exogenous cannabinoids. J Neurosci 25:9746–9751.

Herkenham M, Lynn AB, de Costa BR, Richfield EK (1991) Neuronal localization of cannabinoid receptors in the basal ganglia of the rat. Brain Res 547:267–274.

Hestrin S, Galarreta M (2005) Electrical synapses define networks of neocortical GABAergic neurons. Trends Neurosci 28:304–309.

Hillard CJ, Jarrahian A (2000) The movement of N-arachidonoylethanolamine (anandamide) across cellular membranes. Chem Phys Lipids 108:123–134.

Hirasawa M, Schwab Y, Natah S, Hillard CJ, Mackie K, Sharkey KA, Pittman QJ (2004) Dendritically released transmitters cooperate via autocrine and retrograde actions to inhibit afferent excitation in rat brain. J Physiol 559:611–624.

Hoffman AF, Oz M, Caulder T, Lupica CR (2003) Functional tolerance and blockade of long-term depression at synapses in the nucleus accumbens after chronic cannabinoid exposure. J Neurosci 23:4815–4820.

Hofmann ME, Nahir B, Frazier CJ (2006) Endocannabinoid-mediated depolarization-induced suppression of inhibition in hilar mossy cells of the rat dentate gyrus. J Neurophysiol 96:2501–2512.

Hoffman AF, Oz M, Yang R, Lichtman AH, Lupica CR (2007) Opposing actions of chronic Δ^9-tetrahydrocannabinol and cannabinoid antagonists on hippocampal long-term potentiation. Learn Mem 14:63–74.

Howlett AC, Barth F, Bonner TI, Cabral G, Casellas P, Devane WA, Felder CC, Herkenham M, Mackie K, Martin BR, Mechoulam R, Pertwee RG (2002) International Union of Pharmacology. XXVII. Classification of cannabinoid receptors. Pharmacol Rev 54:161–202.

Ikeda SR (1996) Voltage-dependent modulation of N-type calcium channels by G-protein beta gamma subunits. Nature 380:255–258.

Isokawa M, Alger BE (2005) Retrograde endocannabinoid regulation of GABAergic inhibition in the rat dentate gyrus granule cell. J Physiol 567:1001–1010.

Isokawa M, Alger BE (2006) Ryanodine receptor regulates endogenous cannabinoid mobilization in the hippocampus. J Neurophysiol 95:3001–3011.

Ito M (2001) Cerebellar long-term depression: characterization, signal transduction, and functional roles. Physiol Rev 81:1143–1195.

Iverson LL (2000) The Science of Marijuana. New York: Oxford.

Kaeser PS, Sudhof TC (2005) RIM function in short- and long-term synaptic plasticity. Biochem Soc Trans 33:1345–1349.

Karmarkar UR, Buonomano DV (2002) A model of spike-timing dependent plasticity: one or two coincidence detectors? J Neurophysiol 88:507–513.

Katona I, Sperlagh B, Sik A, Köfalvi A, Vizi ES, Mackie K, Freund TF (1999) Presynaptically located CB_1 cannabinoid receptors regulate GABA release from axon terminals of specific hippocampal interneurons. J Neurosci 19:4544–4558.

Katona I, Sperlagh B, Magloczky Z, Santha E, Köfalvi A, Czirjak S, Mackie K, Vizi ES, Freund TF (2000) GABAergic interneurons are the targets of cannabinoid actions in the human hippocampus. Neuroscience 100:797–804.

Katona I, Urban GM, Wallace M, Ledent C, Jung KM, Piomelli D, Mackie K, Freund TF (2006) Molecular composition of the endocannabinoid system at glutamatergic synapses. J Neurosci 26:5628–5637.

Kawamura Y, Fukaya M, Maejima T, Yoshida T, Miura E, Watanabe M, Ohno-Shosaku T, Kano M (2006) The CB_1 cannabinoid receptor is the major cannabinoid receptor at excitatory presynaptic sites in the hippocampus and cerebellum. J Neurosci 26:2991–3001.

Kearn CS, Blake-Palmer K, Daniel E, Mackie K, Glass M (2005) Concurrent stimulation of cannabinoid CB_1 and dopamine D_2 receptors enhances heterodimer formation: a mechanism for receptor cross-talk? Mol Pharmacol 67:1697–1704.

Kettunen P, Kyriakatos A, Hallen K, El Manira A (2005) Neuromodulation via conditional release of endocannabinoids in the spinal locomotor network. Neuron 45:95–104.

Kim J, Isokawa M, Ledent C, Alger BE (2002) Activation of muscarinic acetylcholine receptors enhances the release of endogenous cannabinoids in the hippocampus. J Neurosci 22:10182–10191.

Kishimoto Y, Kano M (2006) Endogenous cannabinoid signaling through the CB_1 receptor is essential for cerebellum-dependent discrete motor learning. J Neurosci 26:8829–8837.

Klausberger T, Marton LF, O'Neill J, Huck JH, Dalezios Y, Fuentealba P, Suen WY, Papp E, Kaneko T, Watanabe M, Csicsvari J, Somogyi P (2005) Complementary roles of cholecystokinin- and parvalbumin-expressing GABAergic neurons in hippocampal network oscillations. J Neurosci 25:9782–9793.

Köfalvi A, Pereira MF, Rebola N, Rodrigues RJ, Oliveira CR, Cunha RA (2007) Anandamide and NADA bi-directionally modulate presynaptic Ca^{2+} levels and transmitter release in the hippocampus. Br J Pharmacol 151:551–563.

Kreitzer AC, Malenka RC (2005) Dopamine modulation of state-dependent endocannabinoid release and long-term depression in the striatum. J Neurosci 25:10537–10545.

Kreitzer AC, Malenka RC (2007) Endocannabinoid-mediated rescue of striatal LTD and motor deficits in Parkinson's disease models. Nature 445:643–647.

Kreitzer AC, Regehr WG (2001a) Retrograde inhibition of presynaptic calcium influx by endogenous cannabinoids at excitatory synapses onto Purkinje cells. Neuron 29:717–727.

Kreitzer AC, Regehr WG (2001b) Cerebellar depolarization-induced suppression of inhibition is mediated by endogenous cannabinoids. J Neurosci 21:RC174.

Kreitzer AC, Carter AG, Regehr WG (2002) Inhibition of interneuron firing extends the spread of endocannabinoid signaling in the cerebellum. Neuron 34:787–796.

Kushmerick C, Price GD, Taschenberger H, Puente N, Renden R, Wadiche JI, Duvoisin RM, Grandes P, von Gersdorff H (2004) Retroinhibition of presynaptic Ca^{2+} currents by endocannabinoids released via post-synaptic mGluR activation at a calyx synapse. J Neurosci 24:5955–5965.

Lauckner JE, Hille B, Mackie K (2005) The cannabinoid agonist WIN55,212-2 increases intracellular calcium via CB_1 receptor coupling to $G_{q/11}$ G proteins. Proc Natl Acad Sci USA 102:19144–19149.

Leterrier C, Laine J, Darmon M, Boudin H, Rossier J, Lenkei Z (2006) Constitutive activation drives compartment-selective endocytosis and axonal targeting of type 1 cannabinoid receptors. J Neurosci 26:3141–3153.

Levenes C, Daniel H, Soubrie P, Crepel F (1998) Cannabinoids decrease excitatory synaptic transmission and impair long-term depression in rat cerebellar Purkinje cells. J Physiol 510:867–879.

Llano I, Leresche N, Marty A (1991) Calcium entry increases the sensitivity of cerebellar Purkinje cells to applied GABA and decreases inhibitory synaptic currents. Neuron 6:565–574.

Losonczy A, Biro AA, Nusser Z (2004) Persistently active cannabinoid receptors mute a subpopulation of hippocampal interneurons. Proc Natl Acad Sci USA 101:1362–1367.

Mackie K, Stella N (2006) Cannabinoid receptors and endocannabinoids: evidence for new players. AAPS J 8:E298–E306.

Maejima T, Hashimoto K, Yoshida T, Aiba A, Kano M (2001) Presynaptic inhibition caused by retrograde signal from metabotropic glutamate to cannabinoid receptors. Neuron 31:463–475.

Maejima T, Oka S, Hashimotodani Y, Ohno-Shosaku T, Aiba A, Wu D, Waku K, Sugiura T, Kano M (2005) Synaptically driven endocannabinoid release requires Ca^{2+}-assisted metabotropic glutamate receptor subtype 1 to phospholipase $C\beta_4$ signaling cascade in the cerebellum. J Neurosci 25:6826–6835.

Malenka RC, Bear MF (2004) LTP and LTD: an embarrassment of riches. Neuron 44:5–21.

Marcaggi P, Attwell D (2005) Endocannabinoid signaling depends on the spatial pattern of synapse activation. Nat Neurosci 8:776–781.

Marcaggi P, Attwell D (2007) Short- and long-term depression of rat cerebellar parallel fibre synaptic transmission mediated by synaptic crosstalk. J Physiol 578:545–550.

Marinelli S, Di Marzo V, Berretta N, Matias I, Maccarrone M, Bernardi G, Mercuri NB (2003) Presynaptic facilitation of glutamatergic synapses to dopaminergic neurons of the rat substantia nigra by endogenous stimulation of vanilloid receptors. J Neurosci 23:3136–3144.

Marsicano G, Wotjak CT, Azad SC, Bisogno T, Rammes G, Cascio MG, Hermann H, Tang J, Hofmann C, Zieglgansberger W, Di Marzo V, Lutz B (2002) The endogenous cannabinoid system controls extinction of aversive memories. Nature 418:530–534.

Martin SJ, Grimwood PD, Morris RG (2000) Synaptic plasticity and memory: an evaluation of the hypothesis. Annu Rev Neurosci 23:649–711.

Mato S, Robbe D, Puente N, Grandes P, Manzoni OJ (2005) Presynaptic homeostatic plasticity rescues long-term depression after chronic Delta 9-tetrahydrocannabinol exposure. J Neurosci 25:11619–11627.

Matyas F, Yanovsky Y, Mackie K, Kelsch W, Misgeld U, Freund TF (2006) Subcellular localization of type 1 cannabinoid receptors in the rat basal ganglia. Neuroscience 137:337–361.

McBain CJ, Fisahn A (2001) Interneurons unbound. Nat Rev Neurosci 2:11–23.

McPartland JM, Matias I, Di Marzo V, Glass M (2006) Evolutionary origins of the endocannabinoid system. Gene 370:64–74.

Melis M, Pistis M, Perra S, Muntoni AL, Pillolla G, Gessa GL (2004) Endocannabinoids mediate presynaptic inhibition of glutamatergic transmission in rat ventral tegmental area dopamine neurons through activation of CB_1 receptors. J Neurosci 24:53–62.

Mittmann W, Koch U, Hausser M (2005) Feed-forward inhibition shapes the spike output of cerebellar Purkinje cells. J Physiol 563:369–378.

Narushima M, Uchigashima M, Fukaya M, Matsui M, Manabe T, Hashimoto K, Watanabe M, Kano M (2007) Tonic enhancement of endocannabinoid-mediated retrograde suppression of inhibition by cholinergic interneuron activity in the striatum. J Neurosci 27:496–506.

Neu A, Foldy C, Soltesz I (2007) Post-synaptic origin of CB_1-dependent tonic inhibition of GABA release at cholecystokinin-positive basket cell to pyramidal cell synapses in the CA1 region of the rat hippocampus. J Physiol 578:233–247.

Nevian T, Sakmann B (2006) Spine Ca^{2+} signaling in spike-timing-dependent plasticity. J Neurosci 26:11001–11013.
Ohno-Shosaku T, Shosaku J, Tsubokawa H, Kano M (2002) Cooperative endocannabinoid production by neuronal depolarization and group I metabotropic glutamate receptor activation. Eur J Neurosci 15:953–961.
Ohno-Shosaku T, Matsui M, Fukudome Y, Shosaku J, Tsubokawa H, Taketo MM, Manabe T, Kano M (2003) Post-synaptic M_1 and M_3 receptors are responsible for the muscarinic enhancement of retrograde endocannabinoid signalling in the hippocampus. Eur J Neurosci 18:109–116.
Oliet SH, Baimoukhametova DV, Piet R, Bains JS (2007) Retrograde regulation of GABA transmission by the tonic release of oxytocin and endocannabinoids governs post-synaptic firing. J Neurosci 27:1325–1333.
Palay SL, Chan-Palay V (1974) Cerebellar Cortex: Cytology and Organization. New York: Springer-Verlag.
Parrish JC, Nichols DE (2006) Serotonin 5-HT_{2A} receptor activation induces 2-arachidonoylglycerol release through a phospholipase c-dependent mechanism. J Neurochem 99:1164–1175.
Patel S, Hillard CJ (2001) Cannabinoid CB_1 receptor agonists produce cerebellar dysfunction in mice. J Pharmacol Exp Ther 297:629–637.
Piomelli D (2003) The molecular logic of endocannabinoid signalling. Nat Rev Neurosci 4:873–884.
Pitler TA, Alger BE (1992) Post-synaptic spike firing reduces synaptic GABAA responses in hippocampal pyramidal cells. J Neurosci 12:4122–4132.
Rancz EA, Hausser M (2006) Dendritic calcium spikes are tunable triggers of cannabinoid release and short-term synaptic plasticity in cerebellar Purkinje neurons. J Neurosci 26:5428–5437.
Reich CG, Karson MA, Karnup SV, Jones LM, Alger BE (2005) Regulation of IPSP theta rhythm by muscarinic receptors and endocannabinoids in hippocampus. J Neurophysiol 94:4290–4299.
Robbe D, Kopf M, Remaury A, Bockaert J, Manzoni OJ (2002) Endogenous cannabinoids mediate long-term synaptic depression in the nucleus accumbens. Proc Natl Acad Sci USA 99:8384–8388.
Robbe D, Montgomery SM, Thome A, Rueda-Orozco PE, McNaughton BL, Buzsaki G (2006) Cannabinoids reveal importance of spike timing coordination in hippocampal function. Nat Neurosci 9:1526–1533.
Ronesi J, Lovinger DM (2005) Induction of striatal long-term synaptic depression by moderate frequency activation of cortical afferents in rat. J Physiol 562:245–256.
Ronesi J, Gerdeman GL, Lovinger DM (2004) Disruption of endocannabinoid release and striatal long-term depression by post-synaptic blockade of endocannabinoid membrane transport. J Neurosci 24:1673–1679.
Safo PK, Regehr WG (2005) Endocannabinoids control the induction of cerebellar LTD. Neuron 48:647–659.
Schultz W, Dickinson A (2000) Neuronal coding of prediction errors. Annu Rev Neurosci 23:473–500.
Shen M, Thayer SA (1998) The cannabinoid agonist Win55,212-2 inhibits calcium channels by receptor-mediated and direct pathways in cultured rat hippocampal neurons. Brain Res 783:77–84.
Singla S, Kreitzer AC, Malenka RC (2007) Mechanisms for synapse specificity during striatal long-term depression. J Neurosci 27:5260–5264.
Sjostrom PJ, Hausser M (2006) A cooperative switch determines the sign of synaptic plasticity in distal dendrites of neocortical pyramidal neurons. Neuron 51:227–238.
Sjostrom PJ, Nelson SB (2002) Spike timing, calcium signals and synaptic plasticity. Curr Opin Neurobiol 12:305–314.
Sjostrom PJ, Turrigiano GG, Nelson SB (2003) Neocortical LTD via coincident activation of presynaptic NMDA and cannabinoid receptors. Neuron 39:641–654.
Sjostrom PJ, Turrigiano GG, Nelson SB (2004) Endocannabinoid-dependent neocortical layer-5 LTD in the absence of post-synaptic spiking. J Neurophysiol 92:3338–3343.

Sjostrom PJ, Turrigiano GG, Nelson SB (2007) Multiple forms of long-term plasticity at unitary neocortical layer 5 synapses. Neuropharmacology 52:176–184.
Slanina KA, Roberto M, Schweitzer P (2005) Endocannabinoids restrict hippocampal long-term potentiation via CB_1. Neuropharmacology 49:660–668.
Soler-Llavina GJ, Sabatini BL (2006) Synapse-specific plasticity and compartmentalized signaling in cerebellar stellate cells. Nat Neurosci 9:798–806.
Stella N, Schweitzer P, Piomelli D (1997) A second endogenous cannabinoid that modulates long-term potentiation. Nature 388:773–778.
Straiker A, Mackie K (2005) Depolarization-induced suppression of excitation in murine autaptic hippocampal neurones. J Physiol 569:501–517.
Sullivan JM (2000) Cellular and molecular mechanisms underlying learning and memory impairments produced by cannabinoids. Learn Mem 7:132–139.
Szabo B, Schlicker E (2005) Effects of cannabinoids on neurotransmission. Handb Exp Pharmacol 327–365.
Szabo B, Urbanski MJ, Bisogno T, Di Marzo V, Mendiguren A, Baer WU, Freiman I (2006) Depolarization-induced retrograde synaptic inhibition in the mouse cerebellar cortex is mediated by 2-arachidonoylglycerol. J Physiol 577:263–280.
Takahashi KA, Castillo PE (2006) The CB_1 cannabinoid receptor mediates glutamatergic synaptic suppression in the hippocampus. Neuroscience 139:795–802.
Tonini R, Ciardo S, Cerovic M, Rubino T, Parolaro D, Mazzanti M, Zippel R (2006) ERK-dependent modulation of cerebellar synaptic plasticity after chronic Delta^9-tetrahydrocannabinol exposure. J Neurosci 26:5810–5818.
Trettel J, Levine ES (2002) Cannabinoids depress inhibitory synaptic inputs received by layer 2/3 pyramidal neurons of the neocortex. J Neurophysiol 88:534–539.
Turrigiano G (2007) Homeostatic signaling: the positive side of negative feedback. Curr Opin Neurobiol 17:318–324.
Twitchell W, Brown S, Mackie K (1997) Cannabinoids inhibit N- and P/Q-type calcium channels in cultured rat hippocampal neurons. J Neurophysiol 78:43–50.
Tzounopoulos T, Rubio ME, Keen JE, Trussell LO (2007) Coactivation of pre- and post-synaptic signaling mechanisms determines cell-specific spike-timing-dependent plasticity. Neuron 54:291–301.
Uchigashima M, Narushima M, Fukaya M, Katona I, Kano M, Watanabe M (2007) Subcellular arrangement of molecules for 2-arachidonoyl-glycerol-mediated retrograde signaling and its physiological contribution to synaptic modulation in the striatum. J Neurosci 27:3663–3676.
van Beugen BJ, Nagaraja RY, Hansel C (2006) Climbing fiber-evoked endocannabinoid signaling heterosynaptically suppresses presynaptic cerebellar long-term potentiation. J Neurosci 26:8289–8294.
Varela F, Lachaux JP, Rodriguez E, Martinerie J (2001) The brainweb: phase synchronization and large-scale integration. Nat Rev Neurosci 2:229–239.
Varma N, Carlson GC, Ledent C, Alger BE (2001) Metabotropic glutamate receptors drive the endocannabinoid system in hippocampus. J Neurosci 21:RC188:1–5.
Wadiche JI, Jahr CE (2005) Patterned expression of Purkinje cell glutamate transporters controls synaptic plasticity. Nat Neurosci 8:1329–1334.
Wang Z, Kai L, Day M, Ronesi J, Yin HH, Ding J, Tkatch T, Lovinger DM, Surmeier DJ (2006) Dopaminergic control of corticostriatal long-term synaptic depression in medium spiny neurons is mediated by cholinergic interneurons. Neuron 50:443–452.
Wettschureck N, van der Stelt M, Tsubokawa H, Krestel H, Moers A, Petrosino S, Schütz G, Di Marzo V, Offermanns S (2006) Forebrain-specific inactivation of $G_{q/G11}$ family G proteins results in age-dependent epilepsy and impaired endocannabinoid formation. Mol Cell Biol 26:5888–5894.
Whalley BJ, Wilkinson JD, Williamson EM, Constanti A (2004) A novel component of cannabis extract potentiates excitatory synaptic transmission in rat olfactory cortex in vitro. Neurosci Lett 365:58–63.
Wilson RI, Nicoll RA (2001) Endogenous cannabinoids mediate retrograde signalling at hippocampal synapses. Nature 410:588–592.
Wilson RI, Nicoll RA (2002) Endocannabinoid signaling in the brain. Science 296:678–682.

Wilson RI, Kunos G, Nicoll RA (2001) Presynaptic specificity of endocannabinoid signaling in the hippocampus. Neuron 31:453–462.

Yao H, Dan Y (2005) Synaptic learning rules, cortical circuits, and visual function. Neuroscientist 11:206–216.

Yazulla S, Studholme KM, McIntosh HH, Fan SF (2000) Cannabinoid receptors on goldfish retinal bipolar cells: electron-microscope immunocytochemistry and whole-cell recordings. Vis Neurosci 17:391–401.

Yoshida T, Fukaya M, Uchigashima M, Miura E, Kamiya H, Kano M, Watanabe M (2006) Localization of diacylglycerol lipase-alpha around post-synaptic spine suggests close proximity between production site of an endocannabinoid, 2-arachidonoyl-glycerol, and presynaptic cannabinoid CB_1 receptor. J Neurosci 26:4740–4751.

Zhu PJ, Lovinger DM (2005) Retrograde endocannabinoid signaling in a post-synaptic neuron/synaptic bouton preparation from basolateral amygdala. J Neurosci 25:6199–6207.

Zhu PJ, Lovinger DM (2007) Persistent synaptic activity produces long-lasting enhancement of endocannabinoid modulation and alters long-term synaptic plasticity. J Neurophysiol 97:4386–4389.

Zucker RS, Regehr WG (2002) Short-term synaptic plasticity. Annu Rev Physiol 64:355–405.

Chapter 12
Endocannabinoid Functions in Neurogenesis, Neuronal Migration, and Specification

Tibor Harkany, Manuel Guzmán, and Yasmin L. Hurd

Abstract Endocannabinoids act as retrograde messengers thus controlling many synapses in the postnatal brain. In contrast, the concept that endocannabinoid functions are pivotal to fundamental developmental processes, including progenitor proliferation and fate specification, lineage segregation, neuronal migration, differentiation and survival, in the embryonic brain has just begun to emerge. Understanding the basic developmental and signaling principles controlled by endocannabinoids is pertinent to defining the molecular mechanisms establishing functional neuronal circuits with particular emphasis on synapse specification and functional diversification. Deciphering the spatial and temporal context of endocannabinoid signaling will also reveal the molecular substrates of permanent modifications to cellular structure and functions imposed by in utero cannabis exposure. Here, we review the ontogeny and recently identified functions of the endocannabinoid system with emphasis on the neuronal lineage during brain development, and discuss how fetal cannabis exposure may modify neuronal networks such that long-term changes to cognitive functions manifest in the affected offspring.

Introduction

Endocannabinoids from a Developmental Perspective

Our understanding of the structural substrates, spatial composition, and functional significance of endocannabinoid signaling has recently undergone rapid expansion, because of the continued identification of multiple endocannabinoid ligands, exogenous cannabinoids, and other lipid mediators, metabolic enzymes and bioactive intermediates, as well as cannabinoid receptors (Piomelli, 2003; Katona et al., 2006; Mackie and Stella, 2006; Harkany et al., 2007). Accumulating evidence indicates that in the central nervous system (CNS) endocannabinoids and Δ^9-tetrahydrocannabinol (Δ^9-THC), the major psychoactive component in cannabis (*Cannabis spp.*), target cannabinoid receptors primarily on neurons (Katona et al., 1999, 2000; Harkany et al., 2003, 2005), but also on glia (Fernandez-Ruiz et al., 2007). It is generally accepted that

A. Kofalvi (ed.), *Cannabinoids and the Brain.*

CB_1 cannabinoid receptors (CB_1 receptors) are selectively recruited to presynaptic terminals of both inhibitory and excitatory neurons and are ideally positioned to sense endocannabinoids released from postsynaptic cells (Freund et al., 2003). The activity-dependent release of endocannabinoids and their tuning of synaptic plasticity at many synapses throughout the brain (Kreitzer and Regehr, 2001; Wilson and Nicoll, 2001; Freund et al., 2003) support the doctrine that endocannabinoids mediate retrograde synaptic signaling in the adult CNS (Llano et al., 1991; Pitler and Alger, 1992).

Endocannabinoids and Neurodevelopment

Besides this well-established neuromodulatory role, endocannabinoid signaling also subserves principal mechanisms of CNS development: this family of lipid mediators is pivotal in controlling the proliferation, migration, lineage commitment, and survival of neural progenitors (Galve-Roperh et al., 2006) with a continued control of neurogenesis in neurogenic niches (e.g., dentate gyrus) postnatally (Galve-Roperh et al., 2007). Furthermore, endocannabinoids contribute to the phenotypic differentiation of lineage-committed neuronal precursors (Berghuis et al., 2005; Berghuis et al., 2007; Harkany et al., 2007) and the onset of synaptic communication during assembly of functional neuronal networks (Bernard et al., 2005; Berghuis et al., 2007) that directly translate into retrograde synaptic signaling once synapse establishment concludes. The importance of endocannabinoid signaling during neuronal development is underscored by the pathogenic impact of maternal cannabis smoking or CB_1 receptor agonist administration during pregnancy, causing cognitive, motor, and social deficits enduring into the adulthood of the affected offspring (Richardson et al., 1995; Fried et al., 2003; Mereu et al., 2003; Antonelli et al., 2005; Huizink and Mulder, 2006). Here, we discuss recent key findings with regard to the ontogeny of the endocannabinoid signaling system and to its functions during embryogenesis. Moreover, we present findings outlining potential cellular targets of prenatal endocannabinoid actions and pre-/perinatal Δ^9-THC exposure. The emerging significance of endocannabinoid signaling during CNS formation will be supported by data from mammalian expression systems and will be limited to well-accepted ligands, metabolic pathways, and receptors in the CNS, as only circumstantial evidence is at present available on the involvement of the CB_2 cannabinoid receptor (Van Sickle et al., 2005; Fernandez-Ruiz et al., 2007) and the endocannabinoid-sensing orphan G protein-coupled receptor GPR55 (Baker et al., 2006) in instructing neuronal specification.

Ontogeny of the Endocannabinoid System

The establishment of endocannabinoid signaling requires the temporal and spatial coincidence of metabolic enzyme and receptor expression during brain development. In spite of vast efforts directed toward understanding endocannabinoid signaling in the postnatal brain, a surprising lack of data exists with regard to detailed developmental studies on the distribution of *sn*-1 diacylglycerol lipases (DAGLs), the prime 2-arachydonoylglycerol (2-AG) synthetic enzymes, monoacylglycerol lipases ($MAGL_{1/2}$) (Dinh et al., 2002; Muccioli et al., 2007) and fatty acid amide hydrolase (FAAH) (Cravatt et al., 1996), enzymes hydrolyzing anandamide (AEA) and 2-AG, respectively, and the by far known cannabinoid receptors in the developing brain. This caveat in our knowledge is in part due to the ensuing definition of the precise metabolic functions of recently discovered enzymatic entities and the lack of available immunoreagents with rigorously proven specificities against their cellular substrates in embryonic tissues.

Temporal and Spatial Localization of Major Endocannabinoids During Neurodevelopment

AEA and 2-AG levels vary substantially throughout prenatal development (Berrendero et al., 1999; Fernandez-Ruiz et al., 2000). At the periimplantation period (between days 4–6 of pregnancy), AEA concentrations in the uterus define uterine receptivity for embryo implantation (Paria et al., 2001): transiently reduced uterine AEA levels with coincident down-regulation of CB_1 and CB_2 receptor expression in the preimplantation embryo synchronize the onset of uterine receptivity and blastocyst activation enabling implantation competence (Paria et al., 2001). Low AEA concentrations are present in the brain at mid-gestation. In contrast, AEA levels gradually increase throughout the perinatal period until adult levels are reached (Berrendero et al., 1999). Strikingly, 2-AG concentrations (2–8 nmol/g tissue) largely exceed those of AEA (3–6 pmol/g tissue) throughout brain development (Berrendero et al., 1999; Fernandez-Ruiz et al., 2000), with fetal 2-AG levels being similar to those in young and adult rodent brains with a remarkably distinct peak in neonates (Berrendero et al., 1999; Fernandez-Ruiz et al., 2000).

The Ontogeny of Enzymes Involved in 2-AG Biosynthesis

Accumulating evidence indicates the existence of several, often temporally and spatially concerting, biosynthetic pathways enabling the generation of 2-AG and AEA. While a consensus exists about the identities of biosynthetic enzymes pivotal for 2-AG synthesis, the metabolic machinery involved in activity-dependent AEA production is much less understood. Recently, a series of elegant experiments in the Doherty and Di Marzo laboratories demonstrated that the α and β DAGL isoforms (DAGLα/β) are required and are sufficient to generate 2-AG both in heterologous expression systems and in vivo (Bisogno et al., 2003). These studies have also revealed a close spatial association between the sites of DAGLα/β and CB_1 receptor expression during mid-gestation in the mouse embryo thus strongly suggesting the existence of an autocrine endocannabinoid signaling loop regulating axon specification and elongation in subcortical and cerebellar projection tracts (Bisogno et al., 2003). In contrast, DAGLα/β exhibit predominant postsynaptic localization in dendrites of, e.g., hippocampal pyramidal cells, Purkinje cells, and striatal medium spiny neurons in the adult (Bisogno et al., 2003; Katona et al., 2006; Uchigashima et al., 2007), underpinning a developmentally regulated, activity-dependent temporal switch in the sites of endocannabinoid production and release in neurons. The concept that *on-demand* endocannabinoid signaling links axonal specification in the early embryonic brain to synaptogenesis and synaptic plasticity during the neonatal period is supported by recent evidence identifying endocannabinoids as a novel class of axon guidance cues as shown in chemotropic and galvanotropic growth cone turning assays in vitro (Berghuis et al., 2007). Moreover, the translocation of DAGLα/β coincides with the local navigation and postsynaptic target selection of local inhibitory afferents during corticogenesis, thus demonstrating that the specification and extension of axons toward postsynaptic target areas may require autocrine endocannabinoid signaling (Bisogno et al., 2003; Williams et al., 2003), while the precise positioning of synapses on postsynaptic targets, the establishment of cell-to-cell contacts, and the onset of synaptic communication within target regions are controlled by the spatially compartmentalized actions of target-derived endocannabinoids (Berghuis et al., 2007; Harkany et al., 2007).

The Temporal and Spatial Localization of Enzymes Involved in AEA Biosynthesis

A series of candidate enzymes with considerable AEA biosynthetic activity has recently been identified, with *N*-acyl-phosphatidylethanolamine-selective phospholipase D (NAPE-PLD) being considered as a prime candidate for AEA production (Piomelli, 2003; see Chap. 2). However, recent evidence supports the contribution of α/β-hydrolase 4, a lyso-NAPE lipase, to form *N*-acyl ethanolamines including AEA (Simon and Cravatt, 2006) in multistep AEA biosynthetic pathways that proceed through the generation of bioactive intermediaries. Additionally, Liu and colleagues (2006) have identified yet another AEA biosynthetic pathway involving phospholipase C (PLC)-mediated cleavage of NAPE to generate a biologically active intermediate, phospho-AEA, which is dephosphorylated by phosphatases, such as PTPN22, to yield AEA. Clearly, the uncertainty over which enzymes participate in AEA biosynthesis has hampered their histochemical mapping during embryogenesis. Recent data (Berghuis et al., 2007) demonstrate prominent NAPE-PLD expression in dendritic spines of neocortical pyramidal cells in the late-gestational mouse brain that is in stark contrasts with the lack of detectable NAPE-PLD immunoreactivity in earlier stages of CNS development. These findings concur with prior neurochemical observations showing low AEA concentrations in the pM range during early to mid-gestation with progressive enhancement of AEA biosynthesis in the perinatal brain (Berrendero et al., 1999; Morishita et al.,

2005). In addition, this evidence argues that 2-AG bioavailability exceeds that of AEA during early stages of CNS specification.

The Ontogeny of Endocannabinoid Degradation

Monoacylglycerol lipases 1/2 ($MAGL_{1/2}$) (Dinh et al., 2002; Muccioli et al., 2007) and fatty acid amide hydrolase (FAAH) (Cravatt et al., 1996; Freund et al., 2003; Piomelli, 2003) have been established as the major catabolic enzymes of 2-AG and AEA, respectively. Whereas the cellular distribution of $MAGL_{1/2}$ during ontogenesis of the CNS is as yet unknown, FAAH has been detected in radial glia during late gestation and throughout the neonatal period (Aguado et al., 2006; Harkany et al., 2007). In neurons, however, FAAH expression may be either transient or permanent: hippocampal interneurons undergoing intralaminar migration transiently express FAAH in vitro (Berghuis et al., 2005) or during the first postnatal week in vivo (Morozov et al., 2004) with dissipating enzyme levels later. In contrast, FAAH immunoreactivity steadily increases first in proximal then distal dendrites of hippocampal and neocortical pyramidal cell dendrites at birth (Tibor Harkany, unpublished data) and throughout the neonatal period (Morozov et al., 2004). Overall, the cellular positioning of DAGLs, NAPE-PLD, and FAAH with coincident CB_1 receptor expression in neurons suggests that endocannabinoid signaling networks are operational in neonates.

CB_1 Receptors in the Developing Brain: Spatial, Temporal, and Functional Considerations

Relatively precise expression patterns are available for the CB_1 receptor in the developing CNS. CB_1 receptors have been detected as early as day 11 of gestation in the murine CNS (comparable to 5–6 weeks in the human embryo) with gradually increasing levels for both mRNA and receptor density throughout the prenatal period in the whole brain (Berrendero et al., 1999; Garcia-Gil et al., 1999; Fernandez-Ruiz et al., 2000; Bisogno et al., 2003; Wang et al., 2003; Bernard et al., 2005; Berghuis et al., 2007). Similar CB_1 receptor mRNA expression patterns were found during human pre- and postnatal CNS development by in situ hybridization (Wang et al., 2003): CB_1 receptors were detected at week 14 of gestation, with selective receptor expression being present in neurons of the CA2-CA3 hippocampal subfields and in the basal nuclear group of the amygdala by week 20. Similar to the rodent brain, gradually increasing CB_1 receptor mRNA levels were noted in the frontal cortex, hippocampus, basal ganglia, and cerebellum between the fetal period and adulthood in humans. A unique feature of CB_1 receptor distribution in the fetal mouse and human brains is its association with several developing axonal trajectories in the white matter. This type of CB_1 receptor localization, widely considered as *atypical* receptor positioning (Romero et al., 1997), has recently been identified as a prerequisite for guiding the elongating axons to their targets, and to achieving proper synapse positioning of postsynaptic target cells (Berghuis et al., 2007). The evolving concept of endocannabinoid-driven synapse specification is further supported by the removal of CB_1 receptors from developing axonal tracts coincident with the conclusion of synaptogenesis and the selection of postsynaptic targets (Fernandez-Ruiz et al., 2000; Berghuis et al., 2005, 2007). Notably, pharmacological studies unequivocally demonstrated the functionality of CB_1 receptors in embryonic neural tissues since WIN55212-2, a cannabinoid receptor agonist, significantly stimulated $[^{35}S]GTP\gamma S$ binding in both the rodent and human brains (Mato et al., 2003; Wang et al., 2003). Overall, neuroanatomical findings furnish the concept that the endocannabinoid system is expressed and positioned during CNS development such that its activity can ideally tune a broad array of developmental processes in both neural progenitors and in lineage-committed neuronal precursors.

Differential Signaling Through the CB_1 Cannabinoid Receptor Couples to Second Messenger Pathways Regulating Neuronal Survival and Differentiation

Cannabinoid receptors belong to the family of G protein-coupled receptors (GPCRs) with preferential coupling to $G_{i/o}$ proteins. Under certain conditions, however, a shift to signaling through $G_{q/11}$ proteins has been reported (Lauckner et al., 2005). Nevertheless, both G proteins can couple CB_1 receptors to signal transduction pathways regulating, among others, ion channels, neurotransmitter transporters, metabolic enzymes, and cytoskeletal integrity (Iyengar, 2005) (Fig. 1). Accordingly, *on-demand* recruitment of second messengers to the CB_1 receptor, e.g., the Src/Stat3 (Jordan et al., 2005; He et al., 2005), extracellular signal-regulate kinase (ERK1/2) (Galve-Roperh et al., 2000; Rueda et al., 2002; Berghuis et al., 2007), and PI_3K/Akt pathways (Molina-Holgado et al., 2002), and the modulation of sphingolipid-derived signaling mediators and cell death pathways (Guzman, 2003) enhance its potential to dynamically regulate fundamental developmental processes, including neural progenitor proliferation and migration, fate decision, survival, and lineage specification, in a spatially and temporally coordinated manner (Harkany et al., 2007; Galve-Roperh et al., 2007). Recent studies showing CB_1 receptor functionality in the developing human fetus (Mato et al., 2003; Wang et al., 2003) highlight the significance of a physiologically adequate endocannabinoid tone during neurodevelopment, and indicate the importance of the selective recruitment of downstream effector pathways in determining the contributions of endocannabinoids to generating neuronal diversity.

Context-Dependent Downstream Signaling

Recent advances in receptor biology have moved beyond the classic depiction of the CB_1 receptor as a solitary GPCR coupling solely to G proteins, and implicate

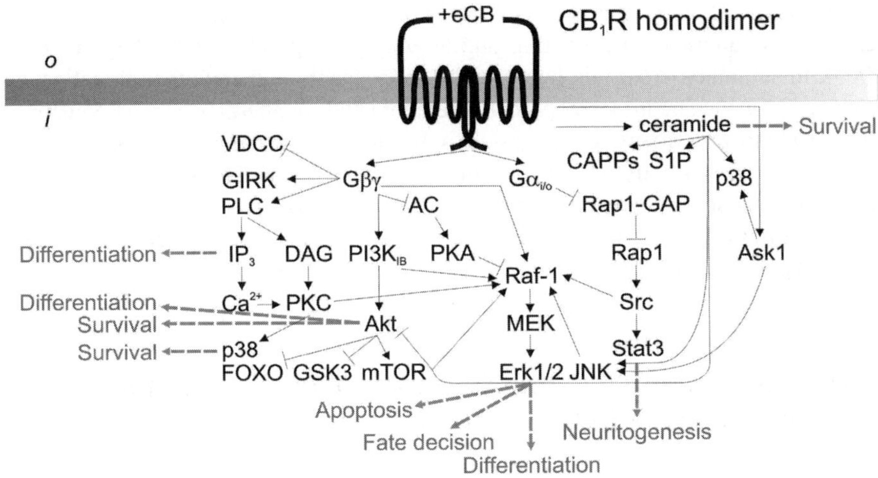

Fig. 1 Second messenger pathways downstream of the $G_{i/o\alpha}$-coupled CB_1 receptor and their physiological output. GPCR signaling relies on receptor homodimers as the functional backbone of signal transduction. Accordingly, agonist binding to CB_1 receptor homodimers recruits distinct second messenger cascades whose recruitment to this receptor is thought to be determined by the actual cellular context. *Grey arrows* point to known biological responses with relevance to CNS development. *AC* adenylyl cyclase; *Ask1* apoptosis signal-regulating kinase 1; *B-Raf* mitogen-activated protein kinase (MAPK) kinase kinase; *CAPP* ceramide-activated protein phosphatase; *DAG* 1,2-diacylglycerol; *eCB* endocannabinoid; *FOXO* forkhead transcription factors; *GAP* GTPase-activating protein; *GIRK* G-protein-gated inwardly rectifying K^+ channel; *GPCRA* GPCR agonist; *GSK3* glycogen synthase kinase-3; *HBEGF* heparin-binding EGF-like growth factor; IP_3 inositol 1,4,5 trisphosphate; *JNK c-Jun* N-terminal kinase; *MEK* ERK kinase; *mTOR* mammalian target of rapamycin; PI_3KIB class IB phosphoinositide 3-kinase; *PKA* protein kinase A; *PKC* protein kinase C; *PLC* phospholipase Cα; *Raf-1* MEK kinase; *S1P* sphingosine 1-phosphate; *Stat3* signal transducer and activator of transcription 3; *VDCC* voltage-dependent Ca^{2+} channel

alternative signaling cascades governing critical events during neurodevelopment (Devi, 2000; Wager-Miller et al., 2002; Hart et al., 2004; Kearn et al., 2005; Rios et al., 2006). CB_1 receptors likely signal as homodimers (Wager-Miller et al., 2002), in agreement with the principle of GPCR signaling that identifies receptor multimers as the key functional signaling units (Devi, 2000). Accumulating evidence suggests that receptor cross-talk may be an essential means to coordinate the coincident actions of multiple ligands, including endocannabinoids, neuropeptides, and neurotransmitters (Pertwee, 2006). Accordingly, signaling interactions of the CB_1 receptor with other developmentally regulated signaling systems, e.g., growth factor signaling pathways, may be essential in controlling progenitor proliferation, precursor migration, and even morphogenesis: basic fibroblast growth factor has been proposed to regulate neural cell growth by increasing the level of 2-AG generation (Williams et al., 2003). Additionally, brain-derived neurotrophic factor production appears essential in cannabinoid-mediated neuroprotection after excitotoxicity (Marsicano et al., 2003;

Khaspekov et al., 2004), and transactivation of TrkB receptors mediates CB_1 receptor regulation of interneuron migration during embryonic development (Berghuis et al., 2005). In summary, cross talk between the endocannabinoid and other signaling systems can influence neurodevelopment. The interaction of alternative endocannabinoid signaling pathways, many of them as yet only partially known, converging on the CB_1 receptor provides a unifying mechanistic perspective explaining the diverse developmental actions of endocannabinoids on various neuron populations. It also emphasizes that overall effects are critically dependent on the balance between diverse signaling cascades and their relative levels of activity.

Endocannabinoid Signaling Controls Neural Progenitor Proliferation and Lineage Commitment

Fate Decision Points

During brain development, the expression of CB_1 and CB_2 receptors, and enzymes associated with endocannabinoid synthesis and degradation coincides with the expansion of neural progenies and their engagement in establishing neuronal diversity (Galve-Roperh et al., 2006; Harkany et al., 2007) (Fig. 2). The presence of functional endocannabinoid signaling networks in neurogenic proliferative zones of the developing brain, and also in neurogenic niches of the adult (see below), suggests that endocannabinoid signals could provide extracellular cues instructing the cellular program of neural progenitors such that they generate appropriate contingents of cell lineages required to build the developing brain. A fine-tuned balance between progenitor cell proliferation and programmed death guarantees the generation of adequate quantities of neural cells during brain development. It is evident that endocannabinoids and related lipid mediators regulate neural progenitor commitment and survival (Guzman

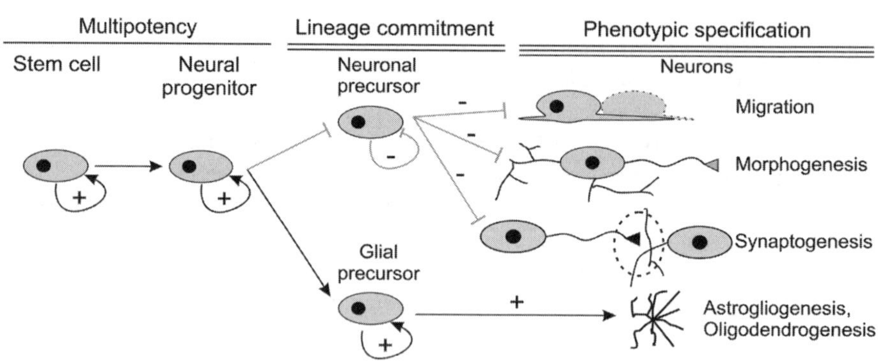

Fig. 2 Endocannabinoid actions during lineage specification in the developing CNS. Note the opposing actions of endocannabinoid signaling on neural progenitor fate decisions and their commitment toward a neuronal or glial lineage. Data were modified from: Guzman et al., 2002; Galve-Roperh et al., 2006, 2007; Harkany et al., 2007

et al., 2002; Guzman, 2003; Aguado et al., 2006). Neural progenitors possess a functional endocannabinoid signaling loop: the capacity to synthesize endocannabinoids, functional CB_1 receptors, and catabolic enzyme(s) (Aguado et al., 2005). Notably, CB_1 receptor activation promotes progenitor cell proliferation (Aguado et al., 2006) and neurosphere generation in vitro. These actions are abrogated in CB_1 receptor knock-out (*CNR1*$^{-/-}$) cells in vitro but increased in FAAH-deficient (*FAAH*$^{-/-}$) neurospheres with robustly elevated AEA levels (Aguado et al., 2005). The endocannabinoid system also plays a role in regulating the primary fate decision point of neural progenitors, e.g., whether neural precursors commit to generate neurons or glia. Consequently, activation of CB_1 receptors on neural progenitors promotes their differentiation into glial cells (Aguado et al., 2006), since in vivo analysis of *CNR1*$^{-/-}$ and *FAAH*$^{-/-}$ mouse brains demonstrated the requirement of an intact endocannabinoid signaling loop to sustain astrogliogenesis (Aguado et al., 2006).

Neurogenesis

In contrast, the impact of CB_1R activation on neurogenesis seems variable. Methanandamide, a nonhydrolizable AEA analogue, significantly decreases neurogenesis, as measured by BrdU incorporation, in the adult dentate gyrus (Rueda et al., 2002). In addition, endocannabinoids decrease the expression of selective markers of early and terminally differentiated neurons, β-III-tubulin and neuron-specific nuclear protein, respectively, and inhibit neurite outgrowth in vitro. Alternatively, SR141716, a selective CB_1 receptor antagonist increases neuronal differentiation of neural progenitors (Rueda et al., 2002; Jin et al., 2004). Collectively, these studies point to the existence of an endogenous cannabinoid tone actively modulating neural progenitor differentiation through the CB_1 receptor. These findings also indicate that the pharmacodynamic and pharmacokinetic characteristics of a given CB_1 receptor ligand, together with the particular (patho)physiological signaling context in which cannabinoid signaling occurs, may determine the extent of neurogenesis modulated by endocannabinoid signaling, and implicate the endocannabinoid system in maintaining the neuron/glia balance during brain development.

AEA-Specific Actions

Data obtained with NG108-15 neuroblastoma cells indicate that AEA, but not 2-AG or WIN55212-2, may also inhibit neuronal differentiation in a CB_1 receptor-independent manner (Galve-Roperh et al., 2006). CB_1 receptor-independent regulation of neurogenesis may however be specific for AEA as a ligand, given its propensity to allosterically regulate the activity of a broad variety of receptors and ion channels affecting neuronal fate (van der Stelt and Di Marzo, 2005). Overall, defining the cellular identities of neuronal precursors in conjunction with identifying the molecular composition of the signal transduction machineries regulated by endocannabinoid signaling in vivo are essential to our understanding of how endo- and phytocannabinoids influence the developmental competence and patterning of neural progenitors.

(Endo)Cannabinoid Regulation of Adult Neurogenesis

Replenishment of neurons continues in neurogenic niches of the postnatal brain: neurons are continuously born from adult subventricular zone and dentate progenitors in the cerebral cortex and hippocampus, respectively, with their progenies generating astroglia and functional neurons integrating in adult neuronal circuitries (Gage, 2002; Kempermann et al., 2004; Harkany et al., 2004; Toni et al., 2007). Initially, Rueda and colleagues (2002) identified a cannabinoid regulatory action on adult neurogenesis that was mediated by CB_1 receptors. This discovery was followed by the identification of endocannabinoid system components in neural progenitor cells (Jin et al., 2004; Aguado et al., 2005): endocannabinoids are produced by neural progenitors and their action on CB_1 receptors of hippocampal and cortical subventricular zone progenitors is required for their proliferation and lineage segregation. Accordingly, $CNR1^{-/-}$ mice show impaired neural progenitor proliferation and self-renewal (Aguado et al., 2005). Conversely, the multifold increased AEA concentrations in $FAAH^{-/-}$ mice generates excess astroglia (Aguado et al., 2006), while pharmacological stimulation of CB_1 receptors triggers neurogenesis (Jiang et al., 2005). Intriguingly, HU-210, a synthetic CB_1 receptor agonist, has been shown to expand hippocampal neurogenesis and exert anxiolytic and antidepressant effects that was attributed to the enhancement of newly born neurons to integrate in corticolimbic circuitries (Jiang et al., 2005). Collectively, these findings suggest that different types of agonists (endogenous vs. synthetic), pathophysiological conditions (e.g., developmental, injury-related, or chronic diseases), and signal bioavailability (locally generated endocannabinoids vs. systemic administration of synthetic agonists) can differentially modify the fate decision points of telencephalic neural progenitors such that either neurons or glial cells will predominate in the newly generated neural cell lineage. The importance of an endocannabinoid regulatory tone on neurogenesis is also depicted by the finding that $CNR1^{-/-}$ mice (Zimmer et al., 1999) suffer from early onset age-related cognitive impairment (Bilkei-Gorzo et al., 2005), a potential consequence of aging-associated decrease in cortical neurogenesis in the absence of the CB_1 receptor (Lie et al., 2004).

Second Messenger Signaling Underpinning Endocannabinoid Actions on Neural Progenitors

Since neural progenitors are endowed with a variety of endocannabinoid system components, including $TRPV_1$, CB_1 and CB_2 receptors, endocannabinoid ligand (AEA, 2-AG) synthesis capacity, and FAAH-mediated metabolism, it is imperative to understand the ligand specificity and divergence/convergence points of the downstream signal transduction cascades brought upon by CB_1 or CB_2 receptor activation. Endocannabinoids can modulate the endogenous differentiation program of neural progenitors either directly or by affecting the production of intermediary mediators in neighboring cells (Rueda et al., 2002; Galve-Roperh et al., 2006). Direct cannabinoid-induced fate decisions of neural progenitors can be attributed to their ability to activate the ERK1/2 pathway (Rueda et al., 2002; Palazuelos et al., 2006): during neocortical neurogenesis, sustained ERK signaling is required to generate neurons on the expense of glia. In this context, CB_1 receptor activation exerts dual effects on ERK1/2 signaling in neurons inasmuch as CB_1 receptor-mediated inhibition of cortical neural progenitor differentiation involves the attenuation of sustained ERK1/2 activation via inhibition of upstream Rap-1/B-Raf signaling (Rueda et al., 2002), whereas neural precursor proliferation, and neuritogenesis in neuronally committed progenitors are reliant on the coincident activation of the ERK pathway (Jordan et al., 2005). The differential control of ERK1/2 activity could also be a likely reason for the different effects of endocannabinoids on neuro- vs. gliogenesis, since glial cells do not express significant amounts of B-Raf, an essential anchor point of this signaling cascade (Galve-Roperh et al., 2006). Recently, Kim and colleagues (2006) have proposed that exogenous cannabinoids can interfere with nitric oxide production, a signaling pathway closely linked with endocannabinoid signaling (Alger, 2005), such that they stimulate CB_1 receptor-dependent adult neurogenesis on the expense of antineurogenic nitric oxide actions. In summary, these data, together with the CB_1 receptor-driven proliferation of human neural progenitors (both the hNSC1 neural stem cell line and a subpopulation of subependimal layer-derived progenitors) (Rueda et al., 2002; Palazuelos et al., 2006; Curtis et al., 2006), demonstrate that endocannabinoid signaling is critical for the maintenance of adult neural progenitor proliferation, self-renewal, and the generation of lineage-committed neuronal precursors, and highlight potential therapeutic implications of endocannabinoid functions with regard to human brain development and disease.

Endocannabinoid Actions Shape Neuronal Phenotypes and Connectivity Patterns

The (endo)cannabinoid-induced switch that commits neural progenitors to gliogenesis at the expense of neurogenesis clearly poses the question whether endocannabinoid effects also extend to the regulation of neuronal migration, and the attainment of particular morphological, physiological, and molecular phenotypes occurring during terminal neuronal differentiation.

Endocannabinoid Actions on Cell Migration

Recent evidence indicates that AEA and WIN55212-2 induce the in vitro migration of late-gestational GABAergic interneurons known to undergo long-distance migration to populate particular neocortical and hippocampal laminae (Berghuis et al., 2005). Notably, endocannabinoid-induced neuronal migration (Song and Zhong, 2000) acts in cooperativity with brain-derived neurotrophic factor (BDNF), a prime migration (Fukumitsu et al., 2006) and prodifferentiation factor (Ventimiglia et al., 1995; Horch and Katz, 2002; Dijkhuizen and Ghosh, 2005), for a variety of CB_1 receptor-expressing neurons, including GABAergic, serotonergic, and dopaminergic cells (Fig. 2). In vivo support for the chemotaxic actions of CB_1 receptor agonists was provided by the finding that prenatal Δ^9-THC increases the density of cholecystokinin-expressing interneurons in the neonatal rat hippocampus (Berghuis et al., 2005). These data, along with the CB_1 receptor agonist-induced migration of HEK-293 cells transfected with a cDNA encoding the CB_1 receptor (Song and Zhong, 2000) indicate that endocannabinoid signaling is instructive and permissive for neuronal migration.

Endocannabinoids and Establishment of Neuronal Connectivity

CB_1 receptor activation also controls neurite outgrowth and synaptogenesis; processes required to generate functionally mature neurons (Fig. 2). AEA and WIN55212-2 inhibit neurite formation and elongation of both GABAergic interneurons (Berghuis et al., 2004, 2005) and excitatory pyramidal cells (J. Mulder and Tibor Harkany, unpublished observations) isolated from the embryonic cerebrum such that AEA even abolishes the morphogenic potential of BDNF. Similarly, cannabinoid ligands, including Δ^9-THC, have been shown to counteract forskolin-induced synaptogenesis in primary hippocampal neurons (Kim and Thayer, 2001). An attractive hypothesis is that autocrine endocannabinoid signaling regulates growth cone differentiation and axon guidance (Bisogno et al., 2003). This concept stems from the finding that 2-AG stimulates neurite outgrowth of cerebellar neurons via a mechanism dependent on intrinsic DAGLα/β activity within axonal growth cones, while CB_1 receptor antagonists abolish N-cadherin and Fgf8-induced neurite extension (Williams et al., 2003). Thus, the question emerges whether the endocannabinoids that differentially control dendrite and axon development originate from within the developing neuron itself or represent paracrine, target-derived morphogens in neural circuits. In this regard, Berghuis and colleagues (2007) have identified endocannabinoids as target-derived axon guidance cues during corticogenesis in the late-gestation mouse embryo: AEA and WIN55212-2 gradients repulsed the axons of GABAergic interneurons and *Xenopus* spinal neurons in chemotropic and galvanotropic growth cone turning assays and this response was dependent on the CB_1 receptor-mediated activation of RhoA guanosine triphosphatases (GTPases) in neuronal growth cones. Intriguingly, AEA and HU-210 were shown to reduce neurogenic differentiation through the recruitment of the Rho family of small GTPases, whose spatially restricted activation controls cytoskeletal integrity (Jaffe and Hall, 2005; Berghuis et al., 2007), thus inducing cell rounding and neurite remodeling in GABAergic neurons and N1E-115 and B103 neuroblastoma cells (Ishii and Chun, 2002;

Galve-Roperh et al., 2006; Berghuis et al., 2007). In contrast, HU-210 promotes neurite outgrowth in Neuro 2A cells by the $G_{i/o\alpha}$-mediated degradation of RapGAPII and subsequent activation of Rap1 (Jordan et al., 2005). Overall, the above evidence together with the finding that conditional genetic deletion of CB_1 receptors in cortical interneurons affects their postsynaptic target selection and affects neuronal connectivity in the neocortex unequivocally define a developmental (morphogenic) niche driven by extracellular endocannabinoid signals.

Prenatal Marijuana Impairs CNS Development

Approximately one-third of Δ^9-THC in the plasma undergoes crossplacental transfer upon cannabis smoking during pregnancy (Hutchings et al., 1989) and affects, besides general fetal growth, development of the CNS. Retrospective longitudinal studies in humans have demonstrated cannabis-related specific long-term abnormalities: from exaggerated startle response and poor habituation to novel stimuli in the infant to cognitive retardation cognition (in tasks requiring visual memory, analysis, and integration), behavioral (e.g., hyperactivity, impulsivity, attention deficit), and social disturbances in adolescent children prenatally exposed to cannabis (Richardson et al., 1995; Fried and Smith, 2001; Fried et al., 2003; Goldschmidt et al., 2004; Antonelli et al., 2005; Gray et al., 2005; Jacobsen et al., 2006; Huizink and Mulder, 2006). The pronounced developmental effects of intrauterine Δ^9-THC exposure are not unexpected considering that CB_1 receptors are preferentially expressed in corticolimbic areas of the human fetal brain where their recruitment to axons participates in neuronal polarization, axon initiation, and postsynaptic target selection (Berghuis et al., 2004, 2005, 2007; Wang et al., 2003, 2004; Spano et al., 2007). Experimental studies clearly substantiate these findings (Mereu et al., 2003; Antonelli et al., 2005; Bernard et al., 2005; Spano et al., 2007) and link behavioral and cognitive deficits and emotional responsiveness to early developmental exposure to cannabis. Notably, prenatal cannabis treatment appears to robustly impact the glutamatergic system since it potently decreases glutamate release from nerve terminals together (Mereu et al., 2003) with diminishing the expression of glutamate transporters, AMPA receptor subunits, and the ability of astroglia to generate glutamine, the major precursor of glutamate in synaptic vesicle pools (Suarez et al., 2002, 2004a,b). More importantly, Δ^9-THC appears to influence the functions of endocannabinoids by increasing their (particularly AEA) synthesis and release in a concentration-dependent manner that is reliant on phospholipase D (PLD) activity (Hunter and Burstein, 1997). The involvement of G protein-coupled cannabinoid receptors in mediating Δ^9-THC-induced AEA synthesis is indicated by the pertussis toxin sensitivity of this response. In addition, the decreased AEA concentrations in the hippocampus of *CNR1*$^{-/-}$ mice (Di Marzo et al., 2000) further supports the concept that CB_1 receptor activity regulates endocannabinoid levels. Furthermore, Δ^9-THC down-regulates the expression and also desensitizes (35–65% of control) CB_1 receptors in a region-specific manner

(Sim et al., 1996; Zhuang et al., 1998; Fernandez-Ruiz et al., 2000), which, in the neonatal hippocampus, could trigger the onset of epileptiform activity (Bernard et al., 2005). While data on Δ^9-THC-induced changes in CB_1 receptor expression patterns during development are lacking, reports indicate no lasting changes in CB_1 receptor mRNA expression or protein levels in adult rats as a consequence of perinatal Δ^9-THC treatment (Garcia-Gil et al., 1999). Nevertheless, recent data show that endocannabinoids released from both interneurons and pyramidal cells in the CA1 region of the hippocampus during the neonatal period activate CB_1 receptors, thereby inhibiting synaptic GABA release and disrupting coherent neuronal network activity (Bernard et al., 2005). The efficacy of CB_1 receptors to form an anchor point for the establishment of information processing in immature neuronal networks (Berghuis et al., 2007; Harkany et al., 2007) suggests the involvement of this neuromodulatory system in cannabis-related developmental impairments and provides a focus for future research to understand the neuronal basis for Δ^9-THC-induced developmental deficits.

Concluding Remarks

Experimental evidence, from the fields of developmental biology, molecular genetics, electrophysiology, neuropharmacology and the neurosciences, together with longitudinal case-controlled human studies demonstrate that (1) endocannabinoid signaling through a variety of cannabinoid-sensing receptors (CB_1 and CB_2 receptors, GPR55, $TRPV_1$ receptors) affects the induction and patterning of the CNS by modulating the phenotypic differentiation program of neural progenitors, the formation of cell–cell contacts, and the onset and efficacy of intercellular synaptic communication. (2) Moreover, endocannabinoid signaling networks are sufficiently organized to evolve into feedback loops underlying retrograde synaptic transmission when immature neuronal networks become operational. (3) Conversely, interference with the precise temporal and spatial coordination of endocannabinoid signaling, through, e.g., in utero cannabis exposure, and genetic (*CNR1, FAAH*) variations (Ujike et al., 2002; Weiser and Noy, 2005), can exert an enduring impact on the establishment and functional specification of synaptic communication in neuronal circuitries subserving learning, memory formation, and motor control. Further identification of endocannabinoid ligands, metabolic enzymes, and receptors, together with defining the cellular context-specific recruitment of second messenger cascades and the affected gene sets, will allow us to understand microenvironmental requirements necessary for physiological endocannabinoid signaling to occur during brain development and will reveal the neural basis of developmental defects imposed by prenatal drug abuse.

Acknowledgments This work was supported by the Swedish Medical Research Council (T.H.), Alzheimer's Association (T.H.), CIBERNED Grant CIBER06/05/0005 (M.G.), and NIH Grants DA15446–04, DA12030–06, DA019350–02, and DA019348 (Y.L.H.).

References

Aguado T, Monory K, Palazuelos J, Stella N, Cravatt B, Lutz B, Marsicano G, Kokaia Z, Guzman M, Galve-Roperh I (2005) The endocannabinoid system drives neural progenitor proliferation. FASEB J 19:1704–1706.

Aguado T, Palazuelos J, Monory K, Stella N, Cravatt B, Lutz B, Marsicano G, Kokaia Z, Guzman M, Galve-Roperh I (2006) The endocannabinoid system promotes astroglial differentiation by acting on neural progenitor cells. J Neurosci 26:1551–1561.

Alger BE (2005) Endocannabinoid identification in the brain: studies of breakdown lead to breakthrough, and there may be NO hope. Sci STKE 2005:e51.

Antonelli T, Tomasini MC, Tattoli M, Cassano T, Tanganelli S, Finetti S, Mazzoni E, Trabace L, Steardo L, Cuomo V, Ferraro L (2005) Prenatal exposure to the CB_1 receptor agonist WIN 55,212-2 causes learning disruption associated with impaired cortical NMDA receptor function and emotional reactivity changes in rat offspring. Cereb Cortex 15:2013–2020.

Baker D, Pryce G, Davies WL, Hiley CR (2006) In silico patent searching reveals a new cannabinoid receptor. Trends Pharmacol Sci 27:1–4.

Berghuis P, Dobszay MB, Ibanez RM, Ernfors P, Harkany T (2004) Turning the heterogeneous into homogeneous: studies on selectively isolated GABAergic interneuron subsets. Int J Dev Neurosci 22:533–543.

Berghuis P, Dobszay MB, Wang X, Spano S, Ledda F, Sousa KM, Schulte G, Ernfors P, Mackie K, Paratcha G, Hurd YL, Harkany T (2005) Endocannabinoids regulate interneuron migration and morphogenesis by transactivating the TrkB receptor. Proc Natl Acad Sci USA 102:19115–19120.

Berghuis P, Rajnicek AM, Morozov YM, Ross RA, Mulder J, Urban GM, Monory K, Marsicano G, Matteoli M, Canty A, Irving AJ, Katona I, Yanagawa Y, Rakic P, Lutz B, Mackie K, Harkany T (2007) Hardwiring the brain: endocannabinoids shape neuronal connectivity. Science 316:1212–1216.

Bernard C, Milh M, Morozov YM, Ben Ari Y, Freund TF, Gozlan H (2005) Altering cannabinoid signaling during development disrupts neuronal activity. Proc Natl Acad Sci USA 102:9388–9393.

Berrendero F, Sepe N, Ramos JA, Di Marzo V, Fernandez-Ruiz JJ (1999) Analysis of cannabinoid receptor binding and mRNA expression and endogenous cannabinoid contents in the developing rat brain during late gestation and early postnatal period. Synapse 33:181–191.

Bilkei-Gorzo A, Racz I, Valverde O, Otto M, Michel K, Sastre M, Zimmer A (2005) Early age-related cognitive impairment in mice lacking cannabinoid CB1 receptors. Proc Natl Acad Sci USA 102:15670–15675.

Bisogno T, Howell F, Williams G, Minassi A, Cascio MG, Ligresti A, Matias I, Schiano-Moriello A, Paul P, Williams EJ, Gangadharan U, Hobbs C, Di Marzo V, Doherty P (2003) Cloning of the first sn1-DAG lipases points to the spatial and temporal regulation of endocannabinoid signaling in the brain. J Cell Biol 163:463–468.

Cravatt BF, Giang DK, Mayfield SP, Boger DL, Lerner RA, Gilula NB (1996) Molecular characterization of an enzyme that degrades neuromodulatory fatty-acid amides. Nature 384:83–87.

Curtis MA, Faull RL, Glass M (2006) A novel population of progenitor cells expressing cannabinoid receptors in the subependymal layer of the adult normal and Huntington's disease human brain. J Chem Neuroanat 31:210–215.

Devi LA (2000) G-protein-coupled receptor dimers in the lime light. Trends Pharmacol Sci 21:324–326.

Dijkhuizen PA, Ghosh A (2005) BDNF regulates primary dendrite formation in cortical neurons via the PI3-kinase and MAP kinase signaling pathways. J Neurobiol 62:278–288.

Di Marzo V, Berrendero F, Bisogno T, Gonzalez S, Cavaliere P, Romero J, Cebeira M, Ramos JA, Fernandez-Ruiz JJ (2000) Enhancement of anandamide formation in the limbic forebrain and reduction of endocannabinoid contents in the striatum of delta9-tetrahydrocannabinol-tolerant rats. J Neurochem 74:1627–1635.

Dinh TP, Carpenter D, Leslie FM, Freund TF, Katona I, Sensi SL, Kathuria S, Piomelli D (2002) Brain monoglyceride lipase participating in endocannabinoid inactivation. Proc Natl Acad Sci USA 99:10819–10824.

Fernandez-Ruiz J, Berrendero F, Hernandez ML, Ramos JA (2000) The endogenous cannabinoid system and brain development. Trends Neurosci 23:14–20.

Fernandez-Ruiz J, Romero J, Velasco G, Tolon RM, Ramos JA, Guzman M (2007) Cannabinoid CB_2 receptor: a new target for the control of neural cell survival? Trends Pharmacol Sci 28:39–45.

Freund TF, Katona I, Piomelli D (2003) Role of endogenous cannabinoids in synaptic signaling. Physiol Rev 83:1017–1066.

Fried PA, Smith AM (2001) A literature review of the consequences of prenatal marihuana exposure. An emerging theme of a deficiency in aspects of executive function. Neurotoxicol Teratol 23:1–11.

Fried PA, Watkinson B, Gray R (2003) Differential effects on cognitive functioning in 13- to 16-year-olds prenatally exposed to cigarettes and marihuana. Neurotoxicol Teratol 25:427–436.

Fukumitsu H, Ohtsuka M, Murai R, Nakamura H, Itoh K, Furukawa S (2006) Brain-derived neurotrophic factor participates in determination of neuronal laminar fate in the developing mouse cerebral cortex. J Neurosci 26:13218–13230.

Gage FH (2002) Neurogenesis in the adult brain. J Neurosci 22:612–613.

Galve-Roperh I, Sanchez C, Cortes ML, del Pulgar TG, Izquierdo M, Guzman M (2000) Anti-tumoral action of cannabinoids: involvement of sustained ceramide accumulation and extracellular signal-regulated kinase activation. Nat Med 6:313–319.

Galve-Roperh I, Aguado T, Rueda D, Velasco G, Guzman M (2006) Endocannabinoids: a new family of lipid mediators involved in the regulation of neural cell development. Curr Pharm Des 12:2319–2325.

Galve-Roperh I, Aguado T, Palazuelos J, Guzman M (2007) The endocannabinoid system and neurogenesis in health and disease. Neuroscientist 13:109–114.

Garcia-Gil L, Romero J, Ramos JA, Fernandez-Ruiz JJ (1999) Cannabinoid receptor binding and mRNA levels in several brain regions of adult male and female rats perinatally exposed to delta^9-tetrahydrocannabinol. Drug Alcohol Depend 55:127–136.

Goldschmidt L, Richardson GA, Cornelius MD, Day NL (2004) Prenatal marijuana and alcohol exposure and academic achievement at age 10. Neurotoxicol Teratol 26:521–532.

Gray KA, Day NL, Leech S, Richardson GA (2005) Prenatal marijuana exposure: effect on child depressive symptoms at ten years of age. Neurotoxicol Teratol 27:439–448.

Guzman M (2003) Cannabinoids: potential anticancer agents. Nat Rev Cancer 3:745–755.

Guzman M, Sanchez C, Galve-Roperh I (2002) Cannabinoids and cell fate. Pharmacol Ther 95:175–184.

Harkany T, Hartig W, Berghuis P, Dobszay MB, Zilberter Y, Edwards RH, Mackie K, Ernfors P (2003) Complementary distribution of type 1 cannabinoid receptors and vesicular glutamate transporter 3 in basal forebrain suggests input-specific retrograde signalling by cholinergic neurons. Eur J Neurosci 18:1979–1992.

Harkany T, Andang M, Kingma HJ, Gorcs TJ, Holmgren CD, Zilberter Y, Ernfors P (2004) Region-specific generation of functional neurons from naive embryonic stem cells in adult brain. J Neurochem 88:1229–1239.

Harkany T, Dobszay MB, Cayetanot F, Hartig W, Siegemund T, Aujard F, Mackie K (2005) Redistribution of CB_1 cannabinoid receptors during evolution of cholinergic basal forebrain territories and their cortical projection areas: a comparison between the gray mouse lemur (Microcebus murinus, primates) and rat. Neuroscience 135:595–609.

Harkany T, Guzman M, Galve-Roperh I, Berghuis P, Devi LA, Mackie K (2007) The emerging functions of endocannabinoid signaling during CNS development. Trends Pharmacol Sci 28:83–92.

Hart S, Fischer OM, Ullrich A (2004) Cannabinoids induce cancer cell proliferation via tumor necrosis factor alpha-converting enzyme (TACE/ADAM17)-mediated transactivation of the epidermal growth factor receptor. Cancer Res 64:1943–1950.

He JC, Gomes I, Nguyen T, Jayaram G, Ram PT, Devi LA, Iyengar R (2005) The G alpha(o/i)-coupled cannabinoid receptor-mediated neurite outgrowth involves Rap regulation of Src and Stat3. J Biol Chem 280:33426–33434.

Horch HW, Katz LC (2002) BDNF release from single cells elicits local dendritic growth in nearby neurons. Nat Neurosci 5:1177–1184.

Huizink AC, Mulder EJ (2006) Maternal smoking, drinking or cannabis use during pregnancy and neurobehavioral and cognitive functioning in human offspring. Neurosci Biobehav Rev 30:24–41.

Hunter SA, Burstein SH (1997) Receptor mediation in cannabinoid stimulated arachidonic acid mobilization and anandamide synthesis. Life Sci 60:1563–1573.

Hutchings DE, Martin BR, Gamagaris Z, Miller N, Fico T (1989) Plasma concentrations of delta-9-tetrahydrocannabinol in dams and fetuses following acute or multiple prenatal dosing in rats. Life Sci 44:697–701.

Ishii I, Chun J (2002) Anandamide-induced neuroblastoma cell rounding via the CB_1 cannabinoid receptors. Neuroreport 13:593–596.

Iyengar R (2005) Teaching resources. Introduction: overview of pathways and networks and GPCR signaling. Sci STKE 2005:tr4.

Jacobsen LK, Pugh KR, Constable RT, Westerveld M, Mencl WE (2006) Functional correlates of verbal memory deficits emerging during nicotine withdrawal in abstinent adolescent cannabis users. Biol Psychiatry 61:31–40.

Jaffe AB, Hall A (2005) Rho GTPases: biochemistry and biology. Annu Rev Cell Dev Biol 21:247–269.

Jiang W, Zhang Y, Xiao L, Van Cleemput J, Ji SP, Bai G, Zhang X (2005) Cannabinoids promote embryonic and adult hippocampus neurogenesis and produce anxiolytic- and antidepressant-like effects. J Clin Invest 115:3104–3116.

Jin K, Xie L, Kim SH, Parmentier-Batteur S, Sun Y, Mao XO, Childs J, Greenberg DA (2004) Defective adult neurogenesis in CB_1 cannabinoid receptor knockout mice. Mol Pharmacol 66:204–208.

Jordan JD, He JC, Eungdamrong NJ, Gomes I, Ali W, Nguyen T, Bivona TG, Philips MR, Devi LA, Iyengar R (2005) Cannabinoid receptor-induced neurite outgrowth is mediated by Rap1 activation through $G_{\alpha o/i}$-triggered proteasomal degradation of Rap1GAPII. J Biol Chem 280:11413–11421.

Katona I, Sperlagh B, Sik A, Köfalvi A, Vizi ES, Mackie K, Freund TF (1999) Presynaptically located CB_1 cannabinoid receptors regulate GABA release from axon terminals of specific hippocampal interneurons. J Neurosci 19:4544–4558.

Katona I, Sperlagh B, Magloczky Z, Santha E, Köfalvi A, Czirjak S, Mackie K, Vizi ES, Freund TF (2000) GABAergic interneurons are the targets of cannabinoid actions in the human hippocampus. Neuroscience 100:797–804.

Katona I, Urban GM, Wallace M, Ledent C, Jung KM, Piomelli D, Mackie K, Freund TF (2006) Molecular composition of the endocannabinoid system at glutamatergic synapses. J Neurosci 26:5628–5637.

Kearn CS, Blake-Palmer K, Daniel E, Mackie K, Glass M (2005) Concurrent stimulation of cannabinoid CB_1 and dopamine D_2 receptors enhances heterodimer formation: a mechanism for receptor cross-talk? Mol Pharmacol 67:1697–1704.

Kempermann G, Wiskott L, Gage FH (2004) Functional significance of adult neurogenesis. Curr Opin Neurobiol 14:186–191.

Khaspekov LG, Brenz Verca MS, Frumkina LE, Hermann H, Marsicano G, Lutz B (2004) Involvement of brain-derived neurotrophic factor in cannabinoid receptor-dependent protection against excitotoxicity. Eur J Neurosci 19:1691–1698.

Kim D, Thayer SA (2001) Cannabinoids inhibit the formation of new synapses between hippocampal neurons in culture. J Neurosci 21:RC146.

Kim SH, Won SJ, Mao XO, Ledent C, Jin K, Greenberg DA (2006) Role for neuronal nitric-oxide synthase in cannabinoid-induced neurogenesis. J Pharmacol Exp Ther 319:150–154.

Kreitzer AC, Regehr WG (2001) Cerebellar depolarization-induced suppression of inhibition is mediated by endogenous cannabinoids. J Neurosci 21:RC174.

Lauckner JE, Hille B, Mackie K (2005) The cannabinoid agonist WIN55,212-2 increases intracellular calcium via CB_1 receptor coupling to $G_{q/11}$ G proteins. Proc Natl Acad Sci USA 102:19144–19149.

Lie DC, Song H, Colamarino SA, Ming GL, Gage FH (2004) Neurogenesis in the adult brain: new strategies for central nervous system diseases. Annu Rev Pharmacol Toxicol 44:399–421.

Liu J, Wang L, Harvey-White J, Osei-Hyiaman D, Razdan R, Gong Q, Chan AC, Zhou Z, Huang BX, Kim HY, Kunos G (2006) A biosynthetic pathway for anandamide. Proc Natl Acad Sci USA 103:13345–13350.

Llano I, Leresche N, Marty A (1991) Calcium entry increases the sensitivity of cerebellar Purkinje cells to applied GABA and decreases inhibitory synaptic currents. Neuron 6:565–574.

Mackie K, Stella N (2006) Cannabinoid receptors and endocannabinoids: evidence for new players. AAPS J 8:E298–E306.

Marsicano G, Goodenough S, Monory K, Hermann H, Eder M, Cannich A, Azad SC, Cascio MG, Gutierrez SO, van der SM, Lopez-Rodriguez ML, Casanova E, Schutz G, Zieglgansberger W, Di Marzo V, Behl C, Lutz B (2003) CB_1 cannabinoid receptors and on-demand defense against excitotoxicity. Science 302:84–88.

Mato S, Del Olmo E, Pazos A (2003) Ontogenetic development of cannabinoid receptor expression and signal transduction functionality in the human brain. Eur J Neurosci 17:1747–1754.

Mereu G, Fa M, Ferraro L, Cagiano R, Antonelli T, Tattoli M, Ghiglieri V, Tanganelli S, Gessa GL, Cuomo V (2003) Prenatal exposure to a cannabinoid agonist produces memory deficits linked to dysfunction in hippocampal long-term potentiation and glutamate release. Proc Natl Acad Sci USA 100:4915–4920.

Molina-Holgado E, Vela JM, Arevalo-Martin A, Almazan G, Molina-Holgado F, Borrell J, Guaza C (2002) Cannabinoids promote oligodendrocyte progenitor survival: involvement of cannabinoid receptors and phosphatidylinositol-3 kinase/Akt signaling. J Neurosci 22:9742–9753.

Morishita J, Okamoto Y, Tsuboi K, Ueno M, Sakamoto H, Maekawa N, Ueda N (2005) Regional distribution and age-dependent expression of N-acylphosphatidylethanolamine-hydrolyzing phospholipase D in rat brain. J Neurochem 94:753–762.

Morozov YM, Ben Ari Y, Freund TF (2004) The spatial and temporal pattern of fatty acid amide hydrolase expression in rat hippocampus during postnatal development. Eur J Neurosci 20:459–466.

Muccioli GG, Xu C, Odah E, Cudaback E, Cisneros JA, Lambert DM, Lopez Rodriguez ML, Bajjalieh S, Stella N (2007) Identification of a novel endocannabinoid-hydrolyzing enzyme expressed by microglial cells. J Neurosci 27:2883–2889.

Palazuelos J, Aguado T, Egia A, Mechoulam R, Guzman M, Galve-Roperh I (2006) Non-psychoactive CB2 cannabinoid agonists stimulate neural progenitor proliferation. FASEB J 20:2405–2407.

Paria BC, Song H, Wang X, Schmid PC, Krebsbach RJ, Schmid HH, Bonner TI, Zimmer A, Dey SK (2001) Dysregulated cannabinoid signaling disrupts uterine receptivity for embryo implantation. J Biol Chem 276:20523–20528.

Pertwee RG (2006) The pharmacology of cannabinoid receptors and their ligands: an overview. Int J Obes (Lond) 1:S13–S18.

Piomelli D (2003) The molecular logic of endocannabinoid signalling. Nat Rev Neurosci 4:873–884.

Pitler TA, Alger BE (1992) Postsynaptic spike firing reduces synaptic $GABA_A$ responses in hippocampal pyramidal cells. J Neurosci 12:4122–4132.

Richardson GA, Day NL, Goldschmidt L (1995) Prenatal alcohol, marijuana, and tobacco use: infant mental and motor development. Neurotoxicol Teratol 17:479–487.

Rios C, Gomes I, Devi LA (2006) Mu opioid and CB1 cannabinoid receptor interactions: reciprocal inhibition of receptor signaling and neuritogenesis. Br J Pharmacol 148:387–395.

Romero J, Garcia-Palomero E, Berrendero F, Garcia-Gil L, Hernandez ML, Ramos JA, Fernandez-Ruiz JJ (1997) Atypical location of cannabinoid receptors in white matter areas during rat brain development. Synapse 26:317–323.

Rueda D, Navarro B, Martinez-Serrano A, Guzman M, Galve-Roperh I (2002) The endocannabinoid anandamide inhibits neuronal progenitor cell differentiation through attenuation of the Rap1/B-Raf/ERK pathway. J Biol Chem 277:46645–46650.

Sim LJ, Hampson RE, Deadwyler SA, Childers SR (1996) Effects of chronic treatment with delta9-tetrahydrocannabinol on cannabinoid-stimulated [^{35}S]GTPgammaS autoradiography in rat brain. J Neurosci 16:8057–8066.

Simon GM, Cravatt BF (2006) Endocannabinoid biosynthesis proceeding through glycerophospho-N-acyl ethanolamine and a role for {alpha}/beta-hydrolase 4 in this pathway. J Biol Chem 281:26465–26472.

Song ZH, Zhong M (2000) CB_1 cannabinoid receptor-mediated cell migration. J Pharmacol Exp Ther 294:204–209.

Spano MS, Ellgren M, Wang X, Hurd YL (2007) Prenatal cannabis exposure increases heroin seeking with allostatic changes in limbic enkephalin systems in adulthood. Biol Psychiatry 61:554–563.

Suarez I, Bodega G, Fernandez-Ruiz JJ, Ramos JA, Rubio M, Fernandez B (2002) Reduced glial fibrillary acidic protein and glutamine synthetase expression in astrocytes and Bergmann glial cells in the rat cerebellum caused by delta^9-tetrahydrocannabinol administration during development. Dev Neurosci 24:300–312.

Suarez I, Bodega G, Fernandez-Ruiz J, Ramos JA, Rubio M, Fernandez B (2004a) Down-regulation of the AMPA glutamate receptor subunits GluR1 and GluR2/3 in the rat cerebellum following pre- and perinatal delta9-tetrahydrocannabinol exposure. Cerebellum 3:66–74.

Suarez I, Bodega G, Rubio M, Fernandez-Ruiz JJ, Ramos JA, Fernandez B (2004b) Prenatal cannabinoid exposure down-regulates glutamate transporter expressions (GLAST and EAAC1) in the rat cerebellum. Dev Neurosci 26:45–53.

Toni N, Teng EM, Bushong EA, Aimone JB, Zhao C, Consiglio A, van Praag H, Martone ME, Ellisman MH, Gage FH (2007) Synapse formation on neurons born in the adult hippocampus. Nat Neurosci 10:727–734.

Uchigashima M, Narushima M, Fukaya M, Katona I, Kano M, Watanabe M (2007) Subcellular arrangement of molecules for 2-arachidonoyl-glycerol-mediated retrograde signaling and its physiological contribution to synaptic modulation in the striatum. J Neurosci 27:3663–3676.

Ujike H, Takaki M, Nakata K, Tanaka Y, Takeda T, Kodama M, Fujiwara Y, Sakai A, Kuroda S (2002) *CNR1*, central cannabinoid receptor gene, associated with susceptibility to hebephrenic schizophrenia. Mol Psychiatry 7:515–518.

van der Stelt M, Di Marzo V (2005) Anandamide as an intracellular messenger regulating ion channel activity. Prostaglandins Other Lipid Mediat 77:111–122.

Van Sickle MD, Duncan M, Kingsley PJ, Mouihate A, Urbani P, Mackie K, Stella N, Makriyannis A, Piomelli D, Davison JS, Marnett LJ, Di Marzo V, Pittman QJ, Patel KD, Sharkey KA (2005) Identification and functional characterization of brainstem cannabinoid CB_2 receptors. Science 310:329–332.

Ventimiglia R, Mather PE, Jones BE, Lindsay RM (1995) The neurotrophins BDNF, NT-3 and NT-4/5 promote survival and morphological and biochemical differentiation of striatal neurons *in vitro*. Eur J Neurosci 7:213–222.

Wager-Miller J, Westenbroek R, Mackie K (2002) Dimerization of G protein-coupled receptors: CB_1 cannabinoid receptors as an example. Chem Phys Lipids 121:83–89.

Wang X, Dow-Edwards D, Keller E, Hurd YL (2003) Preferential limbic expression of the cannabinoid receptor mRNA in the human fetal brain. Neuroscience 118:681–694.

Wang X, Dow-Edwards D, Anderson V, Minkoff H, Hurd YL (2004) In utero marijuana exposure associated with abnormal amygdala dopamine D2 gene expression in the human fetus. Biol Psychiatry 56:909–915.

Weiser M, Noy S (2005) Interpreting the association between cannabis use and increased risk for schizophrenia. Dialogues Clin Neurosci 7:81–85.

Williams EJ, Walsh FS, Doherty P (2003) The FGF receptor uses the endocannabinoid signaling system to couple to an axonal growth response. J Cell Biol 160:481–486.

Wilson RI, Nicoll RA (2001) Endogenous cannabinoids mediate retrograde signalling at hippocampal synapses. Nature 410:588–592.

Zhuang S, Kittler J, Grigorenko EV, Kirby MT, Sim LJ, Hampson RE, Childers SR, Deadwyler SA (1998) Effects of long-term exposure to delta^9-THC on expression of cannabinoid receptor (CB_1) mRNA in different rat brain regions. Brain Res Mol Brain Res 62:141–149.

Zimmer A, Zimmer AM, Hohmann AG, Herkenham M, Bonner TI (1999) Increased mortality, hypoactivity, and hypoalgesia in cannabinoid CB_1 receptor knockout mice. Proc Natl Acad Sci USA 96:5780–5785.

Part II
The Endocannabinoid System in Clinical Neuroscience and Experimental Neuropsychiatry

Chapter 13
Cannabinoids in the Management of Nausea and Vomiting

Linda A. Parker and Cheryl L. Limebeer

Abstract With the discovery of the endocannabinoid system, research investigating the role that this system plays in the control of nausea and vomiting has accelerated. In this chapter, we review some of the evidence in both human clinical trials literature and animal literature demonstrating the potential of cannabinoids to modify nausea and vomiting.

Introduction

Nausea and vomiting are common symptoms reported by patients. They can occur separately and together in many different diseases and are side effects of many drug treatments. Understanding the neurobiological mechanisms responsible for the sensation of nausea and for the reflex of vomiting is important for the development of antiemetic and antinausea treatments. The emetic reflex is conventionally considered to include vomiting, retching, and the more subjective sensation of nausea. However, the organization of the reflex is very complex, because although nausea, retching, and vomiting usually occur in a temporal sequence, they can be separated experimentally (Andrews and Davis, 1995). Although the physiology of vomiting is well understood, the same is not true of nausea (Andrews and Horn, 2006). Chemotherapy treatment for cancer is often accompanied by the serious side effects of nausea and vomiting which may interfere with the completion of treatment. Chemotherapy patients experience three separate types of emetic episodes: (1) acute nausea and/or vomiting that occurs within minutes to hours of receiving a dose of a toxic chemotherapy drug; (2) delayed nausea and/or vomiting that has been arbitrarily defined as emesis that begins or persists more than 24 h after chemotherapy; (3) anticipatory nausea and/or vomiting that occurs when the patient is reexposed to cues associated with the toxin. Anticipatory nausea/vomiting occurs in nearly half of the patients treated, frequently during later cycles of chemotherapy. The more intense the initial acute emetic episode is, the worse is the resultant anticipatory nausea/vomiting (Aapro et al., 1994). A major advance in the control of acute emesis in chemotherapy treatment was the finding that blockade of one subtype of the 5-hydroxytryptamine (5-HT) receptor, the 5-HT_3 receptor, could

A. Kofalvi (ed.), *Cannabinoids and the Brain.*

suppress the acute emetic response (retching and vomiting) induced by cisplatin in the ferret and the shrew (Costall et al., 1986; Miner and Sanger, 1986; Ueno et al., 1987; Matsuki et al., 1988; Torii et al., 1991). In clinical trials with humans, the treatment with 5-HT_3 antagonists often combined with the corticosteroid, dexamethasone, during the first chemotherapy treatment has reduced the incidence of acute vomiting by 70–90% (Reynolds et al., 1991; Tsukada et al., 2001; Bartlett and Koczwara, 2002; Aapro et al., 2003; Ballatori and Roila, 2003; Hickock et al., 2003). If acute vomiting is prevented, the incidence of delayed and anticipatory vomiting is reduced (Aapro et al., 1994). However, the 5-HT_3 antagonists are less effective in suppressing acute nausea than they are in suppressing acute vomiting (Andrews and Davis, 1995; Morrow and Dobkin, 1988; Barlett and Koczwara, 2002; Ballatori and Roila, 2003; Hickok et al., 2003) and they are ineffective in reducing instances of delayed nausea/vomiting (Morrow and Dobkin, 1988; Grelot et al., 1995; Rudd and Naylor, 1996; Rudd et al., 1996; Tsukada et al., 2001; Hesketh et al., 2003) and anticipatory nausea/vomiting (Nesse et al., 1980; Morrow and Dobkin, 1988; Reynolds et al., 1991; Stockhorst et al., 1993; Ballatori and Roila, 2003; Hickok et al., 2003) when they do occur. Therefore, it is likely that another system may be involved in chemotherapy-induced nausea, delayed nausea/vomiting, and anticipatory nausea/vomiting. Two such systems include the Neurokinin 1 (NK_1) tachykinin receptors for substance P (e.g., Rudd and Naylor, 1996; Rudd et al., 1996; Hesketh et al., 2003) and the endocannabinoid system. The effect of cannabinoids on nausea and vomiting is the subject of this review.

Antiemetic Effects of Cannabinoids in Human Clinical Trials

The marijuana plant has been used for several centuries for a number of therapeutic results, including nausea and vomiting. Ineffective treatment of chemotherapy-induced nausea prompted oncologists to investigate the antiemetic properties of cannabinoids in the late 1970s and early 1980s. In these early studies, several clinical trials have compared the effectiveness of Δ^9-THC with placebo or other antiemetic drugs. Comparisons of oral Δ^9-THC with existing antiemetic agents generally indicated that Δ^9-THC was at least as effective as the dopamine antagonist, prochlorperazine (Carey et al., 1983; Ungerleider et al., 1984; Tramer et al., 2001). Three cannabis-based medicines are available: Dronabinol™, Nabilone™, and levonantradol. A systematic review (Tramer et al., 2001) found that oral Nabilone, Dronabinol, and intramuscular levonantradol were more effective than other antiemetics after mild to moderately emetogeneic chemotherapy, but were less effective after highly emetogenic chemotherapy compared with the dopamine antagonist, metoclopramide (Crawford and Buckman, 1986; Cunningham et al., 1988). Withdrawal rates from these trials indicated a narrow therapeutic dose range of effectiveness suggesting a need to carefully titrate the dose. Since these earlier trials, the more effective 5-HT_3 antagonist antiemetic drugs have been developed that reduce acute vomiting during cancer chemotherapy. Additionally, NK_1 receptor antagonists (Aprepitant) have been developed which decrease both acute and delayed emesis from cisplatin-based chemotherapy (Van Belle et al., 2002). To date, clinical trials have not compared the antiemetic effects of cannabinoids with the newer agents. Furthermore, all of the earlier studies involved oral use of cannabinoids which may be less effective than sublingual or inhaled cannabinoids, given the need to titrate the dose (Hall et al., 2005). Because the mechanisms of cannabinoid-induced antiemesis differ from other agents, they may benefit unresponsive patients or may even be found to synergistically facilitate the effects of 5-HT_3 antagonists as is suggested from the animal literature (Kwiatkowska et al., 2004). Cannabinoids produce psychotropic side effects, which partially accounts for their lack of popularity in clinical use (Schwartz and Beveridge, 1994). Patients who have not had any experience with cannabis often find the psychotropic effects unpleasant and disturbing. Most importantly, the development of 5-HT_3 antagonist antiemetic drugs has limited clinical use of cannabis-based medicines. However, 5-HT_3 antagonist antiemetic agents are not as effective in inhibiting nausea as they are in inhibiting vomiting and are ineffective in treating delayed nausea/vomiting or anticipatory nausea/vomiting (e.g., Hickok et al., 2003). There is some evidence that cannabis-based medicines may be effective in treating these more difficult-to-control symptoms. Abrahamov and colleagues (1995) evaluated the antiemetic effectiveness of Δ^8-THC, a close but less psychoactive relative of Δ^9-THC, in children receiving chemotherapy treatment. The children were given Δ^8-THC as oil drops on the tongue or in a bite of food 2 h before the start of each cancer treatment and every 6 h thereafter for 24 h. After a total of 480 treatments, the only side effects reported were slight irritability in two of the youngest children (3.5 and 4 years old); both acute and delayed nausea and vomiting were controlled. More recently, Layeeque and coworkers (2006) evaluated the potential of the combination of Dronabinol and prochlorperazine to reduce postoperative nausea and vomiting following general anesthesia. The rate of nausea and vomiting were improved in patients treated prophylactically with a combination of Dronabinol and prochlorperazine (59% vs. 15% and 29% and 3%, respectively). Many patients have a strong preference for smoked marijuana over the synthetic cannabinoids delivered orally (Tramer et al., 2001). Several reasons have been suggested: (1) advantages of self-titration with the smoked marijuana; (2) difficulty of swallowing the pills while experiencing emesis; (3) faster speed of onset for the inhaled or injected Δ^9-THC than oral delivery; and (4) a combination of the action of other cannabinoids with Δ^9-THC that are found in marijuana. Although many marijuana users have claimed that smoked marijuana is a more effective antiemetic than oral Δ^9-THC, no controlled studies have yet been published that evaluate this possibility. Smoking marijuana may represent a more efficient and rapid route of administration. In addition, Δ^9-THC is only one of over 60 different compounds found in smoked marijuana and some of the additional constituents may contribute to the antiemetic/antinausea effect. Another major cannabinoid found in marijuana is cannabidiol (CBD). Unlike Δ^9-THC, CBD does not produce psychomimetic effects (Mechoulam et al., 2002). In shrews, CBD inhibits cisplatin-induced (Kwiatkowska et al., 2004) and lithium-induced (Parker et al., 2003a) emesis and in rats CBD inhibits nausea as reflected by the conditioned gaping response (Parker et al., 2002b). CBD also interferes with anticipatory retching in shrews (Parker et al., 2006) and with anticipatory gaping (reflective of nausea) in rats (Limebeer et al., 2007). This effect does not appear to be mediated by the action of CBD on CB_1 receptors, because, unlike Δ^9-THC, CBD does not bind to them. It may act by blocking the reuptake of anandamide (an endogenous cannabinoid), or by inhibiting enzymatic hydrolysis of anandamide, or binding with some as of the yet unknown cannabinoid receptor (Mechoulam et al., 2002). Recent evidence indicates that CBD may act as an agonist on the 5-HT_{1A} autoreceptors acting to reduce the availability of 5-HT (Russo et al., 2005). Additionally, CBD has been shown to antagonize the ability of WIN55212-2 to inhibit electrically evoked concentrations of the mouse vas deferens in a manner that appears to be competitive, but does not involve direct competition for CB_1 receptors (Thomas et al., 2004). CBD may also act as an adenosine reuptake inhibitor (Carrier et al., 2006). In mice, CBD is also a highly effective anti-inflammatory agent (Malfait et al., 2000), as well as a neuroprotective antioxidant (Hampson et al., 1998).

Antiemetic Effects of Cannabinoids: Mechanism of Action

The mechanism of action of the suppression of nausea and vomiting produced by cannabinoids has only recently been explored with the discovery of the endocannabinoid system and the development of animal models of nausea and vomiting. Recent reviews on the gastrointestinal effects of cannabinoids have concluded that cannabinoid agonists act mainly via peripheral CB_1 receptors to decrease intestinal motility (Pertwee, 2001), but may act centrally to attenuate emesis (Van Sickle et al., 2001). The dorsal vagal complex (DVC) is involved in the nausea and/or vomiting reactions induced by either vagal gastrointestinal activation or several humoral cytotoxic agents. It is considered the starting point of a final common pathway for the induction of emesis in vomiting species. The DVC consists of the area postrema (AP), nucleus of the solitary tract (NTS), and the dorsal motor nucleus of the vagus (DMNX) in the brainstem of rats, ferrets, and the least shrews. CB_1 receptors, as well as FAAH, have been found in areas of the brain involved in emesis, including the DMNX (Van Sickle et al., 2001). CB_1 receptors in the NTS are activated by Δ^9-THC, and this activation is blocked by the selective CB_1 antagonist/inverse agonists, SR141716A (Rimonabant, Acomplia™; Darmani et al., 2005) and AM251 (Van Sickle et al., 2003). In fact, at higher doses than required to reverse the antiemetic effects of Δ^9-THC, SR141716A produces emesis on its own in the least shrew (Darmani, 2001c) and AM251 potentiates cisplatin-induced emesis in the ferret (Van Sickle et al., 2001). Molecular markers of activation also implicate the role of central CB_1 receptors in the antiemetic effects of Δ^9-THC. Cisplatin pretreatment results in *c-fos* expression in the DMNX, specific subnuclei of the NTS and AP, which is significantly reduced by pretreatment with Δ^9-THC (Van Sickle et al., 2001, 2003). Endogenous cannabinoid ligands, such as anandamide, as well as synthetic cannabinoids, such as WIN55212-2, also act on these receptors (Simoneau et al., 2001). However, Darmani and Johnson (2004) provide evidence that both central and peripheral mechanisms contribute to the antiemetic actions of Δ^9-THC against emesis produced by 5-hydroxytryptophan (5-HTP), the precursor to 5-HT in the least shrew. At lower doses, Δ^9-THC acted centrally as an antiemetic, but at higher doses (10 mg/kg) it acted peripherally. Although anandamide has been reported to have antiemetic properties in the ferret (Van Sickle et al., 2001), 2-AG has emetic properties, most likely via its downstream metabolites (arachidonic acid and prostaglandins), because its emetic activity can be blocked by the COX-2 inhibitor, indomethacin (Darmani, 2002). An evaluation of changes in endocannabinoid levels in response elicited by cisplatin revealed that cisplatin increased levels of 2-AG in the brainstem, but decreased intestinal levels of both 2-AG and AEA (Darmani et al., 2005). Darmani and colleagues (2005) suggested that the central elevation of 2-AG may contribute to the emetic potential of cisplatin (in addition to mobilizing the release of known emetic stimuli such as serotonin, dopamine, and substance P). Most recently, Van Sickle and coworkers (2005) reported the presence of CB_2 receptors on DMNX neurons in both the rat and the ferret that were activated by a CB_2 receptor agonist, 2-arachidonoylglycerol, and by elevated levels of endocannabinoids which also act on CB_1 receptors. The action of 2-AG on these receptors was blocked by pretreatment with the CB_2 antagonist AM630. The authors suggest that "the brainstem receptors are functionally coupled to inhibition of emesis when costimulated with CB_1 receptors by an endogenous cannabinoid capable of activating both receptors." Recent findings indicate that the cannabinoid system interacts with the serotonergic system in the control of emesis. The DVC not only contains CB_1 receptors, but also is densely populated with 5-HT_3 receptors (Himmi et al., 1996, 1998), potentially a site of antiemetic effects of 5-HT_3 antagonists. Anandamide has also been reported to interact with serotonin (Kimura et al., 1998). Cannabinoid receptors are coexpressed with serotonin 5-HT_3 receptors in some neurons in the CNS (Hermann et al., 2002; see Chap. 10) and inhibitory functional interactions have been reported between cannabinoid CB_1 and 5-HT_3 receptors (Fan, 1995; Barann et al., 2002). Additionally, cannabinoids have been shown to reduce the ability of 5-HT_3 agonists to produce emesis (Darmani and Johnson, 2004) and this effect was prevented by pretreatment with SR141716A. Cannabinoids may act at CB_1 presynaptic receptors to inhibit release of newly synthesized serotonin (Schlicker and Kathman, 2001; Howlett et al., 2002; Darmani and Johnson, 2004). Indeed, Darmani and colleagues (2003) report that SR141716A (which produces vomiting in the least shrew) increases brain serotonin and turnover at doses that induced vomiting in the shrew. Furthermore, the antiemetic effects of CBD may be mediated by its ability to act as an agonist on the 5-HT_{1A} autoreceptors reducing the availability of 5-HT (Russo et al., 2005).

Effects of Cannabinoids on Vomiting in Animal Models

To evaluate the antiemetic potential of drug therapies, animal models have been developed. Since rats and mice do not vomit in response to a toxin challenge, it was necessary to develop other animal models on vomiting. As indicated in Table 1 (from Parker et al., 2005), there is considerable evidence that cannabinoids attenuate

Table 1 Effect of cannabinoids on emesis across species

Species	Emetogen	Cannabinoid	Effect on emesis (↓: reduced; –: no effect)
Cat	Cisplatin (7.5 mg/kg, i.v.)	Nabilone (0.025–0.1 mg/kg, i.v.)	↓ McCarthy and Borison (1981)
		N-methyllevonantradol (0.003–0.02 mg/kg, i.v.)	↓ McCarthy and Borison (1981)
Dog	Cisplatin (3 mg/kg, i.v.)	Nabilone™ (0.1 mg/kg, i.v.)	– Gylys et al. (1979)
	Apomorphine (0.05–5 mg/kg, i.v.)	Δ^9-THC (0.003–0.3 mg/kg, i.v.)	– Shannon et al. (1978)
Pigeon	Cisplatin (10 mg/kg, i.v.)	Δ^9-THC (5.0 mg/kg) with $CuCl_2$	↓ Feigenbaum et al. (1989)
	Cisplatin (7.5 mg/kg, i.v.)	HU-211 (2.5 mg/kg) with $CuCl_2$	↓ Feigenbaum et al. (1989)
	Emetine (20 mg/kg, s.c.)	HU-210 (0.012–0.05 mg/kg, s.c.)	↓ Ferrari et al. (1999)
		HU-210 (0.012–0.05 mg/kg, s.c.)	↓ Ferrari et al. (1999)
Ferret	Morphine (1 mg/kg, s.c.)	WIN55212-2 (0.03–0.13 mg/kg, s.c.)	↓ Simoneau et al. (2001)
	Morphine-6-glucuronide (M6G)	Δ^9-THC (1 mg/kg, i.p.)	↓ Van Sickle et al. (2001)
	(0.05 mg/kg, s.c.)	WIN55212-2 (1 mg/kg, i.p.)	↓ Van Sickle et al. (2001)
	Cisplatin (10 mg/kg, i.v.)	methanandamide (3 mg/kg, i.p.)	↓ Van Sickle et al. (2001)
		Δ^9-THC (0.1–1.0 mg/kg, i.p.)	↓ Van Sickle et al. (2003)
Cryptotis parva- (least shrew)	SR141716A (20 mg/kg, i.p.)	CP55940 (1 mg/kg, i.p.)	↓ Darmani (2001a)
	Cisplatin (20 mg/kg, i.p.)	WIN55212-2 (10 mg/kg, i.p.)	↓ Darmani (2001c)
	2-AG (2.5–10 mg/kg, i.p.)	Δ^9-THC (20 mg/kg, i.p.)	↓ Darmani (2001a)
	5-HTP (100 mg/kg, i.p.)	Δ^9-THC (1–10 mg/kg, i.p.)	↓ Darmani (2001b)
	5-HT (5 mg/kg, i.p.)	WIN55212-2 (1–5 mg/kg, i.p.)	↓ Darmani (2001b)
	2-methylserotonin ($5\text{-}HT_3$ agonist)	CP55940 (0.025–0.3 mg/kg)	↓ Darmani et al. (2003)
	(5 mg/kg, i.p.)	CP55940 (0.05–0.1 mg/kg, i.p.)	↓ Darmani (2002)
		WIN55212-2 (1–5 mg/kg, i.p.)	↓ Darmani (2002)
		Δ^9-THC (2.5–5 mg/kg, i.p.)	↓ Darmani (2002)
		CBD (10–20 mg/kg, i.p.)	– Darmani (2002)

(continued)

Table 1 (continued)

Species	Emetogen	Cannabinoid	Effect on emesis (↓: reduced; –: no effect)
		anandamide (5 mg/kg, i.p.)	↓ Darmani (2002)
		methanandamide (10 mg/kg, i.p.)	↓ Darmani (2002)
		SR141716A (2.5–5 mg/kg, i.p.)	↓ Darmani (2002)
		Δ^9-THC (5–20 mg/kg, i.p.)	↓ Darmani and Johnson (2004)
		Δ^9-THC (20 mg/kg, i.p.)	↓ Darmani and Johnson (2004)
		Δ^9-THC (20 mg/kg, i.p.)	
		Δ^9-THC (20 mg/kg, i.p.)	↓ Darmani and Johnson (2004)
			↓ Darmani and Johnson (2004)
Suncus murinus (house musk shrew)	Cisplatin (20 mg/kg, i.p.)	Δ^9-THC (2.5–10 mg/kg, i.p.)	↓ Kwiatkowska et al. (2004)
	Lithium chloride (390 mg/kg, i.p.)	CBD (5–10 mg/kg, i.p.)	↓ Kwiatkowska et al. (2004)
		Δ^9-THC (3–20 mg/kg, i.p.)	↓ Parker et al. (2003a)
		CBD (5–10 mg/kg, i.p.)	↓ Parker et al. (2003a)

vomiting in emetic species. Cannabinoids have been shown to reduce vomiting in cats (McCarthy and Borrison, 1981), pigeons (Feigenbaum et al., 1989; Ferrari et al., 1999), ferrets (Simoneau et al., 2001; Van Sickle et al., 2001, 2003), least shrews, *Cryptotis parva* (Darmani, 2001a,b, 2002; Darmani and Johnson, 2004; Darmani et al., 2005), and the house musk shrews, *Suncus murinus* (Parker et al., 2003a; Kwiatkowska et al., 2004). This data has been reviewed by Parker and colleagues (2005) and is summarized in Table 1.

Effects of Cannabinoids on Nausea in Rats: Conditioned Gaping Model of Nausea

Nausea is more resistant to effective treatment with new antiemetic agents than is vomiting (e.g., Andrews and Horn, 2006) and therefore remains a significant problem in chemotherapy treatment and as a side effect from other pharmacological therapies, such as antidepressants. Even when the cisplatin-induced emetic response is blocked in the ferret by administration of a 5-HT_3 receptor antagonist, *c-fos* activation still occurs in the area postrema, suggesting that an action here may be responsible for some of the other effects of cytotoxic drugs, such as nausea or reduced food intake (Reynolds et al., 1991). In rats, the gastric afferents respond in the same manner to physical and chemical (intragastric copper sulfate and cisplatin) stimulation that precedes vomiting in ferrets (presumably resulting in nausea that precedes vomiting; Hillsley and Grundy, 1998; Billig et al., 2001). Furthermore, 5-HT_3 antagonists that block vomiting in ferrets also disrupt this preceding neural afferent reaction in rats. That is, in the rat, the detection mechanism of nausea is present, but the vomiting response is absent. Nauseogenic doses of CCK and LiCl induce specific patterns of brainstem and forebrain *c-fos* expression in ferrets that are similar to *c-fos* expression patterns in rats (Reynolds et al., 1991; Billig et al., 2001). In a classic review paper, Borrison and Wang (1953) suggest that the rats' inability to vomit can be explained as a species-adaptive neurological deficit and that, in response to emetic stimuli, the rat displays autonomic and behavioral signs corresponding to the presence of nausea, called the prodromata (salivation, papillary dilation, tachypnea, and tachycardia). Over the past number of years, our laboratory has provided considerable evidence that conditioned nausea in rats may be displayed as conditioned rejection reactions (Parker, 1982, 1995, 1998, 2003; Limebeer and Parker, 2000, 2003; Limebeer et al., 2004) using the Taste Reactivity (TR) test (Grill and Norgen, 1978). Rats display a distinctive pattern of rejection reactions (including gaping, chin rubbing, and paw treading) when they are intraorally infused with a bitter tasting quinine solution. This rejection pattern is also displayed to a sweet-tasting solution (that normally elicits hedonic reactions of tongue protrusions) when that solution is paired with a drug that produces vomiting (such as lithium chloride or cyclophosphamide) in species capable of vomiting. Only drugs with emetic properties produce this conditioned gaping reaction when paired with a taste. The most reliable conditioned rejection reaction in the rat is that of gaping (Breslin et al., 1992; Parker, 2003). If conditioned gaping reflects nausea in rats, then antinausea drugs should interfere with this reaction. Limebeer and Parker (2000) demonstrated that when administered prior to a saccharin–lithium pairing, the 5-HT_3 antagonist, ondansetron, prevented the establishment of conditioned gaping in rats, presumably by interfering with lithium-induced nausea. Since ondansetron did not modify unconditioned gaping elicited by bitter quinine solution, the effect was specific to nausea-induced gaping. Subsequently, Limebeer and Parker (2003) demonstrated a very similar pattern following pretreatment with the 5-HT_{1A} autoreceptor antagonist, 8-OH-DPAT, that also reduces serotonin availability and serves as an antiemetic agent in animal models. Most recently, Limebeer and colleagues (2004) report that lesions of the dorsal and median raphe that reduce forebrain serotonin availability interfere with the establishment of conditioned gaping consistent with reports that reduced serotonin availability interferes with nausea. Since rats are incapable of vomiting, we have argued that the gape represents an "incipient vomiting response."

Fig. 1 The rat gape is topographically similar to the shrew retch

As is evident in Fig. 1, the orofacial characteristics of the rat gape are very similar to those of the shrew retch (Parker, 2003). Indeed, Travers and Norgren (1986) suggest that the muscular movements involved in the gaping response mimic those seen in species capable of vomiting. Using the conditioned gaping measure of nausea in rats, we have demonstrated that a low dose (0.5 mg/kg, i.p.) of Δ^9-THC interferes with the establishment and the expression of cyclophosphamide-induced conditioned gaping (Limebeer and Parker, 1999). In addition, the nonintoxicating compound found in marijuana smoke, CBD (5 mg/kg, i.p.), as well as its synthetic dimethylheptyl homolog (5 mg/kg, i.p.), suppressed the establishment and the expression of lithium-induced conditioned gaping (Parker et al., 2002a,b). The potent agonist, HU-210 (0.001–0.01 mg/kg), also suppressed lithium-induced conditioned gaping (Parker and Mechoulam, 2003; Parker et al., 2003b) and this suppression was reversed by the CB_1 antagonist/reverse agonist, SR141716A, suggesting that the effect of HU-210 was mediated by its action at CB_1 receptors. When administered 30 min prior to the conditioning trial, SR141716A did not produce conditioned rejection on its own, but it did potentiate the ability of lithium to produce conditioned gaping. This same pattern has been reported in the emesis literature. Van Sickle and colleagues (2001) reported that although the CB_1 antagonist/reverse agonist AM251 did not produce vomiting on its own, it potentiated the ability of an emetic stimulus to produce vomiting in the ferret. Although low doses of the CB_1 antagonists/inverse agonists SR141716A (Parker et al., 2003b) and AM251 (McLaughlin et al., 2005) did not produce conditioned gaping on their own, higher doses of AM251 (>8 mg/kg, i.p.) produced conditioned gaping reflective of nausea.

This finding suggests that the appetite-suppressant effect of the newly marketed CB_1 antagonist/inverse agonist, Rimonabant, may be partially mediated by the side effect of nausea which is the most commonly reported side effect in human randomized control trials (Pi-Sunyer et al., 2006). Most recently, we evaluated the effect of the silent CB_1 antagonist, AM4113, that does not have inverse agonist properties and found that it did not produce conditioned gaping at doses that produced equivalent feeding suppression as evident with AM251 (Sink et al., 2007). AM251-induced conditioned gaping may thus be mediated by its inverse agonist properties. More compelling evidence that the endocannabinoid system may serve as a regulator of nausea is our recent finding that prolonging the duration of action of anandamide by pretreatment with URB597, a drug that inhibits the enzyme FAAH, also disrupts the establishment of lithium-induced conditioned disgust reactions in rats (Cross-Mellor et al., 2007). Rats pretreated with URB597 (0.3 mg/kg, i.p.) 2 h prior to a saccharin-lithium pairing displayed suppressed conditioned gaping reactions in a subsequent drug-free test. Rats given the combination of URB597 (0.3 mg/kg, i.p.) and anandamide (5 mg/kg, i.p.) displayed even greater suppression of conditioned gaping reactions.

Conditioned Retching in Shrews and Conditioned Gaping in Rats: A Model for Anticipatory Nausea

Anticipatory nausea often develops over the course of repeated chemotherapy sessions (Nesse et al., 1980; Morrow and Dobkin, 1988; Reynolds et al., 1991; Stockhorst et al., 1993; Aapro et al., 1994; Ballatori and Roila, 2003; Hickok et al., 2003). For instance, Nesse and colleagues (1980) described the case of a patient who had severe nausea and vomiting during repeated chemotherapy treatments. After his third treatment, the patient became nauseated as soon as he walked into the clinic building and noticed a "chemical smell," that of isopropyl alcohol. He experienced the same nausea when returning for routine follow-up visits, even though he knew he would not receive treatment. The nausea gradually disappeared over repeated follow-up visits. Nesse and colleagues (1980) reported that about 44% of the patients being treated for lymphoma demonstrated such anticipatory nausea. Anticipatory nausea is best understood as a classically conditioned response (Pavlov, 1927). Control over anticipatory nausea could be exerted at the time of conditioning or at the time of reexposure to the conditioned stimulus (CS). If an antiemetic drug is presented at the time of conditioning, then a reduction in anticipatory nausea would be the result of an attenuated unconditioned response (UCR), that is, reduced nausea produced by the toxin at the time of conditioning thereby attenuating the establishment of the conditioned response (CR). Indeed, when administered during the chemotherapy session, the 5-HT_3 antagonist, granisetron, has been reported to reduce the incidence of anticipatory nausea in repeat cycle chemotherapy treatment (Aapro et al., 1994). On the other hand, if a drug is delivered prior to reexposure to cues previously paired with the toxin-induced nausea, then suppressed anticipatory nausea would be the result of attenuation of the expression of the CR (conditioned nausea); the 5-HT_3 antagonists are ineffective at this stage (Nesse et al., 1980; Morrow and Dobkin, 1988; Reynolds et al., 1991; Stockhorst et al., 1993; Aapro et al., 1994; Ballatori and Roila, 2003; Hickok et al., 2003). Anecdotal evidence suggests that Δ^9-THC alleviates anticipatory nausea in chemotherapy patients (Grinspoon and Bakalar, 1993; Iverson, 2000). Although there has been considerable experimental investigation of unconditioned retching and vomiting in response to toxins, there have been relatively few reports of conditioned retching, that is, emetic reactions elicited by reexposure to a toxin-paired cue (anticipatory nausea). Conditioned retching has been observed to occur in coyotes, wolves, and hawks upon reexposure to cues previously paired with lithium-induced toxicosis (Garcia et al., 1977) and ferrets have been reported to display conditional emetic reactions during exposure to a chamber previously paired with lithium-induced toxicosis (Davey and Biederman, 1998).

Conditioned Retching in the Shrew as a Model of Anticipatory Nausea

We have recently reported that the *Suncus murinus* (house musk shrew) which is capable of vomiting, displays conditioned retching when returned to a chamber previously paired with a dose of lithium that produced vomiting (Parker and Kemp, 2001). Furthermore, this conditioned retching reaction is suppressed by pretreatment with Δ^9-THC. This effect was replicated more recently and extended to demonstrate that the nonpsychoactive compound found in marijuana, CBD, also interfered with the expression of conditioned retching in the shrew, but the 5-HT_3 antagonist ondansetron was completely ineffective (Parker et al., 2006). The doses employed were selected on the basis of their potential to interfere with toxin-induced vomiting in the *Suncus* (Parker et al., 2003a; Kwiatkowska et al., 2004). Therefore, cannabinoids may be potential treatments for anticipatory nausea.

Conditioned Gaping in the Rat as a Model of Anticipatory Nausea

Rodriguez and colleagues (2000) reported that following repeated pairings of a context with lithium, rats will subsequently suppress their consumption of a novel flavored solution when returned to that context. They reasoned that since rats show suppressed consumption of a novel flavored solution when they are ill (Domjan, 1977), the context previously paired with illness must have elicited conditioned nausea promoting suppressed consumption. However, suppressed drinking is not a selective measure of nausea, because rats also suppress consumption as a measure of conditioned fear. If suppressed consumption while in a context previously paired with LiCl is a measure of anticipatory nausea, then rats would also be expected to exhibit conditioned gaping responses when infused with a novel flavored solution in that context. Indeed, a recent study confirmed this expectation (Limebeer et al., 2006). Following four pairings of a distinctive, vanilla odor-laced chamber with LiCl-induced illness, rats were returned to the context for 30 min and received a 1-min intraoral infusion of novel saccharin solution every 5 min. During the infusions, the rats displayed gaping reactions. Surprisingly, the rats also gaped during intervals when they were not being infused with saccharin while in the LiCl-paired context. It was further demonstrated that Δ^9-THC, but not ondansetron, interfered with the conditioned gaping response during both infusion and interinfusion intervals. The finding that rats express conditioned gaping responses when reexposed to a context previously paired with LiCl during interinfusion intervals (Limebeer et al., 2006) suggests that LiCl-paired contextual cues in the absence of the flavor can elicit conditioned nausea. Limebeer and colleagues (2007) recently found that even in the absence of a flavored solution, rats display conditioned gaping reactions during exposure to a distinctive context laced with vanilla odor previously paired with a high dose of lithium, as well as a low dose of lithium and provocative motion. These results are consistent with those of an earlier report by Meachum and Bernstein (1992), that reexposure to a lithium-paired context laced with an odor cue (but not in the absence of an odor cue) elicited gaping reactions in rats. We further demonstrated that pretreatment with cannabidiol (CBD), a nonpsychoactive cannabinoid found in marijuana, prior to reexposure to a lithium-paired distinctive context results in rats expressing fewer gaping responses. These results support the proposal that conditioned gaping is a selective measure of nausea, and this rat model of anticipatory nausea provides a valuable preclinical tool for evaluating the effectiveness of antinausea treatments. Furthermore, cannabinoid compounds may reduce anticipatory nausea.

Concluding Remarks

Since the discovery of the mechanism of action of cannabinoids, our understanding to the role of the endocannabinoid system in the control of nausea and vomiting has greatly increased. In the ferret and shrew models, the site of action has been identified in the emetic area of the brainstem, the dorsal vagal complex. The shrew model, in particular, is cost effective for the evaluation of the antiemetic properties of agents. It is clear that many cannabinoids act on the CB_1 receptors to produce their antiemetic properties; however, it is not known how the nonpsychoactive cannabinoid, cannabidiol, which does not act at the CB_1 receptor, produces antiemetic effects within a limited dose range in the *Suncus murinus* (Parker et al., 2003a; Kwiatkowska et al., 2004). The conditioned gaping response in the rat has provided a glimpse into the antinausea mechanisms of action of cannabinoids, in the absence of a vomiting response. Since nausea is a more difficult symptom to control than vomiting, the gaping model may serve as a useful tool for the development of new antiemetic treatments, as well as for the evaluation of the potential side effects of nausea in newly developed pharmacological treatments. Recent work has also supported anecdotal reports that cannabis may attenuate anticipatory nausea (anticipatory nausea). Using the *Suncus murinus* and the rat models of anticipatory nausea, both Δ^9-THC and CBD effectively prevented conditioned retching and conditioned gaping, respectively, elicited by reexposure to a lithium-paired chamber.

Acknowledgments The authors would like to thank Marion Corrick for care of the shrews. This research was supported by research grants to Linda Parker from the Canadian Institutes of Health Research and the Natural Sciences and Engineering Research Council of Canada.

References

Aapro MS, Kirchner V, Terrey JP (1994) The incidence of anticipatory nausea and vomiting after repeat cycle chemotherapy: the effect of granisetron. Br J Cancer 69:957–960.

Aapro MS, Thuerlimann B, Sessa C, De Pree C, Bernhard J, Maibach R, Swiss Group for Clinical Cancer Research (2003) A randomized double-blind trial to compare the clinical efficacy of granisetron with metoclopramide, both combined with dexamethasone in the prophylaxis of chemotherapy-induced delayed emesis. Ann Oncol 14:291–297.

Abrahamov A, Abrahamov A, Mechoulam R (1995) An efficient new cannabinoid antiemetic in pediatric oncology. Life Sci 56:2097–2102.

Andrews PLR, Davis CJ (1995) The physiology of emesis induced by anti-cancer therapy. In: Reynolds J, Andrews PLR, Davis CJ (eds.), Serotonin and the Scientific Basis of Anti-Emetic Therapy. Oxford: Oxford Clinical Communications, pp. 25–41.

Andrews PLR, Horn CC (2006) Signals for nausea and emesis: implications for models diseases. Auton Neurosci 125:100–115.

Ballatori E, Roila F (2003) Impact of nausea and vomiting on quality of life in cancer patients during chemotherapy. Health Qual Life Outcomes 1:46.

Barann M, Molderings G, Bruss M, Bonisch H, Urban BW, Gothert M (2002) Direct inhibition by cannabinoids of human 5-HT_{3A} receptors: probable involvement of an allosteric modulatory site. Br J Pharm 137:589–96.

Bartlett N, Koczwara B (2002) Control of nausea and vomiting after chemotherapy: what is the evidence? Intern Med J 32:401–407.

Billig I, Yates BJ, Rinaman L (2001) Plasma hormone levels and central c-Fos expression in ferrets after systemic administration of cholecystokinin. Am J Physiol Regul Integr Comp Physiol 281:R1243–R1255.

Borrison HL, Wang SC (1953) Physiology and pharmacology of vomiting. Pharmacol Rev 5:193–230.

Breslin PA, Spector AC, Grill HJ (1992) A quantitative comparison of taste reactivity behaviors to sucrose before and after lithium chloride pairings: a unidimensional account of palatability. Behav Neurosci 106:820–836.

Carey MP, Burish TG, Brenner DE (1983) Delta-9-tetrahydrocannabinol in cancer chemotherapy: research problems and issues. Ann Intern Med 99:106–114.

Carrier EJ, Achampach JA, Hillard CJ (2006) Inhibition of an equilibrative nucleoside transporter by cannabidiol: a mechanism of cannabinoid immunosuppression. Proc Natl Acad Sci USA 103:7895–7900.

Costall B, Domeney AM, Naylor RJ, Tattersall FD (1986) 5-Hydroxytryptamine receptor antagonism to prevent cisplatin-induced emesis. Neuropharmacology 25:959–961.

Crawford SM, Buckman R (1986) Nabilone and metoclopramide in the treatment of nausea and vomiting due to cisplatin: a double blind study. Med Oncol Tumor Pharmacother 3:39–42.

Cross-Mellor SK, Ossenkopp KP, Piomelli D, Parker LA (2007) Effects of the FAAH inhibitor, URB597, and anandamide on lithium-induced taste reactivity responses: a measure of nausea in the rat. Psychopharmacology 190:135–143.

Cunningham D, Bradley CJ, Forrest GJ, Hutcheon AW, Adams L, Sneddon M, Harding M, Kerr DJ, Soukop M, Kaye SB (1988) A randomized trial of oral nabilone and prochlorperazine compared to intravenous metoclopramide and dexamethasone in the treatment of nausea and vomiting induced by chemotherapy regimens containing cisplatin or cisplatin analogues. Eur J Cancer Clin Oncol 24:685–689.

Darmani NA (2001a) Delta-9-tetrahydrocannabinol and synthetic cannabinoids prevent emesis produced by the cannabinoid CB_1 receptor antagonist/inverse agonist SR 141716A. Neuropsychopharmacology 24:198–203.

Darmani NA (2001b) Delta-9-tetrahydrocannabinol differentially suppresses cisplatin-induced emesis and indices of motor function via cannabinoid CB_1 receptor in the least shrew. Pharmacol Biochem Behav 69:239–249.

Darmani NA (2001c) The cannabinoid CB_1 receptor antagonist SR 141716A reverses the antiemetic and motor depressant actions of WIN 55212-2. Eur J Pharm 430:49–58.

Darmani NA (2002) The potent emetogenic effects of the endocannabinoid, 2-AG (2-arachidonoylglycerol) are blocked by Delta^9-tetrahydrocannabinol and other cannabinoids. J Pharmacol Exp Ther 300:34–42.

Darmani NA, Johnson CJ (2004) Central and peripheral mechanisms contribute to the antiemetic actions of delta-9-tetrahydrocannabinol against 5-hydroxytryptophan-induced emesis. Eur J Pharmacol 488:201–212.

Darmani NA, Janoyan JJ, Kumar N, Crim JL (2003) Behaviorally active doses of the CB_1 receptor antagonist SR 141716A increase brain serotonin and dopamine levels and turnover. Pharmacol Biochem Behav 75:777–787.

Darmani NA, McClanahan BA, Trinh C, Petrosino S, Valenti M, Di Marzo V (2005) Cisplatin increases brain 2-arachidonoylglycerol (2-AG) and concomitantly reduces intestinal 2-AG and anandamide levels in the least shrew. Neuropharmacology 49:502–513.

Davey VA, Biederman GB (1998) Conditioned antisickness: indirect evidence from rats and direct evidence from ferrets that conditioning alleviates drug-induced nausea and emesis. J Exp Psychol Anim Behav Process 24:483–491.

Domjan M (1977) Selective suppression of drinking during a limited period following aversive drug treatment in rats. J Exp Psychol Anim Behav Process 3:66–76.

Fan P (1995) Cannabinoid agonists inhibit the activation of 5-HT_3 receptors in rat nodose ganglion neurons. J Neurophysiol 73:907–910.

Feigenbaum JJ, Richmond SA, Weissman Y, Mechoulam R (1989) Inhibition of cisplatin-induced emesis in the pigeon by a non-psychotropic synthetic cannabinoid. Eur J Pharmacol 4:159–165.

Ferrari F, Ottanik A, Giuliani D (1999) Cannabimimetic activity in rats and pigeons of HU-210, a potent antiemetic drug. Pharmacol Biochem Behav 62:75–80.

Garcia J, Rusiniak KW, Brett LP (1977) Conditioning food-illness aversions in wild animals: caveant canonici. In: Davis H, Hurowitz HMB (eds.), Operant Pavlovian Interactions. Hillsdale, NJ: Lawrence Erlbaum, pp. 273–316.

Grelot L, Milano S, LeStunff H (1995) Does 5-HT play a role in the delayed phase of cisplatin-induced emesis? In: Reynolds J, Andrews PLR, Davis CJ (eds.), Serotonin and the Scientific Basis of Anti-Emetic Therapy. Oxford: Oxford Clinical Communications, pp. 181–191.

Grill HC, Norgren R (1978) The taste reactivity test. I. Mimetic responses to gustatory stimuli in neurologically normal rats. Brain Res 143:263–279.

Grinspoon L, Bakalar JB (1993) Marijuana: The Forbidden Medicine. New Haven: Yale University Press.

Gylys JA, Doran KM, Buyniski PJ (1979) Antagonism of cisplatin induced emesis in the dog. Res Commun Chem Pathol Pharmacol 23:61–68.

Hall W, Christie M, Currow D (2005) Cannabinoids and cancer: causation, remediation, and palliation. Lancet Oncol 6:35–42.

Hampson AJ, Grimaldi M, Axelrod J (1998) Cannabidiol and delta-9-tetrahydrocannabinol are neuroprotective antioxidants. Proc Natl Acad Sci USA 95:8268–8273.

Hermann H, Marsicano G, Lutz B (2002) Coexpression of the cannabinoid receptor type 1 with dopamine and serotonin receptors in distinct neuronal subpopulations of the adult mouse forebrain. Neuroscience 109:451–460.

Hesketh PJ, Van Belle S, Aapro M, Tattersall FD, Naylor RJ, Hargreaves R, Carides AD, Evans JK, Horgan KJ (2003) Differential involvement of neurotransmitters through the time course of cisplatin-induced emesis as revealed by therapy with specific receptor antagonists. Eur J Cancer 39:1074–1080.

Hickok JT., Roscoe JA, Morrow GR, King DK, Atkins JN, Fitch TR (2003) Nausea and emesis remain significant problems of chemotherapy despite prophylaxis with 5-Hydroxytryptamine-3 antiemetics. Cancer 97:2880–2886.

Hillsley K, Grundy D (1998) Serotonin and cholecystokinin activate different populations of rat mesenteric vagal afferents. Neurosci Lett 255:63–66.

Himmi T, Dallaporta M, Perrin J, Orsini JC (1996) Neuronal responses to delta9-tetrahyrocannabinol in the solitary tract nucleus. Eur J Pharmacol 312:273–279.

Himmi T, Perrin J, El Ouazzani T, Orsini JC (1998) Neuronal responses to cannabinoid receptor ligands in the solitary tract nucleus. Eur J Pharmacol 359:49–54.

Howlett AC, Barth F, Bonner TI, Cabral P, Casellaa G, Devane WA, Felder CC, Herkenham M, Mackie K, Martin B R, Mechoulam R, Pertwee RG (2002) International Union of Pharmacology. XXVII. Classification of Cannabinoid Receptors. Pharmacol Rev 54:161–202.

Iversen LL (2000) The Science of Marijuana. New York: Oxford University Press.

Kimura T, Ohta T, Watanabe K, Yoshimura H, Yamamoto I (1998) Anandamide, an endogenous cannabinoid receptor ligand, also interacts with 5-hydroxytryptamine (5HT) receptor. Biol Pharm Bull 21:224–226.

Kwiatkowska M, Parker LA, Burton P, Mechoulam R (2004) A comparative analysis of the potential of cannabinoids and ondansetron to suppress cisplatin-induced emesis in the *Suncus murinus* (house musk shrew). Psychopharmacology 174:254–259.

Layeeque R, Siegel E, Kass R, Henry-Tillman RS, Colvert M, Mancino A, Klimberg VS (2006) Prevention of nausea and vomiting following breast surgery. Am J Surg 191:767–772.

Limebeer CL, Parker LA (1999) Delta-9-tetrahydrocannabinol interferes with the establishment and the expression of conditioned disgust reactions produced by cyclophosphamide: a rat model of nausea. Neuroreport 26:371–384.

Limebeer CL, Parker LA (2000) Ondansetron interferes with the establishment and the expression of conditioned disgust reactions: a rat model of nausea. J Exp Psychol Anim Behav Process 26:371–384.

Limebeer CL, Parker LA (2003) The 5-HT_{1A} agonist 8-OH-DPAT dose-dependently interferes with the establishment and the expression of lithium-induced conditioned rejection reactions in rats. Psychopharmacology 166:120–126.

Limebeer CL, Parker, LA, Fletcher, P (2004) 5,7-dihydroxytryptamine lesions of the dorsal and median raphe nuclei interfere with lithium-induced conditioned gaping, but not conditioned taste avoidance, in rats. Behav Neurosci 118:1391–1399.

Limebeer CL, Hall G, Parker LA (2006) Exposure to a lithium-paired context elicits gaping in rats: a model of anticipatory nausea. Physiol Behav 88:398–403.

Limebeer CL, Krohn JP, Rock EM, Cross-Mellor SK, Parker LA, Ossenkopp KP (2007) Exposure to a context previously associated with toxin (LiCl)- or motion-induced sickness elicits conditioned gaping in rats: evidence in support of a model of anticipatory nausea. Behav Brain Res (in press).

Malfait AM, Gallily R, Sumariwalla PF, Malik AS, Andreakos E, Mechoulam R, Feldman M (2000) The non-psychoactive cannabis-constituent cannabidiol is an oral anti-arthritc therapeutic in murine collagen-induced arthritis. Proc Natl Acad Sci USA 97:9561–9566.

Matsuki N, Ueno S, Kaji T, Ishihara A, Wang CH, Saito H (1988) Emesis induced by cancer chemotherapeutic agents in the *Suncus murinus*: a new experimental model. Jpn J Pharmacol 48:303–306.
McCarthy LE, Borison HL (1981) Anti-emetic activity of N-methyllevonantrobil and naboline in cisplatin treated cats. J Clin Pharmacol 21:30S–37S.
McLaughlin PJ, Winston KM, Limebeer CL, Parker LA, Makriyannis A, Salamone JD (2005) The cannabinoid antagonist AM 251 produces food avoidance and behaviors associated with nausea but does not impair feeding efficiency in rats. Psychopharmacology 180:286–293.
Meachum CL, Bernstein IL (1992) Behavioral conditioned responses to contextual and odor stimuli paired with LiCl administration. Physiol Behav 52:895–899.
Mechoulam R, Parker LA, Gallily R (2002) Cannabidiol: an overview of some pharmacological aspects. J Clin Pharmacol 42:11S–19S.
Miner WJ, Sanger GJ (1986) Inhibition of cisplatin-induced vomiting by selective 5-hydroxytryptamine M-receptor antagonism. Br J Pharmacol 88:497–499.
Morrow GR, Dobkin PL (1988) Anticipatory nausea and vomiting in cancer patients undergoing chemotherapy treatment: prevalence, etiology and behavioral interventions. Clin Psychol Rev 8:517–556.
Nesse R M, Carli T, Curtis GC, Kleinman PD (1980) Pretreatment nausea in cancer chemotherapy: a conditioned response? Psychosom Med 42:33–36.
Parker LA (1982) Nonconsummatory and consummatory behavioral CRs elicited by lithium-paired and amphetamine-paired flavors. Learn Motiv 13:281–303.
Parker LA (1995) Rewarding drugs produce taste avoidance, but not taste aversion. Neurosci Biobehav Rev 19:143–151.
Parker LA (1998) Emetic drugs produce conditioned rejection reactions in the taste reactivity test. J Psychophysiol 12:3–13.
Parker LA (2003) Taste avoidance and taste aversion: evidence for two different processes. Learn Behav 31:165–172.
Parker LA, Kemp S (2001) Tetrahydrocannabinol (THC) interferes with conditioned retching in *Suncus murinus*: an animal model of anticipatory nausea and vomiting (ANV). Neuroreport 12:749–751.
Parker LA, Mechoulam R (2003) Cannabinoid agonists and an antagonist modulate conditioned gaping in rats. Integr Physiol Behav Sci 38:134–146.
Parker LA, Corrick ML, Limebeer CL, Kwiatkowska M (2002a) Amphetamine and morphine produce a conditioned taste and place preference in the house musk shrew (*Suncus murinus).* J Exp Psychol Anim Behav Process 28:75–82.
Parker LA, Mechoulam R, Schlievert C (2002b) Cannabidiol, a non-psychoactive component of cannabis, and its dimethylheptyl homolog suppress nausea in an experimental model with rats. Neuroreport 13:567–570.
Parker LA, Kwiatkowska M, Burton P, Mechoulam R (2003a) Effect of cannabinoids on lithium-induced vomiting in the *Suncus murinus*. Psychopharmacology 171:156–161.
Parker LA, Mechoulam R, Shlievert C, Abbott L, Fudge ML, Burton P (2003b) Effects of cannabinoids on lithium-induced conditioned rejection reactions in a rat model of nausea. Psychopharmacology 166:156–162.
Parker LA, Limebeer CL, Kwiatkowska M (2005) Cannabinoids: effects on vomiting and nausea in animal models. In: Mechoulam R (ed.), Cannabinoids as Therapeutics. Basel: Birkhauser Verlag, pp. 183–200.
Parker LA, Kwiatkowska M, Mechoulam R (2006) Delta-9-tetrahydrocannabinol and cannabidiol, but not ondansetron, interfere with conditioned retching reactions elicited by a lithium-paired context in *Suncus murinus*: an animal model of anticipatory nausea and vomiting. Physiol Behav 87:61–71.
Pavlov IP (1927) Conditioned Reflexes (G.V. anrep, trans.).London: Oxford University Press.
Pertwee RG (2001) Cannabinoids and the gastrointestinal tract. Gut 48:859–867.
Pi-Sunyer FX, Aronne LJ, Heshmati HM, Devin J, Rosenstock J, RIO-North American Study Group (2006) Effect of rimonabant, a cannabinoid-1 receptor blocker, on weight and cardiometabolic risk factors in overweight or obese patients. JAMA 295:761–775.

Reynolds DJM, Barber NA, Grahame-Smith DG, Leslie RA (1991) Cisplatin-evoked induction of c-fos protein in the brainstem of the ferret: the effect of cervical vagotomy and the antiemetic 5HT-3 receptor antagonist granisetron. Brain Res 565:321–336.

Rodriguez M, Lopez M, Symonds M, Hall G (2000) Lithium-induced context aversion in rats as a model of anticipatory nausea in humans. Physiol Behav 71:571–579.

Rudd JA, Naylor RJ (1996) An interaction of ondansetron and dexamethasone antagonizing cisplatin-induced acute and delayed emesis in the ferret. Br J Pharmacol 118:209–214.

Rudd JA, Jordan, CC, Naylor, RJ (1996) The action of the NK_1 tachykinin receptor antagonist, CP 99,994, in antagonizing the acute and delayed emesis induced by cisplatin in the ferret. Br J Pharmacol 119:931–936.

Russo EB, Burnett, A, Hall B, Parker KK (2005) Agonist properties of cannabidiol at 5-HT_{1a} receptors. Neurochem Res 30:1037–1043.

Schlicker E, Kathman M (2001) Modulation of transmitter release via presynaptic cannabinoid receptors. TIPS 22:565–571.

Schwartz RH, Beveridge RA (1994) Marijuana as an antiemetic drug: how useful is it today? Opinions from clinical oncologists. J Addict Dis 13:53–65.

Shannon HE, Martin WR, Silcox D (1978) Lack of antiemetic effect of Δ9-tetrahydrocannabinol in apomorphine-induced emesis in the dog. Life Sci 23:49–53.

Simoneau II, Hamza MS, Mata HP, Siegel EM, Vanderah TW, Porreca F, Makriyannis A, Malan P (2001) The cannabinoid agonist WIN 55,212-2 suppresses opioid-induced emesis in ferrets. Anesthesiology 94:882–886.

Sink KS, McLaughlin PJ, Brown C, Xu W, Fan P, Vemuri VK, Wood, JT, Makriyannis A, Parker LA, Salamone JD (2007) The novel cannabinoid CB_1 receptor neutral antagonist AM4113 suppresses food intake and food-reinforced behavior but does not induce signs of nausea in rats. Neuropsychopharmacology (in press).

Stockhorst U, Klosterhalfen S, Klosterhalfen W, Winkelmann M, Steingrueber HJ (1993) Anticipatory nausea in cancer patients receiving chemotherapy: classical conditioning etiology and therapeutical implications. Integr Physiol Behav Sci 28:177–181.

Thomas A, Ross RA, Saha B, Mahadevan A, Razdan RK, Pertwee RJ (2004) 6″-Azidohex-2″-yne-cannabidiol: a potential neutral, competitive cannabinoid CB_1 receptor antagonist. Eur J Pharmacol 487:213–221.

Torii Y, Saito H, Matsuki N (1991) Selective blockade of cytotoxic drug-induced emesis by 5-HT_3 receptor antagonists in *Suncus Murinus*. Jpn J Pharmacol 55:107–113.

Tramer, MR, Carroll D, Campbell FA, Reynolds DJM, Moore RA, McQuay HJ (2001) Cannabinoids for control of chemotherapy induced nausea and vomiting: quantitative systematic review. BMJ 323:1–8.

Travers JB, Norgren R (1986) Electromyographic analysis of the ingestion and rejection of sapid stimuli in the rat. Behav Neurosci 100:544–555.

Tsukada H, Hirose T, Yokoyama A, Kurita Y (2001) Randomized comparison of ondansetron plus dexamethasone with dexamethasone alone for control of delayed cisplatin-induced emesis. Eur J Cancer 37:2398–2404.

Ueno S, Matsuki N, Saito H (1987) *Suncus murinus:* a new experimental model in emesis research. Life Sci 43:513–518.

Ungerleider JT, Andrysiak TA, Fairbanks LA, Tesler AS, Parker RG (1984) Tetrahydrocannabinol vs. prochlorperazine. The effects of two antiemetics on patients undergoing radiotherapy. Radiology 150:598–599.

Van Belle S, Lichinitser M, Navari R, Garin AM, Decramer ML, Riviere A, Thant M, Brestan E, Bui B, Eldridge K, DeSmet M, Michiels N, Reinhardt RR, Carides AD, Evans JK, Gertz BJ (2002) Prevention of cisplatin-induced acute and delayed emesis by the slective neurokinin-1 antagonists, L-758,298 and MK869. Cancer 94:3032–3041.

Van Sickle MD, Oland LD, HO W, Hillard CJ, Mackie K, Davison JS, Sharkey KA (2001) Cannabinoids inhibit emesis through CB_1 receptors in the brainstem of the ferret. Gastroenterology 121:767–774.

Van Sickle MD, Oland LD, Mackie K, Davison JS, Sharkey KA (2003) Δ^9-Tetrahydrocannabinol selectively acts on CB_1 receptors in specific regions of dorsal vagal complex to inhibit emesis in ferrets. Am J Physiol Gastrointest Liver Physiol 285:G566–G576.

Van Sickle MD, Cuncan M, Kingsley PJ, Mouihate A, Urbani P, Mackie K, Stella N, Makriyannis A, Piomelli D, Davison JS, Marnett LJ, Di Marzo V, Pittman QJ, Patel KD, Sharkey KA (2005) Identification and functional characterization of brainstem cannabinoid CB_2 receptors. Science 310:329–332.

Chapter 14
Endocannabinoids in Energy Homeostasis and Metabolic Disorders

Isabel Matias, Vincenzo Di Marzo, and Attila Köfalvi

Abstract It has long been known that marihuana consumption acutely increases appetite (Hollister, 1971) and decreases body temperature (Borgen et al., 1973). These effects predicted that cannabinoids may control energy homeostasis. Later, the endocannabinoid system was discovered, and its role in energy homeostasis moved into focus. The main psychoactive constituent of marihuana, Δ^9-THC, as well as the endocannabinoids, anandamide and 2-arachidonoylglycerol, were shown to stimulate appetite centrally in animal models and in man (Sacks et al., 1990; Williams and Kirkham, 1999; Di Marzo and Matias, 2005). As the ultimate evidence, genetic ablation of the CB_1 receptor was shown to lead to reduced food intake after food deprivation, leanness, resistance to diet-induced obesity, and enhanced leptin sensitivity (Di Marzo et al., 2001; Cota et al., 2003b; Ravinet-Trillou et al., 2004). Meanwhile, obesity and obesity-related complications have put an increasing burden on Western societies' health care; therefore, a large effort has been made to map the role and the therapeutic potential of the endocannabinoid system in obesity and cardiometabolic disorders. This research finally resulted in the introduction of the CB_1 receptor antagonist SR141716A (also known as Rimonabant) in the European market under the name Acomplia™ in the summer of 2006, to treat obesity and its metabolic complications and cardiovascular risk factors (Gelfand and Cannon, 2006). Acomplia™ is the first CB_1 receptor-based medicine. The discussion of the role and clinical impact of endocannabinoids on energy homeostasis, obesity, and diabetes also cannot be limited to the brain. Therefore, although this book is mostly about the CNS, a large part in this chapter will detail peripheral mechanisms as well.

Introduction

Energy homeostasis and metabolism are maintained by multiple mechanisms including the regulation by circulating hormones such as leptin and insulin, for which numerous interactions with the endocannabinoid system have been already documented (Di Marzo et al., 2001; Cota et al., 2003a; Juan-Picó et al., 2006; Matias et al., 2006). Two other classes of molecules that have received increasing

A. Kofalvi (ed.), *Cannabinoids and the Brain.*

attention in energy homeostasis are circulating glucose and fatty acids, which are the main signals of the body's nutritional status. Indeed, the involvement of the endocannabinoid system in glucose and fatty acid homeostasis is still unclear. This review will therefore focus not only on recent findings related to the role of the endocannabinoid system in this context in the central nervous system, and in particular, the hypothalamus, but also on the liver, adipose tissue and skeletal muscle.

Glucose Homeostasis

Oxidizable glucose is the major energy source of the body, and 20–50% of this energy is utilized by the brain under resting condition (Gispen and Biessels, 2000; Fehm et al., 2006). Importantly, the majority of this energy is used to maintain the physicochemical properties of neural membranes. Acute focal or systemic shortage of glucose severely damages body cells because their short-term energy store, glycogen, depletes rapidly. Under such conditions, namely ischemia and infarct (of the heart or brain, and so on) or during hyperinsulinemia, it is critical to reestablish the normal glucose supply of the affected organs, and to localize necrosis. Since the CNS is the most susceptible organ to changes of systemic glucose levels, it is critical to review here recent advances in the role of the endocannabinoid system in glucose homeostasis. Notwithstanding, one of the major glucoregulator systems of the body, namely the HPA axis, is also involved in the control of behavior and hormonal homeostasis. Thus, the impact of the endocannabinoid system on these regulatory mechanisms cannot be entirely separated from glucose homeostasis.

Fatty Acid Homeostasis

Emerging evidence from the last few years indicates that circulating free fatty acids and their derivatives, the long-chain fatty acids (LCFAs), should also be considered as nutritional state sensors (Lam et al., 2005). Therefore, the interplay between de novo lipogenesis and fatty acid oxidation might be the key to control energy homeostasis. In fact, these two phenomena maintain homeostasis by increasing energy expenditure during periods of energy excess and by decreasing it during times of energy deficit. Once the fatty acid is in the cell, a decision must be made whether to direct fatty acid toward mitochondrial oxidation for energy production or toward glycerolipid synthesis for energy storage. Since the endocannabinoid system is upregulated during obesity, and obesity is characterized by atherogenic dyslipidemia, the endocannabinoid system must be involved in fatty acid homeostasis, as will be discussed below.

Physiology and Biochemistry of Energy Homeostasis

Glucose Homeostasis

The majority of ingested carbohydrates, the carbon atoms of catabolized cellular or ingested proteins, and lactate (originated mainly from the skeletal muscle and erythrocytes) can be all converted to glucose in the liver via the biochemical pathways comprised in gluconeogenesis (see Fig. 2). Focal and systemic signals maintain a very complex and delicate balance of systemic glucose level, which has a physiological value of ~5 mM in the plasma and is crucial for the survival of the mammalian organism. The primary effector of these signals is the liver, which takes up or releases glucose, depending on its circulating concentration. Low blood glucose triggers the release of glucagon from pancreatic α-cells and among others, of ACTH and of growth hormone from the pituitary, of glucocorticoids (primarily cortisol) from the adrenal cortex, and of epinephrine from the adrenal medulla, to increase blood glucose either via inhibiting its uptake or by increasing glycogenolysis. Both glucagon and epinephrine, acting on their hepatocyte cell surface receptors, activate glycogene phosphorylase via the cAMP-PKA-mediated cascade, leading to free intracellular glucose-6-phosphate, which is hydrolyzed into glucose and released into the blood. Since muscle and brain cells do not have the enzyme for the latter step (i.e., glucose-6-phosphatase), the glucose-6-phosphate product of hexokinases is retained and oxidized by these tissues. High blood glucose triggers the release of insulin from pancreatic β-cells which in turn stimulates extra-hepatic glucose uptake and glycogen synthesis. Importantly, insulin stimulates the recruitment of glucose transporter complexes on the surface of nonhepatic cells. Glucose transporters comprise a family of an increasing number of members (Joost et al., 2002). Among the most characterized ones, GLUT1 is ubiquitously distributed in various tissues. GLUT2 is found primarily in the intestine, kidney, and liver. GLUT3 is also found in the intestine and GLUT5 in the brain and testis. GLUT5 is also the major glucose transporter present in the membrane of the endoplasmic reticulum (ER) and serves the function of transporting glucose to the cytosol following its dephosphorylation by the ER enzyme glucose-6-phosphatase. Skeletal muscle and adipose tissues contain GLUT4. When the concentration of blood glucose increases in response to food intake, pancreatic glucose uptake is increased by GLUT2, which leads to glucose metabolism and in turn, release of insulin from β-cells, both controlled by the enzyme glucokinase. Hepatocytes, by being virtually freely permeable to glucose, are only marginally affected by insulin. When blood glucose is high, the activity of liver glucokinase elevates. The glucose-6-phosphate product of glucokinase is rapidly converted to glucose-1-phosphate by phosphoglucomutase and then is incorporated into glycogen. Altogether, these data indicate that mainly two peripheral effector organs (namely the liver and the pancreas) and two major peripheral hormones (insulin and glucagon) work hand in hand to maintain normal glucose levels in the plasma. Virtually, more signals and hormones are available to mobilize plasma glucose than to decrease its levels, which corresponds well to the fact that one may live for a relatively long time with untreated hyperglycemia, whereas severe hypoglycemia cannot be managed by the CNS, leading to death in minutes.

Central Regulation of Systemic Glucose Homeostasis

As Fig. 1 indicates, the brain is not a passive participant in glucose homeostasis but actively and bidirectionally communicates with peripheral organs: directly via nerves, and indirectly via hormones. The central regulation of hunger, food seeking, satiation and fatty acid and glucose levels is still ill defined and very complex (Levin, 2006); therefore, its detailed discussion stays beyond the scope of this book. In brief, however, we should mention that the lateral and ventromedial hypothalamic areas play a major role in sensing the levels of nutrients and energy-homeostasis-related signaling molecules (Anand and Brobeck, 1951; Stephens, 1980; Minami et al., 1990; Song and Routh, 2005). Neurons in these areas, together with the proopiomelanocortin- and neuropeptide Y-positive arcuate nucleus neurons and the nucleus tractus solitarius neurons are either excited or inhibited by glucose (Dallaporta et al., 1999; Ibrahim et al., 2003; Kohno et al., 2003; Burdakov et al., 2005; Levin, 2006). The glucosensing mechanism of these neurons employs GLUT3, which is saturated at normal blood glucose levels, and glucokinase, which is the real glucosensor due to its high K_m and the lack of end-product inhibition. When taken up, glucose undergoes glycolysis, which results in elevated ATP levels. Eventually, the increase in ATP levels closes the K_{ATP} potassium channels, thereby depolarizing the axons (as an explanation for the excitatory action) or may hyperpolarize membranes, assumedly through chloride channel activation (inhibitory action). In fact, these mechanisms are quite similar to those in the pancreatic islets that regulate insulin and glucagon release (Kang et al., 2004, Levin, 2006). Interestingly, only dramatic changes in glucose levels activate or inhibit these neurons. Normally, food seeking and ingestive behavior are regulated by several other factors, such as peripheral chemo- and mechanosensor signals converging into the nucleus tractus solitarius through the vagus

Fig. 1 (continued) precursor of α-melanocyte stimulating hormone [αMSH; *αMR*] and adrenocorticotropin [ACTH]); *NPY* neuropeptide Y (*YR*); *CCK* cholecystokinin (*CCKR*); *eCBs* endocannabinoids (*CB_1R*); *GluCs* glucocorticoids (mainly cortisol) (*GluCR*); *CRH* corticotropin-releasing hormone (*CRHR*); *TRH* thyrotropin-releasing hormone (*TRHR*); *TRS* thyroid-stimulating hormone, *AMPK* 5′' AMP-activated protein kinase, *AGRP* agouti-related protein (*αMR* antagonist/inverse agonist); *CART* cocaine/amphetamine-related transcript; *IGF-1* insulin-like growth factor-1 (*IGF1R*); *MCH* melanin-concentrating hormone (*MCHR*); *GHRH* growth-hormone-releasing hormone; *GH* growth hormone; *OX_1R* orexin receptor-1; *OBR* leptin ("obesity") receptor; *OTR* oxytocin receptor; *CatR* catecholamine receptor; *GHSR* growth hormone-secretagogue receptor of ghrelin; *IR* insulin receptor; *ANR* adiponectin receptor. A "pathway inhibiting other pathway" means that the galanine-like peptide-1-positive pathway from the NTSV inhibits the orexigenic action of NPY in the PVNH, whereas AGRP antagonizes αMSH at its receptor

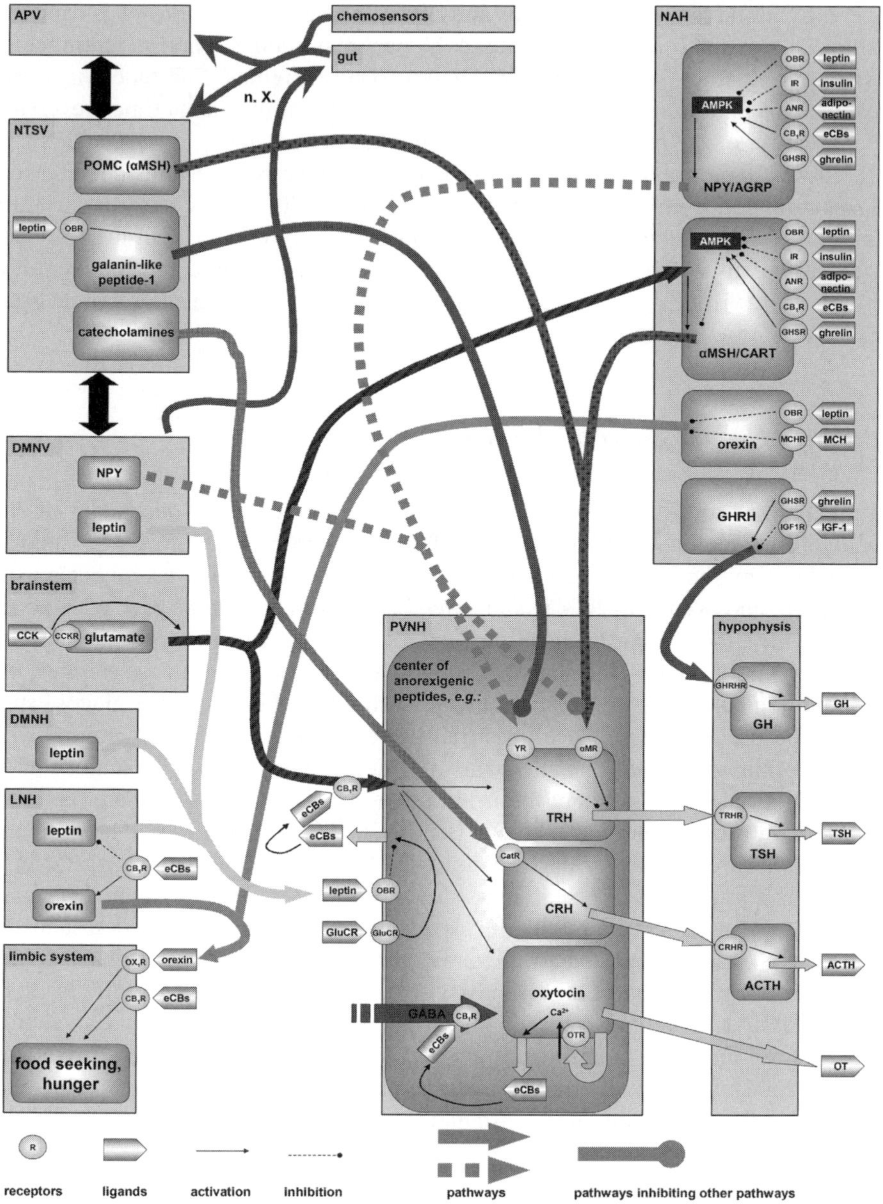

Fig. 1 Schematic Illustration of central regulation of feeding and energy homeostasis, featuring the vagal nuclei, brain stem, hypothalamus, hypophysis, HPA and HPT axes, and the limbic system. For sake of simplicity, only the most important brain areas and pathways are displayed. Ligands without indication of origin are mostly transported by blood to the brain but can be locally released as well. Brain areas: *APV* area postrema of nervus vagus (n. X.); *NTSV* nucleus tractus solitarii of n. X.; *DMNV* dorsal motor nucleus of n. X.; *DMNH* dorsomedial nucleus of hypothalamus; *LNH* lateral nucleus of hypothalamus; *NAH* nucleus arcuatus of hypothalamus; *PVNH* paraventricular nucleus of hypothalamus. Substances, signals, ligands, and their *receptors*: *POMC* proopiomelanocortin (the

nerve, by blood levels of fatty acids and ketone bodies, and by hormones from the fat tissue, the gastrointestinal tract and the pancreas such as ghrelin, leptin, cholecystokinin (CCK), adiponectin or insulin. These signals fine-tune the exact response of glucose-excited and glucose-inhibited neurons, sometimes reversing the direction of the response (Spanswick et al., 2000; Kang et al., 2004; Wang et al., 2004). Once the final direction of the central response is established and further modulated by higher order brain areas such as the limbic system, striatum and cortex, the effector areas such as the lateral hypothalamus and the paraventricular nucleus will be activated and neurohumoral, autonomic and somatomotor information will be passed to the peripheral tissues to regulate energy homeostasis (see below).

Glucose Homeostasis of the Brain

As mentioned above, neuronal activity and function is greatly dependent on the availability of glucose (McCall, 2004; Leybaert, 2005). Therefore, it is understandable why the brain is equipped with specialized glucose uptake mechanisms (McEwen and Reagan, 2004), and why blood flow and transport of glucose through the blood–brain barrier (via the GLUT1), is a function of focal activity (Leybaert, 2005). The brain harvests almost all oxygen and glucose through the blood–brain barrier; therefore, only by increasing blood flow can the glucose supply of the brain increase. Insulin is transported through the blood–brain barrier into the brain with a saturable mechanism, which is thought to be the major source for cerebral insulin. The more the insulin that circulates, the higher are its levels in the brain (Woods and Porte, 1977; Banks, 2004), and most likely, blood is the sole source of cerebral insulin (Woods et al., 2003). Insulin-like growth factor-1 (IGF-1) is homologous to insulin in its structure and amino acid sequence. Its main source is the liver and its circulating levels are very much sustained around the clock. It is also transported to the brain through the blood–brain barrier with higher efficacy (Reinhardt and Bondy, 1994). However, the brain itself also synthesizes IGF-1, which works as an autocrine and/or paracrine anabolic mediator. Both insulin and IGF-1 receptors are widely distributed in the brain (Bondy and Cheng, 2004), yet it is noteworthy that the majority of CNS neurons do not seem to use glucose in function of the circulating insulin level (Kyriaki, 2003). In contrast, in IGF-1 knockout mice, the brain develops to a much smaller size and takes up 30–60% less glucose compared to the brain of wild-type mice (Bondy and Cheng, 2004). This indicates that IGF-1 may be more important in the regulation of glucose uptake than insulin, at least during development. Importantly, both insulin and IGF-1 receptors utilize signaling cascades similar to those activated upon CB_1R activation: (1) the phosphatidylinositol 3-kinase (PI_3K)-PKB/Akt pathway, leading to increased GLUT4 density in the cell membranes, with concomitant increase in glucose uptake (Summers and Birnbaum, 1997) and (2) PKB/Akt-glycogen synthase kinase 3β (GSK3β) pathway, leading to facilitated glycogen and protein synthesis (Cheng et al., 2000; Clodfelder-Miller et al., 2005). These cascades gain importance in neuropsychiatric disorders in which insulin/IGF-1 and CB_1 receptors are implicated in neurodegeneration and hypoplasticity. Apparently, the variability of brain GLUT isoforms is greater than in the rest of the body (McEwen and Reagan, 2004). For instance, GLUT1 and GLUT3 are responsible for the majority of glucose uptake, whereas GLUT2,4,6,8,10 have more specialized distribution and function. In conclusion, the brain is highly dependent on glucose availability; therefore from its part "it does its best" to maintain the continuous glucose supply with high affinity and specialized transporters, with storage of glycogen in resting conditions, with local autocrine and paracrine regulation of glucose utilization, and with active bidirectional neuronal and humoral communication with peripheral organs.

Fatty Acid Homeostasis

Fatty acids are a major source of energy in vertebrates. They can be endogenously synthesized from lipids, carbohydrates or amino acids; however, the main source of lipids is the diet (Fig. 2). In the gut lumen, they are absorbed by intestinal villi cells, a process requiring triglyceride hydrolysis in the lumen and reesterification in epithelial cells. Lipids are then delivered into the general circulation via the lymph as chylomicrons, which are transported to various organs either in the form of triacylglycerols associated with lipoproteins, or as unesterified fatty acids complexed to serum albumin, produced by the adipose tissue after lipolytic stimulation by glucagon and adrenaline. Outside the cell, triacyglycerols are hydrolyzed by lipoprotein lipase to yield again free fatty acids. Fatty acids will then cross the plasma membrane, and also the blood–brain barrier (Wolfgang and Lane, 2006a). Once in the cell, fatty acids are rapidly esterified to fatty acyl-coenzyme A by an acetyl-CoA synthetase. The utilization of fatty acyl-CoA, and particularly of long-chain fatty acyl-CoA (LC-CoA) derived from plasma fatty acids, for either fatty acid oxidation or lipogenesis depends on the body's nutritional status and in particular, on the availability of carbohydrates and on the rate of glycolysis. Importantly, glyceraldehyde-3-phosphate dehydrogenase (GAPDH) is a key enzyme in glycolysis, catalyzing the conversion of D-glyceraldehyde-3-phosphate to 1,3-bisphosphoglycerate in a reaction that is reversible in order to obtain pyruvate. Starting from glucose, phosphofructokinase (PEK), GAPDH, pyruvate kinase (PYK), and lactate dehydrogenase (LDH) transform glucose into pyruvate and lactate. In fact, in fed animals, glucose will be degraded during glycolysis and the resulting acetyl-CoA will be then converted by one of the two isoforms of acetyl-CoA carboxylase, the acetyl-CoA carboxylase 2 (ACC2), into malonyl-CoA, which has been proposed to be a sensor and indicator of the nutritional energy status (Lam et al., 2005). The other isoform of this enzyme, ACC1, is cytosolic and is expressed in different lipogenic cell types and functions in fatty acid synthesis. Both isoforms can be inhibited through their phosphorylation by the AMP-activated protein kinase (AMPK), which is hormonally regulated by the ratio between glucagon and insulin

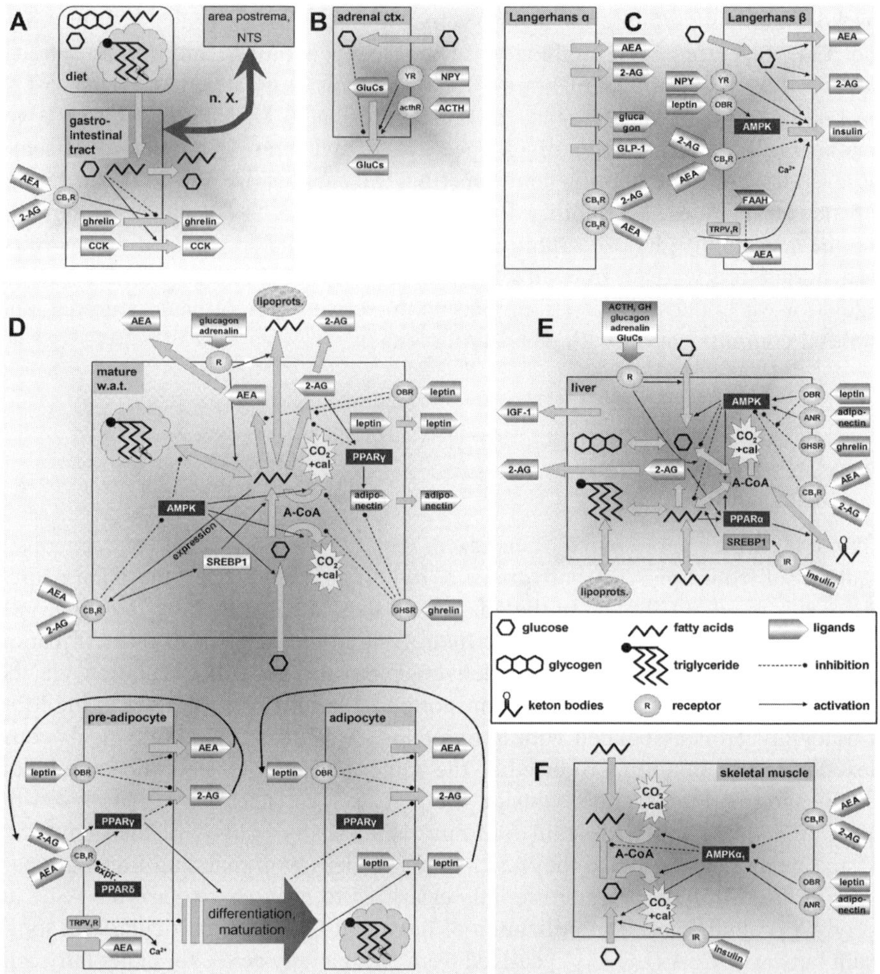

Fig. 2 Major participants in the peripheral regulation of feeding and energy homeostasis, featuring (**a**) the gastrointestinal tract (including the exocrine liver and pancreas), (**b**) the adrenal cortex, (**c**) the endocrine pancreas (α and β cells of the Langerhans' islets), (**d**) the mature adipocytes of the white adipose tissue (w.a.t.) and the maturation of preadipocytes, (**e**) the liver, and (**f**) the skeletal muscle. In contrast to the hypothalamus, in the peripheral tissue, AMPK (5′AMP-activated protein kinase) is activated by leptin and adiponectin, whereas inhibited by ghrelin and endocannabinoids, 2-AG (2-arachidonoylglycerol) and AEA (anandamide). *SREBP1* sterol regulatory element-binding protein; *PPARα,γ* peroxisome proliferator-activated receptor α and γ; *FAAH* fatty acid amidohydrolase, the major anandamide-metabolizing enzyme; *A-CoA* acetyl coenzyme A; *CO_2+cal* complete oxidation into carbon dioxide and heat. Other substances, signals, ligands, and their *receptors*: *NPY* neuropeptide Y (*YR*); *CCK* cholecystokinin; *GluCs* glucocorticoids (mainly cortisol in human) (*GluCR*); *ACTH* adrenocorticotropin (*acthR*); *IGF-1* insulin-like growth factor-1; *$TRPV_1R$* transient release potential receptor vanilloid sybtype-1; *OBR* leptin ("obesity") receptor; *IR* insulin receptor; *ANR* adiponectin receptor; *CB_1R* and *CB_2R* cannabinoid receptor type-1 and -2; GLP-1 glucagon-like polypeptide-1

and will also modulate the intracellular levels of cAMP. AMPK, representing two main isoforms (AMPKα1 and AMPKα2), is considered a cellular fuel gauge and plays a key role in the regulation of energy metabolism. Activated by an increase in the AMP/ATP ratio (ATP depletion), AMPK switches on catabolic pathways such as fatty acid oxidation, and switches off anabolic pathways such as lipogenesis or gluconeogenesis in order to limit further ATP utilization by anabolic pathways and fatty acid synthesis (Hardie et al., 2006). In fact, AMPK regulates glucose and fatty acid homeostasis in the whole body. In the skeletal muscle, activation of AMPK increases glucose uptake and lipid oxidation. Instead, in the liver, activation of AMPK inhibits glucose and lipid synthesis but still increases lipid oxidation. Lipolysis and lipogenesis in the adipose tissue are also reduced following stimulation of AMPK. Activation of pancreatic AMPK is associated with decreased insulin secretion. Collectively, activation of AMPK in muscle, liver, and adipose tissue results in the decrease of circulating glucose and lipids and of ectopic fat accumulation, to maintain the normal homeostasis of glucose and fatty acids and to protect the body against the metabolic syndrome (Rossmeisl et al., 2004). AMPK also integrates hormonal and nutrients signals in multiple tissues. The adipocyte-derived hormones, leptin and adiponectin, activate AMPK in peripheral tissues, including skeletal muscle and liver, thereby increasing energy expenditure and fatty acid oxidation together with glucose uptake (Muoio et al., 1997). In the hypothalamus, AMPK is instead inhibited by leptin and insulin, two hormones that suppress feeding, while ghrelin, a hormone that increases food intake, activates this enzyme (Ahima, 2006; Carling, 2005). This means that in fasted animals, when leptin and insulin circulating levels are low and ghrelin levels are high, increased AMPK activity and intracellular cAMP levels will increase the phosphorylation of ACC thereby decreasing its activity and increasing hypothalamic lipid oxidation.

Fatty Acid β-Oxidation

Fatty acid β-oxidation occurs in the mitochondria once fatty acyl-CoA crosses both inner and outer mitochondrial membranes. The inner mitochondrial membrane is impermeable to fatty acyl-CoAs; therefore, these compounds are carried across this membrane by transmembrane proteins, the carnitine acyltransferases (Wolfgang and Lane, 2006a). These enzymes catalyze the exchange of acyl groups between carnitine and coenzyme A (CoA) and include carnitine acetyltransferases (CATs), carnitine octanoyltransferase (COT), and carnitine palmitoyltransferases (CPTs). CPT-1 and CPT-2 are crucial for the β-oxidation of long-chain fatty acids in the mitochondria by enabling their transport across the mitochondrial membrane into the mitochondrial matrix. It is important to notice that malonyl-CoA is a reversible and effective inhibitor of CPT1. When this compound is present in high levels, namely, as mentioned above, in fed animals, it will block fatty acid β-oxidation, and behave as a key regulator in fatty acid homeostasis (Lam et al., 2005). CTP1, but also the pyruvate dehydrogenase kinase 4 (PDK4), are instead stimulated by a peroxisome proliferators-activated gamma (PPARγ) coactivator-1 (PGC1) to control energy metabolism (Postic et al., 2004; Sugden and Holness, 2006). Substrate competition between fatty acids and glucose occurs at the level of the pyruvate dehydrogenase complex (PDHC), which catalyzes the oxidative decarboxylation of pyruvate to form acetyl-CoA. Thus, it links glycolysis to ATP production and to the tricarboxylic acid cycle (TCA cycle or Krebs' cycle), which depends mostly on the activity of three principal enzymes: fumarase (FUMA), aconitase, and oxoglutarate dehydrogenase (OGDH) (Sugden and Holness, 2006). Inactivation of PDHC by phosphorylation is catalyzed by the PDK4, whose activity depends on the nutritional and endocrine status (Finck and Kelly, 2006). Mitochondrial β-oxidation contributes to energy production via oxidative phosphorylation generating ATP. Fatty acyl-CoA will enter an oxidative spiral, beginning with the dehydrogenation of the fatty acyl-CoA by an acyl-CoA dehydrogenase. Three different acyl-CoA dehydrogenases function in this process, according to the size of the fatty acyl-CoA: the short, medium, and long chain acyl-CoA dehydrogenases which are under the regulation of the peroxisome proliferator-activated receptor alpha (PPARα; see also Chap. 9). Instead, the enoyl CoA hydratase (crotonase) catalyzes the second step of oxidation. The resulting acetyl-CoA units produced at each cycle of fatty acid β-oxidation have three possible destinies: (1) Acetyl-CoA condenses with oxaloacetate, normally provided by the glycolytic pathway via pyruvate, to form citrate which can enter the TCA cycle for complete oxidation to CO_2 and ATP generation. (2) Acetyl-CoA-derived citrate can also be exported to the cytosol for the synthesis of fatty acids, if necessary. (3) In the case of fasting or diabetes, acetyl-CoA can be converted into ketone bodies by the main enzyme involved in this process, the hydroxymethylglutaryl-CoA synthase (HMG-CoAS).

Fatty Acid Synthesis

Fatty acid synthesis starts from acetyl-CoA and malonyl-CoA and involves a sequence of six reactions for each two carbons added to the acyl chain. Fatty acid synthase (FAS) generates saturated fatty acids that can be further modified by the addition of a double bond catalyzed by acyl-CoA desaturases, or stored as triglycerides (Wolfgang and Lane, 2006a,b). FAS is a multifunctional protein organized into globular domains encoded by a single gene. This process is also regulated by the adipocyte determination (ADD) and differentiation factors, sterol regulatory element-binding proteins (SREBPs), which are intracellular transcription factors that are activated by sterol depletion to stimulate fatty acid synthesis (Eberle et al., 2004). The SREBP family is composed of three isoforms, SREBP-1a, SREBP-1c, and SREBP-2, which all have different roles in lipid synthesis. SREBP-1c is involved in fatty acid synthesis and insulin-induced glucose metabolism (particularly, during lipogenesis) and seems to be mainly regulated at the transcriptional level by insulin, whereas SREBP-2 is relatively specific for cholesterol synthesis. In contrast, SREBP-1a seems to be implicated in both pathways (Eberle et al., 2004). The skeletal muscle lacks FAS and serves mostly as a regulatory tissue since it regulates the levels of malonyl-CoA through the activity of ACC2, which converts acetyl-CoA to malonyl CoA, and of malonyl-CoA decarboxylase (MCD), which regenerates acetyl-CoA by decarboxylating malonyl CoA. Again, both enzymes are regulated by AMPK. Instead, in the liver and also in the hypothamus, FAS is dominant (Wolfgang and Lane, 2006a,b).

The Endocannabinoid System Controls Glucose Homeostasis

At the Systemic Level

It is more than forty years that scientists have recognized that marihuana extracts and preparations modulate carbohydrate metabolism in animal models (el-Sourogy et al., 1966). Hashish smoke was shown to increase blood glucose level in rats (Mahfouz et al., 1978), and cannabis resin does the same in dogs together with decreasing glucose tolerance (de Pasquale et al., 1978). Sanz and colleagues (1983, 1985) observed that upon both chronic and single Δ^9-THC injection of rats, energetic and detoxifying glucose metabolism increased in liver postmitochondrial fractions, paralleled by a decrease in glycogen levels. Apart from animal experiments, observations in volunteer human subjects indicated that marihuana extracts may control glucose homeostasis: in six hospitalized volunteer male subjects it was shown that 14 days of pretreatment with a 210 mg/day dose of Δ^9-THC reduced insulin-induced hypoglycemia-evoked increase of growth hormone blood levels to one third and that of cortisol to half (Benowitz et al., 1976). Although we should consider the doses of Δ^9-THC relatively high in the above-listed studies, rediscovering the importance of cannabinoid research in glucose homeostasis with finer pharmacological and molecular tools in the last ten years has supplied us with data overall consistent with those from the 1970s and 1980s. A recent report has accordingly demonstrated that CB_1 receptor activation induces glucose intolerance in rats that were injected with 2 g/kg glucose i.p. (Bermúdez-Siva et al., 2006). When animals were injected with the selective CB_1 receptor agonist ACEA or the endogenous agonist anandamide 30 min prior to glucose administration, significantly higher and prolonged glucose levels were observed compared to control animals. The selective CB_1 receptor antagonist AM251 not only prevented the effect of anandamide and ACEA but also per se facilitated the clearance of glucose from the blood. The capability of CB_1 receptor antagonists to increase glucose tolerance stretches beyond antagonizing the effects of externally given CB_1 receptor agonists. High fat diet-induced obesity (DIO) also results in elevated blood glucose levels in animals, and SR141716A (10 mg/kg for 10 weeks) was shown to reduce leptin, insulin and glucose levels by 67–81% in mice with DIO (Poirier et al., 2005), concomitantly with favorable changes in serum cholesterol levels and HDL/LDL cholesterol ratios. Likewise, SR141716A, at the same dose and already after 14 days of treatment, decreased feeding, blood glucose and insulin levels, and increased insulin sensitivity and, possibly, thermogenesis in the brown adipose tissue, all assessed in lean vs. obese Zucker rats (Doyon et al., 2006). As it will be discussed below, these beneficial actions of CB_1 receptor antagonism in obese animals appear to be effected both peripherally and centrally.

At the Neurohumoral Level

Several lines of evidence indicate the involvement of the endocannabinoid system in the central regulation of systemic glucose homeostasis (Fig. 1). A compelling one is that direct injection of anandamide into the ventromedial hypothalamus induces significant hyperphagia which can be diminished with pretreatment with SR141716A (Jamshidi and Taylor, 2001). In the paraventricular nucleus, acute fast glucocorticoid feedback on the glutamatergic afferents that stimulate the release of several peptide hormones was shown to be mediated by endocannabinoid retrograde actions on glutamate release (Di et al., 2003). This process involves a rapid stimulation of endocannabinoid synthesis through a $G_{s\alpha}$–cAMP–PKA pathway upon stimulation of the plasma membrane glucocorticoid receptor. This synthesis and release of endocannabinoids is counteracted by the peripheral antiobesity hormone leptin via the stimulation of phosphodiesterase-3B and in turn, a decrease in cAMP levels (Malcher-Lopes et al., 2006). This mechanism might explain in part why defective leptin signaling is associated with elevated hypothalamic levels of endocannabinoids in obese *db/db* and *ob/ob* mice and Zucker rats (Di Marzo et al., 2001). As for ACTH and glucocorticoids, it has been shown that CB_1 receptor activation increases their plasma levels and the mRNA levels in the hypothalamus for proopiomelanocortin (POMC) and corticotrophin releasing hormone (CRH), and that the effect of Δ^9-THC on glucocorticoid levels is counteracted by SR141716A (Weidenfeld et al., 1994; Wenger et al., 1997; Corchero et al., 1999; Manzanares et al., 1999; Fig. 1). In CB_1 receptor-knockout mice, a widespread dysregulation of the HPA axis has been recently discovered. These animals display increased corticosterone levels at the circadian peak and increased CRH mRNA expression in the paraventricular nucleus, as well as a hyperresponsiveness of ACTH release to CRH and forskolin at the pituitary level in vitro, and a paradox increase in the release of ACTH and corticosterone upon treatment with low dose of dexamethasone (Cota et al., 2007). The CB_1 receptor antagonist SR141716A has also been shown to stimulate basal circulating corticosterone levels and the activity of the HPA axis in food-deprived obese rats (Doyon et al., 2006). Although these recent data on the effect of CB_1 receptor inactivation on the HPA axis sound somewhat unexpected as they appear to be in contradiction with earlier data on similar effects of CB_1 stimulation, they indicate that one of the central anorexic mechanisms of CB_1 receptor antagonism might exploit the anorexic effects of CRH and, perhaps, melanocortins (Fig. 1). Ghrelin is one of the major orexigenic hormones that are secreted both centrally and peripherally, and regulates ingestive behavior and energy homeostasis (Gil-Campos et al., 2006). Ghrelin is the endogenous agonist for the growth hormone secretagogue receptor. It stimulates appetite when intracerebroventricularly injected, through stimulating the synthesis of neuropeptide Y and agouti-related protein in the arcuate nucleus and hindbrain. It controls the release of proopiomelanocortin, insulin and leptin, and, in turn, the latter two control ghrelin release (Gil-Campos et al., 2006). When fasted rats were injected with SR141716A (5 mg/kg), diminished food intake was observed in the first 20 min after injection, together with a smaller increase in blood ghrelin levels. In fed rats, however, SR141716A injection diminished blood ghrelin level by 35% vs. vehicle-injected animals (Cani et al., 2004; Fig. 2). A subanorectic dose of SR141716A was shown to prevent intrahypothalamic ghrelin from stimulating appetite (Tucci et al., 2004). Later, it was discovered that both ghrelin and the CB_1 receptor agonists, 2-AG and Δ^9-THC, either administered peripherally or centrally, increase AMPK phosphorylation and activity by 50–70% in the hypothalamus (Kola et al., 2005; Fig. 1). At the cellular level, AMPK is an integrator of feeding- and energy-regulator hormonal signals (Hardie, 2004; Xue and Kahn, 2006). Increase in phospho-AMPK, for instance, diminishes the synthesis and increases the oxidation of fatty acids, increases glucose uptake and oxidation and mitochondrial biogenesis. This is what happens when AMPK is stimulated in peripheral tissues (e.g., in the skeletal muscle and adipose tissues) by leptin and adiponectins (Hardy et al., 2004; Fig. 2). Ghrelin and Δ^9-THC strongly suppress AMPK in adipose tissues and in the liver and, instead, they stimulate it in the hypothalamus (Kola et al., 2005; Fig. 1). By doing so, they suppress glucose uptake and glucose oxidation in the liver and, in parallel, set free from inhibitory control the enzymes of gluconeogenesis. Although the role of CB_1 receptors was not investigated by the authors, this mechanism might explain why anandamide and ACEA reduce glucose tolerance in vivo in rats (Bermudez-Siva et al., 2006). Intriguingly, this central vs. peripheral dichotomy also exists for leptin, which inhibits AMPK activity in the hypothalamus but increases it in the peripheral tissues (Hardie, 2004; Figs. 1,2). What could be the underlying mechanism for the opposing central and peripheral effects for endocannabinoids, ghrelin and leptin on AMPK, and how could SR141716A prevent the central action of ghrelin (Tucci et al., 2004)? Although several mechanisms can be hypothesized, one interesting and challenging idea is the heterodimerization between CB_1 receptors and receptors for ghrelin or leptin. The CB_1 receptor and the receptor for another orexigenic peptide, the orexin-1 receptor, also can form a heterodimer which results in a novel functional entity (Ellis et al., 2006), and may be sensitive to cross-desensitization between ligands. Such a hypothetical heterodimer of CB_1 ghrelin/leptin receptors may therefore exist in the hypothalamus but not in the periphery, which would explain the opposing central vs. peripheral effect of endocannabinoids, leptin and ghrelin. Further information on CB_1 receptor heterodimers can be found in Chap. 9.

At the Adipocyte Level

As mentioned above, both ghrelin and Δ^9-THC suppress AMPK activity in the adipose tissue, resulting in adipogenesis and the suppression of fatty acid and glucose oxidation. In view of the fact that increased endocannabinoid levels have been measured in visceral white adipose tissue removed from obese patients (Matias et al., 2006), it is possible that this phenomenon contributes to the increasing fat deposits (see below and Fig. 2). At the cellular level in the white adipose tissue, systemic CB_1 receptor antagonism has been shown to exert additional beneficial effects on glucose metabolism. In high-fat diet-fed C57BL/6J mice, SR141716A (10 mg/kg orally for 40 days) induced upregulation of both glycogen phosphorylase and glycogen synthase. The former enzyme facilitates the breakdown of glycogen and, thus, feeds glycolysis, whereas the latter is a rate-determining enzyme for glycogen synthesis. The authors hypothesized that this would represent an enhanced "futile cycle" in the white adipose tissue cells with concomitant and constant energy dissipation (Jbilo et al., 2005; Fig. 2). The same study enriched further our knowledge on the impact of CB_1 receptor blockade on glucose metabolism in the white adipose tissue. SR141716A induced an upregulation of the insulin-responsive glucose transporter, GLUT4, which suggests a facilitated glucose transport and consequently, increased glycolysis. In contrast to this finding, however, Maccarrone and coworkers have demonstrated that anandamide stimulates insulin-induced glucose uptake in a white adipocyte cell line, the mouse 3T3L1 adipocytes, and that this effect is blocked by SR141716A (Gasperi et al., 2007). Upregulation by SR141716A of glycolytic enzymes, namely the phosphofructokinase, the glyceraldehyde-3-phosphate dehydrogenase, the phosphoglycerate mutase, and the β-enolase, was, however, also demonstrated (Jbilo et al., 2005). In conclusion, blockade of CB_1 receptors might increase glucose uptake and metabolism in the white adipose tissue cells, resulting in an increased energy loss, which might at least in part contribute to the observed 18–27% weight loss in mice with DIO (Ravinet-Trillou et al., 2003; Jbilo et al., 2005; Poirier et al., 2005).

At the Level of Langerhans Islets 1. Cannabinoid Receptors

Juan-Picó and colleagues (2006) observed that CB_1 receptors are not present in insulin-releasing β-cells of pancreatic islets, whereas both β-cells and non-β-cells are equipped with the CB_2 receptor. Pharmacologically, the CB_2 receptor also proved to be the major regulator of insulin secretion. Although they concluded that anandamide and 2-AG decrease intracellular Ca^{2+} oscillations and insulin release via CB_2 receptor activation, paracrine regulation of β-cells by CB_1 receptors in α-cells cannot be excluded; otherwise there would be a mismatch with the findings showing that CB_1 receptor antagonism increases glucose tolerance and decreases blood insulin levels (Bermudez-Siva et al., 2006; Doyon et al., 2006). Two of us recently reported, for the first time in the same study (Starowicz et al., submitted), the expression of endocannabinoid receptors and metabolic enzymes in intact mouse pancreatic islets. The general scenario that emerges from the results that we obtained using both DAB and immuno-fluorescence staining is that endocannabinoid biosynthesizing enzymes are mostly localized in glucagon-secreting β-cells, together with CB_1 and CB_2 receptors, whereas degrading enzymes appear to be mostly localized in insulin-secreting β-cells, where, staining of CB_2, but not CB_1, receptors was also localized (Fig. 2), in agreement with the work by Juan-Pico and colleagues (2006). However, in the same study we also found that CB_1 receptors are expressed in a small population of rat β-cells. Therefore, a cautious conclusion that can be reached from these findings is that, although endocannabinoids are certainly produced from β-cells to act mostly on cannabinoid receptors on these cells, they might, depending on the animal species, also act in a paracrine way on both cannabinoid receptor types expressed in some β-cells, and hence regulate insulin release. Anandamide might also regulate insulin release from α-cells by activating $TRPV_1$ receptors expressed in these cells (see below). These findings provide a potential additional explanation to the reduced CB_1 receptor-induced glucose intolerance observed in vivo (Bermudez-Siva et al., 2006). In fact, autocrine stimulation of CB_1 receptors in α-cells by endocannabinoids might affect glucagon or glucagon-like peptide 1-release from α-cells and, hence, the stimulatory effects of these hormones on insulin release (Moens et al., 1998; Dyachok et al., 2006). Therefore, future studies will have to be aimed at investigating the effect of CB_1 stimulation on hormone release from α-cells. Interestingly, in cultured rat insulinoma β-cells, where both CB_1 and CB_2 receptors are expressed, conditions mimicking hyperglycemia cause elevation of the levels of anandamide and 2-AG (Matias et al., 2006), which implies that in hyperglycemic patients, a dysregulation of pancreatic endocannabinoid levels might represent, depending on the yet-to-be-fully-understood effect of endocannabinoids on insulin release, either a maladaptive cause or an adaptive change for hyperglycemia.

At the Level of Langerhans Islets 2. $TRPV_1$ (Vanilloid) Receptors

As emphasized in the pharmacology of the vanilloid system, endocannabinoids such as anandamide and *N*-arachidonoyl dopamine (NADA) are also capable of activating the $TRPV_1$ vanilloid receptor thereby causing Ca^{2+} and Na^+ entry and depolarization of the cell membrane, with a facilitated exocytosis. In this case, these endogenous ligands are expected to work like capsaicin. Notably, systemic capsaicin administration increases glucose-stimulated insulin level in the dog (Tolan et al., 2001), which prompted Akiba and coworkers (2004) to test the presence of the $TRPV_1$ receptor in β-cells and the effect of capsaicin on insulin release. Both $TRPV_1$ receptor immunoreactivity and mRNA were detected in rat pancreas in cells corresponding to β-cells and in the RIN β-cell line; capsaicin induced the release of insulin from RIN cells, and when systemically administered, elevated insulin levels in fasting rats. One should consider, however, that afferent nerve terminals in the pancreatic islets are also equipped with $TRPV_1$ receptors, and activation of these receptors induces the release of peptide neurohormones which may differently affect the function of the endocrine pancreas. In line with this evidence, neonatal ablation of capsaicin-sensitive afferents in the mouse pancreas increased glucose tolerance in these animals in their adult life (Karlsson et al., 1994).

Furthermore, the $TRPV_1$ receptor can be either desensitized or sensitized by direct interaction with the CB_1 receptor upon coactivation cells coexpressing the two receptors, depending on whether or not the cAMP-dependent protein kinase A is activated (Hermann et al., 2003; Oshita et al., 2005), which may happen in the islets, thereby resulting in a net inhibitory or stimulatory effect, respectively, of anandamide on insulin release. Finally, $TRPV_1$ activation can cause cell toxicity due to the heavy Ca^{2+} and Na^+ load (Kim et al., 2005). In inflamed tissues, activators of the $TRPV_1$ are released from surrounding immune or necrotic cells (Nagy et al., 2004), and ethanol potentiates the action of agonists at the receptor (Trevisani et al., 2002). Therefore it would be wise to investigate whether or not β-cell loss in pancreatitis, especially when exacerbated with ethanol consumption, is a direct consequence of $TRPV_1$ receptor overactivation.

At the Neuronal and Astrocytic Level

Several in vivo studies have addressed the impact of systemic cannabinoid ligand treatment on local cerebral glucose utilization. One of the early studies investigated the effects of low to high doses of Δ^9-THC on radiolabelled 2-deoxy-D-glucose uptake using autoradiography in male rats. Interestingly, a low (0.2 mg/kg) dose of Δ^9-THC increased 2-deoxy-D-glucose uptake in all examined cortical and limbic structures, but not in the examined diencephalic and brainstem areas. In contrast, Δ^9-THC at 2.0 and 10.0 mg/kg inhibited 2-deoxy-D-glucose uptake in most of these regions (Margulies and Hammer, 1991). The anatomic localization of Δ^9-THC effects is therefore consistent with the distribution of CB_1 receptors in the brain (see Chap. 10). However, the biphasic nature of this effect, namely activation at low and inhibition at high concentrations, although typical of many actions of cannabinoids, is not easy to explain. One may assume that CB_1 receptors in GABAergic neurons are more sensitive to cannabinoid agonists, perhaps because of their higher density. Therefore, a low Δ^9-THC concentration would inhibit GABAergic inhibition leading to a net neuronal excitation and higher glucose utilization. By contrast, at higher Δ^9-THC concentrations, glutamate release is also inhibited, resulting in a net inhibition of neuronal excitation and glucose utilization. However, our recent comprehensive work undermines this hypothesis, because we demonstrated that WIN55212-2, a full and potent CB_1 receptor agonist, inhibits presynaptic release of GABA and glutamate with equal potency and similar efficacy (EC_{50}, ~60 nM; Köfalvi et al., 2007). Another early study utilizing positron emission tomography in eight normal human subjects reported that Δ^9-THC increases 18F-2-fluoro-2-deoxyglucose metabolism (Volkow et al., 1991). This increase was only observed in the cerebellum, whereas global cerebral glucose metabolism in response to Δ^9-THC was variable. In this study, a relatively low amount (2 mg) of Δ^9-THC was injected intravenously to the subjects, and furthermore, during a PET scan, the majority of CB_1 receptor-positive forebrain neurons (those regulating motor functions and cognition) are expected to be idle using glucose at baseline level. As we discuss below, glucose uptake in an idle neuronal network is likely not subject to modulation by CB_1 receptors, questioning the usefulness of a PET scan in this type of research. Returning to the rat model, Pontieri and colleagues (1999) have reported that intravenously injected low doses of WIN55212-2 modulated 2-deoxy-D-glucose uptake in selected brain areas of awaken rats, and yet failed to affect behavior. At 0.15 mg/kg, WIN55212-2 elevated 2-deoxy-D-glucose uptake in the shell of the nucleus accumbens by 23%, which was interestingly not observed at the 0.3 mg/kg dose. This bell-shaped dose–response curve can be interpreted as a possible outcome of the interaction at the network level between CB_1 receptor-positive and -negative neurons. At the 0.3 mg/kg dose, however, WIN55212-2 decreased 2-deoxy-D-glucose uptake in the range of 19–33% in the ventromedial thalamus and in all subareas of the hippocampus, whereas other brain areas were still unaffected (Pontieri et al., 1999). In contrast to the findings of Margulies and Hammer (1991) and Pontieri and colleagues (1999), Freedland and colleagues (2002) reported that the low (0.25) dose of Δ^9-THC failed to affect 2-deoxy-D-glucose uptake in the rat brain. They found that only moderate (1.0–2.5 mg/kg) doses of Δ^9-THC (i.p.) inhibited, dose dependently, 2-deoxy-D-glucose uptake in the rat brain 15 min after the injection of the tracer. At the highest Δ^9-THC dose tested, most (28 out of 38) brain areas were affected (in the range of −25 to −42%), and all effects were prevented by pretreatment with SR141716A. Brain areas of the limbic and sensory systems were affected to the highest extent. The same laboratory also reported that a single i.p. injection of Δ^9-THC (2.5 and 10 mg/kg) caused an inhibition of 2-deoxy-D-glucose uptake–an effect lasting for hours depending on the brain area. For instance, at 2.5 mg/kg, significantly reduced 2-deoxy-D-glucose uptake was observed in the auditory and infralimbic cortices, in the superior colliculus, in the amygdala, in the shell of nucleus accumbens, and in the ventral caudate nucleus, 24 h after injection (Whitlow et al., 2002). Paradoxically, Δ^9-THC at 10 mg/kg caused less profound inhibition 24 h after its injection, and only in a limited number of areas, namely in the infralimbic cortex and the central amygdala. Although this observation clearly indicates the involvement of CB_1 receptor desensitization by the high Δ^9-THC dose, the site of action (neuronal, extraneuronal, cerebral or systemic) is unclear. This question is supported by the following observation of the same laboratory. After 7 or 21 days of repeated single injections of Δ^9-THC, the last injection of Δ^9-THC still induced a decrease of 2-deoxy-D-glucose uptake in a few brain areas encompassing the nucleus accumbens, mediodorsal thalamus, basolateral amygdala, portions of the hippocampus and median raphe (Whitlow et al., 2003). Although 2-deoxy-D-glucose uptake was similar to vehicle treatment in the majority of brain areas that were affected by a single dose of 10 mg/kg Δ^9-THC in the previous study, it is notable that the areas showing a reduced 2-deoxy-D-glucose uptake were not restricted only to the infralimbic cortex and the central amygdala – the only two areas that were still affected by Δ^9-THC (10 mg/kg) 24 h after its injection (Whitlow et al., 2002). In other words, it is likely that desensitization of cannabinoid receptors and tolerance to cannabinoid agonists include extraneuronal/extraglial components as well. Even more intriguing was the following report of the same laboratory with the CB_1 receptor antagonist/inverse agonist SR141716A.

SR141716A dose-dependently inhibited 2-deoxy-D-glucose uptake in the thalamus, hippocampus, and limbic system of operant behaving rats; in other words, it induced changes in the same direction as the agonist in the previous tests. More surprisingly, this inhibition was more prominent and covered a larger number of brain areas in Δ^9-THC-tolerant animals, namely in animals in which CB_1 receptor desensitization (in other words, lower CB_1 receptor functionality) is expected. Therefore, what one can conclude from in vivo studies is that both the agonist and the antagonist inhibit glucose uptake, though the antagonist prevents the effect of the agonist (a phenomenon that would then be an occlusion rather than a classical antagonism); and furthermore, tolerance toward these ligands seems unlikely to be solely a local cerebral phenomenon. The studies listed above, however, carry several limitations: (1) first of all, usually it is concluded that if glucose uptake is inhibited then glucose utilization/metabolism is also inhibited. We will discuss below how this is not necessarily the case. The term "metabolic mapping" used in these studies also should be avoided since 2-deoxy-D-glucose can only be "metabolized" to 2-deoxy-D-glucose-6-phosphate as an end-product; therefore, it cannot be a tool for metabolic mapping. (2) Second, systemic injection with cannabinoid agonists decreases cerebral blood flow (Goldman et al., 1975; Bloom et al., 1997) and concomitantly, the level of available cerebral glucose. Cannabinoids via several target organs and receptors also change systemic levels of glucose and different types of hormones (see above), as well as core body temperature, in other words, systemic energy expenditure, at higher concentrations. (3) Last but not least, cannabinoid ligands may affect behavior, and neurotransmission, consequently – neuronal activity. However, the endogenous modulator role of endocannabinoids in brain cells and circuitries encompasses a mere local (autocrine or paracrine) effect on cell somas, and does not necessarily affect the activity of a larger network or, to the extreme end, the levels of blood glucose. To understand this in detail, we should mention that, when resting, neurons take up and use glucose themselves, but when the circuitry is active, astrocytes take up and metabolize glucose at a higher rate than neurons, and then pass lactate to neurons as an alternative fuel and substrate for biosynthetic pathways. Indeed, astrocytes store the majority of the brain energy store, glycogen, which has a rapid turnover corresponding to synaptic activity (Magistretti and Pellerin, 1996; Wiesinger et al., 1997). Altogether, changes in astrocytic glucose metabolism may influence neuronal functions. Earlier investigations in astrocytic cultures have shown that Δ^9-THC and the potent CB_1 receptor agonist HU-210 increase glucose oxidation into CO_2 and glucose incorporation into phospholipids and glycogen, in an SR141716A-dependent manner (Sánchez et al., 1998). The underlying mechanism suggested consisted of sphingomyelin breakdown into ceramide, which in turn stimulated p42/44 MAPKs, leading to Raf-1 phosphorylation and translocation. In summary, acute experimental hyperglycemia, elicited with intravenous or intracerebrovetricular glucose injection, increases neuronal network functions, whereas hypoglycemia obviously induces the opposite effect (Watson and Craft, 2004). Thus, a decrease in cerebral and, more specifically, hippocampal glucose supply attenuates higher order brain functions including memory processes. The above-listed findings together with those showing that both CB_1 receptor agonists and inverse agonists (SR141716A or AM251) decrease glucose uptake in vivo, call for further, carefully planned experiments to unravel the exact contribution of the modulation of glucose uptake and metabolism to the observed – and sometimes controversial – effects of cannabinoid ligands on synaptic transmission, memory, cognition, behavior, as well as in ischemia and neuroprotection.

Possible Roles of CB_1 Receptors in Diabetic Encephalopathy

Diabetes mellitus, the most common metabolic disorder in man, affects a plethora of tissues and organs, including the brain. "Diabetic encephalopathy" encompasses characteristic biochemical, electrophysiological, morphological, and cognitive deficits in diabetic patients (Trudeau et al., 2004). Type-1 diabetic rats also display a compromised long term potentiation (LTP), cognitive deficits and marked changes in the density of pre- and postsynaptic synaptic markers, decreased cell proliferation, and apoptosis in certain cortical regions (Chabot et al., 1997; Jackson-Guilford et al., 2000; Li et al., 2002; Grillo et al., 2005). As discussed above and in several chapters of this book, CB_1 receptors are intricately involved in this spectrum of physiological and pathological mechanisms. This prompted one of us to test if the density and the expression of CB_1 receptors are changed in the streptozotocin-induced classic animal model of type-1 diabetes. We found that in the diabetic animals, CB_1 receptor immunoreactivity was increased in the nerve terminals and in total hippocampal cell membranes as well (Duarte et al., 2007). Binding experiments utilizing [^{3}H]SR141716A have additionally demonstrated (1) that the majority of binding sites is located to the nerve terminals, indicating a major neuromodulator role for the CB_1 receptors; and (2) that in diabetic animals, the binding density is increased – notably by 2–3 times compared to the increases in total nerve terminals and membrane protein densities. This finding might suggest that type-1 diabetes increases not only protein density but also receptor activity. Real-time quantitative analysis of CB_1 receptor expression revealed a *decrease* in the hippocampal level of CB_1 receptor mRNA in the diabetic animals. This suggests an exhaustive (accelerated) translation from mRNA rather than a decreased CB_1 receptor recycling as the underlying mechanism for the increased CB_1 receptor density observed. It also indicates that the expression of CB_1 receptors (namely, the level of CB_1 mRNA) does not always change in a way parallel to that of protein density and function. Indeed, an increase in protein translation was already reported to occur in the hippocampus after traumatic brain injury (Chen et al., 2007), upon activation of a key rate-limiting translational (the mTOR) pathway. It is also notable that our results have been confirmed by Zhang and colleagues (2006): in rat pheochromocytoma (PC12) and human neuroblastoma (SH-SY5Y) cell lines, hyperglycemia induced a decrease in CB_1 receptor expression (mRNA levels), associated with a reduction in total neurite length. The increased density of presynaptic CB_1 receptors may reflect an elevated neuromodulator power to restore normal LTP functions or itself contribute to the impairment of synaptic plasticity – a question to be answered by future functional studies. Elevated CB_1 receptor density in these fractions may reflect a role also in the control of metabolism, cell survival, and neurogenesis. Indeed, CB_1 receptors, forming heterodimers with TrkB BDNF receptors, are responsible for correct migration and axonal arborization of cortical neurons (Berghuis et al., 2005; see Chap. 12). Therefore, impairment of this "cosignaling" results in impaired neuritogenesis, similar to that observed in cultured cells by Zhang and colleagues (2006). The level of phosphorylated (active) Akt is also increased in the streptozotocin model of diabetes, with a concomitant increase in levels of phosphorylated (inactive) glycogen synthase kinase 3β (GSK3β) (Clodfelder-Miller et al., 2005). Since CB_1 receptors can activate the survival factor Akt (Gómez del Pulgar et al., 2002), and thus indirectly inhibit GSK3β, they can promote survival, as well as dendritic arborization, which are normally controlled by the active form of GSK3β. Controlling glucose utilization of hippocampal neurons is another aspect whereby CB_1 receptors may either worsen or improve neuronal functions and survival chances in the absence of insulin, namely, in type-1 diabetes.

The Endocannabinoid System Controls Fatty Acid Homeostasis

At the Systemic Level

Cota and coworkers have demonstrated that CB_1 receptor null mutant (knockout) mice exhibit a slightly albeit significantly lower body weight than wild-type littermates during a period of 12 weeks from birth, starting from week 3 (Cota et al., 2003b). This decrease in body weight is accompanied by a decrease in fat mass and by a corresponding increase of lean mass. Since CB_1 receptor knockout and wild-type mice show similar circadian variations in body temperature and locomotor activity, and only a trend toward higher energy expenditure (which corresponds to the energy combustion and fat and carbohydrates oxidation), the weight loss observed in CB_1 receptor knockout mice must be directly connected to the absence of CB_1 receptors and to their role not only in ingestive behavior but also in fat mass accumulation and in fatty acid synthesis (Cota et al., 2003b). Moreover, the blockade of the CB_1 receptor in a 40-day regimen with SR141716A in the same animal model reduced body weight also persistently in a dose-dependent way, and this effect was accompanied by a decrease of white adipose tissue in epidydimal, perirenal, and lumbar tissues (Cota et al., 2003b).

At the Neurohumoral Level

Several lines of evidence indicate the involvement of the endocannabinoid system in the central regulation of fatty acid homeostasis (Fig. 1). Since leptin modulates the activity of AMPK and fatty acid homeostasis, it was interesting to observe that hypothalamic endocannabinoid levels are decreased after systemic leptin administration in rats, and increased in rodent models of congenital hyperphagia and obesity, such as *db/db* mice, and Zucker rats, where leptin signaling is defective, as well as in *ob/ob* mice, where leptin biosynthesis is defective and exogenous leptin can restore the normal (low) levels of endocannabinoids (Di Marzo et al., 2001). Furthermore, it was shown that hypothalamic endocannabinoid levels are increased in rats deprived of food for a short period, whose plasma leptin levels are normally low, and tend to decrease during food consumption, i.e., when plasma leptin levels are higher (Kirkham et al., 2002; Hanus et al., 2003). These data clearly indicate that endocannabinoids play a role in the control of food intake, but has this control anything to do with the effects of leptin on fatty acid homeostasis in the hypothalamus? Kunos and coworkers were the first to suggest the direct involvement of CB_1 receptors in the hypothalamic stimulation of the lipogenic enzyme, FAS. In fact, treatment of mice with HU-210 induced an increase of both SREBP-1c and FAS expression in this brain area (Osei-Hyiaman et al., 2005), an effect that, in view of the stimulatory action of hypothalamic free fatty acids on food intake, might mediate the appetite-inducing action of the cannabinoid receptor agonist. SR141716A was able to prevent the effect of HU-210 on FAS, even though the CB_1 antagonist did not exhibit any effect per se. However, SR141716A did reduce both the food intake and the expression of hypothalamic SREBP-1c and FAS induced by a cycle of fasting and refeeding constituted by 24-h fasting, followed by a 3-h period of either continued fasting or refeeding with a high-carbohydrate diet. Therefore, SR141716A might inhibit food intake, among other things, also by reducing the expression of SREBP-1c and FAS, in fasted/refed but not in free-feeding mice (Osei-Hyiaman et al., 2005). At the hypothalamic level, another study carried out in rats proposed instead an interaction between cannabinoids and AMPK, another actor involved in fatty acid homeostasis, which stimulates fatty acid oxidation according to the body's hormonal and nutritional status. Kola and coworkers (2005) demonstrated that cannabinoids act at AMPK at both central and peripheral levels with opposing effects, and suggested that AMPK might mediate the orexigenic effect of cannabinoids in the rat hypothalamus but also their lipogenic effects in peripheral tissues, as indicated by the previous study of Osei-Hyiaman and colleagues (2005; see Figs. 1 and 2). Kola and coworkers observed that 2-AG administration to rats increased the total activity of AMPK in the hypothalamus due to an increase of its phosphorylation. Furthermore, and possibly subsequent to AMPK activation, the phosphorylation of ACC1 and ACC2 appeared to be also increased after central injection of cannabinoids. The inactivation of the two ACC isoforms would result in an inhibition of fatty acid synthesis and stimulation of fatty acid oxidation in the hypothalamus. The authors suggested that cannabinoids could potentially increase appetite by central AMPK stimulation or by facilitating the restorative actions of AMPK as the hypothalamus senses fuel deprivation. These data seem to be in contrast with the results of Osei-Hyiaman and coworkers (2005), who observed instead a hypothalamic increase of the expression of FAS. However, (1) these two studies were realized in two different animal species (mouse and rat); (2) in the study of Osei-Hyiaman and colleagues, SR141716A inhibited food intake only in fasted/refed animals but not in free-feeding animals; and (3) no CB_1 antagonist was used by Kola and colleagues to demonstrate the involvement of CB_1 receptors. Last but not least, it is important to keep in mind that the participation and the role of AMPK and also of malonyl-CoA in the hypothalamus is far from being understood and is still under investigation.

At the Adipose Tissue Level

Several studies, carried out either in adipose tissue or in isolated adipocytes, seem to favor the idea that the endocannabinoid system is involved in fatty acid homeostasis in this tissue and that the activation of CB_1 receptors increases de novo lipogenesis. Cota and coworkers (2003b) suggested for the first time the role of the endocannabinoid system in peripheral lipogenesis by showing that CB_1 receptors are expressed in mouse epidydimal fat pads and in a primary epidydimal-derived adipocyte cell line. Furthermore, lipoprotein lipase activity in the primary adipocyte cell line was increased after treatment with the CB_1 agonist WIN55212, and this effect was blocked by SR141716A. Bensaid and coworkers (2003) also advanced a hypothesis that could possibly explain the effect of SR141716A on peripheral lipogenesis. They showed that the expression of one of the major adipocyte-derived hormones, adiponectin, was enhanced by the CB_1 antagonist SR141716A (FIG. 2). They showed that CB_1 is expressed in rat adipose tissue and in the mouse adipocyte 3T3F442A cell line. In addition, they also found that CB_1 receptor expression is upregulated in the adipose tissue of obese Zucker (*fa/fa*) rats in comparison to lean rats, and in differentiated mouse 3T3F442A adipocytes in comparison to undifferentiated adipocytes. SR141716A-induced increase of adiponectin expression in the adipose tissue of obese Zucker (*fa/fa*) rats that was significantly more pronounced that in lean rats. SR141716A also induced adiponectin overexpression in the mouse adipocyte cell line but not on adipocytes from CB_1 receptor knockout mice (Bensaid et al., 2003). In agreement with these results, Matias and coworkers (2006), using the same adipocyte cell line, demonstrated that the activation of CB_1 receptors with HU210 inhibits adiponectin expression in mature/hypertrophic adipocytes and instead stimulates preadipocyte differentiation and lipogenesis. In fact, the chronic stimulation of cannabinoid receptors with HU-210 accelerated the appearance of PPARγ – an early marker of adipocyte differentiation (see Chap. 9 for further reference) – during adipocyte differentiation. Under the same conditions of incubation, HU-210 also stimulated the accumulation of lipid droplets as assessed by Oil Red O-staining. An almost twofold stimulation with HU-210 was observed, also at day 4, when not all preadipocytes are fully differentiated. All these effects were attenuated or reversed by SR141716A (Matias et al., 2006), pointing to the direct role of CB_1 receptors in increasing lipid accumulation in adipocytes (Fig. 2). Stimulation of these receptors is known to be coupled to inhibition of adenylyl cyclase and of cAMP formation, an intracellular event coupled to lipolysis and inhibition of lipogenesis in adipocytes. To assess if the effect of HU-210 on lipid droplets could be due to inhibition of cAMP formation and, hence, inhibition of lipolysis or stimulation of lipogenesis, the authors studied the effect of the compound on forskolin-induced cAMP formation in mature 3T3F442A adipocytes and found that HU-210 dose-dependently inhibited cAMP formation in a way that was significantly attenuated by SR141716A but not by a CB_2 receptor antagonist (Matias et al., 2006). One can then hypothesize that the CB_1 receptor-induced lipogenesis or inhibition of lipolysis might be due to inhibition of cAMP formation. It is important to note that, also in fat pads, the activation of CB_1 receptors stimulates the expression of the important transcription factor SREBP-1c and of its targets, the ACC1 and FAS, suggesting that CB_1 receptors might increase lipid levels by increasing fatty acid synthesis (Osei-Hyiaman et al., 2005). These data, together with the observation that Δ^9-THC inhibits AMPK and, therefore, fatty acid synthesis in the adipose tissue (Kola et al., 2005), and that CB_1 receptor blockade inhibits adipocyte proliferation (Gary-Bobo et al., 2006), suggest that the endocannabinoid system has several potential mechanisms to increase fatty acid storage into white adipocytes and, hence, the mass of adipose tissue (Fig. 2, Table 1). In both visceral and subcutaneous adipose tissue, Δ^9-THC decreases AMPK activity although this effect only in the visceral fat was accompanied by a decrease of AMPK phosphorylation on threonine 172 (Kola et al., 2005). In mouse 3T3F442A adipocytes, SR141716A inhibits cell proliferation in a concentration-dependent manner and stimulates the expression not only of adiponectin, but also

Table 1 Unfavorable cardiometabolic changes in obesity and their reversal by the CB_1 receptor antagonist, SR141716A (Acomplia™)

High-fat (and -sugar) diet (or leptin receptor defect) induces	
Serum	Increase of: the LDL/HDL cholesterol ratio; the levels of glucose, FA, cholesterol, insulin, anandamide (in women) and 2-AG (in men and women)
White adipose tissue	Increase of: adipocyte proliferation and maturation, lipid droplet accumulation, lipogenesis; the level of anandamide (epidydimal fat), anandamide and 2-AG (visceral fat); the expression of PPARγ, SREBP-1c, ACC1, FAS, CB_1 receptor (mainly in visceral fat)
	Decrease of: AMPK function, lipolysis, glycolysis, adiponectin release; the expression of phosphofructokinase, glyceraldehyde-3-phosphate dehydrogenase, phosphoglycerate mutase, β-enolase, CAT, CPT2, crotonase, fumarase, aconitase, oxoglutarate dehydrogenase, cysteine dioxygenase, methylcrotonoyl-CoA carboxylase, methylmalonyl CoA mutase, β_3 adrenergic receptors, GH receptors
Liver	Increase of: the expression of SREBP-1c; the level of anandamide
	Decrease of: glucose oxidation, FA oxidation, AMPK activation
Long-term treatment with SR141716A (Acomplia™) results in	
The reversal of the above detailed unfavorable changes both in animal models and in obese, obese type II diabetic and nonobese type II diabetic patients; and in general, body weight and waist circumference reduction, reduced food-intake, a decrease in cardiometabolic risk factors	

of GAPDH, which is an enzyme involved in lipid and glucose metabolism (Gary-Bobo et al., 2006). However, in agreement with the results by Matias and colleagues (2006), the accumulation of lipids droplets was not affected. The intracellular and secreted levels of adiponectin were increased, respectively, 6.5- and 2.5-fold. Regarding GAPDH expression, the levels of GAPDH mRNA but also the level of cellular GAPDH protein, were increased in a concentration-dependent manner by approximately threefold, also according to the time of exposition to SR141716A (Gary-Bobo et al., 2006). Furthermore, also in whole white adipose tissue, SR141716A was shown to increase GAPDH and other glycolytic and lipolytic enzymes (Jbilo et al., 2005; Table 1). Using DNA chip technology, the comparison between the transcriptional profiles of the white adipose tissues of mice with DIO and lean mice, treated or not with SR141716A, revealed that this CB_1 receptor antagonist enhances lipolysis and energy expenditure in DIO mice possibly by upregulating the expression of the enzymes involved in fatty acid oxidation such as CAT, CPT2, and crotonase, and of enzymes involved in the TCA cycle, such as fumarase, aconitase, and oxoglutarate dehydrogenase (Table 1). Interestingly, the expression of the adenine nucleotide carrier (ANT), which is a carrier protein that exports ATP from mitochondria, together with a component of the respiratory chain, the cytochrome C oxidase subunit Via, were also increased by SR141716A. Lipolysis regulators such as the β_3 adrenergic and growth hormone receptors were also upregulated. Since fatty acids can be endogenously synthesized from amino acids to increase energy expenditure, Jbilo and coworkers also looked at the expression of three enzymes involved in amino acid degradation and showed that SR141716A in DIO mice increased the expression of the cysteine dioxygenase responsible for the oxidative degradation of the cysteine, and of the methylcrotonoyl-CoA carboxylase and methylmalonyl CoA mutase responsible for valine, leucine, and isoleucine degradation (Table 1). Also in the brown adipose tissue, SR141716A increased the genes involved in energy storage and expenditure and in the regulation of mitochondrial functions (Jbilo et al., 2005). Overall, the authors observed how chronic treatment of DIO mice with SR141716A restores a phenotype, in terms of the enzymes involved in glucose and lipid metabolism and in energy expenditure, similar to that of lean mice. Importantly, however, these studies were carried out following the systemic administration of the CB_1 receptor antagonist. As discussed above and below, SR141716A causes weight loss via other central and peripheral mechanisms; therefore, it is not possible to conclude that the changes observed by Jbilo and colleagues (2005) were only due to direct effects of SR141716A on adipocytes. Since leptin and PPARγ are known to participate in the regulation of fatty acid homeostasis in adipocytes, it seemed useful to understand the action of these compounds on endocannabinoids levels. Recently, using again the mouse 3T3F442A adipocyte cell line, we observed that endocannabinoid levels are controlled by these two important regulators of metabolism (Matias et al., 2006). As previously reported in the rodent hypothalamus (Di Marzo et al., 2001), both anandamide and 2-AG levels in mature adipocytes were decreased after either acute or prolonged stimulation with leptin. By contrast, in partially differentiated preadipocytes, leptin decreased anandamide levels but not

2-AG levels. Ciglitazone, a selective agonist of PPARγ, decreased levels of 2-AG but not of anandamide in partially differentiated preadipocytes, but not in mature adipocytes. Given the fact that leptin and PPARγ levels increase and decrease, respectively, when passing from preadipocytes to hypertrophic adipocytes, it is reasonable to expect that, whereas the extent of the inhibitory effect of leptin on endocannabinoid levels increases with differentiation, that of PPARγ instead decreases. Therefore, these regulatory events probably explain why, during 3T3F442A cell differentiation into adipocytes, the levels of 2-AG, which are under the negative control of PPARγ first and leptin later, peak before maturation (when leptin levels are low and PPARγ levels are not high yet) and remain elevated in hypertrophic adipocytes (when PPARγ levels are starting to decrease) (Fig. 2). By contrast, anandamide levels, which are under the negative control of leptin only, peak before maturation (when leptin levels are low) and decrease to levels similar to those observed in preadipocytes after maturation (when leptin levels are the highest) (Matias et al., 2006). A recent study suggested that CB_1 receptors in the adipose tissue are involved in the increase of adipocyte size in visceral and subcutaneous tissue, induced in rats by a prolonged high fat diet, and also participate in the decrease of adipocyte size, induced in these rats by exercise (Yan et al., 2007). In fact, the authors observed how high fat diet-induced adipocyte fattening and exercise-induced adipocyte size reduction are both accompanied by changes in PPARδ levels opposite to those of CB_1 receptors, in the two types of white adipose tissue but not in brown adipose tissue. This prompted the authors to suggest an inverse correlation between PPARδ and CB_1 receptor levels in white adipocytes. Accordingly, when 3T3-L1 adipocytes were submitted to selective silencing of PPARδ mRNA, a significant increase in both CB_1 receptor expression and adipocyte differentiation was observed, whereas adenovirus-mediated overexpression of PPARδ significantly reduced both CB_1 expression and adipocyte differentiation. These findings suggest that adipocyte differentiation is inhibited by PPARδ via actions on CB_1 receptor expression (Yan et al., 2007). Finally, in view of the fact that anandamide, as discussed above, is a full agonist at $TRPV_1$ receptors, it is important to highlight a recent study in which these channels were detected in 3T3-L1-preadipocytes and in visceral adipose tissue from mice and humans (Li Zhang et al., 2007). The authors reported how, in vitro, the $TRPV_1$ agonist capsaicin dose-dependently induces calcium influx and prevents adipogenesis in stimulated 3T3-L1-preadipocytes. RNA interference knockdown of $TRPV_1$ in 3T3-L1-preadipocytes attenuated capsaicin-induced calcium influx, and adipogenesis in stimulated 3T3-L1-preadipocytes was no longer prevented. During regular adipogenesis, the authors observed that $TRPV_1$ channels are downregulated, and this phenomenon was accompanied by a significant and time-dependent reduction of calcium influx. A reduced $TRVP_1$ expression as compared with lean counterparts was found in visceral adipose tissue from obese *db/db* and *ob/ob* mice, and from obese human male subjects. The reduced $TRPV_1$ expression in visceral adipose tissue from obese humans was accompanied by reduced capsaicin-induced calcium influx. Finally, oral administration of capsaicin for 120 days prevented obesity in male wild type mice but not in $TRPV_1$ receptor knockout mice assigned to high fat diet. The authors concluded that the activation of $TRPV_1$ channels by capsaicin prevents adipogenesis and obesity (Li Zhang et al., 2007; Fig. 2). The peak of anandamide levels previously observed to precede 3T3 preadipocyte differentiation into mature adipocytes (Matias et al., 2006) can be interpreted in the light of these results also as an endogenous mechanism contributing to adipocyte differentiation via cessation of $TRPV_1$ receptor-mediated signaling.

At the Liver Level

The first study concerning the role of the endocannabinoid system in fatty acid homeostasis at the level of the liver demonstrated, in rat hepatocytes, that anandamide, via its metabolite acid arachidonic, but not Δ^9-THC, inhibits ACC and a subsequent decrease of malonyl-CoA levels, leading to inhibition of de novo fatty acid synthesis (Fig. 2). Furthermore, fatty acid oxidation was also increased since CPT1 and the rate of ketogenesis from palmitate were found to be stimulated. Both effects on fatty acid synthesis and oxidation were prevented by phenylmethylsulfonyl fluoride (PMSF), an inhibitor of the anandamide enzymatic hydrolysis via FAAH (Guzmán et al., 1995). Demonstrating that it is arachidonic acid, originating from anandamide metabolism, which stimulates the oxidation of fatty acids and inhibits their synthesis, this pioneer study is of great importance and is only in apparent contrast with results obtained later by other laboratories (for instance, Kola et al., 2005; Osei-Hyiaman et al., 2005). In fact, Osei-Hyiaman and coworkers showed that the activation of CB_1 receptors, localized in Kupffer cells and in hepatocytes around perivascular areas, simulate de novo fatty acid synthesis (Osei-Hyiaman et al., 2005). Measurement of the incorporation of tritium into fatty acids in the liver following intrahepatic injection of ^3H_2O after pretreatment of mice with the cannabinoid receptor agonist HU-210 revealed a twofold increase of fatty acid synthesis. Furthermore, in hepatocytes isolated from mice and treated with HU-210, an increase of fatty acids synthesis was also observed. Increase in fatty acids synthesis was found neither in the liver nor in isolated hepatocytes from SR141716A-pretreated or CB_1 receptor knockout ($CB_1^{-/-}$) mice. Furthermore, diet-induced obesity resulted in decreased liver FAAH activity and in increased liver anandamide levels (and hence probably decreased arachidonic acid levels). Subsequently, an increase in fatty acid synthesis was observed – a phenomenon that was significantly decreased by blocking CB_1 receptors by chronic treatment with SR141716A during the high fat diet. The authors also showed that activation of CB_1 receptors stimulates the expression of the transcription factor SREBP-1c and of its targets, the ACC1 and FAS. This finding was similar to some extent to the effects discussed above for the white adipose tissue and the hypothalamus (Osei-Hyiaman et al., 2005). Accordingly, in the same study, the authors showed that in $CB_1^{-/-}$ mice, the level of expression of SREBP-1c in the liver (as well as in the adipose tissue) was lower as compared to that found in wild-type ($CB_1^{+/+}$) mice. These mechanisms seem to explain the lipogenesis-stimulating action of CB_1 receptors in these cells.

The study by Kola and coworkers (2005) instead described an inhibitory effect of Δ^9-THC on AMPK phosphorylation and, hence, on its activity – similar to the action of ghrelin. Since AMPK activation increases fatty acid oxidation and inhibits fatty acid synthesis in hepatocytes by decreasing the activity of ACC, this action of cannabinoids is in agreement with the observation that the activation of CB_1 receptors increases ACC and FAS (Osei-Hyiaman et al., 2005). Notably, the authors never reported if Δ^9-THC acted through CB_1 receptors. This is not a trivial issue since Δ^9-THC is a putative PPARγ activator (O'Sullivan et al., 2005; and see Chap. 9) and a CB_2 receptor agonist as well.

At the Skeletal Muscle Level

Little is known about the presence of the endocannabinoid system in the skeletal muscle and of its involvement in fatty acid homeostasis in this tissue. Recently, the CB_1 receptor antagonist SR141716A was shown to directly affect glucose uptake in the isolated soleus muscle of genetically obese mice (Liu et al., 2005). It has also been shown that the expression of CB_1 receptors in the mouse soleus muscle increases in mice on high-fat diet in comparison to the tissue derived from mice on normal chow (Pagotto et al., 2006). Additional investigations have been carried out by Cavuoto and coworkers (2007) using a primary culture of myotubes from both lean and obese patients. The authors first confirmed the presence of CB_1 receptors in this in vitro model of skeletal muscle and then demonstrated that, in cultures from both lean and obese patients, the CB_1 receptor antagonist AM251 increases the expression of AMPKα1 and decreases that of PDK4 (Cavuoto et al., 2007). Interestingly, anandamide blocked the effect of the antagonist only on the expression of AMPKα1 and not of PDK4, suggesting that the effect of the CB_1 receptor antagonist is CB_1 receptor-mediated at least as far as AMPKα1 activation, and the subsequent activation of lipolysis and fatty acid oxidation, are concerned (Fig. 2). No effect on AMPKα2 was observed with the CB_1 antagonist, which instead inhibited PGC1α expression in myotubes from lean subjects. Therefore, since PGC1α is an activator of PDK4, the inhibitory effect of AM251 on PDK4 might be due to its inhibition of PGC1α only in lean subjects. Accordingly, anandamide stimulated both PDK4 and PGC1α only in cultures from lean patients but never significantly affected the levels of AMPK1α. In view of the fact that PDK4 is an important inhibitor of the pyruvate dehydrogenase complex (PDHC), which links glycolysis to ATP production and to the TCA cycle, its inhibition by AM251 will stimulate glucose flux into the mitochondria and subsequently cause an increase of glucose oxidation in the muscle, and the opposite is likely to occur with anandamide (Cavuoto et al., 2007; Fig. 2). Surprisingly, Kola and coworkers did not observe previously any effect of cannabinoids on AMPK in the skeletal muscle even though administration of either 2-AG or Δ^9-THC did increase cardiac AMPK activity (Kola et al., 2005). This discrepancy is probably due to the fact that Kola and coworkers looked at the whole AMPK activity and not at the expression of the different isoforms of this lipid homeostasis regulator.

Effect of the Diet on Tissue Endocannabinoid Levels and Overactivity of the Endocannabinoid System

Since the body's nutritional status depends on the diet and also on the type of food consumed, it has been suggested that the type of diet and, particularly, of fatty acids contained in the diet might directly influence the levels of the endocannabinoids, possibly by causing a remodeling of the amounts of their biosynthetic phospholipid precursors. In particular, diets rich in ω6-polyunsaturated fatty acids (ω6-PUFAs) and poor in ω3-PUFA – as it is typically the case of many "high fat" diets – have been shown to significantly enhance the levels of anandamide (Berger et al., 2001) or 2-AG (Watanabe et al., 2003) in the postnatal and adult brain, respectively. On the other hand, long-term food deprivation, with the subsequent shortage of essential ω6-fatty acid precursors, was found to reduce the levels of hypothalamic endocannabinoids both in adult rodents (Hanus et al., 2003) and in pups from undernourished dams (Matias et al., 2003). In the latter study, a linear correlation was found between hypothalamic anandamide levels in the pups and their body weight at weaning. Two of us recently observed that the levels of ω3-PUFAs and ω6-PUFAs can also influence the endocannabinoid levels not only in the brain but also directly in the mouse 3T3F442A adipocyte cell line. A strong increase in 2-AG levels following 3-day incubations with arachidonic acid (a ω6-PUFA) and a decrease with docosahexaenoic and eicosapentanoic acids (two ω3-PUFAs) were observed (I. Matias and V. Di Marzo, unpublished data). That the availability of certain biosynthetic precursors, rather than the activity of the biosynthesizing enzymes, might be responsible for changes in endocannabinoid levels was elegantly demonstrated in a recent study by Petersen and colleagues (2006), who investigated the differential changes in the levels of anandamide and other *N*-acyl-ethanolamines (NAEs) in the small intestine of rats following food deprivation and refeeding. Whereas intestinal anandamide levels were found to increase following brief food deprivation, those of other NAEs, which are produced by the action of the same biosynthetic enzymes as anandamide but on different precursors, were found to decrease. The authors showed that the total levels of precursor *N*-acyl-phosphatidyl-ethanolamines (NAPEs) for all NAEs were decreased upon food deprivation, whereas the level of only the anandamide precursor, *N*-arachidonoyl-phosphatidy-ethanolamide (NarPE), increased, with no changes in the activities of the NAPE-PLD and acyl-transferase enzymes, which catalyze NAE biosynthesis, nor of FAAH, which recognizes other NAEs also. The authors concluded that the remodeling of the amide-linked fatty acids of NAPEs (which is likely to be diet dependent) is responsible for the opposite effects of food deprivation and refeeding on the small intestine levels of anandamide and other NAEs (Petersen et al., 2006). Since obesity is in most cases the consequence of sedentary life, in which an increase of food consumption over energy expenditure leads to fat accumulation, and since high fat diets result in an increase of endocannabinoids levels, one can hypothesize that the endocannabinoid system will be overactive in obesity states in both laboratory animals and in man. Indeed, as anticipated in the previous sections, the endocannabinoid system does become overactive during obesity and hyperglycemia (Table 1). Defective leptin signaling is the most likely cause of the permanently elevated endocannabinoid levels found in the hypothalamus of *ob/ob* and *db/db* mice and Zucker rats (Di Marzo et al., 2001). The latter animals also exhibit permanently upregulated CB_1 receptors in the adipose tissue (Bensaid et al., 2003). In mice with DIO, elevated endocannabinoid levels have been reported so far in the epididymal fat (2-AG only) and in the pancreas (both anandamide and 2-AG) (Matias et al., 2006) as well as in the liver (anandamide only) (Osei-Hyiaman et al., 2005). In this latter organ, upregulation of CB_1 receptors was also observed. Also in the visceral and subcutaneous adipose tissue of rats fed with a high fat diet, but not of mice, CB_1 receptors were found to be upregulated (Matias et al., 2006; Yan et al., 2007). In visceral, but not subcutaneous, fat of obese patients, 2-AG, but not anandamide, levels are elevated (Matias et al., 2006), but the same does not seem to always apply to the expression of CB_1 receptors (Engeli et al., 2005; Bluher et al., 2006; Matias et al., 2006; Löfgren et al., 2007), which however are more abundant in tissue and adipocytes from omental vs. subcutaneous fat (Roche et al., 2006; Löfgren et al., 2007). In the blood of women that became obese because of binge eating or menopausa, the levels of anandamide or of both endocannabinoids, respectively, are also significantly higher than in age-matched nonobese controls (Engeli et al., 2005; Monteleone et al., 2005; Table 1), whereas in the blood of obese, particularly if male, patients, the levels of 2-AG, but not anandamide, correlate with intra-abdominal adiposity and with all the cardiometabolic risks associated with ectopic visceral fat (Bluher et al., 2006; Coté et al., 2007; Table 1). Also in the blood of nonobese type II diabetes patients, levels of both anandamide and 2-AG are permanently elevated with respect to those of age- and gender-matched controls (Matias et al., 2006). Finally, a single aminoacid polymorfism in the CB_1 encoding gene (*CNR1*), potentially causing CB_1 malfunctioning, was recently found to be associated to leanness in a special Italian population (Gazzerro et al., 2007). Despite the increasing evidence for an overactive endocannabinoid system in obesity, still very little is known about the possible biochemical mechanisms underlying this phenomenon. Two of us have recently participated in a study (Starowicz et al., submitted) investigating, in mice fed for different periods of time with either a standard diet (STD) or a high fat diet (HFD), the expression and localization of cannabinoid CB_1 and CB_2 receptors and of endocannabinoid metabolizing enzymes (NAPE-PLD and DAGLα, for anandamide and 2-AG biosynthesis, respectively; FAAH and MAGL, for anandamide and 2-AG hydrolysis, respectively) in Langerhans' islets. Already 3 weeks, and also 8 and 14 weeks following a HFD, both DAGLα and NAPE-PLD, which in lean mice are expressed only in α-cells, became expressed also in β-cells, whereas FAAH expression in these latter cells significantly decreased starting with 8 weeks of HFD. We observed that these changes in endocannabinoid metabolic enzyme expression are accompanied by increases of both 2-AG and, particularly, anandamide levels at both 3 and 8 weeks after the beginning of the HFD (Starowicz et al., submitted). Interestingly, however, 14 weeks after HFD, despite the changes in the expression of endocannabinoid metabolic enzymes, no change in endocannabinoid levels was observed in pancreatic islets at this time point, thus underlying again the fact that expression of biosynthetic and degrading enzymes might not be the only factor leading to changes in endocannabinoid levels. At any rate, these findings suggest that following a HFD, endocannabinoids are made also by β-cells, whereupon they might act as autocrine mediators to regulate cell function and insulin secretion via cannabinoid receptors (Juan-Picó et al., 2006; Matias et al., 2006) in an aberrant way. Importantly, also in the liver of DIO mice, increased levels of anandamide were accompanied by decreased expression of FAAH (Osei-Hyiaman et al., 2005), whereas in the visceral fat of obese patients, the increase of 2-AG levels (Matias et al., 2006) is often accompanied by the reduction of the expression of FAAH and MAGL (Engeli et al., 2005; Löfgren et al., 2007). Contrasting results have appeared in the literature regarding the possibility that a polymorphism in the FAAH encoding gene (*FAAH*) is associated with obesity (Sipe et al., 2005; Jensen et al., 2007). These data, taken together, indicate that changes in the activity of biosynthesising or degrading enzymes caused by the diet or genetic factors can be one of the causes, but certainly not the only one, of permanently elevated endocannabinoid levels in peripheral organs of obese individuals. In conclusion, many biochemical mechanisms, including (1) up- or downregulation of biosynthetic and degrading enzymes, respectively, due perhaps to changes in the functional activity of hormones like leptin, insulin, ghrelin, and glucocorticoids, which might control their expression, or (2) the dietary abundance of certain fatty acids instead of others, might explain the elevated endocannabinoid levels during obesity and hyperglycemia (Table 1). Regarding the levels of the expression of CB_1 receptors, it is not known how these can be upregulated (Bensaid et al., 2003; Yan et al., 2007) or downregulated (Engeli et al., 2005; Bluher et al., 2006) during obesity. It has been shown that CCK reduces, and ghrelin enhances, CB_1 levels in a plastic way in the nodose ganglion (Burdyga et al., 2004, 2006), but whether these effects involve any indirect interaction of CCK and ghrelin receptors with the *CNR1* promoter region has not been yet investigated. As outlined in this chapter, endocannabinoids appear to control homeostatic regulation by stimulating the central orexigenic system, inhibiting peripheral lipolysis and modulating glucose metabolism. On the other hand, endocannabinoid levels, and hence the extent of food intake stimulation and fat accumulation, might directly or indirectly depend on eating habits and on certain nutritional regimes, thus originating a "vicious circle" causing more and more food intake and ectopic fat accumulation (Table 1).

Clinical Impact of Cannabinoids in Energy Homeostasis

As anticipated above, blockade of CB_1 receptors with specific antagonists seems to be a fruitful and relatively safe pharmacological strategy to reduce fasting glucose and triglycerides in obese humans. So far, four phase III "Rimonabant In Obesity" (RIO) clinical trials have been completed with SR141716A (Van Gaal et al., 2005;

Despres et al., 2005; Pi-Sunyer et al., 2006; Scheen et al., 2006) and another one (SERENADE, i.e., "Study Evaluating Rimonabant Efficacy in Drug-Naïve Diabetic Patients") has been reported at conferences. It is quite impressive to see how three of the five studies carried out in similar populations of obese patients and also the remaining two carried out in diabetic patients reported overlapping results in terms of safety, body weight, and waist circumference reduction and of amelioration of metabolic parameters, such as fasting glucose levels, insulinemia, HDL cholesterol, and triglyceride levels (Table 1). These beneficial metabolic effects had been predicted from animal studies. In particular, the results of a two-year study, known as Rimonabant in Obesity (RIO)-North America, with over 3,400 patients subject to a mild low calory diet (Pi-Sunyer et al., 2006), can be summarized as follows: (1) a 1-year administration with a 20-mg/day oral dose of SR141716A causes in treated patients weight losses ≥5% and ≥10% in over 62% and 32% of the completers, respectively, vs. 33% and 16% in placebos, respectively. The average weight loss and waist reduction were ~8.8 kg and 8.4 cm, vs. 2.9 kg and 4 cm in placebos, respectively. (ii) After 1-year treatment, the blood triglyceride levels in completers dropped by ~8.5% (vs. a ~4.5% *increase* in placebos) and the HDL cholesterol levels increased by ~17.5% (vs. ~6.3% in placebos). Fasting insulin levels decreased by ~2.7 μIU/ml vs. placebos. (3) Following randomization into placebo or drug continuation at one year, the patients that were kept on SR141716A for another year did not lose further weight, but continued to significantly increase their HDL cholesterol levels, whereas the previously treated patients now taking placebo slowly regained weight to eventually become undistinguishable from the "placebo–placebo" group only at the end of the trial. Results identical to the 1-year results of the RIO-North America study were obtained in two other studies, i.e., the RIO-Lipids, a 1-year study in which a high percent of patients with metabolic syndrome was selected (Després et al., 2005), and the RIO-Europe trials (Van Gaal et al., 2005). In the RIO-Lipids study, it was possible to dissociate ~50% of the beneficial metabolic effects (i.e., increase of adiponectin or HDL-cholesterol levels) from the observed decrease in body weight, in support of the possible direct action of an overactive endocannabinoid system on peripheral cells and tissues involved in these effects, e.g., the adipose tissue and the liver. The three pooled RIO studies (5,580 patients) at one year yielded reassuring results on the safety side, with a 3.6% increase in patients with any adverse events between treated and placebos, and a 5.9% difference between the two groups in patients who discontinued due to adverse events. These events consisted mostly of nausea (1.3%), diarrhea (1.3%), dizziness (0.6%), depression (1.4%), and anxiety (0.7%), and in most cases showed tolerance after the first weeks of treatment, in agreement with results in animal models. The fourth 1-year study was carried out with 1,045 obese patients with type-2 diabetes cotreated with either metformin or sulfonylureas, and in this case changes in glycosylated hemoglobin (HbA1C) were also monitored (Scheen et al., 2006). Apart from reduction in body weight and waist circumference, an increase of HDL cholesterol and a decrease in triglyceride levels were observed with 20 mg/day of SR141716A, which also caused a further reduction of about 0.7% of HbA1C on top of that induced by metformin or sulfonylureas – an effect which was ~50%

independent from weight loss. Finally, the SERENADE trial was a six-month randomized placebo controlled study carried out in 281 participants to confirm the efficacy of SR141716A in type-2 diabetes. The primary objective of this study was to validate the effect of SR141716A on blood glucose measured by its indicator HbA1C in newly diagnosed type-2 diabetes patients not adequately controlled by the diet. After six-month administration with a 20-mg/day oral dose, apart from a reduction in body weight (−6.7 kg vs. −2.7 in placebo) and waist circumference (−6.1 cm vs. −2.4 in placebo), an increase of HDL cholesterol (+10.1% vs. +3.2 in placebo) and a decrease in triglyceride levels (−16.3% vs. −4.4 in placebo), SR141716A also caused a further reduction of about 0.5% of glycosylated hemoglobin (−0.8% vs. −0.3 in placebo). SR141716A (commercial name: Acomplia™, Sanofi-Aventis) is the first CB_1 receptor antagonist/inverse agonist to be approved for therapeutic use in Europe. This compound might be followed in future by other CB_1 receptor antagonists, developed by other companies, now in Phase I and II clinical trials. Like SR141716A, these compounds might not only reduce food intake and body weight in obese patients, but also significantly ameliorate the signs of the metabolic syndrome in overweight/viscerally obese and/or type-2 diabetes patients only partly via weight loss, in other words, via *directly* targeting a potentially overactive endocannabinoid system in peripheral cells and organs (Table 1).

Concluding Remarks

CB_1 receptor activation increases blood glucose levels via several potential mechanisms, including the inhibition of insulin release and of glucose utilization by peripheral tissue and brain cells, whose actual occurrence still needs to be substantiated by future studies. It also stimulates appetite and ingestive behavior through central mechanisms concerning primarily the hypothalamus, but possibly also other brain areas involved in the control of food intake (see Matias and Di Marzo, 2007, for review). Furthermore, CB_1 receptor activation facilitates the growth of fat deposits rather than burning fat as a fuel for cells. As a malignant factor, circulating and fat tissue endocannabinoid levels are increased in overweight and especially abdominally adipose patients, thus giving rise to a vicious circle. These effects can all be counteracted by the CB_1 receptor antagonist SR141716A (Rimonabant, Acomplia™) as well as other compounds with similar activity. Since, at the therapeutically used concentration, SR141716A does not cause major side effects, possibly because it does not block CB_1 receptors throughout the body completely, this compound will likely represent a very effective and promising drug to fight metabolic disorders. On the other hand, under conditions of normal activation, CB_1 receptor activation is certainly beneficial to ensure the optimal energy homeostasis necessary to compensate for the loss of energy occurring following stressful conditions (in which, as discussed in other chapters of this book, the endocannabinoid system plays a major role). In fact, in view of the stress recovery function proposed for the endocannabinoid system, helping with energy replenishment represents one of the crucial physiological functions played by CB_1 receptors. Additional studies are required to understand if CB_1 receptor upregulation in the diabetic brain contributes or counteracts encephalopathy, and hence if SR141716A worsens or attenuates this disorder.

References

Ahima RS (2006) Adipose tissue as an endocrine organ. Obesity (Silver Spring) 14:242S–249S.

Akiba Y, Kato S, Katsube K, Nakamura M, Takeuchi K, Ishii H, Hibi T (2004) Transient receptor potential vanilloid subfamily 1 expressed in pancreatic islet beta cells modulates insulin secretion in rats. Biochem Biophys Res Commun 321:219–225.

Anand BK, Brobeck JR (1951) Hypothalamic control of food intake in rats and cats. Yale J Biol Med 24:123–146.

Banks WA (2004) The source of cerebral insulin. Eur J Pharmacol 490:5–12.

Benowitz NL, Jones RT, Lerner CB (1976) Depression of growth hormone and cortisol response to insulin-induced hypoglycemia after prolonged oral delta-9-tetrahydrocannabinol administration in man. J Clin Endocrinol Metab 42:938–941.

Bensaid M, Gary-Bobo M, Esclangon A, Maffrand JP, Le Fur G, Oury-Donat F, Soubrie P (2003) The cannabinoid CB_1 receptor antagonist SR141716 increases Acrp30 mRNA expression in adipose tissue of obese *fa/fa* rats and in cultured adipocyte cells. Mol Pharmacol 63:908–914.

Berger A, Crozier G, Bisogno T, Cavaliere P, Innis S, Di Marzo V (2001) Anandamide and diet: inclusion of dietary arachidonate and docosahexaenoate leads to increased brain levels of the corresponding N-acylethanolamines in piglets. Proc Natl Acad Sci USA 98:6402–6406.

Berghuis P, Dobszay MB, Wang X, Spano S, Ledda F, Sousa KM, Schulte G, Ernfors P, Mackie K, Paratcha G, Hurd YL, Harkany T (2005) Endocannabinoids regulate interneuron migration and morphogenesis by transactivating the TrkB receptor. Proc Natl Acad Sci USA 102:19115–191120.

Bermudez-Siva FJ, Serrano A, Diaz-Molina FJ, Sanchez Vera I, Juan-Pico P, Nadal A, Fuentes E, Rodriguez de Fonseca F (2006) Activation of cannabinoid CB_1 receptors induces glucose intolerance in rats. Eur J Pharmacol 531:282–284.

Bloom AS, Tershner S, Fuller SA, Stein EA (1997) Cannabinoidinduced alterations in regional cerebral blood flow in the rat. Pharmacol Biochem Behav 57:625–631.

Bluher M, Engeli S, Kloting N, Berndt J, Fasshauer M, Batkai S, Pacher P, Schon MR, Jordan J, Stumvoll M (2006) Dysregulation of the peripheral and adipose tissue endocannabinoid system in human abdominal obesity. Diabetes 55:3053–3060.

Bondy CA, Cheng CM (2004) Signaling by insulin-like growth factor 1 in brain. Eur J Pharmacol 490:25–31.

Borgen LA, Lott GC, Davis WM (1973) Cannabis-induced hypothermia: a dose-effect comparison of crude marihuana extract and synthetic 9-tetrahydrocannabinol in male and female rats. Res Commun Chem Pathol Pharmacol 5:621–626.

Burdakov D, Gerasimenko O, Verkhratsky A (2005) Physiological changes in glucose differentially modulate the excitability of hypothalamic melanin-concentrating hormone and orexin neurons in situ. J Neurosci 25:2429–2433.

Burdyga G, Lal S, Varro A, Dimaline R, Thompson DG, Dockray GJ (2004) Expression of cannabinoid CB_1 receptors by vagal afferent neurons is inhibited by cholecystokinin. J Neurosi 24:2708–2715.

Burdyga G, Varro A, Dimaline R, Thompson DG, Dockray GJ (2006) Ghrelin receptors in rat and human nodose ganglia: putative role in regulating CB-1 and MCH receptor abundance. Am J Physiol Gastrointest Liver Physiol 290:G1289–G1297.

Cani PD, Montoya ML, Neyrinck AM, Delzenne NM, Lambert DM (2004) Potential modulation of plasma ghrelin and glucagon-like peptide-1 by anorexigenic cannabinoid compounds, SR141716A (rimonabant) and oleoylethanolamide. Br J Nutr 92:757–761.

Carling D (2005) AMP-activated protein kinase: balancing the scales. Biochimie 87:87–91.

Cavuoto P, McAinch AJ, Hatzinikolas G, Cameron-Smith D, Wittert GA (2007) Effects of cannabinoid receptors on skeletal muscle oxidative pathways. Mol Cell Endocrinol 267:63–69.

Chabot C, Massicotte G, Milot M, Trudeau F, Gagne J (1997) Impaired modulation of AMPA receptors by calcium-dependent processes in streptozotocin-induced diabetic rats. Brain Res 768:249–256.

Chen S, Atkins CM, Liu CL, Alonso OF, Dietrich WD, Hu BR (2007) Alterations in mammalian target of rapamycin signaling pathways after traumatic brain injury. J Cereb Blood Flow Metab 27:939–949.

Cheng CM, Reinhardt RR, Lee WH, Joncas G, Patel SC, Bondy CA (2000) Insulin-like growth factor 1 regulates developing brain glucose metabolism. Proc Natl Acad Sci USA 97:10236–10241.

Clodfelder-Miller B, De Sarno P, Zmijewska AA, Song L, Jope RS (2005) Physiological and pathological changes in glucose regulate brain Akt and glycogen synthase kinase-3. J Biol Chem 280:39723–39731.

Corchero J, Fuentes JA, Manzanares J (1999) Chronic treatment with CP-55,940 regulates corticotropin releasing factor and proopiomelanocortin gene expression in the hypothalamus and pituitary gland of the rat. LifeSci 64:905–911.

Cota D, Marsicano G, Lutz B, Vicennati V, Stalla GK, Pasquali R, Pagotto U (2003a) Endogenous cannabinoid system as a modulator of food intake. Int J Obes Relat Metab Disord 27:289–301.

Cota D, Marsicano G, Tschop M, Grubler Y, Flachskamm C, Schubert M, Auer D, Yassouridis A, Thone-Reineke C, Ortmann S, Tomassoni F, Cervino C, Nisoli E, Linthorst AC, Pasquali R, Lutz B, Stalla GK, Pagotto U (2003b) The endogenous cannabinoid system affects energy balance via central orexigenic drive and peripheral lipogenesis. J Clin Invest 112:423–431.

Cota D, Steiner MA, Marsicano G, Cervino C, Herman JP, Grubler Y, Stalla J, Pasquali R, Lutz B, Stalla GK, Pagotto U (2007) Requirement of cannabinoid receptor type 1 for the basal modulation of hypothalamic-pituitary-adrenal axis function. Endocrinology 148:1574–1581.

Cote M, Matias I, Lemieux I, Petrosino S, Almeras N, Despres JP, Di Marzo V (2007) Circulating endocannabinoid levels, abdominal adiposity and related cardiometabolic risk factors in obese men. Int J Obes (Lond) 31:692–699.

Dallaporta M, Himmi T, Perrin J, Orsini JC (1999) Solitary tract nucleus sensitivity to moderate changes in glucose level. Neuroreport 10:2657–2660.

Despres JP, Golay A, Sjostrom L, and the Rimonabant in Obesity-Lipids Study Group (2005) Effects of rimonabant on metabolic risk factors in overweight patients with dyslipidemia. N Engl J Med 353:2121–2134.

Di S, Malcher-Lopes R, Halmos KC, Tasker JG (2003) Nongenomic glucocorticoid inhibition via endocannabinoid release in the hypothalamus: a fast feedback mechanism. J Neurosci 23:4850–4857.

Di Marzo V, Matias I (2005) Endocannabinoid control of food intake and energy balance. Nat Neurosci 8:585–589.

Di Marzo V, Goparaju SK, Wang L, Liu J, Batkai S, Jarai Z, Fezza F, Miura GI, Palmiter RD, Sugiura T, Kunos G (2001) Leptin-regulated endocannabinoids are involved in maintaining food intake. Nature 410:822–825.

Doyon C, Denis RG, Baraboi ED, Samson P, Lalonde J, Deshaies Y, Richard D (2006) Effects of rimonabant (SR141716) on fasting-induced hypothalamic-pituitary-adrenal axis and neuronal activation in lean and obese Zucker rats. Diabetes 55:3403–3410.

Duarte JM, Nogueira C, Mackie K, Oliveira CR, Cunha RA, Köfalvi A (2007) Increase of cannabinoid CB_1 receptor density in the hippocampus of streptozotocin-induced diabetic rats. Exp Neurol 204:479–484.

Dyachok O, Isakov Y, Sagetorp J, Tengholm A (2006) Oscillations of cyclic AMP in hormone-stimulated insulin-secreting beta-cells. Nature 439:349–352.

Eberle D, Hegarty B, Bossard P, Ferre P, Foufelle F (2004) SREBP transcription factors: master regulators of lipid homeostasis. Biochimie 86:839–848.

Ellis J, Pediani JD, Canals M, Milasta S, Milligan G (2006) Orexin-1 receptor-cannabinoid CB1 receptor heterodimerization results in both ligand-dependent and -independent coordinated alterations of receptor localization and function. J Biol Chem 281:38812–38824.

Engeli S, Bohnke J, Feldpausch M, Gorzelniak K, Janke J, Batkai S, Pacher P, Harvey-White J, Luft FC, Sharma AM, Jordan J (2005) Activation of the peripheral endocannabinoid system in human obesity. Diabetes 54:2838–2843.

Fehm HL, Kern W, Peters A (2006) The selfish brain: competition for energy resources. Prog Brain Res 153:129–140.

Finck BN, Kelly DP (2006) PGC-1 coactivators: inducible regulators of energy metabolism in health and disease. J Clin Invest 116:615–622.

Freedland CS, Whitlow CT, Miller MD, Porrino LJ (2002) Dose-dependent effects of Delta^9-tetrahydrocannabinol on rates of local cerebral glucose utilization in rat. Synapse 45:134–142.

Gary-Bobo M, Elachouri G, Scatton B, Le Fur G, Oury-Donat F, Bensaid M (2006) The cannabinoid CB_1 receptor antagonist rimonabant (SR141716) inhibits cell proliferation and increases markers of adipocyte maturation in cultured mouse 3T3 F442A preadipocytes. Mol Pharmacol 69:471–478.

Gasperi V, Fezza F, Pasquariello N, Bari M, Oddi S, Agro AF, Maccarrone M (2007) Endocannabinoids in adipocytes during differentiation and their role in glucose uptake. Cell Mol Life Sci 64:219–229.

Gazzerro P, Caruso MG, Notarnicola M, Misciagna G, Guerra V, Laezza C, Bifulco M (2007) Association between cannabinoid type-1 receptor polymorphism and body mass index in a southern Italian population. Int J Obes (Lond) 31:908–912.

Gelfand EV, Cannon CP (2006) Rimonabant: a cannabinoid receptor type 1 blocker for management of multiple cardiometabolic risk factors. J Am Coll Cardiol 47:1919–1926.

Gil-Campos M, Aguilera CM, Canete R, Gil A (2006) Ghrelin: a hormone regulating food intake and energy homeostasis. Br J Nutr 96:201–226.

Gispen WH, Biessels GJ (2000) Cognition and synaptic plasticity in diabetes mellitus. Trends Neurosci 23:542–549.

Goldman H, Dagirmanjian R, Drew WG, Murphy S (1975) Delta^9-tetrahydrocannabinol alters flow of blood to subcortical areas of the conscious rat brain. Life Sci 17:477–482.

Gómez del Pulgar T, De Ceballos ML, Guzmán M, Velasco G (2002) Cannabinoids protect astrocytes from ceramide-induced apoptosis through the phosphatidylinositol 3-kinase/protein kinase B pathway. J Biol Chem 277:36527–36533.

Grillo CA, Piroli GG, Wood GE, Reznikov LR, McEwen BS, Reagan LP (2005) Immunocytochemical analysis of synaptic proteins provides new insights into diabetes-mediated plasticity in the rat hippocampus. Neuroscience 136:477–486.

Guzmán M, Fernández-Ruiz JJ, Sánchez C, Velasco G, Ramos JA (1995) Effects of anandamide on hepatic fatty acid metabolism. Biochem Pharmacol 50:885–888.

Hanus L, Avraham Y, Ben-Shushan D, Zolotarev O, Berry EM, Mechoulam R (2003) Short-term fasting and prolonged semistarvation have opposite effects on 2-AG levels in mouse brain. Brain Res 983:144–151.

Hardie DG (2004) The AMP-activated protein kinase pathway – new players upstream and downstream. J Cell Sci 117:5479–5487.

Hardie DG, Hawley SA, Scott JW (2006) AMP-activated protein kinase-development of the energy sensor concept. J Physiol 574:7–15.

Hermann H, De Petrocellis L, Bisogno T, Schiano Moriello A, Lutz B, Di Marzo V (2003) Dual effect of cannabinoid CB_1 receptor stimulation on a vanilloid VR_1 receptor-mediated response. Cell Mol Life Sci 60:607–616.

Hollister LE (1971) Hunger and appetite after single doses of marihuana, alcohol, and dextroamphetamine. Clin Pharmacol Ther 12:44–49.

Ibrahim N, Bosch MA, Smart JL, Qiu J, Rubinstein M, Ronnekleiv OK, Low MJ, Kelly MJ (2003) Hypothalamic proopiomelanocortin neurons are glucose responsive and express K_{ATP} channels. Endocrinology 144:1331–1340.

Jackson-Guilford J, Leander JD, Nisenbaum LK (2000) The effect of streptozotocin-induced diabetes on cell proliferation in the rat dentate gyrus. Neurosci Lett 293:91–94.

Jamshidi N, Taylor DA (2001) Anandamide administration into the ventromedial hypothalamus stimulates appetite in rats. Br J Pharmacol 134:1151–1154.

Jbilo O, Ravinet-Trillou C, Arnone M, Buisson I, Bribes E, Peleraux A, Penarier G, Soubrie P, Le Fur G, Galiegue S, Casellas P (2005) The CB_1 receptor antagonist rimonabant reverses the diet-induced obesity phenotype through the regulation of lipolysis and energy balance. FASEB J 19:1567–1569.

Jensen DP, Andreasen CH, Andersen MK, Hansen L, Eiberg H, Borch-Johnsen K, Jorgensen T, Hansen T, Pedersen O (2007) The functional Pro129Thr variant of the FAAH gene is not associated with various fat accumulation phenotypes in a population-based cohort of 5,801 whites. J Mol Med 85:445–449.

Juan-Picó P, Fuentes E, Bermudez-Silva FJ, Javier Diaz-Molina F, Ripoll C, Rodriguez de Fonseca F, Nadal A (2006) Cannabinoid receptors regulate Ca^{2+} signals and insulin secretion in pancreatic beta-cell. Cell Calcium 39:155–162.

Kang L, Routh VH, Kuzhikandathil EV, Gaspers L, Levin BE (2004) Physiological and molecular characteristics of rat hypothalamic ventromedial nucleus glucosensing neurons. Diabetes 53:549–559.

Karlsson S, Scheurink AJ, Steffens AB, Ahren B (1994) Involvement of capsaicin-sensitive nerves in regulation of insulin secretion and glucose tolerance in conscious mice. Am J Physiol 267:1071–1077.

Kim SR, Lee DY, Chung ES, Oh UT, Kim SU, Jin BK (2005) Transient receptor potential vanilloid subtype 1 mediates cell death of mesencephalic dopaminergic neurons *in vivo* and *in vitro*. J Neurosci 25:662–671.

Kirkham TC, Williams CM, Fezza F, Di Marzo V (2002) Endocannabinoid levels in rat limbic forebrain and hypothalamus in relation to fasting, feeding and satiation: stimulation of eating by 2-arachidonoylglycerol. Br J Pharmacol 136:550–557.

Köfalvi A, Pereira MF, Rebola N, Rodrigues RJ, Oliveira CR, Cunha RA (2007) Anandamide and NADA bi-directionally modulate presynaptic Ca^{2+} levels and transmitter release in the hippocampus. Br J Pharmacol 151:551–563.

Kohno D, Gao H-Z, Muroya S, Kikuyama S, Yada T (2003) Ghrelin directly interacts with neuropeptide-Y-containing neurons in the rat arcuate nucleus: Ca^{2+} signaling via protein kinase A and N-type channel-dependent mechanisms and cross-talk with leptin and orexin. Diabetes 52:948–956.

Kola B, Hubina E, Tucci SA, Kirkham TC, Garcia EA, Mitchell SE, Williams LM, Hawley SA, Hardie DG, Grossman AB, Korbonits M (2005) Cannabinoids and ghrelin have both central and peripheral metabolic and cardiac effects via AMP-activated protein kinase. J Biol Chem 280:25196–25201.

Kyriaki G (2003) Brain insulin: regulation, mechanisms of action and functions. Cell Mol Neurobiol 23:1–25.

Lam TK, Schwartz GJ, Rossetti L (2005) Hypothalamic sensing of fatty acids. Nat Neurosci 8:579–584.

Levin BE (2006) Metabolic sensing neurons and the control of energy homeostasis. Physiol Behav 89:486–489.

Leybaert L (2005) Neurobarrier coupling in the brain: a partner of neurovascular and neurometabolic coupling? J Cereb Blood Flow Metab 25:2–16.

Li ZG, Zhang W, Grunberger G, Sima AAF (2002) Hippocampal neuronal apoptosis in type 1 diabetes. Brain Res 946:221–231.

Liu YL, Connoley IP, Wilson CA, Stock MJ (2005) Effects of the cannabinoid CB_1 receptor antagonist SR141716 on oxygen consumption and soleus muscle glucose uptake in Lep(ob)/Lep(ob) mice. Int J Obes (Lond) 29:183–187.

Li Zhang L, Yan Liu D, Qun Ma L, Dan Luo Z, Bing Cao T, Zhong J, Cheng Yan Z, Juan Wang L, Gang Zhao Z, Jun Zhu S, Schrader M, Thilo F, Ming Zhu Z, Tepel M. (2007) Activation of Transient Receptor Potential Vanilloid Type-1 Channel Prevents Adipogenesis and Obesity. Circ Res 100:1063–1070.

Lofgren P, Sjolin E, Wahlen K, Hoffstedt J (2007) Human adipose tissue cannabinoid receptor 1 gene expression is not related to fat cell function or adiponectin level. J Clin Endocrinol Metab 92:1555–1559.

Magistretti PJ, Pellerin L (1996) Cellular bases of brain energy metabolism and their relevance to functional brain imaging. Cereb Cortex 6:50–61.

Mahfouz M, Makar AB, Ghoneim MT, Mikhail MM (1975) Effect of hashish on brain gamma aminobutyric acid system, blood fibrinolytic activity and glucose and some serum enzymes in the rat. Pharmazie 30:772–774.

Malcher-Lopes R, Di S, Marcheselli VS, Weng FJ, Stuart CT, Bazan NG, Tasker JG (2006) Opposing crosstalk between leptin and glucocorticoids rapidly modulates synaptic excitation via endocannabinoid release. J Neurosci 26:6643–6650.

Manzanares J, Corchero J, Fuentes JA (1999) Opioid and cannabinoid receptor-mediated regulation of the increase in adrenocorticotropin hormone and corticosterone plasma concentrations induced by central administration of delta^9-tetrahydrocannabinol in rats. Brain Res 839:173–179.

Margulies JE, Hammer Jr RP (1991) Delta 9-tetrahydrocannabinol alters cerebral metabolism in a biphasic, dose-dependent manner in rat brain. Eur J Pharmacol 202:373–378.

Matias I, Di Marzo V (2007) Endocannabinoids and the control of energy balance. Trends Endocrinol Metab 18:27–37.

Matias I, Leonhardt M, Lesage J, De Petrocellis L, Dupouy JP, Vieau D, Di Marzo V (2003) Effect of maternal under-nutrition on pup body weight and hypothalamic endocannabinoid levels. Cell Mol Life Sci 60:382–389.

Matias I, Gonthier MP, Orlando P, Martiadis V, De Petrocellis L, Cervino C, Petrosino P, Pasquali R, Maj M, Pagotto U, Di Marzo V (2006) Regulation, function and dysregulation of endocannabinoids in models of adipose and β-pancreatic cells and in obesity and hyperglicemia. J Clin Endocrinol Metab 91:3171–3180.

McCall AL (2004) Cerebral glucose metabolism in diabetes mellitus. Eur J Pharmacol 490:147–158.

McEwen BS, Reagan LP (2004) Glucose transporter expression in the central nervous system: relationship to synaptic function. Eur J Pharmacol 490:13–24.

Minami T, Shimizu N, Duan S, Oomura Y (1990) Hypothalamic neuronal activity responses to 3-hydroxybutyric acid, an endogenous organic acid. Brain Res 509:351–354.

Moens K, Flamez D, Van Schravendijk C, Ling Z, Pipeleers D, Schuit F (1998) Dual glucagon recognition by pancreatic beta-cells via glucagon and glucagon-like peptide 1 receptors. Diabetes 47:66–72.

Monteleone P, Matias I, Martiadis V, De Petrocellis L, Maj M, Di Marzo V (2005) Blood levels of the endocannabinoid anandamide are increased in anorexia nervosa and in binge-eating disorder, but not in bulimia nervosa. Neuropsychopharmacology 30:1216–1221.

Muoio DM, Dohm GL, Fiedorek Jr FT, Tapscott EB, Coleman RA (1997) Leptin directly alters lipid partitioning in skeletal muscle. Diabetes 46:1360–1363.

Nagy I, Santha P, Jancso G, Urban L (2004) The role of the vanilloid (capsaicin) receptor ($TRPV_1$) in physiology and pathology. Eur J Pharmacol 500:351–369.

Osei-Hyiaman D, DePetrillo M, Pacher P, Liu J, Radaeva S, Batkai S, Harvey-White J, Mackie K, Offertaler L, Wang L, Kunos G (2005) Endocannabinoid activation at hepatic CB_1 receptors stimulates fatty acid synthesis and contributes to diet-induced obesity. J Clin Invest 115:1298–1305.

Oshita K, Inoue A, Tang HB, Nakata Y, Kawamoto M, Yuge O (2005) CB_1 cannabinoid receptor stimulation modulates transient receptor potential vanilloid receptor 1 activities in calcium influx and substance P Release in cultured rat dorsal root ganglion cells. J Pharmacol Sci 97:377–385.

O'Sullivan SE, Tarling EJ, Bennett AJ, Kendall DA, Randall MD (2005) Novel time-dependent vascular actions of Δ^9-tetrahydrocannabinol mediated by peroxisome proliferator-activated receptor gamma. Biochem Biophys Res Commun 337:824–831.
Pagotto U, Marsicano G, Cota D, Lutz B, Pasquali R (2006) The emerging role of the endocannabinoid system in endocrine regulation and energy balance. Endocr Rev 27:73–100.
de Pasquale A, Costa G, Trovato A (1978) The influence of cannabis on glucoregulation. Bull Narc 30:33–41.
Petersen G, Sorensen C, Schmid PC, Artmann A, Tang-Christensen M, Hansen SH, Larsen PJ, Schmid HH, Hansen HS (2006) Intestinal levels of anandamide and oleoylethanolamide in food-deprived rats are regulated through their precursors. Biochim Biophys Acta 1761:143–150.
Pi-Sunyer FX, Aronne LJ, Heshmati HM, Devin J, Rosenstock J, RIO-North America Study Group (2006) Effect of rimonabant, a cannabinoid-1 receptor blocker, on weight and cardiometabolic risk factors in overweight or obese patients: RIO-North America: a randomized controlled trial. JAMA 295:761–775.
Poirier B, Bidouard JP, Cadrouvele C, Marniquet X, Staels B, O'Connor SE, Janiak P, Herbert JM (2005) The anti-obesity effect of rimonabant is associated with an improved serum lipid profile. Diabetes Obes Metab 7:65–72.
Pontieri FE, Conti G, Zocchi A, Fieschi C, Orzi F (1999) Metabolic mapping of the effects of WIN 55212–2 intravenous administration in the rat. Neuropsychopharmacology 21:773–776.
Postic C, Dentin R, Girard J (2004) Role of the liver in the control of carbohydrate and lipid homeostasis. Diabetes Metab 30:398–408.
Ravinet-Trillou C, Arnone M, Delgorge C, Gonalons N, Keane P, Maffrand JP, Soubrie P (2003) Anti-obesity effect of SR141716, a CB_1 receptor antagonist, in diet-induced obese mice. Am J Physiol Regul Integr Comp Physiol 284:R345–R353.
Ravinet-Trillou C, Delgorge C, Menet C, Arnone M, Soubrie P (2004) CB_1 cannabinoid receptor knockout in mice leads to leanness, resistance to diet-induced obesity and enhanced leptin sensitivity. Int J Obes Relat Metab Disord 28:640–648.
Reinhardt RR, Bondy CA (1994) Insulin-like growth factors cross the blood-brain barrier. Endocrinology 135:1753-1761.
Roche R, Hoareau L, Bes-Houtmann S, Gonthier MP, Laborde C, Baron JF, Haffaf Y, Cesari M, Festy F (2006) Presence of the cannabinoid receptors, CB_1 and CB_2, in human omental and subcutaneous adipocytes. Histochem Cell Biol 126:177–187.
Rossmeisl M, Flachs P, Brauner P, Sponarova J, Matejkova O, Prazak T, Ruzickova J, Bardova K, Kuda O, Kopecky J (2004) Role of energy charge and AMP-activated protein kinase in adipocytes in the control of body fat stores. Int J Obes Relat Metab Disord 28:S38–44.
Sacks N, Hutcheson Jr JR, Watts JM, Webb RE (1990) Case report: the effect of tetrahydrocannabinol on food intake during chemotherapy. J Am Coll Nutr 9:630–632.
Sánchez C, Galve-Roperh I, Rueda D, Guzmán M (1998) Involvement of sphingomyelin hydrolysis and the mitogen-activated protein kinase cascade in the Delta^9-tetrahydrocannabinol-induced stimulation of glucose metabolism in primary astrocytes. Mol Pharmacol 54:834–843.
Sanz P, Rodriguez-Vicente C, Repetto M (1985) Alteration of glucose metabolism in liver by acute administration of cannabis. Bull Narc 37:31–35.
Sanz P, Villar P, Repetto M (1983) Effect of cannabis on enzyme induction by phenobarbital. Arch Toxicol Suppl 6:115–120.
Scheen AJ, Finer N, Hollander P, Jensen MD, Van Gaal LF, RIO-Diabetes Study Group (2006) Efficacy and tolerability of rimonabant in overweight or obese patients with type 2 diabetes: a randomised controlled study. Lancet 368:1660–1672.
Sipe JC, Waalen J, Gerber A, Beutler E (2005) Overweight and obesity associated with a missense polymorphism in fatty acid amide hydrolase (FAAH). Int J Obes (Lond) 29:755–759.
Song Z, RouthVH (2005) Differential effects of glucose and lactate on glucosensing neurons in the ventromedial hypothalamic nucleus. Diabetes 54:15–22.
el-Sourogy M, Malek AY, Ibrahim HH, Farag A, el-Shihy A (1966) The effect of Cannabis indica on carbohydrate metabolism in rabbits. J Egypt Med Assoc 49:626–628.

Spanswick D, Smith MA, Mirshamsi S, Routh VH, Ashford ML (2000) Insulin activates ATP-sensitive K^+ channels in hypothalamic neurons of lean, but not obese rats. Nat Neurosci 3:757–758.
Starowicz K, Cristino L, Matias I, Capasso R, Racioppi A, Izzo AA, Di Marzo V (2007) Endocannabinoid dysregulation in the pancreas and adipose tissue of mice fed a high fat diet. Obesity submitted.
Stephens DB (1980) The effects of alimentary infusions of glucose, amino acids, or neutral fat on meal size in hungry pigs. J Physiol 299:453–463.
Sugden MC, Holness MJ (2006) Mechanisms underlying regulation of the expression and activities of the mammalian pyruvate dehydrogenase kinases. Arch Physiol Biochem 112:139–149.
Summers SA, Birnbaum MJ (1997) A role for the serine/threonine kinase, Akt, in insulin-stimulated glucose uptake. Biochem Soc Trans 25:981–988.
Tolan I, Ragoobirsingh D, Morrison EY (2001) The effect of capsaicin on blood glucose, plasma insulin levels and insulin binding in dog models. Phytother Res 15:391–394.
Trevisani M, Smart D, Gunthorpe MJ, Tognetto M, Barbieri M, Campi B, Amadesi S, Gray J, Jerman JC, Brough SJ, Owen D, Smith GD, Randall AD, Harrison S, Bianchi A, Davis JB, Geppetti P (2002) Ethanol elicits and potentiates nociceptor responses via the vanilloid receptor-1. Nat Neurosci 5:546–551.
Trudeau F, Gagnon S, Massicotte G (2004) Hippocampal synaptic plasticity and glutamate receptor regulation: influences of diabetes mellitus. Eur J Pharmacol 490:177–186.
Tucci SA, Rogers EK, Korbonits M, Kirkham TC (2004) The cannabinoid CB_1 receptor antagonist SR141716 blocks the orexigenic effects of intrahypothalamic ghrelin. Br J Pharmacol 143:520–523.
Van Gaal LF, Rissanen AM, Scheen AJ, Ziegler O, Rossner S (2005) Effects of the cannabinoid-1 receptor blocker rimonabant on weight reduction and cardiovascular risk factors in overweight patients: 1-year experience from the RIO-Europe study. Lancet 365:1389–1397.
Volkow ND, Gillespie H, Mullani N, Tancredi L, Grant C, Ivanovic M, Hollister L (1991) Cerebellar metabolic activation by delta-9-tetrahydro-cannabinol in human brain: a study with positron emission tomography and ^{18}F-2-fluoro-2-deoxyglucose. Psychiatry Res 40:69–78.
Wang R, Liu X, Dunn-Meynell AA, Levin BE, Routh VH (2004) The regulation of glucose-excited (GE) neurons in the hypothalamic arcuate nucleus by glucose and feeding relevant peptides. Diabetes 53:1959–1965.
Watanabe S, Doshi M, Hamazaki T (2003) n-3 Polyunsaturated fatty acid (PUFA) deficiency elevates and n-3 PUFA enrichment reduces brain 2-arachidonoylglycerol level in mice. Prostaglandins Leukot Essent Fatty Acids 69:51–59.
Watson GS, Craft S (2004) Modulation of memory by insulin and glucose: neuropsychological observations in Alzheimer's disease. Eur J Pharmacol 490:97–113.
Weidenfeld J, Feldman S, Mechoulam R (1994) Effect of the brain constituent anandamide, a cannabinoid receptor agonist, on the hypothalamopituitary-adrenal axis in the rat. Neuroendocrinology 59:110–112.
Wenger T, Jamali KA, Juaneda C, Leonardelli J, Tramu G (1997) Arachidonyl ethanolamide (anandamide) activates the parvocellular part of hypothalamic paraventricular nucleus. Biochem Biophys Res Commun 237:724–728.
Whitlow CT, Freedland CS, Porrino LJ (2002) Metabolic mapping of the time-dependent effects of delta 9-tetrahydrocannabinol administration in the rat. Psychopharmacology (Berl) 161:129–136.
Whitlow CT, Freedland CS, Porrino LJ (2003) Functional consequences of the repeated administration of Delta^9-tetrahydrocannabinol in the rat. Drug Alcohol Depend 71:169–177.
Wiesinger H, Hamprecht B, Dningen R (1997) Metabolic pathways for glucose in astrocytes. Glia 21:22–34.
Williams CM, Kirkham TC (1999) Anandamide induces overeating: mediation by central cannabinoid (CB_1) receptors. Psychopharmacology (Berl) 143:315–317.

Wolfgang MJ, Lane MD (2006a) Control of energy homeostasis: role of enzymes and intermediates of fatty acid metabolism in the central nervous system. Annu Rev Nutr 26:23–44.

Wolfgang MJ, Lane MD (2006b) The role of hypothalamic malonyl-CoA in energy homeostasis. J Biol Chem 281:37265–37269.

Woods SC, Porte D Jr (1977) Relationship between plasma and cerebrospinal fluid insulin levels of dogs. Am J Physiol 233:E331–334.

Woods SC, Seeley RJ, Baskin DG, Schwartz MW (2003) Insulin and the blood–brain barrier. Curr Pharm Des 9:795–800.

Xue B, Kahn BB (2006) AMPK integrates nutrient and hormonal signals to regulate food intake and energy balance through effects in the hypothalamus and peripheral tissues. J Physiol 574:73–83.

Yan ZC, Liu DY, Zhang LL, Shen CY, Ma QL, Cao TB, Wang LJ, Nie H, Zidek W, Tepel M, Zhu ZM (2007) Exercise reduces adipose tissue via cannabinoid receptor type 1 which is regulated by peroxisome proliferator-activated receptor-delta. Biochem Biophys Res Commun 354:427–433.

Zhang F, Stone V, Smith PJW (2006) Expression of cannabinoid CB_1 receptors in cell culture models of diabetic neuropathy. FENS Abstr 3:A004.20.

Chapter 15
Cannabinoids and Neuroprotection

Veronica A. Campbell and Eric J. Downer

Abstract The majority of neurodegenerative diseases are associated with excessive glutamatergic transmission, oxidative stress and/or inflammatory changes that lead to activation of the apoptotic cascade and subsequent neuronal demise. Cannabinoids have been demonstrated to confer neuroprotection both in vitro and in a number of in vivo paradigms of neurodegeneration including cerebral ischemia, hypoxia, seizures and experimental autoimmune encephalitis. The molecular mechanisms underlying cannabinoid-mediated protection involve both CB_1 receptor-dependent and receptor-independent events. Anti-oxidant activities and the proclivity to reduce excessive glutamatergic synaptic activity underlie some of the neuroprotective effects of cannabinoids. The attenuation of pro-inflammatory signalling coupled with an induction of pro-survival growth factors and enhanced mitogen-activated protein kinase (MAPK) and phosphatidylinositol 3-kinase (PI_3K) activities have also been implicated in the ability of exogenous and endogenous cannabinoids to provide neuroprotection.

Introduction

Loss of neurons is a common feature of a number of neurodegenerative conditions including Alzheimer's disease (AD), Parkinson's disease (PD), epilepsy and stroke. The neuronal loss may be triggered by oxidative stress and subsequent lipid peroxidation, accumulation of misfolded proteins, such as β-amyloid in AD or α-synuclein in PD, or the excessive activation of glutamate receptors. These neuronal insults induce the demise of the cell via a programmed cascade of cellular events that involve, but are not limited to dysregulation of intracellular calcium homeostasis, activation of stress-activated protein kinases, translocation of mitochondrial cytochrome-*c*, which in turn triggers the caspase cascade that kills the cell. There has been substantial interest in the ability of cannabinoids to circumvent the apoptotic (programmed cell death) cascade and offer neuroprotection against a number of insults including ischemia, excitotoxicity, neuroinflammation and experimental models of Alzheimer's disease and multiple sclerosis. Such neuroprotective properties may be elicited pharmacologically, or via manipulation of the endocannabinoid system to confer physiological

A. Kofalvi (ed.), *Cannabinoids and the Brain.*

neuroprotection. Thus, while it is now apparent that cannabinoids have neuroprotective properties, both in vitro and in vivo, the mechanisms underlying their protective effects are complex and will be reviewed herewith.

Cannabinoids and Neuroprotection

The bulk of experimental evidence suggests that cannabinoids act as neuroprotectants in a number of in vitro and in vivo models of neurodegeneration (Guzmán et al., 2002). The molecular mechanisms underlying the protective effects of cannabinoids have been determined principally in vitro. Endogenous cannabinoids protect cultured cortical neurons from oxygen and glucose deprivation independently of CB_1 and CB_2 receptor activation (Nagayama et al., 1999; Sinor et al., 2000). In addition, Δ^9-tetrahydrocannabinol (Δ^9-THC), the principal psychoactive ingredient of marijuana, and the non-psychoactive cannabinoid, cannabidiol, decrease glutamate toxicity in rat cortical neuronal cultures in a manner that is not blocked by CB_1 receptor antagonists, an observation suggestive of a receptor-independent process (Hampson et al., 1998). Cannabidiol also protects cultured rat PC12 cells from toxic beta-amyloid (Aβ)-induced toxicity by virtue of its anti-oxidative and anti-apoptotic properties that are not coupled to CB_1 receptor activation (Iuvone et al., 2004). In contrast, Abood and co-workers (2001) have shown that Δ^9-THC protects mouse spinal neurons against kainate toxicity via activation of the CB_1 receptor. Similarly, the synthetic cannabinoid, WIN55212-2, has been shown to protect cultured hippocampal neurons from glutamate-induced excitotoxicity by a mechanism involving CB_1 receptor activation (Shen and Thayer, 1998) and protective effects mediated through the CB_1 receptor have been observed in a mouse hippocampal cell line and in primary cerebellar cell cultures (Marsicano et al., 2002). The protective effects of some cannabinoids may also be related to the regulation of the NMDA receptor, since the non-psychotropic cannabinoid, HU-211, acts as a stereoselective inhibitor of the NMDA receptor and protects rat forebrain cultures (Nadler et al., 1993) and cortical neuronal cultures (Eshhar et al., 1993) from NMDA-induced neurotoxicity. Neuroprotective effects of cannabinoids have also been found in vivo. Nagayama and co-workers (1999) have shown that WIN55212-2, acting via the CB_1 receptor, decreases hippocampal loss in adult rats following transient global cerebral ischemia. Δ^9-THC has also been shown to protect the neocortex and striatum, but not the hippocampus, from an ischemic insult in adult rats (Louw et al., 2000). Furthermore, Δ^9-THC acting through the CB_1 receptor reduces the neuronal injury that is evoked in neonatal rats following injection of ouabain (van der Stelt et al., 2001). The endocannabinoid, 2-AG, reduces brain oedema, infarct volume and hippocampal damage via the CB_1 receptor in mice following closed head injury (Panikashvili et al., 2001, 2005, 2006). It is also interesting to note that in infant rat models of in vivo neurodegeneration, anandamide precursors, anandamide concentrations and CB_1 receptor density are increased in the cortex (Hansen et al., 2001a,b), which may represent a putative neuroprotective response.

The finding that WIN55,212-2 protects hippocampal astrocytes from the toxic effects of focal administration of ceramide (Gómez del Pulgar et al., 2002) suggests that the neuroprotective effects of the cannabinoid system may be due to an impact on glial cells; thus maintaining glial support of neurons. The ability of cannabinoids to confer neuroprotection are exerted through a variety of mechanisms, including the scavenging of reactive oxygen species (Hampson et al., 1998; Marsicano et al., 2002; Iuvone et al., 2004), inhibition of caspase-3 processing (Iuvone et al., 2004) or by the closing of voltage-sensitive calcium channels and the reduction of calcium influx into the cell (Shen and Thayer, 1996). More recently, the synthetic cannabinoid receptor agonist, WIN55212-2, has been shown to induce morphological changes consistent with neuronal sprouting in vivo (Taglioferro et al., 2006). Taken together, these experimental findings suggest that cannabinoids may have potential therapeutic value to reverse the cellular changes that contribute to neurodegeneration and also promote brain repair.

Cannabinoids and Excitotoxicity

A sustained release of glutamate from glutamatergic nerve terminals results in chronic activation of post-synaptic glutamate receptors (NMDA and AMPA/kainate subtypes), which in turn evoke a prolonged Ca^{2+}influx and consequent disruption of intracellular calcium homeostasis. It is generally accepted that a sustained elevation in intracellular calcium concentration ($[Ca^{2+}]_i$) induces apoptosis, in part by stimulating calcineurin which activates pro-apoptotic caspase-3 (Polster and Fiskum, 2004). Such glutamate receptor overactivation leads to the excitotoxicity that is implicated in both acute conditions, such as stroke, and chronic neurodegenerative conditions, such as amyotrophic lateral sclerosis and Alzheimer's disease. A number of in vitro and in vivo approaches lend support to the neuroprotective properties of cannabinoids being due to an ability to reduce excitotoxic damage. Thus, in hippocampal cultures the excitotoxicity evoked by aberrant patterns of glutamatergic activity is abrogated by activation of CB_1 receptors, although the protection diminishes following long-term treatment with cannabinoid agonists because of receptor desensitization (Gilbert et al., 2006). Similarly, in cultured spinal cord neurons, the toxicity evoked by kainic acid is inhibited by Δ^9-THC in a CB_1 receptor-dependent manner (Abood et al., 2001). Part of this neuroprotective property may be related to the ability of the CB_1 receptor to suppress glutamatergic synaptic activity (Shen et al., 1996; Takahashi and Castillo, 2006) via inhibition of presynaptic Ca^{2+}entry through N- and P/Q-type voltage-dependent Ca^{2+}channels (Mackie and Hille, 1992; Twitchell et al., 1997) and a subsequent prevention of excessive glutamate release. The evidence for a Ca^{2+}-dependent synthesis of AEA and 2-AG Di Marzo et al., 1994; Stella et al., 1997) would suggest that endocannabinoids are generated in response to an intracellular Ca^{2+}load in an attempt to provide feedback inhibition of the excitotoxicity. In this regard it is notable that endocannabinoid upregulation is a feature of a number of neurotoxic paradigms that are associated with elevated intracellular Ca^{2+}concentration (Hansen et al., 1998). Alternative mechanisms of protection against excitotoxicity include inhibition of protein kinase A and reduced nitric oxide generation (Kim et al., 2006a,b) and inhibition of $[Ca^{2+}]_i$ by reducing calcium release from ryanodine-sensitive stores (Zhuang et al., 2005). Manipulation of the endocannabinoid system is also quite likely to be pertinent in mediating protection against excitotoxicity since inhibition of the endocannabinoid transporter or degrading enzyme, fatty acid amide hydrolase (FAAH), enhanced extracellular regulated kinase (ERK) signalling and afforded protection against excitotoxicity both in hippocampal slices and in vivo (Karanian et al., 2005a,b). Of potential clinical relevance are the findings that cannabinoids mediate neuroprotection from excitotoxicity in vivo. In mice lacking the CB_1 cannabinoid receptor ($CB_1^{-/-}$), kainic acid-induced seizures were much more severe than those experienced by wild type animals, inferring that the presence of CB_1 exerted a protective influence (Marsicano et al., 2003). The elegant studies by Monory and colleagues (2006) indicate that the CB_1 receptors located on cortical glutamatergic cells are key elements in the defence against kainic acid-induced excitotoxicity. The exact nature of the endocannabinoid responsible for conferring neuroprotection against excitotoxicity remains to be fully resolved. Stella and co-workers (1997) have demonstrated that glutamate stimulates production of 2-AG, but not anandamide, in the hippocampus, whilst kainic acid-induced seizures are associated with an upregulation of anandamide, but not 2-AG (Marsicano et al., 2003). In an experimental closed head injury model, 2-AG is elevated and exogenous administration of 2-AG reduced the brain oedema and hippocampal cell death, which is a feature of this injury model (Panikashvili et al., 2001, 2005), thereby further supporting a neuroprotective role of endocannabinoids in vivo. The phytocannabinoid Δ^9-THC is also neuroprotective in an in vivo model of excitotoxicity (van der Stelt et al., 2001). In that study, Δ^9-THC reduced neuronal injury elicited by the inhibition of the Na^+/K^+-ATPase in 7- to 8-day-old neonatal rats. This effect was CB_1 receptor-dependent and occurred following a 30 min exposure to Δ^9-THC prior to toxin injection. Thus, the excitotoxicity that is evoked in a number of experimental paradigms is attenuated by cannabinoids and this may be relevant to the development of novel therapies for disorders in which excitotoxicity is a feature.

Cannabinoids and Oxidative Stress

There has been substantial interest in the role of cannabinoids in controlling ischemia-induced damage and there has been intense discussion as to whether cannabinoids act as neuroprotectants or even worsen neuronal damage subsequent to cerebral ischemia. In an experimental stroke model, the CB_1 antagonist, SR141716A, has been found to reduce infarct volume in spite of failing to downregulate excitotoxic NMDA receptors in the ischemic penumbra (Sommer et al., 2006). Following oxygen and glucose deprivation of forebrain slices, an in vitro model for hypoxic-ischemic brain damage in newborn rats, an increase in glutamate release, cell damage and upregulation of pro-inflammatory cytokines and iNOS was observed (Fernandez-Lopez et al., 2006). The finding that WIN55212-2 abrogated these effects provides evidence of a neuroprotective role for CB_1/CB_2 in this system. Furthermore, occlusion of the middle cerebral artery induces a focal cerebral ischemia that is reduced in volume by WIN55212-2 (Bonfils et al., 2006). This latter form of neuroprotection is dependent upon the WIN55212-2-induced hypothermia suggesting that cannabinoids are candidates for a drug-induced hypothermia that may have therapeutic potential in stroke. In vitro studies also support a role for cannabinoids as protectants against oxidative injury. The neurodegeneration evoked by oxidative stress is prevented by anandamide and WIN55212-2, in a manner that is dependent upon inhibition of protein kinase A, and mimicked by antioxidants (Kim et al., 2006a). In addition to protecting cells from damage induced by reactive oxygen species, cannabinoids also confer protection against reactive nitrogen species. Thus, in retinal neurons, the excitotoxicity evoked by glutamate is associated with excessive formation of peroxynitrite that is proposed to lead to the demise of the retinal ganglion cells. The formation of peroxynitrite and subsequent lipid peroxidation and apoptosis is attenuated by both THC and cannabidiol (El-Remessy et al., 2003). The role of cannabinoid receptors in mediating the neuroprotective properties of cannabinoids against oxidative stress is complex. The oxidative stress studies performed in cerebellar granule neurons prepared from $CB_1^{-/-}$ mice would suggest that the CB_1 receptor is not involved in the cellular antioxidant neuroprotective effects of cannabinoids (Marsicano et al., 2002), and recent reports suggest that the antioxidant capacity of cannabinoids is via an ability to chelate $Fe^{2+}$and thus limit Fe^{2+}-induced brain lipid peroxidation (Kessiova et al., 2006). In vivo, cannabidiol offers similar protection against alcohol-induced cell death as that offered by the common anti-oxidant α-tocopherol and suggests that cannabinoid-like lipophilic antioxidant agents may have value in preventing binge ethanol-induced neuronal damage (Hamelink et al., 2005). Antioxidant properties of cannabinoids have also been reported to involve CB_1 receptor activation (Kim et al., 2005).

Cannabinoids and Growth Factors

A number of interesting studies have emerged, which indicate that cannabinoids regulate neurotrophin signaling. Brain-derived neurotrophic factor (BDNF) is responsible for the interneuron migration and morphogenesis via activation of the TrkB receptor. Berghuis and colleagues (2005) have shown that endocannabinoids regulate interneuron migration and morphogenesis by transactivation of TrkB, and their work suggests that prenatal exposure to Δ^9-THC may disrupt that accurate interneuron placement and integration during corticogenesis (see Chap. 12). Furthermore, chronic administration of Δ^9-THC upregulates BDNF expression in the nucleus accumbens, prefrontal cortex and paraventricular nucleus (Butovsky et al., 2005), which may be important in inducing neuroadaptation to cannabinoid exposure. Such cannabinoid-mediated regulation of neurotrophic pathways may be pertinent in the neuroprotection exerted by cannabinoids. In support of this contention, genetic ablation of CB_1 receptors abolishes induction of BDNF that is observed following kainic acid-induced excitotoxicicty and exacerbates the neuronal loss, while application of exogenous BDNF rescues the cells from kainic-acid neurotoxicity (Khaspekov et al., 2004). Thus, BDNF is a critical mediator in the CB_1 receptor-dependent protection against excitotoxicity. In non-neuronal cells the induction of nerve growth factor is also facilitated by cannabinoids, acting through the PI_3K/PKB pathway (Sanchez et al., 2003). The activation of the CB_1 receptor by the endocannabinoid, 2-AG, can also couple to the activated fibroblast growth factor (FGF) receptor to induce an axonal growth response, whilst CB_1 receptor antagonists inhibit axonal growth stimulated by FGF or *N*-cadherin (Williams et al., 2003). The ability of cannabinoids to confer neuroprotection may also be related to their role in the regulation of neurogenesis. The synthetic cannabinoid, WIN55212-2, stimulates adult neurogenesis by opposing the anti-neurogenic effect of nitric oxide (NO) (Kim et al., 2006b) and adult neurogenesis is defective in mice lacking CB_1 receptors (Jin et al., 2004) (see Chap. 12 for further readings). Also, the endocannabinoids have been demonstrated to regulate neurogenesis and neural differentiation (Rueda et al., 2002; Galve-Roperh et al., 2006). Thus, the neuroprotective effects of cannabinoids may involve short-term adaptation to neuronal stress, such as inhibition of glutamate release and oxidative stress, as well as longer-term adaptations associated with de novo neuronal formation and differentiation.

Cannabinoid Signalling Associated with Neuroprotection

The signalling molecules involved in mediating the neuroprotective attributes of cannabinoids are multi-faceted. The control of cell survival is intimately linked to the balance between the activity patterns of members of the mitogen-activated protein kinase (MAPK) family, notably extracellular-regulated protein kinases (ERK) and the stress-activated kinase, p38. The p38 MAPK family mediates cellular responses to stress, such as inflammatory and osmotic insults (Herlarr and Brown, 1999). In contrast, the ERK family members are involved in regulating cell growth and differentiation in response to growth factors and other intracellular messengers (Derkinderen et al., 1999). However, there are also data showing that ERK mediates growth arrest and apoptosis under some circumstances (Grewal et al., 1999). Δ^9-THC and endocannabinoids activate p38 MAPK in hippocampal slices (Derkinderen et al., 2001, 2003) and in PC12 cells (Sarker et al., 2003). Both Δ^9-THC and endocannabinoids have been shown to activate ERK in hippocampal slices (Derkinderen et al., 2003) and in Chinese hamster ovary cells transfected with CB_1 receptor cDNA (Bouaboula et al., 1995). Overall, these findings show that MAPK signalling has a role in the cannabinoid-induced intracellular cascades that may be pertinent in the control of cell fate. In neuronal systems, the endocannabinoid system maintains excitatory synapses in the hippocampus through activation of ERK and integrin-related focal adhesion kinase (FAK) signalling, whereby disruption of either of those kinases has a detrimental affect on the integrity of the synapse (Karanian et al., 2005a). Upregulation of the endocannabinoid system is also associated with enhanced ERK activation and subsequent protection from excitotoxicity (Karanian et al., 2005b). In contrast, the downregulation of p38 MAP kinase is responsible for the neuroprotective and anti-inflammatory effects of cannabidiol in retinal degeneration (El-Remessy et al., 2006) and Δ^9-THC confers protection from NMDA-induced excitotoxicity by virtue of its ability to inhibit p38 MAP kinase (Chen et al., 2005). Another kinase pathway implicated in the neuroprotective role of cannabinoids is the phosphatidylinositol 3-kinase (PI_3K)/Akt pathway whereby the neuroprotection from excitotoxicity offered by the synthetic cannabinoid, HU-210, was associated with activation of Akt, and was reversed by LY294002, an inhibitor of PI_3K (Molina-Holgado et al., 2005). The pro-survival action of HU-210 has been demonstrated to require CB_1 receptor-induced ERK activation downstream of PI_3K/Akt (Galve-Roperh et al., 2002). The consequences of the cannabinoid-mediated regulation of these kinases are quite likely to include a dampening down of key components of the apoptotic cascade (Gomez Del Pulgar et al., 2002). Exposure of telencepthalon cultures to IFNγ is associated with induction of the pro-apoptotic protease, caspase-3, which is lacking in cultures prepared from CB_1 receptor-deficient mice (Jackson et al., 2004) and a number of other studies have demonstrated that cannabinoids inhibit the activation of the apoptotic cascade (Iuvone et al., 2004). A number of neurodegenerative situations, particularly those with a neuroinflammatory component, impact on activity of the transcription factor, NFκB. In closed head injury the increase in 2-AG correlated with a reduction in NFκB transactivation and inhibition of intracellular inflammatory signaling pathways (Panikashvili et al., 2005). Such cannabinoid-mediated inhibition of NFκB may be related to the stabilization of the endogenous inhibitor of NFκB, IKB and prevention of NFκB translocation to the nucleus or inhibition of the transactivation potential of NFκB in a CB_1 receptor-independent manner (Curran et al., 2005). The ability of cannabinoids to regulate neuroinflammatory signalling is quite likely to be critical factor in their ability to prevent AD-like pathology, since cannabinoids have been shown to block the microglial activation and the subsequent cognitive impairment and neuronal loss that is associated with β-amyloid (Ramirez et al., 2005). Also, in PC12 cells exposed to neurotoxic β-amyloid, cannabidiol acts to decrease p38 MAP kinase and NFκB activity, as well as nitrosative stress, to confer neuroprotection (Esposito et al., 2006). Furthermore, cannabinoids provide significant neuroprotection from the consequences of the neuroinflammation that is a feature of experimental allergic encephalomyelitis (EAE), an experimental model of multiple sclerosis (Pryce et al., 2003) in which increased activation of NFκB is observed (Pahan and Schmid, 2000). Thus, the neuroprotective mechanisms of cannabinoids are most likely to include a downregulation in activity of the transcription factors that are pertinent in induction of the pro-inflammatory proteins that serve as key players in neurodegenerative disease.

Neuroprotection vs. Neurotoxicity

The bulk of the experimental evidence indicates that cannabinoids may protect neurons from toxic insults both in vitro (Eshhar et al., 1993; Nadler et al., 1993; Hampson et al., 1998; Shen and Thayer, 1998; Nagayama et al., 1999; Abood et al., 2001; Marsicano et al., 2002; Iuvone et al., 2004) and in vivo (Panikashvili et al., 2000; Hansen et al., 2001a; van der Stelt et al., 2001). The evidence for cannabinoid neurotoxicity is limited to some studies in primary neurons (Chan et al., 1998, Campbell, 2001, Downer et al., 2003), where Δ^9-THC has been found to evoke apoptosis through generation of reactive oxygen species and activation of the stress-activated kinase, *c-Jun* N-terminal kinase via CB_1 receptor. Also, in transformed neural cells a number of studies have identified a pro-apoptotic role for cannabinoids (Sánchez et al., 1998; Jacobsson et al., 2000; Maccarrone et al., 2000; Sarker et al., 2003). In particular, anandamide has been shown to induce apoptotic body formation and DNA fragmentation in cultured neuroblastoma cells (Maccarrone et al., 2000). This neurotoxic effect is independent of CB_1 receptor activation and involves an increase in intracellular calcium concentration, cytochrome-*c* release from the mitochondria and caspase-3 activation. Anandamide also induce apoptosis in cultured PC12 cells via a similar mechanism (Sarker et al., 2003). The observation that anandamide inhibits adult neurogenesis and prevents cortical progenitor cell, neural stem cell and PC12 cell differentiation to the mature neuronal phenotype also highlights the impact of anandamide on neural fate (Rueda et al., 2002). Chronic exposure studies have revealed that THC produces morphological changes in brain structures that are indicative of toxicity (Scallet et al., 1987; Scallet, 1991; Lawston et al., 2000). Specifically, chronic exposure to Δ^9-THC or marijuana extracts alters the structure of the rat hippocampus with decreased mean volume of neurons and number of synapses per unit volume in the hippocampal CA3 region; these structural changes persisted up until at least 7 months after treatment (Scallet et al., 1987). Although apoptotic parameters were not identified in those studies, the authors imply that the morphological changes in the hippocampus are indicative of cannabinoid neurotoxicity. Furthermore, Lawston and co-workers (2000) have shown that adult rats injected twice daily with the synthetic cannabinoid, WIN55212-2 ($2\,mg\ kg^{-1}$), for 21 days exhibit dendritic degradation in CA1 of the hippocampus and retraction from the pyramidal cells. Such morphoregulatory features of cannabinoids in the hippocampus may contribute to cannabinoid-induced memory deficits as changes in the strength of connections between neurons is thought to underlie memory formation (Bolshakov et al., 1997). More recently, the application of functional magnetic resonance imaging (fMRI) has demonstrated a reduction in frontal white-matter volume in substance abusers who abused heroin, cocaine and cannabis (Schlaepfer et al., 2006). Heavy marijuana users were found to have reduced grey matter in the parahippocampal gyrus and reduced white matter in the left parietal lobe, as well as other structural changes (Matochik et al., 2005; see Chap. 22). While the mechanisms responsible for these structural changes have not been identified, the authors suggest a neurotoxic effect, in part mediated through inhibition of myelination. The recurrent transient cerebral ischemic attacks that occur with cannabis use may also cause detrimental effects on the brain (Haubrich et al., 2005). In contrast, Hollister (1986) and Tzilos and colleagues (2005) have found that marijuana does not cause structural damage to the brains of laboratory animals or long-term damage to the human brain. These findings, in conjunction with many in vitro and in vivo experiments describing differential effects of cannabinoids on neuronal viability (Chan et al., 1998; Nagayama et al., 1999; Galve-Roperh et al., 2000; Maccarrone et al., 2000; Sinor et al., 2000; van der Stelt et al., 2001; Zhou and Song, 2001; Iuvone et al., 2004), highlight the complexities associated with cannabinoids and the control of cell survival/death signals in the brain.

Concluding Remarks

Overall, the role of cannabinoids in controlling neuronal cell fate is a complex issue that is influenced by the nature of the toxic insult, the cell type and the particular cannabinoid under investigation. Although the effect of chronic cannabis use on neuronal viability remains to be fully resolved, it is evident that both synthetic and endogenous cannabinoids have the proclivity to confer neuroprotection against a range of insults that are pertinent in excitotoxicity, neuroinflammation, oxidative damage and Alzheimer's pathology. Future therapeutic strategies for neurodegenerative disease may target the endocannabinoid system to offer a cannabinoid-based approach with which to circumvent neurodegeneration.

References

Abood ME, Rizvi G, Sallapudi N, McAllister S (2001) Activation of the CB_1 cannabinoid receptor protects cultured mouse spinal neurons against excitotoxicity. Neuroscience Lett 309:197–201.

Bolshakov VY, Golan H, Kandel ER, Siegelbaum SA (1997) Recruitment of new sites of synaptic transmission during the cAMP-dependent late phase of LTP at CA3-CA1 synapses in the hippocampus. Neuron 19:635–651.

Berghuis P, Dobszay MB, Wang X, Spano S, Ledda F, Sousa KM, Schulte G, Ernfors P, Mackie K, Paratcha G, Hurd YL, Harkany T (2005) Endocannabinoids regulate interneuron migration and morphogenesis by transactivating the TrkB receptor. Proc Natl Acad Sci USA 102:19115–19120.

Bonfils PK, Reith J, Hasseldam H, Johansen FF (2006) Estimation of the hypothermic component in neuroprotection provided by cannabinoids following cerebral ischemia. Neurochem Int 49:508–518.

Bouaboula M, Poinot-Chazel C, Bourrie B, Canat, X Calandra, B Rinaldi-Carmona, M Le Fur G, Casellas P (1995) Activation of mitogen-activated protein kinases by stimulation of the central cannabinoid receptor CB_1. Biochem J 312:637–641.

Butovsky E, Juknat A, Goncharov I, Elbaz J, Eilam R, Zangan A, Vogel Z (2005) *In vivo* upregulation of brain-derived neurotrophic factor in specific brain areas by chronic exposure to tetrahydrocannabinol. J Neurochem 93:802–811.

Campbell VA (2001) Tetrahydrocannabinol-induced apoptosis in cultured cortical neurones is associated with cytochrome c release and caspase-3 activation. Neuropharmacol 40:702–709.

Chan GC, Hinds TR, Impey S, Storm DR (1998) Hippocampal neurotoxicity of Δ^9-Tetrahydrocannabinol. J Neurosci 18:5322–5332.

Chen J, Lee CT, Errico S, Deg X, Cadet JL, Freed, WJ (2005) Protective effects of tetrahydrocannabinol against NMDA-induced AF5 cell death. Brain Res Mol Brain Res 134:215–225.

Curran NM, Griffin BD, O'Toole D, Brady K, Fitzgerald SN, Moynagh PN (2005) the synthetic cannabinoid, R(+)WIN 55, 212-2 inhibits the interleukin-1 signaling pathway in human astrocytes in a cannabinoid receptor-independent manner. J Biol Chem 280:35797–35806.

Derkinderen P, Enslen H, Girault JA (1999) The ERK/MAP-kinases cascade in the nervous system. Neuroreport 10:24–34.

Derkinderen P, Ledent C, Parmentier M, Girault JA (2001) Cannabinoids activate p38 mitogen-activated protein kinases through CB_1 receptors in hippocampus. J Neurochem 77:957–960.

Derkinderen P, Valjent E, Toutant M, Corvol JC, Enslen H, Ledent C, Trzaskos J, Caboche J, Girault JA (2003) Regulation of Extracellular Signal-Regulated Kinase by cannabinoids in hippocampus. J Neurosci 23:2371–2382.

Di Marzo V, Fontana A, Cadas H, Schinell S, Clmlno G, Schwartz J, Piomell D (1994) Formaton and inactivation of endogenous cannabinoid anandamide in central neurones. Nature 372:686–691.

Downer E, Fogarty MP, Campbell VA (2003) Tetrahydrocannabinol-induced neurotoxicity depends on CB1 receptor-mediated c-Jun N-terminal kinase activation in cultured cortical neurons. Br J Pharmacol 140:547–557

El-Remessy AB, Khalil IE, Matragoon S, Abou-Mohamed G, Tsai NJ, Roon P, Caldwell RB, Caldwell RW, Green K, Liou GI (2003) Neuroprotective effect of tetrahydrocannabinol and cannabidiol in NMDA-induced retinal neurotoxicity: involvement of peroxynitrite. Am J Pathol 163:1997–2008.

Eposito G, De Filippis D, Maiuri MC, De Stefano D, Carnuccio R, Iuvone T (2006) Cannabidiol inhibits inducible NOS expression and NO production in β-amyloid-stimulated PC12 neurons through p38 MAP kinase and NFκB involvement. Neurosci Lett 399:91–95.

Eshhar N, Streim S, Biegon A (1993) HU-211, a non-psychotropic cannabinoid, rescues cortical neurones from excitatory amino acid toxicity in culture. Neuroreport 5:237–240.

Fernandez-Lopez D, Martinez-Orgado J, Nunez E, Romero J, Lorenzo P, Moro MA, Lizasoain I (2006) Characterization of the neuroprotective effect of the cannabinoid agonist WIN-55212 in an *in vitro* model of hypoxic-ischemic brain damage in newborn rats. Pediatr Res 60:169–173.

Galve-Roperh I, Rued D, Gómez del Pulgar T, Velasco G, Guzmán M (2002) Mechanism of extracellular signal-regulated kinase activation by the CB_1 cannabinoid receptor. Mol Pharmacol 62:1385–1392.

Galve-Roperh I, Aguado T, Rueda D, Velasco G, Guzman M (2002) Endocannabinoids: a new family of lipid mediators involved in the regulation of neural cell development. Curr Pharm Des 12:2319–2325.

Gilbert GL, Kim HJ, Waataja JJ, Thayer SA (2006) Tetrahydrocannabinol protects hippocampal neurons from excitotoxicity. Brain Res 128:61–69.

Gómez del Pulgar T, de Ceballos ML, Guzmán M, Velasco G (2002) Cannabinoids protect astrocytes from ceramide-induced apoptosis through the phosphatidylinositol 3-kinase/Protein Kinse B pathway. J Biol Chem 277:36527–36533.

Grewal SS, York RD, Stork PJS (1999) Extracellular-signal-regulated kinase signalling in neurons. Curr Opin Neurobiol 9:544–553.

Guzmán M, Sánchez C, Galve-Roperh I (2002) Cannabinoids and cell fate. Pharmacol Ther 95:175–184.

Hamelink C, Hampson A, Wink DA, Eiden LE, Eskay Rl (2005) Comparison of cannabidiol, antioxidants, and diuretics in reversing binge ethanol-induced neurotoxicity. J Pharmacol Exp Ther 314:780–788.

Hampson AJ, Grimaldi M, Axelrod J, Wink D (1998) Cannabidiol and (−)Δ^9-tetrahydrocannabinol are neuroprotective antioxidants. Proc Natl Acad Sci USA 95:8268–8273.

Hansen HH, Ikonomidou C, Bittigau P, Hansen SH, Hansen HS (2001a) Accumulation of the anandamide precursor and other *N*-acylethanolamine phospholipids in infant rat models of *in vivo* necrotic and apoptotic neuronal death. J Neurochem 76:39–46.

Hansen HH, Schmid PC, Bittigau P, Lastres-Becker I, Berrendero F, Manzanare J, Ikonomidou C, Scmid HHO, Fernandez-Ruiz JJ, Hansen HS (2001b) Anandamide, but not 2-arachidonoylglycerol, accumulates during *in vivo* neurodegeneration. J Neurochem 78:1415–1427.

Hansen HS, Lauritzen L, Moesgaard B, Strand AM, Hansen HH (1998) Formation of N-acyl-phosphatidylethanolamines and N-acetylethanolamines proposed role in neurotoxicity. Biochem Pharmacol 55:719–725.

Haubrich C, Diehl R, Donges M, Schiefer J, Loos M, Kosinski C (2005) Recurrent transient ischemic attacks in a cannabis smoker. J Neurol 252:369–370.

Herlarr E, Brown Z (1999) p38 MAPK signaling cascades in inflammatory disease. Mol Med Today 5: 439–447.

Hollister LE (1986) Health aspects of cannabis. Pharmacol Rev 38:1–20.

Iuvone T, Esposito G, Esposito R, Santamaria R, Di Rosa M, Izzo AA (2004) Neuroprotective effect of cannabidiol, a non-psychoactive component from Cannabis sativa, on β-amyloid-induced toxicity in PC12 cells. J Neurochem 89:134–141.

Jackson SJ, Pryce G, Diemel LT, Cuzner ML, Baker D (2005) Cannabinoid-receptor 1 null mice are susceptible to neurofilament damage and caspase-3 activation. Neuroscience 134:261–268.

Jacobsson SOP, Rongard E, Stridh M, Tiger G, Fowler CJ (2000) Serum-dependent effects of tamoxifen and cannabinoids upon C6 glioma cell viability. Biochem Pharmacol 69:1807–1813.

Jin K, Xie L, Kim SH, Parmentier-Batteur S, Sun Y, Mao XO, Childs J, Greenberg DA (2004) Defective adult neurogenesis in CB1 caanbinoid receptor knockout mice. Mo Pharmacol 66:204–208.

Karanian DA, Brown QB, Makriyannis A, Bahr BA (2005a) Blocking cannabinoid activation of FAK and ERK1/2 compromises synaptic integrity in hippocampus. Eur J Pharmacol 508:47–56.

Karanian DA, Brown QB, Makriyannis A, Kosten TA, Bahr BA (2005b) Duel modulation of endocannabinoid transport and fatty acid amide hydrolase protects against excitotoxicity. J Neurosci 25:7813–7820.

Kessiova M, Alexandrova A, Gergieva A, Kirkova M, Todorov S (2006) *In vitro* effects of CB_1 receptor ligands on lipid peroxidation and antioxidant defense systems in the rat brain. Pharmacol Rep 58:870–875.

Khaspekov LG, Brenz Verca MS, Frumkina LE, Hermann H, Marsicano G, Lutz B (2004) Involvement of brain-derived neurotrophic factor in cannabinoid receptor-dependent protection against excitotoxicity. Eur J Neurosci 19:1691–1698.

Kim SH, Won SJ, Mao XA, Jin K, Greenberg DA (2005) Involvement of protein kinase A in cannabinoid receptor-mediated protection from oxidative neuronal injury. J Pharmacol Exp Ther 313:88–94.

Kim SH, Won SJ, Mao XA, Jin K, Greenberg DA (2006a) Molecular mechanisms of cannabinoid protection from neuronal excitotoxicity. Mol Pharmacol 69:691–696.

Kim SH, Won SJ, Mao XA, Ledent C, Jin K, Greenberg DA (2006b) Role for neuronal nitric-oxide synthase in cannabinoid-induced neurogenesis. J Pharmacol Exp Ther 319:150–154.

Lawston J, Borella A, Robinson JK, Whitaker-Azmitia PM (2000) Changes in hippocampal morphology following chronic treatment with the synthetic cannabinoid WIN 55,212-2. Brain Res 877:407–410.

Louw DF, Yang FW, Sutherland GR (2000) The effect of δ–9-tetrahydrocannabinol on forebrain ischemia in rat. Brain Res 857:183–187.

Mackie K, Hille B (1992) Cannabinoids inhibit N-type calcium channels in neuroblastoma-glioma cell lines. Proc Natl Acad Sci USA 89:3825–3829.

Maccarrone M, Lorenzon T, Bari M, Melino G, Finazzi-Agro A (2000) Anandamide induces apoptosis in human cells via vanilloid receptors. J Biol Chem 275:31938–31945.
Marsicano G, Moosmann B, Hermann H, Lutz B, Behl C (2002) Neuroprotective properties of cannabinoids against oxidative stress: role of the cannabinoid receptor CB_1. J Neurochem 80:448–456.
Marsicano G, Goodenough S, Monory K, Hermann H, Eder M, Cannich A, Azad SC, Cascio MG, Gutierrez SO, van der Stelt M, Lopez-Rodriguez ML, Casanova E, Schutz G, Zieglgansberger W, Di Marzo V, Behl C, Lutz B (2003) CB_1 receptors and on-demand defense against excitotoxicity. Science 302:84–88
Matochik JA, Eldreth DA, Cadet JL, Bolla KI (2005) Altered brain tissue composition in heavy marijuana users. Drug Alcohol Depend 77:23–30
Molina-Holgado F, Pinteaux E, Heenan L, Moore JD, Rothwell NJ, Gibson RM (2005) Neuroprotective effects of the synthetic cannabinoid HU-210 in primary cortical neurons are mediated by phosphatidylinositol 3-kinase/Akt signaling. Mol Cell Neurosci 28:189–194.
Monory K, Massa F, Egertova M, Eder M, Blaudzun H, Westenbroek R, Kelsch W, Jacob W, Marsch R, Ekker M, Long J, Rubenstein JL, Goebbels S, Nave KA, During M, Klugmann M, Wolfel B, Dodt HU, Zieglgansberger W, Wotjak CT, Mackie K, Elphick MR, Marsicano G, Lutz B (2006) Endocannabinoid system controls key epileptogenic circuits in the hippocampus. Neuron 51:455–466.
Nadler V, Mechoulam R, Sokolovsky M (1993) The non-psychotropic cannabinoid (+)-3S,4S)-7-hydroxy-Δ^6-tetrahydrocannabinol 1,1-dimethylheptyl (HU-211) attenuates *N*-methyl-D-aspartate receptor-mediated neurotoxicity in primary cultures of rat forebrain. Neurosci Lett 162:43–45.
Nagayama T, Sinor AD, Simon RP, Chen J, Graham SH, Jin K, Greenberg DA (1999) Cannabinoids and neuroprotection in global and focal cerebral ishcemia and in neuronal cultures. J Neurosci 19:2987–2995.
Panikashvili D, Simeonidou C, Ben-Shabat S, Hanus L, Breuer A, Mechoulam R, Shohami E (2001) An endogenous cannabinoid (2-AG) is neuroprotective after brain injury. Nature 413:527–531.
Panikashvili D, Mechoulam R, Beni SM, Alexandrovich A, Shohami E (2005) CB1 cannabinoid receptors are involved in neuroprotection via NFκB. Blood Flow Metab. 25:477–484.
Panikashvili D, Shein NA, Mechoulam R, Trembolver V, Kohen R, Alexandrovich A, Shohami E (2006) The endocannabinoid, 2-AG, protects the blood-brain barrier after closed head injury and inhibits mRNA expression of proinflammatory cytokines. Neurobiol Dis 22:257–264.
Pahan K, Schmid, M (2000) Activation of NFκB in the spinal cord of experimental allergic encephalomyelitis. Neurosci Lett 287:17–20.
Polste BM, Fiskum G (2004) Mitochondrila mechanisms of neural cell apoptosis. J Neurochem 90:1281–1289.
Pryce G, Ahmed Z, Hankey DJ, Jackson SJ, Croxford JL, Pocock JM, Ledent C, Petzold A, Thompson AJ, Giovannoni G, Cuzner ML, Baker D (2003) Cannabinoids inhibit neurodegeneration in animal models of multiple sclerosis. Brain 126:2191–2102.
Ramirez BG, Blazquez C, Gomez del Pulgar T, Guzman M, de Caballos ML (2005) Prevention of Alzheimer's disease pathology by cannabinoids: neuroprotection mediated by blockade of microglial activation. J Neurosci 25:1904–1913.
Rueda D, Navarro B, Martinez-Serrano A, Guzmán M, Galve-Roperh I (2002) The endocannabinoid anandamide inhibits neuronal progenitor cell differentiation through attenuation of the Rap1/B-Raf/ERK pathway. J Biol Chem 277:46645–46650.
Sánchez C, Galve-Roperh I, Canova C, Brachet P, Guzmán M (1998) Δ^9-Tetrahydrocannabinol induces apoptosis in C6 glioma cells. FEBS Lett 436:6–10.
Sánchez MG, Ruiz-Llorente L, Sánchez AM, Díaz-Laviada I (2003) Activation of phosphoinositide 3-kinase/PKB pathway by CB_1 and CB_2 cannabinoid receptors expressed in prostate PC-3 cells. Involvement in Raf-1 stimulation and NGF induction. Cell Signal 15:851–859.
Sarker KP, Biswas KK, Yamakuchi M, Lee K-Y, Hahiguchi T, Krach M, Kitajima I, Maruyama I (2003) ASK1-p38 MAPK/JNK signaling cascade mediates anandamide-induced PC12 cell death. J Neurochem 85:50–61.

Scallet AC, Uemura E, Andrews A, Ali SF, McMillan DE, Paule MG, Brown RM, Slikker W (1987) Morphometric studies of the rat hippocampus following chronic delta-9-tetrahydrocannabinol (THC). Brain Res 436:193–198.

Scallet AC (1991) Neurotoxicology of cannabis and THC: a review of chronic exposure studies in animals. Pharm Biochem Behav, 40:671–676.

Schlaepfer TE, Lancaste E, Heidbreder R, Strain EC, Kosel M, Fisch HU, Pearlson GD (2006) Decreased frontal white-matter volume in chronic substance abuse. Int J Neuropsychopharmacol 9:147–153.

Shen M, Piser TM, Seybold VS, Thayer SA (1996) Cannabinoid receptor agonists inhibit glutamatergic synaptic transmission in rat hippocampal cultures. J Neurosci 16:4322–4334.

Shen M, Thayer SA (1998) The cannabinoid agonist Win55,212-2 inhibits calcium channels by receptor-mediated and direct pathways in cultured rat hippocampal neurons. Brain Res 783:77–84.

Sinor AD, Irvin SM, Greenberg DA (2000) Endocannabinoids protect cerebral cortical neurons from *in vitro* ischemia in rats. Neurosci Lett 278:157–160.

Sommer C, Schomacher M, Berger C, Kuhnert K, Muller HD, Schwab S, Schabitz WR (2006) Neuroprotective cannabinoid receptor antagonist SR141716A prevents downregulation of excitotoxic NMDA receptors in the ischemic penumbra. Acta Neuropathol 112:277–286.

Stella N, Schweitzer P, Piomelli D (1997) A second endogenous cannabinoid that modulates long-term potentiation. Nature 388:773–778.

Tagliaferro P, Javier Ramos A, Onaivi ES, Evrard SG, Lujilde J, Brusco A (2006) Neuronal cytoskeleton and synaptic densities are altered after a chronic treatment with the cannabinoid receptor agonist WIN 55, 212-2. Brain Res 1085:163–176.

Takahashi KA, Castillo PE (2006) CB_1 cannabinoid receptor mediates glutamatergic synaptic suppression in the hippocampus. Neuroscience 139:795–802.

Twitchell W, Brown S, Mackie K (1997) Cannabinoids inhibit N- and P/Q-type calcium channels in cultured rat hippocamapl neurons. J Neurophysiol 78:43–50.

Tzilos GK, Cintron CB, Wood JB, Simpson NS, Young AD, Pope HG, Yurgelun-Todd DA (2005) Lack of hippocampal volume change in long-term cannabis users. Am J Addict 14:64–72.

Van Der Stelt M, Veldeheis WB, Bar PR, Veldink GA, Vliegenthart JFG, Nicolay K (2001) Neuroprotection by Δ^9-Tetrahydrocannabinol, the main active compound in marijuana, against ouabain-induced *in vivo* excitotoxicity. J Neurosci, 21:6475–6479.

Williams EJ, Walsh FS, Doherty P (2003) The FGF receptor uses the endocannabinoid signalling system to couple to an axonal growth response. J Cell Biol 160:481–486.

Zhunag S-Y, Bridges D, Grigoenko E, McCloud S, Boon A, Hampson RE, Deadwyler SA (2005) Cannabinoids produce neuroprotection by reducing intracellular calcium release from ryanodine-sensitive stores. Neuropharmacol 28:1086–1096.

Chapter 16
Neuroinflammation and the Glial Endocannabinoid System

Cristina Benito, Rosa María Tolón, Estefanía Núñez, María Ruth Pazos, and Julián Romero

Abstract The remarkable density and wide distribution of cannabinoid CB_1 receptors in the central nervous system served to explain many of the well-known pharmacological effects of natural, synthetic and endogenous cannabinoids. This receptor type is one of the most abundant cerebral receptors so far described. Its presynaptic location in neurons allows its participation in a myriad of cerebral functions, such as those controlling motor activity or memory and to mediate cannabinoid-induced neuroprotection. At the same time, the psychoactive effects derived from CB_1 activation limited the development of novel therapeutic approaches on the use of cannabinoids. However, recent data have raised the possible interest of the endocannabinoid system in neuroinflammation. These new perspectives can be summarized mostly at two levels: (1) the participation of other components of the endocannabinoid system, mainly CB_2 receptors and fatty acid amide hydrolase (FAAH), in neuroinflammatory processes; and (2) the predominance of the *glial* endocannabinoid system over the *neuronal* endocannabinoid system under pathological conditions. We now know that dramatic changes take place in the endocannabinoid system in the human brain, suggesting its possible involvement in several prevalent diseases, such as Alzheimer's disease, multiple sclerosis or viral encephalitis. This is the subject of the present review.

Introduction

Recent progress in our understanding of neuroinflammatory processes has evidenced the critical role of microglia. These cells are derived from bone marrow and are considered as the resident macrophages of the central nervous system (CNS). It is currently well known that they participate in the first reaction to injury or infection and microglia activation generally precedes any other response (Streit, 2005). They are thus involved in a wide number of acute and chronic pathological states affecting the CNS. In their resting state, microglia serve the role of immune surveillance and host defense, patrolling the cerebral parenchyma and responding against even subtle changes of their microenvironment (Nimmerjahn et al., 2005). When activated, microglial cells experience a series of morphologic and phenotypic

A. Köfalvi (ed.), *Cannabinoids and the Brain.*

changes that allow them to respond more efficiently against the damaging stimulus. Among these changes, microglia up-regulate cell surface receptors, such as complement and major histocompatibility complex receptors, and secrete a number of pro-inflammatory molecules that ultimately collaborate in the neurodegenerative process (Streit, 2005).

The Glial Endocannabinoid System

Endocannabinoid System and Microglia

The possible relevance of the endocannabinoid system in microglial function is an expanding field of research. Tables 1 and 2 summarize much of the current literature regarding cannabinoids and microglia, including in vitro and in vivo studies, respectively. In vitro experiments must face the obvious limitation established by the almost immediate activation of microglial cells in primary culture, but are very useful in providing valuable information on, for instance, cell migration, proliferation or cytokine production. An anti-inflammatory action has been classically attributed to cannabinoids acting on CB_1 and CB_2 receptors, and thus their action on microglia is gaining increasing attention. It must be also noted that non-receptor mediated effects of cannabinoids on microglial function have also been reported. The existence of other type(s) of cannabinoid receptors is still an open issue and might explain some of these observations. Seminal studies by Cabral and co-workers first established that cannabinoids were indeed able to modulate the production of certain molecules by microglia in culture (Waksman et al., 1999; Puffenbarger et al., 2000; Carlisle et al., 2002). Specifically, nitric oxide (NO) production and mRNAs for IL-1α, IL-1β, IL-6 and TNF-α were decreased after exposure of rat microglial cells to Δ^9-THC. Interestingly, while NO production inhibition was a CB_1 receptor-mediated process, cytokine mRNAs modulation was a CB_1 receptor-independent event. These results were somewhat contradictory to those from Stefano and colleagues (1996), who reported increases in NO production by invertebrate microglia and human monocytes after exposure to cannabinoid agonists through a CB_1-mediated mechanism. Cabral's group was also the first to focus their research on the possible role of CB_2 receptors in macrophage/microglia function. Thus, activation of CB_2 receptors present in murine macrophages was shown to decrease antigen processing and subsequent T cell activation (McCoy et al., 1999). Remarkably, CB_2 receptor level of expression could be linked to the cellular activation status, while CB_1 expression levels were constantly low and independent of macrophage activation state. Taken together, these data were highly suggestive of a CB_2 receptor predominant role over CB_1 receptor in macrophages and microglia. Subsequent work from different laboratories has confirmed this suggestion and CB_2 receptor has become a principal target when studying the relationships between the endocannabinoid system and microglial function (see Tables 1 and 2). It seems now

Table 1 Cannabinoids and microglial function (in vitro data)

Reference	Insult/challenge	Ligand(s)	Receptor mediation	Effects of cannabinoid agonists	Involved signalling pathways	Cell type
Stefano et al. (J Biol Chem 271,19238–19242, 1996)	None	AEA	CB_1 mediated	↑ No release		*Mytilus edulis* immunocytes and microglia
		CP55940 WIN55212–2 SR141716A		↑ Cell rounding	Not studied	Human monocytes
Waksman et al. (J Pharm Exp Ther 288, 1357–1366, 1999)	LPS + IFN-γ	CP55940	CB_1 mediated	↓ No release	Gi/Go	Rat microglia
		CP56667		↓ NADPH-diaphorase activity		
		SR141716A				
Puffenbarger et al. (Glia 29, 58–69,2000)	LPS	THC	Non-CB_1	↓ Cytokine mRNAs (IL-1β, IL-1α, IL-6, TNF-α)	Not studied	Rat microglia
		AEA Levonantradol CP55940 CP56667 WIN55212–2 Methanandamide SR141716A SR144528	Non-CB_2			
Carlisle et al. (Int Immunopharmacol 2, 69–82,2002)	Thioglycolate	None	CB_2 mediated	↑ CB_2 expression with cell activation	Not studied	Murine and rat peritoneal macrophages
	LPS					Murine RAW264.7
	IFN-γ					Murine P388D1
						Rat microglia

(continued)

Table 1 (continued)

Reference	Insult/challenge	Ligand(s)	Receptor mediation	Effects of cannabinoid agonists	Involved signalling pathways	Cell type
Facchinetti et al. (Glia 41, 161–168,2003	LPS	AEA	Non-CB_1	↓ TNFα release	Non $G_{i/o}$ mediated	Rat microglia
		2-AG CP55940 WIN55212–2 HU210 AM251 SR141716A SR144528	Non-CB_2			
Walter et al. (J Neurosci 23, 1398–1405, 2003)	ATP	AEA	CB_2 mediated	↑ Cell migration	ERK1/2	BV-2
		2-AG PEA SR141716A SR144528 Cannabinol Cannabidiol O-1918	abn-CBD mediated			Mouse microglia
Klegeris et al. (Br J Pharmacol 139, 775–786, 2003)	LPS + IFN-γ	JWH-015	CB_2 mediated	↓ IL-1β	Not studied	THP-1
		SR141716A SR144528		↓ TNF-α		Human microglia
Franklin and Stella (Eur J Pharmacol 474, 195–198, 2003)	None	ACPA	CB_2 mediated	↑ Cell migration	G_i/G_o	BV-2
		Cannabinol Cannabidiol O-1918 SR141716A	abn-CBD mediated			

Franklin et al. (J Neurosci 23, 7767–7775,2003)	LPS + IFN-γ	PEA	Non-CB_1	↑ Cell migration	G_i/G_o	BV-2
			Non-CB_2 Non-abn-CBD Non-WINr			
Peterson et al. (J Neuroimmunol 147, 123–126, 2004)	HIV-1_{SF162}	WIN55212–2	Not studied	↓ HIV expression (WIN55212-2) No effect (THC)	Not studied	Human fetal microglia
		THC SR141716A SR144528				
Cabral and Marciano-Cabral (J Neuroimmunol 147, 127–130, 2004)	*Acanthamoeba castellanii*	THC	Not studied	↓ Cytokine mRNAs (IL-1β, IL-1α, TNF-α)	Not studied	Rat microglia
Gongora et al. (Immunol Lett 91, 11–16, 2004)	IFN-γ	CP55,940	CB_1 mediated	↓ MHC-II	Not studied	EOC 20
				↓ Class II transactivator		
Witting et al. (PNAS 101, 3214–3219, 2004)	ATP	None	Purinergic $P2X_7$	↑ 2-AG	PI-PLC	Mouse microglia
					DG-lipase ERK1	RTMGL1
Carrier et al. (Mol Pharmacol 65, 999–1007, 2004)	M-CSF	2-AG	CB_2 mediated	↑ Proliferation	Not studied	Mouse microglia
		AEA JWH133 SR144528				
Ramírez et al. (J Neurosci 25, 1904–1913, 2005)	Aβ25–35 and Aβ1–40	HU-210	CB_1 mediated	↓ Morphological changes		
		WIN55212–2 JWH-133	CB_2 mediated	↓ MTT ↓ TNT-α ↑ Neuronal survival		

(continued)

Table 1 (continued)

Reference	Insult/challenge	Ligand(s)	Receptor mediation	Effects of cannabinoid agonists	Involved signalling pathways	Cell type
Ortega-Gutierrez et al. (FASEB J 19, 1338–1340, 2005)	LPS	UCM707	CB_1 mediated	↓ No	Not studied	Mouse microglia
		OMDM1 AEA Methanandamide SR141716A SR144528	CB_2 mediated	↓ iNOS ↓ Cytokines (IL-1β, IL-6, TNF-α)		
Ehrhart et al. (J Neuroinflm 2, 29–42, 2005)	Aβ1–42	JWH-015	CB_2 mediated	↓ IFN-γ-mediated CD40 expression ↓ TNF-α ↓ NO ↑ Phagocytosis	JAK/STATI	Mouse microglia
Maresz et al. (J Neurochem 95, 437–445, 2005)	GM-CSF IFN-γ LPS	None	Not studied	↑ CB_2	Not studied	Mouse microglia
Eljaschewitsch et al. (Neuron 49, 4–8, 2006)	NMDA	AEA	CB_1 mediated	↑ MKP-1	ERK-1/2	Rat microglia
	OGD	WIN55212-2 AM251 AM630	CB_2 mediated	↓ No ↓ iNOS		BV-2 OHSCs
Witting et al. (PNAS 103, 6362–6367, 2006)	IFN-γ	None	Not studied	↓ $P2X_7$–mediated 2-AG	DG lipase-α	Mouse microglia

Carrier et al. (PNAS 103, 7895–7900, 2006)	None	THC	Not studied	↓ Nucleoside uptake	ENT1 nucleoside transporter	EOC-20
		CBD			Adenosine A_{2A}	RAW264.7
Kreutz et al. (Exp Neurol 203, 246–257, 2007)	NMDA	THC	CB_2 mediated (THC, 2-AG)	↓ Number of microglial cells	Not studied	OHSCs
		AEA	Non-CB_2 mediated (AE)	↓ Number of degenerating neurons (2-AG)		
		2-AG				
		AM630				
Mukhopadhyay et al. (J Neuroimmunol 181, 82–92, 2006)	LPS	None	Not studied	↑ CB_2	NF-κB	RAW264.7
					PKA	
					PKC	

Table 2 Cannabinoids and microglial function (in vivo data)

Reference	Insult/disease model	Ligand(s)	Receptor mediation	Molecular effects of cannabinoid agonists	Symptomatic effects of cannabinoid agonists	Animal species
Arevalo-Martin et al. (J Neurosci 23, 2511–2516, 2003)	Theiler's virus	WIN55212	CB_1 mediated	¯ Microglial activation	Motor recovery	Mouse
		ACEA	CB_2 mediated	¯ MHC-II expression	Remyelination	
		JWH-015		¯ CD4+ infiltration		
Zhang et al. (Eur J Neurosci 17, 2750–2754, 2003)	Chronic constriction injury	None	↓ CB_2 mRNA	Not studied	Not studied	Rat
	Freund's complete adjuvant injection					
	Spinal nerve ligation					
Franklin et al. (J Neurosci 23, 7767–7775, 2003)	Focal cerebral ischemia	PEA	Non-CB_1	↓ AEA-induced microglial migration	Not studied	Mouse
			Non-CB_2 Non-abn-CBD Non-WINr			
Cabral and Marciano-Cabral (J Neuroimmunol 147, 127–130, 2004)	Acanthamoeba castellanii	THC	Not studied	↓ Brain accumulation of macrophages	↓ Mortality	Mouse
				¯ Cytokine mRNAs (IL-1β, IL-1α, TNF-α)	↓ Severity of disease	

Ramírez et al. (J Neurosci 25, 1904–1913, 2005)	Aβ25–35 and Aβ1–40	HU-210	Not studied	¯ Microglial activation	¯ Aβ-induced cognitive impairment	Rat
		WIN55212		¯ Aβ-induced neuronal damage	¯ ¯ Aβ-induced cognitive impairment	
		JWH-133				
Mestre et al. (J Neurochem 92, 1327–1339, 2005)	Theiler's virus	AEA	Not studied	¯ Microglial activation	↓ Motor function	Mouse
		OMDM1		¯ MHC-II expression	↓ Motor function	
		OMDM2				
Ortega-Gutierrez et al. (FASEB J 19, 1338–1340, 2005)	Theiler's virus	UCM707	Not studied	¯ Microglial activatioin	↓ Motor function	Mouse
				¯ MHC-II expression		

clear that the expression of microglial CB_1 receptor remains unaltered after cell activation, while that of CB_2 receptor is dramatically increased when microglial cells are exposed to many different types of injuries or challenges. Furthermore, CB_2 receptor activation triggers a decrease in the expression and secretion of pro-inflammatory molecules, such as IL-1 or TNF-α. It must be also considered, however, that other reports suggest a pro-inflammatory CB_2-mediated action on microglia. Thus, Walter and co-workers (2003) and Franklin and co-workers (2003) reported that CB_2 receptor activation leads to an enhancement of microglial migration. In addition, Carrier and colleagues (2004) showed a CB_2 receptor-mediated increase in cell proliferation. These responses are classically considered as pro-inflammatory and could then be paradoxical with the previously described anti-inflammatory actions of cannabinoids through their binding to CB_2 receptors. Nevertheless, it may be argued that these responses are not mutually exclusive, as microglia could proliferate and migrate more rapidly after CB_2 receptor activation to participate in several processes (such as encapsulation of the site of injury, phagocytosis, etc.) and, at the same time, exhibit a lower ability to produce pro-inflammatory molecules. In this line of reasoning, it can be speculated that up-regulated microglial CB_2 receptors could represent an extraordinarily useful target for the development of new therapeutics in neuroinflammation. Finally, it is important to note that microglial cells are a major source of endocannabinoids in the brain. According to results obtained by Stella's group (Stella, 2004), these cells produce almost 20 times more endocannabinoids than astrocytes or neurons do. Furthermore, these authors suggest that microglial cells may be the main producers of endocannabinoids in the inflammatory foci with a subsequent enhancement of local cell recruitment. Interestingly, microglial cells do not exhibit FAAH activity (Franklin et al., 2003).

Endocannabinoid System and Astrocytes

Astrocytes are the most numerous non-neuronal cell type in the CNS (Chen and Swanson, 2003). A well-known function of these cells is to physically structure the brain. A second function is to provide neurons with nutrients such as glucose. The astrocyte end-feet encircling endothelial cells form part of the blood–brain barrier, as they contribute to the formation of tight intercellular junctions between capillary endothelial cells and regulate the expression and function of several endothelial transporters. They perform several essential functions for normal neuronal activity including transmitter's uptake and release, or modulation of synaptic transmission (Piet et al., 2004). They also serve as intermediaries in neuronal regulation of blood flow (Parri and Crunelli, 2003), glucose metabolism and promotion of the myelinating activity of oligodendrocytes. This suggests that astrocytes have an executive-co-ordinating role in the brain (Ishibashi et al., 2006). The presence of CB_1 receptors on astrocytes is well documented. CB_1 receptor immunoreactivity has been described in astrocytes of the nucleus accumbens of Sprague–Dawley rats

(Rodriguez et al., 2001) and also in astrocytes located in the cingulate cortex, the medial forebrain bundle, the amygdala, nucleus accumbens and laminae I and II of the hippocampal dorsal horn of Wistar rats (Moldrich and Wenger, 2000). Perivascular glial fibres have shown moderate to high density of CB_1 protein in olfactory and limbic structures (Salio et al., 2002) (for further information see Chap. 10). In vitro data have raised some conflicts between findings from different laboratories. Cultures from human astrocytomas tumors of different grades and normal rat astrocytes express the CB_1 receptor (Bouaboula et al., 1995; Sánchez et al., 1998a). Furthermore, Molina-Holgado and colleagues (2002a) found that cultured astrocytes from CD1 mice express CB_1 receptors, while Walter and Stella (2003) did not in astrocytes from C57BL/6 mice. Moreover, two studies performed in Swiss-Webster mice astrocytes have shown contradictory results (Sagan et al., 1999; Abood et al., 2001). These discrepancies may reflect differences in the CNS structures used to prepare the astrocytes in culture, differences among species, discrepancies in culture systems or conditions, or ages of cultures. It is important to note that CB_1 expression change depending on the differentiation state of cells in culture (Daaka et al., 1996; Noe et al., 2000). It is still controversial whether astrocytes express CB_2 receptors. Bouaboula and colleagues (1995) and Walter and Stella (2003) found that primary rat and mouse astrocytes do not express CB_2 receptor. On the contrary, Carlisle and colleagues (2002) and Sheng and colleagues (2005) found CB_2 mRNA on cortical primary rat astrocytes and primary human astrocytes, respectively. As mentioned for microglial cells, these paradoxical results may be linked to the fact that levels or expression of CB_2 receptor can change according to cell activation (Carlisle et al., 2002). Immunohistochemical and enzymological data show FAAH expression in human astrocytes. We (Romero et al., 2002) showed FAAH expression in perivascular human astrocytes of the grey and white matter of the human cortex and basal ganglia. In addition, Stella and co-workers (1997) found that mouse astrocytes in culture show FAAH and MAGL protein expression, as well as elicited corresponding enzymatic activities. On the other side, Beltramo and colleagues (1997) found that human astrocytoma cells seem to rapidly metabolize 2-AG. Cannabinoids induce remarkably different effects on astrocytes. When injected into the brain of developing rats, Δ^9-THC may interfere with astroglial differentiation in a sex-dependent manner (Suárez et al., 2000, 2002). Interestingly, WIN55212-2 and HU-210 protect primary astrocytes from ceramide-induced apoptosis via activation of PI_3K/Akt and ERK (Gómez del Pulgar et al., 2002a). It is important to note that, unlike this protective effect on astrocytes, cannabinoids induce apoptosis of glioma cells (Galve-Roperh et al., 2000; Sánchez et al., 2001a; Gómez del Pulgar et al., 2002b). This opposite response to cannabinoids of glioma cells and astrocytes could be based on differences in the regulation of the pathway of de novo ceramide synthesis (Carracedo et al., 2004). Activation of astrocytic CB_1 receptors increases the rate of glucose oxidation to CO_2 as well as the rate of glucose incorporation into phospholipids and glycogen, two phenomena involved in energy supply to the brain (Sánchez et al., 1998b; Blázquez et al., 1999). As perivascular astrocytes are located between cerebral arteries and neurons and regulate energy supply to neighbouring neurons

(Magistretti, 2000; Voutsinos-Porche et al., 2003), CB_1 receptors present at the end-feet of astrocytes may regulate energy supply from blood to neurons. In agreement with this hypothesis, anandamide and Δ^9-THC enhanced the energetic brain metabolism in the rat, probably via the cannabinoid CB_1 receptor (Costa et al., 2004). Additional data show that CB_1 receptor may signal independently of $G_{i/o}$ proteins (see Chap. 9 for detailed discussion). Sánchez and colleagues (2001b) reported that Δ^9-THC, acting at CB_1 receptors, induced sphingomyelin hydrolysis in primary astrocytes, an effect not blocked by pertussis toxin. Other reports show that some effects of cannabinoids on astrocytes are not mediated by CB_1 or CB_2 receptors (Venance et al., 1995; Shivachar et al., 1996; Pertwee, 1999; Sagan et al., 1999; Pearlman et al., 2003; Curran et al., 2005). Whether astrocytes produce endocannabinoids has remained unknown until recently, mainly because of limited sensitivity of the methods used (Di Marzo et al., 1994; Di Tomaso et al., 1997; Stella et al., 1997). However, Walter and colleagues (2002, 2003) recently developed a chemical ionization gas chromatography/mass spectrometry (CI-GC/MS) method that allowed femtomole detection and quantification of anandamide and other acylethanolamides in biological samples. Using this method, these authors detected and quantified anandamide production by mouse astrocytes in culture. Interestingly, endothelin-1 enhanced the production of AEA and 2-AG only (Walter et al., 2002; Walter and Stella, 2003), indicating that activation of different receptor subtypes may selectively increase the production of individual endocannabinoids. Finally, there are a lot of evidence that astrocytes and astrocytoma cell lines inactivate endocannabinoids by uptake and hydrolysis (Beltramo et al., 1997; Deutsch et al., 2000; Muthian et al., 2000; Bisogno et al., 2001; Jonsson et al., 2001).

Endocannabinoid System and Oligodendrocytes

Oligodendrocytes, as myelin-forming cells in the CNS, are responsible for producing the myelin sheath that allows electrical signals to propagate more efficiently. To that end, oligodendroglial processes extend from the cell soma to make contact with axons (Butt and Ransom, 1993). Oligodendrocytes express Ca^{2+}-permeable glutamate receptors and have low resistance to oxidative stress, two factors that make them particularly susceptible to injury (Back et al., 1998). Oligodendrocyte damage compromises brain function, and their injury or death is a prominent feature in demyelinating and neurodegenerative disorders, such as multiple sclerosis. Although the origin of oligodendrocytes capable of remyelinating naked axons is not clear (Levine et al., 2001), oligodendrocyte progenitors exist in the CNS and are recruited to the demyelinated areas to perform the remyelinating process (Keirstead and Blakemore, 1999; Chang et al., 2002). However, survival of proliferating oligodendrocyte progenitors and their successful differentiation to myelinating oligodendrocytes require an appropriate axon–oligodendrocyte contact (Fernandez et al., 2000) and trophic factors released by neurons and astrocytes (Barres et al., 1992; Gard et al., 1995). Molina-Holgado and co-workers (2002b) were able to detect CB_1 receptor immunoreactivity in all of the different developmental stages of rat oligodendrocytes in vivo and in vitro. In primary culture, oligodendrocyte progenitors and differentiated oligodendrocytes also expressed CB_2 receptor, whereas CB_2 expression was absent in vivo. In addition, they reported that CB_1 and CB_2 receptor activation is involved in protecting oligodendrocyte progenitors from apoptosis, via a mechanism dependent on the PI_3K/Akt signaling pathway (Molina-Holgado et al., 2002b). These results supported data obtained by the same group in an animal model of multiple sclerosis in which cannabinoid treatment reduced the MHC class II-restricted CD4 + T cell response rendering a significant increase in the capacity of remyelinating naked axons (Arévalo-Martin et al., 2003). These authors suggested that cannabinoids may favour myelin repair directly because of both anti-inflammatory actions and effects on oligodendrocyte survival and differentiation. A comparative immunohistochemical study carried out in mouse brain on the distribution of FAAH and CB_1 receptors revealed that only FAAH is present in fibre tracts, identified as oligodendrocytes (Egertová et al., 2003). In the mid-brain, the expression of FAAH by oligodendrocytes was particularly striking and particularly abundant in white matter surrounding the cerebellar nuclei. To date, the functional significance of FAAH expression in these glial cells is unknown. The association of FAAH with oligodendrocytes in fibre tracts is of particular interest considering the reported reduction in FAAH activity in the striatum in a rat model of Parkinson's disease (Gubellini et al., 2002). Egertová and colleagues (2003) speculate that the loss of neuronal inputs may also lead to the loss of associated FAAH-expressing oligodendrocytes, which could account for the reduced levels of striatal FAAH in these experimental rats. Nothing is known about whether oligodendrocytes produce endocannabinoids under basal or stimulated conditions or whether are able to take them up (Witting and Stella, 2004).

Cannabinoids and Neuroinflammation

"Neuroinflammation" may be defined as "chronic, sustained cycles of injury and response, in which the cumulative ill effects of immunological microglial and astrocytic activation contribute to and expand the initial neurodestructive effects, thus maintaining and worsening the disease process through their actions" (Streit et al., 2004). This notion originated in the field of Alzheimer's disease (Rogers et al., 1988; Griffin et al., 1989), where it revolutionized our understanding of this disease. From then, it has been also applied to other neurodegenerative diseases such as Parkinson's (McGeer et al., 2001) and Huntington's diseases (Sapp et al., 2001), HIV encephalopathy (Gendelman et al., 1994), multiple sclerosis (Martino et al., 2002), ischemia (Chopp et al., 1994), traumatic brain injury (Dusart and Schwab, 1994), tumor biology (Graeber et al., 2002) and even to normal brain development. The release of pro-inflammatory and neurotoxic mediators (TNF-α, IL-1β, IL-6, eicosanoids, NO and reactive oxygen species) may induce or aggravate brain damage. These factors are predominantly produced by glial cells (mainly reactive microglia) and can be deleterious to neurons (Boje and Arora, 1992; Chao et al., 1992; McGuire et al., 2001; Liu and Hong, 2003). Neuroinflammation incorporates a wide spectrum of complex cellular responses that include activation of microglia and astrocytes and elaboration of cytokines and chemokines, complement proteins, acute phase proteins and related molecular processes. These events may have detrimental effects on neuronal function, leading to neuronal injury with further glial activation and, ultimately, neurodegeneration. As described in the previous section, cell types involved in this process express components of the cannabinoid signalling system that can be endogenously or pharmacologically controlled. Cannabinoid agonists are able to reduce the inflammation that occurs in these diseases. This effect is possibly caused by local effects on glial cells, exerted by either reducing the release of cytotoxic factors or increasing the production of pro-survival molecules (Grundy et al., 2001; Grundy, 2002; Mechoulam et al., 2002; Fowler, 2003; Walter and Stella, 2004). Interestingly, endogenous cannabinoids are also released under neuroinflammation including brain injury (Hansen et al., 2001; Panikashvili et al., 2001; Franklin et al., 2003; Marsicano et al., 2003; Mechoulam and Lichtman, 2003) and are believed to attenuate neuronal damage by binding to CB_1 receptors and protecting against excitotoxicity. These sustained increases in endocannabinoid production constitute a defense mechanism preventing the propagation of neuroinflammation and also of cell damage. The neuroprotective actions of cannabinoids are thought to be mediated through a variety of mechanisms, including antioxidative actions (Hampson et al., 1998), inhibition of NMDA-mediated calcium influx (Mackie and Hille, 1992; Nadler et al., 1993) and inhibition of glutamate release (Shen and Thayer, 1998; Köfalvi et al., 2007). Cannabinoids act on glia and neurons to inhibit the release of proinflammatory molecules, including IL-1, TNF-α and NO (Molina-Holgado et al., 1997,2002a; Shohami et al., 1997; Puffenbarger et al., 2000; Cabral et al., 2001), and enhance the release of the anti-inflammatory cytokines IL-4, IL-10 (Klein et al., 2000), IL-6 (Molina-Holgado et al., 1998) and IL-1ra (Molina-Holgado et al., 2003). Particularly interesting is the inhibitory effect of cannabinoids on the production of TNF-α since this is a major contributor to the pathophysiology of brain injury (Klein et al., 2000). Croxford and Miller (2003) found that WIN55212-2 decreased CNS mRNA encoding for TNF-α in mice infected with Theiler's murine encephalomyelitis virus. In another rat model of brain injury, HU-211 was shown to suppress brain levels of TNF-α directly as well as to reduce mortality and improve clinical outcomes (Shohami et al., 1997). Endogenous and synthetic cannabinoids have the ability to ablate the release of TNF-α elicited by LPS in rat primary cortical microglial and astroglial cells (Facchinetti et al., 2003; Ortega-Gutierrez et al., 2005). This effect does not appear to be mediated by either CB receptor type 1 or type 2 (Facchinetti et al., 2003). Another important inflammation-related mediator is NO, which is produced in response to immune-mediated cellular toxicity playing a role in neurodegeneration (Guzmán et al., 2001; Walter and Stella, 2004). Different cannabinoid chemicals inhibit the release of NO in LPS- or TMEV-stimulated astrocytes, microglia cells or glioma cell line C6 (Molina-Holgado et al., 1997, 2002a, Waksman et al., 1999; Esposito et al., 2001; Ortega-Gutierrez et al., 2005). The effects on NO levels are, at least in part, due to a direct influence on iNOS expression (Ortega-Gutierrez et al., 2005). However, using BV-2 cells, a mouse microglial cell line, WIN55212-2 did not affect basal release of NO, or modulated the LPS/INF-gamma-induced production of NO (Franklin et al., 2003). IL-1 has been identified as an important mediator of diverse forms of experimentally induced brain damage and is expressed rapidly in response to many forms of experimental brain injury, initially by microglia and later by astrocytes (Davies et al., 1999). Cannabinoid agonists (UCM707 and HU-210) induce a significant reduction in the IL-1β levels produced by LPS-stimulated astrocytes (Ortega-Gutierrez et al., 2005). In mouse mixed glial cultures, Molina-Holgado and colleagues (2003) found that HU-210 and CP55940 increase LPS-induced production of IL-1ra, and SR141716A and SR144528 lowered this response. Interestingly, cannabinoid receptor activation failed to do so in knockout mice for these anti-inflammatory cytokines (Molina-Holgado et al., 2003). IL-1ra is a potent endogenous antagonist of all IL-1 actions in the brain (Dinarello and Thompson, 1991), protecting against ischemic, excitotoxic and traumatic brain insults (Allan and Rothwell, 2001). Furthermore, inhibition or deletion of endogenous IL-1ra enhances ischemic brain injury (Loddick et al., 1997) and increases inflammatory responses (Josephs et al., 2000). Anandamide enhances the release of IL-6 from astrocytes infected with TMEV, the virus that elicits a mouse model of multiple sclerosis, an effect blocked by SR141716A (Molina-Holgado et al., 1998). However, it is opposed to the effect observed in LPS-stimulated astrocytes (Ortega-Gutierrez et al., 2005) where UCM707 diminishes IL-6 levels. This result probably may be in relation with the dual character of this cytokine, which can exhibit either pro- or anti-inflammatory properties depending on different factors such as the simultaneous presence of other cytokines. The production of IL-6 by astrocytes could then be related to the anti-inflammatory and/or neuroprotective roles of this cytokine considering that, for example, astrocytes secrete nerve growth factor in response to IL-6 (Frei et al., 1989). Recent studies have implicated CB_2 receptors in the neuroprotective activity of cannabinoids, mainly through a series of glia-dependent anti-inflammatory actions (Fernández-Ruíz et al., 2005). Several studies show that CB_2 receptor activation decreases the production of proinflammatory molecules in several neural cell types such as rat microglial cells (Puffenbarger et al., 2000; Facchinetti et al., 2003), human microglial and THP-1 cells (Stella, 2004), and human astrocytes (Sheng et al., 2005). Activation of CB_2 receptors also reduces the release of proinflammatory factors in animal models of perinatal hypoxia-ischemia (Fernández-López et al., 2006) and Huntington's disease (Fernández-Ruiz and Gonzalez, 2005). The most relevant pro-inflammatory molecules that seem to be under control of the CB_2 receptor include NO, TNF-α, IL-1 and IL-6. In addition, CB_2 receptor activation induce an increase in the release of some anti-inflammatory molecules such as IL-1ra, this molecule may negatively regulate IL-1β (Molina-Holgado et al., 2003). All this data suggest that microglia, astrocytes and oligodendrocytes are sensitive to cannabinoid agonists in different ways, accounting for their anti-inflammatory action. It may be postulated that the beneficial effects on neuroinflammation might be related to several events: inhibition of proinflammatory mediator production, enhancement of anti-inflammatory factor production, inhibition of microglial recruitment and enhancement of astrocyte or oligodendrocyte survival.

The Glial Endocannabinoid System in Human Neurodegenerative Disorders

The distribution of the different components of the endocannabinoid system in the CNS has been extensively studied during the last 15 years. Once the first reports on the existence of a specific receptor protein for cannabinoids were published, the study on its localization was faced. In this sense, elegant autoradiographic studies by Herkenham and co-workers were pioneer (Herkenham et al., 1990, 1991). Thus, by the use of potent and specific radioligands synthesized at Pfizer, these authors obtained evidence on the extensive distribution of cannabinoid CB_1 receptors in the CNS of several animal species, including humans. Basal ganglia structures, cerebellar cortex and hippocampus accounted for the most CB_1 receptor-enriched areas of the brain, explaining some of the most prototypical effects of exogenous cannabinoids (Herkenham et al., 1991; Glass et al., 1997). Once cloned, mRNA distribution studies further confirmed these data and showed that CB_1 receptors were among the most abundant ones in the CNS and exhibited a presynaptic neuronal distribution (Mailleux and Vanderhaegen, 1992). Herkenham and co-workers were also the first to analyze the distribution of CB_2 receptor, revealing its preferential distribution in immune cells and tissues and confirming its absence from the CNS under normal conditions (Lynn and Herkenham, 1994). When the first specific antibodies became available, immunohistochemical studies were performed to describe the precise cellular localization of cannabinoid receptors and FAAH in the CNS. Tsou and colleagues (1998a,b) reported immunohistochemical evidence confirming previous autoradiographic data and describing the predominantly neuronal distribution of these elements of the endocannabinoid system in the murine brain. The first immunohistochemical observations performed in the human brain seemed to corroborate data obtained in other animal species. Both CB_1 receptor and FAAH were abundantly expressed by neurons throughout the brain, with special relevance in cortical neurons, basal ganglia, cerebellar cortex and large neurons of the spinal cord (see Chap. 10). Interestingly, no glial cells showed CB_1 receptor immunoreactivity, while only scarce white matter astrocytes were positive for FAAH enzyme. Several years later, the preferential expression of FAAH on astrocytes has been extensively corroborated. As mentioned earlier, cannabinoid CB_2 receptors seemed to be absent from the CNS (Galiegue et al., 1995). Recent data, however, have raised substantial controversy, with some groups reporting a massive presence of this receptor in neuronal elements of the mouse brain (Gong et al., 2006) and others showing a selective, restricted expression of these receptors in neuronal elements of the brainstem of several animal species (Van Sickle et al., 2005). Concerning the human CNS, our group provided immunohistochemical evidence suggesting the presence of CB_2 receptors in a microglial cell subtype, as the perivascular microglia

(Nuñez et al., 2004). These cells exhibit important differences in respect to other types of microglia and are known to play critical roles in blood–brain barrier homeostasis as well as in pathological states of the CNS (Williams and Hickey, 2002). The selective presence of CB_2 receptors in these cells matched well with the participation of these receptors in immune-related functions and expanded the field to novel approaches like their possible role in the viral infection of the brain.

Alzheimer's Disease

Alzheimer's disease (AD) is one of the most important health challenges in western countries (for details, see Chap. 19). The analysis of human postmortem brain samples from AD patients has provided information on the neuropathology of the endocannabinoid system that raises new hypothesis on the possible role of this system in the prevention and/or treatment of AD (Benito et al., 2003). Of special relevance may be the induction of the expression of CB_2 receptors in microglial cells. It has only been recently accepted that this type of cannabinoid receptor may be present in the CNS, as previous work circumscribed its presence to peripheral cells and tissues of the immune system (Howlett et al., 2002). Although the functions of these receptors in the CNS are far from clear, they may be now considered as diagnostic markers for microglial activation and as relevant candidates for the development of anti-Aβ therapies. The amount of in vitro data in the literature on the anti-inflammatory effects of CB_2 activation, for instance, raises appealing possibilities, as anti-inflammatory compounds are under intense study as putative useful agents for the treatment of AD patients. The modulation of FAAH expression and activity also constitutes an interesting approach. Our current hypothesis suggests that FAAH inhibition may provide benefits for dampening the local inflammatory process triggered by Aβ as it would render more endocannabinoids available for interaction with their receptors. Together with the observed decrease in CB_1 receptor binding and functional coupling in human AD samples (Westlake et al., 1994; Ramirez et al., 2005), the putative psychoactive effects derived from the potentiation of the endogenous cannabinoid tone could have a lower impact. In addition, a decrease in FAAH expression and/or activity would also affect the local levels of arachidonic acid, one of the products of FAAH enzymatic activity on AEA, and precursor of a series of potent pro-inflammatory mediators.

HIV-1-Associated Dementia

The syndrome of cognitive and motor dysfunction observed after infection with human HIV-1 has been designated as HIV-associated dementia (HAD). Many experts believe that HAD is now the most common cause of dementia worldwide among people under 40 years of age (Ellis et al., 1997). There is an incomplete understanding of how HIV infection causes neuronal injury and apoptosis. The principal pathway for HIV entry into the CNS is through infected monocytes, being perivascular macrophages and not the parenchymal microglia the primary cell productively infected (Williams et al., 2001). Some non-productive infection of astrocytes can also occur but it seems well established that neurons are not directly infected. Neuronal injury has been attributed either to a release of neurotoxic factors by HIV-infected microglia/macrophages (and possibly astrocytes) or to neurotoxic HIV proteins (Kaul et al., 2001). HAD is associated pathologically with HIV-1 encephalitis (HIVE). HIVE is characterized by the formation of multinucleated giant cells (through the fusion of inflammatory cells), microglial nodules, infiltration of macrophages from the periphery, widespread astrocytosis, myelin pallor and neuronal loss (Persidsky and Gendelman, 2003). We have recently addressed the question on the status of the endocannabinoid system after HIV-1 infection of the brain by the analysis of human brain tissue samples from HIVE patients and from macaques infected with the simian variant of this virus (the SIVE model) (Benito et al., 2005). Interestingly, these models allow the direct analysis of inflammation in the brain, as samples from infected individuals but without encephalitis are also included. The analysis of human and macaque samples allowed us to conclude that, with little exceptions, a common pattern of inflammation-linked changes in the pattern of expression of cannabinoid receptors and FAAH take place. For the first time, we observed CB_1 receptor positive astrocytes and microglial cells in HIVE samples. This is an important difference with previous observations in AD tissue samples, where CB_1 receptors remained unchanged in their pattern of expression. Although the presence of CB_1 receptor in astrocytes has been previously reported in the rat brain (Rodríguez et al., 2001), their induction in astrocytes in the human brain as a consequence of the viral-triggered inflammatory process deserves to be highlighted. In addition, infiltrated T lymphocytes also exhibited high levels of CB_1 receptor immunoreactivity. In contrast, no glial expression of CB_1 receptors could be noticed in the inflamed macaq ue brain. Possible differences in the inflammatory response of the human vs. macaque respect CB_1 expression could be explained by a possible strain-derived virus phenotype. Similar to what seen in AD human samples, the most dramatic changes occur to CB_2 receptors and FAAH. Both elements of the endocannabinoid system are up-regulated in the inflamed brains of macaques and humans with viral infection. Microglial nodules and infiltrated T lymphocytes exhibited highest levels of CB_2 expression. Interestingly, perivascular microglial cells also exhibited elevated levels of CB_2 immunoreactivity as a consequence of the viral infection of the brain. These cells are known to play a specific and crucial role in the process of viral entry into the CNS and, thus, it seems reasonable to think that selective CB_2 receptor activation could modify this process. FAAH-positive astrocytes were found predominantly in perivascular regions and specifically in areas of cellular infiltration. As astrocytes are known to play a regulatory role in HIV-1 encephalitis by dampening the overexpression of eicosanoids, platelet-activating factor, and TNFα by activated HIV-1 monocytes (Minagar et al., 2002), FAAH overexpressed in glial cells could partially counteract some of these beneficial processes (Weber et al., 2004).

Multiple Sclerosis

Multiple sclerosis (MS) is an inflammatory and demyelinating disease of the CNS, with unknown ethiology (for details, see Chap. 18). The endocannabinoid system is a current target for the treatment of several diseases, including MS (Pryce and Baker, 2005). Clinical evidence from trials confirms the therapeutic potential of cannabinoids in the treatment of multiple sclerosis symptoms (Zajicek et al., 2003, 2005; Rog et al., 2005). These data are supported by previous results obtained in different animal models of MS that show a relief in certain symptoms of this disease. So far, the treatment of MS has focused only in CB_1 receptor activation. However, other results (Arevalo-Martin et al., 2003; Yiangou et al., 2006; Benito et al., 2007; Docagne et al., 2007) postulate that other elements, such as CB_2 receptors and FAAH enzyme, are potential therapeutic targets for the treatment of MS. Moreover, the attractive possibility of finding cannabinoid-based therapies for diseases devoid of undesired CB_1 receptor-mediated psychotropic side effects is also opened. Studies performed in human spinal cord and brain MS samples detected strong CB_2 receptor immunoreactivity in microglia/macrophages in white matter areas, usually within active plaques or in the periphery of chronic lesions (Yiangou et al., 2006; Benito et al., 2007). These results confirm CB_2 expression in glial cells in the human CNS as previously reported in other neuroinflammatory conditions (Benito et al., 2003, 2005). In the brain lesions of MS donors, abundant CB_1 expression was also detected in macrophages located within active plaques. Several in vitro studies documented that microglia/macrophages are involved in phagocytosis of myelin debris in MS lesions, and as a result the process triggers release of pro-inflamatory cytokines and NO (Williams et al., 1994; Mosley and Cuzner, 1996; van der Laan et al., 1996). Although little is known on the effects of cannabinoids on myelin phagocytosis, previous reports have shown that the activation of the endocannabinoid system decrease the production of pro-inflammatory cytokines and NO levels in macrophages/microglia, thus accounting for an anti-inflammatory effect that seems to potentiate the neuroprotection induced by cannabinoids (Mestre et al., 2005; Ortega-Gutierrez et al., 2005). The immunohistochemical study carried out by Benito and colleagues (2007) also revealed the expression of the cannabinoid receptors and the enzyme FAAH in other glial cells. Interestingly, CB_1 receptors were also present in adult oligodendrocytes and OPCs located within MS plaques. These cells are known to be essential to neuroprotection and brain repair since are a key part in the re-myelination process that takes place during the course of the disease (Levine et al., 2001). Previous studies have shown CB_1 and CB_2 expression in the different developmental stages of rat oligodendrocytes in vivo and in vitro (Molina-Holgado et al., 2002b). The activation of these receptors promoted oligodendrocyte survival, via a PI_3K/Akt-dependent mechanism, and thereby enhanced axonal re-myelination in a MS animal model (Molina-Holgado et al., 2002b; Arevalo-Martin et al., 2003). In addition, CB_1 and CB_2 receptors were expressed by perivascular T lymphocytes. The myelin-reactive T lymphocytes are thought to be involved in the demyelinating process and to cause inflammation

(Frohman et al., 2006). Thus, the presence of both type of receptors in T lymphocyte is suggestive of a possible role of the endocannabinoid system in MS-linked, T cell-mediated neuroinflammation, since T cells are known to participate in the pathogenesis of MS (Frohman et al., 2006) and cannabinoids decrease $CD4^+$ infiltration into the spinal cord in an animal model of MS through CB_1 and CB_2 receptor activation (Arevalo-Martin et al., 2003). In contrast to previous data obtained in other pathologies such as Alzheimer's disease (Benito et al., 2003; Ramirez et al., 2005), CB_2 expression was also detected in white matter astrocytes, being the first observation in that type of glial cells in situ in human. There are little data about the role of CB_2 receptors in astrocytes, although in vitro studies suggest that they may modulate the production of different inflammatory mediators (Ortega-Gutierrez et al., 2005; Sheng et al., 2005). More recently, Docagne and co-workers (2007) have proposed a neuroprotective effect in a MS animal model as a result of the concomitant activation of CB_1 receptor in neurons and CB_2 receptor in astrocytes. As previously reported in other neurinflammatory pathologies (Benito et al., 2003, 2005), the endocannabinoid-degrading enzyme FAAH was overexpressed in reactive astrocytes within MS plaques; therefore, this seems to be a strikingly constant feature of this enzyme. Importantly, other arachidonic acid-related enzymes, such as COX-2 or phospholipase-A_2 are also known to be selectively overexpressed in astrocytes under inflammatory stimuli (Sun et al., 2005). FAAH inhibition could have beneficial effects during inflammation because of decreased local production of arachidonic acid and enhanced endogenous cannabinoid tone (Benito et al., 2003; Karanian et al., 2005).

Concluding Remarks

It is now accepted that the endocannabinoid system is an endogenous neuromodulator system that participates in many important processes of the CNS and that acts as a physiological neuroprotectant, both under acute as well as chronic insults. Recent literature shows an increasing attention to in vitro and in vivo animal models of injury in which the activation of the endocannabinoid system results in neuroprotection. In addition, the study of the neuropathology of the endocannabinoid system in the human brain suggests that it may be involved in the neuroinflammation that usually takes place in several diseases, such as AD, HIV-1-encephalitis or MS. Table 3 summarizes some of these findings that confirm a change in the pattern of expression of cannabinoid receptors and FAAH in the chronically damaged human brain. These disease-related modifications suggest a less prominent role for the neuronal CB_1 receptor, the main cannabinoid receptors in the brain, while are indicative of an emerging role for glial CB_2 receptor and FAAH. The relevance of these findings lies on the possible interest of these elements of the endocannabinoid system as new diagnostic markers as well as possible targets for the development of novel therapies by using cannabinoid chemicals, without undesired psychoactive effects.

Table 3 Endocannabinoid system in neurodegenerative diseases (details and abbreviations in the text)

	Control	Alzheimer's disease	Down's syndrome	Sive	Hive	Multiple sclerosis
CB_1	Neurons	Neurons	Neurons	Neurons	Neurons Astrocytes Microglia	neurons Macrophages Oligodendrocytes T lymphocytes
CB_2	Perivascular microglia	Activated microglia	Activated microglia	Activated microglia Perivascular microglia T lymphocytes	Activated microglia Perivascular microglia T lymphocytes	Activated microglia Macrophages Astrocytes
FAAH	Neurons Astrocytes	Neurons Reactive astrocytes	Neurons Reactive astrocytes	Neurons Reactive astrocytes	Neurons Reactive astrocytes	Neurons Reactive astrocytes

References

Abood ME, Rizvi G, Sallapudi N, McAllister S (2001) Activation of the CB_1 cannabinoid receptor protects cultured mouse spinal neurons against excitotoxicity. Neurosci Lett 309:197–201.

Allan SM, Rothwell NJ (2001) Cytokines and acute neurodegeneration. Nat Rev Neurosci 2:734–744.

Arevalo-Martin A, Vela JM, Molina-Holgado E, Borrell J, Guaza C (2003) Therapeutic action of cannabinoids in a murine model of multiple sclerosis. J Neurosci 23:2511–2516.

Back SA, Gan X, Li Y, Rosenberg PA, Volpe JJ (1998) Maturation-dependent vulnerability of oligodendrocytes to oxidative stress-induced death caused by glutathione depletion. J Neurosci 18:6241–6253.

Barres BA, Hart IK, Coles HS, Burne JF, Voyvodic JT, Richardson WD, Raff MC (1992) Cell death and control of cell survival in the oligodendrocyte lineage. Cell 70:31–46.

Beltramo M, Stella N, Calignano A, Lin SY, Makriyannis A, Piomelli D (1997) Functional role of high-affinity anandamide transport, as revealed by selective inhibition. Science 277:1094–1097.

Benito C, Nunez E, Tolon RM, Carrier EJ, Rabano A, Hillard CJ, Romero J (2003) Cannabinoid CB2 receptors and fatty acid amide hydrolase are selectively overexpressed in neuritic plaque-associated glia in Alzheimer's disease brains. J Neurosci 23:11136–11141.

Benito C, Kim WK, Chavarria I, Hillard CJ, Mackie K, Tolon RM, Williams K, Romero J (2005) A glial endogenous cannabinoid system is upregulated in the brains of macaques with simian immunodeficiency virus-induced encephalitis. J Neurosci 25:2530–2536.

Benito C, Romero JP, Tolón RM, Clemente D, Docagne F, Hillard CJ, Guaza C, Romero J (2007) Cannabinoid CB_1 and CB_2 receptors and fatty acid amide hydrolase are specific markers of plaque cell subtypes in human multiple sclerosis. J Neurosci 27:2396–2402.

Bisogno T, Maccarrone M, De Petrocellis L, Jarrahian A, Finazzi-Agrò A, Hillard C, Di Marzo V (2001) The uptake by cells of 2-arachidonylglycerol, an endogenous agonist of cannabinoid receptors. Eur J Biochem 268:1982–1989.

Boje KM, Arora PK (1992) Microglial-produced nitric oxide and reactive nitrogen oxides mediate neuronal cell death. Brain Res 587:250–256.

Bouaboula M, Bourrié B, Rinaldi-Carmona M, Shire D, Le fur G, Casellas P (1995) Stimulation of cannabinoid receptor CB_1 induces krox-24 espression in human astrocytoma cells. J Biol Chem 270:13973–13980.

Butt AM, Ransom BR (1993) Morphology of astrocytes and oligodendrocytes during development in the intact rat optic nerve. J Comp Neurol 338:141–158.

Cabral GA, Harmon KN, Carlisle SJ (2001) Cannabinoid-mediated inhibition of inducible nitric oxide production by rat microglial cells: evidence for CB_1 receptor participation. Adv Exp Med Biol 493:207–214.

Carlisle SJ, Marciano-Cabral F, Staab A, Ludwick C, Cabral GA (2002) Differential expression of the CB_2 cannabinoid receptor by rodent macrophages and macrophage-like cells in relation to cell activation. Int Immunopharmacol 2:69–82.

Carracedo A, Geelen MJ, Diez M, Hanada K, Guzman M, Velasco G (2004) Ceramide sensitizes astrocytes to oxidative stress: protective role of cannabinoids. Biochem J 380:435–440.

Carrier EJ, Kearn CS, Barkmeier AJ, Breese NM, Yang W, Nithipatikom K, Pfister SL, Campbell WB, Hillard CJ (2004) Cultured rat microglial cells synthesize the endocannabinoid 2-arachidonylglycerol, which increases proliferation via a CB2 receptor-dependent mechanism. Mol Pharmacol 65:999–1007.

Chang A, Tourtellotte WW, Rudick R, Trapp BD (2002) Premyelinating oligodendrocytes in chronic lesions of multiple sclerosis. N Engl J Med 346:165–173.

Chao CC, Hu S, Molitor TW, Shaskan EG, Peterson PK (1992) Activated microglia mediate neuronal cell injury via a nitric oxide mechanism. J Immunol 149:2736–2741.

Chen Y, Swanson RA 2003. Astrocytes and brain injury. J Cereb Blood Flow Metab 23:137–149.

Chopp M, Zhang RL, Chen H, Li Y, Jiang N, Rusche JR (1994) Postischemic administration of an anti-Mac-1 antibody reduces ischemic cell damage after transient middle cerebral artery occlusion in rats. Stroke 25:869–875.

Costa B, Colleoni M, Conti S, Parolaro D, Franke C, Trovato AE, Giagnoni G (2004) Oral antiinflammatory activity of cannabidiol, a non-psychoactive constituent of cannabis, in acute carrageenan-induced inflammation in the rat paw. Nauyn-Schmiedebergs Arch Pharmacol 369:74–79.

Croxford JL, Miller SD (2003) Immunoregulation of a viral model of multiple sclerosis using the synthetic cannabinoid R + WIN55,212. J Clin Invest 111:1231–1240.

Curran NM, Griffin BD, O'Toole D, Brady KJ, Fitzgerald SN, Moynagh PN (2005) The synthetic cannabinoid R(+)WIN55,212–2 inhibits the interleukin-1 signaling pathway in human astrocytes in a cannabinoid receptor-independent manner. J Biol Chem 280:35797–35805.

Daaka Y, Friedman H, Klein TW (1996) Cannabinoid receptor proteins are increased in jurkat, human T-cell line after mitogen activation. J Pharmacol Exp Ther 276:776–783.

Davies CA, Loddick SA, Toulmond S, Stroemer RP, Hunt J, Rothwell NJ (1999) The progression and topographic distribution of interleukin-1beta expression after permanent middle cerebral artery occlusion in the rat. J Cereb Blood Flow Metab 19(1):87–98.

Deutsch DG, Glaser ST, Howell JM, Kunz JS, Puffenbarger RA, Hillard CJ, Abumrad N (2000) The cellular uptake of anandamide is coupled to its breakdown by fatty acid amide hydrolase (FAAH). J Biol Chem 276:6967–6973.

Di Marzo V, Fontana A, Cadas H, Shimelli S, Cimino G, Schwart JC, Piommelli D (1994) Formation and inactivation of endogenous cannabinoid anandamide in central neurons. Nature 372:686–619.

Dinarello CA, Thompson RC (1991) Interleukin-1 receptor antagonist competitively inhibits the binding of interleukin-1 to the type II interleukin-1 receptor. J Biol Chem 266:14147–14150.

Di Tomaso E, Cadas H, Gaillet S, Beltramo M, Desarnaud F, Venance L, Piomelli D (1997) Endogenous lipids that activate cannabinoid receptors. Formation and inactivation. Adv Exp Med Biol 407:335–340.

Docagne F, Muneton V, Clemente D, Ali C, Loria F, Correa F, Hernangomez M, Mestre L, Vivien D, Guaza C (2007) Excitotoxicity in a chronic model of multiple sclerosis: neuroprotective effects of cannabinoids through CB_1 and CB_2 receptor activation. Mol Cell Neurosci 34:551–561.

Dusart I, Schwab ME (1994) Secondary cell death and the inmflammatory reaction after dorsal hemisction of the rat spinal cord. Eur J Neurosci 6:712–724.

Egertova M, Cravatt BF, Elphick MR (2003) Comparative analysis of fatty acid amide hydrolase and cb_1 cannabinoid receptor expression in the mouse brain: evidence of a widespread role for fatty acid amide hydrolase in regulation of endocannabinoid signaling. Neuroscience 119:481–496.

Ellis RJ, Deutsch R, Heaton RK, Marcotte TD, McCutchan JA, Nelson JA, Abramson I, Thal LJ, Atkinson JH, Wallace MR, Grant I (1997) Neurocognitive impairment is an independent risk factor for death in HIV infection. Arch Neurol 54:416–424.

Esposito G, Izzo AA, Di Rosa M, and Iuvone T (2001) Selective cannabinoid CB_1 receptor-mediated inhibition of inducible nitric oxide synthase protein expression in C6 rat glioma cells. J Neurochem 78:835–841.

Facchinetti F, Del Giudice E, Furegato S, Passarotto M, Leon A (2003) Cannabinoids ablate release of TNFalpha in rat microglial cells stimulated with lypopolysaccharide. Glia 41:161–168.

Fernandez PA, Tang DG, Cheng L, Prochiantz A, Mudge AW, Raff MC (2000) Evidence that axon-derived neuregulin promotes oligodendrocyte survival in the developing rat optic nerve. Neuron 28:81–90.

Fernandez-Lopez D, Martinez-Orgado J, Nunez E, Romero J, Lorenzo P, Moro MA, Lizasoain I (2006) Characterization of the neuroprotective effect of the cannabinoid agonist WIN-55212 in an *in vitro* model of hypoxic-ischemic brain damage in newborn rats. Pediatr Res 60:169–73.

Fernandez-Ruiz J, Gonzalez S (2005) Cannabinoid control of motor function at the basal ganglia. Handb Exp Pharmacol 168:479–507.

Fernandez-Ruiz JJ, Gonzalez S, Romero J, Ramos JA (2005) Cannabinoids in neurodegeneration and neuroprotection. In Cannabinoids as Therapeutics, Mechoulam R, ed., Birkhäuser Verlag/ Switzerland, pp. 79–109.

Fowler CJ (2003) Plant-derived, synthetic and endogenous cannabinoids as neuroprotective agents. Non-psychoactive cannabinoids, "entourage" compounds and inhibitors of *N*-acyl ethanolamine brakdown as therapeutic strategies to avoid pyschotropic effects. Brain res Rev 41:26–43.

Franklin A, Parmentier-Batteur S, Walter L, Greenberg DA, Stella N (2003) Palmitoylethanolamide increases after focal cerebral ischemia and potentiates microglial cell motility. J Neurosci 23:7767–7775.

Frei K, Malipiero Uv, Leist TP, Zinkernagel RM, Schwab ME, Fontana A (1989) On the cellular source and function of interleukin 6 produced in the central nervous system in viral diseases. Eur J Immunol 19:689–694.

Frohman EM, Racke MK, Raine CS (2006) Multiple Sclerosis. The plaque and its pathogenesis. New Eng J Med 354:942–955.

Galiegue S, Mary S, Marchand J, Dussossoy D, Carriere D, Carayon P, Bouaboula M, Shire D, Le Fur G, Casellas P (1995) Expression of central and peripheral cannabinoid receptors in human immune tissues and leukocyte subpopulations. Eur J Biochem 232:54–61.

Galve-Roperh I, Sánchez C, Cortés M, Gómez del Pulgar T, Izquierdo M, Guzmán M (2000) Anti-tumoral action of cannabinoids: involvement of sustained ceramide accumulation and extracellular signal-regulated kinase activation. Nat Med 6:313–319.

Gard AL, Burrell MR, Pfeiffer SR, Rudge JS, Williams II WC (1995) Astroglial control of oligodendrocyte survival mediated by PDGF and leukemia inhibitory factor-like protein. Development 121:2187–2197.

Gendelman HE, Genis P, Jett M, Zhai QH, Nottet HS (1994) An experimental model sustem for HIV-1-inudced brain injury. Adv Neuroimmunol 4:189–193.

Glass M, Dragunow M, Faull RL (1997) Cannabinoid receptors in the human brain: a detailed anatomical and quantitative autoradiographic study in the fetal, neonatal and adult human brain. Neuroscience 77:299–318.

Gómez del Pulgar T, de CeballoS ML, Guzmán M, Velasco G (2002a) Cannabinoids protect astrocytes from ceramide-induced apoptosis through the phosphatidylinositol 3-kinase B pathway. J Biol Chem 277:36527–36533.

Gómez del Pulgar T, Velasco G, Sánchez C, Haro A, Gumán M (2002b) De novo-synthesized ceramide is involved in cannabinoid-induced apoptosis. Biochem J 363:183–188.

Gong JP, Onaivi ES, Ishiguro H, Liu QR, Tagliaferro PA, Brusco A, Uhl GR (2006) Cannabinoid CB_2 receptors: immunohistochemical localization in rat brain. Brain Res 1071:10–23.

Graeber MB, Scheithauer BW, Kreutzberg GW (2002) Microglia in brain tumors. Glia 40:252–259.

Griffin WS, Stanley LC, Ling C, White L, MacLeod V, Perrot LJ, Whote CL, Araoz C (1989) Brain interleukin 1 and S-100 immunoreactivity are elevated in Down syndrome and Alzheimer disease. Proc Natl Acad Sci 86:7611–7615.

Grundy RI (2002) The therapeutic potential of the cannabinoids in neuroprotection. Expert Opin Investig Drugs 11:1–10.

Grundy RI, Rabuffetti M, Beltramo M (2001) Cannabinoids and neuroprotection. Mol Neurobiol 24:29–51.

Gubellini P, Picconi B, Bari M, Battista N, Calabresi P, Centonze D, Bernardi G, Finazzi-Agro A, Maccarrone M (2002) Experimental parkinsonism alters endocannabinoid degradation: implications for striatal glutamatergic transmission. J Neurosci 22:6900–6907.

Guzmán M, Sanchez C, Galve-Roperh I (2001) Control of the cell survival/death decision by cannabinoids. J Mol Med 78:613–625.

Hampson AJ, Grimaldi M, Axelrod J, Wink D (1998) Cannabidiol and (−)Δ^9-tetrahydrocannabinol are neuroprotective antioxidants. Proc Natl Acad Sci USA 95:8268–8273.

Hansen HH, Schmid PC, Bittigau P, Lastres-Becker I, Berrendero F, Manzanares J, Ikonomidou C, Schmid HH, Fernandez-Ruiz JJ, Hansen HS (2001) Anandamide, but not 2-arachidonoylglycerol, accumulates during *in vivo* neurodegeneration. J Neurochem 78:1415–1427.

Herkenham M, Lynn AB, Little MD, Johnson MR, Melvin LS, De Costa BR, Rice KC (1990) Cannabinoid receptor localization in brain. Proc Natl Acad Sci 87:1932–1936.

Herkenham M, Lynn AB, Johnson MR, Melvin LS, De Costa BR, Rice KC (1991) Characterization and localization of cannabinoid receptors in rat brain: a quantitative *in vitro* autoradiographic study. J Neurosci 11:563–583.

Howlett AC, Barth F, Bonner TI, Cabral G, Casellas P, Devane WA, Felder CC, Herkenham M, Mackie K, Martin BR, Mechoulam R and Pertwee RG (2002) International Union of Pharmacology. XXVII. Classification of cannabinoid receptors. Pharmacol Rev 54:161–202.

Ishibashi T, Dakin K, Stevens B, Lee P, Kozlov S, Stewart C, Fields R (2006) Astrocytes promote myelination in response to electrical impulses. Neuron 49:823–832.

Jonsson K-O, Vandevoorde S, Lambert DM, Tiger G, Fowler CJ (2001) Effects of homologues and analogues of pamitoylethanolamide upon the inactivation of the endocannabinoid anandamide. Br J Pharmacol 133:1263–1275.

Josephs MD, Solorzano CC, Taylor M, Rosenberg JJ, Topping D, Abouhamze A, Mackay SL, Hirsch E, Hirsh D, Labow M, Moldawer LL (2000) Modulation of the acute phase response by altered expression of the IL-1 type 1 receptor or IL-1ra. Am J Physiol Regul Integr Comp Physiol 278:R824–R830.

Karanian DA, Brown QB, Makriyannis A, Kosten TA, Bahr BA (2005) Dual modulation of endocannabinoid transport and fatty acid amide hydrolase protects against excitotoxicity. J Neurosci 25:7813–7820.

Kaul M, Garden GA, Lipton SA (2001) Pathways to neuronal injury and apoptosis in HIV-associated dementia. Nature 410:988–994.

Keirstead HS, Blakemore WF (1999) The role of oligodendrocytes and oligodendrocyte progenitors in CNS remyelination. Adv Exp Med Biol 468:183–197.

Klein TW, Lane B, Newton CA, Friedman H (2000) The cannabinoid system and cytokine network. Proc Soc Exp Biol Med 225:1–8.
Köfalvi A, Pereira MF, Rebola N, Rodrigues RJ, Oliveira CR, Cunha RA (2007) Anandamide and NADA bi-directionally modulate presynaptic Ca^{2+} levels and transmitter release in the hippocampus. Br J Pharmacol doi:10.1038/sj.bjp.0707252.
Levine JM, Reynolds R, Fawcett JW (2001) The oligodendrocyte precursor cell in health and disease. Trends Neurosci 24:39–47.
Liu B, Hong JS (2003) Role of microglia in inflammation-mediated neurodegenerative deseases: mechanisms and strategies for therapeutic intervention. J Pharmacol Exp Ther 304:1–7.
Loddick SA, Wong ML, Bongiorno PB, Gold PW, Licinio J, Rothwell NJ (1997) Endogenous interleukin-1 receptor antagonist is neuroprotective. Biochem Biophys Res Commun 234:211–215.
Lynn AB, Herkenham M (1994) Localization of cannabinoid receptors and nonsaturable high-density cannabinoid binding sites in peripheral tissues of the rat: implications for receptor-mediated immune modulation by cannabinoids. J Pharmacol Exp Ther 268:1612–1623.
Mackie K, Hille B (1992) Cannabinoids inhibit N-type calcium channels in neuroblastoma-glioma cells. Proc Natl Acad Sci USA 89:3825–3829.
Magistretti P (2000) Cellular bases of functional bran imaging: insights from neuron-glia metabolic coupling. Brain Res 886:108–112.
Mailleux P, Vanderhaeghen JJ (1992) Distribution of neuronal cannabinoid receptor in the adult rat brain: a comparative receptor binding radioautography and in situ hybridization histochemistry. Neuroscience 48:655–668.
Marsicano G, Goodenough S, Monory K, Hermann H, Eder M, Cannich A, Azad SC, Cascio MG, Gutierrez SO, van der Stelt M, Lopez-Rodriguez ML, Casanova E, Schutz G, Zieglgansberger W, Di Marzo V, Behl C, Lutz B (2003) CB_1 cannabinoid receptors and on-demand defense against excitotoxicity. Science 302:84–88.
Martino G, Adorini L, Rieckmann P, Hillert J, Kallmann B, Comi G, Filippi M. (2002) Inflammation in multiple sclerosis the good, the bad, and the complex. Lancet Neuro 1:499–509.
McCoy KL, Matveyeva M, Carlisle SJ, Cabral GA (1999) Cannabinoid inhibition of the processing of intact lysozyme by macrophages: evidence for CB_2 receptor participation. J Pharmacol Exp Ther 289:1620–1625.
McGeer PL, Yasojima K, McGeer EG (2001) Inflammation in Parkinson's disease. Adv Neuro 86:83–89.
Mc Guire SO, Ling ZD, Lipton JW, Sortwell CE, Collier TJ, Carvey P (2001) Tumor necrosis factor alpha is toxic to embryonic mesencephalic dopamine neurons. Exp Neurol 169:219–230.
Mechoulam R, Lichtman AH (2003) Neuroscience. Stout guards of the central nervous system. Science 302:65–67.
Mechoulam R, Panikashivili A, Shohami E (2002) Cannabinoids and brain injury: therapeutic implications. Trend Mol Med 8:58–61.
Mestre L, Correa F, Arevalo-Martin A, Molina-Holgado E, Valenti M, Ortar G, Di Marzo V, Guaza C (2005) Pharmacological modulation of the endocannabinoid system in a viral model of multiple sclerosis. J Neurochem 92:1327–1339.
Minagar A, Shapshak P, Fujimura R, Ownby R, Heyes M, Eisdorfer C (2002) The role of macrophage/microglia and astrocytes in the pathogenesis of three neurologic disorders: HIV-associated dementia, Alzheimer disease, and multiple sclerosis. J Neurol Sci 202:13–23.
Moldrich G, Wenger T (2000) Localization of the CB_1 cannabinoid recptor in the rat brain. An immunohistochemical study. Peptides 21:1735–1742.
Molina-Holgado F, Lledo A, Guaza C (1997) Anandamide suppresses nitric oxide and TNF-alpha responses to Theiler's virus or endotoxin in astrocytes. Neuroreport 8:1929–1933.

Molina-Holgado F, Molina-Holgado E, Guaza C (1998) The endogenous cannabinoid anandamide potentiates interleukin-6 production by astrocytes infected with Theiler's murine encephalomyelitis virus by a receptor-mediated pathway. FEBS Lett 433:139–142.

Molina-Holgado F, Molina-Holgado E, Guaza C, Rothwell NJ (2002a) Role of CB1 and CB_2 receptors in the inhibitory effects of cannabinoids on lipopolysaccharde-induced nitric oxide release in astrocytes cultures. J Neurosci Res 67:829–836.

Molina-Holgado E, Vela JM, Arevalo-Martin A, Almazan G, Molina-Holgado F, Borrell J, Guaza C (2002b) Cannabinoids promote oligodendrocyte progenitor survival: involvement of cannabinoid receptors and phosphatidylinositol-3 kinase/Akt signaling. J Neurosci 22:9742–9753.

Molina-Holgado F, Pinteaux E, Moore JD, Molina-Holgado E, Guaza C, Gibson RM, Rothwell NJ (2003) Endogenous interleukin-1 receptor antagonist mediates anti-inflammatory and neuroprotective actions of cannabinoids in neurons and glia. J Neurosci 23:6470–6474.

Mosley K, Cuzner ML (1996) Receptor-mediated phagocytosis of myelin by macrophages and microglia: effect of opsonization and receptor blocking agents. Neurochem Res 21:481–487.

Muthian S, Nithipatikom K, Campbell WB, Hillard CJ (2000) Synthesis and characterization of a fluorescent substrate for the *N*-arachidonoylethanolamine (anandamide) transmembrane carrier. J Pharmacol Exp Ther 293:289–295.

Nadler V, Mechoulam R, Sokolovsky M (1993) Blockade of $^{45}Ca^{2+}$ influx through the *N*-methyl-D-aspartate receptor ion channel by the non-psychoactive cannabinoid HU-211. Brain Res 622:79–85.

Nimmerjahn A, Kirchhoff F, Helmchen F (2005) Resting microglial cells are highly dynamic surveillants of brain parenchyma *in vivo*. Science 308:1314–1318.

Noe SN, Newton C, Widen R, Friedman H, Klein TW (2000) Anti-CD40, anti-CD3, and IL-2 stimulation induce contrasting changes in CB_1 mRNA expression in mouse splenocytes. J Neroimmunol 110:161–167.

Nunez E, Benito C, Pazos MR, Barbachano A, Fajardo O, Gonzalez S, Tolon RM, Romero J (2004) Cannabinoid CB_2 receptors are expressed by perivascular microglial cells in the human brain: an immunohistochemical study. Synapse 53:208–213.

Ortega-Gutierrez S, Molina-Holgado E, Guaza C (2005) Effect of anandamide uptake inhibition in the production of nitric oxide and in the release of cytokines in astrocyte cultures. Glia 52:163–168.

Panikashvili D, Simeonidou C, Ben Shabat S, Hanus L, Breuer A, Mechoulam R, and Shohami E (2001) An endogenous cannabinoid (2-AG) is neuroprotective after brain injury. Nature 413:527–531.

Parri R, Crunelli V (2003) An astrocyte bridge from synapse to blood flow. Nat Neurosci 6:5–6.

Pearlman RJ, Aubrey KR, Vandenberg RJ (2003) Arachidonic acid and anandamide have opposite modulatory actions at the glycine transporter, $GLYT_{1a}$. J Neurochem 84:592–601.

Persidsky Y, Gendelman HE (2003) Mononuclear phagocyte immunity and the neuropathogenesis of HIV-1 infection. J Leukoc Biol 74:691–701.

Pertwee RG (1999) Evidence for the presence of CB_1 cannabinoid receptors on peripheral neurones and for the existence of neuronal non-CB_1 cannabinoid receptors. Life Sciences 65:597–605.

Piet R, Vargová L, Syková E, Poulain D, Oliet S (2004) Physiological contribution of the astrocytic environment of neurons to intersynaptic crosstalk. Proc Natl Acad Sci USA 101:2151–2155.

Pryce G, Baker D (2005) Emerging properties of cannabinoid medicines in management of multiple sclerosis. Trends Neurosci 28:272–276.

Puffenbarger RA, Boothe AC, Cabral GA (2000) Cannabinoids inhibit LPS-inducible cytokine mRNA expression in rat microglial cells. Glia 29:58–69.

Ramirez BG, Blazquez C, Gomez GP, Guzman M, De Ceballos ML (2005) Prevention of Alzheimer's disease pathology by cannabinoids: neuroprotection mediated by blockade of microglial activation. J Neurosci 25:1904–1913.

Rodriguez JJ, Mackie K, Pickel VM (2001) Ultrastructural localization of the CB_1 cannabinoid recptor in µopioid recptor patches of the rat caudate putamen nucleus. J Neurosci 21:823–833.

Rog DJ, Nurmikko TJ, Friede T, Young CA (2005) Randomized, controlled trial of cannabis-based medicine in central pain in multiple sclerosis. Neurology 65:812–819.
Rogers J, Luber-Narod J, Styren SD, Civin WH (1988) Expression of immune system-associated antigens by cells of the human central nervous system: relationship to the pathology of Alzheimer's disease. Neurobiol Aging 9:339–349.
Romero J, Hillard CJ, Calero M, Rabano A (2002) Fatty acid amide hydrolase localization in the human central nervous system: an immunohistochemical study. Brain Res Mol Brain Res 100:85–93.
Sagan S, Venance L, Torrens Y, Cordier J, Glowinski J, Giaume C (1999) Anandamide and WIN 55212–2 inhibit cyclic AMP formation through G-protein-coupled recptors distinct from CB_1 cannabinoid receptors in cultured astrocytes. Eur J Neurosci 11:691–699.
Salio C, Doly S, Fischer J, Franzoni MF, Conrath M (2002) Neuronal and astrocytic localization of the cannabinoid recptor-1 in the dorsal horn of the rat spinal cord. Neurosci Lett 329:13–16.
Sánchez C, Galve-Roperh I, Canova C, Brachet P, Guzmán M (1998a) Delta 9-tetrahydrocannabinol induces apoptosis in a C6 glioma cells. FEBS Lett 436:6–10.
Sánchez C, Galve-Roperh I, Rueda D, Guzmán M (1998b) Involvement of sphingomyelin hydrolysis and the mitogen-activated protein kinase cascade in the delta^9-tetrahydrocannabinol-induced stimulation of glucose metabolism in primary astrocytes. Mol Pharmacol 54:834–843.
Sánchez C, de Ceballos ML, Gómez del Pulgar T, Rueda D, Corbacho C, velasco G, Galve-Roperh I, Huffman JWH, Ramón y Cajal S, Guzmán M (2001a) Inhibition of glioma growth *in vivo* by selective activation of the CB_2 cannabinoid receptor. Cancer Res 61:5784–5789.
Sánchez C, Rueda D, Segui B, Galve-Roperh I, Levade T, Guzmán M (2001b) The CB1 cannabinoid receptor of astrocytes in coupled to sphingomyelin hydrolysis through the adaptor protein Fan. Mol Pharmacol 59:955–959.
Sapp E, kegel KB, Aronin N, Hashikawa T, Uchiyama Y, Tohyama K, Bhide PG, Vonsattel JP, DiFiglia M (2001) Early and progressive accumulation of reactive microglia in the Huntington disease brain. J Neuropathol Exp Neurol 60:161–172.
Shen M, Thayer SA (1998) The cannabinoid agonist Win55,212–2 inhibits calcium channels by receptor-mediated and direct pathways in cultured rat hippocampal neurons. Brain Res 783:77–84.
Sheng WS, Hu S, min X, Cabral GA, Lokensgard JR, Peterson P (2005) Synthetic cannabinoid WIN55,212-2 inhibits generation of inflammatory mediators by IL-1beta stimulated human astrocytes. Glia 49:211–219.
Shivachar AC, Martin BR, Ellis EF (1996) Anandamide- and delta9-tetrahydrocannabinol-evoked arachidonic acid mobilization and blockade by SR141716A [*N*-(Piperidin-1-yl)-5-(4-chlorophenyl)-1-(2,4-dichlorophenyl)-4-methyl-1H-pyrazole-3-carboximide hydrochloride]. Biochem Pharmacol 51:669–676.
Shohami E, Gallily R, Mechoulam R, Bass R, Ben-Hur T (1997) Cytokine production in the brain following closed head injury: dexanabinol (HU-211) is a novel TNF-alpha inhibitor and an effective neuroprotectant. J Neuroimmuno 72:169–177.
Stefano GB, Liu Y, Goligorsky MS (1996) Cannabinoid receptors are coupled to nitric oxide release in invertebrate immunocytes, microglia, and human monocytes. J Biol Chem 271:19238–19242.
Stella N (2004) Cannabinoid signaling in glial cells. Glia 48:267–277.
Stella N, Schweitzer P, Piomelli D (1997) A second endogenous cannabinoid that modulates long-term potentiation. Nature 388:773–778.
Streit WJ (2005) Microglia and neuroprotection: implications for Alzheimer's disease. Brain Res Rev 48:234–239.
Streit WJ, Mrak RE, Griffin WS (2004) Microglia and neuroinflammation: a pathological perspective. J Neuoinflammation 1:14.

Suárez I, Bodega G, Ramos JA, Fernández-Ruiz JJ, Fernández B (2000) Neuronal and astroglial response to pre- and perinatal exposure to delta-9-tetrahydrocannabinol in the rat substantia nigra. Dev Neurosci 22:253–263.

Suárez I, Bodega G, Fernández-Ruiz JJ, Ramos JA, Rubio M, Fernández B (2002) Reduced glial fibrillary acidic protein and glutamine synthase expression in astrocytes and Bergmann glial cells in the rat cerebellum caused by delta-9-tetrahydrocannabinol administration during development. Dev Neurosci 24:300–312.

Sun GY, Xu J, Jensen MD, Yu S, Wood WG, Gonzalez FA, Simonyi A, Sun AY, Weisman GA. (2005) Phospholipase A_2 in astrocytes: responses to oxidative stress, inflammation, and G protein-coupled receptor agonists. Mol Neurobiol 31:27–41.

Tsou K, Brown S, Sanudo-Pena MC, Mackie K, Walker JM (1998a) Immunohistochemical distribution of cannabinoid CB_1 receptors in the rat central nervous sytem. Neuroscience 83:393–411.

Tsou K, Nogueron MI, Muthian S, Sanudo-Pena MC, Hillard CJ, Deutsch DG, Walker JM (1998b) Fatty acid amide hydrolase is located preferentially in large neurons in the rat central nervous system as revealed by immunohistochemistry. Neurosci Lett 254:137–140.

Van der Laan LJ, Ruuls SR, Weber KS, Lodder IJ, Dopp EA, Dijkstra CD (1996) Macrophage phagocytosis of myelin *in vitro* determined by flow cytometry: phagocytosis is mediated by CR3 and induces production of tumor necrosis factor-alpha and nitric oxide. J Neuroimmunol 70:145–152.

Van Sickle MD, Duncan M, Kingsley PJ, Mouihate A, Urbani P, Mackie K, Stella N, Makriyannis A, Piomelli D, Davison JS, Marnett LJ, Di Marzo V, Pittman QJ, Patel KD, Sharkey KA (2005) Identification and functional characterization of brainstem cannabinoid CB_2 receptors. Science 310:329–332.

Venance L, Piomelli D, Glowinski J, Giaume C (1995) Inhibition by anandamide of gap junctions and intercellular calcium signalling in striatal astrocytes. Nature 376:590–594.

Voutsinos-Porche B, Bonvento G, Tanaka K, Steiner P, Welker E, Chatton J-Y, Magistretti PJ, Pellerin L (2003) Glial glutamate transporters mediate a functional metabolic crosstalk between neurons and astrocytes in the mouse developing cortex. Neuron 37:275–286.

Waksman Y, Olson J.M, Carlisle SJ, Cabral GA (1999) The central cannabinoid receptor (CB_1) mediates inhibition of nitric oxide production by rat microglial cells. J Pharmacol Exp Ther 288:1357–1366.

Walter L, Stella N (2003) Endothelin-1 increases 2-arachidonyl glycerol (2-AG) production in astrocytes. Glia 44:85–90.

Walter L, Franklin A, Witting A, Moller T, Stella N (2002) Astrocytes in culture produce anandamide and other acylethanolamides. J Biol Chem 277:20869–20876.

Walter L, Franklin A, Witting A, Wade C, Xie Y, Kunos G, Mackie K, Stella N (2003) Nonpsychotropic cannabinoid receptors regulate microglial cell migration. J Neurosci 23:1398–1405.

Weber A, Ni J, Ling KH, Acheampong A, Tang-Liu DD, Burk R, Cravatt BF, Woodward D (2004) Formation of prostamides from anandamide in FAAH knockout mice analyzed by HPLC with tandem mass spectrometry. J Lipid Res 45:757–763.

Westlake TM, Howlett AC, Bonner TI, Matsuda LA, Herkenham M (1994) Cannabinoid receptor binding and messenger RNA expression in human brain: an *in vitro* receptor autoradiography and in situ hybridization histochemistry study of normal aged and Alzheimer's brains. Neuroscience 63:637–652.

Williams K, Ulvestad E, Waage A, Antel JP, McLaurin J (1994) Activation of adult human derived microglia by myelin phagocytosis *in vitro*. J Neurosci Res 38:433–443.

Williams KC, Hickey WF (2002) Central nervous system damage, monocytes and macrophages, and neurological disorders in AIDS. Annu Rev Neurosci 25:537–562.

Witting A, Stella N (2004) Cannabinoid signaling in glial cells in health and disease. Current Neuropharmacol 2.

Yiangou Y, Facer P, Durrenberger P, Chessell IP, Naylor A, Bountra C, Banati RR, Anand P (2006) COX-2, CB_2 and $P2X_7$-immunoreactivities are increased in activated microglial cells/macrophages of multiple sclerosis and amyotrophic lateral sclerosis spinal cord. BMC Neurol 6:12.

Zajicek J, Fox P, Sanders H, Wright D, Vickery J, Nunn A, Thompson A (2003) Cannabinoids for treatment of spasticity and other symptoms related to multiple sclerosis (CAMS study): multicentre randomised placebo-controlled trial. Lancet 362:1517–1526.
Zajicek JP, Sanders HP, Wright DE, Vickery PJ, Ingram WM, Reilly SM, Nunn AJ, Teare LJ, Fox PJ, Thompson AJ (2005) Cannabinoids in multiple sclerosis (CAMS) study: safety and efficacy data for 12 months follow up. J Neurol Neurosurg Psychiatry 76:1664–1669.

Chapter 17
Targeting Cannabinoid Receptors in Brain Tumors

Guillermo Velasco, Arkaitz Carracedo, Cristina Blázquez, Mar Lorente, Tania Aguado, Cristina Sánchez, Ismael Galve-Roperh, and Manuel Guzmán

Abstract Cannabinoids, the active components of *Cannabis sativa* L., act in the body by mimicking endogenous substances – the endocannabinoids – that activate specific cell surface receptors. Cannabinoids exert various palliative effects in cancer patients. In addition, cannabinoids inhibit the growth of different types of tumor cells, including glioma cells, in laboratory animals. They do so by modulating key cell signaling pathways, mostly the endoplasmic reticulum stress response, thereby inducing antitumoral actions such as the apoptotic death of tumor cells and the inhibition of tumor angiogenesis. Of interest, cannabinoids seem to be selective antitumoral compounds as they kill glioma cells but not their nontransformed astroglial counterparts. On the basis of these preclinical findings, a pilot clinical study of Δ^9-tetrahydrocannabinol (Δ^9-THC) in patients with recurrent glioblastoma multiforme has been recently run. The fair safety profile of Δ^9-THC, together with its possible growth-inhibiting action on tumor cells, may set the basis for future trials aimed at evaluating the potential antitumoral activity of cannabinoids.

Introduction

Gliomas are defined as those tumors that display histological, immunohistochemical, and ultrastructural evidence of glial differentiation. The World Health Organization classifies gliomas according to their cellular features (i.e., resembling astrocytes, oligodendrocytes, or ependymal cells) and their grade of malignancy (from I to IV) (Kleihues et al., 2002). Glioblastoma multiforme (GBM), or grade IV astrocytoma, is the most frequent class of malignant primary brain tumors and one of the most aggressive forms of cancer. As a consequence, survival after diagnosis is normally just 6–12 months (Kleihues et al., 2002; Reardon and Wen, 2006). This dramatic behavior is mainly due to the high invasiveness and proliferation rate of GBM. In addition, GBM exhibits a high resistance to common chemotherapy and radiotherapy. These malignant features may be related to the varying mutations frequently found in these tumors that impact different key pathways involved in the control of cell proliferation, survival, differentiation, and DNA repair (Maher et al., 2001; Kleihues et al., 2002; Reardon and Wen, 2006). Current standard therapeutic strategies for the

A. Köfalvi (ed.), *Cannabinoids and the Brain.*

treatment of GBM are only palliative, and include surgical resection and focal radiotherapy. A large number of chemotherapeutic agents (e.g., alkylating agents such as temozolomide and nitrosureas such as carmustine) have also been tested, but no remarkable improvement on patient survival has been achieved as yet (Lonardi et al., 2005; Reardon and Wen, 2006). Likewise, although dendritic cell- and peptide-based immunotherapy strategies appear promising as a safe approach to induce an antitumor immune response (Yamanaka, 2006), no immunotherapy or gene therapy trial performed to date has been significantly successful. It is therefore essential to develop new therapeutic strategies for the management of GBM, which will most likely require a combination of therapies to obtain significant clinical results. Here we summarize the current knowledge on how a new family of compounds, the cannabinoids, exerts antiglioma actions in laboratory animals, and how a potential cannabinoid-based therapy for GBM might be envisaged.

Cannabinoids and Their Receptors

The hemp plant *Cannabis sativa* L. produces approximately 70 unique compounds known as cannabinoids, of which Δ^9-THC is the most important owing to its high potency and abundance in cannabis (Gaoni and Mechoulam, 1964). Δ^9-THC exerts a wide variety of biological effects by mimicking endogenous substances – the endocannabinoids, anandamide, and 2-arachidonoylglycerol (2-AG) – that bind to and activate specific cannabinoid receptors (see Chaps. 2 and 7). So far, two types of cannabinoid-specific $G_{i/o}$ protein-coupled receptors, CB_1 and CB_2, have been cloned and characterized from mammalian tissues (Howlett et al., 2002). Most of the effects of cannabinoids rely on CB_1 receptor activation. CB_1 receptors are particularly abundant in discrete areas of the brain and peripheral nerve terminals, where they mediate endocannabinoid-dependent neuromodulation (Piomelli, 2003), but are also expressed in many extraneural sites. In contrast, CB_2 receptors were first described in cells and tissues of the immune system and have been long believed to be absent from the brain. Recent data, however, question this notion and support the existence of CB_2 receptors in the central nervous system, specifically in microglial cells, astrocytes, some neuron subpopulations, and glioma cells (Fernández-Ruiz et al., 2007; see Chap. 10). Extensive molecular and pharmacological studies have demonstrated that cannabinoids inhibit adenylyl cyclase through CB_1 and CB_2 receptors. The CB_1 receptor also modulates ion channels, inducing, for example, inhibition of N- and P/Q-type voltage-sensitive Ca^{2+} channels and activation of G protein-coupled K^+ channels (Howlett et al., 2002). Besides these well-established cannabinoid receptor-coupled signaling events, cannabinoid receptors also modulate several pathways that are more directly involved in the control of cell proliferation and survival, including extracellular signal-regulated kinase (ERK) (Bouaboula et al., 1995), *c-Jun* N-terminal kinase and p38 mitogen-activated protein kinase (Liu et al., 2000; Rueda et al., 2000), phosphatidylinositol 3-kinase (PI_3K)/Akt (Gómez del Pulgar et al., 2000), focal adhesion kinase (Derkinderen et al., 1996), and the sphingomyelin cycle (Sanchez et al., 2001).

Antitumoral Activity of Cannabinoids

Cannabinoids have been known for several decades to exert palliative effects in cancer patients, and nowadays capsules of Δ^9-THC (Marinol™) and its synthetic analogue nabilone (Cesamet™) are approved to treat nausea and emesis associated with cancer chemotherapy (Tramer et al., 2001; see Chap. 13). In addition, several clinical trials are testing other potential palliative properties of cannabinoids in oncology such as appetite stimulation and analgesia (Guzmán, 2003; Hall et al., 2005). Besides these palliative actions, cannabinoids have been proposed as potential antitumoral agents on the basis of experiments performed both in cultured cells and in animal models of cancer. These antiproliferative properties of cannabis compounds were first reported 30 years ago, when it was shown that Δ^9-THC inhibits lung adenocarcinoma cell growth in vitro and after oral administration in mice (Munson et al., 1975). Although these observations were promising, further studies in this area were not performed until the late 1990s, mostly by Di Marzo's group (Bifulco and Di Marzo, 2002) and Guzmán's group (Guzmán, 2003). A number of plant-derived (for example, Δ^9-THC and cannabidiol), synthetic (for example, WIN55212-2 and HU-210), and endogenous cannabinoids (for example, anandamide and 2-AG) are now known to exert antiproliferative actions on a wide spectrum of tumor cells in culture (Guzmán, 2003). More importantly, cannabinoid administration to nude mice curbs the growth of various types of tumor xenografts, including lung carcinoma (Munson et al., 1975), glioma (Galve-Roperh et al., 2000), thyroid epithelioma (Bifulco et al., 2001), lymphoma (McKallip et al., 2002), skin carcinoma (Casanova et al., 2003), pancreatic carcinoma (Carracedo et al., 2006a), and melanoma (Blázquez et al., 2006). The requirement of cannabinoid receptors for this antitumoral activity has been revealed by various biochemical and pharmacological approaches, in particular by determining cannabinoid receptor expression in the tumors and by using selective cannabinoid receptor agonists and antagonists.

Antitumoral Activity of Cannabinoids in Gliomas

Most of our research on cannabinoid antitumoral action has focused on gliomas. Initial experiments in cultured glioma cells showed that incubation with cannabinoids induces cell death by an apoptotic mechanism (Sánchez et al., 1998). Further studies with animal models showed that local administration of Δ^9-THC or WIN55212-2 reduced the size of tumors generated by intracranial inoculation of C6 glioma cells in rats, leading to complete eradication of gliomas and increased survival in one third of the treated rats (Galve-Roperh et al., 2000). Additional studies used tumor xenografts generated by subcutaneous injection of glioma cells in the flank of immune-deficient mice. Local administration of Δ^9-THC, WIN55212-2, or the selective CB_2 cannabinoid receptor agonist JWH133 decreased the growth of tumors derived not only from the rat glioma C6 cell line, but also from GBM cells

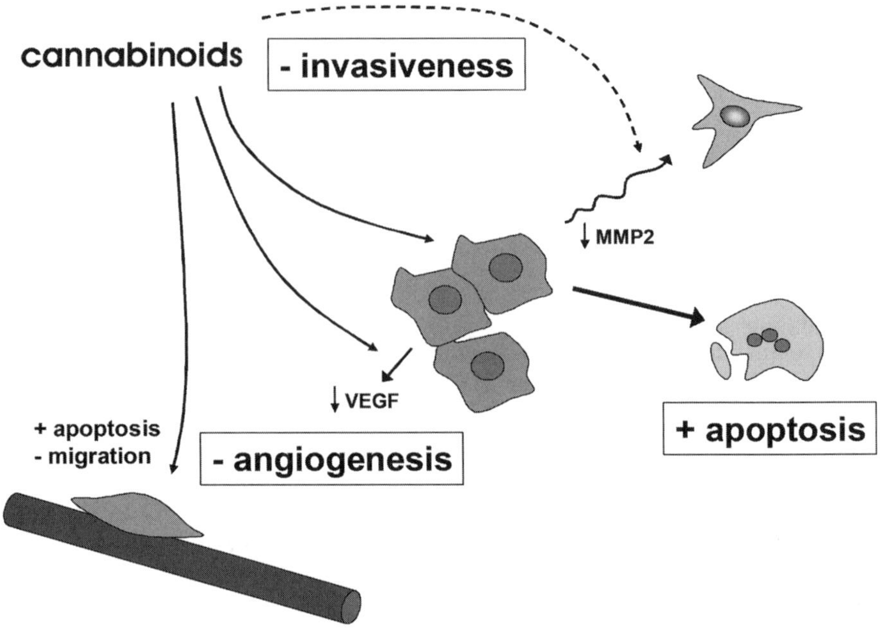

Fig. 1 Antitumoral effect of cannabinoids in gliomas. Cannabinoid administration to mice decreases the growth of gliomas by several mechanisms, including at least (i) reduction of tumor angiogenesis, (ii) induction of tumor cell apoptosis, and perhaps (iii) inhibition of tumor cell migration and invasiveness

obtained from tumor biopsies of patients (Galve-Roperh et al., 2000; Sánchez et al., 2001a,b). These and other studies also showed that cannabinoid receptor activation on glioma cells modulates key signaling pathways involved in cell proliferation and survival. Although the downstream events by which cannabinoids exert their antitumoral action in gliomas are not completely unraveled, there is substantial evidence for the implication of at least two mechanisms: induction of apoptosis of tumor cells and inhibition of tumor angiogenesis (Fig. 1).

Induction of Apoptosis

Cannabinoids induce apoptosis of cultured glioma cells (Sanchez et al., 1998; Galve-Roperh et al., 2000). Different studies have shown that this effect relies on the activation of cannabinoid receptors and the accumulation of the proapoptotic sphingolipid ceramide (Galve-Roperh et al., 2000; Gómez del Pulgar et al., 2002a,b; Ogretmen and Hannun, 2004). However, the molecular mechanisms involved in the triggering of the apoptotic signal by cannabinoids have started to be unraveled only very recently. By using a DNA array approach, we have identified a series of genes that are selectively upregulated in cannabinoid-sensitive but not cannabinoid-resistant glioma cells upon Δ^9-THC treatment (Carracedo et al., 2006b). One of these genes was the stress-regulated protein p8 (also designated candidate of metastasis 1 – Com-1), that belongs to the family of HMG-I/Y transcription factors and was originally described as a gene induced in acute pancreatitis (Mallo et al., 1997). Different experimental approaches confirmed that p8 upregulation is essential for the proapoptotic and antitumoral action of cannabinoids in gliomas and pancreatic tumors (Carracedo et al., 2006a,b). The acute increase of p8 levels after cannabinoid treatment triggers a cascade of events that involves the upregulation of the activating transcription factor 4 (ATF-4) and the C/EBP-homologous protein (CHOP, also called DDIT3 and GADD153). These two transcription factors cooperate in the induction of the tribbles homologue 3 (TRB3, also called TRIB3), a pseudokinase that has been implicated in the induction of apoptosis of tumor cells and neurons (Ohoka et al., 2005). In line with this observation, selective knock-down of ATF-4 and TRB3 prevented cannabinoid-induced apoptosis indicating that this signaling route also operates in glioma cells after treatment with cannabinoids (Carracedo et al., 2006b) (Fig. 2). ATF-4, CHOP, and TRB3 (together with other genes selectively induced upon Δ^9-THC treatment of glioma cells) (Carracedo et al., 2006b) participate in the endoplasmic reticulum (ER) stress response. A series of ER alterations such as calcium depletion, protein

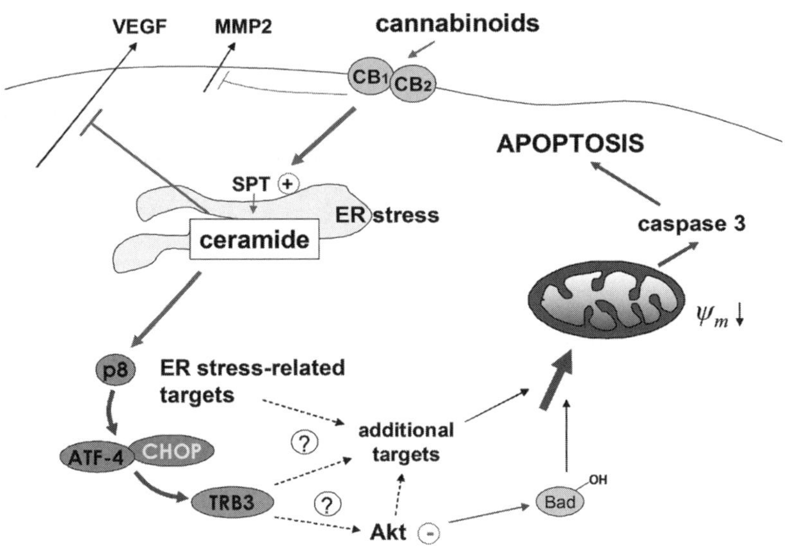

Fig. 2 Mechanism of cannabinoid proapoptotic action in glioma cells. Cannabinoid treatment induces apoptosis of glioma cells via ceramide accumulation and activation of an ER stress-related pathway. The stress-regulated protein p8 plays a key role in this effect by controlling the expression of ATF-4, CHOP, and TRB3. This cascade of events triggers the activation of the mitochondrial intrinsic apoptotic pathway through mechanisms that have not been unraveled as yet. Cannabinoids also decrease the expression of various tumor-progression molecules such as VEGF and MMP2

misfolding, and impairment of protein trafficking to the Golgi triggers this response, which involves attenuation of protein synthesis and selective transcription and translation of a series of genes, mainly involved in favoring correct protein folding (Schroder and Kaufman, 2005). When these ER alterations cannot be repaired by the ER stress response, the damaged cells undergo apoptosis. Several stimuli, including ischemia (Tajiri et al., 2004), viral infection (Li and Holbrook, 2004; Benali-Furet et al., 2005), and drugs such as tunicamycin (Ohoka et al., 2005) or cisplatin (Mandic et al., 2003), induce apoptosis through this pathway. Of interest, cannabinoid-induced ceramide accumulation and ER stress induction seem to be closely linked. Thus, inhibition of ceramide synthesis de novo prevents Δ^9-THC-induced p8, ATF-4, CHOP, and TRB3 upregulation (Carracedo et al., 2006b) as well as ER dilation (authors' unpublished observations), indicating that ceramide accumulation is an early event in cannabinoid-triggered ER stress and apoptosis in glioma cells. Unlike this proapoptotic action of cannabinoids on transformed cells, treatment of primary cultured astrocytes with these compounds triggers neither ceramide accumulation (Carracedo et al., 2004) nor the induction of the aforementioned ER stress-related genes (Carracedo et al., 2006b) (Fig. 3). Furthermore, cannabinoids promote the survival of astrocytes (Gómez del Pulgar et al., 2002a,b), oligodendrocytes (Molina-Holgado et al., 2002), and neurons (Mechoulam et al.,

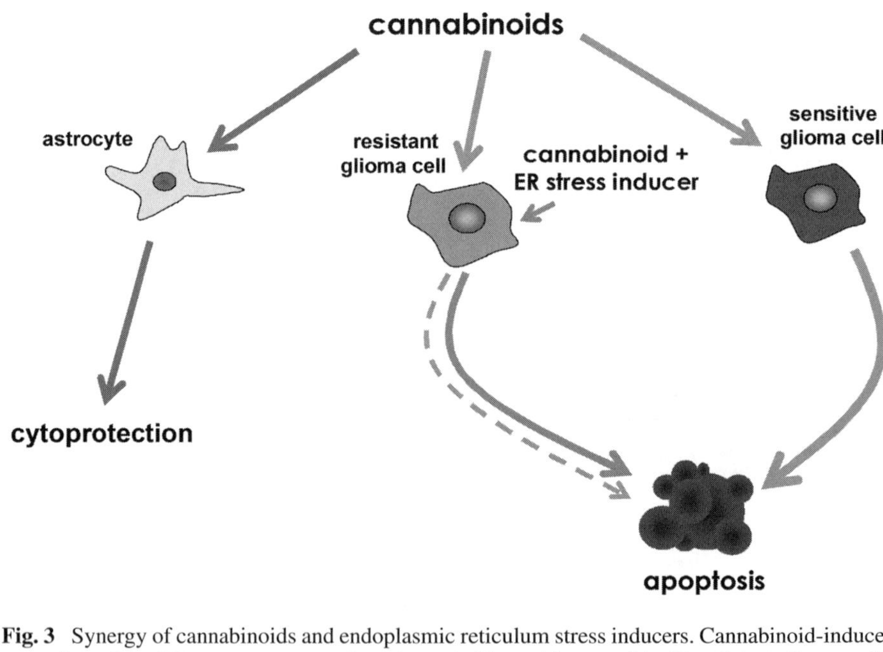

Fig. 3 Synergy of cannabinoids and endoplasmic reticulum stress inducers. Cannabinoid-induced activation of the ER stress proapoptotic pathway is blunted in cannabinoid-resistant glioma cells. This resistance can be overcome by cotreatment with cannabinoids and ER stress-inducing drugs. Cannabinoids protect astrocytes from different proapoptotic stimuli rather than activate the ER stress pathway in them

2002) in different models of injury, suggesting that the antiproliferative effect of cannabinoids is selective for brain tumor cells, the viability of normal brain cells being unaffected or even favored by cannabinoid challenge. The processes downstream of ER stress activation involved in the execution of cannabinoid-induced apoptosis of glioma cells are only partially understood. Decreased mitochondrial membrane potential and caspase 3 activation are observed in cannabinoid-treated glioma cells (Ellert-Miklaszewska et al., 2005; Carracedo et al., 2006b), suggesting that execution of apoptosis occurs via activation of the mitochondrial intrinsic pathway (Fig. 2), a mechanism that is involved in the induction of apoptosis by cannabinoids in other types of tumor cells (Lombard et al., 2005; Herrera et al., 2006). Cannabinoid treatment induces loss of mitochondrial membrane potential in $p8^{+/+}$ but not p8-deficient mouse embryonic fibroblasts, suggesting that the p8-regulated pathway described above is required for the activation of the mitochondrial proapoptotic pathway. On the other hand, cannabinoids inhibit Akt in glioma cells, an effect that is prevented by pharmacological blockade of ceramide synthesis de novo (Gómez del Pulgar et al., 2002a,b). In addition, cannabinoids lead to decreased phosphorylation of the BH3-only protein Bad (Ellert-Miklaszewska et al., 2005), an Akt and extracellular signal-regulated protein kinase (ERK) cascade target which phosphorylation inhibits apoptosis via the intrinsic pathway. These observations suggest that regulation of Akt could be involved in the connection between the ceramide/p8-regulated pathway and the activation of the mitochondrial proapoptotic route (Fig. 2). Modulation of ERK, as well as of the other mitogen-activated protein kinases, could also participate in the induction of apoptosis by cannabinoids in gliomas (Galve-Roperh et al., 2000). Intriguingly, both inhibition (e.g., Ellert-Miklaszewska et al., 2005) and activation (e.g., Galve-Roperh et al., 2000) of ERK have been proposed to participate in this effect. Further research is therefore necessary to clarify the involvement of this signaling cascade in cannabinoid-induced apoptosis.

Inhibition of Tumor Angiogenesis

To grow beyond minimal size, tumors must generate a new vascular supply (angiogenesis) for purposes of cell nutrition, gas exchange, and waste disposal, and therefore blocking the angiogenic process constitutes one of the most promising antitumoral approaches currently available. Immunohistochemical analyses in mouse models of glioma (Blázquez et al., 2003), skin carcinoma (Casanova et al., 2003), and melanoma (Blázquez et al., 2006) have shown that cannabinoid administration turns the vascular hyperplasia characteristic of actively growing tumors to a pattern of blood vessels characterized by small, differentiated, and impermeable capillaries. This is associated with a reduced expression of vascular endothelial growth factor (VEGF) and other proangiogenic cytokines such as angiopoietin-2 and placental growth factor (Blázquez et al., 2003; Casanova et al., 2003; Portella et al., 2003), as well as of type 1 (Portella et al., 2003) and type 2 (Blázquez et al.,

2004) VEGF receptors, in cannabinoid-treated tumors. Pharmacological inhibition of ceramide synthesis de novo abrogates the antitumoral and antiangiogenic effect of cannabinoids in vivo and decreases VEGF production by glioma cells in vitro and by gliomas in vivo (Blázquez et al., 2004), indicating that ceramide plays a general role in cannabinoid antitumoral action. Other reported effects of cannabinoids might be related with the inhibition of tumor angiogenesis and invasiveness by these compounds (Fig. 1). Thus, activation of cannabinoid receptors on vascular endothelial cells in culture inhibits cell migration and survival (Blázquez et al., 2003). Endothelial cell apoptosis was also potently triggered by cannabinoid quinonoid derivatives, although this action seems to be cannabinoid receptor-independent (Kogan et al., 2006). In addition, cannabinoid administration to glioma-bearing mice decreases the activity and expression of matrix metalloproteinase-2 (MMP2), a proteolytic enzyme that allows tissue breakdown and remodeling during angiogenesis and metastasis (Blázquez et al., 2003). In line with this notion, cannabinoid intraperitoneal injection reduces the number of metastatic nodes produced from paw injection of lung (Portella et al., 2003), breast (Grimaldi et al., 2006), and melanoma (Blázquez et al., 2006) cancer cells in mice.

Other Potential Targets of Cannabinoid Action

The identification of the cell(s) of origin of gliomas is still a matter of debate. Although neoplastic transformation of differentiated glial cells was for many years the most accepted hypothesis to explain the origin of gliomas, recent findings support the existence of a stem cell-derived origin for different types of cancers such as gliomas, hematopoietic, breast, and prostate tumors (Jordan et al., 2006). In particular, glioma-derived stem-like cells, which may represent the consequence of transformation of the normal neural stem cell compartment, seem to be crucial for the malignancy of gliomas (Vescovi et al., 2006). We have recently shown that glioma stem-like cells derived from GBM biopsies and glioma cell lines express CB_1 and CB_2 receptors, which activation promotes cell differentiation and inhibit gliomagenesis (Aguado et al., 2007). Interestingly, gene array experiments indicated that cannabinoid receptor activation on glioma stem-like cells downregulates epidermal growth factor (EGF) and fibroblast growth factor (FGF) receptors, in line with the suggestion that cannabinoids mediate at least part of their apoptotic actions on skin and prostate cancer cells by attenuating EGF receptor expression (Casanova et al., 2003; Mimeault et al., 2003) and/or tyrosine kinase activity (Casanova et al., 2003). In addition, the antiproliferative action of cannabinoids in breast, prostate, and thyroid cancer cells may involve a decrease in the activity and/or expression of prolactin (De Petrocellis et al., 1998), nerve growth factor (Melck et al., 2000), and type 1 VEGF receptors (Portella et al., 2003). Furthermore, cannabinoids inhibit type 2 VEGF receptor activation in glioma cells (Blázquez et al., 2004). Taken together, these results indicate that attenuation of the signaling through tyrosine kinase receptors may constitute a common mechanism of cannabinoid growth-inhibiting action.

Cannabinoids as Potential Therapeutic Agents for the Treatment of Gliomas

On the basis of the aforementioned preclinical findings, we have recently conducted a pilot phase I clinical trial in which nine patients with actively growing recurrent GBM were administered Δ^9-THC intratumorally (Guzmán et al., 2006). The patients had previously failed standard therapy (surgery and radiotherapy) and had clear evidence of tumor progression. The primary endpoint of the study was to determine the safety of intracranial Δ^9-THC administration. Δ^9-THC action on length of survival and various tumor cell parameters was also evaluated. A dose escalation regime for Δ^9-THC administration was assessed. The initial dose of Δ^9-THC delivered to the patients was 20–40 μg at day 1, increasing progressively for 2–5 days up to 80–180 μg/day. The median duration of Δ^9-THC administration was 15 days. Under these conditions, cannabinoid delivery was safe and could be achieved without significant psychoactive effects. Median survival of the cohort from the beginning of cannabinoid administration was 24 weeks (95% CI: 15–33). Δ^9-THC decreased tumor cell proliferation (as determined by Ki67 immunostaining; (Guzmán et al., 2006)) and increased tumor cell apoptosis (as determined by active-caspase 3 immunostaining; (Carracedo et al., 2006b)) when administered to two patients. The fair safety profile observed for Δ^9-THC, together with its possible antiproliferative action on tumor cells, may set the basis for future trials aimed at evaluating the potential antitumoral activity of cannabinoids. These possible new trials could involve one or more of the following modifications:

Patients with Newly Diagnosed Tumors

Pilot placebo-controlled trials for recurrent glioblastoma multiforme with temozolomide, a DNA-damaging agent that constitutes the current benchmark for the management of malignant gliomas, showed a very slight impact on overall length of survival (median survival = 24 weeks; 6-month survival = 46–60%) (Dinnes et al., 2002). Further trials in patients with newly diagnosed tumors allowed a clear improvement in the therapeutic efficacy of temozolomide through the development of various administration regimes (Lonardi et al., 2005; Stupp et al., 2005; Reardon and Wen, 2006). It is therefore conceivable that better outcomes could also be obtained with cannabinoid-based therapies in newly diagnosed gliomas.

Δ^9-THC in Combination with Temozolomide

Glioblastoma multiforme – particularly when relapse occurs – is an extremely lethal disease. The success of potential treatments is usually hampered by factors such as the rapid growth, remarkable heterogeneity, high degree of infiltration, and extreme resistance to chemotherapy displayed by these tumors. It is therefore conceivable that combined therapies could provide better results than single-agent therapies. For example, by synergizing via complementary signaling pathways, Δ^9-THC plus temozolomide might exert a more potent clinical impact than either Δ^9-THC or temozolomide alone.

Noninvasive Administration Route

Although intratumoral delivery may allow a high local concentration of the drug in situ, in the case of large tumors, such as actively growing recurrent glioblastoma multifome, the local perfusion through a catheter placed at one point of the tumor constitutes an obvious limitation of the technique. In addition, a noninvasive, less traumatic route would be more desirable in clinical practice. Alternative or complementary options for Δ^9-THC administration would include oral capsules and oro-mucosal sprays.

Other Cannabinoid Ligands

Although the use of cannabinoids in medicine may be limited by their well-known psychotropic effects, it is generally believed that cannabinoids display a fair drug safety profile and that their potential adverse effects are within the range of those accepted for other medications, especially in cancer treatment (Guzmán, 2003; Hall et al., 2005). In line with this idea, Δ^9-THC delivery in the aforementioned clinical study was safe and could be achieved without overt psychoactive effects. As the possible antitumoral action of nabilone has never been evaluated preclinically, Δ^9-THC remains as the unique cannabinoid receptor agonist currently available for cancer clinical trials. Nonetheless, most likely, Δ^9-THC is not the most appropriate cannabinoid agonist for future antitumoral strategies owing to its high hydrophobicity, relatively weak agonistic potency, and ability to elicit CB_1 receptor-mediated psychoactivity. Unfortunately, the current synthetic cannabinoid agonists that have been reported to exert antitumoral actions in animal models and that could theoretically circumvent – at least in part the pharmacokinetic and pharmacodynamic limitations of Δ^9-THC, e.g., WIN55212-2, a more potent and less hydrophobic mixed CB_1/CB_2 receptor agonist (Galve-Roperh et al., 2000), and JWH133, a more potent CB_2 receptor-selective agonist (Sánchez et al., 2001a,b) – are still very far from the clinical application owing to the lack of thorough preclinical toxicology studies. [Au5]

Other Types of Tumors

As mentioned above, we and others have shown that Δ^9-THC and synthetic cannabinoids, besides their antiglioma activity, inhibit the growth of different types of tumor xenografts in mice (see above). Trials on these and other types of tumors might also be run to test the antitumoral activity of cannabinoids in these malignant diseases.

Concluding Remarks

One of the most striking features of gliomas is their high resistance to conventional chemotherapy. Nowadays, it is widely believed that strategies aimed at reducing the mortality caused by these tumors should consist of targeted therapies capable of providing the most efficacious treatment for each individual patient and tumor. This new therapeutic approach would require not only the utilization of new cocktails of chemotherapeutic drugs but, more importantly, the identification of the markers associated with the resistance of tumor cells to these new therapies. The significant antiproliferative action of cannabinoids in animal models of gliomas, together with their low toxicity compared with other chemotherapeutic agents, might make these compounds promising new tools for the management of GBM. Studies performed in our laboratory suggest that resistance of glioma cells to cannabinoid treatment correlates with the ability of these cells to block the activation of the ER stress pathway (authors' unpublished observations). In addition, we have observed that agents that induce ER stress exert a synergic action when administered with cannabinoids (Carracedo et al., 2006b). Likewise, overexpression of p8 or TRB3 sensitizes resistant glioma cells to a further treatment with cannabinoids (Carracedo et al., 2006b). These observations suggest that activation of this route may be investigated as a potential strategy to enhance the response of gliomas to chemotherapy. Research to be performed during the next few years should help to clarify which are the optimal conditions of cannabinoid utilization by identifying the factors that confer resistance to cannabinoid treatment as well as the most efficient approaches for enhancing their antitumoral activity.

Acknowledgments We are indebted to the rest of our lab colleagues for their collaboration and encouragement. Research in our laboratory is financially supported by Ministerio de Educación y Ciencia, Comunidad Autónoma de Madrid, Ministerio de Sanidad y Consumo, Fundación de Investigación Médica Mutua Madrileña Automovilística, and Santander/Complutense.

References

Aguado T, Carracedo A, Julien B, Velasco G, Milman G, Mechoulam R, Álvarez L, Guzmán M, Galve-Roperh I (2007) Cannabinoids inhibit glioma stem-like cell differentiation and inhibit gliomagenesis. J Biol Chem 282:6854–6862.

Benali-Furet NL, Chami M, Houel L, De Giorgi F, Vernejoul F, Lagorce D, Buscail L, Bartenschlager R, Ichas F, Rizzuto R, Paterlini-Brechot P (2005) Hepatitis C virus core triggers apoptosis in liver cells by inducing ER stress and ER calcium depletion. Oncogene 24:4921–4933.

Bifulco M, Di Marzo V (2002) Targeting the endocannabinoid system in cancer therapy: a call for further research. Nat Med 8:547–550.

Bifulco M, Laezza C, Portella G, Vitale M, Orlando P, De Petrocellis L, Di Marzo V (2001) Control by the endogenous cannabinoid system of ras oncogene-dependent tumor growth. FASEB J 15:2745–2747.

Blázquez C, Casanova ML, Planas A, Gómez del Pulgar T, Villanueva C, Fernandez-Acenero MJ, Aragones J, Huffman JW, Jorcano JL, Guzmán M (2003) Inhibition of tumor angiogenesis by cannabinoids. FASEB J 17:529–531.

Blázquez C, Gonzalez-Feria L, Alvarez L, Haro A, Casanova ML, Guzman M (2004) Cannabinoids inhibit the vascular endothelial growth factor pathway in gliomas. Cancer Res 64:5617–5623.

Blázquez C, Carracedo A, Barrado L, Real PJ, Fernandez-Luna JL, Velasco G, Malumbres M, Guzman M (2006) Cannabinoid receptors as novel targets for the treatment of melanoma. FASEB J 20:2633–2635.

Bouaboula M, Poinot-Chazel C, Bourrie B, Canat X, Calandra B, Rinaldi-Carmona M, Le Fur G, Casellas P (1995) Activation of mitogen-activated protein kinases by stimulation of the central cannabinoid receptor CB_1. Biochem J 312:637–641.

Carracedo A, Geelen MJ, Diez M, Hanada K, Guzman M, Velasco G (2004) Ceramide sensitizes astrocytes to oxidative stress: protective role of cannabinoids. Biochem J 380:435–440.

Carracedo A, Gironella M, Lorente M, Garcia S, Guzman M, Velasco G, Iovanna JL (2006a) Cannabinoids induce apoptosis of pancreatic tumor cells via endoplasmic reticulum stress-related genes. Cancer Res 66:6748–6755.

Carracedo A, Lorente M, Egia A, Blázquez C, Garcia S, Giroux V, Malicet C, Villuendas R, Gironella M, Gonzalez-Feria L, Piris MA, Iovanna JL, Guzman M, Velasco G (2006b) The stress-regulated protein p8 mediates cannabinoid-induced apoptosis of tumor cells. Cancer Cell 9:301–312.

Casanova ML, Blazquez C, Martinez-Palacio J, Villanueva C, Fernandez-Acenero MJ, Huffman JW, Jorcano JL, Guzman M (2003) Inhibition of skin tumor growth and angiogenesis *in vivo* by activation of cannabinoid receptors. J Clin Invest 111:43–50.

De Petrocellis L, Melck D, Palmisano A, Bisogno T, Laezza C, Bifulco M, Di Marzo V (1998) The endogenous cannabinoid anandamide inhibits human breast cancer cell proliferation. Proc Natl Acad Sci USA 95:8375–8380.

Derkinderen P, Toutant M, Burgaya F, Le Bert M, Siciliano JC, de Franciscis V, Gelman M, Girault JA (1996) Regulation of a neuronal form of focal adhesion kinase by anandamide. Science 273:1719–1722.

Dinnes J, Cave C, Huang S, Milne R (2002) A rapid and systematic review of the effectiveness of temozolomide for the treatment of recurrent malignant glioma. Br J Cancer 86:501–505.

Ellert-Miklaszewska A, Kaminska B, Konarska L (2005) Cannabinoids down-regulate PI_3K/Akt and Erk signalling pathways and activate proapoptotic function of Bad protein. Cell Signal 17:25–37.

Fernández-Ruiz J, Romero J, Velasco G, Tolon R, Ramos J, Guzmán M (2007) Cannabinoid CB_2 receptor: a new target for controlling neural cell survival? Trends Pharmacol Sci 28:39–45.

Galve-Roperh I, Sanchez C, Cortes ML, Gómez del Pulgar T, Izquierdo M, Guzmán M (2000) Anti-tumoral action of cannabinoids: involvement of sustained ceramide accumulation and extracellular signal-regulated kinase activation. Nat Med 6:313–319.

Gaoni Y, Mechoulam R (1964) Isolation, structure and partial synthesis of an active constituent of hashish. J Am Chem Soc 86:1646–1647.

Gómez del Pulgar T, Velasco G, Guzman M (2000) The CB_1 cannabinoid receptor is coupled to the activation of protein kinase B/Akt. Biochem J 347:369–373.

Gómez del Pulgar T, De Ceballos ML, Guzmán M, Velasco G (2002a) Cannabinoids protect astrocytes from ceramide-induced apoptosis through the phosphatidylinositol 3-kinase/protein kinase B pathway. J Biol Chem 277:36527–36533.

Gómez del Pulgar T, Velasco G, Sánchez C, Haro A, Guzman M (2002b) De novo-synthesized ceramide is involved in cannabinoid-induced apoptosis. Biochem J 363:183–188.

Grimaldi C, Pisanti S, Laezza C, Malfitano AM, Santoro A, Vitale M, Caruso MG, Notarnicola M, Iacuzzo I, Portella G, Di Marzo V, Bifulco M (2006) Anandamide inhibits adhesion and migration of breast cancer cells. Exp Cell Res 312:363–373.

Guzmán M (2003) Cannabinoids: potential anticancer agents. Nat Rev Cancer 3:745–755.

Guzmán M, Duarte MJ, Blazquez C, Ravina J, Rosa MC, Galve-Roperh I, Sanchez C, Velasco G, Gonzalez-Feria L (2006) A pilot clinical study of delta-9-tetrahydrocannabinol in patients with recurrent glioblastoma multiforme. Br J Cancer 95:197–203.

Hall W, Christie M, Currow D (2005) Cannabinoids and cancer: causation, remediation, and palliation. Lancet Oncol 6:35–42.

Herrera B, Carracedo A, Diez-Zaera M, Gomez del Pulgar T, Guzmán M, Velasco G (2006) The CB_2 cannabinoid receptor signals apoptosis via ceramide-dependent activation of the mitochondrial intrinsic pathway. Exp Cell Res 312:2121–2131.

Howlett AC, Barth F, Bonner TI, Cabral G, Casellas P, Devane WA, Felder CC, Herkenham M, Mackie K, Martin BR, Mechoulam R, Pertwee RG (2002) International Union of Pharmacology. XXVII. Classification of cannabinoid receptors. Pharmacol Rev 54:161–202.

Jordan CT, Guzmán ML, Noble M (2006) Cancer stem cells. N Engl J Med 355:1253–1261.

Kleihues P, Louis DN, Scheithauer BW, Rorke LB, Reifenberger G, Burger PC, Cavenee WK (2002). The WHO classification of tumors of the nervous system. J Neuropathol Exp Neurol 61:215–225.

Kogan NM, Blázquez C, Alvarez L, Gallily R, Schlesinger M, Guzman M, Mechoulam R (2006) A cannabinoid quinone inhibits angiogenesis by targeting vascular endothelial cells. Mol Pharmacol 70:51–59.

Li J, Holbrook NJ (2004) Elevated gadd153/chop expression and enhanced c-Jun N-terminal protein kinase activation sensitizes aged cells to ER stress. Exp Gerontol 39:735–744.

Liu J, Gao B, Mirshahi F, Sanyal AJ, Khanolkar AD, Makriyannis A, Kunos G (2000) Functional CB_1 cannabinoid receptors in human vascular endothelial cells. Biochem J 346:835–840.

Lombard C, Nagarkatti M, Nagarkatti PS (2005) Targeting cannabinoid receptors to treat leukemia: role of cross-talk between extrinsic and intrinsic pathways in Delta9-tetrahydrocannabinol (THC)-induced apoptosis of Jurkat cells. Leuk Res 29:915–922.

Lonardi S, Tosoni A, Brandes AA (2005) Adjuvant chemotherapy in the treatment of high grade gliomas. Cancer Treat Rev 31:79–89.

Maher EA, Furnari FB, Bachoo RM, Rowitch DH, Louis DN, Cavenee WK, DePinho RA (2001) Malignant glioma: genetics and biology of a grave matter. Genes Dev 15:1311–1333.

Mallo GV, Fiedler F, Calvo EL, Ortiz EM, Vasseur S, Keim V, Morisset J, Iovanna JL (1997) Cloning and expression of the rat p8 cDNA, a new gene activated in pancreas during the acute phase of pancreatitis, pancreatic development, and regeneration, and which promotes cellular growth. J Biol Chem 272:32360–32369.

Mandic A, Hansson J, Linder S, Shoshan MC (2003) Cisplatin induces endoplasmic reticulum stress and nucleus-independent apoptotic signaling. J Biol Chem 278:9100–9106.

McKallip RJ, Lombard C, Fisher M, Martin BR, Ryu S, Grant S, Nagarkatti PS, Nagarkatti M (2002) Targeting CB_2 cannabinoid receptors as a novel therapy to treat malignant lymphoblastic disease. Blood 100:627–634.

Mechoulam R, Spatz M, Shohami E (2002) Endocannabinoids and neuroprotection. Sci STKE 2002:RE5.

Melck D, De Petrocellis L, Orlando P, Bisogno T, Laezza C, Bifulco M, Di Marzo V (2000) Suppression of nerve growth factor Trk receptors and prolactin receptors by endocannabinoids leads to inhibition of human breast and prostate cancer cell proliferation. Endocrinology 141:118–126.

Mimeault M, Pommery N, Wattez N, Bailly C, Henichart JP (2003) Anti-proliferative and apoptotic effects of anandamide in human prostatic cancer cell lines: implication of epidermal growth factor receptor down-regulation and ceramide production. Prostate 56:1–12.

Molina-Holgado E, Vela JM, Arevalo-Martin A, Almazan G, Molina-Holgado F, Borrell J, Guaza C (2002) Cannabinoids promote oligodendrocyte progenitor survival: involvement of cannabinoid receptors and phosphatidylinositol-3 kinase/Akt signaling. J Neurosci 22:9742–9753.

Munson AE, Harris LS, Friedman MA, Dewey WL, Carchman RA (1975) Antineoplastic activity of cannabinoids. J Natl Cancer Inst 55:597–602.

Ogretmen B, Hannun YA (2004). Biologically active sphingolipids in cancer pathogenesis and treatment. Nat Rev Cancer 4:604–616.

Ohoka N, Yoshii S, Hattori T, Onozaki K, Hayashi H (2005) TRB3, a novel ER stress-inducible gene, is induced via ATF4-CHOP pathway and is involved in cell death. EMBO J 24:1243–1255.

Piomelli D (2003) The molecular logic of endocannabinoid signalling. Nat Rev Neurosci 4:873–884.

Portella G, Laezza C, Laccetti P, De Petrocellis L, Di Marzo V, Bifulco M (2003) Inhibitory effects of cannabinoid CB_1 receptor stimulation on tumor growth and metastatic spreading: actions on signals involved in angiogenesis and metastasis. FASEB J 17:1771–1773.

Reardon DA, Wen PY (2006) Therapeutic advances in the treatment of glioblastoma: rationale and potential role of targeted agents. Oncologist 11:152–164.

Rueda D, Galve-Roperh I, Haro A, Guzman M (2000) The CB_1 cannabinoid receptor is coupled to the activation of c-Jun N-terminal kinase. Mol Pharmacol 58:814–820.

Sánchez C, Galve-Roperh I, Canova C, Brachet P, Guzmán M (1998) Delta-9-tetrahydrocannabinol induces apoptosis in C6 glioma cells. FEBS Lett 436:6–10.

Sánchez C, de Ceballos ML, del Pulgar TG, Rueda D, Corbacho C, Velasco G, Galve-Roperh I, Huffman JW, Ramon y Cajal S, Guzmán M (2001a) Inhibition of glioma growth *in vivo* by selective activation of the CB_2 cannabinoid receptor. Cancer Res 61:5784–5789.

Sánchez C, Rueda D, Segui B, Galve-Roperh I, Levade T, Guzmán M (2001b) The CB_1 cannabinoid receptor of astrocytes is coupled to sphingomyelin hydrolysis through the adaptor protein fan. Mol Pharmacol 59:955–959.

Schroder M, Kaufman RJ (2005). The mammalian unfolded protein response. Annu Rev Biochem 74:739–789.

Stupp R, Mason WP, van den Bent MJ, Weller M, Fisher B, Taphoorn MJ, Belanger K, Brandes AA, Marosi C, Bogdahn U, Curschmann J, Janzer RC, Ludwin SK, Gorlia T, Allgeier A, Lacombe D, Cairncross JG, Eisenhauer E, Mirimanoff RO (2005) Radiotherapy plus concomitant and adjuvant temozolomide for glioblastoma. N Engl J Med 352:987–996.

Tajiri S, Oyadomari S, Yano S, Morioka M, Gotoh T, Hamada JI, Ushio Y, Mori M (2004) Ischemia-induced neuronal cell death is mediated by the endoplasmic reticulum stress pathway involving CHOP. Cell Death Differ 11:403–415.

Tramer MR, Carroll D, Campbell FA, Reynolds DJ, Moore RA, McQuay HJ (2001) Cannabinoids for control of chemotherapy induced nausea and vomiting: quantitative systematic review. Br Med J 323:16–21.

Vescovi AL, Galli R, Reynolds BA (2006) Brain tumour stem cells. Nat Rev Cancer 6:425–436.

Yamanaka R (2006) Novel immunotherapeutic approaches to glioma. Curr Opin Mol Ther 8:46–51.

Chapter 18
Cannabinoids for the Control of Multiple Sclerosis

Gareth Pryce, Sam J. Jackson, and David Baker

Abstract In response to patient perceptions that cannabis can control some of the symptoms of multiple sclerosis (MS), scientific studies in experimental models and clinical trials in MS have been undertaken. These studies and recent understanding of the biology of the cannabinoid system and MS have provided a rationale and objective evidence to support these perceptions. Indeed, the first cannabis-based medicine for the treatment of signs has been recently licensed for use in MS. Although most clinical studies have focused on symptom control, experimental evidence also indicates a potential action for cannabinoids in the control of autoimmune and neurodegenerative processes. These drive the underlying disease pathology that cause the varied symptomatology, for which cannabis-based medicines may currently be used. In the future it may possible to harness the medical benefits that the cannabis system has to offer to control MS, whilst limiting the adverse effects, both physical and psycho-social, associated with smoking cannabis.

Introduction

There has been much recent interest in the potential use of cannabis or cannabinoid reagents for the control of multiple sclerosis (MS). In the absence of effective disease control, people with MS have been willing to seek alternative therapies. They have self-medicated and perceived benefit from taking cannabis (Consroe et al., 1997); little did they know that cannabinoid biology was about to blossom and shed light on these opinions. Multiple sclerosis is a major demyelinating disease of the central nervous system (CNS) that can affect up to 1:500 people in areas of high incidence (Compston and Coles, 2002). The disease is typically associated with relapsing–remitting episodes of neurological dysfunction with varying degrees of clinical recovery prior to the development of progressive accumulation of increasing disability. As a consequence of nerve damage, people with MS accumulate a variety of additional signs such as tremor, spasticity, pain, bladder and sexual dysfunction that greatly diminish "quality of life" of the affected individual (Compston and Coles, 2002; Confavreux and Vukusic, 2006). These clinical features can be observed in experimental autoimmune encephalomyelitis (EAE), which is an autoimmune model of MS

A. Köfalvi (ed.), *Cannabinoids and the Brain.*

(Pryce et al., 2005; Baker and Jackson, 2007). This and viral models of MS have been used to investigate the potential function of cannabinoids for the control of MS. However, some people fail to appreciate the biology of MS and group all data on cannabinoids from animal models, which largely focus on immune response, into one context and relate this to symptom control in human disease. It is very important that one appreciates that both human disease and its models are both complex diseases and that their pathologies and signs result from distinct processes (Compston and Coles, 2002; Bjartmar and Trapp, 2003; Confavreux and Vukusic, 2006; Baker and Jackson, 2007).

Immune Responses

Immune responses cause the formation of CNS lesions and relapsing clinical attacks in MS and EAE (Compston and Coles, 2002; Pryce et al., 2005; Coles et al., 2006; Polman et al., 2006; Baker and Jackson, 2007). These inflammatory, mononuclear cell lesions lead to a loss in CNS homeostasis and if located in clinically eloquent locations can result in clinical disease (Compston and Coles, 2002). Loss of motor function is due to the transient loss of nerve function resulting from oedema and consequent conduction block. In some instances, damage to the oligodendrocyte and the myelin sheaths, which is the pathological hallmark of MS, occurs also (Compston and Coles, 2002). The conduction block alone may explain the paralysis that develops in many acute EAE models as paralysis can occur in the absence of demyelination. Immunosuppressive agents can block the formation of lesions and the development of paralytic relapses (Compston and Coles, 2002; Pryce et al., 2005; Coles et al., 2006; Polman et al., 2006). Cannabinoids have been reported to inhibit the development of paralysis in models of MS (Lyman et al., 1989). However, it is imperative to realize that this inhibition of paralysis is not because the drug is directly affecting symptoms and restoring nerve conduction to the affected limbs, but because it is modulating the immune response. The immune attack fails to materialize in the CNS and so does not cause the initial inflammatory reaction. Therefore, conduction block, the loss of neurotransmission and the consequent development paralysis due to loss of muscle control do not follow.

Progressive Disability

Progressive disability appears due to neurodegenerative processes that start early in the disease course. Nerves are lost due to inflammatory attack and this can be accommodated at least initially by compensation mechanisms such as redundancy and plasticity of affected neural pathways (Bjartmar and Trapp, 2003). The immune attack creates a damaged environment, containing chronic demyelination and low level glial cell activation, which appears to trigger a slow neurodegenerative process (Compston and Coles, 2002; Bjartmar and Trapp, 2003; Confavreux and Vukusic, 2006). For example, toxic ionic imbalances due to the redistribution and function of ion channels and metabolic failure of demyelinated nerves or glutamate excitotoxicity following influences of inflammatory cells or relative loss of inhibitory, GABAergic, circuits can lead to nerve death and the development of chronic, irreversible disability (Bjartmar and Trapp, 2003; Kapoor et al., 2003; Bolton and Paul, 2006; Dutta et al., 2006). This neurodegeneration is the substrate for progressive MS and is not responsive to immunosuppressive drugs (Pryce et al., 2005; Coles et al., 2006; Confavreux and Vukusic, 2006; Metz et al., 2007). Normal neurotransmission is affected and as these lesions occur in sensory or motor pathways they can cause a wide variety of signs (Compston and Coles, 2002). Thus, it is important to realize that symptom control agents will positively influence neurotransmission and thus have a relatively rapid effect. In contrast, immunomodulatory and neuroprotective agents aim to affect the underlying causes of the symptomatology and thus the agents will need to be administered long term early in the disease process and will require time for an effect to be manifest.

Clinical Experience

Symptom Control

Coupling knowledge of the problems of MS with the recent understanding of the biology of the cannabinoid system in regulating neurotransmission suggests that cannabinoids may have a potential beneficial role in symptom control through control of neurotransmission (Fig. 1, Howlett et al., 2002; Wilson and Nicholl, 2002). This supports patient perceptions that cannabis may offer benefit (Consroe et al., 1997; Schnelle et al., 1999; Page et al., 2003; Clark et al., 2004; Ware et al., 2005; Chong et al., 2006; Page and Verhoef, 2006). Symptoms in MS are quite varied and reflect problems related to lesion location within the CNS (Compston and Coles, 2002). However, it is evident that the perceived beneficial effects are not universal across all the varied symptoms of MS and they would suggest that cannabis has value in controlling notably spasms, spasticity, and pain and sleep disturbances (Consroe et al., 1997).

Spasticity

Spasticity is an inappropriate increase in muscle stretch reflexes due to amplified reactivity of motor segments compared to sensory input that leads to limb stiffness. Spasticity is common in MS and becomes more prevalent as nerves are lost

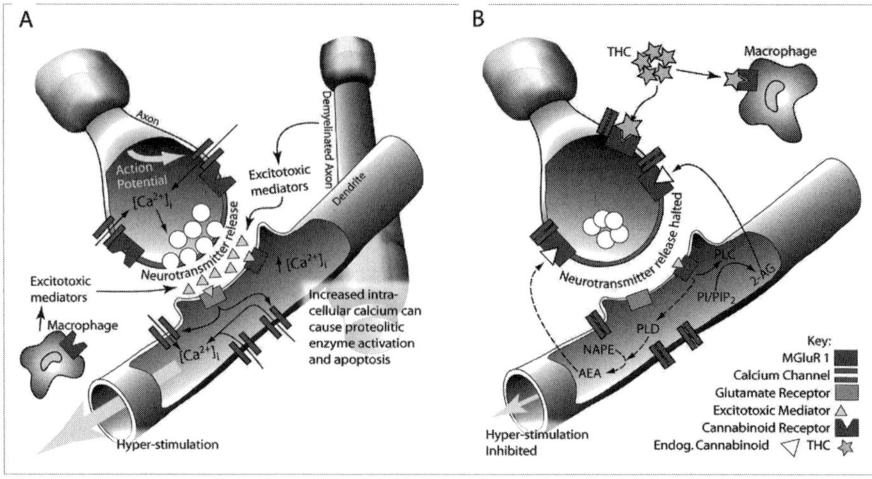

Fig. 1 Cannabinoid control of aberrant neurotransmission and excitotoxicity. (**a**) Symptoms result from aberrant neurotransmission due to demyelination and loss of elements in neural control circuits. This coupled with neuroinflammation can lead to toxic levels of calcium ions, following oxidative and excitotoxic insults, which cause the nerve death that leads to progressive MS. In addition, alterations in neurotransmission, typically glutamatergic transmission, can result in loss of motor control leading to persistent symptoms of disease. (**b**) These elements may be controlled both by exogenous and endogenous cannabinoids, which may involve stimulation of 2-AG and notably anandamide production to reduce excessive neurotransmission and neurotoxic events

and disease progresses, and is in part associated with insufficient GABAergic inhibition of neurotransmission (Kesselring and Thompson, 1997). The majority (>90%) of people with MS who use cannabis for symptom control believe that cannabis can alleviate spasticity (Consroe et al., 1997). Indeed, experimentally induced spasticity, as seen as the residual limb stiffness that develops due to nerve loss following repeated neurological, paralytic attacks caused by immune attack, was tonically controlled by the cannabinoid system (Baker et al., 2000, 2001). Further studies indicate that the CB_1 receptor is the major, although perhaps not exclusive, target for the muscle relaxing effect of cannabis (Baker et al., 2000; Brooks et al., 2002; Wilkinson et al., 2003; Pryce and Baker, 2007). There is limited evidence to suggest that the CB_2 receptor is involved in neuronal function or that it could be involved in the control of spasticity (Howlett et al., 2002; Van Sickle et al., 2005; Pryce and Baker, 2007). This indicates that optimal therapeutic effect of cannabis will invariably be associated with side effects as the CB_1 receptor expressed in different areas of the brain will be controlling the aberrant neurotransmission as well as inducing psychoactive effects. Thus, there may be a small therapeutic window for cannabis-based drugs as suggested from recent clinical trials (Table 1). The perception that cannabis can alleviate spasticity has been largely supported by clinical trials using dose-titrated Δ^9-THC or cannabis extracts

Table 1 Clinical trials of cannabinoid use in multiple sclerosis

Reference	*n*	Spasticity		Pain
		Objective outcome	Subjective outcome	Subjective outcome
Petro and Ellenberger, 1981	8	+ve	n.d.	n.d.
Ungerleider et al., 1987	8	−ve	+ve	n.d.
Killestein et al., 2002	16	−ve	−ve	n.d.
Wade et al., 2003	18	−ve	+ve	+ve
Zajicek et al., 2003	660	−ve	+ve	+ve
Wade et al., 2004	160	−ve	+ve	+ve
Brady et al., 2004	15	n.d.	+ve	+ve
Wade et al., 2006	137	n.d.	+ve	n.d.
Nottcutt et al., 2004	34	n.d.	n.d.	+ve
Svendsen et al., 2005	24	n.d.	n.d.	+ve
Rog et al., 2005	66	n.d.	n.d.	+ve
Vaney et al., 2004	57	−ve	+ve (spasms)	n.d.

Clinical trials in MS have used objective (physician-assessed) and subjective (patient-assessed) outcome measures although in some instances these were not determined (n.d.) and report either positive (+ve) or lack (−ve) of efficacy

containing Δ^9-THC and cannabidiol such as Sativex™ and Cannador™ (Table 1). However, clinical trials using doses that have been titrated to not induce significant psychoactive effects have largely failed to show objective benefit (Table 1), although benefits in walking were apparent (Zajicek et al., 2003). Furthermore long-term treatment with Δ^9-THC appeared to result in significant change in objective measures of spasticity (Zajicek et al., 2005). Medical cannabis is currently undergoing regulatory assessment for use of cannabis extracts in the control of spasticity.

Pain

Pain occurs in the majority of people with MS (Ehde et al., 2006). The types of pain manifest are varied, such as painful spasm and trigeminal neuralgia, but it is often intractable, chronic neuropathic pain, which may be particularly severe (Compston and Coles, 2002; Ehde et al., 2006). Studies show that cannabinoids inhibit experimental pain in animals and neuropathic pain in MS responds to cannabis treatment (Table 1; Walker and Hohmann, 2005; Iskedjian et al., 2007).

Tremor

Tremor, which results from aberrant neurotransmission, is a sign of MS that is difficult to treat (Alusi et al., 2001; Compston and Coles, 2002). Although people with MS have claimed benefit of smoked cannabis on tremor, the only large scale study aimed at assessing the effect of oral cannabis (Cannador™) on MS-associated tremor has failed to detect any of the objective improvement of upper limb tremor compared to placebo (Consroe et al., 1997; Fox et al., 2004). However, there was also some suggestion of a subjective improvement of tremor of oral cannabinoids, most noticeable with long-term use (Zajicek et al., 2003, 2005). However, there have been reports that some tremors do appear to respond to smoked cannabis or oral Δ^9-THC (Clifford, 1983; Meinck et al., 1989; Schon et al., 1999). Although we have reported that cannabinoids can inhibit some tremors in rodents with EAE, the same agent may even exacerbate other tremors (Baker et al., 2000; Baker and Pryce, 2004). Thus, as tremors may develop due to effects in different neuropathological routes, then the cannabinoid may cause different outcomes depending on whether the neurotransmission circuit is intact and whether the cannabinoid receptor is expressed in a pathway that controls stimulatory or inhibitory neural circuits.

Bladder Dysfunction

Bladder dysfunction occurs in the majority of patients with MS, which is associated with degree of spinal cord involvement (Kalsi and Fowler, 2005). Bladder hyperactivity can be caused by nerve damage that influences central inhibitory mechanisms, central sensory or motor pathways or damage that promotes the reorganization of spinal reflex pathways (de Groat, 1998; Kalsi and Fowler, 2005). Experimental studies in rodents suggest that cannabinoids have some potential to limit neurological contractions of bladder and thus the cannabinoid receptors in the bladder, the spinal roots and nervous system are potential pharmacological targets to control, and anecdotal reports suggest that cannabis may alleviate lower urinary tract symptoms (Consroe et al., 1997; Martin et al., 2000. This is supported by results from symptom control trials of Sativex™ in MS and an open label study that reported an inhibitory effect on Δ^9-THC and Sativex™ on incontinence (Wade et al., 2003; Brady et al., 2004). Whilst a study on spasticity failed to detect a significant influence on bladder dysfunction in a substudy aimed at addressing the influence on bladder function both oral Cannador™ and Marinol™ reduced urge incontinence compared to placebo (Zajicek et al., 2003; Freeman et al., 2006). This suggests that cannabinoids may have some utility in the control of bladder dysfunction and support first line treatments such as anti-cholinergic agents (Kalsi and Fowler, 2005).

Cannabis in Symptom Control

The clinical reality of effects of cannabis in symptom control in recent large scale clinical trials, using cannabis extracts that have been dose-titrated to limit psychoactive influences, have as yet failed to show remarkable differences compared to the effects of cannabis and placebo. Thus the efficacy of cannabis, although generally well tolerated, is at best, modest. However, primary outcomes notably in spasticity are based on relatively insensitive, objective scales that have largely been unresponsive to treatment with cannabis-based medicines (Table 1). Furthermore, lack of efficacy may in part relate to route of delivery in some trials such as the oral route, which may limit bioavailability of the drug due to first pass metabolism and compartmentalizing of the active cannabinoids in dietary and body fat due to their hydrophobicity. This makes dosing difficult and reduces the therapeutic window. Drugs such as Sativex™ have yet to gain universal, regulatory approval, despite almost universal, positive effects on spasticity in subjective, patient-oriented scales. Therefore, the millions being spent on trials have merely confirmed what was indicated from surveys performed over a decade ago. By contrast, primary outcome measures in pain are based on subjective scales (Table 1). Although cannabis has not been the panacea in pain relief, some have accepted that it has a place in the medical armoury against disease and Sativex™ has recently been licensed in some North American and European countries for the use relief of chronic pain associated with MS. Furthermore, there has been variation of the response to therapy and must be expected as the location and extent of the lesion load, and expression of cannabinoid receptors within excitatory, inhibitory or disinhibitory circuits will vary between individual and may in part account for seemingly contradictory accounts of the effects of cannabis to suppress or induce a number of signs.

Relapse Rate and Progression

In surveys, there was a perception amongst cannabis smokers that cannabis could have a beneficial effect on the incidence of relapsing disease (Consroe et al., 1997). However, the clinical course of MS within an individual is notoriously difficult, if not impossible to predict. The "field of MS" is full of these unsubstantiated anecdotes, but they have invariably failed to deliver any useful therapies when properly tested in controlled trials. Furthermore, when looking at a generalized population, it appears that relapses are less frequent in later stages of MS, which would correspond to the time when people may be taking cannabis for symptom control, compared to earlier disease (Compston and Coles, 2002). At present, no studies have been undertaken that have been aimed at detecting an effect of cannabis on relapse rate. Most of the published trials in symptom control using cannabis had too few participants, were short in duration and not designed or sufficiently powered to address the issue of an effect of cannabis on relapse rate. Whilst a recent symptom control trial of oral cannabis and Δ^9-THC, using people selected to have relatively stable disease, hinted that there may be a reduced relapse rate during treatment, this was not substantiated upon longer follow-up and no affect on relapse rate was detected (Zajicek et al., 2003, 2005). Furthermore, assessment of a subgroup of people taking cannabis failed to detected significant immune alterations (Killestein et al., 2003; Katona et al., 2005). This suggests that cannabis does not exert significant immunosuppressive effects. However, an effect on progression is suggested from follow-up of long-term oral THC (Zajicek et al., 2005). There appeared to be significant improvements for the subjective and objective measures of control of symptoms not evident during the trial aimed at detecting effects on symptoms (Zajicek et al., 2003, 2005). This suggests that THC can have either a neuroprotective effect by slowing the accumulation of disability or that it is promoting synaptic plasticity that can compensate for the damaging effects of the disease (Tagliaferro et al., 2006). This contrasts to potential worsening of MS following CB_1 receptor antagonism with rimonabant (van Oosten et al., 2004). These ideas of roles for cannabinoids in addition to symptom control have come from clinical experiences, understanding of biology of cannabis and the disease and from evidence produced in experimental studies in animals (Lyman et al., 1989; Pryce et al., 2003; Maresz et al., 2007). These animal-based studies are advanced compared to the human studies and have implications for the application of cannabinoid-based medicines for the treatment of MS.

Implications for the Therapy of Multiple Sclerosis

Symptom Control

The experimental findings that Δ^9-THC is the major component in cannabis that mediates control of spasticity and the adverse effects via CB_1 receptors, notably inhibiting GABAergic signals in cognitive centres, indicates that it will not be possible to truly dissociate positive from adverse events using cannabis (Howlett et al., 2002; Wachtel et al., 2002; Wilkinson et al., 2003; Varvel et al., 2005a,b; Pryce and Baker, 2007). In the long term this will hamper drug development. Δ^9-THC exhibits partial agonist activity at CB_1 receptors and is generally well tolerated for human use (Howlett et al., 2002). The pharmaceutical industry usually aims to produce synthetic ligands with very high affinity and agonism potential as part of their drug development programs. Currently only Nabilone/Cesamet™ has been licensed for the treatment of human disease (Howlett et al., 2002). However, given the wide variability in the ability of humans to tolerate cannabinoids, high affinity agents will only serve to narrow the therapeutic window and will make dosing difficult and increase the chances of adverse events (Zajicek et al., 2003; Wade et al., 2004; Brady et al., 2004). Low affinity agents are preferable candidates for clinical development. Some agents with lowest affinity for the cannabinoid receptors are the endocannabinoids themselves (Howlett et al., 2002). Endocannabinoids are produced "on demand" and are degraded by endogenous mechanisms and are future targets for control of spasticity (Howlett et al., 2002; see Chaps. 2, 3, 11). Indeed in EAE, it was possible to demonstrate that CB_1 receptor antagonism transiently worsened experimental spasticity and that endocannabinoids appeared to be upregulated in areas of damage in spasticity, possibly as a means to control neurotransmission or in response to nerve loss that has accumulated in spastic animals (Baker et al., 2000, 2001; Cabranes et al., 2006). However, the dissection of which endocannabinoid pathway is best to target is currently hampered by the lack of availability of specific agents, which target the different endocannabinoid synthetic and degradation pathways. Although, there is evidence to suggest that anandamide, 2-arachidonoylglycerol and noladin ether can inhibit experimental spasticity (Baker et al., 2000, 2001, and unpublished data), currently only the anandamide degradation pathway is amenable to study and validation. However, most data would suggest that the phospholipase C and diacylglycerol lipase pathway, which produces 2-AG, is probably the most important pathway in normal conditions, in providing the retrograde inhibitory signal in controlling synaptic neurotransmission (Fig. 1; Howlett et al., 2002; Makara et al., 2005; Hashimotodani et al., 2007; Szabo et al., 2006). This would suggest that targeting monoglycerol lipase (see Chap. 3) would be a useful target for therapy. However, as anandamide has a greater affinity for the CB_1 receptor than 2-AG but is hundred times less abundant (Howlett et al., 2002), we suspect that the 2-AG largely serves to maintain the tonicity and underlying signalling of the CB_1 receptor and that it is anandamide, which provides the overriding control of neurotransmission during pathological conditions. Indeed, agents which slow anandamide re-uptake (Baker et al., 2001; De Lago et al., 2004, 2006; Ligresti et al., 2006) and hydrolysis via blockage of fatty aid amide hydrolase (Baker et al., 2001, unpublished) inhibit experimental spasticity. These provide tools with which to develop some form of selective targeting to lesions, which are concentrated in the spinal cord at least in animals, where the endocannabinoid system appears to be dysregulated in contrast to the limbic system which is relatively unaffected in MS (Baker et al., 2001). Furthermore inhibition of excessive neurotransmission not only serves to limit disease symptoms, but can also serve to stop conditions leading to excessive calcium fluxes, which are ultimately lethal to nerves (Fig. 1) and cause the underlying substrate for the development of progressive disability.

Neuroprotection

Nerves are destroyed as a consequence of neuroinflammation that is triggered by immune attack, and also by (auto)immune-independent mechanisms (Compston and Coles, 2002; Bjartmar and Trapp, 2003). Therefore, one route to protect nerves from damage is to induce immunosuppression that prevents neuroinflammation from developing (see below). Another route is to protect the nerves from the damaging effects of the neuroinflammation. Glutamate excitotoxicity, calcium influxes and oxidative stress in excess can cause neurodegeneration in a variety of neurological diseases, including MS. Furthermore hyperexcitation of nerves from electrical activity within an inflammatory environment can cause neurodegeneration so that factors promoting control of excessive neurotransmission and excessive calcium fluxes may also be neuroprotective (Kapoor et al., 2003; Bolton and Paul, 2006; Dutta et al., 2006; Fig. 1). During MS, there may be a relative loss of GABAergic inhibitory signals in motor control and excessive glutamatergic signalling leading to symptoms of disease and/or nerve damage (Bolton and Paul, 2006; Dutta et al., 2006). There is evidence that cannabinoids can inhibit such glutamate-induced damage (Pryce et al., 2003; Docagne et al., 2007; Howlett et al., 2002). In addition to evidence using exogenous cannabinoid receptor agonists, endocannabinoid levels are altered during the course of disease and may be neuroprotective (Howlett et al., 2002; Schabitz et al., 2002; Eljaschewitsch et al., 2006; Witting et al., 2006). In contrast to the relative increase of endocannabinoids in the spinal cord of spastic animals during remission from paralytic attacks (Baker et al., 2001), decreases in endocannabinoid levels have been reported in the brain during periods of paralytic, immune attack (Cabranes et al., 2005; Witting et al., 2006). It has been suggested that such a decrease, possibly due to γ-interferon-induced reduction of 2-arachidonoylglycerol production by microglia, may induce a loss of neuroprotection leading to nerve damage as a consequence of immune attack (Witting et al., 2006). However, the significance of these changes is not completely clear, as the chief pathology in EAE is in the spinal cord and not the brain, which is relatively unaffected during EAE in rodents (Cabranes et al., 2005; Witting et al., 2006). In response to the paralysis, there are dynamic changes in expression levels and importantly, changes in the signalling potential in cannabinoid receptors in different brain regions in the brain (Cabranes et al., 2006). Thus, the observed changes in endocannabinoid levels during immune attack may be reflective of the lack of neurotransmission during paralysis. Although neuronal endocannabinoid expression in the grey matter of people with MS has not been addressed, there have been studies in the white matter of MS tissue, which may relate to more axonal protection or the immune response (Yiangou et al., 2006; Eljaschewitsch et al., 2006; Benito et al., 2007). This demonstrates changes in the cannabinoid receptor expression, notably enhanced glial CB_2 receptor expression (Yiangou et al., 2006). Although it is known that there is significant postmortem generation of endocannabinoids (Schmid et al., 1995; Felder et al., 1996; Kempe et al., 1996), examination of postmortem CNS white matter tissue, sampled hours after death, suggests that anandamide levels are elevated in active lesions compared to normal-appearing white matter in MS/non-MS tissue (Eljaschewitsch et al., 2006). Elevated anandamide levels may stimulate neuroprotection through inhibition of microglial neurotoxicity (Eljaschewitsch et al., 2006). In contrast, some studies have implicated neuroprotection via 2-archidonoyl glycerol-dependent mechanisms, which could be mediated via CB_2 receptor stimulation (Walter et al., 2003; Witting et al., 2006). CB_1 receptor-deficient animals poorly tolerate glutamate excitotoxicity and the inflammatory insult during EAE and show enhanced nerve loss and clinical deficits (Pryce et al., 2003; Jackson et al., 2005). Furthermore, cannabinoid receptor stimulation with Δ^9-THC or synthetic cannabinoids can slow the rate of nerve loss in the spinal cords and slow the accumulation of clinical disability, independent of stopping immune-mediated clinical attacks (Pryce et al., 2003; Croxford et al., submitted). Although cannabinoid control of autoimmune-independent neurodegeneration (Pryce et al., 2005) has yet to be established, cannabinoids have been shown to inhibit autoimmune-independent neurodegeneration by both CB_1 and CB_2 receptor-dependent mechanisms in models of amyotrophic lateral sclerosis (Ramen et al., 2004; Weydt et al., 2005; Bilsland et al., 2006; Kim et al., 2006; Shoemaker et al., 2007). Following the study indicating beneficial effects on controlling progression in long-term follow-up in symptom control trials (Zajicek et al., 2005), the potential of Δ^9-THC to affect progression in MS is currently being investigated in a 3-year trial (http://www.pms.ac.uk/cnrg/cupid.php).

Immunomodulation

Immunomodulation was shown to be a properly of high-dose Δ^9-THC in EAE, even before the discovery and identification of the cannabinoid receptors (Lyman et al., 1989). This efficacy in slowing clinical disease was associated with inhibition of the inflammatory response to be recruited into the CNS (Lyman et al., 1989; Ni et al., 2004; Croxford et al., submitted) and thus shows an activity that is different from an effect on spasticity (Baker et al., 2000, 2001). There are an increasing number of studies in EAE which suggest that cannabinoids could have an immunmodulatory role. This could be via both CB_1 and CB_2 receptor-dependent pathways (Fig. 2; Maresz et al., 2007) as also suggested in viral models of MS (Arevalo-Martin et al., 2003; Croxford and Miller, 2003). Cannabinoid receptor stimulation blocks T cell function and the conditions that lead to microglial activation and migration (Arevalo-Martin et al., 2003; Franklin and Stella, 2003; Walter et al., 2003; Maresz et al., 2007). Recent studies suggest that human T cells do not express significant levels of cannabinoid receptors until activated (Borner et al., 2007; Coopman et al., 2007). However, it is possible to show that CB_2 receptor regulates T cell apoptosis and that may be mediated by CNS-derived endocannabinoids (Sanchez et al., 2006; Lombard et al., 2007; Maresz et al., 2007). Thus, stimulation of CB_2 receptors reduces pathogenic Th1/Th17 responses, including inhibition of gamma interferon production, which can be reflected by the failure to upregulate major histocompatibility complex class II antigens on glial cells, thus reducing their capacity to present antigen to T cells (Arevalo-Martin et al., 2003). Likewise, it has been reported that WIN55212-2 inhibits leucocyte migration into brains of mice with EAE by partially a CB_2 receptor-dependent mechanism (Ni et al., 2004). In contrast, CB_2 receptor inverse agonists (see Chap. 7) have been reported to inhibit leucocyte diapedesis into tissues (Lunn et al., 2006; Oka et al., 2006) and other studies have failed to find evidence for immunomodulatory effects of CB_2 receptor agonists or antagonists (Croxford et al., submitted) and indicate that cannabinoid-mediated immunmodulation by Δ^9-THC in EAE is mediated largely via the CB_1 receptor (Fujiwara and Egashira, 2004; Croxford et al., submitted; Maresz et al., 2007).

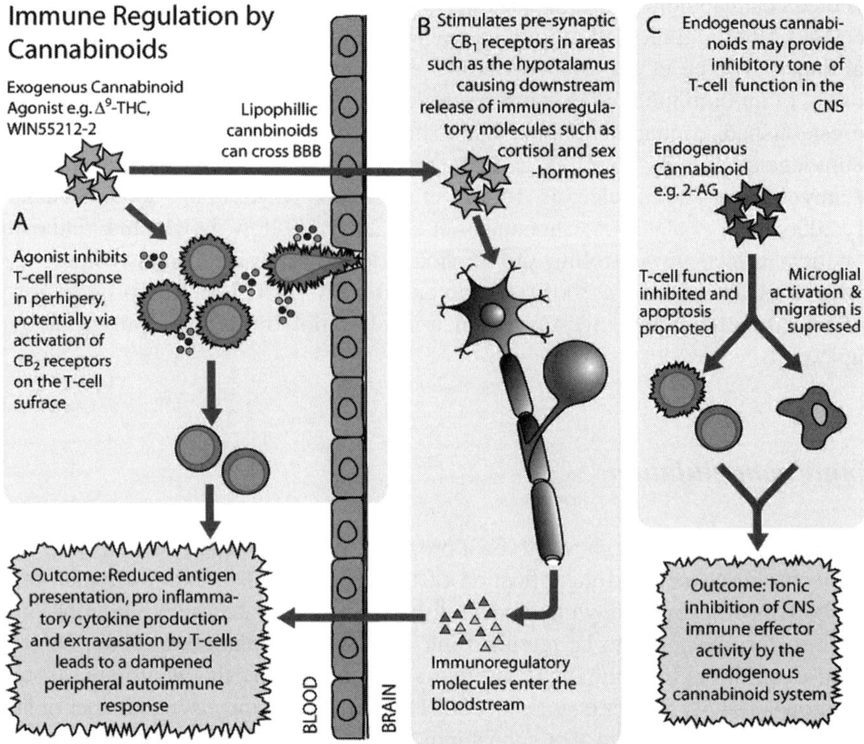

Fig. 2 Immunoregulation by cannabinoids. Immunoregulation by cannabinoids could influence peripheral (**a**, **b**) or (**c**) central immune responses by either (**a**, **c**) direct or (**b**) indirect effector mechanisms

However, this form of immunosuppression does not appear to be due to cannabinoid stimulation on T cells, but is secondary to neuronal CB_1 receptor stimulation (Maresz et al., 2007) resulting in the downstream production of immunosuppressive molecules such as glucocorticosteroids that are known to tonically control the neuroinflammatory response in EAE (Pertwee, 1974; Wirguin et al., 1994; Bolton et al., 1997; Murphy et al., 1998). Furthermore, immunosuppression shown in vitro occurs only at high micromolar cannabinoid concentrations that are unlikely to be reached under normal physiological conditions (Kraft and Kress, 2004) and in vivo at high doses that induce significant cannabimimetic effects, which would be unlikely to be used in the clinic (Wirguin et al., 1994; Croxford and Miller, 2003; Croxford et al., submitted; Maresz et al., 2007). Likewise, arvanil is a cannabinoid with mild cannabinoid receptor affinity, which is such a potent $TRPV_1$ vanillloid receptor agonist that is not effectively inhibited in mice by the $TRPV_1$ antagonist, capsazepine (Brooks et al., 2002; Correll et al., 2004; Pryce and Baker, 2007). This agent can induce modest immunosuppression of acute EAE via $TRVP_1$ responses (Malfitano et al., 2006; Marquez et al., 2006), but again at doses which will probably induce immunosuppressive stress responses in vivo, due to the notably noxious/cannabimimetic effects via $TRPV_1$ receptor responses that this molecule produces (Brooks et al., 2002; Pryce and Baker, 2007). Inhibitors of endocannabinoid degradation have been reported to inhibit the immune-mediated disease in viral and autoimmune models (Cabranes et al., 2005; Mestre et al., 2005); however, in EAE at least this was again associated with their activity at $TRPV_1$ receptors rather than an action on cannabinoid receptors (Cabranes et al., 2005). Further work is needed to clarify the role of cannabinoids in immunoregulation. Synthetic Δ^9-THC is currently licensed for treatment of chemotherapy-induced nausea and wasting associated with acquired immunodeficiency syndrome (AIDS). If cannabis were markedly immunosuppressive, as some experimental studies would lead us to believe, then it would be unlikely that Marinol™ would be considered useful or desirable for use in AIDS. Therefore the immunomodulatory effect of cannabinoids shown in EAE may represent an artefact of dose, which has limited relevance to human use. Cannabis smokers are not overtly immunosuppressed and as already mentioned, evidence for marked immunomodulatory effect that would represent in a reduction in relapse rate has not yet been detected in cannabis trials in MS (Killestein et al., 2003; Katona et al., 2005; Zajicek et al., 2005). Although cannabinoids may have some limited potential for modulating immune responses, this is probably of limited clinical significance and the value of cannabinoids for MS is more likely to be in the control of symptoms and progression than in influencing relapsing immune-mediated disease as an immune modulator.

Concluding Remarks

Current data indicate that whilst cannabis may not be the "wonder drug" that was initially hoped, non-smoked cannabis extracts have a small but significant impact on some symptoms of MS and it is likely that cannabis will enter the pharmacopoeia to a greater extent in the not too distant future. Currently much has been made of the value of mixing the weak CB_1 receptor antagonist/anandamide uptake inhibitor/weak $TRPV_1$ receptor agonist phytocannabinoid, cannabidiol (CBD) with Δ^9-THC for medicinal cannabis (Russo and Guy, 2006). Whilst it is argued that CBD positively influences the pharmacokinetics and reduces the psychoactive potential of Δ^9-THC (Russo and Guy, 2006), the reasons for mixing plant cannabinoids compared to using pure synthetic cannabinoids are perhaps more based on the need to identify product novelty and a marketing niche than on scientific merit. Both British and American cannabis users with MS claim benefit from smoking cannabis (Consroe et al., 1997), yet cannabis in the USA tends to have a low/marginal CBD content (ElSohly et al., 2000). This supports experimental studies that Δ^9-THC is the major active component of cannabis, where CBD by itself has no demonstrable activity in experimental spasticity (Baker et al., 2000; Wilkinson et al., 2003). Furthermore, it is reported that CBD exhibits minimal influence on the induction of adverse events on Δ^9-THC action or pharmacokinetics in the ratios used clinically (Nadulski et al., 2005; Varvel et al., 2006). Surprisingly it has been reported that

CBD acts as a CB receptor antagonist (Thomas et al., 2007) and may be consistent with the observations that Δ^9-THC may exhibit enhanced therapeutic benefit over cannabis extracts containing CBD (Brady et al., 2004; Zajicek et al., 2005). Although it appears that medical cannabis is well tolerated, with time we will be able to assess whether the consequences of long-term use are acceptable (Iverson, 2005; Wade et al., 2006). Once efficacy is accepted and a market is generated, the real value of cannabinoid therapy will be identified. Agents that target the endocannabinoid system are likely to be the future for cannabinoid therapeutics. However, at present only experimental agents are available and we are not aware of an endocannabinoid degradation inhibitor that has been shown to be safe for human use. These agents may be able to harness the medical benefits that the cannabis system has to offer, whilst limiting the adverse effects, both physical and psycho-social, associated with smoking cannabis.

Acknowledgements The authors would like to thank the support of the Multiple Sclerosis Society of Great Britain and Northern Ireland, the National Multiple Sclerosis Society and Aims2cure.

References

Alusi SH, Worthington J, Glickman S, Bain PG (2001) A study of tremor in multiple sclerosis. Brain 124:720–730.

Arevalo-Martin A, Vela JM, Molina-Holgado E, Borrell J, Guaza C (2003) Therapeutic action of cannabinoids in a murine model of multiple sclerosis. J Neurosci 23:2511–2516.

Baker D, Jackson SJ (2007) Models of multiple sclerosis. Adv Clin Neurosci Rehab 6:10–12.

Baker D, Pryce G (2004) The potential role of the endocannabinoid system in the control of multiple sclerosis. Curr Med Chem 4:195–202.

Baker D, Pryce G, Croxford JL, Brown P, Huffman JW, Pertwee RG, Layward L (2000) Cannabinoids control spasticity and tremor in an animal model of multiple sclerosis. Nature 404:84–87.

Baker D, Pryce G, Croxford JL, Brown P, Makryiannis A, Pertwee R, Layward L, Di Marzo V (2001) Endocannabinoids control spasticity in a multiple sclerosis model. FASEB J 15:300–302.

Benito C, Romero JP, Clemente D, Docagne F, Hillard C, Guaze C, Tolon RM, Romero J (2007) Cannabinoid CB_1 and CB_2 receptors and FAAH are specific markers of plaque cell subtypes in human multiple sclerosis. J Neurosci 27:2396–2402.

Bilsland LG, Dick JRT, Pryce G, Petrosino S, Di Marzo V, Baker D, Greensmith L (2006) Manipulation of the endocannabinoid system ameliorates disease symptoms in the $SOD1^{G93A}$ mouse model of ALS. FASEB J 20:1003–1005.

Bjartmar C, Trapp BD (2003) Axonal degeneration and progressive neurologic disability in multiple sclerosis. Neurotox Res 5:157–164.

Bolton C, Paul C (2006) Glutamate receptors in neuroinflammatory demyelinating disease. Mediators Inflamm 2006:1–12.

Bolton C, O'Neill JK, Allen SJ, Baker D (1997) Regulation of chronic relapsing experimental allergic encephalomyelitis by endogenous and exogenous glucocorticoids. Int Arch Allergy Immunol 114:74–80.

Borner C, Hollt V, Sebald W, Kraus J (2007) Transcriptional regulation of the cannabinoid receptor type 1 gene in T cells by cannabinoids. J Leukoc Biol 81:336–343.

Brady CM, DasGupta R, Dalton C, Wiseman OJ, Berkley KJ, Fowler CJ (2004) An open-label pilot study of cannabis-based extracts for bladder dysfunction in advanced multiple sclerosis. Mult Scler 10:425–433.

Brooks JW, Pryce G, Bisogno T, Jagger SI, Hankey DJR, Brown P, Bridges D, Ledent C, Bifulco M, Rice AS, Di Marzo V, Baker D (2002) Arvanil-induced inhibition of spasticity and persistent pain: further evidence for additional therapeutic non-CB_1 cannabinoid receptors. Eur J Pharmacol 439:83–92.

Cabranes A, Venderova K, de Lago E, Fezza F, Sanchez A, Mestre L, Valenti M, Garcia-Merino A, Ramos JA, Di Marzo V, Fernandez-Ruiz J (2005) Decreased endocannabinoid levels in the brain and beneficial effects of agents activating cannabinoid and/or vanilloid receptors in a rat model of multiple sclerosis. Neurobiol Dis 20:207–217.

Cabranes A, Pryce G, Baker D, Fernández-Ruiz J (2006) Changes in CB_1 receptors in motor-related brain structures of chronic relapsing experimental allergic encephalomyelitis mice. Brain Res 1107:199–205.

Chong MS, Wolff K, Wise K, Tanton C, Winstock A, Silber E (2006) Cannabis use in patients with multiple sclerosis. Mult Scler 12:646–651.

Clark AJ, Ware MA, Yazer E, Murray TJ, Lynch ME (2004) Patterns of cannabis use among patients with multiple sclerosis. Neurology 62:2098–2100.

Clifford DB (1983) Tetrahydrocannabinol for tremor in multiple sclerosis. Ann Neurol 13:669–671.

Coles AJ, Cox A, Le Page E, Jones J, Trip SA, Deans J, Seaman S, Miller DH, Hale G, Waldmann H, Compston DA (2006) The window of therapeutic opportunity in multiple sclerosis: evidence from monoclonal antibody therapy. J Neurol 253:98–108.

Compston A, Coles A (2002) Multiple sclerosis. Lancet 359:1221–1231.

Confavreux C, Vukusic S (2006) Accumulation of irreversible disability in multiple sclerosis: from epidemiology to treatment. Clin Neurol Neurosurg 108:327–332.

Consroe P, Musty R, Rein J, Tillery W, Pertwee R (1997). The perceived effects of smoked cannabis on patients with multiple sclerosis. Eur Neurol 38:44–48.

Coopman K, Smith LD, Wright KL, Ward SG (2007) Temporal variation in CB_2R levels following T lymphocyte activation: evidence that cannabinoids modulate CXCL12-induced chemotaxis. Int Immunopharmacol 7:360–371.

Correll CC, Phelps PT, Anthes JC, Umland S, Greenfeder S (2004) Cloning and pharmacological characterization of mouse $TRPV_1$. Neurosci Lett 370:55–60.

Croxford JL, Miller SD (2003) Immunoregulation of a viral model of multiple sclerosis using the synthetic cannabinoid R + WIN55,212. J Clin Invest 111:1231–1240.

Croxford JL, Pryce G, Jackson SJ, Ledent C, Giovannoni G, Pertweee RG, Yamamura T, Baker D (2007) Cannabinoid-mediated neuroprotection, not immunosuppression, may be more relevant to multiple sclerosis. Submitted for publication.

De Groat WC (1998) Anatomy of the central neural pathways controlling the lower urinary tract. Eur Urol 34:2–5.

De Lago E, Ligresti A, de Lago E, Ortar G, Morera E, Cabranes A, Pryce G, Bifulco M, Baker D, Fernadez-Ruiz J, Di Marzo V (2004) *In vivo* pharmacological actions of two novel inhibitors of anandamide cellular uptake. Eur J Pharmacol 484:249–257.

De Lago E, Fernández-Ruiz J, Ortega-Gutiérrez S, Cabranes A, Pryce G, Baker D, López-Rodríguez, Ramos JA (2006) UCM707, an inhibitor of anandamide uptake, behaves as a symptom control agent in models of Huntington's disease and multiple sclerosis, but fails to delay/arrest the progression of different motor-related disorders. Eur Neuropsychopharmacol 16:7–18.

Docagne F, Muneton V, Clemente D, Ali C, Loria F, Correa F, Hernangomez M, Mestre L, Vivien D, Guaza C (2007) Excitotoxicity in a chronic model of multiple sclerosis: neuroprotective effects of cannabinoids through CB_1 and CB_2 receptor activation. Mol Cell Neurosci 34:551–561.

Dutta R, McDonough J, Yin X, Peterson J, Chang A, Torres T, Gudz T, Macklin WB, Lewis DA, Fox RJ, Rudick R, Mirnics K, Trapp BD (2006) Mitochondrial dysfunction as a cause of axonal degeneration in multiple sclerosis patients. Ann Neurol 59:478–489.

Ehde DM, Osborne TL, Hanley MA, Jensen MP, Kraft GH (2006) The scope and nature of pain in persons with multiple sclerosis. Mult Scler 12:629–638.

Eljaschewitsch E, Witting A, Mawrin C, Lee T, Schmidt PM, Wolf S, Hoertnagl H, Raine CS, Schneider-Stock R, Nitsch R, Ullrich O (2006) The endocannabinoid anandamide protects neurons during CNS inflammation by induction of MKP-1 in microglial cells. Neuron 49:67–79.

ElSohly MA, Ross SA, Mehmedic Z, Arafat R, Yi B, Banahan III BF (2000) Potency trends of delta^9-THC and other cannabinoids in confiscated marijuana from 1980–1997. J Forensic Sci 45:24–30.

Felder CC, Nielsen A, Briley EM, Palkovits M, Priller J, Axelrod J, Nguyen DN, Richardson JM, Riggin RM, Koppel GA, Paul SM, Becker GW (1996) Isolation and measurement of the endogenous cannabinoid receptor agonist, anandamide, in brain and peripheral tissues of human and rat. FEBS Lett 393:231–235.

Fox P, Bain PG, Glickman S, Carroll C, Zajicek J (2004) The effect of cannabis on tremor in patients with multiple sclerosis. Neurology 62:1105–1109.

Franklin A, Stella N (2003) Arachidonylcyclopropylamide increases microglial cell migration through cannabinoid CB_2 and abnormal-cannabidiol-sensitive receptors. Eur J Pharmacol 474:195–198.

Freeman RM, Adekanmi O, Waterfield MR, Waterfield AE, Wright D, Zajicek J (2006) The effect of cannabis on urge incontinence in patients with multiple sclerosis: a multicentre, randomised placebo-controlled trial (CAMS-LUTS). Int Urogynecol J Pelvic Floor Dysfunct 17:636–641.

Fujiwara M, Egashira N (2004) New perspectives in the studies on endocannabinoid and cannabis: abnormal behaviors associate with CB_1 cannabinoid receptor and development of therapeutic application. J Pharmacol Sci 96:362–366.

Hashimotodani Y, Ohno-Shosaku T, Kano M (2007) Presynaptic monoacylglycerol lipase activity determines basal endocannabinoid tone and terminates retrograde endocannabinoid signaling in the hippocampus. J Neurosci 27:1211–1219.

Howlett AC, Barth F, Bonner TI, Cabral G, Casellas P, Devane WA, Felder CC, Herkenham M, Mackie K, Martin BR, Mechoulam R, Pertwee RG (2002) International Union of Pharmacology. XXVII. Classification of cannabinoid receptors. Pharmacol Rev 54:161–202.

Iskedjian M, Bereza B, Gordon A, Piwko C, Einarson TR (2007) Meta-analysis of cannabis based treatments for neuropathic and multiple sclerosis-related pain. Curr Med Res Opin 23:17–24.

Iversen L (2005) Long-term effects of exposure to cannabis. Curr Opin Pharmacol 5:69–72.

Jackson SL, Pryce G, Diemel DT, Baker D (2005) Cannabinoid receptor null mice are susceptible to neurofilament damage and caspase 3 activation. Neuroscience 134:261–268.

Kalsi V, Fowler CJ (2005) Therapy insight: bladder dysfunction associated with multiple sclerosis. Nat Clin Pract Urol 2:492–501.

Kapoor R, Davies M, Blaker PA, Hall SM, Smith KJ (2003) Blockers of sodium and calcium entry protect axons from nitric oxide-mediated degeneration. Ann Neurol 53:174–80.

Katona S, Kaminski E, Sanders H, Zajicek J (2005) Cannabinoid influence on cytokine profile in multiple sclerosis. Clin Exp Immunol 140:580–585.

Kempe K, Hsu FF, Bohrer A, Turk J (1996) Isotope dilution mass spectrometric measurements indicate that arachidonylethanolamide, the proposed endogenous ligand of the cannabinoid receptor, accumulates in rat brain tissue post mortem but is contained at low levels in or is absent from fresh tissue. J Biol Chem 271:17287–17295.

Kesselring J, Thompson AJ (1997) Spasticity, ataxia and fatigue in multiple sclerosis. Baillieres Clin Neurol 6: 429–445.

Killestein J, Hoogervorst EL, Reif M, Kalkers NF, Van Loenen AC, Staats PG, Gorter RW, Uitdehaag BM, Polman CH (2002) Safety, tolerability, and efficacy of orally administered cannabinoids in MS. Neurology 58:1404–1407.

Killestein J, Hoogervorst EL, Reif M, Blauw B, Smits M, Uitdehaag BM, Nagelkerken L, Polman CH (2003) Immunomodulatory effects of orally administered cannabinoids in multiple sclerosis. J Neuroimmunol 137:140–143.

Kim K, Moore DH, Makriyannis A, Abood ME (2006) AM1241, a cannabinoid CB2 receptor selective compound, delays disease progression in a mouse model of amyotrophic lateral sclerosis. Eur J Pharmacol 542:100–105.

Kraft B, Kress HG (2004) Cannabinoids and the immune system. Of men, mice and cells. Schmerz 18:203–210.

Ligresti A, Cascio MG, Pryce G, Kulasegram S, Beletskaya I, De Petrocellis L, Saha B, Mahadevan A, Visintin C, Baker D, Wiley J, Martin BR, Razdan RK, Di Marzo V (2006). New potent and selective inhibitors of anandamide re-uptake with anti-spastic activity in a mouse model of multiple sclerosis. Br J Pharmacol 147:83–91.

Lombard C, Nagarkatti M, Nagarkatti P (2007) CB_2 cannabinoid receptor agonist, JWH-015, triggers apoptosis in immune cells: potential role for CB_2-selective ligands as immunosuppressive agents. Clin Immunol 122:259–270.

Lunn CA, Fine JS, Rojas-Triana A, Jackson JV, Fan X, Kung TT, Gonsiorek W, Schwarz MA, Lavey B, Kozlowski JA, Narula SK, Lundell DJ, Hipkin RW, Bober LA (2006) A novel cannabinoid peripheral cannabinoid receptor-selective inverse agonist blocks leukocyte recruitment *in vivo*. J Pharmacol Exp Ther 316:780–788.

Lyman WD, Sonett JR, Brosnan CF, Elkin R, Bornstein MB (1989) Delta 9-tetrahydrocannabinol: a novel treatment for experimental autoimmune encephalomyelitis. J Neuroimmunol 23:73–81.

Makara JK, Mor M, Fegley D, Szabo SI, Kathuria S, Astarita G, Duranti A, Tontini A, Tarzia G, Rivara S, Freund TF, Piomelli D (2005) Selective inhibition of 2-AG hydrolysis enhances endocannabinoid signaling in hippocampus.Nat Neurosci 8:1139–1141.

Malfitano AM, Matarese G, Pisanti S, Grimaldi C, Laezza C, Bisogno T, Di Marzo V, Lechler RI, Bifulco M (2006) Arvanil inhibits T lymphocyte activation and ameliorates autoimmune encephalomyelitis. J Neuroimmunol 171:110–119.

Maresz K, Pryce G, Ponomarev ED, Marsicano G, Croxford JL, Shriver LP, Ledent C, Cheng X, Carrier E, Mann MK, Giovannoni G, Pertwee RG, Yamamura T, Buckley NE, Hillard CJ, Lutz B, Baker D, Dittel BN (2007) Direct suppression of CNS autoimmune inflammation via the cannabinoid receptor CB_1 on neurons and CB_2 on autoreactive T Cells. Nat Med 13:492–497.

Marquez N, De Petrocellis L, Caballero FJ, Macho A, Schiano-Moriello A, Minassi A, Appendino G, Munoz E, Di Marzo V (2006) Iodinated N-acylvanillamines: potential "multiple-target" anti-inflammatory agents acting via the inhibition of t-cell activation and antagonism at vanilloid $TRPV_1$ channels. Mol Pharmacol 69:1373–1382.

Martin RS, Luong LA, Welsh NJ, Eglen RM, Martin GR, MacLennan SJ (2000) Effects of cannabinoid receptor agonists on neuronally evoked contractions of urinary bladder tissues isolated from rat, mouse, pig, dog, monkey and human. Br J Pharmacol 129:1707–1715.

Meinck HM, Schonle PW, Conrad B (1989) Effect of cannabinoids on spasticity and ataxia in multiple sclerosis. J Neurol 236:120–122.

Mestre L, Correa F, Arevalo-Martin A, Molina-Holgado E, Valenti M, Ortar G, Di Marzo V, Guaza C (2005) Pharmacological modulation of the endocannabinoid system in a viral model of multiple sclerosis. J Neurochem 92:1327–1339.

Metz I, Lucchinetti CF, Openshaw H, Garcia-Merino A, Lassmann H, Freedman MS, Atkins HL, Azzarelli B, Kolar OJ, Bruck W (2007). Autologous haematopoietic stem cell transplantation fails to stop demyelination and neurodegeneration in multiple sclerosis. Brain 130:1254–1262.

Murphy LL, Munoz RM, Adrian BA, Villanua MA (1998) Function of cannabinoid receptors in the neuroendocrine regulation of hormone secretion. Neurobiol Dis 6:432–446.

Nadulski T, Pragst F, Weinberg G, Roser P, Schnelle M, Fronk EM, Stadelmann AM (2005). Randomized, double-blind, placebo-controlled study about the effects of cannabidiol (CBD) on the pharmacokinetics of Delta9-tetrahydrocannabinol (THC) after oral application of THC versus standardized cannabis extract. Ther Drug Monit 27:799–810.

Ni X, Geller EB, Eppihimer MJ, Eisenstein TK, Adler MW, Tuma RF (2004) Win 55212-2, a cannabinoid receptor agonist, attenuates leukocyte/endothelial interactions in an experimental autoimmune encephalomyelitis model. Mult Scler 10:158–164.

Notcutt W, Price M, Miller R, Newport S, Phillips C, Simmons S, Sansom C (2004) Initial experiences with medicinal extracts of cannabis for chronic pain: results from 34 'N of 1' studies. Anaesthesia 59:440–452.

Oka S, Wakui J, Gokoh M, Kishimoto S, Sugiura T (2006) Suppression by WIN55212-2, a cannabinoid receptor agonist, of inflammatory reactions in mouse ear: interference with the actions of an endogenous ligand, 2-arachidonoylglycerol. Eur J Pharmacol 538:154–162.

Page SA, Verhoef MJ (2006) Medicinal marijuana use: experiences of people with multiple sclerosis. Can Fam Physician 52:64–65.

Page SA, Verhoef MJ, Stebbins RA, Metz LM, Levy JC (2003) Cannabis use as described by people with multiple sclerosis. Can J Neurol Sci 30:201–205.

Pertwee RG (1974) Tolerance to the effect of delta1-tetrahydrocannabinol on corticosterone levels in mouse plasma produced by repeated administration of cannabis extract or delta1-tetrahydrocannabinol. Br J Pharmacol 51:391–397.

Petro DJ, Ellenberger Jr C (1981) Treatment of human spasticity with delta 9-tetrahydrocannabinol. J Clin Pharmacol 21:413S–416S.

Polman CH, O'Connor PW, Havrdova E, Hutchinson M, Kappos L, Miller DH, Phillips JT, Lublin FD, Giovannoni G, Wajgt A, Toal M, Lynn F, Panzara MA, Sandrock AW, AFFIRM Investigators (2006) A randomized, placebo-controlled trial of natalizumab for relapsing multiple sclerosis. N Engl J Med 354:899–910.

Pryce G, Baker D (2007) Control of spasticity in a multiple sclerosis model is CB_1, not CB_2, cannabinoid receptors. Br J Pharmacol 150:519–525.

Pryce G, Ahmed Z, Hankey DRJ, Jackson SL, Croxford JL, Pocock JM, Ledent C, Petzold A, Thompson AJ, Giovannoni G, Cuzner ML, Baker D (2003) Cannabinoids inhibit neurodegeneration in multiple sclerosis models. Brain 126:2191–2202.

Pryce G, O'Neill JKA, Croxford JL, Amor S, Hankey DRJ, Giovannoni G, Baker D (2005) Immunological tolerance that eliminates relapses, fails to halt secondary progression in a chronic multiple sclerosis model. J Neuroimmunol 165:41–52.

Raman C, McAllister SD, Rizvi G, Patel SG, Moore DH, Abood ME (2004) Amyotrophic lateral sclerosis: delayed disease progression in mice by treatment with a cannabinoid. Amyotroph Lateral Scler Other Motor Neuron Disord 5:33–39.

Rog DJ, Nurmikko TJ, Friede T, Young CA (2005) Randomized, controlled trial of cannabis-based medicine in central pain in multiple sclerosis. Neurology 65:812–819.

Russo E, Guy GW (2006) A tale of two cannabinoids: the therapeutic rationale for combining tetrahydrocannabinol and cannabidiol. Med Hypotheses 66:234–246.

Sanchez AJ, Gonzalez-Perez P, Galve-Roperh I, Garcia-Merino A (2006) R-(+)-[2,3-Dihydro-5-methyl-3-(4-morpholinylmethyl)-pyrrolo-[1,2,3-de]-1,4-benzoxazin-6-yl]-1-naphtalenyl-methanone ameliorates experimental autoimmune encephalomyelitis and induces encephalitogenic T cell apoptosis: partial involvement of the CB_2 receptor. Biochem Pharmacol 72:1697–1706.

Schabitz WR, Giuffrida A, Berger C, Aschoff A, Schwaninger M, Schwab S, Piomelli D (2002) Release of fatty acid amides in a patient with hemispheric stroke: a microdialysis study. Stroke 33:2112–2114.

Schmid P C, Krebsbach R J, Perry S R, Dettmer TM, Maasson JL, Schmid HH (1995) Occurrence and postmortem generation of anandamide and other long-chain N-acylethanolamines in mammalian brain. FEBS Lett 375:117–120.

Schnelle M, Grotenhermen F, Reif M, Gorter RW (1999) Results of a standardized survey on the medical use of cannabis products in the German-speaking area. Forsch Komplementarmed 6:28–36.

Schon F, Hart PE, Hodgson TL, Pambakian AL, Ruprah M, Williamson EM, Kennard C (1999) Suppression of pendular nystagmus by smoking cannabis in a patient with multiple sclerosis. Neurology 53:2209–2210.

Shoemaker JL, Seely KA, Reed RL, Crow JP, Prather PL (2007) The CB2 cannabinoid agonist AM-1241 prolongs survival in a transgenic mouse model of amyotrophic lateral sclerosis when initiated at symptom onset. J Neurochem 101:87–98.

Svendsen KB, Jensen TS, Hansen HJ, Bach FW (2005) Sensory function and quality of life in patients with multiple sclerosis and pain. Pain 114:473–481.

Szabo B, Urbanski MJ, Bisogno T, Di Marzo V, Mendiguren A, Baer WU, Freiman I (2006) Depolarization-induced retrograde synaptic inhibition in the mouse cerebellar cortex is mediated by 2-arachidonoylglycerol. J Physiol 577:263–280.

Tagliaferro P, Javier Ramos A, Onaivi ES, Evrard SG, Lujilde J, Brusco A (2006) Neuronal cytoskeleton and synaptic densities are altered after a chronic treatment with the cannabinoid receptor agonist WIN 55,212-2. Brain Res 1085:163–176.

Thomas A, Baillie GL, Phillips AM, Razdan RK, Ross RA, Pertwee RG (2007) Cannabidiol displays unexpectedly high potency as an antagonist of CB_1 and CB_2 receptor agonists in vitro. Br J Pharmacol 150:613–823.

Ungerleider JT, Andyrsiak T, Fairbanks L, Ellison GW, Myers LW (1987) Delta-9-THC in the treatment of spasticity associated with multiple sclerosis. Adv Alcohol Subst Abuse 7:39–50.

Vaney C, Heinzel-Gutenbrunner M, Jobin P, Tschopp F, Gattlen B, Hagen U, Schnelle M, Reif M (2004) Efficacy, safety and tolerability of an orally administered cannabis extract in the treatment of spasticity in patients with multiple sclerosis: a randomized, double-blind, placebo-controlled, crossover study. Mult Scler 10:417–424.

van Oosten BW, Killestein J, Mathus-Vliegen EM, Polman CH (2004) Multiple sclerosis following treatment with a cannabinoid receptor-1 antagonist. Mult Scler 10:330–331.

van Sickle MD, Duncan M, Kingsley PJ, Mouihate A, Urbani P, Mackie K, Stella N, Makriyannis A, Piomelli D, Davison JS, Marnett LJ, Di Marzo V, Pittman QJ, Patel KD, Sharkey KA (2005) Identification and functional characterization of brainstem cannabinoid CB2 receptors. Science 310:329–332.

Varvel SA, Anum E, Niyuhire F, Wise LE, Lichtman AH (2005a) Delta^9-THC-induced cognitive deficits in mice are reversed by the GABA(A) antagonist bicuculline. Psychopharmacology 178:317–327.

Varvel SA, Bridgen DT, Tao Q, Thomas BF, Martin BR, Lichtman AH (2005b) Delta9-tetrahydrocannbinol accounts for the antinociceptive, hypothermic, and cataleptic effects of marijuana in mice. J Pharmacol Exp Ther 314:329–337.

Varvel SA, Wiley JL, Yang R, Bridgen DT, Long K, Lichtman AH, Martin BR (2006) Interactions between THC and cannabidiol in mouse models of cannabinoid activity. Psychopharmacology 186:226–234.

Wachtel SR, ElSohly MA, Ross SA, Ambre J, de Wit H (2002) Comparison of the subjective effects of Delta^9-tetrahydrocannabinol and marijuana in humans. Psychopharmacology 161:331–339.

Wade DT, Robson P, House H, Makela P, Aram J (2003) A preliminary controlled study to determine whether whole-plant cannabis extracts can improve intractable neurogenic symptoms. Clin Rehabil 17:21–29.

Wade DT, Makela P, Robson P, House H, Bateman C (2004) Do cannabis-based medicinal extracts have general or specific effects on symptoms in multiple sclerosis? A double-blind, randomized, placebo-controlled study on 160 patients. Mult Scler 10:434–441.

Wade DT, Makela PM, House H, Bateman C, Robson P (2006) Long-term use of a cannabis-based medicine in the treatment of spasticity and other symptoms in multiple sclerosis. Mult Scler 12:639–645.

Walker JM, Hohmann AG (2005) Cannabinoid mechanisms of pain suppression. Handb Exp Pharmacol 168:509–554.

Walter L, Franklin A, Witting A, Wade C, Xie Y, Kunos G, Mackie K, Stella N (2003) Nonpsychotropic cannabinoid receptors regulate microglial cell migration. J Neurosci 23:1398–1405.

Ware MA, Adams H, Guy GW (2005) The medicinal use of cannabis in the UK: results of a nationwide survey. Int J Clin Pract 59:291–295.

Weydt P, Hong S, Witting A, Moller T, Stella N, Kliot M (2005) Cannabinol delays symptom onset in SOD1 (G93A) transgenic mice without affecting survival. Amyotroph Lateral Scler Other Motor Neuron Disord 6:182–184.

Wilkinson JD, Whalley BJ, Baker D, Pryce G, Constanti A, Gibbons S, Williamson EM (2003) Medicinal cannabis: is delta9-tetrahydrocannabinol necessary for all its effects? J Pharm Pharmacol 55:1687–1694.

Wilson RI, Nicoll RA (2002) Endocannabinoid signalling in the brain. Science 296:678–682.

Wirguin I, Mechoulam R, Breuer A, Schezen E, Weidenfeld J, Brenner T (1994) Suppression of experimental autoimmune encephalomyelitis by cannabinoids. Immunopharmacology 28:209–214.

Witting A, Chen L, Cudaback E, Straiker A, Walter L, Rickman B, Moller T, Brosnan C, Stella N (2006) Experimental autoimmune encephalomyelitis disrupts endocannabinoid-mediated neuroprotection. Proc Natl Acad Sci USA 103:6362–6367.

Yiangou Y, Facer P, Durrenberger P, Chessell IP, Naylor A, Bountra C, Banati RR, Anand P (2006) COX-2, CB_2 and $P2X_7$-immunoreactivities are increased in activated microglial cells/macrophages of multiple sclerosis and amyotrophic lateral sclerosis spinal cord. BMC Neurol 6:12.

Zajicek J, Fox P, Sanders H, Wright D, Vickery J, Nunn A, Thompson A, UK MS Research Group (2003) Cannabinoids for treatment of spasticity and other symptoms related to multiple sclerosis (CAMS study): multicentre randomised placebo-controlled trial. Lancet 362:1517–1526.

Zajicek JP, Sanders HP, Wright DE, Vickery PJ, Ingram WM, Reilly SM, Nunn AJ, Teare LJ, Fox PJ, Thompson AJ (2005) Cannabinoids in multiple sclerosis (CAMS) study: safety and efficacy data for 12 months follow up. J Neurol Neurosurg Psychiatry 76:1664–1669.

Chapter 19
Endocannabinoids in Alzheimer's Disease

María L. de Ceballos

Abstract Alzheimer's disease (AD), the major cause of dementia, is a chronic neurodegenerative disorder. Although our understanding of the cellular and molecular events involved in the pathophysiology of the disease has greatly advanced, few effective therapies had been introduced into the clinic. The characterization of the cannabinoid system has been defined during the last few years and cannabinoid-based therapies are beginning to be recognized for the treatment of different diseases. According to recent evidence, cannabinoid receptors are localized to senile plaques in AD brain, in particular in activated microglial cell clusters. On the other hand, cortical CB_1 positive neurons are lost and CB_1 receptor expression and functioning are markedly decreased in the neurologic disorder. Furthermore, in AD models, in vivo cannabinoids prevent the cognitive impairment, while reducing the loss of neuronal markers and of markers of gliosis. The beneficial effects of cannabinoids in preventing neurotoxicity induced by β-amyloid (Aβ) may rely on the anti-inflammatory properties of cannabinoids, given that they reduce the effects Aβ on microglial cells and on astrocytes, as judged by in vitro experiments, and can be brought about by both cannabinoid receptor-dependent and -independent mechanisms. These findings may set the basis for the use of these compounds, that combine both anti-inflammatory and neuroprotective actions, as a therapeutic approach for AD.

Introduction

Alzheimer's disease (AD) is a chronic and progressive neurodegenerative disease, which is the most frequent cause of dementia. Indeed it is considered to be responsible for 50% of cases of dementia. In particular, AD currently affects around 15 million people worldwide. The risk of suffering AD varies with age; thus, the incidence increases from 0.5% per year at the age of 65 to 8% per year after the age of 85. In the opinion of some authors, this high prevalence of the disease may be considered as an epidemic and all the efforts should be made to prevent the disease so as to delay the onset of symptoms and to alleviate those symptoms in already diagnosed patients. This represents a major challenge to researchers in Biomedicine, whether medicinal chemists, pharmacologists, cellular/molecular biologists or clinical neurologists. AD patients

A. Köfalvi (ed.), *Cannabinoids and the Brain.*

have problems in recent memory and abstract thinking, spatial and temporal disorientation, altered judgement, problems with speech and alterations in behaviour. Mental function and the activities of daily living progressively worsen. In this disorder, brain regions involved in learning and memory processes are reduced in size. They present the pathological alterations characteristic of the disease and show loss of particular subsets of neurons. Those areas are neocortical regions, the hippocampus, nuclei of the amygdala, the nucleus basalis of Meynert and several monoaminergic nuclei of the brain stem. The pathological lesions which characterize the disease are localized to those brain areas, namely the senile plaques, vascular β-amyloid (Aβ) accumulation (amyloid angiopathy) and neurofibrillary tangles (La-Ferla and Odo, 2005). Neuritic plaques are extracellular deposits comprising a core of aggregated Aβ, surrounded by activated microglia and reactive astrocytes. Aβ, a 4kDa peptide, is generated from a large molecule, the Aβ precursor protein (APP), through sequential cleavage by β- and γ-secretases. The 40-amino acid form of Aβ predominates in brain, but longer peptides are thought to be the major pathogenic forms in AD. Neurofibrillary tangles are intraneuronal aggregates of paired helical filaments of hyperphosphorylated tau, a microtubule-associated protein. In general, the disorder is sporadic and not inherited, while in other cases is caused by mutations in different genes, that results in familial cases with early onset of the disease. Mutations in three different genes have been found in familial AD. The genes are present in chromosome 21, 14 and 1, and encode APP (Goate et al., 1991), presenilin-1 (PS1; Sherrington et al., 1995) and presenilin-2 (PS2; Levy-Lahad et al., 1995). The origin of sporadic AD, accounting for more than 90% of the cases, appears to be multifactorial and indeed several risk factors have been identified such as advanced age, female gender, diet (high calorie and fat diet), and prior cardiovascular disease or brain trauma. Genetic variants may also confer risk for suffering AD such as apolipoprotein E (ApoE ε4), alpha2 macroglobulin and several mitochondrial genes. An invariant feature of AD is the marked microglial activation in the vulnerable regions of the brain. Several studies have shown that in AD, microglia is attracted and associates to the plaques surrounding them (McGeer et al., 1987; Dickson et al., 1988; Haga et al., 1989). The role played by activated microglia is under debate, given that they may be either neurotoxic or neuroprotective under different circumstances. Cultured microglial cells, activated by Aβ, release toxic species such as proinflammatory cytokines which can cause neurodegeneration. Therefore, neurotoxicity has been observed in co-cultures of microglia with neurons (Combs et al., 1999, 2001; Tan et al., 2000; Xie et al., 2002). The neuroprotective role of microglia is supported by immunization studies with Aβ peptides or with antibodies directed against Aβ in transgenic mouse models. These manipulations effectively reduce Aβ burden in the brain restoring cognitive functions (Bard et al., 2000; Janus et al., 2000). This strategy has been investigated in patients, although severe undesired side effects made the trial to be discontinued. The development of different experimental animal models of AD has greatly helped our understanding of the pathophysiology of the disease and the search for new therapeutic strategies. Several decades ago, analogues of glutamic acid were used to destroy particular subsets of neurons known to degenerate in the neurologic disorders (e.g. cholinergic neurons in the nucleus basalis magnocellularis). After Aβ purification and characterization, a great number of studies have demonstrated that the peptide was neurotoxic to cultured cells and some of the biochemical and pathological changes were mimicked in animals intraventricularly or focally injected with the peptide (Frautschy et al., 1991; Kowall et al., 1991; Delobette et al., 1997). Finally, the recognition of familial AD due to different mutations supported the development of transgenic mice bearing them (Kobayashi and Chen, 2005). Thus, different APP transgenics with different human mutations have been described, Aβ levels are markedly increased compared with wild type mice and they show plaques, which progress from diffuse to mature plaques (Games et al., 1995). These plaques are surrounded by activated microglia and reactive astrocytes and have markers of inflammation (Frautschy et al., 1998; Jantzen et al., 2002). More importantly, the APP transgenic mice have pronounced impairment in learning and memory processes, as judged by their performance in different tests, in particular in those involving spatial navigation (Müller et al., 1994; Chen et al., 2000). Although PS1 and PS2 transgenic mice do not have an overt phenotype, whether pathological or behavioural, double mutants APP/PS1 (Holcomb et al., 1998) or APP/PS2 show significant changes earlier than the single transgenics. However, these animals do not show one of the hallmarks of AD, namely tau pathology. This has prompted the generation of triple transgenics APP/PS1/tau to mimick more faithfully the pathology observed in AD patients. During the last decades, in light of the high incidence of AD, the research effort made has increased our understanding on the cellular and molecular events associated with the pathology. Animal models, human post-mortem material and genetic analyses have all provided important cues to the etiology of AD and indeed the present search for effective therapies is based on these findings (Martin, 1999; Michaelis, 2003; Mattson, 2004). These may fall into two different kinds of therapies: palliative, i.e. drugs aimed to symptom relief, and disease-modifying agents, which prevent/delay its onset or slow the course of the disease. Drugs aimed at treating AD belong to very different pharmacological subclasses. To maintain ACh levels, acetylcholinesterase inhibitors have been developed to block the major ACh-degrading enzyme. Indeed, these drugs are presently approved and introduced into the clinic. Aβ production and deposition – markedly increased in AD – are targets for other series of compounds. Therefore, several drugs which inhibit the production of Aβ (secretase inhibitors), or favour its clearance (including vaccination) or block its aggregation are under active investigation. Other strategies aimed at decreasing neurodegeneration and favouring neuroprotection have been sought (Scorer, 2001; Benson, 2005). In this section we may include antioxidants, glutamate antagonists and trophic factors. Antinflammatory drugs are actively studied as well, since they decrease the risk of suffering the disease (Broe et al., 2000) and the impact of inflammation in AD presently recognized (Akiyama et al., 2000).

Changes in the Endocannabinoid System in Alzheimer's Disease

The first work addressing the study of cannabinoid receptors in AD was that of Westlake and colleagues (1994). By means of autoradiography with the mixed CB_1/CB_2 cannabinoid agonist CP55940 they reported decreases in binding in several hippocampal subfields and basal ganglia regions, but no changes in cortical regions. They found no actual correlation between the changes in cannabinoid receptors in AD with the pathological alterations, whether senile plaques or neurofibrillary tangles. However, it should be mentioned that autoradiography lacks cellular resolution that may explain those results. Therefore the localization of cannabinoid receptors in AD brain has been investigated by immunohistochemistry. Benito and colleagues (2003) observed the presence of CB_2 receptors in cells with microglial morphology and co-labelled with CD68, a microglial marker, surrounding plaques, in enthorrinal cortex and hippocampus in AD. These findings were confirmed by other authors (Ramírez et al., 2005) that found CB_1 and CB_2 localized to plaques along with microglial activation markers. Interestingly, CB_1 positive neurons, present in high numbers in frontal cortex in control subjects, according to the results in primates (Ong and Mackie, 1999), were markedly reduced in AD brains, in particular in areas showing enhanced microglial activation (Ramírez et al., 2005), suggesting their vulnerability to the toxic species released by activated microglial cells. The presence of CB_2 receptors in neurons has been recently described in brain stem of several species (van Sickle et al., 2005), in the cerebellum (Ashton et al., 2006) and in other brain regions of the rat (Gong et al., 2006), although in previous studies it was not documented (see Chap. 10 for *pro* and *contra* arguments). Therefore, it was an unexpected observation to find CB_2 positive neurons with similar morphology to tangle-like neurons and in dystrophic neurites in AD cortex (Ramírez et al., 2005), a finding deserving further study. The localization and expression of fatty acid amide hydrolase (FAAH), one of the major endocannabinoid hydrolysing enzymes, has been studied in AD as well. While activated microglia in senile plaques express both types of cannabinoid receptors, FAAH positive hypertrophic astrocytes have been described adjacent to plaques in AD brain (Benito et al., 2003), although in controls the labelling was predominantly neuronal. Furthermore, FAAH enzyme activity was detected in individual plaques, but not in similar tissue pieces of healthy subjects. Apart from the above-described overt changes in the cellular localization of cannabinoid receptors in brains from patients with AD, cannabinoid receptor density and function have been found to be markedly altered. Indeed, we observed that CB_1 receptor immunoreactivity was significantly decreased (Ramírez et al., 2005). More importantly, their function has been compromised, as judged by the pronounced reduction in G protein-coupling, observed by experiments with WIN55212-2-stimulated GTPγS binding (Ramírez et al., 2005). The fact that protein nitration was found in AD plaques along with CB_1 and CB_2 receptor immunoreactivity and that this biochemical modification inactivates some proteins (Aoyama et al., 2000), prompted us to study if protein nitration of cannabinoid receptors could occur in this neurological disorder. Indeed, both receptors were markedly nitrated in comparison with control subjects – a change that may explain their decreased G protein-coupling. In summary, cannabinoid receptors present a different localization in AD brain compared to that in normal brain and CB_1 receptor density and function is compromised, suggesting that these changes may be important in the ethiopathology of the disease.

A Possible Therapeutic Role of Endocannabinoids in Alzheimer's Disease

Aβ is toxic to different neural and cell lines in culture. After the seminal work by Yankner (1996), overwhelming evidence has been gathered on the characteristics of this toxicity, and it has determined that several mechanisms are responsible for it. In fact, the peptide promotes oxidative stress (Behl et al., 1994; Butterfield et al., 2002), disrupts calcium homeostasis (Mattson et al., 1993, 2000), enhances extracellular levels of glutamate due to glutamate transport inhibition, as well as induces apoptotic or necrotic cell death, depending on the experimental conditions or the cell system employed. Cannabinoids are neuroprotective agents as shown in both in vitro and in vivo experiments (see Chap. 15) modelling acute brain damage. In that neuroprotection, different mechanisms appear to be involved such as decreased glutamate release, modulation of calcium channel activity, reduced release of inflammatory molecules (e.g. TNF-α), antioxidant effects or enhancement of trophic factor support. Therefore this type of drugs may modulate the neurodegeneration occurring in models of AD, an issue only recently investigated. Several recent works have studied the effects of different cannabinoids on the in vitro effects of Aβ (Table 1). Milton (2002) reported that the toxicity of high concentrations (25 μM) of Aβ peptides was prevented by endocannabinoids in NT-2 cells, a human teratocarcinoma cell line that can be differentiated into neuronal phenotype, and in myeloma cells. Thus, anandamide and noladin ether protected the cells by a CB_1 and a mixed CB_1/CB_2 receptor-mediated effect, respectively, depending on the cell line used (Milton 2002). These results obtained with cell lines were not reproduced in our neuronal cultures. Thus, different synthetic cannabinoids (HU-210, WIN55212-2

Table 1 Effects of cannabinoids on the cellular/molecular actions of β-amyloid

Peptide	Model	Drug	Effect	Reference
$Aβ_{1-40}$ (fib)	primary microglia	WIN, HU-210, JWH133	↓TNF-α	Ramírez et al., 2005
	microglia/neurons	WIN, JWH133	↓neurotoxicity	
$Aβ_{1-42}$ (fib)	N9 + CD40L	JWH015	↓NO_2^- ↓TNF-α	Erhart et al., 2005
$Aβ_{1-42}$ (sol)	C6 C6/PC12	ACEA, WIN	↓NO_2^-,↓iNOS ↓phospho tau	Esposito et al., 2006c
$Aβ_{1-42}$ (sol)	PC12	CBD	↓GSK3β, ↑β-catenin ↓phospho tau	Esposito et al., 2006a
$Aβ_{1-42}$ (sol)	PC12	CBD	↓NO_2^-, ↓iNOS, ↓p38 MAPK, ↓NFκB	Esposito et al., 2006b
Aβ	AchE + THC		↓aggregation	Eubanks et al., 2006

AChE acetilcholinesterase; *fib* fibrillar; *GSK3β* glycogen synthase kinase 3β; *sol* soluble; *THC* $Δ^9$-THC; *TNF-α* tumour necrosis factor-α; *WIN* WIN55212-2

or JWH133) did not prevent direct toxicity of Aβ on primary cortical neurons in culture (Ramírez et al., 2005). Cannabinoids modulate glial activity; in particular, they diminish the reactive phenotype of astrocytes and microglial cells (see Chap. 16). The anti-inflammatory properties of cannabinoids can be observed in Aβ-stimulated microglial cells in culture (Table 1). Aβ challenge induced a morphological activated phenotype and an increase in mitochondrial respiration which was blocked by HU-210. The increased release of the cytokine TNF-α following Aβ addition to the microglia cultures, both at 4 h and at 24 h, was prevented by HU-210, and the effect was mimicked by WIN55212-2, devoid of antioxidant properties (Marsicano et al., 2002), and by the CB_2 selective agonist JWH133 (Ramírez et al., 2005). These results have been confirmed and extended in primary microglial cell cultures and in the N9 microglial cell line (Ehrhart et al., 2006). Indeed, the CB_2 receptor-selective agonist JWH015 reduced microglial TNF-α and nitrite generation induced either by IFN-γ or Aβ peptide challenge in the presence of CD40 ligation. As previously mentioned, in our work we found no protection when Aβ toxicity was studied either on neurons or astrocytes in culture. However a significant neuroprotection was observed when we investigated the effects of cannabinoid on the neurotoxicity mediated by Aβ microglial activation. Indeed, co-addition of either WIN55212-2 or JWH133 with Aβ to microglial cells cultured on inserts, to avoid the peptide direct toxicity to neurons, prevented neuron cell death, and the effect was mediated by CB_1 and/or CB_2 receptors (Ramírez et al., 2005; Table 1). Aβ-induced iNOS expression and nitrite generation by C6 glioma cells in culture has been found to be counteracted by the synthetic cannabinoid WIN55212-2 and the endocannabinoid analogue ACEA in a dose-dependent manner (Esposito et al., 2006c). CB_1 receptors appear to be responsible of the effects, since the CB_2 selective agonist JWH015 was ineffective and the CB_1 selective antagonist rimonabant blocked the ACEA action. Importantly, Aβ-induced tau hyperphospholyration – one of the hallmarks of AD – in differentiated PC12 co-cultured with C6 cells, and this change was prevented by ACEA but not JWH015 treatment (Esposito et al., 2006c). Therefore, the inflammatory response consequence of Aβ stimulation of C6 cells is effectively counteracted by cannabinoids being beneficial to neuron-like cells. Interesting results have been reported with the non-psychotropic constituent of marijuana cannabidiol (Table 1), that has no or very low affinity for CB_1 or CB_2 receptors, but that shows prominent antioxidant properties. In fact this cannabinoid molecule showed antiapoptotic activity in cultured PC12 cells when added prior Aβ exposure and significantly elevated cell survival while it decreased ROS production, lipid peroxidation, caspase 3 levels, DNA fragmentation and intracellular calcium (Iuvone et al., 2004). The inhibition of nitrite generation and iNOS expression induced by Aβ in PC12 cells was accompanied by a reduction of p38 MAP kinase and in the redox-sensitive transcription factor NFκB (Esposito et al., 2006b). The drug also effectively counteracted the increase in phosphorylation of tau and GSK3β and the decrease in expression of β-catenin observed upon Aβ challenge of PC12 cells (Esposito et al., 2006a). Taken together, these results suggest that cannabidiol, which has anti-inflammatory and antioxidant properties (Hampson et al., 1998), may represent an interesting compound that blocks Aβ toxicity and the molecular pathways leading to it. Cannabinoids are also able to prevent different biochemical and behavioural changes observed in the rat model of AD following its administration in vivo. This model recapitulates many of the pathological and neurochemical alterations found in AD brain. Interestingly, a reduction in CB_1 receptor expression similar to that in AD was also found. In rats repeatedly injected with Aβ for 7 days, WIN55212-2 prevented the microglial activation observed at day 8 and the astrogliosis existing at day 15. More importantly, the cognitive impairment and the loss of neuronal markers at 2 months following Aβ administration were counteracted by cannabinoid treatment (Ramírez et al., 2005). Van der Stelt and colleagues (2006) reported an increase in the hippocampus in 2-AG levels, but not anandamide, at 12 days after a single cortical injection of Aβ, concomitant with a reduction in neuronal markers and an increase in markers of gliosis. Increased endocannabinoid levels in this condition resembles that found after other kind of brain injuries (Panikashvili et al., 2001), and may subserve a neuroprotective role. VDM11, an inhibitor of endocannabinoid reuptake, enhanced rat hippocampal and mouse brain endocannabinoid levels when administered subchronically, starting either at 3 or 7 days after Aβ administration until the 12th to 14th day. The drug restored neuronal or inflammatory markers to basal levels and the loss of memory retention in the passive avoidance task in mice, but only when administered at the third day after Aβ injection (van der Stelt et al., 2006). Taken together, the results available so far in the rodent models of AD (e.g. rats or mice injected with Aβ either intraventricularly or focally) suggest an involvement of the cannabinoid system in the effects of Aβ (decreased CB_1 protein expression, increased endocannabinoid levels) and a beneficial action of both endocannabinoids (brought about by reuptake inhibition) or synthetic molecules on the AD-like pathology and symptoms of the animals. The amelioration of the learning and memory impairment by cannabinoids is of note, and appears to contradict the well-known memory-disrupting effects of acute administration of this type of molecules (see Chap. 22). In fact, CB_1 antagonists show memory-enhancing effects in different paradigms and it has been reported that rimonabant administered before the retention trial counteracted the amnesia induced by Aβ injected 7 days prior to the learning trial (Mazzola et al., 2003), but ineffective when given at other time points. However, cannabinoids may not be detrimental for memory processes when administered chronically, given the tolerance to the effects upon repeated exposure to cannabinoids, including their memory-disruptive effects (Hampson et al., 2003).

Concluding Remarks

Cannabinoids have several pharmacologic properties that make them suitable for the treatment of AD, aimed either at preventing or slowing the progress of AD. Indeed some of them have antioxidant capacity, and oxidative stress is an early process induced by Aβ in different experimental conditions in vitro and in vivo, which also exists in AD. Another interesting feature is their anti-inflammatory actions (Klein, 2005; Stella, 2004), which may reduce the ongoing inflammatory process in the neurological disease considered to be a consequence of glial activation, preventing its negative effects on neuronal processes. In this regard, CB_2 receptor-selective agonists may be a better choice compared with mixed cannabinoids given that the former are devoid of their psychoactivity (Hanus et al., 1999; Sánchez et al., 2001) induced by CB_1 receptor activation, which may cast concern when translating this kind of therapy to the clinic. Nevertheless, the modulator effects of cannabinoids on immune function may be a double-edged sword that should be taken into account. However, it would be possible to obtain neuroprotection with no accompanying psychoactivity by using low doses of mixed cannabinoids or CB_1 selective agonists, an issue that should be addressed as well. In addition, Δ^9-THC has been recently shown to competitively inhibit the activity of AChE in vitro as well as preventing AChE-induced Aβ aggregation (Eubanks et al., 2006) and both mechanisms would positively impact on the progression of AD. This mechanism is independent of cannabinoid receptors and relies on the chemical structure of the molecule, as revealed by computational modelling of the Δ^9-THC–AChE interaction, but may be well shared by other cannabinoids with a similar chemical structure. Finally, cannabinoids may be useful for the management of anorexia and disturbed behaviour, major problems in advanced AD, as shown in a clinical trial report (Volicer et al., 1997). In summary, cannabinoids are showing very interesting effects on AD-like pathology in vivo and their effects in vitro are beginning to delineate the mechanism that underlie the beneficial effects of those drugs (de Ceballos and Guzmán, 2005). Several cellular targets appear to be involved in the actions of cannabinoids against Aβ neurotoxicity, namely microglial cells, astrocytes and the neurons itself. Undoubtedly, present and future studies will contribute to the possible endorsement of cannabinoids in the treatment of AD, a devastating disease.

Acknowledgements Dr. B.G. Ramírez is acknowledged for her collaboration and M.E. Fernández de Molina for editorial help. The work of María L. de Ceballos is funded by the Ministerio de Educación y Ciencia (SAF2002–01566 and SAF2005–02845) and Comunidad de Madrid (S-BIO/0170/2006).

References

Akiyama H, Barg, S, Barnum S, Bradt B, Bauer J, Cole GM, Cooper NR, Eikelenboom P, Emmerling M, Fiebich BL, Finch CE, Frautschy SA, Griffin WST, Hampel H, Hull M, Landreth G, Lue L-F, Mrak R, Mackenzie IR, McGeer PL, O'Banion K, Pachter J, Pasinetti G, Plata-Salaman C, Rogers J, Rydell R, Shen Y, Sreit W, Strohmeyer R, Tooyoma I, Van Muiswinkel FL, Veerhuis R, Walker D, Webster S, Wegrzyniak B, Wenk G, Wyss-Coray T (2000) Inflammation and Alzheimer's disease. Neurobiol Aging 21:383–421.

Aoyama K, Matsubara K, Fujikawa Y, Nagahiro Y, Shimizu K, Umegae N, Hayase N, Shiono H, Kobayashi S (2000) Nitration of manganese superoxide dismutase in cerebrospinal fluids is a marker for peroxynitrite-mediated oxidative stress in neurodegenerative diseases. Ann Neurol 47:524–527.

Ashton JC, Friberg D, Darlington CL, Smith PF (2006) Expression of the cannabinoid CB_2 receptor in the rat cerebellum: an immunohistochemical study. Neurosci Lett 396:113–116.
Bard F, Cannon C, Barbour R, Burke R-L, Games D, Grajeda H, Guido T, Hu K, Huang J, Johnson-Wood K, Khan K, Kholodenko D, Lee M, Lieberbug I, Motter R, Nguyen M, Soriano F, Vasquez N, Weiss K, Welch B, Seubert P, Schenk D, Yednock T (2000) Peripherally administered antibodies against amyloid β-peptide enter the central nervous system and reduce pathology in a mouse model of Alzheimer disease. Nat Med 6:916–919.
Behl C, Davis JB, Lesley R, Schubert D (1994) Hydrogen peroxide mediates amyloid β protein toxicity. Cell 77:817–827.
Benito C, Núñez E, Tolón RM, Carrier EJ, Rábano A, Hillard CJ, Romero J (2003) Cannabinoid CB_2 receptors and fatty acid amide hydrolase are selectively overexpressed in neuritic plaque-associated glia in Alzheimer's disease brains. J Neurosci 23:11136–11141.
Benson A (2005) Alzheimer's disease: a tangled issue. Drug Discov Today 10:749–751.
Broe GA, Grayson DA, Creasey HM, Waite LM, Casey BJ, Bennett HP, Brooks WS, Halliday GM (2000) Anti-inflammatory drugs protect against Alzheimer disease at low doses. Arch Neurol 57:1586–1591.
Butterfield DA, Castegna A, Lauderback CM, Drake J (2002) Evidence that amyloid beta-peptide-induced lipid peroxidation and its sequelae in Alzheimer's disease brain contributes to neuronal death. Neurobiol Aging 23:655–664.
Chen G, Chen KS, Knox J, Inglis J, Bernard A, Martin SJ, Justice A, McConlogue L, Games D, Freedman SB, Morris RG (2000) A learning deficit related to age and beta-amyloid plaques in a mouse model of Alzheimer's disease. Nature 408:975–979.
Combs CK, Johnson DE, Cannady SB, Lehman TM, Landreth GE (1999) Identification of microglial signal transduction pathways mediating a neurotoxic response to amyloidogenic fragments of beta-amyloid and prion proteins. J Neurosci 19:928–939.
Combs CK, Karlo JC, Kao SC, Landreth GE (2001) Beta-amyloid stimulation of microglia and monocytes results in TNFalpha-dependent expression of inducible nitric oxide synthase and neuronal apoptosis. J Neurosci 21:1179–1188.
de Ceballos ML, Guzmán M (2005) The role of cannabinois in preventing the neurodegenerative process occurring in Alzheimer's disease. Drugs Fut 30:807–814.
Delobette S, Privat A, Maurice T (1997) *In vitro* aggregation facilitates β-amyloid peptide-(25–35)-induced amnesia in the rat. Eur J Pharmacol 319:1–4.
Dickson DW, Farlo J, Davies P, Crystal H, Fuld P, Yen S-H (1988) Alzheimer's disease: a double-labeling immunohistochemical study of senile plaques. Am J Pathol 132:86–101.
Ehrhart J, Obregón D, Mori T, Hou H, Sun N, Bai Y, Klein T, Fernández F, Tan J (2005) Stimulation of cannabinoid receptor 2 (CB_2) supresses microglial activation. J Neuroinflam 2:1–13.
Esposito G, De Filipis D, Carnuccio R, Izzo AA, Iuvone T (2006a) The marijuana component cannabidiol inhibits β-amyloid-induced tau protein hyperphosphorylation through Wnt/β-catenin pathway rescue PC12 cells. J Mol Med 84:253–258.
Esposito G, De Filipis D, Mauiuri MC, De Stefano D, Carnuccio R, Iuvone T (2006b) Cannabidiol inhibits nitric oxide synthase protein expression and nitric oxide production in β-amyloid stimulated PC12 neurons through p38 MAP kinase and NFκB involvement. Neurosci Lett 399:91–95.
Esposito G, De Filipis D, Steardo L, Scuderi C, Savani C, Cuomo V, Iuvone T (2006c) CB_1 receptor selective activation inhibits β-amyloid-induced iNOS protein expression in C6 cells and subsequently blunts tau protein hyperphosphorylation in co-cultured neurons. Neurosci Lett 404:342–346.
Eubanks LM, Rogers CJ, Beuscher IV AE, Koob GF, Olson AJ, Dickerson TJ, Janda KD (2006) A molecular link between the active component of marijuana and Alzheimer's disease pathology. Mol Pharm 3:773–777.
Frautschy SA, Baird A, Cole GM (1991) Effects of injected Alzheimer beta-amyloid cores in rat brain. Proc Natl Acad Sci USA 88:8362–8366.
Frautschy SA, Cole GM, Baird A (1998) Microglial response to amyloid plaques in APPsw transgenic mice. Am J Pathol 152:307–317.

Games D, Adams D, Alessandrini R, Barbour R, Berthelette P, Blackwell C, Carr T, Clemens J, Donaldson T, Gillespie F (1995) Alzheimer-type neuropathology in transgenic mice overexpressing V717F beta-amyloid precursor protein. Nature 373:523–527.

Goate A, Chartier-Harlin MC, Mullan M, Brown J, Crawford F, Fidani L, Giuffra L, Haynes A, Irving N, James L, Mant R, Newton P, Rooke K, Roques P, Talbot C, Pericak-Vance M, Roses A, Williamson R, Rossor M, Owen M, Hardy J (1991) Segregation of a missense mutation in the amyloid precursor protein gene with familial Alzheimer's disease. Nature 349:704–706.

Gong JP, Onaivi ES, Ishiguro H, Liu QR, Tagliaferro PA, Brusco A, Uhl GR (2006) Cannabinoid CB_2 receptors: immunohistochemical localization in rat brain. Brain Res 1071:10–23.

Haga S, Akai K, Ishii T (1989) Demonstration of microglial cells in and around senile (neuritic) plaques in the Alzheimer brain. An immunohistochemical study using a novel monoclonal antibody. Acta Neuropathol 77:569–575.

Hampson AJ, Grimaldi M, Axelrod J, Wink D (1998) Cannabidiol and (-)Delta9-tetrahydrocannabinol are neuroprotective antioxidants. Proc Natl Acad Sci USA 95:8268–8273.

Hampson RE, Simeral JD, Kelly EJ (2003) Tolerance to the memory disruptive effects of cannabinoids involves adaptation by hippocampal neurons. Hippocampus 13:543–556.

Hanus L, Breuer A, Tchilibon S, Shiloah S, Goldenberg D, Horowitz M, Pertwee RG, Ross RA, Mechoulam R, Fride E (1999) HU-308: a specific agonist for CB_2, a peripheral cannabinoid receptor. Proc Natl Acad Sci USA 96:14228–14233.

Holcomb L, Gordon MN, McGowan MC, Yu X, Benkovic S, Jantzen P, Wright K, Saad I, Mueller R, Morgan D, Sanders S, Zehr C, O'Campo K, Hardy J, Prada C, Eckman C, Younkin S, Hsiao K, Duff K (1998) Accelerated Alzheimer-type phenotype in transgenic mice carrying both mutant amyloid precursor protein and presenilin 1 transgenes. Nat Med 4:97–100.

Iuvone T, Esposito G, Esposito R, Santamaria R, Di Rossa M, Izzo AA (2004) Neuroprotective effects of cannabidiol, a non-psychoactive component from Cannabis sativa, on β-amyloid-induced toxicity in PC12 cells. J Neurochem 89:134–141.

Jantzen PT, Connor KE, DiCarlo G, Wenk GL, Wallace GL, Rojiani AM, Coppola D, Morgan D, Gordo MN (2002) Microglial activation and β-amyloid deposit reduction caused by nitric oxide-releasing nonsteroidal anti-inflammatory drug in amyloid precursor protein plus presenilin-1 transgenic mice. J Neurosci 22:2246–2254.

Janus C, Pearson J, McLaurin J, Mathews PM, Jiang Y, Schmidt SD, Chishti MA, Horne P, Heslin D, French J, Mount HT, Nixon RA, Mercken M, Bergeron C, Fraser PE, St George-Hyslop P, Westaway D (2000) Abeta peptide immunization reduces behavioural impairment and plaques in a model of Alzheimer's disease. Nature 408:979–982.

Klein TW (2005) Cannabinoid-based drugs and anti-inflammatory therapeutics. Nat Rev Immunol 5:400–411.

Kobayashi DM, Chen KS (2005) Behavioural phenotypes of amyloid-based genetically modified mouse models of Alzheimer's disease. Genes Brain Behav 4:173–196.

Kowall NW, Beal MF, Busciglio J, Duffy LK, Yankner BA (1991) An *in vivo* model for the neurodegenerative effects of beta amyloid and protection by substance P. Proc Natl Acad Sci USA 88:7247–7251.

LaFerla FM, Oddo S (2005) Alzheimer's disease: Abeta, tau and synaptic dysfunction. Trends Mol Med 11:170–176.

Levy-Lahad E, Wijsman EM, Nemens E, Anderson L, Goddard KA, Weber JL, Bird TD, Schellenberg GD (1995) A familial Alzheimer's disease locus on chromosome 1. Science 269:970–973.

Marsicano G, Moosmann B, Hermann HM, Lutz B, Behl C (2002) Neuroprotective properties of cannabinoids against oxidative stress: role of the cannabinoid receptor CB_1. J Neurochem 80:448–456.

Martin JB (1999) Molecular basis of the neurodegenerative disorders. N Engl J Med 340:1970–1980.

Mattson MP (2004) Pathways towards and away from Alzheimer's disease. Nature 5:631–639.

Mattson MP, Barger SW, Cheng B, Lieberburg I, Smith-Swintosky VL, Rydel RE (1993) Beta-amyloid precursor protein metabolites and loss of neuronal Ca2+ homeostasis in Alzheimer's disease. Trends Neurosci 16:409–414.

Mattson MP, La Ferla FM, Chan SL, Leissring MA, Shepel PN, Geiger JD (2000) Calcium signaling in the ER: its role in neuronal plasticity and neurodegenerative disorders. Trends Neurosci 23:222–229.
Mazzola C, Micale V, Drago F (2003) Amnesia induced by β-amyloid fragments is counteracted by cannabinoid CB_1 receptor blockade. Eur J Pharmacol 477:219–225.
McGeer PL, Itagak S, Tago H, McGeer EG (1987) Reactive microglia in patients with senile dementia of the Alzheimer type are positive for the histocompatibility glycoprotein HLA-DR. Neurosci Lett 79:195–200.
Michaelis ML (2003) Drugs targeting Alzheimer's disease: some things old and some things new. J Pharmacol Exp Ther 304:897–904.
Milton NGN (2002) Anandamide and noladin ether prevent neurotoxicity of the human amyloid-β peptide. Neurosci Lett 332:127–130.
Müller U, Cristina N, Li Z-W, Wolfer DP, Lipp H-P, Rulicke T, Brandner S, Aguzzi T, Weissmann C (1994) Behavioral and anatomical deficits in mice homozygous for a modified β-amyloid precursor protein. Cell 79:755–765.
Ong WY, Mackie K (1999) A light and electron microscopic study of the CB_1 cannabinoid receptor in primate brain. Neuroscience 92:1177–1191.
Panikashvili D, Simeonidou C, Ben-Shabat S, Hanus L, Breue A, Mechoulam R, Shohami E (2001) An endogenous cannabinoid (2-AG) is neuroprotective after brain injury. Nature 413:527–531.
Ramírez BG, Blázquez C, Gómez del Pulgar T, Guzmán M, de Ceballos ML (2005) Prevention of Alzheimer's disease pathology by cannabinoids: neuroprotection mediated by blockade of microglial activation. J Neurosci 25:1904–1913.
Sánchez C, de Ceballos ML, Gómez del Pulgar T, Rueda D, Corbacho C, Velasco G, Galve-Roperh I, Huffman JW, Ramón y Cajal S, Guzmán M (2001) Inhibition of glioma growth *in vivo* by selective activation of the CB_2 cannabinoid receptor. Cancer Res 61:5784–5789.
Scorer CA (2001) Preclinical and clinical challenges in the development of disease-modifying therapies for Alzheimer's disease. Drug Discov Today 6:1207–1219.
Sherrington R, Rogaev EI, Liang Y, Rogaeva EA, Levesque G, Ikeda M, Chi H, Lin C, Li G, Holman K, et al. (1995) Cloning of a gene bearing missense mutations in early-onset familial Alzheimer's disease. Nature 375:754–760.
Stella N (2004) Cannabinoid signalling in glial cells. Glia 48:267–277.
Tan J, Town T, Mori T, Saxe Y, Crawford F, Mullan M (2000) CD45 opposes β-amyloid peptide-induced microglial activation via inhibition of p44/42 mitogen-activated protein kinase. J Neurosci 20:7587–7594.
van der Stelt M, Mazzola C, Esposito G, Matias I, Petrosino S, De Filipis D, Micale V, Steardo L, Drago F, Iuvone T, Di Marzo V (2006) Endocannabinoids and β-amyloid-induced nerurotoxicity *in vivo*: effect of pharmacological elevation of endocannabinoid levels. Cell Mol Life Sci 63:1410–1424.
Van Sickle MD, Duncan M, Kingsley PJ, Mouihate A, Urbani P, Mackie K, Stella N, Makriyannis A, Piomelli D, Davison JS, Marnett LJ, Di Marzo V, Pittman QJ, Patel KD, Sharkey KA (2005) Identification and functional characterization of brainstem cannabinoid CB_2 receptors. Science 310:329–332.
Volicer L, Stelly M, Morris J, McLaughlin J, Volicer BJ (1997) Effects of dronabinol on anorexia and disturbed behaviour in patients with Alzheimer's disease. Int J Geriatr Psychiatry 12:913–919.
Westlake TM, Howlett AC, Bonner TI, Matsuda LA, Herkenham M (1994) Cannabinoid receptor binding and messenger RNA expression in human brain: an *in vitro* receptor autoradiography and *in situ* hybridization histochemistry study of normal aged and Alzheimer's brains. Neuroscience 63:637–652.
Xie Z, Wei M, Morgan TE, Fabrizio P, Han D, Finch CE, Longo VD (2002) Peroxynitrite mediates neurotoxicity of amyloid β-peptide_{1-42} and lipopolysaccharide-activated microglia. J Neurosci 22:3484–3492.
Yankner BA (1996) Mechanisms of neuronal degeneration in Alzheimer's disease. Neuron 16:921–932.

Chapter 20
The Endocannabinoid System as a Therapeutic Target in Epilepsy

Krisztina Monory and Beat Lutz

Abstract A tightly regulated balance between excitatory and inhibitory neurotransmission is required for proper functioning of the brain in the long term. Although a strong excitatory drive is necessary for processes such as learning and memory, exaggerated levels of excitation may lead to the neuronal system getting out of control, possibly leading to pathological processes. They may range from epileptiform seizures to neurodegenerative disorders, finally resulting in a massive neuronal cell death. The present chapter will discuss the function of the endocannabinoid system in the maintenance of the balance between excitation and inhibition in the brain, and its possible therapeutic exploitation as a target against epilepsy.

Introduction

In the forebrain, large interconnected neuronal networks are able to generate synchronized activities. However, if excessive synchronized bursts of activities take place in cortical, hippocampal and thalamocortical networks, respectively, epileptiform seizures may occur, with possible long-lasting changes of network activities and the development of a state of hyperexcitability, a pathology called epilepsy (McCormick and Contreras, 2001). Epilepsy syndromes can be classified into two major categories, regarding the sites of hyperexcitability. Partial seizures occur within localized forebrain regions, while generalized seizures happen in the entire forebrain. Generalized seizures fall into two categories: A patient with a *grand mal* seizure becomes unconscious and experiences strong muscular convulsions all over the body, while a *petit mal* seizure (also called absence seizure) is much milder and leads only to a temporary loss in consciousness, often lasting only for a few seconds. But in any case, regardless whether the lapse of consciousness is long or only a few seconds, such seizures can be life threatening in particular circumstances. Epilepsy is a major health burden, affecting about 1% of the human population, with a cumulative lifetime incidence of up to 3%. The incidence is particularly high during the first years of life and in elderly persons (LaRoche and Helmers, 2004). There are more than

A. Köfalvi (ed.), *Cannabinoids and the Brain.*

40 recognized types of epileptic syndromes (McCormick and Contreras, 2001). Although several drugs are available, the current medical treatments are not fully satisfactory, as the drugs are ineffective or not fully effective in a considerable number of patients (up to 30–40%). This problem may be due to the fact that epilepsy constitutes a complex disease of multiple origins. More than 70 epilepsy susceptibility genes have been found in humans to date (Noebels, 2003), and numerous small nucleotide polymorphisms (SNPs) have been associated with epilepsy (Clancy and Kass, 2003). Therefore, the quest for novel therapeutic targets is an important line of research.

Cannabis and Epilepsy

The use of *Cannabis* as a possible treatment of a variety of diseases has a long history, dating back about 5,000 years (Iversen, 2000; Mechoulam, 1986; see Chap. 1). Since the identification of the psychoactive compound in *Cannabis sativa* L., Δ^9-tetrahydrocannabinol (Δ^9-THC) (Gaoni and Mechoulam, 1964), and the discovery of the endocannabinoid system (Matsuda et al., 1990; Piomelli, 2003), the mechanisms underlying the reported effects in numerous *Cannabis* applications have been started to be elucidated (Di Marzo et al., 2004; Mackie, 2006). However, the possible beneficial effects of *Cannabis* extracts or of distinct cannabinoid compounds, including Δ^9-THC and the non-psychoactive cannabinoid cannabidiol, in the treatment against epilepsy have not yet been validated by well designed clinical trials that are of a quality which would meet the current standard. In addition, few patients' reports are available, and altogether, the data available to date is insufficient to make firm conclusions (Keeler and Reifler, 1967; Consroe et al., 1975; Cunha et al., 1980; Ames and Cridland, 1986; Ellison et al., 1990; Ng et al., 1990; Gordon and Devinsky, 2001; Schwenkreis and Tegenthoff, 2003; Lutz, 2004; Pertwee, 2004). Nevertheless, it is important to note that the reported cases and the small clinical trials hint to a possible therapeutic application of *Cannabis* and its derivatives in the treatment against epilepsy. The most disturbing observation is that cannabinoids may be protective against seizures or they may worsen seizures, and in many cases they had no effects. This is true both for humans with epilepsy, but also in animal models, in particular in rodent models of epilepsy. Therefore, further research aimed at understanding this dichotomy of the action of cannabinoids. In a first step, it is important to understand the protective mechanisms of the endocannabinoid system in the control of excitability and epileptiform seizures in animal models. It is hoped that these investigations are able to lead to a more rational design of strategies using the endocannabinoid system as a therapeutic target in appropriate animal models of human epilepsy and, afterwards, in well designed clinical studies in humans. Nevertheless, clearly, it is still a long path to reach this goal. In this chapter, it is discussed how the endocannabinoid system protects the brain from excessive neuronal activities and epileptic seizures.

The Endocannabinoid System in the Control of Neuronal Excitability and Epileptiform Seizures

The Control of Neuronal Excitability

In neurons, the typical intracellular effects after binding of agonists to CB_1 receptors are (1) inhibition of adenylyl cyclase, leading to decreased levels of intracellular cAMP, (2) stimulation of potassium channels (A-type and inwardly rectifying potassium channels), leading to an increased efflux of potassium ions, and (3) inhibition of voltage-dependent calcium channels (N- and P/Q-type), leading to a decreased calcium ion influx. Collectively, CB_1 receptor agonists render neurons less excitable. At the level of synaptic transmission, the activation of CB_1 receptors at the presynaptic site leads to a transient decrease of neurotransmitter release from the presynaptic site (Chevaleyre et al., 2006). Thus, based on these general characteristics of CB_1 receptor signalling, it becomes evident that the type of synapse where CB_1 receptors are expressed and activated is an extremely important aspect to consider. In fact, the dichotomy of cannabinoid effects on seizure frequency has its counterpart at the level of expression sites. It has clearly been established that CB_1 receptors are present both on inhibitory synapses (i.e. γ-aminobutyric acid-containing, GABAergic terminals) (Freund et al., 2003) as well as on excitatory synapses (i.e. glutamate-containing terminals) (Marsicano et al., 2003; Piomelli, 2003; Chevaleyre et al., 2006; Monory et al., 2006; see Chap. 10). As a stringent consequence of this observation, the activation of CB_1 receptors on glutamatergic neurons should be anti-convulsive, while the activation of CB_1 receptors on GABAergic neurons should be pro-convulsive. Within this frame, indeed, the specific loss of receptors on glutamatergic terminals was shown to lead to an increased excitability of hippocampal glutamatergic neurons when treated with the excitatory compound kainic acid, which activates glutamate receptors (Marsicano et al., 2003) (Fig. 1). Similarly, at the behavioural level, loss of CB_1 receptors on glutamatergic terminals led to a significant increase of kainic acid-induced seizures (Marsicano et al., 2003; Monory et al., 2006). Surprisingly, however, deleting CB_1 receptor from GABAergic cells did not lead to any changes in seizure behaviour (Monory et al., 2006). This, however, does not contradict the notion that CB_1 activation on GABAergic cells is pro-convulsive, as most probably, in an acute excitatory situation in vivo, these CB_1 receptors do not get activated by endocannabinoids. It was shown several years ago that application of both the hippocampal cultures with CB_1 receptor agonists leads to a decrease of Ca^{2+} spiking and neuronal death, consistent with the reduction of neuronal excitability after CB_1 receptor activation. However, sustained activation of CB_1 receptors by Δ^9-THC in autaptic hippocampal neurons, in which glutamatergic axons project back to their own dendrites, leads to an abolishment of the transient decrease of glutamatergic transmission, called depolarization-induced suppression of excitation (DSE) (Straiker and Mackie, 2007). This is important to consider when systemic application of CB_1 receptor agonists is used in the treatment against seizure. In the following paragraphs, the role of the

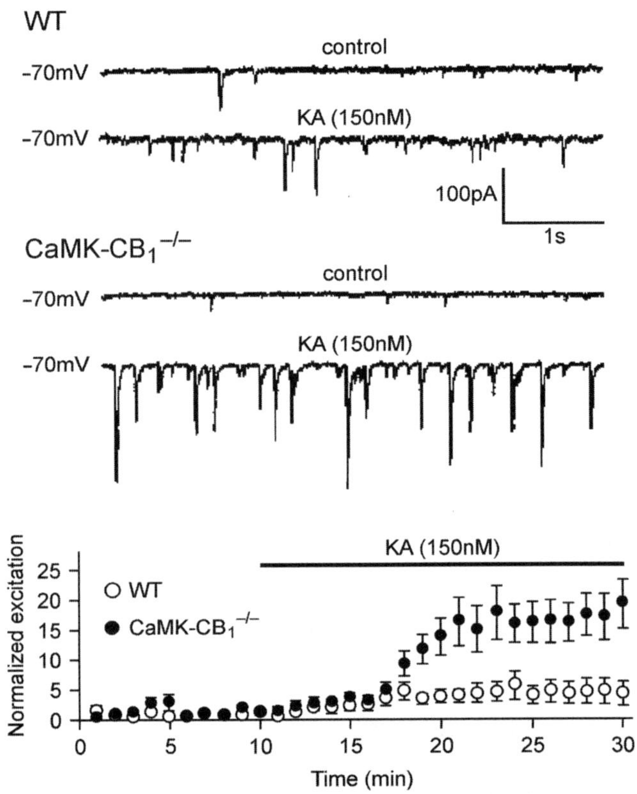

Fig. 1 CB_1 receptors mediate suppression of kainic acid (KA)-induced excitation of CA1 hippocampal glutamatergic (pyramidal) neurons. (**a**) Representative traces of pyramidal neurons in wild-type mice (WT) before (*upper part*) and after (*lower part*) KA treatment. Excitation is non-significantly changed. (**b**) Mice lacking CB_1 receptors in principal neurons of the forebrain, including hippocampal glutamatergic neurons (CaMK-$CB_1^{-/-}$), display excessive neuronal activities after KA treatment. (**c**) Normalized excitation in the course of KA treatment in the two mutant mice

endocannabinoid system in several animal models of epilepsy will be discussed. Using these model systems, it is hoped to understand the "logic" of the endocannabinoid system (Piomelli, 2003) in the protection against epileptiform seizures and to outline possible further investigations in order to finally exploit the endocannabinoid system as a therapeutic target against epilepsy in humans.

Febrile Seizure

Febrile (fever-induced) seizure is the most frequent form of seizure in childhood, affecting about 3–5% of humans between the ages of 6 months and 6 years (Shinnar and Glauser, 2002). Most febrile seizures are benign. However, one-third of them are associated with a risk to subsequent epilepsy, in particular if a prolonged duration of seizures (more than 10–20 min) had occurred (French et al., 1993). A very valuable model for this human disorder has been established by exposing early postnatal rats to an increased temperature, comparable to the human situation. Typically, at postnatal day 10, rat pups are exposed to moderately heated air for 30 min, finally leading to an increased core temperature of 41–42°C. Concomitantly, seizures occur and can be quantified. These febrile seizure events induce also long-term hyperexcitability and a decrease of seizure threshold. Studies revealed that alterations of inhibitory neurotransmission are the hallmarks for the pathological process (Chen et al., 1999). Further studies revealed changes in the endocannabinoid system as a possible mechanism underlying the observed alterations of inhibitory neurotransmission. Increased levels of CB_1 receptors were found on presynaptic terminals of GABAergic interneurons in the hippocampus, leading to an increase in CB_1 receptor-mediated suppression of GABA transmission. This was measured by using the electrophysiological paradigm of depolarization-induced suppression of inhibition (DSI) (Chen et al., 2003). Interestingly, no changes of CB_1 receptor levels in glutamatergic neurons nor changes in glutamate transmission were observed, as DSE was not altered after febrile seizure (Chen et al., 2007). Thus, altogether, these studies clearly revealed a link between the decrease in GABA transmission and the increase in excitability with the GABAergic-specific up-regulation of CB_1 receptor function. In another study, it was found that hyperthermia causes respiratory alkalosis in the immature brain, and the suppression of alkalosis by incubation with 5% CO_2 abolishes seizures immediately during hyperthermia, and, furthermore, CO_2 treatment prevents the up-regulation of CB_1 receptors in GABAergic terminals (Schuchmann et al., 2006). In an elegant set of experiments, the question was raised whether or not blocking of CB_1 receptor signalling during febrile seizure induction is able to prevent the long-term hyperexcitability (Chen et al., 2007). Indeed, application of the CB_1 receptor antagonist rimonabant during hyperthermia blocked the increase in CB_1 receptor protein levels and the concomitant enhancement of DSI. Importantly, the long-term changes in hyperexcitability were also abolished. Six weeks after hyperthermia, seizure thresholds were monitored by acute injection of kainic acid. Rimonabant treatment was able to completely abolish the increased seizure frequency as seen in untreated animals after hyperthermia. Thus, although blockade of CB_1 receptors generally leads to a hyperexcitability of neurons, in this case, the acute CB_1 receptor blockade is beneficial for the prevention of hyperexcitability in the long-term. In future studies, it will be interesting to investigate the effects of CB_1 receptor modulators in animals that have already acquired febrile seizure.

The Pilocarpine Model of Status Epilepticus and Acquired Epilepsy

Status epilepticus refers to a life threatening condition, in which the brain is in a state of persistent seizure, i.e. either in a continuous or in a recurrent seizure activity without gaining consciousness for longer than 30 min. Such a state is a major emergency and is associated with significant long-term damages of the brain or even immediate mortality (up to 10% of the patients). Furthermore, the long-term consequence of a status epilepticus is a process called epileptogenesis, finally resulting in acquired epilepsy, a state of neuronal hyperexcitablity and recurrent seizures (DeLorenzo et al., 2005). These pathological processes can be modelled in rodents. A single injection of pilocarpine, a muscarinic acetylcholine receptor agonist, into rats induces a status epilepticus, which is terminated after 30 min by injections of diazepam to enhance GABA signalling via $GABA_A$ receptor. The resulting neuronal injury in the hippocampus, as well as the persistent spontaneous recurrent seizures, resembles the human pathology (Mello et al., 1993; Wallace et al., 2003). Importantly, the quest for novel therapeutic targets in this model of epilepsy is of high relevance, as it represents a refractory epileptic condition which is not easily treated by conventional anti-convulsive drugs (Leite and Cavalheiro, 1995). After induction of the status epilepticus with pilocarpine, freely moving rats were monitored for several days to weeks by electroencephalography and video monitoring (Wallace et al., 2003). Pharmacological treatments targeting the endocannabinoid system were performed, revealing that blockade of CB_1 receptor function increased seizure frequency, while CB_1 receptor agonist treatment reduced the seizure frequency much more efficiently than the widely used anti-convulsive drugs phenobarbital, a potentiator of $GABA_A$-mediated transmission, and phenytoin, an inhibitor of sodium currents. In addition, an increase of hippocampal levels of the endocannabinoid 2-arachidonoylglycerol (2-AG) and CB_1 receptor mRNA levels were reported (Wallace et al., 2003). A further study investigated the changes of the endocannabinoid system in this model of acquired epilepsy in more detail (Falenski et al., 2007). In contrast to the changes observed in the febrile seizure model, where an up-regulation of CB_1 receptors was reported in a particular neuronal population, i.e. the GABAergic interneurons, the situation in the pilocarpine model is much more complex. CB_1 receptor immunoreactivity, CB_1 receptor agonist binding and CB_1 receptor-mediated G protein activity (by using the [^{35}S]GTPγS assay) were quantified in different subregions of the hippocampus. Both up- and down-regulation were reported. While, e.g. in stratum radiatum at the hippocampal subregion CA3, a clear up-regulation was detected, stratum pyramidale and inner molecular layer of the dentate gyrus contained lower levels of CB_1 receptor activity. Thus, a region-specific redistribution of hippocampal CB_1 receptor protein occurred. These effects are not easy to be interpreted. In particular, in this set of experiments, it was not possible to determine exactly in which neuronal populations the up- and down-regulation, respectively, occurred. This is important as CB_1 receptors are expressed both in glutamatergic and GABAergic neurons (Monory et al., 2006). It is reasonable to suggest that an up-regulation in glutamatergic neurons is anti-convulsive, while an up-regulation in GABAergic neurons is pro-convulsive. Further investigations will have to address these issues. In general, however, the question remains whether or not this redistribution of hippocampal CB_1 receptors is beneficial or harmful for brain function. If it were similar to the case of febrile seizure, the redistribution of CB_1 receptors could constitute a compensatory process which participates in the pathology observed. However, our current knowledge does not yet allow a conclusion. An in vitro model of status epilepticus was recently developed (DeLorenzo et al., 2005; Blair et al., 2006). Early postnatal rat hippocampal neurons are grown on a glial support layer. After two weeks, medium is changed to a saline buffered medium without serum. For induction of the status epilepticus, medium without Mg^{2+} was used. Addition of 1 mM of Mg^{2+} stopped the continuous epileptiform high-frequency bursts. This was typically done after 3 h. This treatment led to a model of acquired epilepsy, as characterized by spontaneous recurrent epileptiform discharges. Thus, pharmacological treatments could be tested during the time period of these spontaneous discharges, monitoring an aggravation or a mitigation of the spontaneous discharges. Altogether, these experiments were able to show that both synthetic CB_1 receptor agonist (Blair et al., 2006) and endocannabinoids (Deshpande et al., 2007b) were able to alleviate acquired epilepsy. In contrary, pharmacological blockade of CB_1 receptor function worsened the frequency of discharges (Deshpande et al., 2007c). In another set of experiments, the effects on the acute status epilepticus were investigated (Deshpande et al., 2007a). As a large fraction of the status epilepticus cases is refractory to commonly used anti-epileptic treatments, the alleviating effects of CB_1 receptor agonists on high-frequency spiking were compared with the effects evoked by the widely used benzodiazepine. Remarkably, while benzodiazepine lost its effectiveness 30 min after induction of the status epilepticus, the CB_1 receptor agonist WIN55212-2 retained its efficacy even after 2 h. Thus, this is a very promising result for the treatment of status epilepticus cases with CB_1 receptor agonists.

Acute Model of Kainic Acid-Induced Seizures

Excessive activation of glutamatergic transmission is considered as a key pathogenic event that can lead to epileptiform seizures. Hippocampal glutamatergic circuits are particularly prone to excessive excitatory activities. The model of acute kainic acid-induced seizures has been shown to be a useful experimental paradigm to understand the mechanisms underlying epileptiform seizures of temporal lobe epilepsies (Ben-Ari and Cossart, 2000), although this model represents properly only the acute phase of the seizures. Kainic acid (KA), a drug originally discovered in a red alga, is able to pass the blood–brain barrier and activates efficiently a particular subtype of glutamate receptors, the KA receptors, leading to excessive excitatory transmission in the brain. The intensity and frequency of the acutely induced seizures can be monitored and quantified over the course of about 2 h (Schauwecker and Steward, 1997). Using this model, CB_1 receptors were shown to be crucial for the protection against seizures (Marsicano et al., 2003). Deficiency of CB_1 receptors in the mouse leads to a highly increased susceptibility to KA-induced seizures as compared to wild-type littermates. The endocannabinoid system is activated during KA-induced seizures, as significantly elevated levels of anandamide (Marsicano et al., 2003; Wettschureck et al., 2006) and 2-AG (Wettschureck et al., 2006) were measured at about the time when first signs of seizure activities are observed, i.e. 20 min after KA injection. This activation requires, at least in part, the function of the heterotrimeric G proteins of the G_q/G_{11} family. Genetic deletion of the α-subunits of G_q and G_{11}, $G_{\alpha q}$ and $G_{\alpha 11}$, in forebrain glutamatergic neurons revealed that the KA-induced up-regulation of 2-AG, but not of anandamide was completely abolished (Wettschureck et al., 2006). The deficit in elevated 2-AG production after KA treatment may explain, at least in part, the increased susceptibility of these mutant mice to KA-induced seizures. In fact, pharmacological treatment of mutant mice with the endocannabinoid transport inhibitor OMDM-2 was able to rescue the phenotype of the mutant mice, and alleviated the increased seizure frequency. The elevation of endocannabinoids after KA treatment appears to be transient, as 1 h after KA injection, anandamide levels returned to basal levels (Marsicano et al., 2003). Thus, a distinct temporal course of the changes in the activity of the endocannabinoid system seems to be an important feature of this endogenous protective system. The concept of such a transient activation of the endocannabinoid system after KA treatment may explain the apparent contrasting results obtained in the analysis of mice deficient in the endocannabinoid degrading enzyme fatty acid amid hydrolase (FAAH) (Clement et al., 2003). FAAH deficient mice show a 10- to 15-fold increase in anandamide levels in the brain. However, these elevated levels do not mediate enhanced protection against KA-induced seizures. In contrary, the mutant mice displayed an increased susceptibility to KA. Thus, continuously elevated levels of endocannabinoids are apparently pro-convulsive and do not provide protection. In addition, systemic application of CB_1 receptor agonists to wild-type mice prior to KA injection were not able to protect against KA-induced seizures, at least with the agonists and doses tested (Clement et al., 2003; Krisztina Monory and Beat Lutz, unpublished results). Thus, it appears that the activation of all CB_1 receptors in the brain *before* the excitotoxic event occurs is not beneficial at all. This might be explained by the notion that CB_1 receptors are prematurely activated and are not able to perceive the endocannabinoid signalling when it is needed and acutely activated after KA-induced seizures. A similar feature is observed in the transient CB_1 receptor-mediated decrease of neurotransmitter release, both at glutamatergic and GABAergic synapses, called DSE and DSI, respectively (Chevaleyre et al., 2006). In these electrophysiological paradigms, the application of CB_1 receptor antagonists disrupts suppression of neurotransmitter release, clearly showing a CB_1 receptor-dependent mechanism. However, CB_1 receptor agonists similarly abolish these processes, suggesting that activation of CB_1 receptors prior to the induction of DSI/DSE occludes the mechanism required for the transient suppression of neurotransmission mediated by endocannabinoids. The protective effects of the endocannabinoid system are characterized not only by a temporal specificity. There is also an important specificity given by the fact that CB_1 receptors are expressed in functionally opposing neuronal subpopulations. It has been established for more than ten years that CB_1 receptors are highly abundant on terminals of GABAergic neurons (Freund et al., 2003). However, it has been a long-lasting quest for showing functional CB_1 receptor proteins on glutamatergic terminals, although there has been compelling evidence from electrophysiological experiments

(Shen et al., 1996) and in situ hybridization studies (Matsuda et al., 1993; Marsicano and Lutz, 1999). To this end, several recent studies convincingly showed the presence and the functional importance of CB_1 receptors on glutamatergic neurons. Using the Cre/loxP technique to generate mouse lines with neuron subtype-specific deficiencies of CB_1 receptors (Fig. 2), the role of this receptor in the protection from KA-induced seizures was established (Marsicano et al., 2003; Monory et al., 2006). Three different mouse lines were analysed in the KA model of epileptiform seizures. One line (named CaMK-$CB_1^{-/-}$) shows a complete loss of CB_1 receptors on all forebrain projecting neurons, including cortical and hippocampal glutamatergic neurons as well as GABAergic striatal medium spiny neurons. The second line has a specific loss of CB_1 receptors on cortical and hippocampal glutamatergic neurons (named Glu-$CB_1^{-/-}$) (Fig. 3). Finally, in the third line, CB_1 receptor expression is lost on all GABAergic neurons (named GABA-$CB_1^{-/-}$) (Fig. 3). The phenotypic analyses of the CaMK-$CB_1^{-/-}$ line gave for the first time conclusive insights into the role of CB_1 receptors on glutamatergic neurons as a "stout guard"

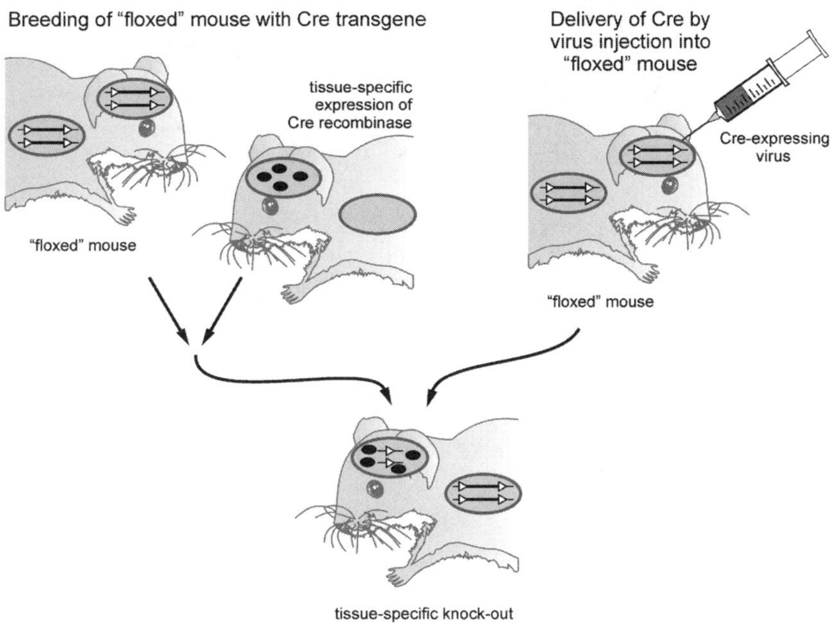

Fig. 2 Cre/loxP technique for the generation of tissue-specific gene inactivation in mouse. In the so-called floxed mouse, two loxP sites (*open arrowheads*) were introduced by homologous recombination in embryonic stem cells. The loxP sites, oriented as head-to-tail, are adjacent to the coding region of the CB_1 gene. The insertion of the loxP sites does not influence the expression of the CB_1 gene, and thus, the "floxed" mouse can be considered as a wild-type control mouse line. The loxP sites are recognized by the Cre recombinase and the sequence between the two loxP sites is deleted. The introduction of the Cre recombinase can occur by two different methods, either by crossing the floxed mouse with a transgenic mouse line expressing Cre recombinase in a tissue-specific manner, or by site-specific injection of virus expressing Cre recombinase. These methods lead to a loss of CB_1 receptors in specific neuronal subpopulation and brain regions, respectively

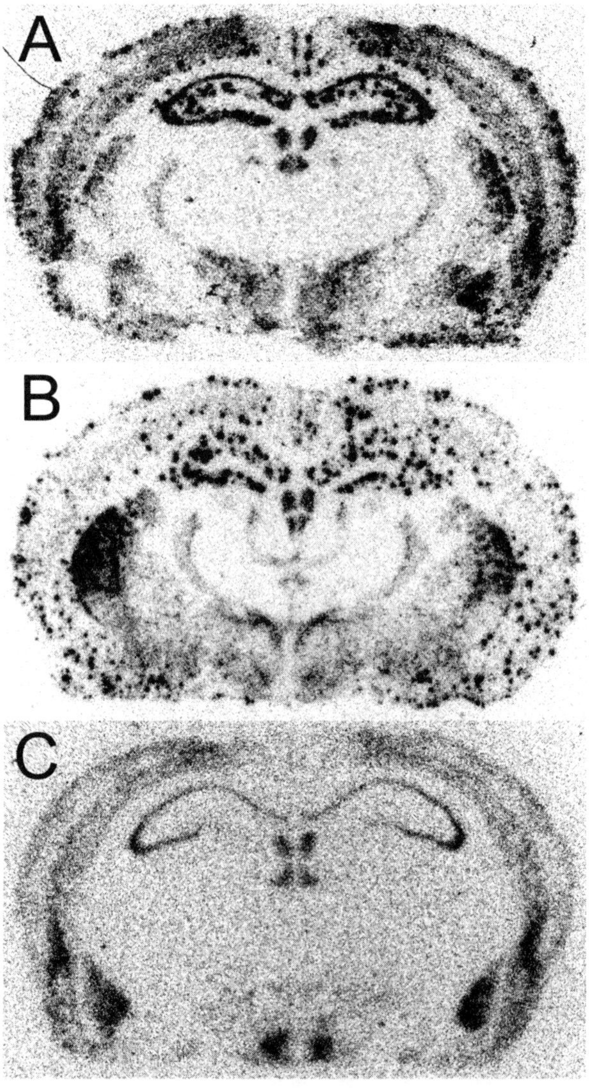

Fig. 3 Deletion of CB_1 receptors in glutamatergic or GABAergic neurons, by applying the Cre/loxP technique. Bright field micrographs of coronal sections of forebrains of wild-type (**a**), Glu-$CB_1^{-/-}$ (**b**) and GABA-$CB_1^{-/-}$ (**c**) mice showing the distribution of CB_1 mRNA, as detected by in situ hybridization with a [^{35}S]-radiolabelled riboprobe for CB_1. Note the complex expression pattern in the wild-type animal with CB_1 mRNA being present in glutamatergic neurons (low level, "blurry" expression in cortical areas) and GABAergic neurons (strong, spot-like expression in cortex and hippocampus, intense staining in striatum) as well. In Glu-$CB_1^{-/-}$ mice (**b**), only the strong spot-like expression remains in the cortex (representing GABAergic interneurons), while in GABA-$CB_1^{-/-}$ mice (**c**), the remaining, widespread but low level of cortical CB_1 expression originates from glutamatergic principal neurons

of the central nervous system (Mechoulam and Lichtman, 2003). As measured by in vitro patch clamp experiments with slices from the CaMK-$CB_1^{-/-}$ line, loss of CB_1 receptors in hippocampal glutamatergic neurons leads to a hyperexcitability of these neurons upon KA treatment (Marsicano et al., 2003) (Fig. 1). Consistently, this mutant line displayed increased susceptibility to KA-induced seizures, to a degree that is comparable with the phenotype observed in mice with a loss of CB_1 receptors in all the cells of the body (Marsicano et al., 2003). As discussed above, endocannabinoids appear to provide a protection against seizures, as their levels transiently increase upon KA treatment. Based on this concept, the pharmacological treatment of KA-injected mice with UCM707, an endocannabinoid transport inhibitor, was able to enhance this endogenous protective mechanism, by prolonging and/or intensifying the endocannabinoid signalling at the time and the location where it is needed to mediate protection. UCM707-treated animals showed a lower seizure frequency than untreated mice. This protective effect was missing in the mutant line CaMK-$CB_1^{-/-}$, consistent with the notion of the importance of CB_1 receptors on glutamatergic terminals. In this mutant line, it was also shown that well established neuroprotective signalling pathways were impaired in its induction after KA treatment. Phosphorylation of p42 and p44 extracellular signal regulated kinases (ERKs), the expression of the transcription factors c-fos and zif268, and the expression of the gene encoding brain-derived neurotrophic factor (BDNF) were not induced to the same extent as observed in wild-type littermate controls. Similar impairments in *c-fos* and zif268 gene induction were observed in mice with a forebrain-specific loss of $G_{q\alpha}$ and $G_{11\alpha}$ (Wettschureck et al., 2006). As discussed above, this mutant mouse line has also some deficits in endocannabinoid synthesis after KA injection and displayed increased susceptibility to seizures. Thus, this congruent phenotype of both mutant mouse lines suggests that *c-fos* and zif268 are important protective gene products downstream of the endocannabinoid signalling. Further experiments aimed at characterizing the neuronal cell population and brain regions which mediate the CB_1-dependent protection from KA-induced seizures (Monory et al., 2006). As CaMK-$CB_1^{-/-}$ mice lost CB_1 receptor expression also in the striatum and in subcortical forebrain regions, it was important to generate another mutant mouse line with a loss of CB_1 receptors more restricted to cortical and hippocampal glutamatergic neurons (Fig. 3). To this end, the line Glu-$CB_1^{-/-}$ was generated. Consistent with the results obtained so far, these mutant mice also displayed increased susceptibility to KA-induced seizures (Monory et al., 2006). Furthermore, it was important to investigate the role of CB_1 receptors on GABAergic neurons. Interestingly, the newly established GABA-$CB_1^{-/-}$ mice (Fig. 3) showed the same responses to KA as littermate wild-type controls. This is in contrast to the expectations, as activation of CB_1 receptors on GABAergic neurons would suppress GABA transmission and, thus, loss of CB_1 receptors on this neuronal population could, in theory, lead to an enhanced protection against KA-induced seizures. Thus, these results may suggest that in the case of a very strong excitatory burst such as a KA treatment does, CB_1 receptors on GABAergic interneurons are not proconvulsive, and the possible CB_1 receptor-mediated depression of GABA transmission is either overridden by the strong glutamatergic drive or the receptors on GABAergic

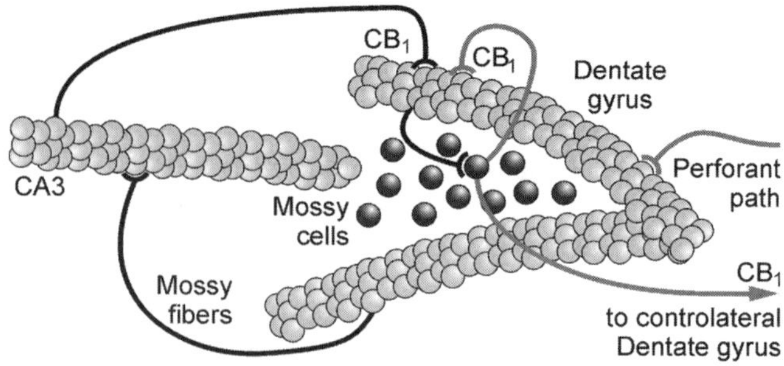

Fig. 4 Proposed circuits involved in the CB_1 receptor-mediated suppression of excitatory drive in the hippocampus. At the inner molecular layer of the dentate gyrus, CB_1 receptor-positive glutamatergic terminals synapse to granule cells of the dentate gyrus. These terminals represent projections from ipsilateral and contralateral mossy cells, located in the hilus of the dentate gyrus, and from the CA3 subregion of the hippocampus. CB_1 receptor-negative projections are from the dentate gyrus to the mossy cells and to the CA3 region. There is evidence that the mossy cells are crucially involved in the control of excitatory drive in the hippocampus. Thus, in the case of CB_1 receptor deficiency on glutamatergic terminals, including those in the inner molecular layer of the dentate gyrus, this endogenous dampening mechanism is impaired

neurons do not get activated during such an insult. In a next step, it was asked which glutamatergic neurons might be relevant for the protection. We found that the inner molecular layer of the dentate gyrus contains abundantly synaptic terminals that co-express CB_1 receptor protein with the vesicular glutamate transporter 1 ($VGlut_1$), a marker for glutamatergic terminals. The molecular layer obtains input from CA3 pyramidal neurons and from the hilus of the hippocampus, where a distinct population of glutamatergic neurons is located, namely the mossy cells. These neurons are considered as central players when excessive neuronal activities may develop to epileptiform seizures (Ratzliff et al., 2002). Further experiments substantiated the importance of CB_1 receptors in the hippocampal region. Cre recombinase-expressing virus was specifically injected into the hippocampus, leading to a hippocampal loss of CB_1 receptors. Consistently, these mice displayed increased seizure responses to KA as compared to control mice, indicating that indeed the CB_1 receptors in the hippocampus mediate the protection from seizures. The current interpretation of all these experiments is depicted in Fig. 4, suggesting a pivotal role of CB_1 receptors in mossy cells.

The Endocannabinoid System and Other Messenger Systems

Recent investigations revealed that targeting different elements of the endocannabinoid system might be a promising direction towards the development of novel therapeutic approaches for seizure-related pathologies. Thus, directly stimulating or blocking CB_1 receptor, blocking endocannabinoid re-uptake or degradation have proved to be beneficial in some or other models of epileptic disorders. Yet, an important notion is that in vivo the neuromodulatory systems always operate in an intricately complex milieu of many other neurotransmitter or neuromodulatory systems. Therefore, they may interact with each other at several levels. Such an interaction has recently been shown to occur with the action of valproate, a rather widely used anti-epileptic drug, probably acting via GABA transmission. The maximal electroshock-induced seizure threshold test was applied in mouse (Luszczki et al., 2006). With this model, the median effective dose ED_{50} was determined. The administration of the specific CB_1 receptor agonist ACEA (arachidonoyl-2′-chloroethylamide) together with the FAAH inhibitor PMSF (phenylmethylsulfonyl fluoride) enhanced the effects elicited by valproate. Thus, the EC_{50} value of valproate was shifted to a lower concentration. This is interesting, as the co-application of different drugs at lower concentrations is more effective than when applied alone. This might be beneficial regarding the reduction of the side effects. The major endocannabinoids are anandamide and 2-AG (Piomelli, 2003). However, several endocannabinoid-like compounds exist, including N-palmitoylethanolamide (PEA). A putative high affinity PEA receptor has not yet been identified (Mackie and Stella, 2006). PEA was tested in several models of seizures. PEA was shown to have potent anti-epileptic effects with EC_{50} comparable to phenytoin, in particular in pentylenetetrazol-induced seizures (Lambert et al., 2001). Another study used the model of kindled amygdaloid seizures, also concluding that PEA produces antiepileptic effects, though it does not suppress them completely (Sheerin et al., 2004).

Concluding Remarks

Recent research on the role of the endocannabinoid system regarding a protective mechanism against seizure and epilepsy in animal model systems has provided several new insights. (1) The long history of *Cannabis* extracts suggests that they are able to alleviate seizure frequency under certain conditions. However, currently, no clinical trials have substantiated these observations, and in various animal models, the effects are controversial. Therefore, it will be important to identify the active compounds in these extracts regarding the alleviation of seizures. Furthermore, clinical trials with various components from *Cannabis* extracts should be performed. (2) The mere application of CB_1 receptor agonists does not reliably alleviate seizures. The effects may depend on the seizure model and the species (e.g. rat or mouse) used. The psychotropic activity of CB_1 receptor agonists is often perceived by patients as a serious side effect. Furthermore, there is a high risk of the development of tolerance by the treatment with direct CB_1 receptor agonists. As an alternative approach, application of endocannabinoid re-uptake inhibitors appears to be promising. However, long-term treatments have not yet been performed with such substances. (3) It is important to note that, depending on the origin of the pathology, CB_1 receptors on both hippocampal glutamatergic neurons and

GABAergic interneurons are centrally involved in the state of hyperexcitability and the emergence of seizures. On glutamatergic neurons, CB_1 receptors provide protection, while it was shown that enhanced expression and activity of CB_1 receptors on GABAergic neurons are pro-convulsive. In further investigations, it will be crucial to understand the biochemical and cell biological properties of CB_1 receptors on glutamatergic and GABAergic neurons, respectively. By this approach, it is expected to find strategies to specifically target CB_1 receptors on glutamatergic neurons, thereby alleviating exaggerated glutamatergic transmission, or to inhibit CB_1 receptors on GABAergic neurons.

References

Ames FR, Cridland S (1986) Anticonvulsant effect of cannabidiol. S Afr Med J 69:14.
Ben-Ari Y, Cossart R (2000) Kainate, a double agent that generates seizures: two decades of progress. Trends Neurosci 23:580–587.
Blair RE, Deshpande LS, Sombati S, Falenski KW, Martin BR, DeLorenzo RJ (2006) Activation of the cannabinoid type-1 receptor mediates the anticonvulsant properties of cannabinoids in the hippocampal neuronal culture models of acquired epilepsy and status epilepticus. J Pharmacol Exp Ther 317:1072–1078.
Chen K, Baram TZ, Soltesz I (1999) Febrile seizures in the developing brain result in persistent modification of neuronal excitability in limbic circuits. Nat Med 5:888–894.
Chen K, Ratzliff A, Hilgenberg L, Gulyas A, Freund TF, Smith M, Dinh TP, Piomelli D, Mackie K, Soltesz I (2003) Long-term plasticity of endocannabinoid signaling induced by developmental febrile seizures. Neuron 39:599–611.
Chen K, Neu A, Howard AL, Foldy C, Echegoyen J, Hilgenberg L, Smith M, Mackie K, Soltesz I (2007) Prevention of plasticity of endocannabinoid signaling inhibits persistent limbic hyperexcitability caused by developmental seizures. J Neurosci 27:46–58.
Chevaleyre V, Takahashi KA, Castillo PE (2006) Endocannabinoid-mediated synaptic plasticity in the CNS. Annu Rev Neurosci 29:37–76.
Clancy CE, Kass RS (2003) Pharmacogenomics in the treatment of epilepsy. Pharmacogenomics 4:747–751.
Clement AB, Hawkins EG, Lichtman AH, Cravatt BF (2003) Increased seizure susceptibility and proconvulsant activity of anandamide in mice lacking fatty acid amide hydrolase. J Neurosci 23:3916–3923.
Consroe PF, Wood GC, Buchsbaum H (1975) Anticonvulsant nature of marihuana smoking. JAMA 234:306–307.
Cunha JM, Carlini EA, Pereira AE, Ramos OL, Pimentel C, Gagliardi R, Sanvito WL, Lander N, Mechoulam R (1980) Chronic administration of cannabidiol to healthy volunteers and epileptic patients. Pharmacology 21:175–185.
DeLorenzo RJ, Sun DA, Deshpande LS (2005) Cellular mechanisms underlying acquired epilepsy: the calcium hypothesis of the induction and maintenance of epilepsy. Pharmacol Ther 105:229–266.
Deshpande LS, Blair RE, Nagarkatti N, Sombati S, Martin BR, DeLorenzo RJ (2007a) Development of pharmacoresistance to benzodiazepines but not cannabinoids in the hippocampal neuronal culture model of status epilepticus. Exp Neurol 204:705–713.
Deshpande LS, Blair RE, Ziobro JM, Sombati S, Martin BR, DeLorenzo RJ (2007b) Endocannabinoids block status epilepticus in cultured hippocampal neurons. Eur J Pharmacol 558:52–59.
Deshpande LS, Sombati S, Blair RE, Carter DS, Martin BR, DeLorenzo RJ (2007c) Cannabinoid CB_1 receptor antagonists cause status epilepticus-like activity in the hippocampal neuronal culture model of acquired epilepsy. Neurosci Lett 411:11–16.

Di Marzo V, Bifulco M, De Petrocellis L (2004) The endocannabinoid system and its therapeutic exploitation. Nat Rev Drug Discov 3:771–784.

Ellison JM, Gelwan E, Ogletree J (1990) Complex partial seizure symptoms affected by marijuana abuse. J Clin Psychiatry 51:439–440.

Falenski KW, Blair RE, Sim-Selley LJ, Martin BR, DeLorenzo RJ (2007) Status epilepticus causes a long-lasting redistribution of hippocampal cannabinoid type 1 receptor expression and function in the rat pilocarpine model of acquired epilepsy. Neuroscience 146:1232–1244.

French JA, Williamson PD, Thadani VM, Darcey TM, Mattson RH, Spencer SS, Spencer DD (1993) Characteristics of medial temporal lobe epilepsy: I. Results of history and physical examination. Ann Neurol 34:774–780.

Freund TF, Katona I, Piomelli D (2003) Role of endogenous cannabinoids in synaptic signaling. Physiol Rev 83:1017–1066.

Gaoni Y, Mechoulam R (1964) Isolation, structure and partial synthesis of an active constituent of hashish. J Am Chem Soc 86:1646–1647.

Gordon E, Devinsky O (2001) Alcohol and marijuana: effects on epilepsy and use by patients with epilepsy. Epilepsia 42:1266–1272.

Iversen LL (2000) The Science of Marijuana. Oxford: Oxford University Press.

Keeler MH, Reifler CB (1967) Grand mal convulsions subsequent to marijuana use. Case report. Dis Nerv Syst 28:474–475.

Lambert DM, Vandevoorde S, Diependaele G, Govaerts SJ, Robert AR (2001) Anticonvulsant activity of N-palmitoylethanolamide, a putative endocannabinoid, in mice. Epilepsia 42:321–327.

LaRoche SM, Helmers SL (2004) The new antiepileptic drugs: scientific review. JAMA 291:605–614.

Leite JP, Cavalheiro EA (1995) Effects of conventional antiepileptic drugs in a model of spontaneous recurrent seizures in rats. Epilepsy Res 20:93–104.

Luszczki JJ, Czuczwar P, Cioczek-Czuczwar A, Czuczwar SJ (2006) Arachidonyl-2 -chloroethylamide, a highly selective cannabinoid CB_1 receptor agonist, enhances the anticonvulsant action of valproate in the mouse maximal electroshock-induced seizure model. Eur J Pharmacol 547:65–74.

Lutz B (2004) On-demand activation of the endocannabinoid system in the control of neuronal excitability and epileptiform seizures. Biochem Pharmacol 68:1691–1698.

Mackie K (2006) Cannabinoid receptors as therapeutic targets. Annu Rev Pharmacol Toxicol 46:101–122.

Mackie K, Stella N (2006) Cannabinoid receptors and endocannabinoids: evidence for new players. AAPS J 8:E298–E306.

Marsicano G, Lutz B (1999) Expression of the cannabinoid receptor CB_1 in distinct neuronal subpopulations in the adult mouse forebrain. Eur J Neurosci 11:4213–4225.

Marsicano G, Goodenough S, Monory K, Hermann H, Eder M, Cannich A, Azad SC, Cascio MG, Gutierrez SO, van der Stelt M, Lopez-Rodriguez ML, Casanova E, Schütz G, Zieglgänsberger W, Di Marzo V, Behl C, Lutz B (2003) CB_1 cannabinoid receptors and on-demand defense against excitotoxicity. Science 302:84–88.

Matsuda LA, Lolait SJ, Brownstein MJ, Young AC, Bonner TI (1990) Structure of a cannabinoid receptor and functional expression of the cloned cDNA. Nature 346:561–564.

Matsuda LA, Bonner TI, Lolait SJ (1993) Localization of cannabinoid receptor mRNA in rat brain. J Comp Neurol 327:535–550.

McCormick DA, Contreras D (2001) On the cellular and network bases of epileptic seizures. Annu Rev Physiol 63:815–846.

Mechoulam R (1986) The pharmacohistory of *Cannabis sativa*. In: Mechoulam R (ed.), Cannabinoids as Therapeutic Agents. Boca Raton, FL: CRC Press, pp. 1–19.

Mechoulam R, Lichtman AH (2003) Neuroscience. Stout guards of the central nervous system. Science 302:65–67.

Mello LE, Cavalheiro EA, Tan AM, Kupfer WR, Pretorius JK, Babb TL, Finch DM (1993) Circuit mechanisms of seizures in the pilocarpine model of chronic epilepsy: cell loss and mossy fiber sprouting. Epilepsia 34:985–995.

Monory K, Massa F, Egertova M, Eder M, Blaudzun H, Westenbroek R, Kelsch W, Jacob W, Marsch R, Ekker M, Long J, Rubenstein JL, Goebbels S, Nave KA, During M, Klugmann M, Wolfel B, Dodt HU, Zieglgansberger W, Wotjak CT, Mackie K, Elphick MR, Marsicano G, Lutz B (2006) The endocannabinoid system controls key epileptogenic circuits in the hippocampus. Neuron 51:455–466.

Ng SK, Brust JC, Hauser WA, Susser M (1990) Illicit drug use and the risk of new-onset seizures. Am J Epidemiol 132:47–57.

Noebels JL (2003) The biology of epilepsy genes. Annu Rev Neurosci 26:599–625.

Pertwee RG (2004) The pharmacology and therapeutic potential of cannabidiol. In: Di Marzo V (ed.), Cannabinoids. New York: Kluwer Academic/Plenum Publishers, pp. 32–42.

Piomelli D (2003) The molecular logic of endocannabinoid signalling. Nat Rev Neurosci 4:873–884.

Ratzliff AH, Santhakumar V, Howard A, Soltesz I (2002) Mossy cells in epilepsy: rigor mortis or vigor mortis? Trends Neurosci 25:140–144.

Schauwecker PE, Steward O (1997) Genetic determinants of susceptibility to excitotoxic cell death: implications for gene targeting approaches. Proc Natl Acad Sci USA 94:4103–4108.

Schuchmann S, Schmitz D, Rivera C, Vanhatalo S, Salmen B, Mackie K, Sipila ST, Voipio J, Kaila K (2006) Experimental febrile seizures are precipitated by a hyperthermia-induced respiratory alkalosis. Nat Med 12:817–823.

Schwenkreis P, Tegenthoff M (2003) Therapeutic use of cannabinoids in neurology. Schmerz 17:367–373.

Sheerin AH, Zhang X, Saucier DM, Corcoran ME (2004) Selective antiepileptic effects of N-palmitoylethanolamide, a putative endocannabinoid. Epilepsia 45:1184–1188.

Shen M, Piser TM, Seybold VS, Thayer SA (1996) Cannabinoid receptor agonists inhibit glutamatergic synaptic transmission in rat hippocampal cultures. J Neurosci 16:4322–4334.

Shinnar S, Glauser TA (2002) Febrile seizures. J Child Neurol 17:S44–S52.

Straiker A, Mackie K (2007) Metabotropic suppression of excitation in murine autaptic hippocampal neurons. J Physiol 578:773–785.

Wallace MJ, Blair RE, Falenski KW, Martin BR, DeLorenzo RJ (2003) The endogenous cannabinoid system regulates seizure frequency and duration in a model of temporal lobe epilepsy. J Pharmacol Exp Ther 307:129–137.

Wettschureck N, van der SM, Tsubokawa H, Krestel H, Moers A, Petrosino S, Schutz G, Di Marzo V, Offermanns S (2006) Forebrain-specific inactivation of G_q/G_{11} family G proteins results in age-dependent epilepsy and impaired endocannabinoid formation. Mol Cell Biol 26:5888–5894.

Chapter 21
The Endocannabinoid System in the Physiology and Pathology of the Basal Ganglia

Gregory L. Gerdeman and Javier Fernández-Ruiz

Abstract Multiple lines of evidence indicate a prominent role for the cannabinoid signaling system in the control of basal ganglia function, exerted by modulating the activity of various classic neurotransmitters, such as GABA, dopamine, or glutamate, that operate within this circuit. Throughout the basal ganglia, the activity-evoked release of endocannabinoids has been found to directly regulate the release and plasticity of both excitatory and inhibitory synapses. These observations, together with the demonstration that different elements (receptors, ligands, enzymes) of the cannabinoid signaling system are markedly disturbed in basal ganglia disorders, namely Parkinson's and Huntington's disease, provide support to the idea that cannabinoid-based medicines, with selectivity for different targets of the cannabinoid signaling system, might have therapeutic benefits in these disorders. These benefits would include the alleviation of specific motor symptoms but they could be also extended to a possible delay or arrest of disease progression based on the neuroprotective and/or neuroregenerative properties of certain cannabinoid compounds. In this chapter we review the anatomical, neurochemical, electrophysiological, and pharmacological bases that sustain the importance of the cannabinoid system in basal ganglia function, attempting also to present current information and future lines for research on the therapeutic potential of this system in basal ganglia disorders.

Introduction

Among the different neurobiological processes in which the cannabinoid system has been implicated, the control of movement deserves special attention. In comparison to other brain structures, the endocannabinoids and their receptors – including not only the CB_1 receptor subtype, but also the CB_2 receptor and the related vanilloid $TRPV_1$ receptor – are abundant in the basal ganglia (Mackie, 2005). In the basal ganglia, the activation or blockade of some of these receptors, in particular the CB_1 receptor, produces motor alterations (Romero et al., 2002). However, the location of these receptors at different sites in the basal ganglia circuit may sometimes produce paradoxical effects. In general, cannabinoid agonists have powerful inhibitory actions on motor activity, although there are differences in magnitude

A. Köfalvi (ed.), *Cannabinoids and the Brain.*

and duration for their effects depending on their differences in receptor affinity, potency, and/or metabolic stability (Fernández-Ruiz and González, 2005). These behavioral consequences following the activation or blockade of CB_1 receptors are certainly related to the capability of these receptors to regulate the activity of several neurotransmitters, including GABA, dopamine, and glutamate, that are importantly involved in basal ganglia function (Lupica and Riegel, 2005; Chevaleyre et al., 2006). Furthermore, recent investigations have found different elements of the cannabinoid system to be significantly altered in different basal ganglia disorders, phenomena demonstrated either in human patients or in different animal models for these diseases. This opens the possibility that cannabinoid compounds, acting selectively at key targets of this signaling system (e.g., enzymes, receptors, transporters), may be used to alleviate specific motor symptoms and/or to delay the progression of different disorders affecting the basal ganglia function, in particular Parkinson's disease (PD) or Huntington's disease (HD). The present chapter will address many lines of research evidence pertinent to this hypothesis, which establish a basis to support the future clinical application of cannabinoid-based medicines in basal ganglia disorders.

The Anatomy and Neurochemistry of the Endocannabinoid System in the Basal Ganglia

That the cannabinoid system is a key modulator of basal ganglia function was first supported by studies dealing with the anatomical identification and the biochemical quantification of the different elements of this signaling system (Romero et al., 2002; Fernández-Ruiz and González, 2005). These studies have strongly demonstrated that endocannabinoids and their receptors, in comparison with other brain structures, are significantly abundant in the basal ganglia. Among the different key proteins (receptors, transporter, enzymes) of the cannabinoid signaling system, most studies have concentrated on the expression of the CB_1 receptor, but also more recently, the functionally related $TRPV_1$ receptor (Fig. 1). Further readings can be found in Chap. 10 on the general distribution of these receptors in the basal ganglia.

Receptors for Endocannabinoids in the Basal Ganglia

In the beginning of the 1990s, several authors conducted exhaustive autoradiographic studies aimed at establishing the anatomical distribution of the CB_1 receptor in the brain of rats and other laboratory species, studies that were later followed by the characterization of this receptor type in human brains. These studies demonstrated that the basal ganglia are among the brain structures that contain highest levels of both binding (Herkenham et al., 1991a) and mRNA expression (Mailleux and Vanderhaeghen, 1992a) for the CB_1 receptor, a fact

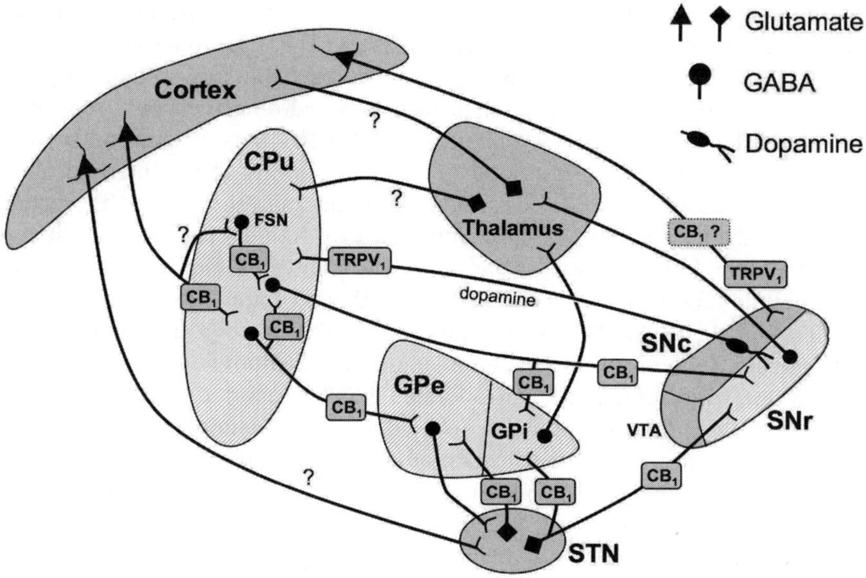

Fig. 1 Simplified circuit anatomy of the basal ganglia, highlighting presynaptic expression of CB_1 and $TRPV_1$ receptors. Neurons originating from hatched areas are GABAergic. All others are glutamatergic except the dopamine projection from substantia nigra pars compacta (SNc). CB_1 receptor function on cortico-nigral projections are postulated based on findings in the ventral tegmental area (VTA) but have not been directly shown in the SNc. Anatomical evidence for CB_1 receptors in the SNc is limited, and so the expression of DSI in this region might be due to prominent dendritic extensions into the substantia nigra pars reticulata (SNr), where CB_1 receptor-positive striatonigral terminals occur. Intracortical CB_1 receptors and other important CB_1 receptor-expressing basal ganglia inputs (such as from the amygdala or involving the VTA) are omitted for clarity, and axons labeled with a question mark are considered possible for CB1 receptor expression. Intrastriatal inhibitory circuitry is simplified; GABAergic inhibition is not specific to the direct striatonigral pathway. *CPu* caudate-putamen (striatum); *GPe* and *Gpi* globus pallidus external and internal; *STN* subthalamic nucleus; *FSN* fast-spiking interneuron

later corroborated by Tsou et al. (1998a) when the development of specific antibodies for the CB_1 receptor permitted immunohistochemical analyses. According to these studies, CB_1 receptors are abundantly distributed in different structures of the rat basal ganglia circuitry, in particular the three nuclei recipient of striatal efferent outputs (external globus pallidus (GPe), internal globus pallidus (GPi, the entopeduncular nucleus in rodents), and substantia nigra pars reticulata (SNr)), which contain high levels of receptor binding (Herkenham et al., 1991a) but do not contain mRNA for this receptor (Mailleux and Vanderhaeghen, 1992a). CB_1 receptor mRNA transcripts are, however, located in the caudate-putamen (striatum) (Mailleux and Vanderhaeghen, 1992a) suggesting that CB_1 receptors should be presynaptically located in striatal neurons projecting to the basal ganglia output nuclei. To validate this hypothesis, Herkenham et al. (1991b) conducted an interesting series of anatomical studies, using lesions of specific neuronal subpopulations in the basal ganglia with different neurotoxins and analyzing the changes in CB_1 receptor levels. These authors found an almost complete disappearance of CB_1 receptors when they lesioned striatal projection neurons with ibotenate, but not when they lesioned other neuronal subpopulations in the basal ganglia (Herkenham et al., 1991b), thus confirming the location of the CB_1 receptor in striatal projection neurons. As mentioned above, these early studies always used autoradiographic analysis of the CB_1 receptor because of the absence of available antibodies for immunolabeling. These tools were available, however, some years later and allowed a more precise analysis of the cellular and subcellular distribution of this receptor type in the basal ganglia (Tsou et al., 1998a; Marsicano and Lutz, 1999; Egertová and Elphick, 2000; Hohman and Herkenham, 2000; Julian et al., 2003; Köfalvi et al., 2005; Matyas et al., 2006), which essentially corroborated the data reported by Herkenham and coworkers. Thus, CB_1 receptors are located in striatal neurons projecting to the functional unit formed by the substantia nigra and the GPi – the so-called direct striatal efferent pathway – and also in those projecting to the GPe – the so-called indirect striatal efferent pathway (see Fig. 1). Both neuronal subpopulations use GABA as neurotransmitter and therefore express glutamic acid decarboxylase, but they can be differentiated by specific markers which are selectively coexpressed with CB_1 receptors in either striatonigral (dynorphin/substance P and D_1 receptor) or striatopallidal (enkephalin and D_2 receptors) pathways (Hohmann and Herkenham, 2000; Julian et al., 2003). Striatal projection neurons are not, however, the only neuronal subpopulations that contain CB_1 receptors in the basal ganglia. The autoradiographic studies mentioned before also allowed the identification of measurable levels of mRNA for this receptor in the subthalamic nucleus (STN), together with the absence of detectable levels of receptor binding in that structure (Mailleux and Vanderhaeghen, 1992a). This is compatible with the presence of CB_1 receptors in subthalamic neurons projecting to the SNr (and also to the GPi), preferably at axonal sites. These neurons are glutamatergic, thus reproducing the classic pattern described for the CB_1 receptor in most brain regions where it is generally located in neurons with GABAergic or glutamatergic phenotype (see Chap. 10). Motor control is therefore influenced by CB_1 receptors at both glutamatergic and GABAergic synapses intrinsic to the basal ganglia circuitry. Intrinsic striatal neurons have also been recently identified as expressing CB_1 receptors. For instance, Hohmann and Herkenham (2000) reported that certain striatal interneurons, containing somatostatin or acetylcholine (ACh), do not express CB_1 receptors (see also Uchigashima et al., 2007). However, these authors and others (Fusco et al., 2004; Uchigashima et al., 2007) demonstrated that some subclasses of striatal interneurons are CB_1 receptor-positive. These included most of the GABAergic interneurons that are labeled with parvalbumin, but also according to Fusco et al. (2004), roughly one-third of cholinergic interneurons. CB_1 receptors are not, however, expressed by nigrostriatal dopaminergic neurons, as several studies which used selective lesions of these neurons with 6-hydroxydopamine (6-OHDA) (Herkenham et al., 1991b) or double labeling (tyrosine hydroxylase vs. CB_1 receptors) methods (Julian et al.,

2003) have strongly demonstrated. According to these data, it was generally assumed that the effects of cannabinoids on dopamine transmission in the basal ganglia would be always indirect and presumably exerted through CB_1 receptors located on GABAergic and/or glutamatergic afferents to the substantia nigra pars compacta (SNc) (Romero et al., 2002; van der Stelt and Di Marzo, 2003; Fernández-Ruiz and González, 2005; and see Chap. 22). However, there is recent evidence indicating that $TRPV_1$ receptors, for which certain endocannabinoids, including anandamide (AEA), are able to act as endogenous ligands, are located in nigrostriatal dopaminergic neurons. This provides an alternative for AEA and other endocannabinoids to act through these receptors to directly control dopamine synthesis and release, as supported by the following observations: (1) $TRPV_1$ receptors colocalize with tyrosine hydroxylase in the basal ganglia (Mezey et al., 2000); (2) $TRPV_1$ receptor density is lowered by treatment with 6-OHDA which damages dopaminergic neurons (Lastres-Becker et al., 2005); (3) $TRPV_1$ receptor binding is enhanced in the striatum of mice deficient in dopamine transporter (Tzavara et al., 2006); and (4) AEA, but not Δ^9-tetrahydrocannabinol (Δ^9-THC), was found to inhibit striatal dopamine release in vitro (de Lago et al., 2004a). In contrast to some of the above observations, however, other authors reported that CB_1 and $TRPV_1$ receptors might colocalize in the caudate-putamen, the GP and the SN (Cristino et al., 2006). In addition, activation of the $TRPV_1$ receptor – a nonselective cation channel (see Chaps. 1 and 8) – may excite, rather than inhibit, SNc neurons through indirect, presynaptic mechanisms (Marinelli et al., 2003; see below). Studies performed during the 1990s indicated that the CB_2 receptor, the other major cannabinoid receptor type, does not appear to be present in the healthy brain (Felder and Glass, 1998), although it might be induced in response to different damaging stimuli (Fernández-Ruiz et al., 2007). However, recent data from different laboratories have found that the normal brain also contains CB_2 receptors located in different cellular elements, including neurons and different types of glial cells, of different species including humans and rodents (Nuñez et al., 2004; Van Sickle et al., 2005; Ashton et al., 2006; Gong et al., 2006) – although see Chaps. 6, 9, 10. This also includes the basal ganglia, in particular the striatum and the SN, where Gong et al. (2006) detected immunoreactivity for this receptor type. Other studies have employed a combination of RT-PCR analysis and immunolabeling techniques to investigate CB_2 receptor expression in the striatum of naive rats, observing that this receptor subtype is indeed expressed in striatal cells, but possibly located in astrocytes rather than in neurons (Fernández-Ruiz et al., 2007). Its role in the control of the basal ganglia function in normal conditions remains to be elucidated, although there is some evidence that it can be involved in the protection of striatal neurons against different types of damaging stimuli (Fernández-Ruiz et al., 2007). There are, in fact, growing indications that in multiple disease states, CB_2 receptors are induced or upregulated in glial cells (activated astrocytes, reactive microglia) to regulate the protective and/or cytotoxic influences that these cells exert on neuronal homeostasis (Benito et al., 2003, 2005, 2007; Pazos et al., 2004). Lastly, some studies have suggested the possibility that a novel receptor type, different from CB_1, CB_2, or $TRPV_1$ receptors but active for certain cannabinoids, may be present in the basal ganglia and participate in motor effects of these compounds under certain circumstances. In general, the pharmacological profile of this unknown receptor would be relatively similar to the CB_1 receptor, yet its distinct identity is supported by pharmacological experiments conducted in mice with genetic deletions of classic cannabinoid receptors (Di Marzo et al., 2000a,b).

Endocannabinoid Ligands in the Basal Ganglia

Endocannabinoid ligands, AEA and 2-arachidonoylglycerol (2-AG), are also abundant in the basal ganglia (Bisogno et al., 1999; Di Marzo et al., 2000c; Fernández-Ruiz and González, 2005), showing concentrations that are in general higher than those measured in other brain regions. Endocannabinoids are particularly abundant in the basal ganglia output structures, the GP and SN (Di Marzo et al., 2000c), which parallels the high densities for CB_1 receptors reported in these two structures (Herkenham et al., 1991a; Mailleux and Vanderhaeghen, 1992a). In the case of AEA, the GP and the SN are the two regions where the levels of this endocannabinoid are the highest (Di Marzo et al., 2000c). The existence of in situ synthesis for this endocannabinoid can be confirmed by the detection of its metabolic precursor *N*-arachidonoylphosphatidylethanolamine (NAPE; see Chap. 2) in the basal ganglia (Di Marzo et al., 2000d) and of the enzyme responsible for this process, NAPE-phospholipase D (NAPE-PLD), in the whole brain (Okamoto et al., 2004). One important issue is to determine the neuronal phenotype in which this process occurs. There is indirect evidence indicating that AEA would be generated in striatal (Ronesi et al., 2004; Ade and Lovinger, 2007) and nigral (Wallmichrath and Szabo, 2002a,b) neurons, in both cases aimed at controlling CB_1 receptor function in cortical afferents to the striatum, and striatal afferents to the SNr, respectively (see below). It is also important to mention that the levels of AEA are significantly elevated in the striatum after the stimulation of dopaminergic D_2 receptors (Giuffrida et al., 1999; Beltramo et al., 2000). A priori this increase would reflect that the generation of AEA would be D_2 receptor-dependent, but it can also reflect a low rate of degradation, since a recent study reported both increased NAPE-PLD and reduced fatty acid amide hydrolase (FAAH), the enzyme that degrades AEA, in striatal slices following D_2 receptor activation (Centonze et al., 2004). In any case, the elevation of AEA levels following D_2 receptor activation would be compatible with the idea that the cannabinoid system may serve as an inhibitory feedback mechanism controlling dopamine-induced motor stimulation (Giuffrida et al., 1999). Striatal medium spiny neurons (MSNs) also express high levels of diacylglycerol lipase α (see Chap. 2), allowing for the activity-dependent generation of 2-AG (Uchigashima et al., 2007). In the basal ganglia, as in other brain areas, 2-AG might be more prominent than AEA as a rapid neuromodulator (see Chap. 11), yet AEA clearly mediates some aspects of striatal function. Most findings of 2-AG activity have been connected to electrophysiological studies, and will be discussed below.

Endocannabinoid Transport in the Basal Ganglia

It is generally assumed that the anandamide membrane transporter (AMT) should be highly concentrated in the basal ganglia (Fernández-Ruiz and González, 2005). However, the evidence supporting this view is indirect and generated mainly by pharmacological data, since to date no AMT has been isolated or cloned (see Chap. 3). Nonetheless, data obtained with several AEA analogues that behave in vitro as putative AMT inhibitors (Giuffrida et al., 2001) produced important effects in several bioassays aimed at functionally detecting changes in the control of movement in laboratory animals (Fernández-Ruiz and González, 2005). This is the case with compounds such as AM404 (González et al., 1999; Beltramo et al., 2000), VDM11 (de Lago et al., 2004b), UCM707 (de Lago et al., 2002), or OMDM2 (de Lago et al., 2004b). These compounds generally inhibited motor activity in rodents, with different potencies depending on the type of inhibitor, but the most important observation was that they were able to potentiate the motor inhibition exerted by subeffective doses of AEA (Fernández-Ruiz and González, 2005). Based on these pharmacological data, there is a general consensus that the AMT is present, and possibly abundantly concentrated, in the basal ganglia. However, further experiments are certainly necessary to demonstrate the molecular identity of the AMT activity and how it is regulated in the brain. Such experiments should receive significant benefit from the recent development of covalent AMT inhibitors (Moriello et al., 2006), which might serve as novel tools to isolate and quantify the protein involved in this process.

Endocannabinoid Metabolism in the Basal Ganglia

FAAH is the enzyme that catalyzes the hydrolysis of AEA and related *N*-acylethanolamines, and so it plays a key role in the regulation of brain levels of this endocannabinoid (see Chap. 3). In general, FAAH is located in cell bodies and dendrites of neurons that are postsynaptic to axonal terminals containing CB_1 receptors (Egertová et al., 2003; see Chap. 10), thus showing a pattern complementary to these receptors and supportive of the role of endocannabinoids as retrograde signaling molecules. FAAH enzyme is present in numerous brain structures, including structures of the basal ganglia (Desarnaud et al., 1995; Tsou et al., 1998b; Egertová et al., 2003), which is concordant with the subtle motor anomalies found in FAAH null mutant mice (Cravatt et al., 2001; Lichtman et al., 2002). However, while some authors detected high or moderate levels of FAAH in the GP and SN (Desarnaud et al., 1995; Tsou et al., 1998b), more recent studies found that the abundance of CB_1 receptors typical of these two regions correlated with very little FAAH expression (Egertová et al., 2003). By contrast, these authors detected significant expression of FAAH in the striatum, preferentially located in neurons but also in glial cells, in particular in oligodendrocytes (Egertová et al., 2003), an observation that these authors related to the reduction of FAAH levels reported in the striatum of parkinsonian rats (Gubellini et al., 2002).

Finally, monoacylglycerol lipase (MAGL), the enzyme involved in the degradation of 2-AG, and also other related monoacylglycerols (see Chap. 3), has been also detected in the basal ganglia (Dinh et al., 2002). Its pattern is somewhat different than FAAH, which suggests that AEA and 2-AG might subserve different functional roles in the basal ganglia circuitry (see also discussion in the next section). However, as stated above, both enzymes accept as substrates various *N*-acylethanolamines or monoacylglycerols, respectively; therefore, they should not be considered as definitive markers specific for cannabinoid signaling. In any case, the identification of these two enzymes, FAAH and MAGL, in the basal ganglia supports the notion that endocannabinoids generated in the different neuronal subpopulations of these structures are also degraded locally as a mechanism to efficiently inactivate these signaling molecules.

Endocannabinoids Control Neurotransmitter Release and Synaptic Plasticity in the Basal Ganglia

Physiological studies over the last several years have demonstrated that a principal and widespread function of endocannabinoids in the brain is to modulate fast synaptic neurotransmission (see Chap. 11). Both short- and long-lasting plasticity of synaptic transmission are mediated by activity-dependent release of endocannabinoids, which act in a retrograde manner at presynaptic sites to regulate the vesicular release of classical neurotransmitters (Freund et al., 2003; Chevaleyre et al., 2006). Given the distribution and dense expression levels of the CB_1 receptor within the basal ganglia, it is not surprising that endocannabinoid signaling has now been found to modulate synaptic function extensively within this network, indicating a prominent and complex role in modulating the physiology of behavioral output (see Fig. 1). Experimentally, the "on-demand" release of endocannabinoids is evoked either by the direct electrical stimulation of afferent synaptic pathways, pharmacological manipulations (such as the activation of certain $G_{q/11}$-coupled receptors), or by direct depolarization of a neuron using whole-cell voltage-clamp electrophysiology. In the latter case, retrograde signaling by endocannabinoids is measured as depolarization-evoked suppression of inhibition or excitation (DSI or DSE), when studying inhibitory or excitatory synapses, respectively. The mechanisms underlying these phenomena have been studied extensively – especially in the hippocampus and cerebellum – and are the subject of several excellent reviews (Alger, 2002; Freund et al., 2003; Diana and Marty, 2004; Hashimotodani et al., 2007; see also Chap. 11). As of preparing this chapter, however, there has not yet been a careful review of the many recently elaborated mechanisms of endocannabinoid-mediated synaptic plasticity in the basal ganglia. As such processes may be highly relevant to basal ganglia function in health and disease, we will discuss the topic at length in this section.

Glutamate

As the primary input nucleus of the basal ganglia, the striatum receives a massive convergence of glutamatergic axons from all areas of the cerebral cortex, as well as from the thalamus (Tepper, 2006). Coincident excitatory input from cortical and/or thalamic afferents largely controls striatal output, driving MSNs into a depolarized "up-state" from which these cells fire action potentials (Wilson and Kawaguchi, 1996; Tepper, 2006). The synaptic integration of excitatory inputs in the striatum is therefore a major means by which cortical activity can be translated into motor patterns through the basal ganglia (Graybiel et al., 1994). A number of labs have recently confirmed an important role of endocannabinoid signaling in modulating the presynaptic function of glutamatergic synapses in both dorsal and ventral striatum (Chevaleyre et al., 2006). The finding that striatal glutamate release is acutely inhibited by CB_1 receptor activation (Gerdeman and Lovinger, 2001; Robbe et al., 2001; Huang et al., 2001; Gubellini et al., 2002; Köfalvi et al., 2005) was initially confusing, because the loss of CB_1 receptor binding following excitotoxic lesioning of the striatum suggested the postsynaptic expression of these receptors (Herkenham et al., 1991a). It is important to note, however, that the methods employed to quantify these findings focused on medial striatal areas where CB_1 receptor expression is the lowest (Herkenham et al., 1991b; Glass et al., 1997; Tsou et al., 1998a), and binding was assayed at a time point that could allow for significant degeneration of afferent cortical axons from the site of the lesion (Herkenham et al., 1991a). Nonetheless, immunohistochemical detection techniques using CB_1 receptor-directed antibodies have yielded inconsistent results, with some authors failing to see labeling of asymmetric, excitatory synapses (Matyas et al., 2006). Others have found CB_1 receptor expression levels that, although modest in comparison to the very dense expression of this receptor at other synapses, nonetheless support the findings of physiological experiments (Robbe et al., 2001; Rodriguez et al., 2001; Köfalvi et al., 2005; Uchigashima et al., 2007). Efforts to detect CB_1 receptor mRNA in corticostriatal projection neurons using in situ hybridization techniques were also initially unsuccessful (Marsicano and Lutz, 1999), but recent experiments designed to specifically investigate this neuronal population have yielded positive results (Uchigashima et al., 2007). Considering all the evidence, it is likely that only very low levels of CB_1 receptors are required for functioning in striatal glutamatergic axon terminals of mature animals.

Striatal Long-Term Depression (LTD)

Evidence for the physiological activity of presynaptic striatal CB_1 receptors emerged with the finding that endocannabinoids act as retrograde messengers critical for the induction of striatal long-term synaptic depression (LTD) (Gerdeman et al., 2002; Robbe et al., 2002; Kreitzer and Malenka, 2005). LTD is a form of synaptic plasticity that is usually induced by high-frequency stimulation of afferent corticostriatal axons, and is thought to play important roles in striatal mnemonic function, including the development of behavioral habits (Gerdeman et al., 2003, 2006; Yin and Knowlton, 2006). In concert with mechanisms of long-term potentiation (LTP), striatal LTD may powerfully sculpt the learning and execution of motor repertoires involving the basal ganglia (Graybiel,

2005; Yin and Knowlton, 2006), and has received considerable attention in models of basal ganglia pathology (Calabresi et al., 1996, 2000; Graybiel and Rauch, 2000; Gubellini et al., 2002; Gerdeman et al., 2003; Kreitzer and Malenka, 2007). Importantly, endocannabinoids have now been found as necessary for the induction of striatal LTD via several different methods of induction (Chevaleyre et al., 2006), including a protocol meant to mimic the naturally occurring, depolarized up-state transitions of striatal MSNs in vivo (Kreitzer and Malenka, 2005). In dorsal striatum, endocannabinoid-dependent LTD is also dependent on D_2 dopamine receptors (Calabresi et al., 1992; Tang et al., 2001) and $mGluRs_{1/5}$ (Gubellini et al., 2001; Sung et al., 2001; Kreitzer and Malenka, 2005). Endocannabinoid synthesis and release appears to represent a coincidence detector in striatal MSNs, signaling not only convergent excitatory inputs releasing glutamate (a likely requirement for sufficient postsynaptic depolarization and $mGluR_{1/5}$ activation), but also the coincident release of dopamine from nigrostriatal boutons (Gerdeman et al., 2003; Yin and Lovinger, 2006). Such an associative role of CB_1 and D_2 receptors in mediating the complex synaptic mechanisms of striatal LTD may have important implications for striatal pathologies and for the function of dopamine as a learning signal critical for motivated behaviors (Graybiel, 2005; Schultz, 2006). As noted above, activation of D_2 receptors has been found to stimulate release of AEA in the rat striatum in vivo, especially under depolarizing conditions of high external K^+ (Giuffrida et al., 1999). Recent findings have led to disagreement regarding the mechanisms by which D_2 receptors may promote endocannabinoid release during the induction of LTD (Wang et al., 2006; Kreitzer and Malenka, 2007). Wang et al. (2006) reported a mechanism to explain the observation that D_2 receptor-dependent LTD can be regularly induced in striatal MSNs that do not themselves express D_2 receptors (see Wilson, 2006, for commentary). These authors demonstrated that regardless of the verified expression of D_1 or D_2 receptors in recorded MSNs, striatal LTD was highly reproducible and involved a D_2 receptor-mediated inhibition of tonically active cholinergic interneurons. According to this model, therefore, the presence or lack of D_2 receptors on the MSNs themselves are largely irrelevant to the pathways leading to LTD. Rather, D_2 receptor activation on interneurons leads to a decrease in synaptic ACh, which then facilitates postsynaptic endocannabinoid production by deactivating a signaling cascade involving M_1 muscarinic ACh receptors (mAChRs) and $Ca_v1.3$ (L-type) voltage-gated Ca^{2+} channels (Wang et al., 2006) (Fig. 2). In other words, the requisite generation of AEA for striatal LTD, driven by influx of extracellular Ca^{2+}, is under a tonic inhibition by M_1 mAChRs, and the populations of D_2 receptors responsible for enhancing endocannabinoid release are located on cholinergic interneurons, where they induce a pause in the release of ACh (Watanabe and Kimura, 1998; Wang et al., 2006). Kreitzer and Malenka (2007) have provided evidence for a different hypothesis. Using transgenic mice to identify fluorescently labeled medium spiny neurons as either D_1 or D_2 receptor expressing (Wang et al. (2006) also employed this technique), these authors reported that only D_2 receptor-expressing neurons of the indirect striatal outflow pathway readily express endocannabinoid-dependent LTD. This was attributed

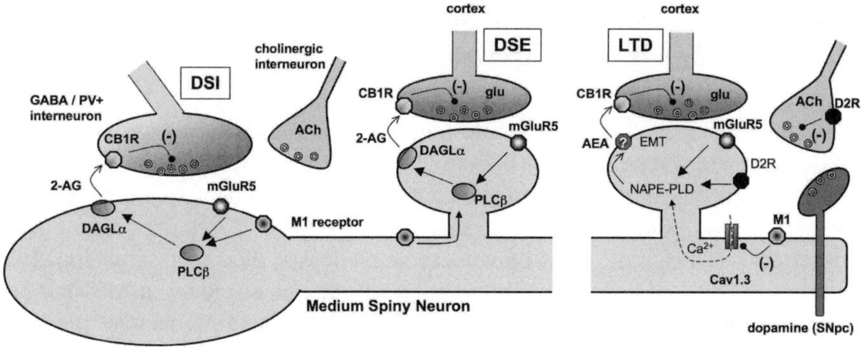

Fig. 2 Model of synaptic endocannabinoid signaling in striatal medium spiny neurons. In all cases, retrograde activation of presynaptic CB_1 receptors leads to decreased neurotransmitter release. At somatodendritic GABAergic synapses formed by parvalbumin-expressing (PV+) interneurons, DSI is induced by $mGluR_5$ and M_1 mAChRs paired with postsynaptic depolarization, involving the activation of DAGLα to generate 2-AG. $mGluR_5$ and DAGLα are also critically required to elicit DSE at glutamatergic synapses, where the M_1 mAChR plays only a facilitating role. Corticostriatal LTD is mechanistically distinct, and involves anandamide (AEA) rather than 2-AG. LTD induction requires activation of postsynaptic $mGluRs_{1/5}$, hypothesized to promote AEA synthesis via elevation of intracellular Ca^{2+} levels (not shown) to activate *n*-acyl-phosphatidylethanolamine specific PLD (NAPE-PLD). Dopamine signaling at D_2 receptors is required to transiently inhibit ACh release, thereby removing a tonic blockade of postsynaptic $Ca_v1.3$ Ca^{2+} channels. Postsynaptic D_2 receptors (when present) may also enhance AEA generation or its carrier-facilitated efflux through an endocannabinoid membrane transporter (EMT). Mechanisms for DSE and LTD may coexist in dendritic spines, but are depicted separately for the sake of clarity. Presynaptic effectors controlling vesicular release probability also may vary among synapses, and the presynaptic mechanisms underlying the persistent expression of LTD are not well understood. This scheme is an integration of numerous references described in the text

to a greater intrinsic excitability of indirect pathway neurons. Thus, these neurons are expected to more readily release AEA in response to synaptically driven Ca^{2+} influx and $mGluR_{1/5}$ activation, augmented in some way by postsynaptic D_2 receptors (Kreitzer and Malenka, 2007; see also Chap. 11). These novel results are difficult to reconcile with the level of reproducibility seen for the induction of striatal LTD in prior studies (e.g., Gerdeman et al., 2002; Ronesi et al., 2004; Kreitzer and Malenka, 2005), and are in direct disagreement with the article just described (Wang et al., 2006). Regardless of how this controversy is resolved, it is compelling to presume that endocannabinoid release may at least occur more readily from striatal neurons of the indirect pathway under physiological conditions, where afferent activity is not as robust as during stimuli used to evoke LTD in striatal slices. These findings imply that in the striatum, inhibitors of endocannabinoid reuptake or enzymatic degradation may have markedly different effects than direct CB_1 receptor agonists, which inhibit glutamate release at neurons of both pathways (Gerdeman and Lovinger, 2001; Kreitzer and Malenka, 2007; Uchigashima et al., 2007).

Short-Term Plasticity of Glutamate Release via Endocannabinoids in the Basal Ganglia

Endocannabinoid-mediated retrograde inhibition of glutamate release also occurs in the striatum on a rapid and transient timeframe, and can be evoked by stimulation of afferent corticostriatal synapses in a frequency-dependent manner (Yin and Lovinger, 2006). Similarly to LTD, this phenomenon also requires elevation of internal Ca^{2+} via $mGluR_{1/5}$ and coincident activation of the D_2 class of dopamine receptors. Frequency-dependent suppression of glutamate release in MSNs therefore represents another form of endocannabinoid-mediated associative plasticity that may regulate motor behaviors or procedural learning. DSE has also been recently demonstrated in striatal MSNs, and likewise appears to be strongly dependent on direct activation of $mGluR_5$ (Narushima et al., 2006a; Uchigashima et al., 2007), closely linking endocannabinoid signaling to synaptic input. The extent of a physiological role for DSE is not entirely clear, and optimal parameters for inducing the phenomenon are not conditions expected to occur in vivo (5 s at a holding potential of 0 mV; Narushima et al., 2006a). Nonetheless, even small influences on glutamate release might have significant influence on synaptic integration, and can be underestimated in distal dendrites using voltage clamp techniques. DSE could function as a purely synapse-specific negative feedback loop, or it might additionally represent a heterosynaptic mechanism by which spillover of glutamate can inhibit transmission at neighboring synapses (see Chap. 11). Such a question will be difficult to determine in the striatum and throughout the basal ganglia, where axo-dendritic arrangements are not nearly as predictable as in hippocampus or cerebellum (Tepper, 2006). An intriguing picture emerging from recent studies is that different endocannabinoids appear to mediate DSE and LTD in the striatum. DSE is prevented by inhibitors of DAGL, demonstrating a critical involvement of 2-AG (Uchigashima et al., 2007). This physiological finding was presented alongside highly consistent anatomical data demonstrating postsynaptic distribution patterns of DAGLα and $mGluR_5$, physically apposed to presynaptic CB_1 receptors (Uchigashima et al., 2007). In contrast, CB_1 receptor-dependent striatal LTD is not blocked by DAGL inhibitors. Rather, the ability to induce striatal LTD in brain slices has been correlated to a developmental increase in striatal AEA content, and is greatly facilitated by exogenously applied AEA (Ade and Lovinger, 2007). LTD can also be rescued in striatal neurons, either by blocking extracellular AMT activity (Ronesi et al., 2004; Ronesi and Lovinger, 2005), or by extracellular application of URB-597, a specific FAAH inhibitor (Kreitzer and Malenka, 2007). As FAAH is the primary enzymatic pathway for AEA hydrolysis, this latter finding is consistent with the observation that striatal levels of AEA, but not 2-AG, are enhanced by local application of the D_2 receptor agonist quinpirole either in vivo (Giuffrida et al., 1999) or in acute brain slices (Ade et al., 2003) – and quinpirole also facilitates LTD (Kreitzer and Malenka, 2005, 2007; Wang et al., 2006). Thus, mechanisms of AEA signaling – including an apparent requirement for transport-mediated efflux (Ronesi et al., 2004) – might be specialized for mediating enduring depression of synaptic efficacy in response to convergent input (from both glutamate and dopamine-releasing afferents) (see Fig. 2). It is not known which endocannabinoid is primarily responsible for transient, frequency-dependent suppression of excitation in these neurons (Yin and Lovinger, 2006), but similarities with LTD are notable. These include a dependence on D_2 receptors and elevations of internal Ca^{2+} (which has not been directly demonstrated for either DSE or DSI in striatum), suggesting that perhaps AEA is also the retrograde messenger for this form of CB_1 receptor-mediated synaptic plasticity. Interestingly, whereas AEA-dependent LTD has been more easily induced in D_2 receptor-expressing MSNs (Kreitzer and Malenka, 2007), Uchigashima et al. (2007) reported immunolabeling of DAGLα to appear more intensely in neuronal processes expressing the D_1 receptor. While more research will surely refine these models, the distinction of separate functional pathways of retrograde endocannabinoid signaling in striatal MSNs highlights the possibility that specific pharmacological targeting of these pathways might have differential effects on the extrapyramidal motor system, distinct also from direct activation of CB_1 receptors. Glutamatergic synapses in other nuclei of the basal ganglia are also sensitive to modulation by endocannabinoids. These include projections from the STN to both the GPe and SNr. Using electrophysiological techniques either in vivo (Sañudo-Peña and Walker, 1997) or in brain slices (Szabo et al., 2000), excitation of SNr neurons by subthalamonigral axons was inhibited by cannabinoids, as suggested by earlier behavioral studies (Sañudo-Peña et al., 1996, 1999). Similarly, direct electrical stimulation of the STN evokes glutamatergic excitatory postsynaptic currents (EPSCs) in GPe neurons, and the CB_1 receptor agonist WIN55212-2 inhibits these EPSCs by a presynaptic mechanism (Freiman and Szabo, 2005). WIN55212-2 also was shown to decrease the spontaneous firing of GP neurons measured in anesthetized rats, consistent with a decrease in glutamate release from tonically active subthalamopallidal inputs (Miller and Walker, 1996, 1998). Thus, in agreement with anatomical data, STN neurons projecting to either the GP or SNr express CB_1 receptors preferentially at axonal sites (Freiman and Szabo, 2005), where cannabinoids inhibit the release of glutamate. Glutamatergic synapses onto the dopaminergic neurons of the ventral tegmental area (VTA) are modulated by endocannabinoids as well (Melis et al., 2004a,b; Riegel and Lupica, 2004), where it has been elegantly demonstrated that 2-AG acts as a retrograde messenger mediating DSE (Melis et al., 2004b). Moreover, Melis et al. provided evidence that electrical stimulations of the PFC, reflective of in vivo firing, are sufficient to elicit 2-AG release and negative feedback to suppress glutamatergic transmission. Riegel and Lupica (2004) demonstrated that endocannabinoid release is enhanced by inducing burst firing of dopamine neurons – activating CB_1 receptors at both excitatory and inhibitory synapses within the VTA. Space does not permit further discussion of endocannabinoid modulation of VTA activity (Lupica and Riegel, 2005), although it is likely to have numerous implications for the neurobiology of drug abuse, as well as motor function and the sensitivities of DA neurons to excitotoxicity (Melis et al., 2006). Lastly, there is some evidence that CB_1 receptor agonists can inhibit glutamate reuptake, either in striatal brain slices (Brown et al., 2003) or synaptosomal preparations (Köfalvi et al., 2005; see Chap. 9). Although these effects are more likely to be CB_1 receptor-independent (see Chap. 9), alterations in the transporter-mediated reuptake of glutamate can exert pronounced influence on the excitation of neurons and on the stimulation of 2-AG by extrasynaptic mGluRs (see Chap. 11). Whether or not endocannabinoids inhibit glutamate reuptake in physiological settings remains to be seen. There is, however, evidence that endocannabinoids can enhance synaptic glutamate in the SNc by activating $TRPV_1$ receptor cation channels on presynaptic sites, which would be expected to depolarize the axon terminal (Marinelli et al., 2003).

Inhibition of Striatal GABA Release

Pharmacological studies have long demonstrated interactions between cannabinoids and GABA signaling within the extrapyramidal motor system (Pertwee and Wickens, 1991; Sañudo-Peña et al., 1999). As mentioned above, CB_1 receptors are highly expressed at symmetric, GABA-releasing synapses throughout the basal ganglia. Accordingly, some of the first uses of brain slice electrophysiology to investigate CB_1 receptor signaling found cannabinoid modulation of GABA release in the striatum (Szabo et al., 1998) and SNr (Chan and Yung, 1998; Chan et al., 1998). In the striatum, GABAergic inputs to MSNs arise from axon collaterals from other MSNs, and from at least three types of interneuron (Tepper, 2006). Recent studies have indicated that synapses formed by the parvalbumin-expressing (PV+), fast spiking class of interneurons (FSNs) are physiologically regulated by retrograde endocannabinoids (Freiman et al., 2006; Narushima et al., 2006b, 2007; Matyas et al., 2006; Uchigashima et al., 2007). In technically demanding experiments, paired recordings have been used to analyze FSN-MSN and MSN-MSN synaptic connections (Freiman et al., 2006; Narushima et al., 2006b). Both of these synapses were found to be sensitive to the CB_1 receptor agonist WIN55212-2 (Freiman et al., 2006), consistent with a role for endocannabinoid signaling at these synapses. Accordingly, DSI can be induced in MSNs, but only with concomitant activation of either $mGluR_{1/5}$ or M_1 mAChRs, and this receptor-driven endocannabinoid release was only detected at synapses from PV + FSNs in paired recordings (Freiman et al., 2006; Narushima et al., 2006b). These interneurons are believed to be highly important to information processing in the striatum, providing a source of strong feed-forward inhibition to MSNs (Koos and Tepper, 1999; Tepper, 2006). There is also evidence that short-term synaptic depression of these synapses can significantly influence the likelihood of MSNs to fire action potentials in response to afferent excitatory inputs (Fitzpatrick et al., 2001). Endocannabinoid-mediated suppression of GABA release at FSN-MSN synapses may therefore play an important role in the modulation of striatal output. Uchigashima et al. (2007) found that DAGLα activation is necessary for DSI, and that this enzyme is localized close to CB_1 receptor-expressing synaptic sites (including PV + GABAergic terminals) and to $G_{q/11}$-coupled receptors that activate 2-AG signaling (see Fig. 2). This builds a compelling story that 2-AG is the retrograde messenger responsible for both DSE and DSI in the striatum (Uchigashima et al., 2007), as also appears to be the case for DSI in the SNr (Szabo et al., 2006), cerebellum (Galante and Diana, 2004; Szabo et al., 2006), and hippocampus (Hashimotodani et al., 2007). Release of 2-AG has also been measured from striatal slice cultures following electrical stimulation (Jung et al., 2005), although other protocols have yielded primarily AEA from acutely prepared slices (Ade et al., 2003), or from the striatum in vivo (Giuffrida et al., 1999). Striatal DSI appears to be tightly under the control of the giant aspiny cholinergic interneurons (Narushima et al., 2007). Although these cells represent only 5% of the total striatal neuronal population, the tonically active cholinergic interneurons extend dense axonal arbors throughout a large region of the striatum, endowing it with the largest concentration of ACh in the mammalian brain (Tepper, 2006). Narushima et al. (2007) recently showed elegantly that DSI in striatal MSNs was enhanced by increasing ambient ACh or by evoking spikes within individual nearby cholinergic interneurons, and this effect required the activation of postsynaptic M_1 mAChRs (see Fig. 2). It is fascinating then that ACh release may play contrasting endocannabinoid-mediated roles between excitatory and inhibitory neurotransmission in the striatum. M_1 mAChR activation may chronically inhibit AEA synthesis at striatal spines (Wang et al., 2006), yet tonically enhance 2-AG synthesis at somatodendritic GABAergic synapses (Narushima et al., 2007). ACh has not previously been highly implicated in striatal cannabinoid function, given that at least most cholinergic interneurons do not express CB_1 receptors (see above), and in striatal brain slices, evoked release of ACh is not modulated by CB_1 receptor ligands (Gifford et al., 1997). Yet, based on these recent investigations of DSI and LTD, spontaneous firing activities of cholinergic interneurons may play critical roles in regulating endocannabinoid synthesis and release in MSNs. This could have important mechanistic implications for models of striatal contributions to adaptive behavior, and striatal function in disease states. It is known that the tonically active cholinergic neurons recorded in vivo exhibit a distinct pause in firing in response to reward-associated events encountered by the animal (Graybiel et al., 1994; Schultz, 2006). This pause is correlated to dopamine cell firing and activation of D_2 receptors (Watanabe and Kimura, 1998), and is thought to play important roles in striatal processes of associative learning (Schultz, 2006). The recent findings reviewed here suggest that dopamine-induced pauses in interneuron activity and ACh release might act as a cellular switch that regulates endocannabinoid-mediated synaptic plasticity and its control over striatal output (Wang et al., 2006; Narushima et al., 2007; see Fig. 2).

GABA Release in GP and SNr

Experiments measuring rat behavior and neuronal activity in vivo first provided evidence that GABA release in the GP and SNr is inhibited by presynaptic CB_1 receptors densely expressed on the axon terminals of striatal afferent projections (Miller and Walker, 1995, 1996, 1998; Sañudo-Peña et al., 1996, 1999; Tersigni and Rosenberg,

1996). Studies using whole-cell patch-clamp methods in brain slice preparations have corroborated this evidence. Thus, in the SNr, activation of CB_1 receptors by WIN55212-2 decreased spontaneous or locally evoked IPSCs by a presynaptic mechanism (Chan and Yung, 1998; Chan et al., 1998). This finding was supported by studies that more specifically stimulated striatonigral afferent axons (Wallmichrath and Szabo, 2002a,b). DSI was also found to occur in both SNr (Wallmichrath and Szabo, 2002b; Yanovsky et al., 2003) and SNc neurons (Yanovsky et al., 2003), and there is some evidence that endocannabinoids may be tonically released by SNr neurons to inhibit striatonigral inputs (Wallmichrath and Szabo, 2002a). The effects of exogenous cannabinoid agonists on motor behaviors therefore involve influencing the balance of excitatory and inhibitory synapses in the SNr (Sañudo-Peña et al., 1999). In normal situations, the most overt motor effects are likely due to inhibition of the spontaneously active STN inputs releasing glutamate, but this may change according to interactions with neighboring dopamine neurons (Sañudo-Peña et al., 1998a) or following dopamine depletion in a PD model (Sañudo-Peña et al., 1998b). The effects of cannabinoids on GP output are likewise dependent upon the relative activities of excitatory (subthalamopallidal) and inhibitory (striatopallidal) inputs, as both are inhibited by presynaptic CB_1 receptors (Miller and Walker, 1996, 1998; Sañudo-Peña et al., 1999; see Fig. 1). In a slice preparation preserving inputs to the GPe from the caudate/putamen, intrastriatal stimulation evoked CB_1 receptor-sensitive IPSCs in GP neurons, and this striatopallidal synapse was found to exhibit a modest DSI mediated by endocannabinoids and associated with robust elevations in intracellular Ca^{2+} (Engler et al., 2006). It is compelling to speculate that DSI would be enhanced by activation of $G_{q/11}$-coupled mGluRs, which could be relevant to associative processes between STN and striatal inputs to the GP. For example, a recent study found that repetitive activation of subthalamopallidal axons causes an $mGluR_1$-mediated postsynaptic depolarization and a subsequent enhancement in postsynaptic spiking that lasts much longer than the depolarization itself, over 25 s (Kaneda et al., 2007). Although it has not yet been tested, an endocannabinoid-mediated DSI may contribute to this phenomenon – especially if enhanced by $mGluR_{1/5}$ as in multiple other brain areas (see Chap. 11) – thereby allowing STN inputs to more strongly drive GP activity. Such an interaction might suggest an increase in GP endocannabinoid contents in PD, in which STN neurons can exhibit enhanced burst firing that likely correlates to impaired movement and sensory motor processing (Bevan et al., 2002). This is consistent with findings in multiple animal PD models where elevations in GP endocannabinoid tone were observed (Di Marzo et al., 2000c; van der Stelt et al., 2005). Whether this relates directly to PD symptoms – or is perhaps reflective of a hyperactivated homeostatic response – is presently unknown.

I-LTD in the Basal Ganglia?

In addition to DSI and other transient endocannabinoid effects on GABAergic synapses, LTD of inhibitory inputs (I-LTD) has been observed in multiple brain areas

(Chevaleyre et al., 2006). A recent finding suggests that I-LTD may occur within inhibitory networks of the basal ganglia, particularly in the superior colliculus (SC), where CB_1 receptors can inhibit motor behavior (Sañudo-Peña et al., 2000a). Using a tissue culture system, mGluR- and CB_1 receptor-dependent suppression of inhibition was observed in SC neurons innervated by cortical explants (Henneberger et al., 2007). This transient effect developed into a lasting, presynaptic I-LTD following high frequency activation of excitatory cortical axons and postsynaptic mGluRs. Such endocannabinoid-mediated heterosynaptic plasticity within the SC could contribute to proper sensorimotor visual processing (Henneberger et al., 2007), which is disrupted by Δ^9-THC in humans (Ploner et al., 2002). It remains to be seen if this I-LTD occurs in mature SC tissue, or elsewhere among the many collateral GABAergic synapses in the basal ganglia.

GABA Reuptake Transporters

Some authors have found evidence for an inhibition of [^{3}H]GABA uptake by CB_1 receptors in the basal ganglia (Maneuf et al., 1996a,b; Romero et al., 1998a), which could provide a homeostatic function to balance the inhibition of GABA release by endocannabinoids. Interpretation of these studies is hampered, however, by the use of drug concentrations that are many times higher than those required to fully activate CB_1 receptors (see Chap. 9). Other reports have indeed contradicted some of these findings (Köfalvi et al., 2005; Venderova et al., 2005; Engler et al., 2006), with one study indicating that CB_1 receptors inhibit GABA uptake in the SNr, but not the GP (Romero et al., 1998a). In either nucleus, studies utilizing whole-cell voltage-clamp electrophysiology have not shown evidence for alterations in the decay of GABA-mediated currents following CB_1 receptor activation, which would be expected if there were an inhibition of GABA reuptake (Chan and Yung, 1998; Chan et al., 1998; Yanovsky et al., 2003; Engler et al., 2006; Wallmichrath and Szabo, 2002a). In the GP, Engler et al. (2006) directly tested this hypothesis, finding that WIN55212-2 does not prolong the decay of GABAergic IPSCs in a manner similar to a specific blocker of GABA uptake transporters. Therefore, while an endocannabinoid mechanism for modulating amino acid transporters cannot be ruled out, definitive support for such a model awaits the use of expression systems to describe such a function in molecular detail.

Dopamine

Systemically delivered cannabinoids can influence dopamine cell firing (French et al., 1997; Gessa et al., 1998) and the regulation of tyrosine hydroxylase (Romero et al., 1995a,b, 2002). In accordance with many of the mechanisms just discussed, there is good evidence that cannabinoid agonists can disinhibit SNc cell firing by acting on striatonigral terminals (Yanovsky et al., 2003), which would lead to elevations of striatal dopamine observed following Δ^9-THC administration (Castaneda et al., 1991; Ng Cheong Ton et al., 1988; see Chap. 22). Some of these terminals preferentially target postsynaptic metabotropic $GABA_B$ receptors, which contribute to cell firing in complicated ways (Lupica and Riegel, 2005). This is consistent with the observation that some behavioral effects of cannabinoids are distinctly modulated by $GABA_B$ receptors (Romero et al., 1996a). As mentioned, however, endocannabinoids could also excite SNc neurons via $TRPV_1$ receptors (Marinelli et al., 2003), and the influence of CB_1 receptors on cell firing may vary in context-dependent ways (Gueudet et al., 1995; Melis et al., 2004b; Lupica and Riegel, 2005; see also Chap. 11 for a broader discussion of this concept). A disinhibition model of cannabinoid effects on dopamine cell function is supported by anatomical data (see above). It is therefore not expected that endocannabinoids within the striatum could influence dopamine release through direct actions at dopamine release sites (see Chap. 22). There are nonetheless conflicting reports in the literature in this regard. Notably, Cadogan et al. (1997) found an inhibitory effect of cannabinoids on electrically stimulated release of [^{3}H]dopamine in striatal slices, which has been widely interpreted as suggesting a direct inhibition of dopaminergic transmission by striatal endocannabinoids. This careful study used concentrations of CP55940 and AEA that are generally considered specific for CB_1 receptors in ex vivo brain slice experiments (rife with nonspecific binding opportunities for hydrophobic ligands). Inhibition by CB_1 receptor agonists on the release of preloaded [^{3}H]dopamine was blocked by the CB_1 receptor antagonist Rimonabant at concentrations as low as 10 nM (Cadogan et al., 1997). However, methods of whole-slice electrical stimulation employed by these authors to evoke the release of dopamine lack both specificity and physiological basis. It is likely that neuronal elements other than the nigrostriatal dopaminergic axons were direct targets of cannabinoids in this study. As mentioned, CB_1 receptors inhibit corticostriatal glutamate release, and glutamate is likely to directly enhance striatal dopamine release via activation of ionotropic receptors on nigrostriatal afferents (Borland and Michael, 2004). Polysynaptic effects of exogenous *n*-methyl-D-aspartate (NMDA) may also explain the results of Kathmann et al. (1999), rather than a cannabinoid effect on dopamine release. For instance, neither WIN55212-2 nor CP55940 were found to inhibit the release of endogenous dopamine induced by single, localized electrical pulses (Szabo et al., 1999; see also Köfalvi et al., 2005). There may, however be CB_1 receptor-independent effects of high-dose cannabinoid agonists on dopamine transporter function in the striatum (Price et al., 2007; see Chap. 9). It is not clear how these findings apply to endocannabinoid function in the basal ganglia in vivo. Price et al. (2007) used bit high ligand concentrations, and curiously saw similar effects with both AM251, a CB_1 receptor antagonist, and AM404, an AMT inhibitor. In contrast, Fernandez-Espejo et al. (2004) used AM251 to block the effects of AM404 on dopamine-mediated turning behaviors when both drugs were injected directly into the striatum. Other authors failed to see direct effects of either Rimonabant or CB_1 receptor agonists on uptake of radiolabeled extracellular dopamine (Cadogan et al., 1997; Köfalvi et al., 2005), further indicating that any effects of these compounds on dopamine transporters are CB_1 receptor independent and require high concentrations that are not likely to mimic endocannabinoid signaling (Price et al., 2007). Considering multiple lines of study, effects of endocannabinoids on dopamine release in vivo are most likely the result of alterations in SNc cell firing, through presynaptic modulation of GABA release or basal ganglia network activity, rather than a direct effect on dopamine release or reuptake. In summary, given that CB_1 receptors can modulate the release of both glutamate and GABA in multiple basal ganglia nuclei, it is not surprising that cannabinoid agonists such as Δ^9-THC would have complex and dose-dependent effects on motor output (Sañudo-Peña et al., 1999, 2000b; Fernandez-Ruiz et al., 2002). It should also be expected that systemic application of CB_1 receptor antagonists might preferentially target excitatory or inhibitory transmission, depending on dosage. This highlights the need to develop and characterize pharmacological agents that act specifically on endocannabinoid uptake or metabolism. Nonetheless, even regarding endocannabinoids, it is not yet clear whether these molecules exert similar, or imbalanced effects on excitatory vs. inhibitory synapses under normal conditions, and how this might change in basal ganglia pathologies. There is further complexity added by the apparent formation of heteromeric receptor assemblies involving CB_1 receptors, which may allow other transmitter systems to regulate cannabinoid effects on motor output through allosteric receptor interactions or other cooperative mechanisms (Kearn et al., 2005; Schoffelmeer et al., 2006; Carriba et al., 2007). Endocannabinoid signals appear to be an important component of regulating brain responses to particular contexts or patterns of afferent neuronal activity (see Chap. 11). Neuronal firing throughout the basal ganglia is correlated to behavioral activation, and a great many of the synaptic connections are sensitive to cannabinoids (see Fig. 1). Behavioral consequences of exogenous, systemically delivered cannabinoids may thus vary depending on the situational context and recent history of an individual. As this field advances rapidly, increasingly diverse techniques are being applied to test such hypotheses. It is important to recognize that the widespread role of endocannabinoid signaling in the basal ganglia – sometimes at functionally opposing synaptic inputs to the same neurons – reflects a complex physiology that should be interpreted through multiple approaches to understand systems-level effects of cannabinoid-based medicines. For example, some authors have described cannabinoid signaling as a "brake" on dopaminergic activity (Rodriguez de Fonseca et al., 1998), whereas others have emphasized the integral role of endocannabinoids as part of signaling cascades downstream of D_2 receptor activation, which may functionally occlude dopamine effects at the same synapses (Chevaleyre et al., 2006; and see above). These differences in emphasis are not exclusive, but largely reflect the particular historical perspectives of various methodologies, and how these are applied to a systems-level of interpretation. Ideally, increasing sophistication in neuroscience allows for multidisciplinary findings to be integrated and reconciled; it is our hope that the overview provided by this chapter is helpful in this regard.

Pharmacological Effects of Cannabinoids on Motor Behavior

It is evident from the pronounced activity of the cannabinoid system in the basal ganglia circuitry that this signaling system plays an important role in the control of movement. Accordingly, mice deficient in the gene encoding for FAAH, which exhibit a high and permanent elevation of brain AEA levels, develop a series of subtle motor disturbances, in particular in the response to different pharmacological stimuli (Cravatt et al., 2001; Lichtman et al., 2002). It is quite likely that the effects found in FAAH-deficient mice are mediated by the activation of CB_1 receptors (Cravatt et al., 2001), and indeed CB_2 receptor knockout mice have not been reported to exhibit any motor disturbances (Buckley et al., 2000). By contrast, CB_1 receptor-deficient mice have exhibited important motor alterations (Ledent et al., 1999; Zimmer et al., 1999), despite the fact that the two models developed so far exhibited apparently opposite motor phenotype, i.e., hyperlocomotion was observed in the CB_1 knockout mouse model developed by Ledent et al. (1999), whereas hypoactivity was evident in the strain developed by Zimmer et al. (1999). Methodological differences, varying anxiety levels in experimental populations, and other reasons have been argued to explain the differences between these two models. It is possible that these differences are related to the multiplicity of sites where these receptors modulate synaptic transmission in the basal ganglia circuitry, subserving a multifactorial regulation of network activity and behavior. As will be detailed below, this could be also associated with the few paradoxical results obtained in pharmacological studies, in particular when experiments employed cannabinoid compounds with notably different pharmacodynamic or pharmacokinetic properties, or used very different doses or times of treatment (Fernández-Ruiz and González, 2005). However, these paradoxical results are a minor component within the literature published so far, where there is a general consensus that the activation of CB_1 receptors is usually followed by an inhibition of motor activity. By contrast, the blockade of CB_1 receptors attenuates the effects of agonists and is sometimes even associated with hyperlocomotion, due perhaps to the inverse agonist properties displayed by most CB_1 receptor antagonists in heterologous expression systems (see Chap. 7). In laboratory animals, the administration of cannabinoid agonists produces dose-dependent impairments in the open-field, ring test, actimeter, rotarod, or other tests frequently employed to record motor activity (Sañudo-Peña et al., 1999; Romero et al., 2002; Fernández-Ruiz and González, 2005). This is the case of Δ^9-THC, the prototypical tricyclic cannabinoid derived from *Cannabis sativa*, AEA, and a variety of synthetic agonists (Crawley et al., 1993; Fride and Mechoulam, 1993; Wickens and Pertwee, 1993; Smith et al., 1994; Romero et al., 1995a,b). The data in humans reinforce the same idea, since individuals that consume cannabis can experience diverse psychomotor effects reflected by a global impairment of motor performance (especially in complex and demanding tasks) associated with incoordination, ataxia, tremulousness, and weakness (Dewey, 1986; Consroe, 1998; Kalant, 2004).

Effects of Plant-Derived, Synthetic, or Endogenous Cannabinoid Agonists

Much information on the involvement of the cannabinoid system in the control of movement has been obtained in studies testing phytocannabinoid effects in laboratory animals. For instance, the administration of Δ^9-THC reduced the spontaneous motor activity and provoked catalepsy by itself or enhanced the cataleptic activity of muscimol in mice (Pertwee et al., 1988). In rats, Δ^9-THC also reduced the spontaneous activity and the frequency of stereotypic movements (Navarro et al., 1993; Romero et al., 1995a; Jarbe et al., 1998), increased the inactivity (Rodríguez de Fonseca et al., 1994; Romero et al., 1995a; Jarbe et al., 1998), and disrupted fine motor control (McLaughlin et al., 2000). Δ^9-THC also affected the motor effects of classic hypo- and hyperkinetic substances in rats; for example, it potentiated reserpine-induced motor inhibition (Moss et al., 1981) and muscimol-induced catalepsy (Wickens and Pertwee, 1993), but reduced amphetamine-induced hyperlocomotion (Gorriti et al., 1999). Other plant-derived cannabinoids, like cannabinol (CBN) and cannabidiol (CBD), also produced motor inhibition (Hiltunen et al., 1988), although their effects were weaker than those caused by Δ^9-THC in accordance with their lower affinity for CB_1 receptors, in particular, in the case of CBD (see Chap. 9). Synthetic but somewhat nonspecific cannabinoid agonists, developed in an attempt to improve either pharmacokinetic or pharmacodynamic properties of classic phytocannabinoids, produced powerful inhibitory effects in a variety of motor tests and animal models (Consroe, 1998; Sañudo-Peña et al., 1999; Romero et al., 2002). Selective agonists for the CB_1 receptor, such as the metabolically stable arachidonoyl-chloroethylamide (ACEA), also impaired motor function in laboratory animals (Schuster et al., 2002), although the effects found in that study were relatively quite small. By contrast, selective agonists for the CB_2 receptor, such as HU308 or JWH133, failed to reproduce these effects (Hanus et al., 1999; Huffman, 2005), thus supporting the possible favorability of CB_2 selective ligands for certain neurological pathologies (Fernández-Ruiz et al., 2007). The inhibitory effects reported for plant-derived or synthetic cannabinoids have been, in general, reproduced by endocannabinoids, in particular by AEA, which has been the endogenous cannabinoid ligand most studied for its effects on the control of movement. Thus, one year after the discovery of AEA, Fride and Mechoulam (1993) reported that this endocannabinoid decreases rearing behavior and causes immobility in mice, results that were subsequently reproduced by Crawley et al. (1993) and Smith et al. (1994) in rats. In addition, Wickens and Pertwee (1993) found that muscimol-induced catalepsy in rats was potentiated by AEA as well as Δ^9-THC. The group of Fernandez-Ruiz has also contributed to characterize the motor effects of AEA by measuring the dose-dependent and time-course responses exhibited by this endocannabinoid in rats subjected to open-field analysis. We found that AEA inhibited motor and stereotypic behaviors in a dose-related manner as did Δ^9-THC (Romero et al., 1995a), but, compared with the time-course response exhibited by this phytocannabinoid, AEA showed a biphasic pattern that is related to its lower metabolic stability

(Romero et al., 1995b). This was confirmed after repeating the same type of experiments with methanandamide, a more stable analogue of AEA (Romero et al., 1996b; Jarbe et al., 1998). It is possible, however, that the differences between the motor effects of AEA and those caused by Δ^9-THC might be originated by the existence of other targets available for AEA, but not for classic cannabinoids, to influence motor behavior (Di Marzo et al., 2000a,b; de Lago et al., 2004a; see Chap. 9). However, most of the literature published in relation with the motor inhibitory effects of the different cannabinoid agonists indicates that these are CB_1 receptor mediated. It is true that authors report subtle differences in magnitude and duration of these motor effects, but they can be attributed to the use of different compounds with differences in receptor affinity, potency, and/or metabolic stability. As already discussed, another key factor is that the CB_1 receptor modulates a diversity of synapses within the basal ganglia circuitry – which are likely to be recruited in different context- and experience-dependent ways. Variable effects of cannabinoids on excitatory vs. inhibitory synapses, for example – or in functionally oppositional basal ganglia nuclei – may also relate to the biphasic effects of cannabinoids in some studies reporting increased motor behavior in mice (Souilhac et al., 1995) or rats (Sañudo-Peña et al., 2000b) following very low doses.

Effects of Inhibitors of Endocannabinoid Inactivation

Hypokinetic effects may also be achieved by indirect activation of the CB_1 receptor through the modulation of different elements that prolong endocannabinoid activity, such as inhibitors of the AMT, FAAH, or MAGL. Inhibition of these activities produced generally equivalent effects to those observed with direct CB_1 receptor agonists, with the advantage that they might be used with comparatively minimal side effects (Fernández-Ruiz and González, 2005). A guiding hypothesis is that such compounds will not produce motor effects by themselves, or that these will be small, but they can enhance the motor effects of endocannabinoids. Again, this may be an oversimplification, and such effects may be somewhat context-dependent, as suggested, for example, by conflicting results with the FAAH inhbitor URB597. In some studies, this inhibitor did not produce hypomotility or catalepsy when administred alone (Jayamanne et al., 2006), whereas in others, URB597 was either sedating (Van Sickle et al., 2005), or it reduced movement in normal rodents but not in those previously treated with reserpine (Lee et al., 2006). Other "indirect agonists," interesting for their hypokinetic effects, are compounds characterized as inhibitors of the putative AMT. The most important compounds in this category are AM404 (González et al., 1999; Beltramo et al., 2000), VDM11 (de Lago et al., 2004b), OMDM2 (de Lago et al., 2004b), or UCM707 (de Lago et al., 2002). This last compound is the most potent and selective AMT inhibitor developed so far. Compared with transport inhibitors like AM404 that have hypokinetic activity by themselves – perhaps due to activity at other targets (González et al., 1999; see Chap. 9) – UCM707 does not produce any motor effects when administered alone, but it causes a significant potentiation of hypokinetic effects of subeffective doses of AEA (de Lago et al., 2002). This enables this compound to be an interesting pharmacological tool for those diseases, such as HD or other hyperkinetic disorders, where a hypofunction of the cannabinoid signaling has been documented (see below).

Effects of Cannabinoid Receptor Antagonists

As mentioned before, the motor effects of most of cannabinoid agonists were usually prevented by blockade of CB_1 receptors with Rimonabant or other selective antagonists (Souilhac et al., 1995; Di Marzo et al., 2001). However, Rimonabant and other CB_1 receptor blockers have in some reports induced stereotypies and hyperlocomotion in laboratory animals (Compton et al., 1996), which is similar to those data obtained in CB_1 receptor knockout mice that exhibited equivalent motor disturbances (Ledent et al., 1999). The motor effects of these compounds might be related to their reported inverse agonist properties. This raises an interesting therapeutic possibility since it suggests CB_1 receptor inverse agonists to be useful for the treatment of hypokinetic signs associated with overactivity of the cannabinoid system, which seems to include PD and related disorders (discussed in detail below). By the above rationale, a similar therapeutic benefit might be reached by pharmacological inhibition of endocannabinoid synthetic enzymes, such as NAPE-PLD or DAGL. These enzymes have been identified and characterized only recently, however, and selective pharmacological tools to study their potential therapeutic application are still lacking. One interesting compound, however, may be the DAGL inhibitor O3841 (Bisogno et al., 2006), whose motor effects are presently under investigation.

Involvement of $TRPV_1$ Receptors in Motor Effects of Certain Cannabinoids

Recent data have suggested that certain cannabinoids may also produce motor inhibition through the activation of vanilloid $TRPV_1$ receptors (de Lago et al., 2004a). This suggestion is supported by the identification of these receptors in nigrostriatal dopaminergic neurons within the basal ganglia circuitry (Mezey et al., 2000), and also by some pharmacological studies indicating that classic vanilloid agonists or antagonists affected movement. For instance, although some studies, conducted in the 1980s using intranigral injection of capsaicin, described an enhancement of motor activity by this $TRPV_1$ receptor agonist (Dawbarn et al., 1981; Hajos et al., 1988), more recent data indicate that the administration of capsaicin was followed by a strong reduction in locomotor activity in rodents and that these effects are reversed by capsazepine (Di Marzo et al., 2001; Lee et al., 2006). In contrast to CB_1 receptor agonists, capsaicin did not enhance the hypokinetic action of reserpine (Lee et al., 2006). With the above information in mind, several studies have tried to demonstrate that the $TRPV_1$ receptors located within the basal ganglia might represent an alternative target for certain cannabinoid agonists (eicosanoid-derived cannabinoids with vanilloid-like activity, such as AEA or AM404, but not classic cannabinoids) to improve movement in basal ganglia disorders (Lastres-Becker et al., 2002a, 2003a; de Lago et al., 2004a). Thus, in the case of AEA, there exists recent data that demonstrate that the motor inhibition caused by this endocannabinoid is reversed by the blockade of $TRPV_1$ receptors with capsazepine but not by the blockade of CB_1 receptors with Rimonabant (de Lago et al., 2004a). In the same line, other studies described that the increase of $TRPV_1$ receptor binding observed in the striatum of mice deficient in the dopamine transporter (DAT) would be a mechanism aimed at compensating the spontaneous hyperactivity and low striatal AEA levels exhibited by these mice (Tzavara et al., 2006). The importance of the $TRPV_1$ receptor is also evident in studies conducted with AM404 to reduce hyperkinesia in rat models of HD (Lastres-Becker et al., 2002a, 2003a) or with other putative AMT inhibitors that reduced spontaneous hyperlocomotion in DAT knockout mice (Tzavara et al., 2006). In both studies, the reduction in motor hyperactivity was primarily dependent on the capability of these compounds to activate $TRPV_1$ receptors, either directly or indirectly through the elevation of AEA levels.

Endocannabinoids and Parkinson's Disease

The wealth of preclinical research we have described supports the idea that manipulations of the cannabinoid signaling system could be a fruitful therapeutic approach to different disorders affecting the function of the basal ganglia. Additional compelling evidence derives from a large number of studies conducted in postmortem tissue or biological fluids obtained from patients with basal ganglia disorders – or using well-described animal models of these conditions – in which disease symptomology was correlated to significant changes in markers of endocannabinoid function within the basal ganglia. This evidence provides definitive support to the possible use of cannabinoid-based medicines to alleviate symptoms and/or provide neuroprotection in basal ganglia disorders. This includes prominent hypokinetic or hyperkinetic disorders affecting specific neuronal subpopulations within the basal ganglia, such as PD (Lastres-Becker and Fernández-Ruiz, 2006) and HD (Lastres-Becker et al., 2003b, Maccarone et al., 2007), respectively, but also other disorders of the basal ganglia, such as primary dystonias or dyskinesias of different origins, and compulsive tic disorders such as Tourette's Syndrome (Table 1). PD is the most prevalent disorder directly affecting basal ganglia function (de Lau and Breteler, 2006) and there is now considerable interest in the therapeutic possibilities of

Table 1 Alleviation of motor symptoms and/or delay of the disease progression with cannabinoids in patients or animal models of different basal ganglia disorders

Neurological disorder	Symptom relief	Disease progression
Huntington's disease	$TRPV_1$ receptor agonists reduce hyperkinesia (3NP-lesioned rats)	CB_1 receptor agonists reduce excitotoxicity (quinolinate-lesioned rats)
	CB_1 receptor agonists produce only modest effects (3NP-lesioned rats)	CB_2 receptor agonists attenuate microglial toxicity (malonate-lesioned rats)
	Inhibitors of the endocannabinoid uptake are effective only if they also bind $TRPV_1$ receptors (3NP-lesioned rats)	CBD and Δ^9-THC reduce oxidative injury by mechanisms independent of cannabinoid receptors (3NP-lesioned rats)
Parkinson's disease	CB_1 receptor antagonists reduce bradykinesia and restore locomotion in parkinsonian rats, but they do not work in patients	CBD, Δ^9-THC and AM404 reduce oxidative injury by mechanisms independent of cannabinoid receptors (6-hydroxydopamine-lesioned rats)
	CB_1 receptor agonists reduce tremor in parkinsonian animals	CB_1 agonists and inhibitors of the endocannabinoid transporter are not effective (6-hydroxydopamine-lesioned rats)
	Cannabinoid receptor agonists interact with dopaminergic agonists to improve motor deterioration in parkinsonian animals	CB_2 receptor agonists produce only modest effects (in vitro models)
	CB_1 receptor agonists and antagonists delay and reduce levodopa-induced dyskinesia (patients and various animal models)	
Tourette's syndrome	Plant-derived cannabinoids and analogues reduce tics (patients)	
Dystonia	Classic and nonclassic cannabinoid agonists have antidystonic effects (patients and laboratory animals)	
Dyskinesia	CB_1 agonists and or antagonists delay and reduce levodopa-induced dyskinesia (patients and various animal models)	

cannabinoid-based medicines for either alleviating specific symptoms or delaying the progression of this disease (Lastres-Becker and Fernández-Ruiz, 2006). The major clinical neuropathology in PD includes bradykinesia (slowness of movement), rigidity and tremor, which develop following the progressive degeneration of dopaminergic neurons of the SNc and a subsequent severe dopaminergic denervation of its target structures (Blandini et al., 2000). Although the etiology of PD is presently unknown, there is consensus that both genetic (i.e., mutations in different PD-related genes: α-synuclein, parkin, PINK, dardarin, etc.; Abeliovich and Beal, 2006) and environmental (i.e., pesticides, antidopaminergic drugs; Di Monte, 2003) insults are important factors to trigger the disease. This progresses through a series of concomitant cytotoxic processes (i.e., altered proteolysis, oxidative stress, excitotoxicity, mitochondrial failure, and inflammatory stimuli; McGeer et al., 2001; Sherer et al., 2001; Sethi, 2002; Wood-Kaczmar et al., 2006) that synergistically interact to trigger the progressive loss of nigral dopaminergic neurons. Dopaminergic replacement therapy with levodopa represents a useful remedy to release rigidity and bradykinesia in PD patients (Carlsson, 2002; Singh et al., 2007), at least in the early and middle phases of this disease, but patients develop an irreversible state of dyskinesia after 5 or 10 years of prolonged levodopa treatment (Fabbrini et al., 2007). Therefore, major challenges for novel pharmacological therapies in PD are (1) the finding of an alternative symptomatic treatment for those patients that do not respond well to levodopa or to the other treatments enhancing dopaminergic transmission; (2) the development of novel medicines for advanced phases of the disease; (3) the treatment of tremor which is a prominent symptom in a third of patients; (4) the attenuation of levodopa-induced dyskinesia with the use of different types of coadjuvants; and (5) the development of an efficient therapy to arrest or delay the progression of nigral degeneration. It is possible that compounds elevating or inhibiting endocannabinoid activity might provide some benefits in all or part of these therapeutic demands.

Changes in the Endocannabinoid System in PD

Thinking that the activation of the cannabinoid signaling reduces movement, one might expect that this system would become overactive following the dopamine denervation that occurs in PD. Some time ago, it was demonstrated that this hypothesis is correct, as it was observed that the density of CB_1 receptors, as well as the capability of these receptors to activate GTP-binding proteins, were significantly increased in postmortem basal ganglia obtained from PD patients (Lastres-Becker et al., 2001a). In accordance with these data, Pisani et al. (2005) also found an increase in endocannabinoid levels in the cerebrospinal fluid of PD patients. However, a frequent problem with the data obtained in patients is that it is difficult to be precise about whether these increases are exclusively related to the selective degeneration of nigrostriatal dopaminergic neurons, or whether they are a consequence of, or are associated with, the dopaminergic replacement therapy with levodopa that patients receive over the course of several years. For example, in the study conducted by Hurley et al. (2003) in postmortem tissues from normal and parkinsonian human subjects, the authors found a reduction of CB_1 receptor-mRNA levels, assessed by RT-PCR, in some structures (e.g., caudate nucleus, anterior dorsal putamen, GPe) but not in the remaining basal ganglia, and suggested that their data might have resulted from long-term dopamine-increasing treatment received by their patient population prior to death. To determine the relative influence of levodopa treatment in the data found in another study in PD patients (Lastres-Becker et al., 2001a), parallel analyses were conducted in MPTP-lesioned nonhuman primate models of PD, with or without a chronic treatment with the dopamine precursor. Results strongly indicated that the increase in the number and function of CB_1 receptors was directly related to the degeneration of dopaminergic neurons (Lastres-Becker et al., 2001a), a finding also reported by van der Stelt et al. (2005) for the increase in endocannabinoid levels that these authors found in the striatum and GPe, but not in the GPi or SN, in the same PD primate model. Interestingly, in both studies, the chronic administration of levodopa attenuated the increase in the number of CB_1 receptors (Lastres-Becker et al., 2001a) and in the levels of endocannabinoid ligands (van der Stelt et al., 2005). A similar finding emerged from studies conducted in rats lesioned with 6-OHDA by the group of Mauro Macarrone (Gubellini et al., 2002; Macarrone et al., 2003). These authors found that elevated endocannabinoid levels and other abnormalitites of the cannabinoid system observed following 6-OHDA lesions (Gubellini et al., 2002) were markedly reduced by chronic treatment with levodopa (Macarrone et al., 2003). Collectively, these observations support the existence of an imbalance between dopamine and endocannabinoids at the basal ganglia in PD, which is consistent with the conclusions of Mailleux and Vanderhaeghen (1993) that cannabinoid signaling in the basal ganglia is under a negative control exerted by dopamine transmission. Several authors (e.g., van der Stelt et al., 2005) have proposed that the increase in endocannabinoid signaling that parallels dopaminergic denervation of the striatum might represent a compensatory mechanism aimed at reducing an excess of glutamate transmission in this structure, in concordance with the more general suggestion that the reduction of excitatory inputs might serve as an antiparkinsonian therapy (Wu and Frucht, 2005). It is important to consider that enhanced endocannabinoid signaling might also aggravate parkinsonian symptoms due to the hypokinetic profile of direct or indirect agonists of CB_1 receptors (but see Kreitzer and Malenka, 2007). Heightened measures of cannabinoid signaling in the basal ganglia, as reflected in upregulation of CB_1 receptors or increases in endocannabinoid levels, has also been reported in studies using different models of experimental parkinsonism in laboratory animals (Mailleux and Vanderhaeghen, 1993; Romero et al., 2000; Di Marzo et al., 2000c; González et al., 2005). However, the issue retains certain controversy since there are a few studies that reported no changes in endocannabinoid signaling in PD (Herkenham et al., 1991b), or found a reduction (Silverdale et al., 2001) or a dependency on chronic levodopa cotreatment (Zeng et al., 1999). Despite these conflicting data, it is widely evidenced that the cannabinoid signaling becomes overactive in the basal ganglia in PD. This body of evidence is supportive of a view that a general overactivity of endocannabinoid signaling is an event that develops when the degeneration of nigral neurons has progressed to a certain extent and the major parkinsoniam symptoms are already evident. However, some recent studies suggest that losses and/or malfunctioning of the cannabinoid signaling system, mainly at the level of the CB_1 receptor, might be an early event linked to the PD pathogenesis itself. This can be studied in patients affected by incidental Lewy body disease, an early and presymptomatic phase of PD, in which individuals present Lewy bodies and a low degree of nigral pathology, but not neurological symptoms. Using postmortem tissue from a small population of these patients, it was observed that CB_1 receptors already exhibited a trend toward an increase in some basal ganglia structures (Lastres-Becker et al., 2001a). The issue can be also studied in recently developed mouse models of deficiency or mutation in specific genes linked in humans with the development of parkinsonism, such as mice deficient in the PARK-2 (parkin; see Itier et al., 2003) or PARK-1 (α-synuclein; Cabin et al., 2002) genes, or overexpressing a mutated form of α-synuclein (Gispert et al., 2003). The importance of these genetic models is that they can be considered as representative of early stages of parkinsonism when animals only exhibit small disturbances in motor behaviors or markers of dopamine function, but no evidence for protein aggregation or neurodegeneration. Interestingly, these mice show several alterations in the synthesis, density, or function of CB_1 receptors in the SN and other basal ganglia structures (González et al., 2005; Fernández-Ruiz, unpublished results). It is important to emphasize that these receptor changes occurred in a situation where dopaminergic dysfunction rather than neuronal death is the major extant pathological event, thus indicating that these receptor changes might be an early event presumably involved in the neurodegenerative process. These anomalies (losses or malfunctioning) in CB_1 receptors might trigger excitotoxicity, inflammation, or other cytotoxic events that are normally under the control of these receptors, contributing to PD disease progression.

Potential of Cannabinoid-Based Therapies in PD

As mentioned above, the present therapy in PD only allows the alleviation of specific parkinsonian symptoms but fails to delay the progression of nigral degeneration. This consists in the use of the dopamine precursor levodopa which is able to release the rigidity and bradykinesia typical of most parkinsonian patients, in particular, during the early and middle phases of this disease (Carlsson, 2002; Singh et al., 2007). However, not all patients respond to levodopa, and, in those that are well responders, levodopa frequently loses efficacy, and ultimately leads to the appearance of an irreversible dyskinetic state in a period of 5–10 years. Therefore, the treatment of PD patients is demanding the urgent development of novel medicines (1) that are able to reduce parkinsonian symptoms in patients that do not respond to levodopa; (2) that do not develop dyskinesia after prolonged uses; and (3) that serve as neuroprotective molecules able to delay or arrest the progress of nigral degeneration. Cannabinoid-based compounds, either agonists or antagonists, might serve as novel medicines for PD, either used alone or as coadjuvants with classic therapies, as recent preclinical studies and a few clinical trials have indicated. According to these studies, cannabinoid compounds might be used to alleviate parkinsonian symptoms (including the attenuation of levodopa-induced dyskinesia) and to prevent SNc cell death (Brotchie, 2000, 2003; Fernández-Ruiz and González, 2005; Lastres-Becker and Fernández-Ruiz, 2006). However, the type of cannabinoid compound to be used in each of these two aspects might be different. Parkinsonian bradykinesia might be effectively treated by the use of selective CB_1 receptor antagonists, compounds that could also be recommended for delaying the development of dyskinesia associated with long-term levodopa treatment (Brotchie, 2003). By contrast, antioxidant cannabinoid agonists might be the best option to protect dopaminergic nigral cells from death (Lastres-Becker and Fernández-Ruiz, 2006), although this would cover only one component – oxidative injury – of the complex pathophysiology of PD. It is possible that the antiexcitotoxic and/or anti-inflammatory properties of CB_1 and CB_2 agonists, respectively, may be also used to protect nigral cells from death (Fernández-Ruiz et al., 2005).

Alleviation of Parkinsonian Symptoms with Cannabinoid-Based Compounds

Studies carried out almost exclusively in laboratory animals revealed that CB_1 receptor agonists, despite their hypokinetic profile, may be useful in PD under certain circumstances. For instance, they are able to interact with dopaminergic agonists to improve motor impairments (Anderson et al., 1995; Maneuf et al., 1997; Brotchie, 1998; Sañudo-Peña et al., 1998b). In a recent study, Kreitzer and Malenka (2007) proposed that endocannabinoid enhancement, using FAAH or MAGL inhibitors, combined with D_2 receptor agonists may attenuate motor deficits in PD (see striatal LTD discussion above). Cannabinoid agonists might also reduce tremor associated with overactivity of the STN, based on the capability of CB_1 receptors to inhibit glutamate release from subthalamonigral axon terminals (Sañudo-Peña et al., 1998, 1999). Lastly, CB_1 receptor agonists have also been related to a decrease and/or delay in the development of levodopa-induced dyskinesia, as has been described in both laboratory animals (Fox et al., 2002a; Ferrer et al., 2003; Segovia et al., 2003) and human patients (Sieradzan et al., 2001), although a recent clinical trial using a cannabis extract conducted by Carroll et al. (2004) did not replicate the observations in the study by Sieradzan et al. (2001) who used the Δ^9-THC analogue Nabilone™. Despite these therapeutic benefits suggested for CB_1 receptor agonists against certain parkinsonian symptoms, a generally accepted view is that, due to their hypokinetic profile, it is unlikely that these agonists might be useful to alleviate bradykinesia, the major symptom in PD patients. In fact, the few completed studies in humans and MPTP-lesioned primates confirm this view, since the administration of phytocannabinoid agonists was interpreted to worsen motor disability (Consroe, 1998; Müller-Vahl et al., 1999c). By contrast, the blockade of CB_1 receptors has been proposed as a better alternative to reduce bradykinesia, although only in special circumstances which will be described below. CB_1 receptor antagonists, used as coadjuvants with the classic dopaminergic replacement therapy, have been proposed as a better alternative than agonists to delay the appearance and to reduce the severity of levodopa-induced dyskinesia (Brotchie, 2000, 2003), a proposal that has been recently evaluated with positive results in nonhuman primate (van der Stelt et al., 2005) and rat (Segovia et al., 2003) models of PD. The use of CB_1 receptor antagonists in PD is supported by studies describing that cannabinoid signaling becomes overactive in the basal ganglia of PD patients (Lastres-Becker et al., 2001a; Pisani et al., 2005). It is also supported by data obtained in multiple animal models of parkinsonism, including reserpine depletion (Di Marzo et al., 2000c), chronic dopaminergic blockade (Mailleux and Vanderhaeghen, 1993), neurotoxin lesioning with 6-OHDA (Mailleux and Vanderhaeghen, 1993; Romero et al., 2000; Gubellini et al., 2002; Fernández-Espejo et al., 2004), or MPTP (Lastres-Becker et al., 2001a; van der Stelt et al., 2005), or genetic manipulations (González et al., 2005; Fernández-Ruiz, unpublished results). As described above, there are various hypotheses as to how CB_1 receptor antagonists – or agonists – could elicit benefit in a manner specific to different aspects of PD expression (bradykinesia vs. tremor, for example). Given the diverse and synapse-specific effects of CB_1 receptors to regulate neurotransmitter release in the complex circuitry of the basal ganglia (see Fig. 1), it seems premature to espouse any present model too strictly. With that said, although some evaluations of CB_1 receptor blockade on animal PD models have shown positive effects (Di Marzo et al., 2000c; van der Stelt et al., 2005), others have found no benefits (Meschler et al., 2001). The only clinical trial developed so far with the CB_1 receptor antagonist Rimonabant in PD patients showed no improvements (Mesnage et al., 2004). This study was conducted, however, with a population of patients that were all well-responders to classic dopaminergic replacement therapy (Mesnage et al., 2004), and it is possible (but unexamined) that Rimonabant might function better in those patients with poor response to levodopa. Preclinical data indeed suggest that the blockade of CB_1 receptors would require special circumstances to be effective in PD, including attention paid to the following points: (1) the use of low doses of Rimonabant (< 1 mg/kg) in rats (González et al., 2006), although van der Stelt et al. (2005) reported beneficial effects at a dose of 3 mg/kg in MPTP-lesioned primates and (2) that Rimonabant might be effective only at very advanced phases of the disease (Fernández-Espejo et al., 2005). If this hypothesis is correct, it would be possible to have an antiparkinsonian agent for conditions at which classic levodopa therapy generally fails: patients with poor response and/or advanced phases of this disease, which would represent an important pharmacological advantage. On the other hand, the description of $TRPV_1$ receptors in nigrostriatal dopaminergic neurons (Mezey et al., 2000), as well as the neurochemical data indicating that they might play a role in the regulation of dopamine release from nigral neurons (de Lago et al., 2004a), open an additional therapeutic possibility for the endocannabinoid/endovanilloid systems to be used for the alleviation of parkinsonian symptoms. There is evidence that $TRPV_1$ receptors are reduced in the striatum of parkinsonian rats (Lastres-Becker et al., 2005), likely as a consequence of the death of neurons containing this receptor, which would represent a problem for therapeutic manipulation. However, the blockade of remaining receptors in surviving cells might be used to enhance dopamine transmission – assuming an inhibitory function of these receptors in the regulation of dopamine release (de Lago et al., 2004a) – although this requires further verification.

Neuroprotection with Cannabinoid-Based Compounds in PD

Cannabinoids appear to represent a special class of molecules, in which a multifaceted combination of complementary effects exert an overall neuroprotective influence that is mediated by multiple mechanisms (van der Stelt and Di Marzo, 2005; see Chaps. 15 and 16). Thus, cannabinoid agonists are able to reduce excitotoxicity, calcium influx, oxidative injury, cerebro-vasoconstriction, body temperature, and/or local inflammatory events – all of which are effects that could contribute to increasing neuronal survival in this and other acute or chronic neurodegenerative disorders (Fernández-Ruiz et al., 2005; Sarne and Mechoulam, 2005; van der Stelt and Di Marzo, 2005). An interesting aspect of this potential is that it does not include exclusively the activation of CB_1 receptors. Other important targets of the cannabinoid signaling system (e.g., CB_2 receptors) seem to also play key roles. Such diverse neuroprotective potential presents the cannabinoid system as an excellent tool for the treatment of neurodegenerative disorders, because it allows the combination of multiple, supplemental strategies. Therapeutic approaches therefore go beyond the exclusive use of CB_1 receptor agonists, the clinical use of which is confounded by their psychotrophic side effects (Fowler, 2005). Because of their capability to regulate glial influences to neuronal homeostasis, CB_2 receptor agonists are a likely clinical alternative for neuroprotection (Fernández-Ruiz et al., 2007) and possibly, because of its antioxidant and anti-inflammatory properties, CBD may also be of similar use (Mechoulam et al., 2002). Other neuroprotective strategies are being directed toward methods to enhance endocannabinoids, such as targeting the AMT or degradative enzymes (Di Marzo et al., 2004; Fowler et al., 2005). Details on the molecular and cellular pathways underlying the effects of cannabinoids as neuroprotective agents have been discussed in another chapter of this book and will not be addressed here. We will instead concentrate exclusively on the neuroprotective potential of cannabinoids for basal ganglia disorders, for which a schematic diagram of different mechanisms proposed has been outlined in Fig. 3. As regard to PD, recent preclinical studies carried out with Δ^9-THC have provided solid evidence that this plant-derived cannabinoid may reduce the degeneration of nigrostriatal dopaminergic neurons in rats with hemiparkinsonism caused by the unilateral application of 6-OHDA (Lastres-Becker et al., 2005). However, the hypokinetic profile of this phytocannabinoid may represent a disadvantage for any future clinical exploitation of this effect. Interestingly, another phytocannabinoid, CBD, proved to exhibit the same degree of neuroprotection against the toxicity of 6-OHDA in rats (Lastres-Becker et al., 2005). CBD has minimal affinity for CB_1 and CB_2 receptors compared to Δ^9-THC, and this likely represents some important advantages for a possible clinical exploitation of this effect: (1) CBD does not

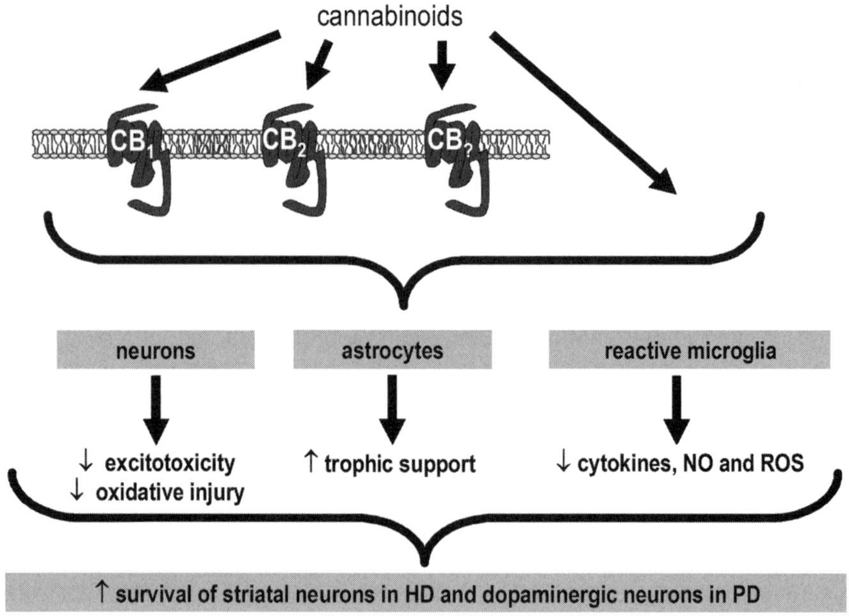

Fig. 3 Mechanisms suggested for the protection exerted by cannabinoids against the damage of specific basal ganglia structures in HD and PD

activate CB_1 receptors (see Chaps. 7 and 9) and so it would not produce any aggravation of motor disability that compounds activating this receptor type (Mechoulam et al., 2002; Russo and Guy, 2006) and (2) CBD does not elicit tolerance compared to Δ^9-THC when used for prolonged treatments (Malfait et al., 2000; Hayakawa et al., 2007). In favor of Δ^9-THC, however, this compound, but not CBD, is able to activate CB_2 receptors, which might represent another important neuroprotective target in PD because of their anti-inflammatory potential; although CBD is also anti-inflammatory by mechanisms that have not yet been identified (Walter and Stella, 2004). On the other hand, the fact that both phytocannabinoids were equally effective in increasing neuronal survival, despite their pharmacodynamic differences, is suggestive that the mechanism providing neuroprotection against 6-OHDA toxicity would be cannabinoid receptor-independent and would emphasize antioxidant properties of these compounds (Lastres-Becker et al., 2005). The same type of antioxidant properties have been proposed for explaining the neuroprotective potential of phytocannabinoids in other in vitro or in vivo models of neurodegeneration (Hampson et al., 1998; Sagredo et al., 2007). Corroborating this hypothesis, recent in vivo studies – investigating the potential of different cannabinoid-based molecules with selectivity for CB receptor types or for the AMT (García-Arencibia et al., 2007) – indicated that only compounds with antioxidant properties, such as AM404, were able to reduce the nigral toxicity of 6-OHDA (García-Arencibia et al., 2007). This effect was likely originated by the antioxidant potential of the phenolic group present in the chemical structure of AM404, rather than its capability to act as an AMT inhibitor, since other AMT inhibitors devoid of this antioxidant potential, like UCM707, failed to protect dopaminergic neurons (García-Arencibia et al., 2007). Selective agonists for the CB_1 receptor, like ACEA, also failed to protect these neurons (García-Arencibia et al., 2007), which is in part concordant with the data published by Kim et al. (2005) indicating that CB_1 receptors enhance rather than reduce the toxic effect of the $TRPV_1$ agonist capsaicin in an in vitro model of PD. As mentioned above, the cause of dopaminergic cell death in PD is still unknown and possibly involves the combination of different pathogenic mechanisms. Alterations in the function of glial cells, including the activation of CB_2 receptor-expressing microglia, might be one of these mechanisms, playing a potentially important role in the initiation and/or early progression of the dopaminergic denervation (Whitton, 2007). Activated microglia and other glial cells, for example, have been described in close proximity to nigral neurons (Kim et al., 2000; Gao et al., 2002; Whitton, 2007). These cells produce several cytotoxic factors, such as TNFα, IL-1β, nitric oxide, reactive oxygen species, and others, that have been reported to be elevated in the SN and the caudate-putamen of PD patients (Nagatsu et al., 2000; Whitton, 2007). Reactive microgliosis has been also described in different animal models of Parkinsonism (Whitton, 2007), and in these animals, the control of the activity of these cells has been reported to provide neuroprotection (Morale et al., 2006). With this idea in mind, some in vitro studies evaluated the importance of glial metabolism in the effects of cannabinoid agonists on 6-OHDA neuronal toxicity. Results indicated that the activation of CB receptors located in glial cells, but not in neurons, was associated with an increase in neuronal survival, presumably exerted by increasing the trophic support exerted by glial cells on neurons (Lastres-Becker et al., 2005). These experiments utilized HU-210, a cannabinoid with antioxidant potential but certainly not selective for CB_2 receptors. However, the importance of CB_2 receptors in the function of glial cells, and in particular microglia (Walter and Stella, 2004), supports a possible role for this receptor type to provide neuroprotection in PD (Fernández-Ruiz et al., 2007). Yet, recent in vivo data only reflected a very modest contribution of CB_2 receptors to the neuroprotective effect of cannabinoids observed in hemiparkinsonian rats (García-Arencibia et al., 2007). Collectively, the different studies conducted so far strongly indicate that the noted neuroprotective properties of certain cannabinoids are mechanistically related to their antioxidant potentials (Lastres-Becker et al., 2005; García-Arencibia et al., 2007). It is important to remark that this fact is of special relevance in a degenerative disorder in which oxidative injury is a major hallmark of pathogenesis (Blandini et al., 2000). CB receptor-independent antioxidant effects are presumably exerted by the phenolic structure of these compounds, which allows them to act as scavengers of reactive oxygen species; although, they might also act by improving the function of endogenous antioxidant enzymes (García-Arencibia et al., 2007). Additionally, cannabinoids may provide neuroprotection through the activation of CB_2 receptors located in glial cells which would regulate the influence of these cells on neuronal homeostasis (Stella, 2004; Walter and Stella, 2004; Fernández-Ruiz et al., 2007). Therefore, the major challenges for future research in this area are essentially two. First, the development of additional preclinical studies aimed at identifying the true potential of CB_2 receptors in PD, in particular: (1) to determine whether these receptors are induced or upregulated in reactive microglia or astrocytes, as happens in other neurodegenerative disorders (Fernández-Ruiz et al., 2007) and (2) to identify the molecular mechanism(s) through which these receptors control the cytotoxic or protective influences exerted by glial cells on neuronal homeostasis, as also described in other neurodegenerative disorders (Fernández-Ruiz et al., 2007). The second challenge would be the clinical validation of the neuroprotective potential of (1) CB_2 receptor agonists – if the above proposed experiments indicate that this receptor type is involved – and in particular (2) the antioxidant cannabinoids CBD and Δ^9-THC, for which preclinical studies have already reported solid and promising possibilities. In this sense, CBD combined with Δ^9-THC is the basis for one of the cannabis-based medicines, Sativex™, which is presently being subjected to clinical testing for its potential in a variety of neurological disorders (Russo and Guy, 2006). This combination appears an excellent option for the clinical exploitation of a cannabinoid-based medicine able to control the progression of PD pathogenesis.

Endocannabinoids and Huntington's Disease

HD is an inherited, progressive, and fatal neurodegenerative disorder caused by the expansion of polyglutamines in the N-terminal of a protein identified in the study of this disease, which was called huntingtin and whose gene (IT15) is located on chromosome 4. The disorder belongs to the family of diseases caused by an excess of CAG repeats in the genes encoding for different proteins, including huntingtin in HD and ataxins in the different spinocerebellar ataxias (Riley and Orr, 2006). In the case of HD, the mode of transmission is autosomal dominant and the limit for CAG repeats is 35. Expansions between 36 and 39 lead to incomplete penetrance, whereas the occurrence of typical adult-onset HD starts with a number of repeats greater than 40. Variables such as an earlier age of onset, the rate of disease progression, and the severity of neuronal damage/neurological deficits mostly correlated with the number of glutamines found in huntingtin (Walker, 2007). Although mutated huntingtin is constitutively expressed, only a few cells are sensitive to its toxic effects. Thus, HD is characterized by a dramatic loss of neurons in the striatum and the cerebral cortex, resulting in motor abnormalities (chorea), cognitive disturbances (dementia), and early death (10–20 years after diagnosed). The mechanism(s) by which the mutated huntingtin causes the progressive loss of striatal and cortical neurons is still pending complete description, but there is consensus that HD is a multifactorial disease, and multiple molecular and cellular mechanisms have been reported to be involved in the pathophysiology (Li and Li, 2006). Part of these pathogenic mechanisms, possibly those that operate in the initiation and during the first steps of disease progression, involve a series of conformational changes caused by polyglutamine expansion in the huntingtin protein, resulting in altered protein–protein interactions, abnormal protein aggregation, and proteolysis (Li and Li, 2004; Borrell-Pagès et al., 2006). These are followed by transcriptional dysregulation affecting some genes involved in neuronal survival (e.g., brain-derived neurotrophic factor, BDNF), excitotoxicity, mitochondrial dysfunction (i.e., complex II deficiency), oxidative stress, and local inflammatory events, and culminating in extensive loss of striatal and cortical neurons (Cattaneo et al., 2005; Borrell-Pagès et al., 2006). At present, there is no specific pharmacotherapy to alleviate symptoms and/or to arrest or delay striatal and cortical degeneration in HD (see below). In this context, cannabinoid-related compounds have been proposed as potential novel medicines for this disorder considering the data generated in a series of preclinical studies (Lastres-Becker et al., 2003b). The rationale for these studies is based on the suspicion that cannabinoid agonists might be used to alleviate hyperkinetic symptomatology in HD because they are hypokinetic substances. In addition, they can also protect striatal neurons from death because of their neuroprotective properties.

Changes in the Endocannabinoid System in HD

Contrarily to the case of PD, the hypokinetic function exerted by cannabinoid signaling would suggest a priori that this system becomes progressively hypofunctional in HD due to the hyperkinetic nature of this disorder (Lastres-Becker et al., 2003b). Studies in postmortem basal ganglia from HD patients have confirmed this hypothesis proving a massive loss of CB_1 receptors in the SN and GPe, and a lesser reduction in the putamen (Glass et al., 1993, 2000; Richfield and Herkenham, 1994). The loss of CB_1 receptors reflects the characteristic neuronal loss observed in the basal ganglia in HD, which predominantly affects GABAergic MSNs. On this level, the loss of CB_1 receptors have been seen as equivalent to the losses of other phenotypic markers of striatal neurons, such as substance P, dynorphin, enkephalin, and adenosine and dopamine receptors (Hersch and Ferrante, 1997). The data obtained in different animal models of HD generally corroborated the findings obtained in human postmortem tissues. Thus, marked losses of CB_1 receptors were also evident in the striatum, GP and SN of rats with lesions of striatal GABA projection neurons, induced with mitochondrial or excitotoxic toxins (Page et al., 2000; Lastres-Becker et al., 2001b, 2002a,b). These toxins, in particular 3-nitropropionic acid (3-NP), reproduce in animals the characteristic mitochondrial complex II deficit described in patients (see Gu et al., 1996) and are associated with the same cytotoxic events that have been proposed for the etiology of the human disease, i.e., failure of energy metabolism, glutamate excitotoxicity, and oxidative stress, leading to progressive neuronal death (Alexi et al., 1998; Brouillet et al., 1999, 2005). As in humans, the losses of CB_1 receptors found in animals lesioned with neurotoxins might represent a mere side effect caused by the progressive and selective destruction of striatal MSNs where these receptors are located. However, there is evidence that these losses happen in early stages where neuronal death does not exist or is still minimal (this important aspect will be addressed below). Interestingly, the losses of CB_1 receptors described in rats lesioned with 3-NP were accompanied by a reduction in the levels of endocannabinoids that was mostly evident in the caudate-putamen (Lastres-Becker et al., 2001b). Therefore, it is possible that endocannabinoid signaling becomes progressively hypofunctional in the basal ganglia in HD. It is likely that this might contribute to some extent to the hyperkinesia typical of this disorder and support a hypothesis that CB_1 receptor agonists might be beneficial to alleviate motor deterioration. In evident contrast with the loss of CB_1 receptors typical of HD, CB_2 receptors seem to be induced or upregulated in the basal ganglia in response to damaging conditions. This seems to occur in reactive microglial cells that are recruited to lesioned sites, and also in astrocytes, in animal models of HD (Fernández-Ruiz et al., 2005, 2007), in similarity to studies with patients affected for other neurodegenerative or neuroinflammatory disorders (Benito et al., 2003, 2005, 2007; see Chap. 16). In these cells, CB_2 receptors would presumably play a protective role by reducing the cytotoxic influences of reactive microglial cells on neuronal homeostasis (Fernández-Ruiz et al., 2007). The occurrence of reactive microgliosis is documented in the striatum of HD patients (Pavese et al., 2006), which would support that they also experience a possible induction or upregulation of CB_2 receptors – this would have interesting pharmacological implications in a disease with a poor therapeutic outcome. In this sense, a recent microarray study conducted in blood samples from HD patients revealed the gene encoding for the CB_2 receptor as one of the genes altered during the development and progression of HD, although this gene was not included among the nine key genes that authors proposed as potential biomarkers of this disease (Borovecki et al., 2005). On the other hand, in HD, as happens in PD, the changes observed in endocannabinoid signaling, rather than being mere secondary consequences of striatal injury, might contribute to the pathogenesis and/or early progression of this disease. Thus, the marked losses of CB_1 receptors described in middle and advanced phases of HD have been already observed at very early phases. They occur in advance of other receptor losses and even before the appearance of major HD symptomatology, therefore in conditions in which cell death is still very low. This has been described by Glass et al. (2000) using postmortem human brains at different grades of HD progression, and further validated in animal models (Lastres-Becker et al., 2003b). For instance, losses of CB_1 receptors were also observed in the basal ganglia of different transgenic mouse models (R6/1, R6/2, and HD94) that express mutated forms of the human huntingtin with different numbers of CAG repeats/polyglutamine expansions (Denovan-Wright and Robertson, 2000; Lastres-Becker et al., 2002c; Naver et al., 2003; McCaw et al., 2004). Importantly, the magnitude of CB_1 receptor down-regulation correlated with the size of the CAG repeats (McCaw et al., 2004) and were attenuated when transgenic animals are housed in an enriched environment (Glass et al., 2004), in parallel to a slow progression of HD symptoms (Hockly et al., 2002). One important aspect of these data is that the reductions of CB_1 receptors were evident at ages of the animals at which cell dysfunction rather than cell death is the major change that takes place. This is absolutely consistent with the results obtained by Glass et al. (2000) in early grades of the human disease, supporting the hypothesis that losses of CB_1 receptors are an early event presumably involved in the initiation and/or first phases of HD pathogenesis. With this notion in mind, Centonze et al. (2005) conducted a series of electrophysiological experiments in the R6/2 transgenic mouse model of HD trying to identify early alterations of endocannabinoid signaling that may relate to the progression of the disease. These authors found a greatly reduced sensitivity of striatal GABAergic synapses to the presynaptic inhibitory effects of CB_1 receptor activation (Centonze et al., 2005; Maccarone et al., 2007). Again, they found that these alterations were not a consequence of striatal degeneration because they were observed at an early stage in the disease progression in R6/2 mice, before the primary initiation of cell death (Turmaine et al., 2000). The mechanism(s) underlying this effect is presently unknown, but there is some evidence that it might be related to impairment in the efficiency of these receptors to activate certain intracellular signals. One of the earliest events following 3-NP intoxication in rats is the occurrence of various anomalies in CB_1 receptor agonist-induced stimulation of GTP-binding proteins in striatal neurons (Lastres-Becker et al., 2004). This occurred in absence of changes in binding and mRNA levels for this receptor and before the first signs of neurodegeneration and neurological deterioration (Lastres-Becker et al., 2004). For some authors, this response and the subsequent losses of CB_1 receptors (Glass et al., 2000; Denovan-Wright and Robertson, 2000; Lastres-Becker et al., 2002c) are interpreted as a compensatory mechanism which might counteract excitoxic damage to MSNs by enhancing GABAergic synaptic function through reduced presynaptic inhibition (see Fig. 2; Maccarrone et al., 2007). An alternative explanation (not necessarily exclusive of the former) could be that the losses or the malfunctioning of CB_1 receptors in specific neuronal subpopulations of the basal ganglia might render these neurons more vulnerable to different cytotoxic stimuli (e.g., oxidative stress, excitotoxicity, inflammation; see Chaps. 15 and 16) that frequently cooperate to produce cell damage in HD (van der Stelt et al., 2002; Fernández-Ruiz et al., 2005). If this were the case, the activation of these receptors might be used as a neuroprotectant strategy in this disease, as will be addressed with more detail in the following section.

Potential of Cannabinoid-Based Therapies in HD

Despite the progression in the elucidation of molecular events involved in the pathogenesis in HD (Cattaneo et al., 2005; Li and Li, 2006; Borrell-Pagès et al., 2006), there is no parallel progression in the development of novel medicines capable of alleviating symptoms and/or delaying degeneration. To date, the only therapeutic tools used in HD include mainly antidopaminergic drugs to reduce the hyperkinesia characteristic of the first phases of the disease (Factor and Friedman, 1997) and antiglutamatergic agents to reduce excitotoxicity (Kieburtz, 1999). There are some novel pharmacological strategies (e.g., unsaturated fatty acids, coenzyme Q10, minocycline, inhibitors of histone deacetylases) that are still under clinical testing (Bonelli and Wenning, 2006; Butler and Bates, 2006;Walker, 2007). In any case, the outcome of these strategies, measured in terms of improving quality of life for HD patients, is still too limited. In this context, cannabinoid agonists might be a reasonable alternative since they can act as antihyperkinetic substances as well as neuroprotective agents.

Alleviation of Hyperkinetic Symptoms with Cannabinoid-Based Compounds

As antihyperkinetic substances, CB_1 receptor ligands may act to acutely recover the neurochemical deficits typical of the hyperkinetic phase of HD (Lastres-Becker et al., 2002a, 2003a). As the disease progresses, this property may be limited by the massive loss of CB_1 receptors and the occurrence of akinesia rather than hyperkinesia as a major symptom during advanced states of this disorder (Lastres-Becker et al., 2002b). Therefore, the activation of the cannabinoid system may serve to alleviate motor disturbances in HD only during the first and intermediate phases of this disease. However, this property seems to be restricted to certain cannabinoids that combine the capability to enhance endocannabinoid signaling but also to directly activate $TRPV_1$ receptors. AM404, for example, was able to reduce hyperkinesia and lead to the recovery from GABAergic and dopaminergic deficits in rats with striatal lesions caused by local application of 3-NP (Lastres-Becker et al., 2002a, 2003a). While activity of this compound as an AMT blocker will enhance the action of endocannabinoids at the CB_1 receptor, the population of CB_1 receptors in the striatum is progressively reduced in parallel to the progression of HD, so that indirect CB_1 receptor activation is likely to be most relevant only in early grades of the disease when cell death is still low (Lastres-Becker et al., 2003b). However, AM404 remained efficacious in more advanced phases of the disease characterized by a moderate degree of neuronal death (Lastres-Becker et al., 2002a), which suggested the contribution of an additional mechanism – possibly the participation of the $TRPV_1$ receptor – in the antihyperkinetic potential of AM404. This was corroborated in a series of experiments demonstrating: (1) that the antihyperkinetic effects of AM404 were reversed by antagonists for the $TRPV_1$ receptor but not for the CB_1 receptor (Lastres-Becker et al., 2003a); (2) that direct agonists of CB_1 receptors, such as CP55940, only produced very modest antihyperkinetic effects (Lastres-Becker et al., 2003a); (3) that inhibitors of endocannabinoid inactivation, devoid of direct capability to activate $TRPV_1$ receptors, such as VDM11 or AM374, were not effective (Lastres-Becker et al., 2003a), and the potent AMT inhibitor UCM707 only produced modest effects (de Lago et al., 2006); and (4) that the $TRPV_1$ receptor agonist capsaicin was also antihyperkinetic (Lastres-Becker et al., 2003a), as was the endocannabinoid/endovanilloid hybrid arvanil, although apparently through distinct mechanisms (de Lago et al., 2005). These data collectively suggest that the $TRPV_1$ receptor may be a relevant target for the treatment of hyperkinesia, the major symptom in HD at least during first grades of the disease. This proposal will need, of course, further clinical validation, especially considering that the only clinical trials conducted to date with cannabinoids to reduce choreic movements in HD patients have failed. It is important to remark that these negative effects could be explained by the lack of $TRPV_1$ receptor activation by the phytocannabinoids (Consroe, 1998) or their synthetic analogues (Müller-Vahl et al., 1999b) used in those clinical trials. The best option for this clinical validation would be the development of "hybrid" compounds with a dual capability to activate both $TRPV_1$ and CB_1 receptors, although the relative contribution of both targets would ideally vary along the course of the disease due to the progressive reduction reported for CB_1 receptors (but with no changes reported in $TRPV_1$ receptor expression) (Lastres-Becker et al., 2003b).

Neuroprotection with Cannabinoid-Based Compounds in HD

As was discussed for PD, the neuroprotective (Romero et al., 2002; Lastres-Becker et al., 2003b; Fernández-Ruiz et al., 2005, 2007) and even neuroregenerative (Galve-Roperh et al., 2007) properties of certain cannabinoids (see Chaps. 15 and 16) may add significantly to the therapeutic utility of these compounds in HD. To date, this neuroprotective potential has been examined only in animal (Lastres-Becker et al., 2003c, 2004; Pintor et al., 2006) and cellular (Aiken et al., 2004; Wang et al., 2005) models of this disease, and although the matter is still far from being completely elucidated, some results have provided promising expectations for a clinical evaluation and future application for patients. It is again valuable to note that glial CB_2 receptors – upregulated in conditions of striatal degeneration (Fernández-Ruiz et al., 2007), in apparent contrast to the loss of neuronal CB_1 receptors in HD – might represent therapeutic targets to attenuate striatal degeneration in this disorder. A recent in vitro study screened a library of 1,040 compounds for their capability to protect cultured PC12 cells from death caused by an expanded polyglutamine form of huntingtin exon 1 and found that various cannabinoids, including CBN, CBD, Δ^8-THC, and Δ^9-THC, showed reproducible protection in this assay (Aiken et al., 2004). However, this was not replicated in a similar study with the same library of compounds (Wang et al., 2005). The issue has been also recently evaluated in vivo using different rat models of striatal damage generated with excitotoxic or mitochondrial toxins that selectively replicate some of the cytotoxic events that cooperatively contribute to HD pathogenesis. For example, Pintor et al. (2006) hypothesized that HD patients (which present low levels of CB_1 receptors in the striatum (Glass et al., 1993, 2000; Richfield and Herkenham, 1994)) would experience a reduction in CB_1 receptor-mediated inhibition of glutamate release in this structure, thus resulting in excitotoxicity. To validate this hypothesis, they used rats lesioned with quinolinate and found that the activation of CB_1 receptors indeed reduced the striatal damage caused by this excitotoxin (Pintor et al., 2006). Cannabinoids may also be effective against other types of neurotoxic events that specifically operate in HD patients. An example of this is the complex II deficiency characteristic of HD patients (Gu et al., 1996) that may be experimentally reproduced by using selective inhibitors of complex II like 3-NP, characteristics of which have been detailed above. Striatal injury in rats lesioned with this toxin progresses by mechanisms that mainly involve nonapoptotic pathways, since neuronal death in this model is reportedly caspase 3-independent and produced instead via the Ca^{2+}-regulated protein calpain (Bizat et al., 2003; Galas et al., 2004). The phytocannabinoids, Δ^9-THC (Lastres-Becker et al., 2004) and CBD (Sagredo et al., 2007), have been found to protect striatal neurons against the in vivo toxicity of 3-NP in rats, and to a similar extent. Importantly, selective agonists of CB_1, CB_2, or $TRPV_1$ receptors failed to provide neuroprotection in this animal model (Sagredo et al., 2007), again implicating antioxidant actions as a critical mechanism. This finding is particularly relevant because the generation of reactive oxygen and nitrogen species seems to be a key process, among others, in the striatal injury provoked by intoxication in rats with 3-NP (Brouillet et al., 2005). CB_2 receptors have also been proposed as an alternative target for the treatment of HD based on data obtained in another rat model of HD (Fernández-Ruiz et al., 2007). This model was generated by unilateral injections of malonate, another complex II inhibitor that, in contrast with 3-NP, is reversible, allowing certain recovery of the mitochondrial function and producing the death of striatal neurons through the activation of apoptotic machinery (via activation of NMDA receptors and caspase-3) (Toulmond et al., 2004). In this model, selective agonists for the CB_2 receptor, but not for the CB_1 receptor, were able to protect striatal neurons from malonate-induced cell death (Fernández-Ruiz et al., 2007). AMT inhibitors like UCM707 (de Lago et al., 2006) or antioxidant cannabinoids like CBD (Fernández-Ruiz et al., 2007) also failed to reproduce this neuroprotective effect, thus stressing the importance of CB_2 receptors. This was also supported by experiments conducted with selective CB_2 receptor antagonists (Fernández-Ruiz et al., 2005, 2007). As mentioned above, CB_2 receptors are present in low concentrations in the healthy striatum, presumably located in astrocytes, but in response to malonate-induced damage, they are induced or upregulated in reactive microglial cells (as several double-labeling immunohistochemical analyses have confirmed), which then become activated and migrate to the lesioned sites (Fernández-Ruiz et al., 2007). In these cells, CB_2 receptors might control the production of cytotoxic factors, such as nitric oxide, reactive oxygen species, and in particular, proinflammatory cytokines released by microglial cells, deteriorating neuronal homeostasis (Stella, 2004; Walter and Stella, 2004; Fernández-Ruiz et al., 2007; see Chaps. 15 and 16). This hypothesis is supported by data obtained in HD patients that show (1) activation of glial cells (i.e., astrocytes, oligodendroglia or microglia; see Rajkowska et al., 1998; Sapp et al., 2001) associated with the toxicity of mutated huntingtin (Singhrao et al., 1999) and (2) higher levels of TNFα and IL-2 (Bonifati and Kishore, 2007). The activation of CB_2 receptors may in fact reduce the malonate-induced increase in the production of TNFα, which behaves as a proinflammatory mediator aggravating the striatal damage generated by the mitochondrial failure caused by the neurotoxin (Fernández-Ruiz et al., unpublished results). On the other hand, it is important to remark that the induction or upregulation of CB_2 receptors in glial cells in response to malonate-induced damage is not a phenomenon exclusive of the striatum and of HD, since it has also been observed in other brain regions and in other disorders (Benito et al., 2003, 2005, 2007). Therefore, it should be interpreted as a part of an endogenous response against neuronal death caused by local inflammatory events (Pazos et al., 2004; Fernández-Ruiz et al., 2005, 2007). Importantly, this endogenous response may be the basis for the design of a novel neuroprotective strategy based on selective CB_2 receptor agonists capable of controlling microglial toxic influences of neurons. This therapeutic option is consistent with the general idea that neuroprotective and anti-inflammatory properties should be combined for the therapy of neurodegenerative disorders, since both neurodegeneration and neuroinflammation are frequently cooperative events in these disorders (Maccarrone et al., 2007). There are therefore three key mechanisms that enable cannabinoids to provide neuroprotection in HD (see Fig. 3 for a schematic overview): (1) their capability to normalize glutamate release processes via CB_1 receptors, which is expected to mitigate excitotoxic events that occur in this pathology; (2) the receptor-independent antioxidant potential of certain cannabinoids, which can reduce the oxidative injury that takes place in HD; and (3) their activity at CB_2 receptors to control microglial influences on neuronal survival, thus reducing the local inflammatory events that are associated with striatal degeneration. The availability of rat models of HD that reproduce these phenomena with certain selectivity has allowed researchers to resolve separable cannabinoid mechanisms that are differentially effective against each of these cytotoxic processes. However, these cytotoxic events occur in a cooperative manner during the pathogenesis of HD in humans (Borrell-Pages et al., 2006). In this context, the use of nonselective or hybrid compounds with relatively broad-spectrum cannabinoid properties might be the best solution. Centonze et al. (2007) have recently reviewed this issue providing excellent ideas that support how different cannabinoids with specific properties may be combined as a rational strategy in HD and in other neurodegenerative disorders, irrespective of the nature of the primary insult. To this point, we have tried to provide all available information to sustain that cannabinoids have neuroprotective potential in HD, exerted by preventing striatal cell death caused by different neurotoxic stimuli that operate in this disease. However, this is not the only way for cannabinoids to delay or arrest the progression of the disease. In this regard, it is important to indicate that cannabinoids have been recently involved in the control of adult neurogenesis (Aguado et al., 2005; see also Chap. 12), which occurs mainly in two forebrain regions, the subventricular zone and the hippocampal dentate gyrus (Taupin and Gage, 2002). Although the evidence accumulated so far on this potential is still very limited, some studies have already suggested that it could serve as a novel therapy for different neurodegenerative disorders, including HD (Galve-Roperh et al., 2006, 2007; Maccarrone et al., 2007). According to such a view, a cannabinoid-sensitive mechanism could allow the replacement of dead neurons through the proliferation of cell progenitors, their differentiation into neurons, and their migration to the damaged striatum, where they might acquire the typical phenotype of the striatal MSNs that are mainly lost in HD. Recent data have actually demonstrated the presence of a population of progenitor cells expressing cannabinoid receptors in the subependymal layer of the adult normal and HD human brain (Curtis et al., 2006), which might represent a suitable source for the replacement of cells lost due to striatal degeneration (Maccarrone et al., 2007). In a general sense, this effect might be comparable to the effects reported for different neurotrophic factors, such as FGF-2 and BDNF in HD models (Barnabe-Heider and Miller, 2003; Curtis et al., 2003; Jin et al., 2005).

Other Disorders Affecting the Basal Ganglia

Cannabinoid-based medicines might also provide certain benefits to alleviate motor or behavioral abnormalities in other disorders affecting directly or indirectly the basal ganglia structures. This is the case for tardive dyskinesia, Gilles de la Tourette's syndrome, primary dystonias, and other disorders (Consroe, 1998; Romero et al., 2002; Müller-Vahl, 2003). In brief, cannabinoid agonists have antidystonic effects reported in humans (Fox et al., 2002b) or animal models (Richter and Löscher, 1994, 2002). Patients with Gilles de la Tourette's syndrome – a compulsive tic disorder of proposed striatal etiology – experienced a reduction in the severity of behavioral tics when they were treated with plant-derived cannabinoids; this includes data obtained from patients who self-medicate with cannabis (Hemming and Yellowlees, 1993; Consroe, 1998; Müller-Vahl et al., 1998, 1999a, 2002; Müller-Vahl, 2003). Cannabinoids might also be useful for the treatment of different types of dyskinesias, in particular, in the case of levodopa-induced dyskinesias where both agonists and antagonists of CB_1 receptors have been reported to be beneficial (see key references in the section corresponding to PD in this chapter). However, this pharmacological evidence has progressed with little indications that the endocannabinoid system is specifically altered during the development of these pathologies. In a similar way, the neurochemical substrates underlying the beneficial effects reported for cannabinoids in these basal ganglia disorders have not yet been identified. There is a growing perspective, however, that disorders of unwanted or habitual behaviors – ranging from dyskinesias and Tourette's syndrome to drug addiction – may be fundamentally related to the "habit-learning" or procedural memory functions of the basal ganglia (Gerdeman et al., 2003; Yin and Knowlton, 2006), and either pathologically "learned" or reinforced through mechanisms of long-term synaptic plasticity (which could be overactivated by genetic variation in key proteins). As discussed at length earlier in this chapter, such synaptic learning functions in the striatum and its target nuclei of the basal ganglia are under a remarkable level of control by retrograde endocannabinoid signaling and presynaptic CB_1 receptors, as well as ACh, which has fundamental roles in striatal mnemonic function. It is proposed that CB_1 receptor-dependent LTD in the striatum may be a substrate for the learning of both adaptive behavioral habits in rats (Gerdeman et al., 2006, unpublished observations) and chronic compulsive behaviors related not only to drug addiction, but also perhaps to disorders of compulsive movement

(Gerdeman et al., 2003). Two other prevalent disorders with notable basal ganglia malfunctioning, and thus deserving some additional comment, are multiple sclerosis (MS) and Alzheimer's disease (AD). Both disorders should not be mentioned here with great detail because they are not originated by a primary degeneration of the basal ganglia, and also because they have been extensively addressed in Chaps. 18 and 19. However, the primary cause of both disorders (e.g., autoimmune activation in the case of MS, and β-amyloid-dependent degeneration of cortical and subcortical areas in the case of AD) secondarily induces malfunctioning of the basal ganglia circuitry triggering the occurrence of a series of notable extrapyramidal symptoms. Studies in laboratory animals have convincingly demonstrated that both direct and indirect cannabinoid receptor agonists are useful to alleviate motor-related symptoms in MS, including spasticity, tremor, and dystonia as the most relevant (Pertwee, 2002; Baker and Pryce, 2003). These preclinical observations have been mostly validated in a series of clinical trials recently completed using some cannabinoid-based medicines (Zajicek et al., 2003, 2005; Vaney et al., 2004; Wade et al., 2004). An important observation for the purpose of the present chapter is that in rodent models of this disease, most changes observed in CB_1 receptors were in large part restricted to basal ganglia structures (Berrendero et al., 2001; Cabranes et al., 2006), which may be related to the fact that motor deterioration is one of the most prominent neurological signs in MS. The changes in receptors were also accompanied by changes in endocannabinoid levels seen in multiple brain structures, including, but not restricted to, the basal ganglia (Baker et al., 2001; Cabranes et al., 2005). Changes found in basal ganglia endocannabinoid markers (Berrendero et al., 2001; Cabranes et al., 2005, 2006) were, however, not accompanied by parallel changes in vesicular transmitters such as GABA, dopamine, serotonin, or glutamate (Cabranes et al., 2005). This is an important observation that may in part explain the greater potential of cannabinoid-related compounds to alleviate the motor deterioration (spasticity, tremor, dystonia) observed in patients with MS. AD is likewise not a disorder of the basal ganglia, but the occurrence of extrapyramidal signs, possibly caused by the degeneration of glutamate cortical afferents to the striatum, is frequent in these patients (Kurlan et al., 2000). Studies in postmortem brain regions of patients affected by this disease have revealed a significant loss of CB_1 receptors within the basal ganglia, in particular the caudate nucleus, and to a lesser extent, the SN and GP (Westlake et al., 1994). However, these changes did not correlate with the histopathology (e.g., location of neuritic plaques or fibrillary tangles) suggesting that these results related more to increasing age (Mailleux and Vanderhaeghen, 1992b; Romero et al., 1998b) rather than to an effect selectively associated with the histopathology characteristic of AD (Westlake et al., 1994). Additional studies have analyzed the expression of CB_2 receptors in postmortem tissue from patients with AD (Benito et al., 2003; Ramírez et al., 2005). These authors described increased CB_2 receptor detection in activated microglia surrounding senile plaques, in parallel to losses in the number of CB_1 receptors located in degenerating neurons (Benito et al., 2003; see details in Chaps. 16 and 19). This observation suggests key relationships of both CB_1 and CB_2 receptors in either the pathogenesis or treatment of AD (Milton, 2002; Iuvone et al., 2004; Ramírez et al., 2005; Esposito et al., 2007).

Concluding Remarks

The cannabinoid signaling system plays a key role in the control of basal ganglia function, as now supported by a large body of multidisciplinary research. We have extensively reviewed this evidence, and used these findings to propose models for cannabinoid receptor function in health and disease of the basal ganglia. We feel that this body of research has established an excellent rationale for the development of novel pharmacotherapies with compounds that selectively target specific elements of the cannabinoid system, thus increasing or reducing the endogenous activity of this signaling system through various mechanisms. These approaches hold promise not only for the alleviation of specific symptoms of multiple disorders (e.g., hyperkinesia in HD, tremor and bradykinesia in PD, tics in Tourette's syndrome, or spasticity and dystonia in MS) but they also could provide a remarkable array of benefits in terms of delaying or arresting the progression of these neurodegenerative diseases, due to the neuroprotectant or neuroregenerative properties described for certain cannabinoids. However, most of the studies which have examined the therapeutic potential of these compounds in basal ganglia disorders have been conducted in animal or cellular models, whereas the number of clinical trials is still too limited. Therefore, we see two major horizons for future research in this field. First – the development of novel compounds with more selectivity for the different proteins (CB_1, CB_2, and $TRPV_1$ receptors, AMT, synthetic enzymes, and the degrading enzymes, FAAH, MAGL) which altogether constitute the cannabinoid system. This would represent an attempt to have therapeutic compounds that minimize the frequent side effects observed when classic cannabinoids are used in patients – yet combinations of different types of cannabinoids might remain the best option in particular cases, and the therapeutic utility of known phytocannabinoids are clearly still being discovered. Second – the development of more clinical studies that validate the symptomatic and/or neuroprotective properties of different cannabinoids for the treatment of basal ganglia pathology. This will allow the promising potentials of cannabinoid medicines to progress from growing clinical evidence to actual clinical utility.

Acknowledgments Studies included in this chapter have been possible by grants from "Ministerio de Educación y Ciencia, Programa Nacional de Biomedicina" (SAF2003-–08269 and SAF2006-–11333), CIBERNED (CB06/05/0089), and "Comunidad de Madrid" (S-SAL-0261/2006) to Javier Fernández-Ruiz. Gregory L. Gerdeman acknowledges support by the National Institute on Drug Abuse (DA14263-–04), and thanks Shinai A. Schindler for critical reading of an earlier draft of the manuscript.

References

Abeliovich A, Beal MF (2006) Parkinsonism genes: culprits and clues. J Neurochem 99:1062–1072.

Ade KK, Lovinger DM (2007) Anandamide regulates postnatal development of long-term synaptic plasticity in the rat dorsolateral striatum. J Neurosci 27:2403–2409.

Ade KK, Kunos G, Lovinger DM (2003) Stimulation of dopamine D_2 receptors induces the formation of anandamide (AEA) but not 2-arachidonoyl glycerol (2-AG) in rat cortical-striatal slices. Symposium on the Cannabinoids International Cannabinoid Research Society (Burlington, VT), p. 88.

Aguado T, Monory K, Palazuelos J, Stella N, Cravatt B, Lutz B, Marsicano G, Kokaia Z, Guzman M, Galve-Roperh I (2005) The endocannabinoid system drives neural progenitor proliferation. FASEB J 19:1704–1706.

Aiken CT, Tobin AJ, Schweitzer ES (2004) A cell-based screen for drugs to treat Huntington's disease. Neurobiol Dis 16:546–555.

Alexi T, Hughes PE, Faull RLM, Williams LE (1998) 3-Nitropropionic acid's lethal triplet: cooperative pathways of neurodegeneration. Neuroreport 9:57–64.

Alger B (2002) Retrograde signaling in the regulation of synaptic transmission: focus on endocannabinoids. Prog Neurobiol 68:247–286.

Anderson LA, Anderson JJ, Chase TN, Walters JR (1995) The cannabinoid agonists WIN55,212-2 and CP55,940 attenuate rotational behaviour induced by a dopamine D_1 but not D_2 agonist in rats with unilateral lesions of the nigrostriatal pathway. Brain Res 691:106–114.

Ashton JC, Friberg D, Darlington CL, Smith PF (2006) Expression of the cannabinoid CB_2 receptor in the rat cerebellum: an immunohistochemical study. Neurosci Lett 396:113–116.

Baker D, Pryce G (2003) The therapeutic potential of cannabis in multiple sclerosis. Expert Opin Investig Drugs 12:561–567.

Baker D, Pryce G, Croxford JL, Brown P, Pertwee RG, Makriyannis A, Khanolkar A, Layward L, Fezza F, Bisogno T, Di Marzo V (2001) Endocannabinoids control spasticity in a multiple sclerosis model. FASEB J 15:300–302.

Barnabe-Heider F, Miller FD (2003) Endogenously produced neurotrophins regulate survival and differentiation of cortical progenitors via distinct signaling pathways. J Neurosci 23:5149–5160.

Beltramo M, Rodríguez de Fonseca F, Navarro M, Calignano A, Gorriti MA, Grammatikopoulos G, Sadile AG, Giuffrida A, Piomelli D (2000) Reversal of dopamine D_2 receptor responses by an anandamide transport inhibitor. J Neurosci 20:3401–3407.

Benito C, Nunez E, Tolon RM, Carrier EJ, Rabano A, Hillard CJ, Romero J (2003) Cannabinoid CB_2 receptors and fatty acid amide hydrolase are selectively overexpressed in neuritic plaque-associated glia in Alzheimer's disease brains. J Neurosci 23:11136–11141.

Benito C, Kim WK, Chavarria I, Hillard CJ, Mackie K, Tolon RM, Williams K, Romero J (2005) A glial endogenous cannabinoid system is upregulated in the brains of macaques with simian immunodeficiency virus-induced encephalitis. J Neurosci 25:2530–2536.

Benito C, Romero JP, Tolon RM, Clemente D, Docagne F, Hillard CJ, Guaza C, Romero J (2007) Cannabinoid CB_1 and CB_2 receptors and fatty acid amide hydrolase are specific markers of plaque cell subtypes in human multiple sclerosis. J Neurosci 27:2396–2402.

Bernardi G, et al. (2004) A critical interaction between dopamine D_2 receptors and endocannabinoids mediates the effects of cocaine on striatal GABAergic transmission. Neuropsychopharmacology 29:1488–1497.

Berrendero F, Sánchez A, Cabranes A, Puerta C, Ramos JA, García-Merino A, Fernández-Ruiz J (2001) Changes in cannabinoid CB_1 receptors in striatal and cortical regions of rats with experimental allergic encephalomyelitis, an animal model of multiple sclerosis. Synapse 41:195–202.

Bevan MD, Magill PJ, Terman D, Bolam JP, Wilson CJ (2002) Move to the rhythm: oscillations in the subthalamic nucleus-external globus pallidus network. Trends Neurosci 25:525–531.

Bisogno T, Berrendero F, Ambrosino G, Cebeira M, Ramos JA, Fernández-Ruiz JJ, Di Marzo V (1999) Brain regional distribution of endocannabinoids: implications for their biosynthesis and biological function. Biochem Biophys Res Commun 256:377–380.

Bisogno T, Cascio MG, Saha B, Mahadevan A, Urbani P, Minassi A, Appendino G, Saturnino C, Martin B, Razdan R, Di Marzo V (2006) Development of the first potent and specific inhibitors of endocannabinoid biosynthesis. Biochim Biophys Acta 1761:205–212.

Bizat N, Hermel JMH, Humbert S, Jacquard C, Créminon C, Escartin C, Saudou F, Krajewski S, Hantraye P, Brouillet E (2003) *In vivo* calpain/caspase cross-talk during 3-nitropropionic acid-induced striatal degeneration: implication of a calpain-mediated cleavage of active caspase-3. J Biol Chem 278:43245–43253.

Blandini F, Nappi G, Tassorelli C, Martignoni E (2000) Functional changes in the basal ganglia circuitry in Parkinson's disease. Prog Neurobiol 62:63–88.

Bonelli RM, Wenning GK (2006) Pharmacological management of Huntington's disease: an evidence-based review. Curr Pharm Des 12:2701–2720.

Bonifati DM, Kishore U (2007) Role of complement in neurodegeneration and neuroinflammation. Mol Immunol 44:999–1010.

Borland LM, Michael AC (2004) Voltammetric study of the control of striatal dopamine release by glutamate. J Neurochem 91:220–229.

Borovecki F, Lovrecic L, Zhou J, Jeong H, Then F, Rosas HD, Hersch SM, Hogarth P, Bouzou B, Jensen RV, Krainc D (2005) Genome-wide expression profiling of human blood reveals biomarkers for Huntington's disease. Proc Natl Acad Sci USA 102:11023–11028.

Borrell-Pages M, Zala D, Humbert S, Saudou F (2006) Huntington's disease: from huntingtin function and dysfunction to therapeutic strategies. Cell Mol Life Sci 63:2642–2660.

Brotchie JM (1998) Adjuncts to dopamine replacement: a pragmatic approach to reducing the problem of dyskinesia in Parkinson's disease. Mov Disord 13:871–876.

Brotchie JM (2000) The neural mechanisms underlying levodopa-induced dyskinesia in Parkinson's disease. Ann Neurol 47:S105–S114.

Brotchie JM (2003) CB_1 cannabinoid receptor signalling in Parkinson's disease. Curr Opin Pharmacol 3:54–61.

Brouillet E, Conde F, Beal MF, Hantraye P (1999) Replicating Huntington's disease phenotype in experimental animals. Prog Neurobiol 59:427–468.

Brouillet E, Jacquard C, Bizat N, Blum D (2005) 3-Nitropropionic acid: a mitochondrial toxin to uncover physiopathological mechanisms underlying striatal degeneration in Huntington's disease. J Neurochem 95:1521–1540.

Brown TM, Brotchie JM, Fitzjohn SM (2003) Cannabinoids decrease corticostriatal synaptic transmission via an effect on glutamate uptake. J Neurosci 23:11073–11077.

Buckley NE, McCoy KL, Mezey E, Bonner T, Zimmer A, Felder CC, Glass M, Zimmer A (2000) Immunomodulation by cannabinoids is absent in mice deficient for the cannabinoid CB_2 receptor. Eur J Pharmacol 396:141–149.

Butler R, Bates GP (2006) Histone deacetylase inhibitors as therapeutics for polyglutamine disorders. Nat Rev Neurosci 7:784–796.

Cabin DE, Shimazu K, Murphy D, Cole NB, Gottschalk W, McIlwain KL, Orrison B, Chen A, Ellis CE, Paylor R, Lu B, Nussbaum RL (2002) Synaptic vesicle depletion correlates with attenuated synaptic responses to prolonged repetitive stimulation in mice lacking alpha-synuclein. J Neurosci 22:8797–8807.

Cabranes A, Venderova K, de Lago E, Fezza F, Sánchez A, Mestre L, Valenti M, García-Merino A, Ramos JA, Di Marzo V, Fernández-Ruiz J (2005) Decreased endocannabinoid levels in the brain and beneficial effects of agents activating cannabinoid and/or vanilloid receptors in a rat model of multiple sclerosis. Neurobiol Dis 20:207–217.

Cabranes A, Pryce G, Baker D, Fernández-Ruiz J (2006) Changes in CB_1 receptors in motor-related brain structures of chronic relapsing experimental allergic encephalomyelitis mice. Brain Res 1107:199–205.

Cadogan AK, Alexander SP, Boyd EA, Kendall DA (1997) Influence of cannabinoids on electrically evoked dopamine release and cyclic AMP generation in the rat striatum. J Neurochem 69:1131–1137.

Calabresi P, Maj R, Mercuri NB, Bernardi G (1992) Coactivation of D_1 and D_2 dopamine receptors is required for long-term synaptic depression in the striatum. Neurosci Lett 142:95–99.

Calabresi P, Pisani A, Mercuri NB, Bernardi G (1996) The corticostriatal projection: from synaptic plasticity to dysfunctions of the basal ganglia. Trends Neurosci 19:19–24.

Calabresi P, Giacomini P, Centonze D, Bernardi G (2000) Levodopa-induced dyskinesia: a pathological form of striatal synaptic plasticity? Ann Neurol 47:60–68; discussion:68–69.

Carlsson A (2002) Treatment of Parkinson's with L-DOPA. The early discovery phase, and a comment on current problems. J Neural Transm 109:777–787.

Carriba P, Ortiz O, Patkar K, Justinova Z, Stroik J, Themann A, Müller C, Woods AS, Hope BT, Ciruela F, Casadó V, Canela EI, Lluis C, Goldberg SR, Moratalla R, Franco R, Ferré S (2007) Striatal adenosine A_{2A} and cannabinoid CB_1 receptors form functional heteromeric complexes that mediate the motor effects of cannabinoids. Neuropsychopharmacology doi:10.1038/sj.npp.1301375.

Carroll CB, Bain PG, Teare L, Liu X, Joint C, Wroath C, Parkin SG, Fox P, Wright D, Hobart J, Zajicek JP (2004) Cannabis for dyskinesia in Parkinson disease: a randomized double-blind crossover study. Neurology 63:1245–1250.

Castaneda E, Moss DE, Oddie SD, Whishaw IQ (1991) THC does not affect striatal dopamine release: microdialysis in freely moving rats. Pharmacol Biochem Behav 40:587–591.

Cattaneo E, Zuccato C, Tartari M (2005) Normal huntingtin function: an alternative approach to Huntington's disease. Nat Rev Neurosci 6:919–930.

Centonze D, Battista N, Rossi S, Mercuri NB, Finazzi-Agro A, Bernardi G, et al. (2004) A critical interaction between dopamine D_2 receptors and endocannabinoids mediates the effects of cocaine on striatal GABAergic transmission. Neuropsychopharmacology 29:1488–1497.

Centonze D, Rossi S, Prosperetti C, Tscherter A, Bernardi G, Maccarrone M, Calabresi P (2005) Abnormal sensitivity to cannabinoid receptor stimulation might contribute to altered gamma-aminobutyric acid transmission in the striatum of R6/2 Huntington's disease mice. Biol Psychiatry 57:1583–1589.

Centonze D, Finazzi-Agro A, Bernardi G, Maccarrone M (2007) The endocannabinoid system in targeting inflammatory neurodegenerative diseases. Trends Pharmacol Sci 28:180–187.

Chan PK, Yung WH (1998) Occlusion of the presynaptic action of cannabinoids in rat substantia nigra pars reticulata by cadmium. Neurosci Lett 249:57–60.

Chan PK, Chan SC, Yung WH (1998) Presynaptic inhibition of GABAergic inputs to rat substantia nigra pars reticulata neurones by a cannabinoid agonist. Neuroreport 9:671–675.

Chevaleyre V, Takahashi KA, Castillo PE (2006) Endocannabinoid-mediated synaptic plasticity in the CNS. Annu Rev Neurosci 29:37–76.

Compton DR, Aceto MD, Lowe J, Martin BR (1996) *In vivo* characterization of a specific cannabinoid antagonis (SR141716A): inhibition of Δ^9-tetrahydrocannabinol-induced responses and apparent agonist activity. J Pharmacol Exp Ther 277:586–594.

Consroe P (1998) Brain cannabinoid systems as targets for the therapy of neurological disorders. Neurobiol Dis 5:534–551.

Cravatt BF, Demarest K, Patricelli MP, Bracey MH, Giang DK, Martin BR, Lichtman AH (2001) Supersensitivity to anandamide and enhanced endogenous cannabinoid signaling in mice lacking fatty acid amide hydrolase. Proc Natl Acad Sci USA 98:9371–9376.

Crawley JN, Corwin RL, Robinson JK, Felder ChC, Devane WA, Axelrod J (1993) Anandamide, an endogenous ligand of the cannabinoid receptor, induces hypomotility and hypothermia *in vivo* in rodents. Pharmacol Biochem Behav 46:967–972.

Cristino L, de Petrocellis L, Pryce G, Baker D, Guglielmotti V, Di Marzo V (2006) Immunohistochemical localization of cannabinoid type 1 and vanilloid transient receptor potential vanilloid type 1 receptors in the mouse brain. Neuroscience 139:1405–1415.

Curtis MA, Penney EB, Pearson AG, van Roon-Mom WM, Butterworth NJ, Dragunow M, Connor B, Faull RL (2003) Increased cell proliferation and neurogenesis in the adult human Huntington's disease brain. Proc Natl Acad Sci USA 100:9023–9027.

Curtis MA, Faull RL, Glass M (2006) A novel population of progenitor cells expressing cannabinoid receptors in the subependymal layer of the adult normal and Huntington's disease human brain.. J Chem Neuroanat 31:210–215.

Dawbarn D, Harmar AJ, Pycock CJ (1981) Intranigral injection of capsaicin enhances motor activity and depletes nigral 5-hydroxytryptamine but not substance P. Neuropharmacology 20:341–346.

de Lago E, Fernandez-Ruiz J, Ortega-Gutierrez S, Viso A, Lopez-Rodriguez ML, Ramos JA (2002) UCM707, a potent and selective inhibitor of endocannabinoid uptake, potentiates hypokinetic and antinociceptive effects of anandamide. Eur J Pharmacol 449:99–103.

de Lago E, de Miguel, Lastres-Becker I, Ramos JA, Fernández-Ruiz J (2004a) Involvement of vanilloid-like receptors in the effects of anandamide on motor behavior and nigrostriatal dopaminergic activity: *in vivo* and *in vitro* evidence. Brain Res 1007:152–159.

de Lago E, Ligresti A, Ortar G, Morera E, Cabranes A, Pryce G, Bifulco M, Baker D, Fernández-Ruiz J, Di Marzo V (2004b) *In vivo* pharmacological actions of two novel inhibitors of anandamide cellular uptake. Eur J Pharmacol 484:249–257.

de Lago E, Urbani P, Ramos JA, Di Marzo V, Fernández-Ruiz J (2005) Arvanil, a hybrid endocannabinoid and vanilloid compound, behaves as an antihyperkinetic agent in a rat model of Huntington's disease. Brain Res 1050:210–216.

de Lago E, Fernández-Ruiz J, Ortega-Gutierrez S, Cabranes A, Pryce G, Baker D, Lopez-Rodriguez M, Ramos JA (2006) UCM707, an inhibitor of the anandamide uptake, behaves as a symptom control agent in models of Huntington's disease and multiple sclerosis, but fails to delay/arrest the progression of different motor-related disorders. Eur Neuropsychopharmacol 16:7–18.

de Lau LM, Breteler MM (2006) Epidemiology of Parkinson's disease. Lancet Neurol 5:525–535.

Denovan-Wright EM, Robertson HA (2000) Cannabinoid receptor messenger RNA levels decrease in subset neurons of the lateral striatum, cortex and hippocampus of transgenic Huntington's disease mice. Neuroscience 98:705–713.

Desarnaud F, Cadas H, Piomelli D (1995) Anandamide amidohydrolase activity in rat brain microsomes. Identification and partial purification. J Biol Chem 270:6030–6035.

Dewey WL (1986) Cannabinoid pharmacology. Pharmacol Rev 38:151–178.

Diana MA, Marty A (2004) Endocannabinoid-mediated short-term synaptic plasticity: depolarization-induced suppression of inhibition (DSI) and depolarization-induced suppression of excitation (DSE). Br J Pharmacol 142:9–19.

Di Marzo V, Breivogel C, Bisogno T, Melck D, Patrick G, Tao Q, Szallasi A, Razdan RK, Martin BR (2000a) Neurobehavioral activity in mice of N-vanillyl-arachidonyl-amide. Eur J Pharmacol 406:363–374.

Di Marzo V, Breivogel CS, Tao Q, Bridgen DT, Razdan RK, Zimmer AM, Zimmer A, Martin BR (2000b) Levels, metabolism, and pharmacological activity of anandamide in CB_1 cannabinoid receptor knockout mice: evidence for non-CB_1, non-CB_2 receptor-mediated actions of anandamide in mouse brain. J Neurochem 75:2434–2444.

Di Marzo V, Hill MP, Bisogno T, Crossman AR, Brotchie JM (2000c) Enhanced levels of endocannabinoids in the globus pallidus are associated with a reduction in movement in an animal model of Parkinson's disease. FASEB J 14, 1432–1438.

Di Marzo V, Berrendero F, Bisogno T, Gonzalez S, Cavaliere P, Romero J, Cebeira M, Ramos JA, Fernandez-Ruiz J (2000d) Enhancement of anandamide formation in the limbic forebrain and reduction of endocannabinoid contents in the striatum of Δ^9-tetrahydrocannabinol-tolerant rats. J Neurochem 74:1627–1635.

Di Marzo V, Lastres-Becker I, Bisogno T, De Petrocellis L, Milone A, Davis JB, Fernandez-Ruiz J (2001) Hypolocomotor effects in rats of capsaicin and two long chain capsaicin homologues. Eur J Pharmacol 420:123–131.

Di Marzo V, Bifulco M, De Petrocellis L (2004) The endocannabinoid system and its therapeutic exploitation. Nat Rev Drug Discov 3:771–784.

Di Monte DA (2003) The environment and Parkinson's disease: is the nigrostriatal system preferentially targeted by neurotoxins? Lancet Neurol 2:531–538.

Dinh TP, Freund TF, Piomelli D (2002) A role for monoglyceride lipase in 2-arachidonoylglycerol inactivation. Chem Phys Lipids 121:149–158.

Egertova M, Elphick MR (2000) Localisation of cannabinoid receptors in the rat brain using antibodies to the intracellular C-terminal tail of CB. J Comp Neurol 422:159–171.

Egertova M, Cravatt BF, Elphick MR (2003) Comparative analysis of fatty acid amide hydrolase and CB_1 cannabinoid receptor expression in the mouse brain: evidence of a widespread role for fatty acid amide hydrolase in regulation of endocannabinoid signaling. Neuroscience 119:481–496.

Engler B, Freiman I, Urbanski M, Szabo B (2006) Effects of exogenous and endogenous cannabinoids on GABAergic neurotransmission between the caudate-putamen and the globus pallidus in the mouse. J Pharmacol Exp Ther 316:608–617.

Esposito G, Iuvone T, Savani C, Scuderi C, De Filippis D, Papa M, Di Marzo V, Steardo L (2007) Opposing control of cannabinoid receptor stimulation on amyloid-β induced reactive gliosis: *in vitro* and *in vivo* evidence. J Pharmacol Exp Ther 322:1144–1152.

Fabbrini G, Brotchie JM, Grandas F, Nomoto M, Goetz CG (2007) Levodopa-induced dyskinesias. Mov Disord 22:1379–1389.

Factor SA, Friedman JH (1997) The emerging role of clozapine in the treatment of movement disorders. Mov Disord 12:483–496.

Felder CC, Glass M (1998) Cannabinoid receptors and their endogenous agonists. Annu Rev Pharmacol Toxicol 38:179–200.

Fernández-Espejo E, Caraballo I, Rodriguez de Fonseca F, Ferrer B, El Banoua F, Flores JA, Galan-Rodriguez B (2004) Experimental parkinsonism alters anandamide precursor synthesis, and functional deficits are improved by AM404: a modulator of endocannabinoid function. Neuropsychopharmacology 29:1134–1142.

Fernández-Espejo E, Caraballo I, de Fonseca FR, El Banoua F, Ferrer B, Flores JA, Galan-Rodriguez B (2005) Cannabinoid CB_1 antagonists possess antiparkinsonian efficacy only in rats with very severe nigral lesion in experimental parkinsonism. Neurobiol Dis 18:591–601.

Fernández-Ruiz J, González S (2005) Cannabinoid control of motor function at the basal ganglia. In: Pertwee RG (ed.), Handbook of Experimental Pharmacology – 168 – Cannabinoids. Heidelberg: Springer-Verlag, pp. 479–507.

Fernandez-Ruiz J, Lastres-Becker I, Cabranes A, Gonzalez S, Ramos JA (2002) Endocannabinoids and basal ganglia functionality. Prostaglandins Leukot Essent Fatty Acids 66:257–267.

Fernández-Ruiz J, González S, Romero J, Ramos JA (2005) Cannabinoids in neurodegeneration and neuroprotection. In: Mechoulam R (ed.), Cannabinoids as Therapeutics (MDT). Switzerland: Birkhaüser Verlag, pp. 79–109.

Fernández-Ruiz J, Romero J, Velasco G, Tolón RM, Ramos JA, Guzmán M (2007) Cannabinoid CB_2 receptor: a new target for the control of neural cell survival? Trends Pharmacol Sci 28:39–45.

Ferrer B, Asbrock N, Kathuria S, Piomelli D, Giuffrida A (2003) Effects of levodopa on endocannabinoid levels in rat basal ganglia: implications for the treatment of levodopa-induced dyskinesias. Eur J Neurosci 18:1607–1614.

Fitzpatrick JS, Akopian G, Walsh JP (2001) Short-term plasticity at inhibitory synapses in rat striatum and its effects on striatal output. J Neurophysiol 85:2088–2099.

Fowler CJ (2005) Pharmacological properties and therapeutic possibilities for drugs acting upon endocannabinoid receptors. Curr Drug Targets CNS Neurol Disord 4:685–696.

Fowler CJ, Holt S, Nilsson O, Jonsson KO, Tiger G, Jacobsson SO (2005) The endocannabinoid signaling system: pharmacological and therapeutic aspects. Pharmacol Biochem Behav 81:248–262.

Fox SH, Henry B, Hill M, Crossman A, Brotchie J (2002a) Stimulation of cannabinoid receptors reduces levodopa-induced dyskinesia in the MPTP-lesioned nonhuman primate model of Parkinson's disease. Mov Disord 17:1180–1187.

Fox SH, Kellett M, Moore AP, Crossman AR, Brotchie JM (2002b) Randomised, double-blind, placebo-controlled trial to assess the potential of cannabinoid receptor stimulation in the treatment of dystonia. Mov Disord 17:145–149.

Freiman I, Szabo B (2005) Cannabinoids depress excitatory neurotransmission between the subthalamic nucleus and the globus pallidus. Neuroscience 133:305–313.

Freiman I, Anton A, Monyer H, Urbanski MJ, Szabo B (2006) Analysis of the effects of cannabinoids on identified synaptic connections in the caudate-putamen by paired recordings in transgenic mice. J Physiol 575:789–806.
French ED, Dillon K, Wu X (1997) Cannabinoids excite dopamine neurons in the ventral tegmentum and substantia nigra. Neuroreport 8:649–652.
Freund TF, Katona I, Piomelli D (2003) Role of endogenous cannabinoids in synaptic signaling. Physiol Rev 83:1017–1066.
Fride E, Mechoulam R (1993) Pharmacological activity of the cannabinoid receptor agonist, anandamide, a brain constituent. Eur J Pharmacol 231:313–314.
Fusco FR, Martorana A, Giampa C, De March Z, Farini D, D'Angelo V, Sancesario G, Bernardi G (2004) Immunolocalization of CB_1 receptor in rat striatal neurons: a confocal microscopy study. Synapse 53:159–167.
Galante M, Diana MA (2004) Group I metabotropic glutamate receptors inhibit GABA release at interneuron-Purkinje cell synapses through endocannabinoid production. J Neurosci 24:4865–4874.
Galas MC, Bizat N, Cuvelier L, Bantubungi K, Brouillet E, Schiffmann SN, Blum D (2004) Death of cortical and striatal neurons induced by mitochondrial defect involves differential molecular mechanisms. Neurobiol Dis 15:152–159.
Galve-Roperh I, Aguado T, Rueda D, Velasco G, Guzman M (2006) Endocannabinoids: a new family of lipid mediators involved in the regulation of neural cell development. Curr Pharm Des 12:2319–2325.
Galve-Roperh I, Aguado T, Palazuelos J, Guzman M (2007) The endocannabinoid system and neurogenesis in health and disease. Neuroscientist 13:109–114.
Gao HM, Jiang J, Wilson B, Zhang W, Hong JS, Liu B (2002) Microglial activation-mediated delayed and progressive degeneration of rat nigral dopaminergic neurons: relevance to Parkinson's disease. J Neurochem 81:1285–1297.
García-Arencibia M, González S, de Lago E, Ramos JA, Mechoulam R, Fernández-Ruiz J (2007) Evaluation of the neuroprotective effect of cannabinoids in a rat model of Parkinson's disease: importance of antioxidant and cannabinoid receptor-independent properties. Brain Res 1134:162–170.
Gerdeman G, Lovinger DM (2001) CB_1 cannabinoid receptor inhibits synaptic release of glutamate in rat dorsolateral striatum. J Neurophysiol 85:468–471.
Gerdeman GL, Ronesi J, Lovinger DM (2002) Postsynaptic endocannabinoid release is critical to long-term depression in the striatum. Nat Neurosci 5:446–451.
Gerdeman GL, Partridge JG, Lupica CR, Lovinger DM (2003) It could be habit forming: drugs of abuse and striatal synaptic plasticity. Trends Neurosci 26:184–192.
Gerdeman GL, Schechter JB, French ED (2006) Inhibition of stimulus-response (habit) learning by striatal injection of the CB_1 antagonist rimonabant. Symposium on the Cannabinoids International Cannabinoid Research Society (Burlington, VT), p. 178.
Gessa GL, Melis M, Muntoni AL, Diana M (1998) Cannabinoids activate mesolimbic dopamine neurons by an action on cannabinoid CB_1 receptors. Eur J Pharmacol 341:39–44.
Gifford AN, Samiian L, Gatley SJ, Ashby Jr CR (1997) Examination of the effect of the cannabinoid receptor agonist, CP 55,940, on electrically evoked transmitter release from rat brain slices. Eur J Pharmacol 324:187–192.
Gispert S, Del Turco D, Garrett L, Chen A, Bernard DJ, Hamm-Clement J, Korf HW, Deller T, Braak H, Auburger G, Nussbaum RL (2003) Transgenic mice expressing mutant A53T human alpha-synuclein show neuronal dysfunction in the absence of aggregate formation. Mol Cell Neurosci 24:419–429.
Giuffrida A, Parsons LH, Kerr TM, Rodriguez de Fonseca F, Navarro M, Piomelli D (1999) Dopamine activation of endogenous cannabinoid signaling in dorsal striatum. Nat Neurosci 2:358–363.
Giuffrida A, Beltramo M, Piomelli D (2001) Mechanisms of endocannabinoid inactivation: biochemistry and pharmacology. J Pharmacol Exp Ther 298:7–14.
Glass M, Faull RLM, Dragunow M (1993) Loss of cannabinoid receptors in the substantia nigra in Huntington's disease. Neuroscience 56:523–527.

Glass M, Dragunow M, Faull RL (1997) Cannabinoid receptors in the human brain: a detailed anatomical and quantitative autoradiographic study in the fetal, neonatal and adult human brain. Neuroscience 77:299–318.

Glass M, Dragunow M, Faull RL (2000) The pattern of neurodegeneration in Huntington's disease: a comparative study of cannabinoid, dopamine, adenosine and GABA-A receptor alterations in the human basal ganglia in Huntington's disease. Neuroscience 97:505–519.

Glass M, van Dellen A, Blakemore C, Hannan AJ, Faull RL (2004) Delayed onset of Huntington's disease in mice in an enriched environment correlates with delayed loss of cannabinoid CB_1 receptors. Neuroscience 123:207–212.

Gong JP, Onaivi ES, Ishiguro H, Liu QR, Tagliaferro PA, Brusco A, Uhl GR (2006) Cannabinoid CB_2 receptors: immunohistochemical localization in rat brain. Brain Res 1071:10–23.

González S, Romero J, de Miguel R, Lastres-Becker I, Villanúa MA, Makriyannis A, Ramos JA, Fernández-Ruiz J (1999) Extrapyramidal and neuroendocrine effects of AM404, an inhibitor of the carrier-mediated transport of anandamide. Life Sci 65:327–336.

González S, Mena MA, Lastres-Becker I, Serrano A, de Yebenes JG, Ramos JA, Fernandez-Ruiz J (2005) Cannabinoid CB_1 receptors in the basal ganglia and motor response to activation or blockade of these receptors in parkin-null mice. Brain Res 1046:195–206.

Gonzalez S, Scorticati C, Garcia-Arencibia M, de Miguel R, Ramos JA, Fernandez-Ruiz J (2006) Effects of rimonabant, a selective cannabinoid CB_1 receptor antagonist, in a rat model of Parkinson's disease. Brain Res 1073–1074:209–219.

Gorriti MA, Rodríguez de Fonseca F, Navarro M, Palomo T (1999) Chronic (−) Δ^9-tetrahydrocannabinol treatment induces sensitization to the psychomotor effects of amphetamine in rats. Eur J Pharmacol 365:133–142.

Graybiel AM (2005) The basal ganglia: learning new tricks and loving it. Curr Opin Neurobiol 15:638–644.

Graybiel AM, Rauch SL (2000) Toward a neurobiology of obsessive-compulsive disorder. Neuron 28:343–347.

Graybiel AM, Aosaki T, Flaherty AW, Kimura M (1994) The basal ganglia and adaptive motor control. Science 265:1826–1831.

Gu M, Gash MT, Mann VM, Javoy-Agid F, Cooper JM, Schapira AH (1996) Mitochondrial defect in Huntington's disease caudate nucleus. Ann Neurol 39:385–389.

Gubellini P, Saulle E, Centonze D, Bonsi P, Pisani A, Bernardi G, Conquet F, Calabresi P (2001) Selective involvement of $mGlu_1$ receptors in corticostriatal LTD. Neuropharmacology 40:839–846.

Gubellini P, Picconi B, Bari M, Battista N, Calabresi P, Centonze D, Bernardi G, Finazzi-Agro A, Maccarrone M (2002) Experimental parkinsonism alters endocannabinoid degradation: implications for striatal glutamatergic transmission. J Neurosci 22:6900–6907.

Gueudet C, Santucci V, Rinaldi-Carmona M, Soubrie P, Le Fur G (1995) The CB_1 cannabinoid receptor antagonist SR 141716A affects A9 dopamine neuronal activity in the rat. Neuroreport 6:1421–1425.

Hajos M, Engberg G, Nissbrandt H, Magnusson T, Carlsson A (1988) Capsaicin-sensitive vasodilatatory mechanisms in the rat substantia nigra and striatum. J Neural Transm 74:129–139.

Hampson AJ, Grimaldi M, Axelrod J, Wink D (1998) Cannabidiol and (−)Δ^9-tetrahydrocannabinol are neuroprotective antioxidants. Proc Natl Acad Sci USA 95:8268–8273.

Hanus L, Breuer A, Tchilibon S, Shiloah S, Goldenberg D, Horowitz M, Pertwee RG, Ross RA, Mechoulam R, Fride E (1999) HU-308: a specific agonist for CB_2, a peripheral cannabinoid receptor. Proc Natl Acad Sci USA 96:14228–14233.

Hashimotodani Y, Ohno-Shosaku T, Kano M (2007) Ca^{2+}-assisted receptor-driven endocannabinoid release: mechanisms that associate presynaptic and postsynaptic activities. Curr Opin Neurobiol 17:360–365.

Hayakawa K, Mishima K, Nozako M, Ogata A, Hazekawa M, Liu AX, Fujioka M, Abe K, Hasebe N, Egashira N, Iwasaki K, Fujiwara M (2007) Repeated treatment with cannabidiol but not Δ^9-

tetrahydrocannabinol has a neuroprotective effect without the development of tolerance. Neuropharmacology 52:1079–1087.

Hemming M, Yellowlees PM (1993) Effective treatment of Tourette's syndrome with marijuana. J Psychopharmacol 7:389–391.

Henneberger C, Redman SJ, Grantyn R (2007) Cortical efferent control of subcortical sensory neurons by synaptic disinhibition. Cereb Cortex 17:2039–2049.

Herkenham M, Lynn AB, de Costa BR, Richfield EK (1991a) Neuronal localization of cannabinoid receptors in the basal ganglia of the rat. Brain Res 547:267–274.

Herkenham M, Lynn AB, Johnson MR, Melvin LS, de Costa BR, Rice KC (1991b) Characterization and localization of cannabinoid receptors in rat brain: a quantitative *in vitro* autoradiographic study. J Neurosci 11:563–583.

Hersch SM, Ferrante RJ (1997) Neuropathology and pathophysiology of Huntington's disease. In: Watts RL, Koller WC (eds.), Movement Disorders. Neurologic Principles and Practice. New York: McGraw-Hill, pp. 503–518.

Hiltunen AJ, Jarbe TU, Wangdahl K (1988) Cannabinol and cannabidiol in combination: temperature, open-field activity, and vocalization. Pharmacol Biochem Behav 30:675–678.

Hockly E, Cordery PM, Woodman B, Mahal A, van Dellen A, Blakemore C, Lewis CM, Hannan AJ, Bates GP (2002) Environmental enrichment slows disease progression in R6/2 Huntington's disease mice. Ann Neurol 51:235–242.

Hohmann AG, Herkenham M (2000) Localization of cannabinoid CB_1 receptor mRNA in neuronal subpopulations of rat striatum: a double-label in situ hybridization study. Synapse 37:71–80.

Huang CC, Lo SW, Hsu KS (2001) Presynaptic mechanisms underlying cannabinoid inhibition of excitatory synaptic transmission in rat striatal neurons. J Physiol 532:731–748.

Huffman JW (2005) CB_2 receptor ligands. Mini Rev Med Chem 5:641–649.

Hurley MJ, Mash DC, Jenner P (2003) Expression of cannabinoid CB_1 receptor mRNA in basal ganglia of normal and parkinsonian human brain. J Neural Transm 110:1279–1288.

Itier JM, Ibanez P, Mena MA, Abbas N, Cohen-Salmon C, Bohme GA, Laville M, Pratt J, Corti O, Pradier L, Ret G, Joubert C, Periquet M, Araujo F, Negroni J, Casarejos MJ, Canals S, Solano R, Serrano A, Gallego E, Sanchez M, Denefle P, Benavides J, Tremp G, Rooney TA, Brice A, Garcia de Yebenes J (2003) Parkin gene inactivation alters behaviour and dopamine neurotransmission in the mouse. Hum Mol Genet 12:2277–2291.

Iuvone T, Esposito G, Esposito R, Santamaria R, Di Rosa M, Izzo AA (2004) Neuroprotective effect of cannabidiol, a non-psychoactive component from Cannabis sativa, on beta-amyloid-induced toxicity in PC12 cells. J Neurochem 89:134–141.

Jarbe TU, Sheppard R, Lamb RJ, Makriyannis A, Lin S, Goutopoulos A (1998) Effects of Δ^9-tetrahydrocannabinol and (R)-methanandamide on open-field behavior in rats. Behav Pharmacol 9:169–174.

Jayamanne A, Greenwood R, Mitchell VA, Aslan S, Piomelli D, Vaughan CW (2006) Actions of the FAAH inhibitor URB597 in neuropathic and inflammatory chronic pain models. Br J Pharmacol 147:281–288.

Jin K, LaFevre-Bernt M, Sun Y, Chen S, Gafni J, Crippen D, Logvinova A, Ross CA, Greenberg DA, Ellerby LM (2005) FGF-2 promotes neurogenesis and neuroprotection and prolongs survival in a transgenic mouse model of Huntington's disease. Proc Natl Acad Sci USA 102:18189–18194.

Julian MD, Martin AB, Cuellar B, Rodriguez De Fonseca F, Navarro M, Moratalla R, Garcia-Segura LM (2003) Neuroanatomical relationship between type 1 cannabinoid receptors and dopaminergic systems in the rat basal ganglia. Neuroscience 119:309–318.

Jung KM, Mangieri R, Stapleton C, Kim J, Fegley D, Wallace M, Mackie K, Piomelli D (2005) Stimulation of endocannabinoid formation in brain slice cultures through activation of group I metabotropic glutamate receptors. Mol Pharmacol 68:1196–1202.

Kalant H (2004) Adverse effects of cannabis on health: an update of the literature since 1996. Prog Neuropsychopharmacol Biol Psychiatry 28:849–863.

Kaneda K, Kita T, Kita H (2007) Repetitive activation of glutamatergic inputs evokes a long-lasting excitation in rat globus pallidus neurons *in vitro*. J Neurophysiol 97:121–133.

Kathmann M, Bauer U, Schlicker E, Gothert M (1999) Cannabinoid CB_1 receptor-mediated inhibition of NMDA- and kainate-stimulated noradrenaline and dopamine release in the brain. Naunyn Schmiedebergs Arch Pharmacol 359:466–470.
Kearn CS, Blake-Palmer K, Daniel E, Mackie K, Glass M (2005) Concurrent stimulation of cannabinoid CB_1 and dopamine D_2 receptors enhances heterodimer formation: a mechanism for receptor cross-talk? Mol Pharmacol 67:1697–1704.
Kieburtz K (1999) Antiglutamate therapies in Huntington's disease. J Neural Transm Suppl 55:97–102.
Kim WG, Mohney RP, Wilson B, Jeohn GH, Liu B, Hong JS (2000) Regional difference in susceptibility to lipopolysaccharide-induced neurotoxicity in the rat brain: role of microglia. J Neurosci 20:6309–6316.
Kim SR, Lee DY, Chung ES, Oh UT, Kim SU, Jin BK (2005) Transient receptor potential vanilloid subtype 1 mediates cell death of mesencephalic dopaminergic neurons *in vivo* and *in vitro*. J Neurosci 25:662–671.
Köfalvi A, Rodrigues RJ, Ledent C, Mackie K, Vizi ES, Cunha RA, Sperlagh B (2005) Involvement of cannabinoid receptors in the regulation of neurotransmitter release in the rodent striatum: a combined immunochemical and pharmacological analysis. J Neurosci 25:2874–2884.
Koos T, Tepper JM (1999) Inhibitory control of neostriatal projection neurons by GABAergic interneurons. Nat Neurosci 2:467–472.
Kreitzer AC, Malenka RC (2005) Dopamine modulation of state-dependent endocannabinoid release and long-term depression in the striatum. J Neurosci 25:10537–10545.
Kreitzer AC, Malenka RC (2007) Endocannabinoid-mediated rescue of striatal LTD and motor deficits in Parkinson's disease models. Nature 445:643–647.
Kurlan R, Richard IH, Papka M, Marshall F (2000) Movement disorders in Alzheimer's disease: more rigidity of definitions is needed. Mov Disord 15:24–29.
Lastres-Becker I, Fernandez-Ruiz J (2006) An overview of Parkinson's disease and the cannabinoid system and possible benefits of cannabinoid-based treatments. Curr Med Chem 13:3705–3718.
Lastres-Becker I, Cebeira M, de Ceballos M, Zeng B-Y, Jenner P, Ramos JA, Fernández-Ruiz J (2001a) Increased cannabinoid CB_1 receptor binding and activation of GTP-binding proteins in the basal ganglia of patients with Parkinson's syndrome and of MPTP-treated marmosets. Eur J Neurosci 14:1827–1832.
Lastres-Becker I, Fezza F, Cebeira M, Bisogno T, Ramos JA, Milone A, Fernández-Ruiz J, Di Marzo V (2001b) Changes in endocannabinoid transmission in the basal ganglia in a rat model of Huntington's disease. Neuroreport 12:2125–2129.
Lastres-Becker I, Hansen HH, Berrendero F, de Miguel R, Pérez-Rosado A, Manzanares J, Ramos JA, Fernández-Ruiz J (2002a) Alleviation of motor hyperactivity and neurochemical deficits by endocannabinoid uptake inhibition in a rat model of Huntington's disease. Synapse 44:23–35.
Lastres-Becker I, Gomez M, De Miguel R, Ramos JA, Fernández-Ruiz J (2002b) Loss of cannabinoid CB_1 receptors in the basal ganglia in the late akinetic phase of rats with experimental Huntington's disease. Neurotox Res 4:601–608.
Lastres-Becker I, Berrendero F, Lucas JJ, Martín-Aparicio E, Yamamoto A, Ramos JA, Fernández-Ruiz J (2002c) Loss of mRNA levels, binding and activation of GTP-binding proteins for cannabinoid CB_1 receptors in the basal ganglia of a transgenic model of Huntington's disease. Brain Res 929:236–242.
Lastres-Becker I, de Miguel R, De Petrocellis L, Makriyannis A, Di Marzo V, Fernández-Ruiz J (2003a) Compounds acting at the endocannabinoid and/or endovanilloid systems reduce hyperkinesia in a rat model of Huntington's disease. J Neurochem 84:1097–1109.
Lastres-Becker I, De Miguel R, Fernández-Ruiz J (2003b) The endocannabinoid system and Huntington's disease. Curr Drug Targets CNS Neurol Disord 2:335–347.
Lastres-Becker I, Bizat N, Boyer F, Hantraye P, Brouillet E, Fernández-Ruiz J (2003c) Effects of cannabinoids in the rat model of Huntington's disease generated by an intrastriatal injection of malonate. Neuroreport 14:813–816.

Lastres-Becker I, Bizat N, Boyer F, Hantraye P, Fernández-Ruiz J, Brouillet E (2004) Potential involvement of cannabinoid receptors in 3-nitropropionic acid toxicity *in vivo*. Neuroreport 15:2375–2379.

Lastres-Becker I, Molina-Holgado F, Ramos JA, Mechoulam R, Fernández-Ruiz J (2005) Cannabinoids provide neuroprotection against 6-hydroxydopamine toxicity *in vivo* and *in vitro*: relevance to Parkinson's disease. Neurobiol Dis 19:96–107.

Lastres-Becker I, Molina-Holgado F, Ramos JA, Mechoulam R, Fernández-Ruiz J (2005) Cannabinoids provide neuroprotection against 6-hydroxydopamine toxicity *in vivo* and *in vitro*: relevance to ParkinsonLedent C, Valverde O, Cossu G, Petitet F, Aubert JF, Beslot F, Böhme GA, Imperato A, Pedrazzini T, Roques BP, Vassart G, Fratta W, Parmentier M (1999) Unresponsiveness to cannabinoids and reduced addictive effects of opiates in CB_1 receptor knockout mice. Science 283:401–404.

Lee J, Di Marzo V, Brotchie JM (2006) A role for vanilloid receptor 1 ($TRPV_1$) and endocannabinnoid signalling in the regulation of spontaneous and L-DOPA induced locomotion in normal and reserpine-treated rats. Neuropharmacology 51:557–565.

Li SH, Li XJ (2004) Huntingtin and its role in neuronal degeneration. Neuroscientist 10:467–475.

Li S, Li XJ (2006) Multiple pathways contribute to the pathogenesis of Huntington disease. Mol Neurodegener 1:19.

Lichtman AH, Hawkins EG, Griffin G, Cravatt BF (2002) Pharmacological activity of fatty acid amides is regulated, but not mediated, by fatty acid amide hydrolase *in vivo*. J Pharmacol Exp Ther 302:73–79.

Lupica CR, Riegel AC (2005) Endocannabinoid release from midbrain dopamine neurons: a potential substrate for cannabinoid receptor antagonist treatment of addiction. Neuropharmacology 48:1105–1116.

Maccarrone M, Gubellini P, Bari M, Picconi B, Battista N, Centonze D, Bernardi G, Finazzi-Agro A, Calabresi P (2003) Levodopa treatment reverses endocannabinoid system abnormalities in experimental parkinsonism. J Neurochem 85:1018–1025.

Maccarrone M, Battista N, Centonze D (2007) The endocannabinoid pathway in Huntington's disease: a comparison with other neurodegenerative diseases. Prog Neurobiol 81:349–379.

Mackie K (2005) Distribution of cannabinoid receptors in the central and peripheral nervous system. In: Pertwee RG (ed.), Handbook of Experimental Pharmacology – 168 – Cannabinoids. Heidelberg: Springer-Verlag, pp. 299–325.

Mailleux P, Vanderhaeghen JJ (1992a) Distribution of neuronal cannabinoid receptor in the adult rat brain: a comparative receptor binding radioautography and *in situ* hybridization histochemistry. Neuroscience 48:655–668.

Mailleux P, Vanderhaeghen JJ (1992b) Age-related loss of cannabinoid of cannabinoid receptor binding sites and mRNA in the rat striatum. Neurosci Lett 147:179–181.

Mailleux P, Vanderhaeghen JJ (1993) Dopaminergic regulation of cannabinoid receptor mRNA levels in the rat caudate-putamen: an in situ hybridization study. J Neurochem 61:1705–1712.

Malfait AM, Gallily R, Sumariwalla PF, Malik AS, Andreakos E, Mechoulam R, Feldmann M (2000) The nonpsychoactive cannabis constituent cannabidiol is an oral anti-arthritic therapeutic in murine collagen-induced arthritis. Proc Natl Acad Sci USA 97:9561–9566.

Maneuf YP, Crossman AR, Brotchie JM (1996a) Modulation of GABAergic transmission in the globus pallidus by the synthetic cannabinoid WIN 55,212-2. Synapse 22:382–385.

Maneuf YP, Nash JE, Croosman AR, Brotchie JM (1996b) Activation of the cannabinoid receptor by Δ^9-THC reduces GABA uptake in the globus pallidus. Eur J Pharmacol 308:161–164.

Maneuf YP, Croosman AR, Brotchie JM (1997) The cannabinoid receptor agonist WIN 55,212-2 reduces D_2, but not D_1, dopamine receptor-mediated alleviation of akinesia in the reserpine-treated rat model of Parkinson's disease. Exp Neurol 148:265–270.

Marinelli S, Di Marzo V, Berretta N, Matias I, Maccarrone M, Bernardi G, Mercuri NB (2003) Presynaptic facilitation of glutamatergic synapses to dopaminergic neurons of the rat substantia nigra by endogenous stimulation of vanilloid receptors. J Neurosci 23:3136–3144.

Marsicano G, Lutz B (1999) Expression of the cannabinoid receptor CB_1 in distinct neuronal subpopulations in the adult mouse forebrain. Eur J Neurosci 11:4213–4225.
Matyas F, Yanovsky Y, Mackie K, Kelsch W, Misgeld U, Freund TF (2006) Subcellular localization of type 1 cannabinoid receptors in the rat basal ganglia. Neuroscience 137:337–361.
McCaw EA, Hu H, Gomez GT, Hebb AL, Kelly ME, Denovan-Wright EM (2004) Structure, expression and regulation of the cannabinoid receptor gene (CB_1) in Huntington's disease transgenic mice. Eur J Biochem 271:4909–4920.
McGeer PL, Yasojima K, McGeer EG (2001) Inflammation in Parkinson's disease. Adv Neurol 86:83–89.
McLaughlin PJ, Delevan CE, Carnicom S, Robinson JK, Brener J (2000) Fine motor control in rats is disrupted by Δ^9-tetrahydrocannabinol. Pharmacol Biochem Behav 66:803–809.
Mechoulam R, Parker LA, Gallily R (2002) Cannabidiol: an overview of some pharmacological aspects. J Clin Pharmacol 42:11S–19S.
Melis M, Pistis M, Perra S, Muntoni AL, Pillolla G, Gessa GL (2004a) Endocannabinoids mediate presynaptic inhibition of glutamatergic transmission in rat ventral tegmental area dopamine neurons through activation of CB_1 receptors. J Neurosci 24:53–62.
Melis M, Perra S, Muntoni AL, Pillolla G, Lutz B, Marsicano G, Di Marzo V, Gessa GL, Pistis M (2004b) Prefrontal cortex stimulation induces 2-arachidonoyl-glycerol-mediated suppression of excitation in dopamine neurons. J Neurosci 24:10707–10715.
Melis M, Pillolla G, Bisogno T, Minassi A, Petrosino S, Perra S, Muntoni AL, Lutz B, Gessa GL, Marsicano G, Di Marzo V, Pistis M (2006) Protective activation of the endocannabinoid system during ischemia in dopamine neurons. Neurobiol Dis 24:15–27.
Meschler JP, Howlett AC, Madras BK (2001) Cannabinoid receptor agonist and antagonist effects on motor function in normal and 1-methyl-4-phenyl-1,2,5,6-tetrahydropyridine (MPTP)-treated non-human primates. Psychopharmacology 156:79–85.
Mesnage V, Houeto JL, Bonnet AM, Clavier I, Arnulf I, Cattelin F, Le Fur G, Damier P, Welter ML, Agid Y (2004) Neurokinin B, neurotensin, and cannabinoid receptor antagonists and Parkinson's disease. Clin Neuropharmacol 27:108–110.
Mezey E, Toth ZE, Cortright DN, Arzubi MK, Krause JE, Elde R, Guo A, Blumberg PM, Szallasi A (2000) Distribution of mRNA for vanilloid receptor subtype 1 (VR1), and VR1-like immunoreactivity, in the central nervous system of the rat and human. Proc Natl Acad Sci USA 97:3655–3660.
Miller AS, Walker JM (1995) Effects of a cannabinoid on spontaneous and evoked neuronal activity in the substantia nigra pars reticulata. Eur J Pharmacol 279:179–185.
Miller AS, Walker JM (1996) Electrophysiological effects of a cannabinoid on neural activity in the globus pallidus. Eur J Pharmacol 304:29–35.
Miller AS, Walker JM (1998) Local effects of cannabinoids on spontaneous activity and evoked inhibition in the globus pallidus. Eur J Pharmacol 352:199–205.
Milton NG (2002) Anandamide and noladin ether prevent neurotoxicity of the human amyloid-beta peptide. Neurosci Lett 332:127–130.
Morale MC, Serra PA, L'episcopo F, Tirolo C, Caniglia S, Testa N, Gennuso F, Giaquinta G, Rocchitta G, Desole MS, Miele E, Marchetti B (2006) Estrogen, neuroinflammation and neuroprotection in Parkinson's disease: glia dictates resistance versus vulnerability to neurodegeneration. Neuroscience 138:869–878.
Moriello AS, Balas L, Ligresti A, Cascio MG, Durand T, Morera E, Ortar G, Di Marzo V (2006) Development of the first potential covalent inhibitors of anandamide cellular uptake. J Med Chem 49:2320–2332.
Moss DE, McMaster SB, Rogers J (1981) Tetrahydrocannabinol potentiates reserpine-induced hypokinesia. Pharmacol Biochem Behav 15:779–783.
Müller-Vahl KR (2003) Cannabinoids reduce symptoms of Tourette's syndrome. Expert Opin Pharmacother 4:1717–1725.
Müller-Vahl KR, Kolbe H, Schneider U, Emrich HM (1998) Cannabinoids: posible role in the pathophysiology of Gilles de la Tourette-syndrome. Acta Psychiatr Scand 98:502–506.

Müller-Vahl KR, Schneider U, Kolbe H, Emrich HM (1999a) Treatment of Tourette-syndrome with Δ^9-tetrahydrocannabinol. Am J Psychiatry 156:495.

Müller-Vahl KR, Schneider U, Emrich HM (1999b) Nabilone increases choreatic movements in Huntington's disease. Mov Disord 14:1038–1040.

Müller-Vahl KR, Kolbe H, Schneider U, Emrich HM (1999c) Cannabis in movement disorders. Forsch Komplementärned 6:23–27.

Müller-Vahl KR, Schneider U, Koblenz A, Jobges M, Kolbe H, Daldrup T, Emrich HM (2002) Treatment of Tourette's syndrome with Δ^9-tetrahydrocannabinol (THC): a randomized crossover trial. Pharmacopsychiatry 35:57–61.

Nagatsu T, Mogi M, Ichinose H, Togari A (2000) Changes in cytokines and neurotrophins in Parkinson's disease. J Neural Transm 60:277–290.

Narushima M, Hashimoto K, Kano M (2006a) Endocannabinoid-mediated short-term suppression of excitatory synaptic transmission to medium spiny neurons in the striatum. Neurosci Res 54:159–164.

Narushima M, Uchigashima M, Hashimoto K, Watanabe M, Kano M (2006b) Depolarization-induced suppression of inhibition mediated by endocannabinoids at synapses from fast-spiking interneurons to medium spiny neurons in the striatum. Eur J Neurosci 24:2246–2252.

Narushima M, Uchigashima M, Fukaya M, Matsui M, Manabe T, Hashimoto K, et al. (2007) Tonic enhancement of endocannabinoid-mediated retrograde suppression of inhibition by cholinergic interneuron activity in the striatum. J Neurosci 27:496–506.

Navarro M, Fernández-Ruiz J, de Miguel R, Hernández ML, Cebeira M, Ramos JA (1993) Motor disturbances induced by an acute dose of Δ^9-tetrahydrocannabinol: possible involvement of nigrostriatal dopaminergic alterations. Pharmacol Biochem Behav 45:291–298.

Naver B, Stub C, Moller M, Fenger K, Hansen AK, Hasholt L, Sorensen SA (2003) Molecular and behavioral analysis of the R6/1 Huntington's disease transgenic mouse. Neuroscience 122:1049–1057.

Ng Cheong Ton JM, Gerhardt GA, Friedemann M, Etgen AM, Rose GM, Sharpless NS, Gardner EL (1988) The effects of delta 9-tetrahydrocannabinol on potassium-evoked release of dopamine in the rat caudate nucleus: an *in vivo* electrochemical and *in vivo* microdialysis study. Brain Res 451:59–68.

Nuñez E, Benito C, Pazos MR, Barbachano A, Fajardo O, González S, Tolón R, Romero J (2004) Cannabinoid CB_2 receptors are expressed by perivascular microglial cells in the human brain: an immunohistochemical study. Synapse 53:208–213.

Okamoto Y, Morishita J, Tsuboi K, Tonai T, Ueda N (2004) Molecular characterization of a phospholipase D generating anandamide and its congeners. J Biol Chem 279:5298–5305.

Page KJ, Besret L, Jain M, Monaghan EM, Dunnett SB, Everitt BJ (2000) Effects of systemic 3-nitropropionic acid-induced lesions of the dorsal striatum on cannabinoid and mu-opioid receptor binding in the basal ganglia. Exp Brain Res 130:142–150.

Pavese N, Gerhard A, Tai YF, Ho AK, Turkheimer F, Barker RA, Brooks DJ, Piccini P (2006) Microglial activation correlates with severity in Huntington disease: a clinical and PET study. Neurology 66:1638–1643.

Pazos MR, Nuñez E, Benito C, Tolón R, Romero J (2004) Role of the endocannabinoid system in Alzheimer's disease: new perspectives. Life Sci 75:1907–1915.

Pertwee RG (2002) Cannabinoids and multiple sclerosis. Pharmacol Ther 95:165–174.

Pertwee RG, Wickens AP (1991) Enhancement by chlordiazepoxide of catalepsy induced in rats by intravenous or intrapallidal injections of enantiomeric cannabinoids. Neuropharmacology 30:237–244.

Pertwee RG, Greentree SG, Swift PA (1988) Drugs which stimulate or facilitate central GABAergic transmission interact synergistically with Δ^9-tetrahydrocannabinol to produce marked catalepsy in mice. Neuropharmacology 27:1265–1270.

Pintor A, Tebano MT, Martire A, Grieco R, Galluzzo M, Scattoni ML, Pezzola A, Coccurello R, Felici F, Cuomo V, Piomelli D, Calamandrei G, Popoli P (2006) The cannabinoid receptor agonist WIN 55,212-2 attenuates the effects induced by quinolinic acid in the rat striatum. Neuropharmacology 51:1004–1012.

Pisani A, Fezza F, Galati S, Battista N, Napolitano S, Finazzi-Agro A, Bernardi G, Brusa L, Pierantozzi M, Stanzione P, Maccarrone M (2005) High endogenous cannabinoid levels in the cerebrospinal fluid of untreated Parkinson's disease patients. Ann Neurol 57:777–779.

Ploner CJ, Tschirch A, Ostendorf F, Dick S, Gaymard BM, Rivaud-Pechoux S, et al. (2002) Oculomotor effects of delta-9-tetrahydrocannabinol in humans: implications for the functional neuroanatomy of the brain cannabinoid system. Cereb Cortex 12:1016–1023.

Price DA, Owens WA, Gould GG, Frazer A, Roberts JL, Daws LC, et al. (2007) CB_1-independent inhibition of dopamine transporter activity by cannabinoids in mouse dorsal striatum. J Neurochem 101:389–396.

Rajkowska G, Selemon LD, Goldman-Rakic PS (1998) Neuronal and glial somal size in the prefrontal cortex: a postmortem morphometric study of schizophrenia and Huntington disease. Arch Gen Psychiatry 55:215–224.

Ramirez BG, Blazquez C, Gomez del Pulgar T, Guzman M, de Ceballos ML (2005) Prevention of Alzheimer's disease pathology by cannabinoids: neuroprotection mediated by blockade of microglial activation. J Neurosci 25:1904–1913.

Richfield EK, Herkenham M (1994) Selective vulnerability in Huntington's disease: preferential loss of cannabinoid receptors in lateral globus pallidus. Ann Neurol 36:577–584.

Richter A, Loscher W (1994) (+)-WIN 55,212-2, a novel cannabinoid receptor agonist, exerts antidystonic effects in mutant dystonic hamsters. Eur J Pharmacol 264:371–377.

Richter A, Loscher W (2002) Effects of pharmacological manipulations of cannabinoid receptors on severity of dystonia in a genetic model of paroxysmal dyskinesia. Eur J Pharmacol 454:145–151.

Riegel AC, Lupica CR (2004) Independent presynaptic and postsynaptic mechanisms regulate endocannabinoid signaling at multiple synapses in the ventral tegmental area. J Neurosci 24:11070–11078.

Riley BE, Orr HT (2006) Polyglutamine neurodegenerative diseases and regulation of transcription: assembling the puzzle. Genes Dev 20:2183–2192.

Robbe D, Alonso G, Duchamp F, Bockaert J, Manzoni OJ (2001) Localization and mechanisms of action of cannabinoid receptors at the glutamatergic synapses of the mouse nucleus accumbens. J Neurosci 21:109–116.

Robbe D, Kopf M, Remaury A, Bockaert J, Manzoni OJ (2002) Endogenous cannabinoids mediate long-term synaptic depression in the nucleus accumbens. Proc Natl Acad Sci USA 99:8384–8388.

Rodríguez de Fonseca F, Gorriti MA, Fernández-Ruiz J, Palomo T, Ramos JA (1994) Downregulation of rat brain cannabinoid binding sites after chronic Δ^9-tetrahydrocannabinol treatment. Pharmacol Biochem Behav 47:33–40.

Rodriguez de Fonseca F, Del Arco I, Martin-Calderon JL, Gorriti MA, Navarro M (1998) Role of the endogenous cannabinoid system in the regulation of motor activity. Neurobiol Dis 5:483–501.

Rodriguez JJ, Mackie K, Pickel VM (2001) Ultrastructural localization of the CB_1 cannabinoid receptor in mu- opioid receptor patches of the rat Caudate putamen nucleus. J Neurosci 21:823–833.

Romero J, García L, Cebeira M, Zadrozny D, Fernández-Ruiz J, Ramos JA (1995a) The endogenous cannabinoid receptor ligand, anandamide, inhibits the motor behaviour: role of nigrostriatal dopaminergic neurons. Life Sci 56:2033–2040.

Romero J, de Miguel R, García-Palomero E, Fernández-Ruiz J, Ramos JA (1995b) Time-course of the effects of anandamide, the putative endogenous cannabinoid receptor ligand, on extrapyramidal function. Brain Res 694:223–232.

Romero J, Garcia-Palomero E, Fernandez-Ruiz JJ, Ramos JA (1996a) Involvement of $GABA_B$ receptors in the motor inhibition produced by agonists of brain cannabinoid receptors. Behav Pharmacol 7:299–302.

Romero J, García-Palomero E, Lin SY, Ramos JA, Makriyannis A, Fernández-Ruiz J (1996b) Extrapyramidal effects of methanandamide, an analog of anandamide, the endogenous CB_1 receptor ligand. Life Sci 58:1249–1257.

Romero J, de Miguel R, Ramos JA, Fernández-Ruiz J (1998a) The activation of cannabinoid receptors in striatonigral neurons inhibited GABA uptake. Life Sci 62:351–363.
Romero J, Berrendero F, García-Gil L, de la Cruz P, Ramos JA, Fernández-Ruiz J (1998b) Loss of cannabinoid receptor binding and messenger RNA levels and cannabinoid agonist-stimulated [35S]-GTPγS binding in the basal ganglia of aged rats. Neuroscience 84:1075–1083.
Romero J, Berrendero F, Pérez-Rosado A, Manzanares J, Rojo A, Fernández-Ruiz J, de Yébenes JG, Ramos JA (2000) Unilateral 6-hydroxydopamine lesions of nigrostriatal dopaminergic neurons increased CB_1 receptor mRNA levels in the caudate-putamen. Life Sci 66:485–494.
Romero J, Lastres-Becker I, de Miguel R, Berrendero F, Ramos JA, Fernandez-Ruiz J (2002) The endogenous cannabinoid system and the basal ganglia. Biochemical, pharmacological, and therapeutic aspects. Pharmacol Ther 95:137–152.
Ronesi J, Lovinger DM (2005) Induction of striatal long-term synaptic depression by moderate frequency activation of cortical afferents in rat. J Physiol 562:245–256.
Ronesi J, Gerdeman GL, Lovinger DM (2004) Disruption of endocannabinoid release and striatal long-term depression by postsynaptic blockade of endocannabinoid membrane transport. J Neurosci 24:1673–1679.
Russo E, Guy GW (2006) A tale of two cannabinoids: The therapeutic rationale for combining tetrahydrocannabinol and cannabidiol. Med Hypotheses 66:234–246.
Sagredo O, Ramos JA, Decio A, Mechoulam R, Fernández-Ruiz J (2007) Cannabidiol reduced the striatal atrophy caused 3-nitropropionic acid *in vivo* by mechanisms independent of the activation of cannabinoid receptors. Eur J Neurosci Eur J Neuro Sci 26:843–851.
Sañudo-Peña MC, Walker JM (1997) Role of the subthalamic nucleus in cannabinoid actions in the substantia nigra of the rat. J Neurophysiol 77:1635–1638.
Sañudo-Peña MC, Patrick SL, Patrick RL, Walker JM (1996) Effects of intranigral cannabinoids on rotational behavior in rats: interactions with the dopaminergic system. Neurosci Lett 206:21–24.
Sañudo-Peña MC, Force M, Tsou K, Miller AS, Walker JM (1998a) Effects of intrastriatal cannabinoids on rotational behavior in rats: interactions with the dopaminergic system. Synapse 30:221–226.
Sañudo-Peña MC, Patrick SL, Khen S, Patrick RL, Tsou K, Walker JM (1998b) Cannabinoid effects in basal ganglia in a rat model of Parkinson's disease. Neurosci Lett 248:171–174.
Sañudo-Peña MC, Tsou K, Walker JM (1999) Motor actions of cannabinoids in the basal ganglia output nuclei. Life Sci 65:703–713.
Sañudo-Peña MC, Tsou K, Romero J, Mackie K, Walker JM (2000a) Role of the superior colliculus in the motor effects of cannabinoids and dopamine. Brain Res 853:207–214.
Sañudo-Peña MC, Romero J, Seale GE, Fernandez-Ruiz JJ, Walker JM (2000b) Activational role of cannabinoids on movement. Eur J Pharmacol 391:269–274.
Sapp E, Kegel KB, Aronin N, Hashikawa T, Uchiyama Y, Tohyama K, Bhide PG, Vonsattel JP, DiFiglia M (2001) Early and progressive accumulation of reactive microglia in the Huntington disease brain. J Neuropathol Exp Neurol 60:161–172.
Sarne Y, Mechoulam R (2005) Cannabinoids: between neuroprotection and neurotoxicity. Curr Drug Targets CNS Neurol Disord 4:677–684.
Schoffelmeer AN, Hogenboom F, Wardeh G, De Vries TJ (2006) Interactions between CB_1 cannabinoid and mu opioid receptors mediating inhibition of neurotransmitter release in rat nucleus accumbens core. Neuropharmacology 51:773–781.
Schultz W (2006) Behavioral theories and the neurophysiology of reward. Annu Rev Psychol 57:87–115.
Schuster J, Ates M, Brune K, Guhring H (2002) The cannabinoids R(-)-7-hydroxy-delta-6-tetrahydrocannabinol-dimethylheptyl (HU-210), 2-O-arachidonoylglycerylether (HU-310) and arachidonyl-2-chloroethylamide (ACEA) increase isoflurane provoked sleep duration by activation of cannabinoids 1 (CB_1)-receptors in mice. Neurosci Lett 326:196–200.
Segovia G, Mora F, Crossman AR, Brotchie JM (2003) Effects of CB_1 cannabinoid receptor modulating compounds on the hyperkinesia induced by high-dose levodopa in the reserpine-treated rat model of Parkinson's disease. Mov Disord 18:138–149.
Sethi KD (2002) Clinical aspects of Parkinson disease. Curr Opin Neurol 15:457–460.

Sherer TB, Betarbet R, Greenamyre JT (2001) Pathogenesis of Parkinson's disease. Curr Opin Investig Drugs 2:657–662.

Sieradzan KA, Fox SH, Hill M, Dick JP, Crossman AR, Brotchie JM (2001) Cannabinoids reduce levodopa-induced dyskinesia in Parkinson's disease: a pilot study. Neurology 57:2108–2111.

Silverdale MA, McGuire S, McInnes A, Crossman AR, Brotchie JM (2001) Striatal cannabinoid CB_1 receptor mRNA expression is decreased in the reserpine-treated rat model of Parkinson's disease. Exp Neurol 169:400–406.

Singhrao SK, Neal JW, Morgan BP, Gasque P (1999) Increased complement biosynthesis by microglia and complement activation on neurons in Huntington's disease. Exp Neurol 159:362–376.

Singh N, Pillay V, Choonara YE (2007) Advances in the treatment of Parkinson's disease. Prog Neurobiol 81:29–44.

Smith PB, Compton DR, Welch SP, Razdan RK, Mechoulam R, Martin BR (1994) The pharmacological activity of anandamide, a putative endogenous cannabinoid, in mice. J Pharmacol Exp Ther 270:219–227.

Souilhac J, Poncelet M, Rinaldi-Carmona M, Le-Fur G, Soubrie P (1995) Intrastriatal injection of cannabinoid receptor agonists induced turning behavior in mice. Pharmacol Biochem Behav 51:3–7.

Stella N (2004) Cannabinoid signaling in glial cells. Glia 48:267–277.

Sung KW, Choi S, Lovinger DM (2001) Activation of group I mGluRs is necessary for induction of long-term depression at striatal synapses. J Neurophysiol 86:2405–2412.

Szabo B, Dorner L, Pfreundtner C, Norenberg W, Starke K (1998) Inhibition of GABAergic inhibitory postsynaptic currents by cannabinoids in rat corpus striatum. Neuroscience 85:395–403.

Szabo B, Muller T, Koch H (1999) Effects of cannabinoids on dopamine release in the corpus striatum and the nucleus accumbens *in vitro*. J Neurochem 73:1084–1089.

Szabo B, Wallmichrath I, Mathonia P, Pfreundtner C (2000) Cannabinoids inhibit excitatory neurotransmission in the substantia nigra pars reticulata. Neuroscience 97:89–97.

Szabo B, Urbanski MJ, Bisogno T, Di Marzo V, Mendiguren A, Baer WU, et al. (2006) Depolarization-induced retrograde synaptic inhibition in the mouse cerebellar cortex is mediated by 2-arachidonoylglycerol. J Physiol 577:263–280.

Tang K, Low MJ, Grandy DK, Lovinger DM (2001) Dopamine-dependent synaptic plasticity in striatum during *in vivo* development. Proc Natl Acad Sci USA 98:1255–1260.

Taupin P, Gage FH (2002) Adult neurogenesis and neural stem cells of the central nervous system in mammals. J Neurosci Res 69:745–749.

Tepper JMaP, D (2006) Microcircuits in the striatum: striatal cell types and their interaction. In: Grillner SaG, A.M., (ed.), Microcircuits: The Interface between Neurons and Global Brain Function. Cambridge, MA: The MIT Press.

Tersigni TJ, Rosenberg HC (1996) Local pressure application of cannabinoid agonists increases spontaneous activity of rat substantia nigra pars reticulata neurons without affecting response to iontophoretically-applied GABA. Brain Res 733:184–192.

Toulmond S, Tang K, Bureau Y, Ashdown H, Degen S, O'Donnell R, Tam J, Han Y, Colucci J, Giroux A, Zhu Y, Boucher M, Pikounis B, Xanthoudakis S, Roy S, Rigby M, Zamboni R, Robertson GS, Ng GY, Nicholson DW, Fluckiger JP (2004) Neuroprotective effects of M826, a reversible caspase-3 inhibitor, in the rat malonate model of Huntington's disease. Br J Pharmacol 141:689–697.

Tsou K, Brown S, Sañudo-Peña MC, Mackie K, Walker JM (1998a) Immunohistochemical distribution of cannabinoid CB_1 receptors in the rat central nervous system. Neuroscience 83:393–411.

Tsou K, Nogueron MI, Muthian S, Sañudo-Peña M, Hillard CJ, Deutsch DG, Walker JM (1998b) Fatty acid amide hydrolase is located preferentially in large neurons in the rat central nervous system as revealed by immunohistochemistry. Neurosci Lett 254:137–140.

Turmaine M, Raza A, Mahal A, Mangiarini L, Bates GP, Davies SW (2000) Nonapoptotic neurodegeneration in a transgenic mouse model of Huntington's disease. Proc Natl Acad Sci USA 97:8093–8097.

Tzavara ET, Li DL, Moutsimilli L, Bisogno T, Di Marzo V, Phebus LA, Nomikos GG, Giros B (2006) Endocannabinoids activate transient receptor potential vanilloid 1 receptors to reduce hyperdopaminergia-related hyperactivity: therapeutic implications. Biol Psychiatry 59:508–515.

Uchigashima M, Narushima M, Fukaya M, Katona I, Kano M, Watanabe M (2007) Subcellular arrangement of molecules for 2-arachidonoyl-glycerol-mediated retrograde signaling and its physiological contribution to synaptic modulation in the striatum. J Neurosci 27:3663–3676.

van der Stelt M, Di Marzo V (2003) The endocannabinoid system in the basal ganglia and in the mesolimbic reward system: implications for neurological and psychiatric disorders. Eur J Pharmacol 480:133–150.

van der Stelt M, Di Marzo V (2005) Cannabinoid receptors and their role in neuroprotection. Neuromol Med 7:37–50.

van der Stelt M, Veldhuis WB, Maccarrone M, Bar PR, Nicolay K, Veldink GA, Di Marzo V, Vliegenthart JF (2002) Acute neuronal injury, excitotoxicity, and the endocannabinoid system. Mol Neurobiol 26:317–346.

van der Stelt M, Fox SH, Hill M, Crossman AR, Petrosino S, Di Marzo V, Brotchie JM (2005) A role for endocannabinoids in the generation of parkinsonism and levodopa-induced dyskinesia in MPTP-lesioned non-human primate models of Parkinson's disease. FASEB J 19:1140–1142.

Vaney C, Heinzel-Gutenbrunner M, Jobin P, Tschopp F, Gattlen B, Hagen U, Schnelle M, Reif M (2004) Efficacy, safety and tolerability of an orally administered cannabis extract in the treatment of spasticity in patients with multiple sclerosis: a randomized, double-blind, placebo-controlled, crossover study. Mult Scler 10:417–424.

Van Sickle MD, Duncan M, Kingsley PJ, Mouihate A, Urbani P, Mackie K, Stella N, Makriyannis A, Piomelli D, Davison JS, Marnett LJ, Di Marzo V, Pittman QJ, Patel KD, Sharkey KA (2005) Identification and functional characterization of brainstem cannabinoid CB_2 receptors. Science 310:329–332.

Venderova K, Brown TM, Brotchie JM (2005) Differential effects of endocannabinoids on [^{3}H]-GABA uptake in the rat globus pallidus. Exp Neurol 194:284–287.

Wade DT, Makela P, Robson P, House H, Bateman C (2004) Do cannabis-based medicinal extracts have general or specific effects on symptoms in multiple sclerosis? A double-blind, randomized, placebo-controlled study on 160 patients. Mult Soler 10:434–441.

Walker FO (2007) Huntington's disease. Lancet 369:218–228.

Wallmichrath I, Szabo B (2002a) Analysis of the effect of cannabinoids on GABAergic neurotransmission in the substantia nigra pars reticulata. Naunyn Schmiedebergs Arch Pharmacol 365:326–334.

Wallmichrath I, Szabo B (2002b) Cannabinoids inhibit striatonigral GABAergic neurotransmission in the mouse. Neuroscience 113:671–682.

Walter L, Stella N (2004) Cannabinoids and neuroinflammation. Br J Pharmacol 141:775–785.

Wang W, Duan W, Igarashi S, Morita H, Nakamura M, Ross CA (2005) Compounds blocking mutant huntingtin toxicity identified using a Huntington's disease neuronal cell model. Neurobiol Dis 20:500–508.

Wang Z, Kai L, Day M, Ronesi J, Yin HH, Ding J, Tkatch T, Lovinger DM, Surmeier DJ (2006) Dopaminergic control of corticostriatal long-term synaptic depression in medium spiny neurons is mediated by cholinergic interneurons. Neuron 50:443–452.

Watanabe K, Kimura M (1998) Dopamine receptor-mediated mechanisms involved in the expression of learned activity of primate striatal neurons. J Neurophysiol 79:2568–2580.

Westlake TM, Howlett AC, Bonner TI, Matsuda LA, Herkenham M (1994) Cannabinoid receptor binding and messenger RNA expression in human brain: an *in vitro* receptor autoradiography and in situ hybridization histochemistry study of normal aged and Alzheimer's brains. Neuroscience 63:637–652.

Whitton PS (2007) Inflammation as a causative factor in the aetiology of Parkinson's disease. Br J Pharmacol 150:963–976.

Wickens AP, Pertwee RG (1993) Δ^9-Tetrahydrocannabinol and anandamide enhance the ability of muscimol to induce catalepsy in the globus pallidus of rats. Eur J Pharmacol 250:205–208.

Wilson CJ (2006) Striatal D_2 receptors and LTD: yes, but not where you thought they were. Neuron 50:347–348.
Wilson CJ, Kawaguchi Y (1996) The origins of two-state spontaneous membrane potential fluctuations of neostriatal spiny neurons. J Neurosci 16:2397–2410.
Wood-Kaczmar A, Gandhi S, Wood NW (2006) Understanding the molecular causes of Parkinson's disease. Trends Mol Med 12:521–528.
Wu SS, Frucht SJ (2005) Treatment of Parkinson's disease: what's on the horizon? CNS Drugs 19:723–743.
Yanovsky Y, Mades S, Misgeld U (2003) Retrograde signaling changes the venue of postsynaptic inhibition in rat substantia nigra. Neuroscience 122:317–328.
Yin HH, Knowlton BJ (2006) The role of the basal ganglia in habit formation. Nat Rev Neurosci 7:464–476.
Yin HH, Lovinger DM (2006) Frequency-specific and D_2 receptor-mediated inhibition of glutamate release by retrograde endocannabinoid signaling. Proc Natl Acad Sci USA 103:8251–8256.
Zajicek J, Fox P, Sanders H, Wright D, Vickery J, Nunn A, Thompson A (2003) Cannabinoids for treatment of spasticity and other symptoms related to multiple sclerosis (CAMS study): multicentre randomised placebo-controlled trial. Lancet 362:1517–1526.
Zajicek JP, Sanders HP, Wright DE, Vickery PJ, Ingram WM, Reilly SM, Nunn AJ, Teare LJ, Fox PJ, Thompson AJ (2005) Cannabinoids in multiple sclerosis (CAMS) study: safety and efficacy data for 12 months follow up. J Neurol Neurosurg Psychiatry 76:1664–1669.
Zeng BY, Dass B, Owen A, Rose S, Cannizzaro C, Tel BC, Jenner P (1999) Chronic L-DOPA treatment increases striatal cannabinoid CB_1 receptor mRNA expression in 6-hydroxydopamine-lesioned rats. Neurosci Lett 276:71–74.
Zimmer A, Zimmer AM, Hohmann AG, Herkenham M, Bonner TI (1999) Increased mortality, hypoactivity, and hypoalgesia in cannabinoid CB_1 receptor knockout mice. Proc Natl Acad Sci USA 96:5780–5785.

Chapter 22
The Endocannabinoid System is a Major Player in Schizophrenia

Attila Köfalvi and Markus Fritzsche

Abstract Converging lines of evidence point to an inextricable role of the endocannabinoid system in schizophrenia. Marijuana consumption (1) elicits psychotic symptoms similar to schizophrenia; (2) precipitates schizophrenia in susceptible individuals; (3) worsens psychosis; and (4) is more prevalent among schizophrenia patients. (5) Genetic linkage studies have mapped a locus around the CB_1 cannabinoid receptor gene (*CNR1*), which potentially confers susceptibility to schizophrenia, and (6) within *CNR1*, several polymorphisms reportedly associate with this disease. (7) The endocannabinoid system controls brain areas and signalling systems implicated in schizophrenia, (8) and is overactive in patients, (9). It correlates with symptom severity and is reversible with certain antipsychotics. Finally, (10) the naturally occurring CB_1 receptor antagonist cannabidiol exhibits a promising antipsychotic profile in pharmacological model-psychoses and schizophrenia. In summarizing the pertinent epidemiological and molecular data, we define schizophrenia as a manifestation of aberrant circuitry formation at the synaptic level and propose that the liability to develop psychosis is driven by imbalanced co-signalling between endocannabinoids and other neuromodulatory pathways already implicated in schizophrenia.

Introduction

Schizophrenia – a break between thought and emotion – is a psychiatric disorder characterized by hallucinations, delusions, reduced attention and motivation, as well as deterioration in social functioning (Kraepelin, 1899; Bleuler, 1911). Although family, twin and adoption studies conclusively demonstrate the importance of genetic risk factors, researchers have found it surprisingly difficult to disentangle its mode(s) of transmission, or to obtain some finding evidence for single genes, whilst others suggest that many genes must act in combination or under environmental constraints. The incomplete concordance rate for schizophrenia in monozygotic twins falls far short of the 100% that would be expected from genetically identical individuals. At the same time, unaffected monozygotic twins and their affected co-twins show an equal proportion of schizophrenic offspring. This strongly implies that an unaffected identical twin possesses, but does not

A. Kofalvi (ed.), *Cannabinoids and the Brain.*

express, a genetic pre-disposition to schizophrenia. Equally diverse are the putative 'schizophrenic' traits. These are transmitted as minor deviations such as decreased brain volume, increased ventricular space, or dysfunctional eye tracking that are relatively innocent in themselves. If an individual is unlucky enough to inherit several of these traits, confounded by prenatal infection or obstetric complications, the cumulative effect of these risk factors may propel the individual over a threshold for the full expression of psychosis. Schizophrenia is one of the most frequent mental disorders affecting people in the range of 7.7–43.0 per 100,000, as recently reassessed by McGrath (2006). The almost sixfold difference between the lowest and the highest value appears to be influenced by complex epigenetic variables including season of birth, place of habitation and life style, among others. Schizophrenia also occurs with significantly higher frequency in men than in women with the ratio of 1.4:1 (McGrath, 2006). We will frequently use the term 'schizophrenia' in singular form for the sake of simplicity, without the intention to refer to it as one well-defined type of disorder. More exactly, schizophrenia is characterized by a multiplicity of signs and symptoms, no single one of which is present in all patients. To complicate matters, schizophrenia is frequently combined with symptoms typical of other disorders such as drug dependence, depression and obsessive compulsive disorder. Nevertheless, as this highly prevalent illness is recognized throughout the world, there must be a clinical pattern which gives schizophrenia conceptual unity. To correctly recognize and treat the disease presents a significant challenge not only to psychiatrists, but also to the molecular biologists and neuroscientists, where it is particularly difficult to adduce plausible explanations for its uniquely human symptomatology, which can hardly be modelled in animals (Boksa, 2007). Here, we comprehensively review the converging lines of molecular evidence which seem to confirm that psychotic mechanisms are in part driven by the endocannabinoid system. Nonetheless, we would also like to add the cautionary remark that the putative modes of action still require further elaborate studies from the cell to the in vivo level, to be conclusively proved.

Neurochemical Factors

Not surprisingly, a ca. twofold increase in D_2 receptor density has been reported in post-mortem brains of untreated schizophrenia patients, since the neuromodulators most commonly implicated in schizophrenia are dopamine and serotonin (Seeman, 1987; Kapur and Remington, 1996, 2001). Among their receptors, blockade of D_2-like and $5\text{-}HT_{2A}$ receptors have proved to be highly efficient in therapy. This term, 'D_2-like', encompasses the pharmacologically hardly distinguishable D_2, D_3 and D_4 receptors. For the sake of simplicity, we use the term D_2 rather than the cumbersome D_2-like when we speak in general. The majority of typical and atypical antipsychotics exhibit considerable affinity for D_2 receptors, whereas some of them (i.e., remoxipiride and amisulpiride) lack comparable affinity for $5\text{-}HT_{2A}$ receptors. Others such as the atypical quetiapine have high $5\text{-}HT_{2A}$ and low D_2 receptor affinity, and still, all of them are effective antipsychotics. Quetiapine displays a short and rapidly declining peak of D_2 occupancy, which may explain that it is free from extrapyramidal and prolactin-associated side effects. This indicates that, rather than a continuous D_2 receptor blockade, which is perhaps not quiescent in the effective treatment of many schizophrenic patients (Kapur et al., 2000), the modulation of phasic dopaminergic signalling may be preferable. This transient dopaminergic signalling is under endocannabinoid control and is also modulated by substances of abuse in a CB_1 receptor-dependent manner in the ventral tegmental area (VTA; Szabo et al., 2002; Cheer et al., 2004, 2007), which is a major area involved in schizophrenia (Boksa, 2007). Phasic modulation of dopamine release in the VTA may contribute to a possible maladaptive signalling circle in the disease (see below). To further illustrate the complexity of the picture, we have to point out that in a common animal model of schizophrenia, classical antipsychotics reverse most effects (hyperlocomotion and stereotypic behaviour) of the dopamine-releasing drug amphetamine, whereas atypical antipsychotics do not necessarily do so. In man, amphetamine psychosis is similar to schizophrenia in terms of clear sensorium, auditory hallucinations and sensitivity to phenothiazines. In contrast, the appearance of strong sexual stimulation, stereotypic compulsive behaviour and the lack of flattened affect and formal thought disorder all indicate that high dopamine levels are responsible only for a part of the symptomatology of schizophrenia. N-methyl-D-aspartate (NMDA) receptor hypofunction is widely accepted to contribute to the pathomechanism of the disease particularly in those areas which are enriched in dopamine and serotonin (Mechri et al., 2001; MacDonald and Chafee, 2006; Mouri et al., 2007). This was originally recognized, because the NMDA blocker phencyclidine induced schizophrenia-like symptoms in animals and humans (Boksa, 2007). As it is suggested, the NMDA-erg hypofunction is due to the dysregulation of the receptor function (indirectly due to mutations in the DTNBP1 and/or NRG1 and/or RGS4 and/or DAOA genes) as well as attributable to a loss of NMDA receptor-positive synaptic contacts (MacDonald and Chafee, 2006; Gu et al., 2007), which all lead to impaired executive processes in the prefrontal cortex. Activation of the D_4 dopamine and the 5-HT_{1A} serotonin receptors down-regulates prefrontal cortical NMDA receptors, and for instance, the blockade of RGS4 function augments this effect of the 5-HT_{1A} receptor (Gu et al., 2007). Furthermore, this 5-HT_{1A} receptor-mediated action on prefrontal cortical NMDA receptors was exaggerated by sub-chronic phencyclidine treatment of rats, which down-regulated the RGS4 (Gu et al., 2007). Accordingly, a decreased RGS4 expression is also observed in the cortex of schizophrenia patients (Mirnics et al., 2001). The NMDA hypothesis of schizophrenia is further strengthened by several other observations concerning the other non-competitive NMDA blocker, ketamine. At sub-anaesthetic doses, ketamine elicits the majority of positive and negative symptoms of schizophrenia. It causes defects in smooth pursuit eye tracking, a typical marker of schizophrenia, and elevates cortical and striatal synaptic dopamine levels (Mechri et al., 2001). The third frequently used NMDA channel blocker, MK-801 or dizocilpine, induces the typical schizophrenia syndromes, hyperlocomotion and stereotypy, in rats, which are fully prevented by the antipsychotics, haloperidol and eticlopride, but clozapine counteracts hyperlocomotion only (Hoffman, 1992). MK-801 also elicits another typical symptom of schizophrenia, the impairment of the neurobehavioural phenomenon 'pre-pulse inhibition' (PPI), which represents the inhibition of a startle reflex by a preceding sensory stimulus of a lower intensity (see below). Importantly, the NMDA subtype of glutamate receptors was shown to be involved in schizophrenia phenotypes. Blockade of the metabotropic subtype $mGluR_5$ by its selective antagonist MTEP dose-dependently induces social isolation in rats without causing stereotypy and hyperlocomotion (Koros et al., 2007). Notably, $mGluR_5$ is a major post-synaptic receptor whose activation is required to induce endocannabinoid release in the brain (see below and Chaps. 11 and 21). Therefore, $mGluR_5$ gains critical importance in CB_1 receptor-mediated synaptic plasticity (Chevaleyre et al., 2006). These findings highlight that neuromodulators, other than dopamine and serotonin and their receptors, D_2 and $5\text{-}HT_{2A}$, also contribute to the manifestation of the disease, which we demonstrate later in relation to the endocannabinoid system.

Developmental Variables

Cognitive, behavioural, emotional and motor anomalies may precede the onset of schizophrenia in a significant number of susceptible individuals (Mäki et al., 2004; Isohanni et al., 2006). It means that the putative anatomical abnormalities in schizophrenia should be observed before the onset of psychosis in childhood. Grey matter thinning, for example, was demonstrated by numerous neuroimaging studies in the frontal, occipital, temporal and parietal heteromodal association cortices, among others (Narr et al., 2002, 2004, 2005a,b; Yamasue et al., 2004), and one can find, in fact, hardly any major brain region where abnormalities are not reported (Shenton et al., 2001). The cause and onset of the deficits is unclear, but they are very likely of genetic or developmental origin (Shenton et al., 2001). The epigenetic background is most evident in close relatives, i.e., in case of discordant monozygotic twins when one of the twins is schizophrenic and the other is healthy (Cannon et al., 2002; Narr et al., 2002). The fact that genes are necessary, but not sufficient to provoke the phenotypic manifestation of schizophrenia, begs a question: Which one of all the numerously reported candidate genes contributes to the developmental brain dysfunction? Essentially, all of them may be involved by some means, but there are common processes during the development, which should be highlighted.

Oddities in Schizophrenia at the Systemic Level

Any gene–environment interaction causes schizophrenia, certain oddities coinciding in the disorder. Compared to matched controls, rheumatoid arthritis and cancer occur rarely, and visible nail-fold capillaries are more common in schizophrenic patients (Hanson and Gottesman, 2005; Kalkman, 2006). This may mean that genetic and epigenetic factors virtually converge to the same pathways, and also suggest that schizophrenia is not only a brain disease. However, a system-wise view of cellular pathways is required to understand it as a whole. Crucial for the survival of neurons in global cerebral and systemic defence against stress and infections, the phosphatidylinositide 3-kinase (PI_3K) pathway is one of the prime candidates (Kalkman, 2006). For instance, hypoactivity of the PI_3K pathway – as observed in schizophrenia – may be responsible for the reduced cancer incidence, and for the higher vulnerability of the brain to stressors and infections (Kalkman, 2006) as well. CB_1 receptors are coupled to PI_3K, but the direction of this control, whether it is negative or positive, depends on the condition of the neuron and on other signalling systems which are directly coupled to the CB_1 receptor (Ellert-Miklaszewska et al., 2005; Kalkman, 2006; Harkány et al., 2007; and see below). Likewise, the systemic oddities may account for the impairment of the immune system. The cannabinoid receptors, CB_1 and CB_2, are widely expressed in the immune system, which has an impact on cell proliferation and immune responses. Through controlling the levels of immune mediators, cannabinoids thereby influence cancer growth, inflammation, pain, autoimmune reactivity, brain injury and haematopoiesis, apart from other more direct modes of action (Gladkevich et al., 2004; Massi et al., 2006; Fig. 1). A simple skin flush analysis, for example, demonstrates that both schizophrenic non-cannabis abusers and healthy cannabis consumers respond to methylnicotinate with a reduced sensitivity compared to matched controls (Smesny et al., 2007). This peripheral marker of disturbed arachidonic acid pathways implies that the endocannabinoid system is co-affected, as being a major signalling system derived from PI_3K-coupled membrane lipids that control a vast array of immune reactions (Massi et al., 2006; and see Chap. 16).

The Endocannabinoid System of the Brain

The anatomy, molecular biology, pharmacology and pathophysiology of the molecules participating in the brain endocannabinoid cascade are extensively reviewed in the previous chapters. Nonetheless, now we just briefly reiterate the most important points. The main cannabinoid-sensing cell surface receptor on neurons is the CB_1 receptor, whereas the CB_2 receptor likely appears only on glia cells under pathological conditions (Pertwee, 2006; Köfalvi et al., 2006b, 2007; see Chaps. 10 and 16). $TRPV_1$ 'vanilloid' receptor is also believed to be present in the brain, but perhaps only post-synaptically (Köfalvi et al., 2006a,b, 2007; see Chaps. 8 and 10). CB_1 receptor is primarily an inhibitory metabotropic receptor. It inhibits adenylyl cyclase activity, especially when signalling as a homodimer coupled to $G_{\alpha i/o}$, and in being negatively coupled to voltage-gated Ca^{2+} channels and positively to K^+ channels, CB_1 receptor stimulation inhibits the release of transmitters. In addition, CB_1 receptor also couple to several intracellular messengers such as PI_3K, PKA, Akt, GSK3β and ERK1/2 (Pertwee, 2006; Harkány et al., 2007; see Chap. 5). CB_1 receptors constitute the most frequently expressed metabotropic receptors in the brain. This density is even more

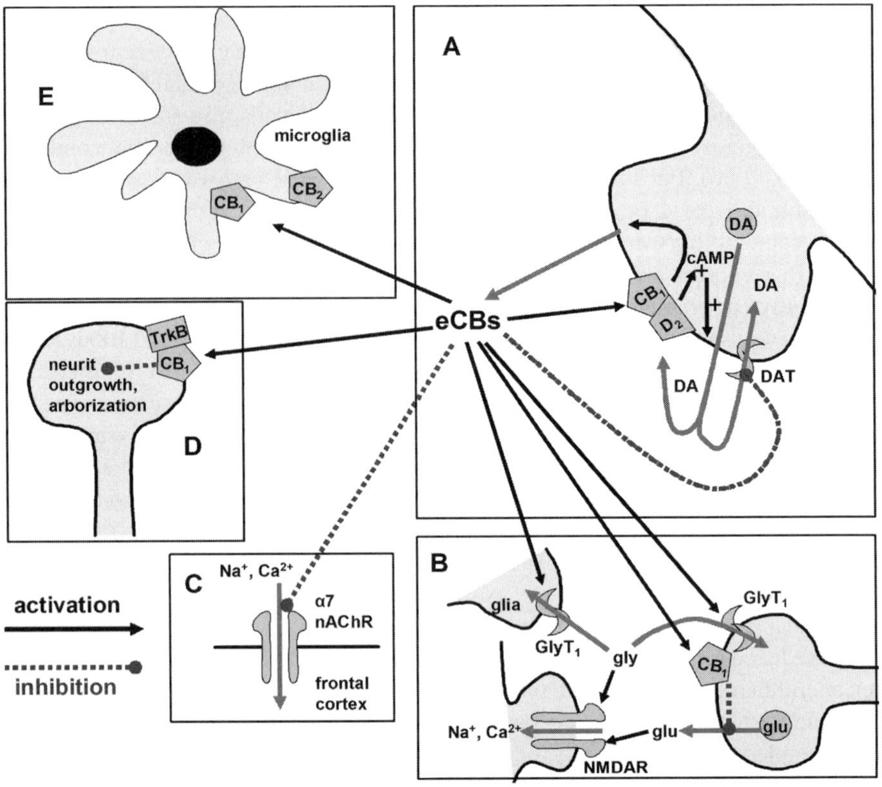

Fig. 1 Endocannabinoids (eCBs) and dopamine (DA) mutually enhance each other's level resulting in a vicious circle. This malignant cycle may be interrupted by CB_1 receptor antagonists (e.g., Sativex™) or D_2 receptor antagonists. Note that for sake of simplicity, we denote the participating dopamine receptor as D_2 but it rather refers to D_2-like receptors (see text). Panel (**a**) demonstrates that dopamine release induces D_2 dopamine receptor-mediated endocannabinoid release. Endocannabinoids in turn, inhibit dopamine transporters (DAT) resulting in an increased synaptic dopamine level (hyperdopaminergic state). Dopamine, together with endocannabinoids, co-activates the CB_1/D_2 heterodimer with a concomitant excitation of the plasma membrane, inducing further endocannabinoid and dopamine release. In panel (**b**), the excess endocannabinoid level exerts inhibition on the release of glutamate through CB_1 receptor activation, as well as facilitates the synaptic clearance of glycine through activating its transporter ($GlyT_1$). These altogether worsen the hypo-NMDA-ergic state and for instance, contribute to memory dysfunction and hypofrontality. (**c**) Excess levels of endocannabinoids directly inhibit the α7 nicotinic acetylcholine receptor (α7 nAChR), contributing to the disruption of phenomena 'sensory gating' and 'smooth pursuit eye movements', as well as to hypofrontality and memory dysfunctions. (**d**) Overactivation of the CB_1 receptors (which, here, is thought to form heterodimers with nerve growth factor receptors, e.g., the TrkB) in axonal growth cones and mature axons terminals hampers normal axonal development and synaptic plasticity, resulting in hypoplasticity, decreased grey and white matter volumes, hypofrontality and memory dysfunctions. (**e**) Microglia are also sensitive to excess endocannabinoid levels, and may enter a continuously activated phase in which they may exert detrimental effects on neurons

pronounced in the prefrontal cortex, basal ganglia and limbic system (Herkenham et al., 1990), in areas which display major developmental deficits in schizophrenia. CB_1 receptors are virtually present on all types of neurons, with highest expression levels GABA-ergic and glutamatergic terminals. Additionally, CB_1 receptors are abundant on hippocampal cholinergic terminals, less abundant on hippocampal dopaminergic nerve terminals and the least abundant on mesolimbic and striatal dopaminergic and striatal cholinergic nerve terminals (Katona et al., 1999, 2000; Köfalvi et al., 2005; Degroot et al., 2006; and see Chaps. 10 and 21). Insurmountable evidence demonstrates that under physiological conditions, CB_1 receptors exert tonic inhibition on hippocampal acetylcholine and dopamine release, as well as phasic inhibition on GABA and glutamate release throughout the brain (Katona et al., 1999, 2000; Köfalvi et al., 2005, 2007; Degroot et al., 2006). By means of phasic modulation of these latter neurotransmitters, endocannabinoids exert a transient indirect facilitatory action on VTA dopaminergic activity (see below). Endocannabinoids, anandamide, 2-arachidonoylglycerol (2-AG) and some other arachidonic acid-derivative substances are released via enzymatic cleavage upon post-synaptic Ca^{2+} entry or upon activation of post-synaptic metabotropic receptors of the D_2 or $mGluR_5$ subtypes, and then return to the pre-synaptic side to exert inhibition. This so-called retrograde endocannabinoid transmission is of great importance in synaptic plasticity (Chevaleyre et al., 2006; see Chap. 11). Consistently, D_2 receptor blockade abrogates CB_1 receptor-dependent synaptic plasticity in the VTA and basal ganglia, whereas D_2 receptor activation enhances it (Melis et al., 2004b; Ronesi and Lovinger, 2005).

Interactions Between the Endocannabinoid and Other Major Signalling Systems

Canonical Interactions

Compared to the relatively simple role of CB_1 receptors in the pre-synaptic inhibition of GABA and glutamate release, the whole endocannabinoid system displays a much more complex interaction with dopaminergic neuromodulation. Our discussion here focuses mainly on the VTA, but several observations are also applicable to the substantia nigra. As we mentioned before, CB_1 receptors do not seem to directly control dopamine release in the VTA and in the striatum because of the low density and frequency of CB_1 receptors in dopaminergic cells – though other factors seem to underlie this phenomenon as well (Köfalvi et al., 2005; Lupica and Riegel, 2005). The primary inputs of VTA neurons are mostly GABA-ergic from several brain areas including the nucleus accumbens (ventral striatum), and glutamatergic from the prefrontal cortex. Conversely, their output is dopaminergic with important innervations to the prefrontal cortex and the nucleus accumbens. Consequently, both inhibitory and excitatory inputs of dopaminergic cells can be inhibited by CB_1 receptor activation in the VTA (Lupica and Riegel, 2005), but the net effect apparently leads to facilitation in the mesolimbic area. First, this comes as no surprise, taking into consideration that illicit drug abuse, including cannabis, can cause dependence, and drug dependence is known to be positively associated with the activity of the VTA (Lupica and Riegel, 2005). Second, SR141716A (Rimonabant, Acomplia™) as well as other systemically administered CB_1 receptor antagonist counteract drug-seeking behaviour, and drug self-administration in animals and significantly heightens the rate of successful smoking cessation in man (Maldonado et al., 2006). Third, cognitive impairment by CB_1 receptor activation and cognitive enhancement by SR141716A administration can be also explained by the modulation of dopamine levels in the prefrontal cortex. One of the earliest studies demonstrated that systemically administered Δ^9-THC increased rat striatal dopamine levels up to 200%. Depending on the timing and site of administration, the serotonin uptake inhibitor fluoxetine could bi-directionally modulate this increase (Malone and Taylor, 1999) predicting a complex interaction between the serotonin, dopamine and endocannabinoid systems. Another preliminary study showed that intravenously administered Δ^9-THC as well as the potent synthetic cannabinoid, WIN55212-2, dose-dependently increased the firing rate and burst firing in the majority of antidromically identified VTA-prefrontal dopaminergic neurons (Diana et al., 1998). Accordingly, intravenous Δ^9-THC dose dependently increased the firing rate of VTA efferens projecting to the nucleus accumbens and neostriatum in an SR141716A-sensitive manner (Melis et al., 2000). As mentioned above, the D_2 receptor antagonist eticlopride inhibits, whereas the D_2 receptor agonist quinpirol enhances CB_1 receptor-dependent synaptic plasticity (Melis et al., 2004b). The same laboratory additionally illustrated that the stimulation of the medial prefrontal cortex of the rat increases spiking and bursting probability, as well as augments frequency within the bursts in the VTA, in the presence of the CB_1 receptor antagonist SR141716A. Slightly unexpectedly, this study implicates that $mGluR_1$ and not $mGluR_5$ or D_2 receptor elicits post-synaptic Ca^{2+} rise in the VTA cells resulting in 2-AG deliberation and retrograde endocannabinoid transmission (Melis et al., 2004a). It would mean that low to moderate glutamatergic activation preferentially induces endocannabinoid release via $mGluR_1$ activation, whereas strong activation may select D_2 receptor-mediated endocannabinoid release (Melis et al., 2004a). Thus in contrast to the mesoaccumbal pathway, the inhibitory mesoprefrontocortical feedback loop is under a phasic negative indirect endocannabinoid tone, as demonstrated previously (Pistis et al., 2001). Regarding the serotonergic system, hitherto, little is known about how CB_1 receptor activation influences the levels of this neuromodulator. It has been shown that a low fraction of serotonergic raphe neurons contain CB_1 receptor mRNA, and their axon terminals are endowed with the CB_1 receptor in the hippocampus and the amygdala (Haring et al., 2007). Therefore, it is presumed that CB_1 receptor may directly control serotonin release in specific brain areas.

Non-Canonical Interactions

Endocannabinoids can increase the level of dopamine in an additional manner. It is quite intriguing that endocannabinoids possess the ability of direct interaction with some transporters and ligand-gated ion channels (see Chap. 9). The dopamine and serotonin transporters, DAT and SERT, can either directly or indirectly, but always CB_1 receptor-independently, be inhibited by endocannabinoids and WIN55212-2, respectively (Chen et al., 2003; Steffens and Feuerstein, 2004; Price et al., 2007). As a result, an increase in the level of anandamide will reduce the clearance of extracellular dopamine. Apart from monoamines, anandamide and 2-AG facilitate the transport of glycine at the glycine transporter 1A ($GlyT_{1A}$) (Pearlman et al., 2003; see Chap. 9). As a result, an increase in endocannabinoid levels would concomitantly result in NMDA-ergic hypofunction, especially if D-serine level is also decreased (Fig. 1). The overall picture of interactions is exemplified by the ability of anandamide, but not of Δ^9-THC, to potentiate the NR1/NR2A NMDA receptors by up to 50% in the hippocampus, cortex and cerebellum (Hampson et al., 1998; see Chap. 9).

CB_1 Receptor Heterodimers

To date, it has not been clarified whether signalling at receptor heterodimers is the rule or the exception. Therefore, we regard it as neither canonical nor non-canonical. As broadly discussed in Chap. 9, the frequent CB_1 receptor heterodimerization is a dynamic process which transiently (or eventually, chronically) couples the endocannabinoid signalling to other signalling pathways. Stimulation of such heterodimers often activates alternative signalling cascades through coupling to optional downstream effectors. The most well known such heterodimer is CB_1/D_2 (Glass and Felder, 1997; Jarrahian et al., 2004; Kearn et al., 2004), which acutely signals with $G_{s\alpha}$, increasing cAMP level and MAPK activation, but chronically may switch back to $G_{i/o\alpha}$-mediated signalling. The CB_1/A_{2A} adenosine receptor heterodimer also stimulates cAMP, and A_{2A} receptor blockade counteracts the motor depressant effects of intrastriatally administered WIN55212-2 (Carriba et al., 2007). This demonstrates that at least in the rat striatum, CB_1 receptor function is highly dependent on A_{2A} receptors. Pharmacological assays propose the existence of a $5\text{-}HT_2/CB_1$ receptor heterodimer, and on this receptor complex, ligands for each receptor enhance binding to the other receptor (Cheer et al., 1999; Devlin and Christopoulos, 2002). Last but not least, CB_1 receptors can form heterocomplexes with some types of receptor tyrosine-kinases (Harkány et al., 2007). CB_1 receptor activation triggers the migration of progenitor neurons and attenuates neurotrophin-induced neuronal differentiation and neurite outgrowth. In the context of schizophrenia and developmental miswiring, the transactivation of the TrkB receptor in the growth cone of developing axons is of utmost importance (Berghuis et al., 2005; see below and Chap. 12 and Fig. 1).

Marijuana Abuse and Schizophrenia

Due to limited space, we were obliged to select only a limited amount from the large number of studies on this topic. We do apologize for omitting many significant publications, which have attracted our attention though.

Similarities Between Marijuana Effects and Schizophrenia Symptoms

As discussed in the previous chapters and this chapter, the CB_1 receptor has a quiescent role in learning and reward in the hippocampus, cortex and mesolimbic area. Specifically, one single dose of Δ^9-THC, the main psychoactive CB_1 receptor agonist constituent of marijuana, impairs synaptic plasticity for three days in the hippocampus and the nucleus accumbens via functional tolerance of the CB_1 receptor, which does not involve down-regulation or uncoupling (Mato et al., 2004). In contrast, one week treatment with Δ^9-THC already induces a significant uncoupling of the CB_1 receptor in cortico-accumbal synapses (Mato et al., 2005). Although less is known about the molecular changes upon cannabis exposure in man, it is clear that acute marijuana consumption elicits symptoms resembling the positive symptoms of schizophrenia, while chronic marijuana consumption results in a phenotype that is highly similar to the core negative (or residual) symptoms of schizophrenia. Altogether, common marijuana abuse and schizophrenia symptoms encompass: impaired attention and cognition, perceptual alterations, reduced binocular depth inversion, avolition and lack of motivation, apathy, psychotic episodes, hallucinations, altered judgement, false beliefs, and psychomotor anomalies (Halikas et al., 1972; Negrete, 1989; Turner and Tsuang, 1990; Chaudry et al., 1991; Mathers and Ghodse, 1992; McGuire et al., 1994; Emrich et al., 1997; Johns, 2001; Semple et al., 2003, 2005; Solowij and Michie, 2007).

Marijuana Abuse as a Risk Factor for Schizophrenia

Associations between marijuana abuse and schizophrenia have been recognized at least for the last 30 years. It was reported several times that prolonged cannabis abuse (ca. ≥ 50 times, and especially in the young) precipitates psychotic symptoms in vulnerable subjects, also triggers the relapse of psychotic symptoms in schizophrenic patients and worsens positive symptoms of schizophrenia (Andreasson et al., 1987; Negrete, 1989; Turner and Tsuang, 1990; Linszen et al., 1994; Arseneault et al., 2002, 2004; van Os et al., 2002; Weiser et al., 2002; Zammit et al., 2002). Curiously, drug abuse – marijuana in 60% of the cases – is significantly more frequent among individuals with schizophrenia than individuals in the general community (Shearn and Fitzgibbons, 1972; Andreasson et al., 1987; Schneier and Siris, 1987; Cuffel, 1992; Linszen et al., 1994; Kovasznay et al., 1997). Perhaps one of the most interesting studies was conducted by McGuire and co-workers (1995), who matched cannabis-positive and -negative acute psychotic patients for urine screening, and then estimated the lifetime morbid risk of schizophrenia among the subjects' first degree relatives. The result was 7.1% vs. 0.7% in favour of cannabis-positive patients indicating a strong genetic liability in cannabis abuse and psychosis.

Is Cannabis a Risk Factor for Schizophrenia?

As John Macleod, George Davey Smith and Matthew Hickman highlighted in their letter to Lancet (2006), caution is required when interpreting cannabis abuse as a risk factor for schizophrenia. The most important evidence to support this view is that there is no major increase in the frequency of schizophrenia occurrence in the population in spite of the fact that there is a significant growth in the number of cannabis abusers in young people (Drewe et al., 2004). In fact, both the rate of incidence and the prevalence of schizophrenia fluctuate in function of decades, ethnicity, geographical location (countries, towns, villages), economy, life style, migration and several other factors (McGrath, 2006). On the contrary, as discussed here, cannabis abuse is more frequent in the pre-schizophrenic life of newly diagnosed schizophrenia patients than in aged matched controls. This virtual mismatch can be explained in the following manner: (1) A so-far healthy subject – who certainly becomes schizophrenic eventually – is more prone to try cannabis (and other drugs of abuse) than the one who never develops schizophrenia; or (2) decades ago other factors precipitated schizophrenia in susceptible subjects, but in the recent decades, it is cannabis that substitutes a large percentage of previously unidentified precipitating factors; or (3) cannabis ultimately accelerates the onset of the disease in susceptible subjects who would undoubtedly develop schizophrenia in the end. These three explanations semantically do not differ much from each other, but at point (1) we proposed a reverse correlation. A similar reverse causality was screened and rejected in a meta-analysis of prospective studies, which also determined the pooled odds as 2.1 for marijuana abuse to provoke schizophrenia (Henquet et al., 2005). This result most closely corresponds to the result of another meta-analysis which calculated the pooled odds ratio as 2.9 from seven studies (Semple et al., 2005). Therefore, we are left with the conclusion that only the precipitating factor – if needed – and/or the speed of onset are different in relation to chronic marijuana consumption. A recent longitudinal study investigated subjects who were identified as at risk to develop psychosis in the frame of the Cognitive Assessment and Risk Evaluation (CARE) Program. From the group of participants who did not meet criteria for cannabis abuse/dependence, 3.1% developed schizophrenia in one year. From the group of cannabis abusers, however, a significantly (ten times) higher rate, 31%, of the subjects converted to psychosis in one year (Kristensen and Cadenhead, 2007). Although the small sample size does not allow drawing a general solid conclusion on the rate of increase due to cannabis abuse, it warns that cannabis itself can indeed provoke transition to schizophrenia from sub-schizophrenic stages in individuals with pre-existing liability. This theory seems to be bolstered by studies demonstrating that schizophrenia debuts significantly earlier in chronic cannabis abusers compared to non-abusers (Jockers-Scherubl et al., 2003, 2004). As further detailed below, it is still an open possibility that heavy marijuana consumption may precipitate schizophrenia in the absence of other susceptibility (genetic) factors. However, all these facts discussed in this paragraph can be substantially modified by environmental factors as well.

Peculiarities of the Endocannabinoid System in Schizophrenia

Increased CB_1 Receptor Density in Schizophrenia Patients

Three studies have demonstrated in post-mortem brain of schizophrenia patients that CB_1 receptor binding density is increased in areas known to be involved in the disorder. The first study was performed in the dorsolateral prefrontal cortex (Brodmann's area 9), which is involved in information processing and planning tasks. Here, a significant 19% increase was found in the binding of the tritium-labelled potent CB_1 receptor agonist CP55940 in schizophrenia cases (Dean et al., 2001). The dorsolateral prefrontal cortex has reciprocal connections with the anterior and posterior cingulate regions. The anterior cingulate cortex (Brodmann's area 24) is involved in cognition, attention and motivation. The impairment of these three higher-order brain functions resembles core negative symptoms of schizophrenia. In this brain area, a significant 64% increase in [^{3}H]SR141716A binding was observed in schizophrenia patients (Zavitsanou et al., 2004). SR141716A is a selective CB_1 receptor antagonist, and binding assay with the radioligand [^{3}H]SR141716A is a useful tool to demonstrate changes in the CB_1 receptor protein density (Duarte et al., 2007). The posterior cingulate cortex (Brodmann's area 23) has also been implicated in the pathomechanism of schizophrenia. This area is the most sensitive to phencyclidine treatment (Sharp et al., 1994; Olney and Farber, 1995), and shows reduced activation during semantic memory tasks in schizophrenia patients (Tendolkar et al., 2004). Furthermore, Newell and co-workers (2006) reported an age-independent 25% increase in CB_1 receptor density, evaluated by [^{3}H]CP55940 binding, in layers I–II of the posterior cingulate cortex of the schizophrenia group, but not in layers III–VI. Interestingly, this brain area displays a reduced metabolic rate in schizophrenia patients (Haznedar et al., 2004), which may be explained by the inhibitory control of CB_1 receptors, exerted on brain glucose metabolism (unpublished observation by one of the authors, and see Chap. 14).

Increased Endocannabinoid Levels in Schizophrenia Patients

In schizophrenia patients, not only the density of CB_1 receptors, but the level of its endogenous agonists such as anandamide and palmitoylethanolamide was shown to be increased in the cerebrospinal fluid (Leweke et al., 1999). Importantly, this finding was independent from age, gender and medication, whereas 2-AG levels were below detection in both healthy controls and patients. The same laboratory additionally observed that in drug-naïve first episodic paranoid-type schizophrenia patients, the level of anandamide is eightfold higher (0.057 pmol/ml) than in healthy controls, negatively correlating with psychotic symptoms (Giuffrida et al., 2004). It is of utmost importance that the authors also found that 'typical' antipsychotics of D_2-like dopamine receptor antagonist activity, but not 'atypical' antipsychotics of preferential 5-HT_{2A} antagonist activity, abolished this increase. In the plasma of acute schizophrenic patients, in average, a threefold higher anandamide level (7.8 pmol/ml) was detected compared to healthy controls (De Marchi et al., 2003) independently of symptom scores. This increase reduced by half upon clinical remission, accompanied by a significant decrease in CB_2 receptor and FAAH, but not CB_1 receptor, mRNA expression. Here, we should note that while De Marchi and colleagues measured blood anandamide levels in the picomolar range (namely, 2.58 pmol/ml on average for healthy subjects), Giuffrida and colleagues (2004) found the serum level of anandamide six times lower in healthy controls. In antipsychotic naïve schizophrenics, this value was only 52% higher, which was not statistically significant (Giuffrida et al., 2004). The reason for this discrepancy may lie in the fact that De Marchi and colleagues assayed plasma and blood cells together. More interestingly, Giuffrida and colleagues (2004) reported a change (a 35% drop vs. control) in palmitoylethanolamide level of the cerebrospinal fluid opposing to that they had published previously (Leweke et al., 1999). But since the same study found control plasma anandamide level 60 times higher than in the cerebrospinal fluid in the same subjects (Giuffrida et al., 2004), the extra blood anandamide in schizophrenia is (1) likely blood-born *or at least* not due to leakage from the brain and (2) may explain the impairment of the immune system in schizophrenic patients, which normalizes upon remission (Muller et al., 2000; Gladkevich et al., 2004; see above). The attenuated skin-flush reaction observed by Smesny and colleagues (2007; see above) in patients and healthy marijuana smokers thus could be explained by the inhibition of specific immune responses upon chronic cannabinoid receptor overactivation by endogenous and exogenous ligands (Massi et al., 2006). Intriguing that the more potent and efficacious endocannabinoid, 2-AG, has not yet been evaluated in blood samples of schizophrenic patient. However, much prior to the recognition of 2-AG as the main endocannabinoid molecule (see Chap. 2), Kaiya and colleagues proposed that the platelet level of diacyl-glycerol (the precursor of 2-AG; see Chap. 2) may predict the outcome of schizophrenia treatment (Kaiya et al., 1989). This assumption later led to the schizophrenia theory of platelet 2-AG release (Pryor, 2000), but still, further studies are invited to confirm it.

Mapping Schizophrenia to the Human CB_1 Receptor Gene CNR1

The most striking evidence for the involvement of the endocannabinoid system in schizophrenia is found at the level of the hCB_1 receptor gene *CNR1* (OMIM114610). In 1997, Cao and colleagues revealed from two independent data sets with a two-stage approach and non-parametric linkage analysis that a new locus in the chromosome 6, the 6q13-q26, confers susceptibility to schizophrenia. Coincidentally, as previously reported by Hoehe and colleagues (1991), the *CNR1* is located to the 6q14–q15 region. This finding by itself still does not seem to be sufficiently substantial, because, for instance, the $5\text{-}HT_{1E}$ receptor gene is also mapped to the human chromosome 6q14–q15 (Levy et al., 1994), and serotonin is also believed to play a role in schizophrenia (see above). Nevertheless, the *CNR1* locus exhibits several single nucleotide polymorphisms and has an $(AAT)_n$ microsatellite at 3′-UTR (Dawson, 1995), 18 kbp away from the exon 4 translational start site (Zhang et al., 2004). This microsatellite is associated with polydrug abuse (Comings et al., 1997), with the P300 event-related potential (Johnson et al., 1997), and with the childhood antecedent of attention deficit and hyperactivity disorder (ADHD) in alcoholics (Ponce et al., 2003). Among the possible variations of this 3′-UTR flanking region, the 9-repeat allele is statistically significantly *positively*, whereas the 17-repeat allele *negatively* associated with a susceptibility to hebephrenic schizophrenia in a surveyed Japanese adult population (Ujike et al., 2002). Notably, the frequency of the 9-repeat allele in the paranoid schizophrenics was also higher (by the factor of 1.8) compared to control subjects, but this failed to reach statistical significance. In addition, the 10-repeat allele was also found to significantly increase susceptibility to schizophrenia, since it was only found in a small number of patients, but not in the control group (Ujike et al., 2002). It is also important to note that hebephrenic schizophrenia is characterized by predominant negative symptoms, which are strikingly similar to chronic cannabinoid psychoses (Weiser and Noy, 2005). This encompasses an amotivational state, decreased information processing and weakened planning tasks. As mentioned above, brain areas which play a crucial role in these mental processes display an increased CB_1 receptor density in schizophrenia. Therefore, although it is unknown how AAT polymorphism affects the expression and function of the CB_1 receptor, it is quite possible that $(AAT)_9$ genotype causes an accelerated transcription from *CNR1*. In contrast to the findings of Ujike and co-workers (2002), Martínez-Gras and colleagues (2006) found in a smaller size Spanish population that the $(AAT)_{10}$ is a protective polymorphism, regarding that 32.9% of healthy controls had this allele, compared to the statistically significantly less 23.5% of schizophrenia patients. Last but not least, it is also of interest that according to another study, no association was found between AAT polymorphism and schizophrenia in a Chinese poplulation (Tsai et al., 2000). These kinds of differences may be due to disease-related issues and statistical inhomogeneity among the studies. However, it must be highlighted that all studies found different allele frequency distributions in the investigated groups of control and schizophrenic patients. For instance, the $(AAT)_{15}$ (34.8%) and the $(AAT)_{16}$ (28.7%) were the most frequent ones in the investigated 296 Japanese controls, whereas the $(AAT)_{10}$ was virtually absent (none of the controls had it). In contrast, $(AAT)_{10}$ prevailed among the tested 111 Spanish people, and $(AAT)_{14}$ followed it with 27.0%. Together with similar differences in the other studies (Dawson, 1995; Tsai et al., 2000) and in other studies not listed here, we can arrive at the conclusion that the allele frequency distribution may substantially vary based on ethnicity. Furthermore, as Martínez-Gras and colleagues (2006) assumed, AAT polymorphism alone is perhaps not a single factor predicting someone's susceptibility to schizophrenia, but it should be in linkage disequilibrium with other functional polymorphisms. As mentioned above, several single point polymorphisms have been described for the *CNR1*. For instance, a homozygous genotype for one of these single base mutations in the first exon is a significant major pre-disposition factor for intravenous substance abuse in a French Caucasian schizophrenic population (Leroy et al., 2001). Altogether, these suggest that a coincidental appearance of two or more specific polymorphisms in the *CNR1* (and maybe in other genes) would render the subject susceptible to the disease.

Altered Endocannabinoid Signalling Behind the Pathomechanisms of Schizophrenia

In conclusion, if any of the plant-, blood- or brain-derived CB_1 receptor agonist is in excess and/or CB_1 receptors transduce the cannabinoid signal in excess, the consequence of these actions points to the direction of ethiopathology of schizophrenia. In the following sections, we demonstrate several possible hypotheses which all can represent a pathomechanism leading to schizophrenia or at least contributing to the symptoms of the disease.

Developmental Hypothesis

As discussed in Chap. 12, cannabinoids can profoundly modulate neuronal development. Yet before disentangling this question of schizophrenia, we need to take a closer look at the major extracellular signalling systems involved in brain development and plasticity.

(a) Transient disconnection of the CA1 and CA2 regions in rat hippocampus in a critical period of development hampers the neurogenesis of the dentate gyrus, which, in turn, impairs the development of prefrontal cortical wiring, reduces levels of brain-derived neurotrophic factor (BDNF), glutamic acid decarboxilase of 67 kDa (GAD67; the major enzyme involved in GABA synthesis), N-acetylaspartate (NAA) and glycogen synthase kinase-3β (GSK3β) ending up in schizophrenia-like symptoms (Lipska, 2004). Repeated administration of phencyclidine and MK-801 (in animal models of schizophrenia) also induces decreased neurogenesis in the dentate gyrus, which is antagonized by clozapine, but not haloperidol (Maeda et al., 2007). Furthermore, the endogenous NMDA receptor enhancers, glycine and D-serine, given exogenously together with phencyclidine also attenuate these impairments. These, altogether, point towards the critical role of NMDA receptors in the developing forebrain; therefore, the gene polymorphisms reported above causing NMDA-ergic deficit may all induce aberrant circuitry development. Importantly, BDNF via the TrkB receptor potentiates glutamate release and glutamatergic transmission (Kang and Schuman, 1995; Pereira et al., 2006), and it also facilitates GABA-ergic neurotransmission (Baldelli et al., 2005). Together with its role in activity-dependent synaptic plasticity, namely, strengthening synaptic contacts, BDNF is primarily implicated in the development of the brain. BDNF is required also for the growth of dopaminergic and serotonergic neurons (Knusel et al., 1991; Mamounas et al., 1995). The single nucleotide Val66Met polymorphism of BDNF does not affect the activity of the protein at the TrkB receptor, but impairs its secretion (Chen et al., 2004). This, for instance, may explain several anatomical and pathophysiological alterations in schizophrenia, but the underlying exact mechanisms are yet to be determined. In schizophrenic patients, decreased BDNF and TrkB (but not TrkC) mRNA expression is observed together with prefrontal cortical decrease in GAD67 and parvalbumin, all of which are thought to underlie the cognitive deficit in patients (Hashimoto et al., 2005). Compared to BDNF, the decrease in TrkB expression correlated better with that of GAD67, and down-regulation of TrkB expression, but not that of BDNF, resulted in a similar down-regulation of prefrontal cortical GAD67 and parvalbumin expression in mice (Hashimoto et al., 2005). This indicates that the Val66Met polymorphism of BDNF is less likely to be involved in the reduction of cortical inhibition, which is confirmed by Hashimoto and Lewis (2006). Altogether, there can be several mechanisms which may point to the fact that mutations could cause decrease in grey matter and neuronal wiring.

(b) As for the bad consequences of prolonged cannabis intake in adolescents, the developmental hypothesis seems to be the most suitable model. This implies that a chronically aberrant CB_1 receptor signalling undermines the correct selection of progenitors, and then their migration, maturation and arborization, which consequently results in hypoplasticity, grey matter deficit and wiring anomalies. First, both embryonic and adult neural stem/progenitor cells are endowed with CB_1 receptors. As a result, chronic treatment of hippocampal stem cell culture with the potent CB_1 receptor agonist HU-210 induce proliferation, while chronic injection of adult rats promotes neurogenesis in the dentate gyrus (Jiang et al., 2005). Conversely, impaired neurogenesis was observed in adult CB_1 receptor knockout mice (about 50% less bromodeoxyuridine-positive cells were found in the dentate gyrus and subventricular zone than in the wild-type littermates (Jin et al., 2004)). In hippocampal culture, WIN55212-2, Δ^9-THC and anandamide nearly abolish stimulation-induced new synapse formation in a short timescale (Kim and Thayer, 2001). Curiously, a 12-day injection with Δ^9-THC at a continuous low or at escalating doses increase the length of the dendrites as well as the number of dendritic branches in the shell of the nucleus accumbens and in the medial prefrontal cortex (Kolb et al., 2006). As a highly conceivable underlying mechanism for these results were observed, the involvement of growth factor receptor (possibly the TrkB) heterodimers with CB_1 receptors is proposed. Namely, both BDNF and endocannabinoids play an important role in synaptic plasticity, but also in the migration of progenitors, maturation of neurons, synapse development and establishment of new synaptic contacts (Berghuis et al., 2005, 2007; Harkány et al., 2007; see Chap. 12). In short, BDNF and endocannabinoids induce chemotaxis in foetal interneurons, but CB_1 receptor activation does the same via trans-activation of the TrkB receptor in the heterodimer, whereas endocannabinoids suppress BDNF-induced morphogenesis of these neurons. Additionally, one of the authors of this review observed that the density of pre-synaptic CB_1 receptors was fairly stable during the post-natal lifespan of the rat, but that of $mGluR_5$ fell to 60% between the first two and eight weeks of post-natal life, and further declined progressively (AK, unpublished observation). Since $mGluR_5$ is implicated in the generation of post-synaptic endocannabinoid release, our data point to the fact that as soon as the vast majority of neuro- and neuritogenesis occur and synaptic contacts are established in the young adult, a down-regulation of $mGluR_5$ density takes place. As detailed above, the impairment of neurogenesis in the developing dentate gyrus at a critical stage results in adult brain anatomical deficits, resembling those in schizophrenia patients (Lipska, 2004; Maeda et al., 2007). Obviously, a chronic overactivation of CB_1 receptors in the developing brain of adolescents by marijuana abuse – even if the endogenous cannabinoid signal is down-regulated – would continuously suppress the morphogenesis of the neurons in the dentate gyrus and in the cortex, contributing to hypoplasticity and brain maldevelopment (Fig. 1). To test this hypothesis, Szeszko and colleagues (2007) examined prefrontal grey and white matter regions in cannabis user and non-cannabis user patients hospitalized with first episode of schizophrenia, and found less anterior cingulate grey matter in cannabis users, compared with non-cannabis user patients and healthy volunteers. As for the nerve growth factors, both serum NGF and BDNF levels were shown to be significantly higher in drug-naïve schizophrenic patients with long-term cannabis abuse than in non-drug-abusing schizophrenic patients or controls (Jockers-Scherubl et al., 2003, 2004), and this increase was normalized upon remission due to antipsychotics (Jockers-Scherubl et al., 2006). The authors concluded that the source of the extra growth factor levels is the brain. Thus it is plausible to assume that extra NGF and BDNF release tries to overcome the impaired functionality of the growth factor receptor/CB_1 receptor heterodimers, but further studies are invited to corroborate this assumption.

Disrupted Sensory Gating and Smooth Pursuit Eye Movements

As we already discussed above, schizophrenia involves anatomical deficits in prefrontal regions involved in eye movement and motor planning and in temporal and parietal areas which support multimodal sensory and perceptual integration, auditory perception and episodic memory (Cannon et al., 2002). Perhaps this is another major reason for those specific deficits observed in schizophrenia patients called sensorimotor gating deficit and voluntary smooth pursuit eye movement deficit. It is very important to underline that for both deficits, an impaired cholinergic signalling at the $\alpha7$ nicotinic acetylcholine receptor ($\alpha7$ nAChR) is proposed as a partial or full underlying mechanism, and cigarette smoking and treatment with nicotinic agonists are known to alleviate or prevent these symptoms in man (Adler et al., 1998; Olincy et al., 2003). Furthermore, such an impaired signalling at the $\alpha7$ nAChR is proposed to participate in the pathomechanisms for cognitive anomalies in schizophrenia and therefore, $\alpha7$ nAChR agonists represent valuable therapeutic tools to recover cognitive impairment (Olincy et al., 2006; Pichat et al., 2007). The sensorimotor gating deficit can be reliably experimented by PPI of the startle response in rodents and man (see above). Cannabidiol, a non-psychoactive constituent of marijuana, possesses weak CB_1 receptor antagonist activity (see Chap. 9). Cannabidiol (5 mg/kg, i.p.) was shown to reverse disruptions in PPI induced by MK-801, and this effect was mimicked by clozapine as well (Long et al., 2006). Similar findings are reported with the potent and selective CB_1 receptor antagonists, AM251 and SR141716A, which reduced the PPI-disruptive effects of dizocilpine, phencyclidine and apomorphine (Ballmaier et al., 2007). Interestingly, NRG1 heterozygotic mice are more sensitive to the effect of Δ^9-THC on PPI than their wild-type littermates (Boucher et al., 2007; note that NRG1 null-mutant mice do not survive and therefore can not be tested in this assay). Altogether, these data provide direct evidence that cannabinoid agonist can impair auditory sensory gating and smooth pursuit eye movements via CB_1 receptor-mediated and non-CB_1 receptor-mediated mechanisms. Although mutations in the $\alpha7$ nAChR gene are also confirmed to underscore these physiological processes, it is important to mention that the endocannabinoids, anandamide and 2-AG, inhibit currents evoked at $\alpha7$ nAChR in the nanomolar range (Oz et al., 2003, 2004). Therefore, high endocannabinoid levels can directly induce impairment in cognition, sensory gating and smooth pursuit eye movements. Inhibitors of CB_1 and D_2 receptors can indirectly attenuate this impairment via reducing endocannabinoid release, as proposed above and as observed in other studies (Fig. 1).

D_2 or D_2/CB_1?

In the VTA and the basal ganglia, post-synaptic D_2 receptors are thought to be involved in eliciting post-synaptic Ca^{2+} rise from intracellular stores, which in turn, activates enzymes which cleave endocannabinoids from their precursors, and then endocannabinoids are released into the synaptic cleft (Melis et al., 2004b; Ronesi and Lovinger, 2005, and see below as well). Furthermore, schizophrenia is a hyperdopaminergic and hypercannabinergic state, as extensively discussed above. Additionally, D_2 receptors are inhibitory receptors, but the D_2/CB_1 receptor heterodimer is coupled to $G_{s\alpha}$, thereby eliciting facilitatory responses, including Ca^{2+} level elevations (Glass and Felder, 1997; Jarrahian et al., 2004; Kearn et al., 2004). This prompts us to ask: "Could it be possible that the D_2-mediated synaptic plasticity is rather a D_2/CB_1 receptor-mediated synaptic plasticity?" If it was so, then high transient dopamine and endocannabinoid levels would continually produce endocannabinoids in excess. Moreover, high endocannabinoid levels reduce the uptake of dopamine and serotonin, whereas facilitate that of glycine (see above), and inhibit the release of glutamate and GABA via CB_1 receptor activation, and then everything comes full circle: this excess dopamine and endocannabinoid level will further elevate the levels of endocannabinoids, dopamine and serotonin, and diminish that of glycine, besides inhibiting the release of glutamate, both contributing to the hypo-NMDA-ergic state, and also inhibiting the release of GABA whereby increasing the post-synaptic activity of mesolimbic dopaminergic cells. Now, if all these assumptions were true then (1) (endo)cannabinoids should be able to increase dopamine levels in the brain – and this was already demonstrated by several studies both in animal models and in schizophrenia patients (Melis et al., 2000, 2004b; Malone and Taylor, 1999; Voruganti et al., 2001; and see above) and (2) antipsychotics should be able to disrupt endocannabinoid signalling to exert their beneficial effects. And indeed, the D_2 receptor antagonist clozapine displaces the binding of the CB_1 receptor agonist [^{3}H]CP55940 in rat nucleus accumbens (Sundram et al., 2005), indicating that clozapine can interrupt CB_1 receptor signalling via D_2 receptor blockade. Additional studies are necessary to strengthen or reject this hypothesis (Fig. 1).

Impaired Working Memory

Anatomical and functional data in rodents and man suggest that the endocannabinoid system modulates cognitive functions, such as working memory, through depolarization-induced suppression of inhibition (DSI) on cholecystokinin (CCK)-positive terminals in the prefrontal cortex, and these terminals are presumed to fine-tune the network oscillation of parvalbumin containing neurons in the gamma frequency range (Lewis and Hashimoto, 2007). Intriguingly, the working memory deficits commonly observed in schizophrenia (Goldman-Rakic, 2005) are associated with both reduced gamma band power and deficient perisomatic afferents to pyramidal neurons from parvalbumin containing GABA neurons. Activation of CB_1 receptors through the use of Δ^9-THC could therefore result in an additional deficit in perisomatic GABAergic input to prefrontal pyramidal neurons in individuals with schizophrenia by inhibiting GABA release from CCK-positive interneurons (Lewis and Hashimoto, 2007). The reported upregulation of CB_1 receptor binding sites in schizophrenic brains (Ujike and Morita, 2004) might further worsen this deleterious effect on working memory. These observations suggest that endocannabinoids play a critical role in the circuitry which subserves cognitive functions such as those which are disturbed in schizophrenia. Likewise, Δ^9-THC may induce a host of perceptual distortions indistinguishable from schizophrenia (Iversen, 2003; D'Souza, 2007), one of which is particularly interesting.

Psychotic Time Distortions

Altered time estimation has been reported with considerable consistency in both cannabis intoxication and schizophrenia (Melges, 1982; Elvevag et al., 2003). The characteristic temporal distortion under the influence of Δ^9-THC is best described by the poet J.R. Anderson:

> The first effect – and this remained true for every subsequent occasion – was the alteration of time values. Time was so immensely lengthened that it practically ceased to exist. But

> this slowing-down … did not apply to the processes of thought. Those, on the contrary, appeared to be very greatly accelerated (cited in Hicks et al., 1984).

In other words, when Δ^9-THC speeds up the physiological processing of time, the subject estimates the passage of physical time to be proportionally longer compared to the clock. This phenomenon is also experienced by schizophrenic patients when subjective time seems to be passing more quickly and physical time more slowly than expected.

> "Time has stopped; there is no time… The past and future have collapsed into the present, and I can't tell them apart". "The world had become timeless." She knew that the "clocks still marched onward," but she was "in a different realm" where "everything is happening at once" (Melges, 1982).

This kind of asynchrony, being counter current to our common sense intuition of time, appears to be directly related to other perceptual alteration including spatial distortions and dissolution of the self.

> A common hallucination induced by large doses of cannabis is time and space distortion: minutes seem like hours, small rooms yawn into caverns, and every activity is imbued with a sense of timeless grandeur…More importantly, in the ecstatic union of the human and the divine represented by this ritual, the sense of self is transcended by both partners. The role of cannabis in Tantric ceremony is thus to enable the worshippers to feel the divinity within and without themselves (Aldrich, 1977).

Accordingly, Melges (1982) cites a patient with schizophrenia who experienced 'mystical awareness' in which she felt she could 'see beyond' ordinary reality. But later, her sense of 'psychic powers' and revelations dissipated, as she entered the 'abyss of timelessness'. Along with this, she had lost her 'grip' on who she was and felt 'pushed and pulled' by 'strange forces and voices' that made her do things against her will. While unconscious processes are characterized by timelessness (Freud, 1915), time distortions with disintegration of the subject–object boundary are fundamental aspects of altered states of consciousness (Ludwig, 1966). Likewise, impaired goal-directed behaviour and loss of self control are among the most striking clinical clues that a patient may be psychotic. Objectively, this can be observed in the disorganization of the patient's thought and actions. Subjectively, when asked about his or her future perspective, the psychotic patient often reports to have completely lost control over what might happen in the future (Melges, 1982). Despite decades of intensive research by psychologists, anatomists and clinicians, the physiological substrate of temporal information processing by the brain (Buhusi and Meck, 2005) has remained incompletely understood. One of the major parsimonious findings concerning time perception in the seconds to minutes range is the ability to accelerate subjective time with CB_1 and D_2 receptor agonists, and to decrease it with the classical antipsychotic dopamine D_2 receptor antagonists (Meck, 1996). Interestingly, genetic polymorphisms within both CB_1 receptor (Johnson et al., 1997) and D_2 receptors (Blum et al., 1994) caused alterations of P300 (see above and below). This event-related potential, known to correlate with attention across time-subsequent goal-directed action (Munson et al., 1984), is among the most robust electroencephalographic indices for schizophrenia

(Blackwood, 2000) and long-term cannabis abuse (Solowij et al., 1991). In addition, there is considerable support for a dissociation of the internal clock affected by dopaminergic manipulation from a memory stage affected by cholinergic manipulation. As mentioned previously, dopaminergic antagonists produce a deceleration of the subjective clock speed in proportion to their affinity for the dopamine D_2 receptor, while the systematic discrepancies in the accuracy of temporal memory are proportional to pharmacological effect of acetylcholine (ACh) in the frontal cortex (Buhusi and Meck, 2005). Based on this functional dichotomy, a biologically plausible model has been developed. It describes timing as an emergent oscillatory property of medium spiny (MS) neurons in the striatum activated by coincidental cortical inputs. Glutamatergic afferents from the motor cortex increase synchrony and coincidental activity on MS neurons, when animals are trained to expect a 'go' signal at a certain point in time, while attention modulates the coincidental activity of somatosensory and visual cortical afferents. These oscillators are assumed to be synchronized at the onset of the trial, and to oscillate at a fixed beta frequency throughout the criterion interval (Meck, 1996; Buhusi and Meck, 2005). Fully in line with this assumption are the reported intrinsic beta2 frequencies (20–30 Hz) of efferent pyramidal neurons in the cortex (Roopun et al., 2006), as well as the 25 Hz timing frequency assessed in humans (Brown et al., 2002). Very intriguing, therefore, are the recently reported findings of electroencephalographic alterations in the beta frequency range in schizophrenic patients (Gross et al., 2006) and cannabis abusers (Skosnik et al., 2006). In addition, the total years of cannabis use positively correlate with their schizotypal profile, and those scoring higher in schizotypy demonstrate larger deficits in neuronal synchronization (Skosnik et al., 2006).

Antagonistic Signal Transduction Converges onto the Calcium Cascade

(a) *Dopamine vs. ACh.* The prefrontal cortex is involved in this process and made up of areas thought to mediate specific aspects of the temporal organization of behaviour (Luria, 1973; Fuster, 1989; Simmons and Richmond, 2007). In functional (Fujii and Graybiel, 2005) and anatomical terms (Eblen and Graybiel, 1995), one of the prime output targets of these neocortical areas is the striatum (nucleus accumbens and caudate-putamen), the main afferent structure of the so-called 'basal ganglia' (Hanero et al., 2002). Overall, there are two major inputs to the striatum: the corticostriatal and thalamostriatal pathways; and these pathways are under direct, mutually antagonistic (Shapovalova, 2000) influence by dopaminergic (Goto and Grace, 2005a,b) and cholinergic neurons (Kimura et al., 2003; Samejima et al., 2005). The effect of these modulatory neurons upon the activity of MS neurons has recently been shown to be critically modulated by endocannabinoid retrograde messengers (Kreitzer and Malenka, 2007; Narushima et al., 2007). Dopamine D_2 receptors and muscarine M_1 receptors thereby play a role as 'coincidence detectors' of dopamine-glutamate and ACh-glutamate co-activation, respectively (Kreitzer and Malenka, 2007; Uchigashima et al., 2007). On MS neurons, several other 'competing' coincidence detectors exist at the cellular level involving CB_1/D_2 (Glass and Felder, 1997; Andersson et al., 2005) and D_1/D_2 receptor heterodimerization (Rashid et al., 2007). Particularly intriguing, in our context, is the fact that in MS neurons, both D_1 and the CB_1/D_2 receptor heterodimers converge on the same target, the G_s protein-coupled-metabotropic receptor subunit of the protein kinase A (PKA) cascade (Andersson et al., 2005). Complementary to this pathway, the D_1/D_2 receptor heterodimers (Rashid et al., 2007), the CB_1 (Lauckner et al., 2005) and M_1 receptors (Narushima et al., 2007), all converge on the $G_{q/11}$ protein-coupled-metabotropic receptor subunit, which in turn is pivotal for the direct and indirect downstream activation of the inositol-triphosphate (IP_3) calcium cascade (Berridge, 1998), including phospholipase C (PLC), the calmodulin-dependent protein kinase II and protein kinase C (PKC) (see Fig. 2).

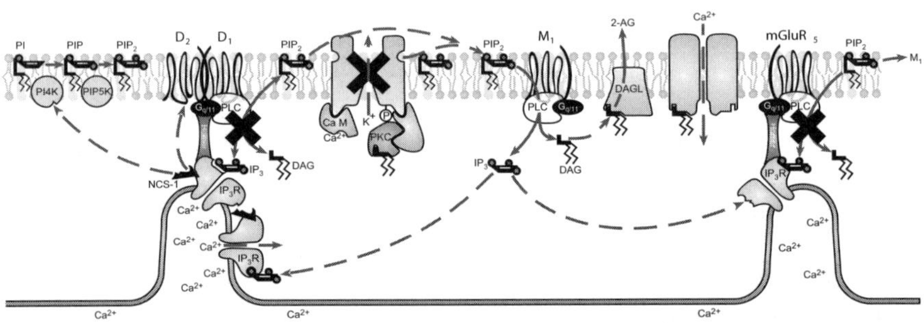

Fig. 2 Dysbalanced calcium influx at the root of psychosis. Integrated in the phosphoinositide-endocannabinoid system, increases in intracellular free Ca^{2+} regulate a wide variety of biological processes during fertilization, immune responses, neuronal migration and synaptic plasticity during development and memory formation. To ensure tight control of the pertinent Ca^{2+} signalling cascades, complementary effector pathways have evolved in terms of spatial and temporal differentiation (Berridge, 1998). Within the post-synaptic striatal neuron, scaffolding of D_1/D_2 and $mGluR_5$ membrane receptors to calcium channels (IP_3R) on the endoplasmatic reticulum (ER) makes sure that dopaminergic and glutamatergic calcium release from these local Homer-coupled microdomains is effectively segregated from the global elevation of Ca^{2+} elicited by cholinergic M_1 receptor activation. To engender synaptic plasticity by means of long term potentiation and depression, calcium signalling depends on temporal patterns of Ca^{2+} depletion from the ER through burst and tonic firing, respectively (Cui et al., 2007). During tonic firing, M_1 receptor-mediated stimulation of PLC leads to a closure of K_{ir2} channels by depleting the membrane bound substrate PIP_2 (Delmas et al., 2004; Carr and Surmeier, 2007; Suh and Hille, 2007). By analogy, M_1 receptor activation could also deprive the two other PLC-dependent receptor complexes of their pivotal substrate PIP_2, resulting in disintegration of D_1/D_2 receptors and the characteristic NCS-1 coupled upregulation of D_2 receptor in schizophrenia (Bergson et al., 2003; Wang and Goldman-Rakic, 2004). Impairment of D_2 receptor-driven behavioural control over M_1 receptor-mediated sensory inputs (Yeomans, 1995; Shergill et al., 2005) may thus reflect the core of psychosis within a broader framework of altered spatiotemporal Ca^{2+} signal transduction at the synaptic level

(b) *D_1/D_2 receptor co-activation of $G_{q/11}$ subunits.* The striatum is not only assumed to play a principal role in psychostimulant addiction (Hyman et al., 2006), striatal dysfunction has also been implicated in several neuropsychiatric disorders, such as schizophrenia (Joyce and Gurevich, 1999). A diminished link between D_1 and D_2 dopamine receptors has been noted in schizophrenic brains, and it has been proposed that altered calcium signalling may be the central molecular factor in schizophrenia (Bergson et al., 2003; Rashid et al., 2007). Associated with this alleged role, heteromeric D_1/D_2 dopamine receptor signalling is required for $G_{q/11}$-coupled activation of calmodulin-dependent protein kinase II and subsequent intracellular calcium release in brain (Rashid et al., 2007). This provides an important hint to the characteristic onset of schizophrenia in late adolescence (Kraepelin, 1899), because the D_1/D_2 signalling-complex can be more readily detected in mice which are eight months of age, compared to younger animals, and explains why both co-activation of D_1 and D_2 receptors as well as activation of calmodulin-dependent protein kinase II are necessary for the post-adolescent induction of behavioural sensitization to psychostimulants such as cocaine (Rashid et al., 2007). Given the fact that endocannabinoids facilitate the effects of commonly abused drugs including cocaine (Gardner, 2002; Cheer et al., 2007), it is also noteworthy that a direct inactivation of the $G_{q/11}$ subunit in mice leads to impaired endocannabinoid formation and increasing seizure susceptibility in adolescence (Wettschureck et al., 2006).

(c) ***Scaffoding of $G_{q/11}$ to intracellular calcium release.*** To understand the full impact of $G_{q/11}$-coupled D_1/D_2 co-activation on striatal brain function and dysfunction, its relation in terms of molecular anatomy has to be viewed from yet another perspective: the coupling of $G_{q/11}$ to group I metabotropic glutamate receptors (mGluRs). Cocaine and other psychostimulants, such as methylphenindate, are known to induce the genetic expression of an ineffective splice variant of Homer (Brakeman et al., 1997; Yano and Steiner, 2005). This splice variant, $Homer_{1a}$, competes with $Homer_{1b/c}$ for binding to mGluRs and uncouples mGluRs from IP_3 receptors. For, by virtue of their ability to form multimers, $Homer_{1b/c}$ assembles mGluRs in large macromolecular complexes directly to the IP_3 receptor on endoplasmatic reticulum (ER), the main source of intracellular calcium influx (Berridge, 1998). $Homer_{1a}$, the short, activity-dependent splice variant of $Homer_{1b/c}$, however, lacks the ability of linking mGluRs to synaptic proteins, and functions as an endogenous negative modulator of the direct $PLC\text{-}IP_3$-mediated intracellular calcium influx. It has also been proposed that $Homer_{1a}$ functions as an endogenous antagonist of the mGluR-signalling pathway. Inefficient interaction of $Homer_{1a}$ with striatal post-synaptic $G_{q/11}$-coupled receptors such as the metabotropic glutamate receptor 5 ($mGluR_5$) (Simonyi et al., 2005) results in major impairment of neural processes involving learning, memory and epileptogenesis. Additionally, as discussed above, $mGluR_1$ and $mGluR_5$ are also critical receptors to induce post-synaptic endocannabinoid release. Complex motor tasks and amphetamine-induced stereotypy are significantly altered in transgenic mouse lines that over-express $Homer_{1a}$ in their striata. Since pharmacologically induced loss of $Homer_{1a}$ in these mice rescues the normal motor phenotype, it can be safely assumed that $Homer_{1a}$, and not other factors such as the genetic background, are responsible for these defects. Furthermore, transgenic mice, which overexpress $Homer_{1a}$ exclusively in strisomes, display stronger defects than mice which overexpress $Homer_{1a}$ in the matrix compartment. This suggests that $Homer_{1a}$-induced modulation of $mGluR_5$ signalling (see above) in striosomal efferents (of the 'direct' pathway to dopaminergic neurons in the brainstem) has a greater impact on function than in the matrix (indirect pathway). Consistent with this notion, critical signalling effectors of $mGluR_5$ such as PLC and the IP_3 receptors have been reported to be enriched in the striosomal compartment, although neither $mGluR_5$ themselves nor endogenous $Homer_{1b/c}$ appear to display selectivity of expression over the patch–matrix domain of the striatum (Gerfen, 2004; Tappe and Kuner, 2006).

(d) *Antagonistic signalling and binary neuronal function*. Several studies in rodents have revealed that the matrix and striosomes are highly specific in terms of afferent–efferent connectivity: the striosomal neurons process inputs from structures associated with dopamine-mediated motivational behaviour, learning and goal-directed action, on the one hand, and inputs from the intralaminar thalamic nuclei associated with ACh-mediated attention, on the other. In primates, this dual mode is reciprocated in the structure of prefrontal cortex: higher 'limbic' efferents from the anterior cingulate and anterior insular cortex project to the striosomes, in addition to intralaminar nuclei of the thalamus and the caudally adjacent tegmental area and periaqueductal grey (PAG). The remaining parts of the prefrontal cortex project preferentially to the matrix compartment which receives inputs associated with somatic locomotor behaviour in response to sensory inputs (Eblen and Graybiel, 1995; Tappe and Kuner, 2006). The anterior cigulate and anterior insular cortices, by contrast, are associated with 'visceral' emotional behaviour (Damasio 1999; Damasio et al., 2000), and the PAG constitutes one of the ultimate effectors of emotional and cardiorespiratory responses to social and environmental challenges (Holstege et al., 2004; Keay and Bandler, 2004; Green et al., 2007). It is, therefore, not surprising that limbic structures have traditionally been implicated in schizophrenia. Functional brain imaging identifies the anterior cingulate gyrus as one of the main dysfunctional regions in schizophrenia. Emotional dysfunction and impaired stress coping, at a phenomenal level (Bleuler, 1911), and diminished heart rate variability (as an indication of anticipatory distress), at a physiological level, are very characteristic signs and symptoms in schizophrenia (Malaspina et al., 2002; Bär et al., 2005; Fritzsche et al., 2006). In summary, a specific convergence of inputs into the striatum is closely associated with the emotional processing of attention and motivation, faculties that are typically impaired in schizophrenia (Kraepelin, 1899; Bleuler, 1911) and chronic cannabis psychosis (D'Souza, 2007). Thereby, the striosomal compartment plays a critical role in the dopaminergic ($G_{q/11}$-$mGluR_5$-$Homer_1$-IP_3 receptor-mediated) trigger of calcium release through the ER membrane of the MS neuron (Berridge, 1998) on the one hand, and the integration of signals as part of the ($G_{q/11}$-mediated) intracellular component of neural calcium release involves ACh, on the other. The dual mode of dopaminergic–cholinergic signalling in this compartment does not only appear to be functional within different afferent structures from the prefrontal 'limbic' and intralaminar thalamic system. The duality also reflects the binary membrane mechanism of neural calcium release (Berridge, 1998). This brings us to the crucial question: "How is the dopaminergic ER-bound calcium signalling mechanism related to cholinergic calcium signalling?" We will see in the sequel that the M_1-$G_{q/11}$ receptor complex constitutes the plasma membrane (PM) component of the binary system and that its function, in terms of signal integration, is inversely related to the dopaminergic modulation of MS neurons.

(e) *Signal integration by depleting membrane-bound substrates*. Let us first start with the basic M_1-mediated mechanism of action onto the potassium channel. In efferents from the rodent limbic cortex, the layer V pyramidal cells, M_1 muscarinic receptor stimulation depolarizes the membrane and reduces constitutively active inwardly rectifying (K_{ir2}) potassium channel. This results in robust membrane depolarization and tonic firing. More specifically, M_1 receptor-mediated stimulation of PLC leads to a depletion of membrane phosphatidylinositol-4,5-bisphosphate (PIP_2) and the closure of K_{ir2} channels, because binding of PIP_2 to the C-terminus of K_{ir2} channel subunits is necessary for channel opening (see Fig. 2). In other words, M_1 receptor-triggered signal integration in the membrane is induced by means of depleting a pivotal, membrane-bound fatty acid substrate through its degrading enzyme PLC (Delmas et al., 2004; Carr and Surmeier, 2007; Suh and Hille, 2007). Consistent with this finding is the effect on retrograde depolarization-induced suppression of inhibition (DSI), which is prompted by M_1-PLC-mediated release of endocannabinoids in the neocortex (Hill et al., 2007) and striatum (Uchigashima et al., 2007). In addition to direct membrane depolarization, PLC-mediated DSI and PLC-mediated closure of largely dendritic K_{ir2} potassium channels dramatically enhances the summation of excitatory synaptic potentials. Complimented by $M_{2/4}$ receptor-mediated inhibitory component of the same terminals, ACh thus reduces the post-synaptic consequences of single afferent volleys, but potentiates the response to temporally coherent bursts of synaptic activity in vitro. Whether this is also true in a natural setting, awaits demonstration in vivo (Carr and Surmeier, 2007). Nevertheless, the conclusions appear to be complementary to those drawn from the effects of M_1 receptor stimulation on glutamatergic afferents (Uchigashima et al., 2007), the striatal recipients of the layer V pyramidal cells. At the membrane level, this effect is engendered, at least in part, by depriving other receptors of their substrate through PLC. In summary, it is tempting to assume that direct activation of PLC does not only deprive K_{ir2} signalling of PIP_2, but could also deprive PLC dependent D_1/D_2-$G_{q/11}$ signalling of its pivotal substrate PIP_2. If this is the case, the latter would disintegrate and uncouple D_1/D_2 membrane-bound heterodimerization. M_1 receptor-dependent enhancement of the inositol pathway is also known to affect downstream calcium-PIP_3 receptors indirectly through intracellular diffusion, because in the case of M_1, $G_{q/11}$ is not directly bound to PIP_3 receptors (Delmas et al., 2004). As a consequence, M_1-dependent activation (or alternatively, D_1/D_2-dependent deactivation) could counter-regulate the down-regulation of D_2 by means of the neuronal calcium sensor-1 (NCS-1), which possesses molecular docking pockets that bind to the D_2 and PIP_3 receptors as well as to a type III phosphatidylinositol-4-kinase (Burgoyne et al., 2004). Irrespective of the precise mechanism involved, NCS-1 and D_2 receptors have been reported to be up-regulated in schizophrenia (Bergson et al., 2003; Wang and Goldman-Rakic, 2004), and NCS-coupled up-regulation of D_2 receptor would, in a vicious circle, further impair the primarily D_1 driven (Rashid et al., 2007) cohesion of the D_1/D_2-$G_{q/11}$ receptor complex. Worse still, NCS-coupled activation of the phosphatidylinositol-4-kinase would further deplete upstream substrates of the inositol pathway (Burgoyne et al., 2004), a scenario that is fully in line with the reported deprivation of membrane-bound lipids and related second messengers in schizophrenia (Arvindakshan et al., 2003) and chronic cannabis abuse (Smesny et al., 2007). Taken together, there is ample evidence that these phenomena may play a role in schizophrenia at the molecular level, but the question arises whether the functional consequence of this scenario would also be consistent with what we know about ACh-dopamine interaction in the striatum.

(f) *Retrograde signalling at centre stage*. Owing to differential cellular distribution of the receptors and their downstream molecular cascades on excitatory and inhibitory synapses on MS neurons and their microcircuits, activation of dopaminergic and cholinergic receptors exert an opposite effect on MS neurons. This anterograde antagonism, which seems to be at the core of striatal function (Shapovalova, 2000; Kimura et al., 2004; Ragozzino and Choi, 2004; Minamimoto et al., 2005), mirrors the dual role of retrograde regulation in the striatum. On excitatory glutamatergic synapses of the indirect pathway, D_2 receptor-mediated enhancement of retrograde endocannabinoid release suppresses excitatory glutamatergic transmission, and this results in depolarization-induced suppression of excitation (DSE) diminishing the activity of the MS neuron (Kreitzer and Malenka, 2007). This stands in contrast to the endocannabinoid-mediated effects on inhibitory synapses of both direct and indirect pathways, where M_1 activation alone strongly enhances 2-AG synthesis by PLC, as reported in the hippocampus and cerebellum. As a result, the retrograde release of 2-AG from the MS neuron suppresses inhibitory GABA transmission onto MS neurons through DSI, which transiently enhances overall striatal output (Narushima et al., 2007; Uchigashima et al., 2007). In the striatum, this is the least we know at present (but see also Surmeier et al., 2007). It is also clear that the site from where neurotransmitters are released at the MS neuron does not always correspond with the location of the respective receptors. ACh released at the shaft of the MS neuron, for example, has to cover a certain distance before it couples to the corresponding M_1 receptor on the spines (Uchigashima et al., 2007). Dopaminergic release from afferent buttons near the middle of the MS dendritic tree is even further apart from the respective D_2 receptor on its apical segments. For simple anatomical reasons, therefore, 'phasic' volume transmission has been postulated for both ACh and dopamine along the MS neuron. Otherwise, the topological facts would make no sense (Saulskaya, 2000). By contrast, dendritic D_1 and D_2 receptors normally co-localize during tonic transmission inside the synaptic space opposite the dopaminergic terminals (Saulskaya, 2000) in close proximity to the adjacent perisynaptic M_1 and $mGluR_5$ receptor sites (Uchigashima et al., 2007). Such sub-cellular proximity stipulates close functional interaction of the D_1/D_2-$G_{q/11}$ signalling complex, not only with the $mGluR_5$-$G_{q/11}$ complex, as discussed in detail, but also with the M_1-$G_{q/11}$ signalling complex. Taking into consideration that the afferents to the cholinergic terminals to the striatum stem from the intralaminar thalamus, and the afferents to the mGluRs exclusively from the dopaminergic limbic cortex – reminding us of Meck's statement that decision and timing processes reflect two sides of the same coin (Buhusi and Meck, 2005) – it is tempting to assume some kind of attentional switch (Kimura et al., 2004, Minamimoto et al., 2005). If this switch is engendered through the cholinergic interface, what would be the behavioural effect? If dysfunctional, the question immediately arises what the neuropsychiatric sequelae would be.

Basic Reaction to Challenge and Dysfunctional Sensorimotor Integration

At the most basic physiological level, the cholinergic switch would trigger an archaic type of information process or choice pattern: stop moving and watch out, as if one type of signal integration takes precedence over another type of signal integration (Berridge, 1998). As evidenced by experiment (Shapovalova, 2000; Ragozzino and Choi, 2004), this duality is deeply rooted in evolutionary biology. It constitutes the two most basic moieties of structure and function in the nerve axis, one for sensing and deciphering incoming data from the environment and the other for acting in it. In the lowest monocellular forms, behaviour is dependent upon direct contact with the environment and is limited to two reactions: moving towards or withdrawal (Stein and Meredith, 1993). As organisms progress up the phylogenetic scale, this innate behavioural pattern is integrated into higher levels of the neuraxis and promoted to a potentially infinite number of exploratory, goal-oriented and aversive responses. Consequently, the binary system of signal integration does not only continue to exist at the cellular level (Stein and Meredith, 1993; Berridge, 1998), the functional dichotomy remains preserved in the basic anatomical organization of the mammalian brain (Swanson, 2000) and can be observed during risk assessment behaviour, when foraging animals leave the safe haven of their territory for feeding and procreation (Misslin, 2003). Betz (1875) first extrapolated to the motor cortex the posterior–anterior dichotomy that has prevailed in the course of evolution along the nerve axis from the spinal cord on upwards. However, at the pinnacle of evolution, in the associative cortex, there is one fundamental difference between perceptual and motor function. The first essentially receptive mode gets dominated by memory and sensory representations, and the latter mode, in addition to being dopaminergic and motivational, also gets operational in subserving sensorimotor integration. This dual mode of neural functioning reflects and extends the widely accepted dichotomy in neuropsychology, which has proved so eminently useful for understanding higher associative motor and memory deficits (Luria, 1973; Fuster, 1989).

(a) *Delusions of Alien Control*. In shedding additional light onto the psychiatric deficits, the serendipitous observations by Giacomo Rizzolatti and colleagues have put the valid scheme of sensorimotor integration into a new perspective. Normally, when individuals observe an action performed by another person, a 'replica' of that action is automatically generated in the so-called mirror neurons of the cortex, recruiting the same neuronal circuits that become active when such action is generated by the observers themselves (Rizzolatti and Gallese, 2003). Being mandatory for understanding the intentions of others and motor learning, this mechanism normally does not lead to confusion. Yet, it is only possible to discriminate self-generated action from action made by others, if there exist signals preceding action initiation as well as signals following movement onset, that is, the representation of an extended time dimension. If this temporal representation is absent or dysfunctional, the mirror neurons produce delusions of alien control by default. Schizophrenic patients have precisely this problem, because incorrect predictions cause the delusion of self-generated actions as externally generated. To be precise, schizophrenics erroneously misattribute their own actions to an external source, because they suffer from a defect in their ability to predict the sensory consequences of their actions (Shergill et al., 2005). In addition to the deficits to anticipate motor sequences, they also have defects in error correction and memory for action, and brain-imaging studies during hallucinatory experiences demonstrate increased activation in associated sensory regions (Frith et al., 2000). Consistent with the model described by Jackson (1932), dysfunctional overactivity in these areas results from a lack of incoming inhibition from the frontal cortex, which normally attenuates activity associated with predicted, self-generated re-afferent stimuli. By default, the physiological role of this inhibitory frontal process is to enhance the salience of sensations that have an external cause: if the predicted trajectory is discordant, as when one's arm is passively deviated by someone else, the respective motion is labelled as foreign. Conversely, if the predicted sensory inputs match the actual sensory consequences of active movement engendered by the frontal lobe, it is labelled as one's own. In schizophrenia, misattribution of the self to the outside or vice versa, which Schneider included among his first rank symptoms, can be best described as actions created, not by the patients themselves, but by some outside forces. At the root of the psychotic disorder, such a dysfunction relates to higher motor control over sensory inputs (Shergill et al., 2005) and reflects the kind of dysbalance in dopamine-driven motor control over ACh-mediated sensory inputs discussed above. This closes the conceptual loop between the phenomenological and molecular level and brings us back to the central tenet of the present chapter – the missing link between schizophrenia and cannabis psychosis.

(b) *Dreams, nightmares and psychosis*. Refashioned in the absence of a direct sensorimotor input, dreaming is widely held to emerge through activated cortical memory networks under the influence of subcortical afferents. Activation of the cholinergic system, with its source in a nuclear complex at the base of the brainstem, is thereby critically implicated; in rapid eye movement (REM) sleep, both the basal forebrain and thalamic corticocopetal projections are stimulated by cholinergic afferents originating mainly from the pedunculopontine and laterodorsal tegmenta. In this context, long-standing speculations about the similarities between dreaming and psychotic conditions are substantiated by the following main arguments. Compared to normal controls, certain patients with schizophrenia show an earlier onset and a decreased latency of REM sleep, as well as a potential increase in the number of cholinergic cells in the tegmentum. There is evidence that administration of antipsychotic drugs attenuates abnormal increases in cortical ACh release and repeated administration of hallucinogenic psychostimulants augments drug-induced increases in cortical ACh efflux. A substantial body of literature also indicates that ACh exerts its role not alone but by interaction with other neurotransmitters, including dopamine, glutamate, GABA and, not least, the endocannabinoids, which are all involved in schizophrenia (Yeomans, 1995; Sarter and Bruno, 2000). The memory activated by ACh in dreams is not only distorted, often beyond recognition by the awakened dreamer, but it lacks a critical attribute of conscious awareness: temporality. Cut off from sensory inputs and context, the subject cannot but project the events in the dream to any time but the present; the cortical neuronal networks, anchored in the present as they are without time tags and references, seem to lack the associative links to a time frame (Freud, 1915). The dream, though replete with past experience, appears in the present and lacks the phenomenal attributes of past and future. Taken together, it comes with no surprise that the typical time distortions in cannabis and schizophrenic psychoses are closely related to dreamlike episodes. During psychedelic cannabis experience, there are frequent momentous transitions from a 'dreamy' state of consciousness to full awareness and vice versa. Typically, the lapses back into the 'dreamy' state go unnoticed and are beyond control, whether you sit, stand or walk. In self-experiments with high dosages of Dronabinol™ and Δ^9-THC, one of the authors of this chapter (MF) and his medical colleagues could reliably reproduce these strikingly asymmetric shifts in consciousness. These can only be noted, i.e., by pressing a button linked to the EEG, at the moment the subjects 'awake', but not in the opposite way. In the classic *Les paradis artificiels*, in which the poet Charles Baudelaire reports about his self-experiments with cannabis, this phenomenon is eloquently brought to the point:

> When the personal self … and the notion of time disappear completely … you may from time to time wake up shortly. It seems that you are stepping out of a marvellous and fantastic world. Yet, you actually preserve the faculty of observing yourself, and tomorrow you will remember part of your impressions Baudelaire (1966).

Bleuler (1911) observed similar episodic mental changes in early schizophrenia, referring to the Kraepelinian (1899) concept of 'blocking' as being of fundamental significance for its initial diagnosis. In his words, the patients exhibited relatively short periods of 'inattention', during which even the most powerful stimuli could not influence their train of thought. Sometimes this involves the entire psyche, speech and motility. Being to some extent aware of the disrupting process, the patients often describe the subjective experiences during such episodes as 'trances', 'attacks', 'dazes', 'blank spells' or 'stoppages of the mind'. Some patients say: "it's a delight – you don't feel anxious until you come out of it". Others are 'breaking up into bits', and they interpret the dissolution of their identity as an impending death of the self (Chapman, 1966). It is also noteworthy that in the build up of these phenomena, before the block occurs, schizophrenic patients have difficulties in co-ordinating simple motor sequences, namely, they suffer from paroxysmal ideokinetic dyspraxia (Chapman, 1966). We do not know what kinds of associations provide cortical networks with the time tags that are missing in dreams, but we do know that the sense of an extended time is not the only one missing in it; olfactory and gustatory representations are practically unheard of. In fact, most dreams have only two major characteristics. According to the magnitude dedicated to visual and motor representations in the primate cortex, they are mostly visual and they contain movement. It is as if the extended visual and frontal regions of the cortex attracted the lion's share of brain stem inputs. In addition to moving visual images, kinaesthetic sensations are thus common in dreams, but auditory ones are rare. This point is of importance, as schizophrenic hallucinations are often auditory: voices perceived as originating from someone else or 'aliens' in one's body. Kraepelin (1899) argued that these symptoms are associated with the temporal lobe, the function or structure of which being most probably altered in schizophrenia (Talbot and Arnold, 2002). It was, therefore, tempting to compare temporal lobe epilepsy and its typical dreamy states (Vignal et al., 2007) with the episodes of altered states of consciousness in schizophrenia (Jackson, 1932). With the advent of depth recording, it was possible to obtain intracerebral EEG abnormalities from schizophrenic patients. These insights have provided a unique contribution to our knowledge of psychosis, since, owing to ethical constraints, studies of this nature will not likely ever be repeated. During periods of acutely psychotic behaviour, Heath (1954) discovered abnormal electrical activity that localized predominantly to the septum and to a lesser extent to the hippocampus proper and amygdala. The particular septal region could be associated with the hippocampal formation, as demonstrated by Heath (1954) in comparative neurophysiological studies with cats and primates. In the meantime, this septal hippocampal system has been disclosed to constitute a functional, mainly cholinergic complex (Swanson, 2000; Risold, 2004) that projects via brainstem tegmental areas and the intralaminar thalamus (Kimura et al., 2004) to the striatum and related 'limbic' areas in the anterior cingulate gyrus and insula (Nieuwenhuys et al., 1988; Jones, 2007). Hippocampal hyperactivity during intoxication with Δ^9-THC is probably induced by DSI. Whether in schizophrenia, septal hyperactivity is caused by developmental malformation, genetic or epigenetic, must for the present remain a matter of conjecture. In any case, Heath (1962) showed that patients with acute psychosis due to clinically established temporal lobe epilepsy exhibited higher amplitude spiking and more slow-wave activity, compared to what was typically seen in schizophrenia. Although the anatomical regions of abnormal depth recordings were the same for the two groups, the pattern of activity was different, and this was so even during periods when the epileptic patients were displaying psychotic features indistinguishable from the schizophrenic. In every schizophrenic case, however, drowsiness or sleep intensified either the amplitude of the characteristic spikes or the frequency of their discharge. Strong or frequent 'spikers' showed the synchronized discharge, whether they were alert or asleep, but invariably the spiking was intensified with the occurrence of sleep. According to Heath (1954), this point was of paramount concern to the electrophysiological analysis of schizophrenia, reminding us of Hughlings Jackson's postulate that you will find out about insanity if you find out about dreams (Jackson, 1932; Gottesmann, 2006).

DARP-32

Downstream to the PLC-IP_3-mediated calcium signalling cascade, psychotic dysfunction may likewise emerge at the genetic level. Application of muscarinic and/or glutamatergic agonists to neurons reportedly evokes Ca^{2+} waves propagating from the dendritic tree into the nucleus. Such rises in nuclear Ca^{2+} may in turn activate dopamine- and cAMP-regulated phosphoprotein of 32 kDa (DARPP-32)-mediated gene transcription (Power and Sah, 2007), complementary to the well-known DARPP-32-mediated phosphorylation of target proteins. A polymorphism of the DARPP-32 gene, *PPP1R1B*, has recently been associated with the risk for schizophrenia in a family-based study (Meyer-Lindenberg et al., 2007), and, upon cannabis consumption, DARPP-32-mediated gene transcription and/or phosphorylation of downstream targets may impact on the state of excitability of striatopallidal neurons. More specifically, DARP-32 is abundantly expressed in the MS neurons of the striatum (Ouimet et al., 1998), and activation of CB_1 receptors induces its phosphorylation through PKA, while inactivation of the PKA phosphorylation site on DARP-32 impairs the psychomotor effects of CB_1 receptor agonists (Andersson et al., 2005). It is of importance that disrupted signalling at the D_2 and A_{2A} receptors also impair the ability of CB_1 receptors to exert motor depressant effects through the phosphorylation of DARP-32. As mentioned above, CB_1 receptors form heterodimers with the D_2 receptors (Glass and Felder, 1997; Jarrahian et al., 2004; Kearn et al., 2004), as well as with the A_{2A} receptors (Carriba et al., 2007), and intact signalling at A_{2A} receptors is a prerequisite for the motor effects of cannabinoids (Carriba et al., 2007). At the crossroads of calcium-dependent PKA and PKC phosporylation (Chen et al., 2007), DARP-32 is considered a major integrator of molecular signalling at the nuclear level, and it may be safe to implicate its dysfunction as an additional causative factor in schizophrenia.

The Antipsychotic Cannabidiol (Sativex™)

The weak CB_1 receptor antagonist phytocannabinoid, cannabidiol, has been recognized as a possible antipsychotic drug more than a decade ago (Zuardi et al., 1995). It reduces apomorphine-induced stereotypic behaviour in rats, but in contrast to haloperidol, neither elevates prolactine level nor induces catalepsy at high doses. Additionally, cannabidiol reduces both ketamine- and amphetamine-induced hyper-locomotion in mice (Moreira and Guimarães, 2005). As for the safety profile, up to the tested highest dose (daily 0.7 g for six weeks), cannabidiol fails to cause toxicity or significant pathological alterations in healthy volunteers and Huntington's disease patients. In healthy volunteers under ketamine-induced psychosis, cannabidiol is also effective in reducing psychosis, particularly delusion of alien control, indicating that cannabidiol can be a safe and well-tolerated antipsychotic. Unfortunately, this conjecture has only been tested in a highly limited number of schizophrenic patients. A 19-year-old female, suffering from serious side effects of conventional antipsychotics, improved significantly after a four-week treatment with cannabidiol, but relapsed when cannabidiol was substituted with haloperidol (Zuardi et al., 1995). Among the other 22–23-year-old male patients, one responded well to cannabidiol, while the remaining two were in a refractory phase and responded neither to cannabidiol nor to clozapine. Due to the small number of observations, a substantial effort is invited to explore if cannabidiol could be an alternative medicine in non-treatment-resistant schizophrenia (Zuardi et al., 2006).

Concluding Remarks

One of the most remarkable mental capacities that nature has bestowed upon human beings is our sense of time in which we exist. We can reflect on our protracted existence that extends from the present back into the past and forward into the distant future. Sometimes we even fathom at novel, previously unlearned things. Recent evidence has started to elucidate the physiological basis of this faculty being so characteristically distorted in both cannabis psychosis and schizophrenia – the fact that timing by the brain is associated with temporal sensorimotor integration and personal identity. Two other lines of evidence merge into conceptual unity: one being rooted in the dopaminergic and the other in the cholinergic neurotransmitter system, with the endocannabinoid retrograde messenger system at centre stage. This corroborates the long-held hypothesis that cannabis abuse and its psychotic manifestation are closely related to schizophrenia and, more specifically, suggests that the epigenetic liability to developing psychosis is driven by imbalanced calcium co-signalling between endocannabinoid and other neuromodulator pathways already implicated in schizophrenia. The dysfunctional information processing and aberrant formation of neuronal circuitry, as a result, shed new light onto the underlying physiological process at the synaptic level, reminding us of Virchov's conclusion that 'diseases represent merely the course of physiological phenomena, yet under altered conditions' (1847). Apart from the epidemiological data gathered in the last decades, this should prompt the society to (1) help adolescents understand that marijuana abuse is a major risk factor for developing schizophrenia, (2) test the therapeutic efficacy of CB_1 receptor antagonists in psychotic patients and (3) raise funds for an in-depth scientific and clinical investigation of the physiological role of endocannabinoids in other neuropsychiatric disorders.

Acknowledgements Attila Köfalvi is grateful for the III/BIO/56/2005 grant and for the Fundação para a Ciência e Tecnologia of the Portuguese Government (POCI2010/SFRH/BPD/18506/2004).

References

Adler LE, Olincy A, Waldo M, Harris JG, Griffith J, Stevens K, Flach K, Nagamoto H, Bickford P, Leonard S, Freedman R (1998) Schizophrenia, sensory gating, and nicotinic receptors. Schizophr Bull 24:189–202.

Aldrich MR (1977) Tantric cannabis use in India. J Psychedelic Drugs 9:227–233.

Andersson M, Usiello A, Borgkvist A, Pozzi L, Dominguez C, Fienberg AA, Svenningsson P, Fredholm BB, Borrelli E, Greengard P, Fisone G (2005) Cannabinoid action depends on phosphorylation of dopamine- and cAMP-regulated phosphoprotein of 32 kDa at the protein kinase A site in striatal projection neurons. J Neurosci 25:8432–8438.

Andreasson S, Allebeck P, Engstrom A, Rydberg U (1987) Cannabis and schizophrenia. A longitudinal study of Swedish conscripts. Lancet 2:1483–1486.

Arseneault L, Cannon M, Poulton R, Murray R, Caspi A, Moffitt TE (2002) Cannabis use in adolescence and risk for adult psychosis: longitudinal prospective study. BMJ 325:1212–1213.

Arseneault L, Cannon M, Witton J, Murray RM (2004) Causal association between cannabis and psychosis: examination of the evidence. Br J Psychiatry 184:110–117.

Arvindakshan M, Sitasawad S, Debsikdar V, Ghate M, Evans D, Horrobin DF, Bennett C, Ranjekar PK, Mahadik SP (2003) Essential polyunsaturated fatty acid and lipid peroxide levels in never-medicated and medicated schizophrenia patients. Biol Psychiatry 53:56–64.

Baldelli P, Hernandez-Guijo JM, Carabelli V, Carbone E (2005) Brain-derived neurotrophic factor enhances GABA release probability and nonuniform distribution of N- and P/Q-type channels on release sites of hippocampal inhibitory synapses. J Neurosci 25:3358–3368.

Ballmaier M, Bortolato M, Rizzetti C, Zoli M, Gessa G, Heinz A, Spano P (2007) Cannabinoid receptor antagonists counteract sensorimotor gating deficits in the phencyclidine model of psychosis. Neuropsychopharmacology 32:2098–2107.

Bär KJ, Letzsch A, Jochum T, Wagner G, Greiner W, Sauer H (2005) Loss of efferent vagal activity in acute schizophrenia. J Psychiatr Res 39:519–527.

Baudelaire C (1966) Les Paradis Artificiels. Paris: Garnier-Flammarion.

Berghuis P, Dobszay MB, Wang X, Spano S, Ledda F, Sousa KM, Schulte G, Ernfors P, Mackie K, Paratcha G, Hurd YL, Harkany T (2005) Endocannabinoids regulate interneuron migration and morphogenesis by transactivating the TrkB receptor. Proc Natl Acad Sci USA 102:19115–191120.

Berghuis P, Rajnicek AM, Morozov YM, Ross RA, Mulder J, Urban GM, Monory K, Marsicano G, Matteoli M, Canty A, Irving AJ, Katona I, Yanagawa Y, Rakic P, Lutz B, Mackie K, Harkany T (2007) Hardwiring the brain: endocannabinoids shape neuronal connectivity. Science 316:1212–1216.

Bergson C, Levenson R, Goldman-Rakic PS, Lidow MS (2003) Dopamine receptor-interacting proteins: the Ca^{2+} connection in dopamine signaling. Trends Pharmacol Sci 24:486–492.

Berridge MJ (1998) Neuronal calcium signaling. Neuron 21:13–26.

Betz VA (1875) Two centers in the human brain cortex.

Blackwood D (2000) P300, a state and a trait marker in schizophrenia. Lancet 355:771–72.

Bleuler E (1911) Handbuch der Psychiatrie. Dementia praecox oder Gruppe der Schizophrenien. Leipzig: Deuticke.

Blum J, Braverman ER, Dinardo MJ, Wood RC, Sheridan PJ (1994) Prolonged P300 latency in a neuropsychiatric population with the D_2 dopamine receptor A1 allele. Pharmacogenetics 4:313–322.

Boksa P (2007) Of rats and schizophrenia. J Psychiatry Neurosci 32:8–10.

Boucher AA, Arnold JC, Duffy L, Schofield PR, Micheau J, Karl T (2007) Heterozygous neuregulin 1 mice are more sensitive to the behavioural effects of Delta^9-tetrahydrocannabinol. Psychopharmacology (Berl) 192:325–336.

Boydell J, van Os J, Caspi A, Kennedy N, Giouroukou E, Fearon P, Farrell M, Murray RM (2006) Trends in cannabis use prior to first presentation with schizophrenia, in South-East London between 1965 and 1999. Psychol Med 36:1441–1446.

Brakeman PR, Lanahan AA, O'Brien R, Roche K, Barnes CA, Huganir RL, Worley PF (1997) Homer: a protein that selectively binds metabotropic glutamate receptors. Nature 368:284–288.

Brown P, Williams D, Aziz T, Mazzone P, Oliviero A, Insola A, Tonali P, Di Lazzaro V (2002) Pallidal activity recorded in patients with implanted electrodes predictively correlates with eventual performance in a timing task. Neurosci Lett 330:188–192.

Buhusi CV, Meck WH (2005) What makes us tick? Functional and neural mechanisms of interval timing. Nat Rev Neurosci 6:755–765.

Burgoyne RD, O'Callaghan DW, Hasdemir B, Haynes LP, Tepikin AV (2004) Neuronal Ca^{2+}-sensor proteins: multitalented regulators of neuronal function. Trends Neurosci 27:203–209.

Cannon TD, Thompson PM, van Erp TG, Toga AW, Poutanen VP, Huttunen M, Lonnqvist J, Standerskjold-Nordenstam CG, Narr KL, Khaledy M, Zoumalan CI, Dail R, Kaprio J (2002) Cortex mapping reveals regionally specific patterns of genetic and disease-specific gray-matter deficits in twins discordant for schizophrenia. Proc Natl Acad Sci USA 99:3228–3233.

Cao Q, Martinez M, Zhang J, Sanders AR, Badner JA, Cravchik A, Markey CJ, Beshah E, Guroff JJ, Maxwell ME, Kazuba DM, Whiten R, Goldin LR, Gershon ES, Gejman PV (1997) Suggestive evidence for a schizophrenia susceptibility locus on chromosome 6q and a confirmation in an independent series of pedigrees. Genomics 43:1–8.

Carr DB, Surmeier DJ (2007) M_1 Muscarinic receptor modulation of k_{ir2} channels enhances temporal summation of excitatory synaptic potentials in prefrontal cortex pyramidal neurons. J Neurophysiol 97:3432–3438.

Carriba P, Ortiz O, Patkar K, Justinova Z, Stroik J, Themann A, Müller C, Woods AS, Hope BT, Ciruela F, Casadó V, Canela EI, Lluis C, Goldberg SR, Moratalla R, Franco R, Ferré S (2007) Striatal adenosine A_{2A} and cannabinoid CB_1 receptors form functional heteromeric complexes that mediate the motor effects of cannabinoids. Neuropsychopharmacol *doi:10.1038/sj.npp.1301375*.

Chapman J (1966) The early symptoms of schizophrenia. Br J Psychiatry 112:225–251.

Chaudry HR, Moss HB, Bashir A, Suliman T (1991) Cannabis psychosis following bhang ingestion. Br J Addict 86:1075–1081.

Cheer JF, Cadogan AK, Marsden CA, Fone KC, Kendall DA (1999) Modification of 5-HT_2 receptor mediated behaviour in the rat by oleamide and the role of cannabinoid receptors. Neuropharmacology 38:533–541.

Cheer JF, Wassum KM, Heien ML, Phillips PE, Wightman RM (2004) Cannabinoids enhance subsecond dopamine release in the nucleus accumbens of awake rats. J Neurosci 24:4393–4400.

Cheer JF, Wassum KM, Sombers LA, Heien ML, Ariansen JL, Aragona BJ, Phillips PE, Wightman RM (2007) Phasic dopamine release evoked by abused substances requires cannabinoid receptor activation. J Neurosci 27:791–795.

Chen N, Appell M, Berfield JL, Reith ME (2003) Inhibition by arachidonic acid and other fatty acids of dopamine uptake at the human dopamine transporter. Eur J Pharmacol 478:89–95.

Chen L, Bohanick JD, Nishihara M, Seamans JK, Yang CR (2007) Dopamine $D_{1/5}$ receptor-mediated long-term potentiation of intrinsic excitability in rat prefrontal cortical neurons: Ca^{2+}-dependent intracellular signaling. J Neurophysiol 97:2448–2464.

Chen ZY, Patel PD, Sant G, Meng CX, Teng KK, Hempstead BL, Lee FS (2004) Variant brain-derived neurotrophic factor(BDNF)(Met66) alters the intracellular trafficking and activity-dependent secretion of wild-type BDNF in neurosecretory cells and cortical neurons. J Neurosci 24:4401–4411.

Chevaleyre V, Takahashi KA, Castillo PE (2006) Endocannabinoid-mediated synaptic plasticity in the CNS. Annu Rev Neurosci 29:37–76.

Comings DE, Muhleman D, Gade R, Johnson P, Verde R, Saucier G, MacMurray J (1997) Cannabinoid receptor gene (CNR1): association with *i.v.* drug use. Mol Psychiatry 2:161–168.

Cuffel BJ (1992) Prevalence estimates of substance abuse in schizophrenia and their correlates. J Nerv Ment Dis 180:589–592.

Cui G, Bernier BE, Harnett MT, Morikawa H (2007) Differential regulation of action potential- and metabotropic glutamate receptor-induced Ca^{2+} signals by inositol 1,4,5-trisphosphate in dopaminergic neurons. J Neurosci 27:4776–4785.

Damasio AR (1999) The Feeling of What Happens – Body and Emotion in the Making of Consciousness. New York: Harcourt Brace.

Damasio AR, Grabowski TJ, Bechara A, Damasio H, Ponto LL, Parvizi J, Hichwa RD (2000) Subcortical and cortical brain activity during the feeling of self-generated emotions. Nat Neurosci 3:1049–1056.

Dawson E (1995) Identification of a polymorphic triplet repeat marker for the brain cannabinoid receptor gene: use in linkage and association studies. Psychiatr Genet 5(S50–S51):850.

De Marchi N, De Petrocellis L, Orlando P, Daniele F, Fezza F, Di Marzo V (2003) Endocannabinoid signaling in the blood of patients with schizophrenia. Lipids Health Dis 2:5.

Dean B, Sundram S, Bradbury R, Scarr E, Copolov D (2001) Studies on [^{3}H]CP-55940 binding in the human central nervous system: regional specific changes in density of cannabinoid-1 receptors associated with schizophrenia and cannabis use. Neuroscience 103:9–15.

Degroot A, Köfalvi A, Wade MR, Davis RJ, Rodrigues RJ, Rebola N, Cunha RA, Nomikos GG (2006) CB_1 receptor antagonism increases hippocampal acetylcholine release: site and mechanism of action. Mol Pharmacol 70:1236–1245.

Delmas P, Crest M, Brown DA (2004) Functional organization of PLC signaling microdomains in neurons. Trends Neurosci 27:41–47.

Devlin MG, Christopoulos A (2002) Modulation of cannabinoid agonist binding by 5-HT in the rat cerebellum. J Neurochem 80:1095–1102.

Drewe M, Drewe J, Riecher-Rossler A (2004) Cannabis and risk of psychosis. Swiss Med Wkly 134:659–663.

D'Souza DC (2007) Cannabinoids and psychosis. Int Rev Neurobiol 78:289–326.

Duarte JM, Nogueira C, Mackie K, Oliveira CR, Cunha RA, Köfalvi A (2007) Increase of cannabinoid CB_1 receptor density in the hippocampus of streptozotocin-induced diabetic rats. Exp Neurol 204:479–484.

Eblen F, Graybiel AM (1995) Highly restricted origin of prefrontal cortical inputs to striosomes in the macaque monkey. J Neurosci 15:5999–6013.

Ellert-Miklaszewska A, Kaminska B, Konarska L (2005) Cannabinoids down-regulate PI_3K/Akt and Erk signaling pathways and activate proapoptotic function of Bad protein. Cell Signal 17:25–37.

Elvevag B, McCormack T, Gilbert A, Brown GD, Weinberger DR, Goldberg TE (2003) Duration judgements in patients with schizophrenia. Psychol Med 33:1249–1261.

Emrich HM, Leweke FM, Schneider U (1997) Towards a cannabinoid hypothesis of schizophrenia: cognitive impairments due to dysregulation of the endogenous cannabinoid system. Pharmacol Biochem Behav 56:803–807.

Freud S (1915) The unconscious. In: Strachey J (ed.), Complete Psychological Works, Vol. 14, translated in 1964. London: Hogarth.

Frith CD, Blakemore S, Wolpert DM (2000) Explaining the symptoms of schizophrenia: abnormalities in the awareness of action. Brain Res Brain Res Rev 31:357–363.

Fritzsche M, Fritzsche LN, Kosidubova SM, Prognimak AB, Mayorov OY (2006) Asymmetric information-processing in development, evolution and psychopathology. In: Hellige JB, Bogen JE (eds.), Cognition, Brain, Behavior, Special Issue. Napoca: CBB, pp. 311–342.

Fujii N, Graybiel AM (2005) Time-varying covariance of neural activities recorded in striatum and frontal cortex as monkeys perform sequential-saccade tasks. Proc Natl Acad Sci USA 102:9032–9037.

Fuster JM (1989) The Prefrontal Cortex, Anatomy, Physiology, and Neuropsychology of the Frontal Lobe. New York: Raven.

Gardner EL (2002) Addictive potential of cannabinoids: the underlying neurobiology. Chem Phys Lipids 121:267–290.

Gerfen CR (2004) Basal ganglia. In: Paxinos G (ed.), The Rat Nervous System. Amsterdam: Elsevier, pp. 455–508.

Giuffrida A, Leweke FM, Gerth CW, Schreiber D, Koethe D, Faulhaber J, Klosterkotter J, Piomelli D (2004) Cerebrospinal anandamide levels are elevated in acute schizophrenia and are inversely correlated with psychotic symptoms. Neuropsychopharmacology 29:2108–2114.

Gladkevich A, Kauffman HF, Korf J (2004) Lymphocytes as a neural probe: potential for studying psychiatric disorders. Prog Neuropsychopharmacol Biol Psychiatry 28:559–576.

Glass M, Felder CC (1997) Concurrent stimulation of cannabinoid CB_1 and dopamine D_2 receptors augments cAMP accumulation in striatal neurons: evidence for a G_s linkage to the CB_1 receptor. J Neurosci 17:5327–5333.

Goldman-Rakic PS (2005) Working memory dysfunction in schizophrenia. In: Salloway SP, Malloy PF, Duffy JD (eds.), The Frontal Lobes and Neuropsychiatric Illness. Washington, DC: American Psychiatric PI, pp. 71–82.

Goto Y, Grace AA (2005a) Dopamine-dependent interactions between limbic and prefrontal cortical plasticity in the nucleus accumbens: disruption by cocaine sensitization. Neuron 47:255–266.

Goto Y, Grace AA (2005b) Dopaminergic modulation of limbic and cortical drive of nucleus accumbens in goal-directed behavior. Nat Neurosci 8:805–812.

Gottesmann C (2006) The dreaming sleep stage: a new neurobiological model of schizophrenia? Neuroscience 140:1105–1115.

Green AL, Wang S, Owen SL, Aziz TZ (2007) The periaqueductal grey area and the cardiovascular system. Acta Neurochir Suppl 97:521–528.

Gross A, Joutsiniemi SL, Rimon R, Appelberg B (2006) Correlation of symptom clusters of schizophrenia with absolute powers of main frequency bands in quantitative EEG. Behav Brain Funct 2:23.

Gu Z, Jiang Q, Yan Z (2007) RGS4 modulates serotonin signaling in prefrontal cortex and links to serotonin dysfunction in a rat model of schizophrenia. Mol Pharmacol 71:1030–1039.

Halikas JA, Goodwin DW, Guze SB (1972) Marihuana use and psychiatric illness. Arch Gen Psychiatry 27:162–165.

Hampson AJ, Bornheim LM, Scanziani M, Yost CS, Gray AT, Hansen BM, Leonoudakis DJ, Bickler PE (1998) Dual effects of anandamide on NMDA receptor-mediated responses and neurotransmission. J Neurochem 70:671–676.

Hanson DR, Gottesman II (2005) Theories of schizophrenia: a genetic-inflammatory-vascular synthesis. BMC Med Genet 6:7.

Häring M, Marsicano G, Lutz B, Monory K (2007) Identification of the cannabinoid receptor type 1 in serotonergic cells of raphe nuclei in mice. Neuroscience 146:1212–1219.

Harkány T, Guzman M, Galve-Roperh I, Berghuis P, Devi LA, Mackie K (2007) The emerging functions of endocannabinoid signaling during CNS development. Trends Pharmacol Sci 28:83–92.

Hashimoto T, Bergen SE, Nguyen QL, Xu B, Monteggia LM, Pierri JN, Sun Z, Sampson AR, Lewis DA (2005) Relationship of brain-derived neurotrophic factor and its receptor TrkB to altered inhibitory prefrontal circuitry in schizophrenia. J Neurosci 25:372–383.

Hashimoto T, Lewis DA (2006) BDNF Val66Met polymorphism and GAD67 mRNA expression in the prefrontal cortex of subjects with schizophrenia. Am J Psychiatry 163:534–537.

Haznedar MM, Buchsbaum MS, Hazlett EA, Shihabuddin L, New A, Siever LJ (2004) Cingulate gyrus volume and metabolism in the schizophrenia spectrum. Schizophr Res 71:249–262.

Heath RG (1954) Studies in Schizophrenia. A Multidisciplinary Approach to Mind-Brain Relationships. Cambridge: Harvard University Press.

Heath RG (1962) Common characteristics of epilepsy and schizophrenia: clinical observation and depth electrode studies. Am J Psychiatry 118:1013–1026.

Henquet C, Murray R, Linszen D, van Os J (2005) The environment and schizophrenia: the role of cannabis use. Schizophr Bull 31:608–612.

Herkenham M, Lynn AB, Little MD, Johnson MR, Melvin LS, de Costa BR, Rice KC (1990) Cannabinoid receptor localization in brain. Proc Natl Acad Sci USA 87:1932–1936.

Herrero MT, Barcia C, Navarro JM (2202) Functional anatomy of thalamus and basal ganglia. Childs Nerv Syst 18:386–404.

Hicks RE, Gualtieri CT, Mayo JP, Perez-Reyes M (1984) Cannabis, atropine, and temporal information processing. Neuropsychobiology 12:229–237.
Hill EL, Gallopin T, Ferezou I, Cauli B, Rossier J, Schweitzer P, Lambolez B (2007) Functional CB1 receptors are broadly expressed in neocortical GABAergic and glutamatergic neurons. J Neurophysiol 97:2580–2589.
Hoehe MR, Caenazzo L, Martinez MM, Hsieh WT, Modi WS, Gershon ES, Bonner TI (1991) Genetic and physical mapping of the human cannabinoid receptor gene to chromosome 6q14–q15. New Biol 3:880–885.
Hoffman DC (1992) Typical and atypical neuroleptics antagonize MK-801-induced locomotion and stereotypy in rats. J Neural Transm Gen Sect 89:1–10.
Holstege G, Mouton LJ, Gerrits N (2004). Emotional motor system. In: Paxinos G (ed.), The Human Nervous System. Amsterdam: Elsevier, pp. 1306–1325.
Hyman SE, Malenka RC, Nestler EJ (2006) Neural mechanisms of addiction: the role of reward-related learning and memory, Annu Rev Neurosci. 2006; 29:565–98.
Isohanni M, Miettunen J, Maki P, Murray GK, Ridler K, Lauronen E, Moilanen K, Alaraisanen A, Haapea M, Isohanni I, Ivleva E, Tamminga C, McGrath J, Koponen H (2006) Risk factors for schizophrenia. Follow-up data from the Northern Finland 1966 Birth Cohort Study. World Psychiatry 5:168–171.
Iversen L (2003) Cannabis and the brain. Brain 126:1252–1270.
Jackson JH (1932) In: Taylor J (ed.), Selected Writings of John Hughlings Jackson. London: Hodder & Stoughton.
Jarrahian A, Watts VJ, Barker EL (2004) D_2 dopamine receptors modulate G_α-subunit coupling of the CB_1 cannabinoid receptor. J Pharmacol Exp Ther 308:880–886.
Jiang W, Zhang Y, Xiao L, Van Cleemput J, Ji SP, Bai G, Zhang X (2005) Cannabinoids promote embryonic and adult hippocampus neurogenesis and produce anxiolytic- and antidepressant-like effects. J Clin Invest 115:3104–3116.
Jin K, Xie L, Kim SH, Parmentier-Batteur S, Sun Y, Mao XO, Childs J, Greenberg DA (2004) Defective adult neurogenesis in CB_1 cannabinoid receptor knockout mice. Mol Pharmacol 66:204–208.
Jockers-Scherubl MC, Danker-Hopfe H, Mahlberg R, Selig F, Rentzsch J, Schurer F, Lang UE, Hellweg R (2004) Brain-derived neurotrophic factor serum concentrations are increased in drug-naive schizophrenic patients with chronic cannabis abuse and multiple substance abuse. Neurosci Lett 371:79–83.
Jockers-Scherubl MC, Matthies U, Danker-Hopfe H, Lang UE, Mahlberg R, Hellweg R (2003) Chronic cannabis abuse raises nerve growth factor serum concentrations in drug-naive schizophrenic patients. J Psychopharmacol 17:439–445.
Jockers-Scherubl MC, Rentzsch J, Danker-Hopfe H, Radzei N, Schurer F, Bahri S, Hellweg R (2006) Adequate antipsychotic treatment normalizes serum nerve growth factor concentrations in schizophrenia with and without cannabis or additional substance abuse. Neurosci Lett 400:262–266.
Johns A (2001) Psychiatric effects of cannabis. Br J Psychiatry 178:116–122.
Johnson JP, Muhleman D, Mc Murray J, Gade R, Verde R, Ask M, Kelley J, Comings DE (1997) Association between the cannabinoid receptor gene (CNR1) and the P300 event-related potential. Mol Psychiatry 2:169–171.
Jones EG (2007) The Thalamus. California: Cambridge University Press.
Joyce JN, Gurevich EV (1999) D_3 receptors and the actions of neuroleptics in the ventral striatopallidal system of schizophrenics. Ann NY Acad Sci 877:595–613.
Kaiya H, Nishida A, Imai A, Nakashima S, Nozawa Y (1989) Accumulation of diacylgylcerol in platelet phosphoinositide turnover in schizophrenia: a biological marker of good prognosis? Biol Psychiatry 26:669–676.
Kalkman HO (2006) The role of the phosphatidylinositide 3-kinase-protein kinase B pathway in schizophrenia. Pharmacol Ther 110:117–134.
Kang H, Schuman EM (1995) Long-lasting neurotrophin-induced enhancement of synaptic transmission in the adult hippocampus. Science 267:1658–1662.
Kapur S, Remington G (2001) Atypical antipsychotics: new directions and new challenges in the treatment of schizophrenia. Annu Rev Med 52:503–517.

Kapur S, Remington G (1996) Serotonin-dopamine interaction and its relevance to schizophrenia. Am J Psychiatry 153:466–476.

Kapur S, Zipursky R, Jones C, Shammi CS, Remington G, Seeman P (2000) A positron emission tomography study of quetiapine in schizophrenia: a preliminary finding of an antipsychotic effect with only transiently high dopamine D_2 receptor occupancy. Arch Gen Psychiatry 57:553–559.

Katona I, Sperlágh B, Magloczky Z, Santha E, Köfalvi A, Czirjak S, Mackie K, Vizi ES, Freund TF (2000) GABAergic interneurons are the targets of cannabinoid actions in the human hippocampus. Neuroscience 100:797–804.

Katona I, Sperlágh B, Sík A, Köfalvi A, Vizi ES, Mackie K, Freund TF (1999) Presynaptically located CB_1 cannabinoid receptors regulate GABA release from axon terminals of specific hippocampal interneurons. J Neurosci 19:4544–4558.

Kearn CS, Blake-Palmer K, Daniel E, Mackie K, Glass M (2004) Concurrent stimulation of cannabinoid CB_1 and dopamine D_2 receptors enhances heterodimer formation: a mechanism for receptor cross-talk? Mol Pharmacol 67:1697–1704.

Keay KA, Bandler R (2004) Periaqueductal gray. In: Paxinos G (ed.), The Rat Nervous System. Amsterdam: Elsevier, pp. 244–257.

Kim D, Thayer SA (2001) Cannabinoids inhibit the formation of new synapses between hippocampal neurons in culture. J Neurosci 21:RC146.

Kimura M, Yamada H, Matsumoto N (2003) Tonically active neurons in the striatum encode motivational contexts of action. Brain Dev Suppl 1:S20–23.

Kimura M, Minamimoto T, Matsumoto N, Hori Y (2004) Monitoring and switching of cortico-basal ganglia loop functions by the thalamo-striatal system. Neurosci Res 48:355–360.

Knusel B, Winslow JW, Rosenthal A, Burton LE, Seid DP, Nikolics K, Hefti F (1991) Promotion of central cholinergic and dopaminergic neuron differentiation by brain-derived neurotrophic factor but not neurotrophin 3. Proc Natl Acad Sci USA 88:961–965.

Köfalvi A, Oliveira CR, Cunha RA (2006a) Lack of evidence for functional $TRPV_1$ vanilloid receptors in rat hippocampal nerve terminals. Neurosci Lett 403:151–156.

Köfalvi A, Pereira MF, Rebola N, Rodrigues RJ, Oliveira CR, Cunha RA (2007) Anandamide and NADA bi-directionally modulate presynaptic Ca^{2+} levels and transmitter release in the hippocampus. Br J Pharmacol 151:551–563.

Köfalvi A, Rebola N, Rodrigues RJ, Pereira MF, Cunha RA (2006b) Evidence for CB_1Rs, but lack of evidence for presynaptic functional CB_2Rs and $TRPV_1$Rs in the hippocampus. Annual International Cannabinoid Research Society Meeting, Tihany, Hungary.

Köfalvi A, Rodrigues RJ, Ledent C, Mackie K, Vizi ES, Cunha RA, Sperlagh B (2005) Involvement of cannabinoid receptors in the regulation of neurotransmitter release in the rodent striatum: a combined immunochemical and pharmacological analysis. J Neurosci 25:2874–2884.

Kolb B, Gorny G, Limebeer CL, Parker LA (2006) Chronic treatment with Delta-9-tetrahydrocannabinol alters the structure of neurons in the nucleus accumbens shell and medial prefrontal cortex of rats. Synapse 60:429–436.

Kovasznay B, Fleischer J, Tanenberg-Karant M, Jandorf L, Miller AD, Bromet E (1997) Substance use disorder and the early course of illness in schizophrenia and affective psychosis. Schizophr Bull 23:195–201.

Kraepelin E (1899) Psychiatrie, *Dementia praecox*. Ein Lehrbuch für Studierende und Aerzte. Leipzig: Barth.

Kreitzer AC, Malenka RC (2007) Endocannabinoid-mediated rescue of striatal LTD and motor deficits in Parkinson's disease models. Nature 445:643–647.

Kristensen K, Cadenhead KS (2007) Cannabis abuse and risk for psychosis in a prodromal sample. Psychiatry Res 151:151–154.

Lauckner JE, Hille B, Mackie K (2005) The cannabinoid agonist WIN55,212-2 increases intracellular calcium via CB_1 receptor coupling to $G_{q/11}$ G proteins. Proc Natl Acad Sci USA 102:19144–19149.

Leroy S, Griffon N, Bourdel MC, Olie JP, Poirier MF, Krebs MO (2001) Schizophrenia and the cannabinoid receptor type 1 (CB_1): association study using a single-base polymorphism in coding exon 1. Am J Med Genet 105:749–752.

Levy FO, Holtgreve-Grez H, Tasken K, Solberg R, Ried T, Gudermann T (1994) Assignment of the gene encoding the 5-HT_{1E} serotonin receptor (S31) (locus HTR1E) to human chromosome 6q14–q15. Genomics 22:637–640.

Leweke FM, Giuffrida A, Wurster U, Emrich HM, Piomelli D (1999) Elevated endogenous cannabinoids in schizophrenia. Neuroreport 10:1665–1669.

Lewis DA, Hashimoto T (2007) Deciphering the disease process of schizophrenia: the contribution of cortical GABA neurons. Int Rev Neurobiol 78:109–131.

Linszen DH, Dingemans PM, Lenior ME (1994) Cannabis abuse and the course of recent-onset schizophrenic disorders. Arch Gen Psychiatry 51:273–279.

Lipska BK (2004) Using animal models to test a neurodevelopmental hypothesis of schizophrenia. J Psychiatry Neurosci 29:282–286.

Long LE, Malone DT, Taylor DA (2006) Cannabidiol reverses MK-801-induced disruption of prepulse inhibition in mice. Neuropsychopharmacology 31:795–803.

Ludwig AM (1966) Altered states of consciousness. Arch Gen Psychiatry 15:225–234.

Lupica CR, Riegel AC (2005) Endocannabinoid release from midbrain dopamine neurons: a potential substrate for cannabinoid receptor antagonist treatment of addiction. Neuropharmacology 48:1105–1116.

Luria AR (1973) The Working Brain: An Introduction to Neuropsychology. New York: Basic Books.

MacDonald III AW, Chafee MV (2006) Translational and developmental perspective on N-methyl-D-aspartate synaptic deficits in schizophrenia. Dev Psychopathol 18:853–876.

Macleod J, Davey Smith G, Hickman M (2006) Does cannabis use cause schizophrenia? Lancet 367:1055.

Maeda K, Sugino H, Hirose T, Kitagawa H, Nagai T, Mizoguchi H, Takuma K, Yamada K (2007) Clozapine prevents a decrease in neurogenesis in mice repeatedly treated with phencyclidine. J Pharmacol Sci 103:299–308.

Mäki P, Veijola J, Rantakallio P, Jokelainen J, Jones PB, Isohanni M (2004) Schizophrenia in the offspring of antenatally depressed mothers: a 31-year follow-up of the Northern Finland 1966 Birth Cohort. Schizophr Res 66:79–81.

Malaspina D, Dalack G, Leitman D, Corcoran C, Amador XF, Yale S, Glassman A, Gorman JM (2002) Low heart rate variability is not caused by typical neuroleptics in schizophrenia patients. CNS Spectr 7:53–57.

Maldonado R, Valverde O, Berrendero F (2006) Involvement of the endocannabinoid system in drug addiction. Trends Neurosci 29:225–232.

Malone DT, Taylor DA (1999) Modulation by fluoxetine of striatal dopamine release following Delta9-tetrahydrocannabinol: a microdialysis study in conscious rats. Br J Pharmacol 128:21–26.

Mamounas LA, Blue ME, Siuciak JA, Altar CA (1995) Brain-derived neurotrophic factor promotes the survival and sprouting of serotonergic axons in rat brain. J Neurosci 15:7929 –7939.

Martinez-Gras I, Hoenicka J, Ponce G, Rodriguez-Jimenez R, Jimenez-Arriero MA, Perez-Hernandez E, Ampuero I, Ramos-Atance JA, Palomo T, Rubio G (2006) $(AAT)_n$ repeat in the cannabinoid receptor gene, *CNR1*: association with schizophrenia in a Spanish population. Eur Arch Psychiatry Clin Neurosci 256:437–441.

Massi P, Vaccani A, Parolaro D (2006) Cannabinoids, immune system and cytokine network. Curr Pharm Des 12:3135–3146.

Mathers DC, Ghodse AH (1992) Cannabis and psychotic illness. Br J Psychiatry 161:648–653.

Mato S, Chevaleyre V, Robbe D, Pazos A, Castillo PE, Manzoni OJ (2004) A single in-vivo exposure to delta ^{9}THC blocks endocannabinoid-mediated synaptic plasticity. Nat Neurosci 7:585–586.

Mato S, Robbe D, Puente N, Grandes P, Manzoni OJ (2005) Presynaptic homeostatic plasticity rescues long-term depression after chronic Delta 9-tetrahydrocannabinol exposure. J Neurosci 25:11619–11627.

McGrath JJ (2006) Variations in the incidence of schizophrenia: data versus dogma. Schizophr Bull 32:195–197.

McGuire PK, Jones P, Harvey I, Bebbington P, Toone B, Lewis S, Murray RM (1994) Cannabis and acute psychosis. Schizophr Res 13:161–167.

McGuire PK, Jones P, Harvey I, Williams M, McGuffin P, Murray RM (1995) Morbid risk of schizophrenia for relatives of patients with cannabis-associated psychosis. Schizophr Res 15:277–281.

Mechri A, Saoud M, Khiari G, d'Amato T, Dalery J, Gaha L (2001) Glutaminergic hypothesis of schizophrenia: clinical research studies with ketamine. Encephale 27:53–59.

Meck WH (1996) Neuropharmacology of timing and time perception. Cogn Brain Res 3:227–242.

Melges FT (1982) Time and the Inner Future – A Temporal Approach to Psychiatric Disorders. New York: Wiley.

Melis M, Gessa GL, Diana M (2000) Different mechanisms for dopaminergic excitation induced by opiates and cannabinoids in the rat midbrain. Prog Neuropsychopharmacol Biol Psychiatry 24:993–1006.

Melis M, Perra S, Muntoni AL, Pillolla G, Lutz B, Marsicano G, Di Marzo V, Gessa GL, Pistis M (2004a) Prefrontal cortex stimulation induces 2-arachidonoyl-glycerol-mediated suppression of excitation in dopamine neurons. J Neurosci 24:10707–10715.

Melis M, Pistis M, Perra S, Muntoni AL, Pillolla G, Gessa GL (2004b) Endocannabinoids mediate presynaptic inhibition of glutamatergic transmission in rat ventral tegmental area dopamine neurons through activation of CB_1 receptors. J Neurosci 24:53–62.

Meyer-Lindenberg A, Straub RE, Lipska BK, Verchinski BA, Goldberg T, Callicott JH, Egan MF, Huffaker SS, Mattay VS, Kolachana B, Kleinman JE, Weinberger DR (2007) Genetic evidence implicating DARPP-32 in human frontostriatal structure, function, and cognition. J Clin Invest 117:672–682.

Minamimoto T, Hori Y, Kimura M (2005) Complementary process to response bias in the centromedian nucleus of the thalamus. Science 308:1798–1801.

Mirnics K, Middleton FA, Stanwood GD, Lewis DA, Levitt P (2001) Disease-specific changes in regulator of G-protein signaling 4 (RGS4) expression in schizophrenia. Mol Psychiatry 6:293–301.

Misslin R (2003) The defense system of fear: behavior and neurocircuitry. Neurophysiol Clin 33:55–66.

Moreira FA, Guimarães FS (2005) Cannabidiol inhibits the hyperlocomotion induced by psychotomimetic drugs in mice. Eur J Pharmacol 512:199–205.

Mouri A, Noda Y, Noda A, Nakamura T, Tokura T, Yura Y, Nitta A, Furukawa H, Nabeshima T (2007) Involvement of a dysfunctional dopamine-D_1/NMDA-NR1 and CaMKII pathway in the impairment of latent learning in a model of schizophrenia induced by phencyclidine. Mol Pharmacol doi:10.1124/mol.106.032961.

Muller N, Riedel M, Gruber R, Ackenheil M, Schwarz MJ (2000) Immune system and schizophrenia. An integrative view. Ann NY Acad Sci 917:456–467.

Munson R, Ruchkin DS, Ritter W, Sutton S, Squires NK (1984) The relation of P3b to prior events and future behavior. Biol Psychol 19:1–29.

Narr KL, Bilder RM, Toga AW, Woods RP, Rex DE, Szeszko PR, Robinson D, Sevy S, Gunduz-Bruce H, Wang YP, DeLuca H, Thompson PM (2005a) Mapping cortical thickness and gray matter concentration in first episode schizophrenia. Cereb Cortex 15:708–719.

Narr KL, Cannon TD, Woods RP, Thompson PM, Kim S, Asunction D, van Erp TG, Poutanen VP, Huttunen M, Lonnqvist J, Standerksjold-Nordenstam CG, Kaprio J, Mazziotta JC, Toga AW (2002) Genetic contributions to altered callosal morphology in schizophrenia. J Neurosci 22:3720–3729.

Narr KL, Thompson PM, Szeszko P, Robinson D, Jang S, Woods RP, Kim S, Hayashi KM, Asunction D, Toga AW, Bilder RM (2004) Regional specificity of hippocampal volume reductions in first-episode schizophrenia. Neuroimage 21:1563–1575.

Narr KL, Toga AW, Szeszko P, Thompson PM, Woods RP, Robinson D, Sevy S, Wang Y, Schrock K, Bilder RM (2005b) Cortical thinning in cingulate and occipital cortices in first episode schizophrenia. Biol Psychiatry 58:32–40.

Narushima M, Uchigashima M, Fukaya M, Matsui M, Manabe T, Hashimoto K, Watanabe M, Kano M (2007) Tonic enhancement of endocannabinoid-mediated retrograde suppression of inhibition by cholinergic interneuron activity in the striatum. J Neurosci 27:496–506.

Negrete JC (1989) Cannabis and schizophrenia. Br J Addict 84:349–351.

Newell KA, Deng C, Huang XF (2006) Increased cannabinoid receptor density in the posterior cingulate cortex in schizophrenia. Exp Brain Res 172:556–560.

Nieuwenhuys R, Voogd J, van Huijzen Ch (1988) The Human Central Nervous System. Berlin: Springer.

Olincy A, Harris JG, Johnson LL, Pender V, Kongs S, Allensworth D, Ellis J, Zerbe GO, Leonard S, Stevens KE, Stevens JO, Martin L, Adler LE, Soti F, Kem WR, Freedman R (2006) Proof-of-concept trial of an alpha7 nicotinic agonist in schizophrenia. Arch Gen Psychiatry 63:630–638.

Olincy A, Johnson LL, Ross RG (2003) Differential effects of cigarette smoking on performance of a smooth pursuit and a saccadic eye movement task in schizophrenia. Psychiatry Res 117:223–236.

Olney JW, Farber NB (1995) Glutamate receptor dysfunction and schizophrenia. Arch Gen Psychiatry 52:998–1007.

Ouimet CC, Langley-Guillion KC, Greengard P (1998) Quantitative immunochemistry of DARPP-32-expressing neurons in the rat caudatoputamen. Brain Res 808:8–12.

Oz M, Ravindran R, Zhang L, Morales M (2003) Endogenous cannabinoid, anandamide inhibits neuronal nicotinic acethylcholine receptor-mediated responses in Xenopus oocytes. J Pharmacol Exp Ther 306:1003–1010.

Oz M, Zhang L, Ravindran A, Morales M, Lupica CR (2004) Differential effects of endogenous and synthetic cannabinoids on alpha7-nicotinic acetylcholine receptor-mediated responses in Xenopus Oocytes. J Pharmacol Exp Ther 310:1152–1160.

Pearlman RJ, Aubrey KR, Vandenberg RJ (2003) Arachidonic acid and anandamide have opposite modulatory actions at the glycine transporter, $GLYT_{1a}$. J Neurochem 84:592–601.

Pereira DB, Rebola N, Rodrigues RJ, Cunha RA, Carvalho AP, Duarte CB (2006) Trkb receptors modulation of glutamate release is limited to a subset of nerve terminals in the adult rat hippocampus. J Neurosci Res 83:832–844.

Pertwee RG (2006) Cannabinoid pharmacology: the first 66 years. Br J Pharmacol 147: S163–S171.

Pichat P, Bergis OE, Terranova JP, Urani A, Duarte C, Santucci V, Gueudet C, Voltz C, Steinberg R, Stemmelin J, Oury-Donat F, Avenet P, Griebel G, Scatton B (2007) SSR180711, a novel selective alpha7 nicotinic receptor partial agonist: (II) efficacy in experimental models predictive of activity against cognitive symptoms of schizophrenia. Neuropsychopharmacology 32:17–34.

Pistis M, Porcu G, Melis M, Diana M, Gessa GL (2001) Effects of cannabinoids on prefrontal neuronal responses to ventral tegmental area stimulation. Eur J Neurosci 14:96–102.

Ponce G, Hoenicka J, Rubio G, Ampuero I, Jimenez-Arriero MA, Rodriguez-Jimenez R, Palomo T, Ramos JA (2003) Association between cannabinoid receptor gene (CNR1) and childhood attention deficit/hyperactivity disorder in Spanish male alcoholic patients. Mol Psychiatry 8:466–467.

Power JM, Sah P (2007) Distribution of IP3-mediated calcium responses and their role in nuclear signaling in rat basolateral amygdala neurons. J Physiol 580:835–857.

Price DA, Owens WA, Gould GG, Frazer A, Roberts JL, Daws LC, Giuffrida A (2007a) CB_1-independent inhibition of dopamine transporter activity by cannabinoids in mouse dorsal striatum. J Neurochem 101:389–396.

Price G, Cercignani M, Parker GJ, Altmann DR, Barnes TR, Barker GJ, Joyce EM, Ron MA (2007) Abnormal brain connectivity in first-episode psychosis: A diffusion MRI tractography study of the corpus callosum. Neuroimage 35:458–466.

Pryor SR (2000) Is platelet release of 2-arachidonoyl-glycerol a mediator of cognitive deficits? An endocannabinoid theory of schizophrenia and arousal. Med Hypotheses 55:494–501.

Ragozzino ME, Choi D (2004) Dynamic changes in acetylcholine output in the medial striatum during place reversal learning. Learn Mem 11:70–77.

Rashid AJ, So CH, Kong MM, Furtak T, El-Ghundi M, Cheng R, O'Dowd BF, George SR (2007) D_1-D_2 dopamine receptor heterooligomers with unique pharmacology are coupled to rapid activation of $G_{q/11}$ in the striatum. Proc Natl Acad Sci USA 104:654–659.
Risold PY (2004) The septal region. In: Paxinos G (ed.), The Rat Nervous System. Amsterdam: Elsevier, pp. 602–636.
Rizzolatti G, Gallese V (2003) Mirror neurons. In: Nadel L (ed.), Encyclopedia of Cognitive Science. London: Nature PG, Vol. III, pp. 37–42.
Ronesi J, Lovinger DM (2005) Induction of striatal long-term synaptic depression by moderate frequency activation of cortical afferents in rat. J Physiol 562:245–256.
Roopun AK, Middleton SJ, Cunningham MO, LeBeau FE, Bibbig A, Whittington MA, Traub RD (2006) A beta2-frequency (20–30 Hz) oscillation in nonsynaptic networks of somatosensory cortex. Proc Natl Acad Sci USA 103:15646–15650.
Samejima K, Ueda Y, Doya K, Kimura M (2005) Representation of action-specific reward values in the striatum. Science 310:1337–1340.
Sarter M, Bruno JP (2000) Cortical cholinergic inputs mediating arousal, attentional processing and dreaming: differential afferent regulation of the basal forebrain by telencephalic and brainstem afferents. Neuroscience 95:933–952.
Saulskaya NB (2000) Volume transmission in the striatum as constituting information processing In: Miller R, Ivanitsky AM, Balaban PM (eds.), Complex Brain Functions – Conceptual Advances in Russian Neuroscience. Singapore: OPA, pp. 1–19.
Schneier FR, Siris SG (1987) A review of psychoactive substance use and abuse in schizophrenia. Patterns of drug choice. J Nerv Ment Dis 175:641–652.
Seeman P (1987) Dopamine receptors and the dopamine hypothesis of schizophrenia. Synapse 1:133–152.
Semple DM, McIntosh AM, Lawrie SM (2005) Cannabis as a risk factor for psychosis: systematic review. J Psychopharmacol 19:187–194.
Semple DM, Ramsden F, McIntosh AM. (2003) Reduced binocular depth inversion in regular cannabis users. Pharmacol Biochem Behav 75:789–793.
Shapovalova KB (2000) The striatal cholinergic system and instrumental behaviour. In: Miller R, Ivanitsky AM, Balaban PM (eds.), Complex Brain Functions – Conceptual Advances in Russian Neuroscience. Singapore: OPA, pp. 263–288.
Sharp FR, Butman M, Koistinaho J, Aardalen K, Nakki R, Massa SM, Swanson RA, Sagar SM (1994) Phencyclidine induction of the hsp 70 stress gene in injured pyramidal neurons is mediated via multiple receptors and voltage gated calcium channels. Neuroscience 62:1079–1092.
Shearn CR, Fitzgibbons DJ (1972) Patterns of drug use in a population of youthful psychiatric patients. Am J Psychiatry 128:1381–1387.
Shenton ME, Dickey CC, Frumin M, McCarley RW (2001) A review of MRI findings in schizophrenia. Schizophr Res 49:1–52.
Shergill SS, Samson G, Bays PM, Frith CD, Wolpert DM (2005) Evidence for sensory prediction deficits in schizophrenia. Am J Psychiatry 162:2384–2386.
Simmons JM, Richmond BJ (2007) Dynamic changes in representations of preceding and upcoming reward in monkey orbitofrontal cortex. Cereb Cortex doi:10.1093/cercor/bhm034.
Simonyi A, Schachtman TR, Christoffersen GR (2005) The role of metabotropic glutamate receptor 5 in learning and memory processes. Drug News Perspect 18:353–361.
Skosnik PD, Krishnan GP, Aydt EE, Kuhlenshmidt HA, O'Donnell BF (2006) Psychophysiological evidence of altered neural synchronization in cannabis use: relationship to schizotypy. Am J Psychiatry 163:1798–1805.
Smesny S, Rosburg T, Baur K, Rudolph N, Sauer H (2007) Cannabinoids influence lipid-arachidonic acid pathways in schizophrenia. Neuropsychopharmacology 32:2067–2073.
Solowij N, Michie PT (2007) Cannabis and cognitive dysfunction: parallels with endophenotypes of schizophrenia? J Psychiatry Neurosci 32:30–52.
Solowij N, Michie PT, Fox AM (1991) Effect of long-term cannabis use on selective attention: an event-related potential study. Pharmacol Biochem Behav 40:683–688.

Steffens M, Feuerstein TJ (2004) Receptor-independent depression of DA and 5-HT uptake by cannabinoids in rat neocortex – involvement of Na^+/K^+-ATPase. Neurochem Int 44:529–538.

Stein BE, Meredith MA (1993) The Merging of the Senses. Cambridge: MIT.

Suh BC, Hille B (2007) Regulation of KCNQ channels by manipulation of phosphoinositides. J Physiol 582:911–916.

Sundram S, Copolov D, Dean B (2005) Clozapine decreases [^{3}H]CP55940 binding to the cannabinoid 1 receptor in the rat nucleus accumbens. Naunyn Schmiedebergs Arch Pharmacol 371:428–433.

Surmeier DJ, Ding J, Day M, Wang Z, Shen W (2007) D_1 and D_2 dopamine-receptor modulation of striatal glutamatergic signaling in striatal medium spiny neurons. Trends Neurosci 30:228–235.

Swanson LW (2000) Cerebral hemisphere regulation of motivated behavior. Brain Res 886:113–164.

Szabo B, Siemes S, Wallmichrath I (2002) Inhibition of GABAergic neurotransmission in the ventral tegmental area by cannabinoids. Eur J Neurosci 15:2057–2061.

Szeszko PR, Robinson DG, Sevy S, Kumra S, Rupp CI, Betensky JD, Lencz T, Ashtari M, Kane JM, Malhotra AK, Gunduz-Bruce H, Napolitano B, Bilder RM (2007) Anterior cingulate grey-matter deficits and cannabis use in first-episode schizophrenia. Br J Psychiatry 190:230–236.

Talbot K, Arnold SA (2002) The parahippocampal region in schizophrenia. In: Witter M, Wouterlood F (eds.), The Parahippocampal Region – Organization and Role in Cognitive Function. Oxford: Oxford University Press, pp. 297–320.

Tappe A, Kuner R (2006) Regulation of motor performance and striatal function by synaptic scaffolding proteins of the Homer1 family. Proc Natl Acad Sci USA 103:774–779.

Tendolkar I, Weis S, Guddat O, Fernandez G, Brockhaus-Dumke A, Specht K, Klosterkotter J, Reul J, Ruhrmann S (2004) Evidence for a dysfunctional retrosplenial cortex in patients with schizophrenia: a functional magnetic resonance imaging study with a semantic-perceptual contrast. Neurosci Lett 369:4–8.

Tsai SJ, Wang YC, Hong CJ (2000) Association study of a cannabinoid receptor gene (CNR1) polymorphism and schizophrenia. Psychiatr Genet 10:149–151.

Turner WM, Tsuang MT (1990) Impact of substance abuse on the course and outcome of schizophrenia. Schizophr Bull 16:87–95.

Uchigashima M, Narushima M, Fukaya M, Katona I, Kano M, Watanabe M (2007) Subcellular arrangement of molecules for 2-arachidonoyl-glycerol-mediated retrograde signaling and its physiological contribution to synaptic modulation in the striatum. J Neurosci 27:3663–3676.

Ujike H, Morita Y (2004) New perspectives in the studies on endocannabinoid and cannabis: cannabinoid receptors and schizophrenia. J Pharmacol Sci 96:376–381.

Ujike H, Takaki M, Nakata K, Tanaka Y, Takeda T, Kodama M, Fujiwara Y, Sakai A, Kuroda S (2002) CNR1, central cannabinoid receptor gene, associated with susceptibility to hebephrenic schizophrenia. Mol Psychiatry 7:515–518.

van Os J, Bak M, Hanssen M, Bijl RV, de Graaf R, Verdoux H (2002) Cannabis use and psychosis: a longitudinal population-based study. Am J Epidemiol 156:319–327.

Vignal JP, Maillard L, McGonigal A, Chauvel P (2007) The dreamy state: hallucinations of autobiographic memory evoked by temporal lobe stimulations and seizures. Brain 130:88–99.

Virchow R (1847) Standpoints in scientific medicine. In: Rather LJ (ed.), Diseases, Life and Man; Selected Essays by Rudolf Virchow. Stanford: Stanford University Press.

Voruganti LN, Slomka P, Zabel P, Mattar A, Awad AG (2001) Cannabis induced dopamine release: an in-vivo SPECT study. Psychiatry Res 107:173–177.

Wang Y, Goldman-Rakic PS (2004) D_2 receptor regulation of synaptic burst firing in prefrontal cortical pyramidal neurons. Proc Natl Acad Sci USA 101:5093–5098.

Weiser M, Knobler HY, Noy S, Kaplan Z (2002) Clinical characteristics of adolescents later hospitalized for schizophrenia. Am J Med Genet 114:949–955.

Weiser M, Noy S (2005) Interpreting the association between cannabis use and increased risk for schizophrenia. Dialogues Clin Neurosci 7:81–85.

Wettschureck N, van der Stelt M, Tsubokawa H, Krestel H, Moers A, Petrosino S, Schutz G, Di Marzo V, Offermanns S (2006) Forebrain-specific inactivation of $G_{q/G11}$ family G proteins results in age-dependent epilepsy and impaired endocannabinoid formation. Mol Cell Biol 26:5888–5894.

Yamasue H, Iwanami A, Hirayasu Y, Yamada H, Abe O, Kuroki N, Fukuda R, Tsujii K, Aoki S, Ohtomo K, Kato N, Kasai K (2004) Localized volume reduction in prefrontal, temporolimbic, and paralimbic regions in schizophrenia: an MRI parcellation study. Psychiatry Res 131:195–207.

Yano M, Steiner H (2005) Methylphenidate (Ritalin) induces Homer 1a and zif 268 expression in specific corticostriatal circuits. Neuroscience 132:855–865.

Yeomans JS (1995) Role of tegmental cholinergic neurons in dopaminergic activation, antimuscarinic psychosis and schizophrenia. Neuropsychopharmacology 12:3–16.

Zammit S, Allebeck P, Andreasson S, Lundberg I, Lewis G (2002) Self reported cannabis use as a risk factor for schizophrenia in Swedish conscripts of 1969: historical cohort study. BMJ 325:1199.

Zavitsanou K, Garrick T, Huang XF (2004) Selective antagonist [^{3}H]SR141716A binding to cannabinoid CB_1 receptors is increased in the anterior cingulate cortex in schizophrenia. Prog Neuropsychopharmacol Biol Psychiatry 28:355–360.

Zhang PW, Isighuro H, Ohtsuki T, Hess J, Carillo F, Walther D, Onaivi ES, Arinami T, Uhl GR (2004) Human cannabinoid receptor 1:5′ exons, candidate regulatory regions, polymorphisms, haplotypes and association with polysusbstance abuse. Mol Psychiatry 9:916–931.

Zuardi AW, Crippa JA, Hallak JE, Moreira FA, Guimaraes FS (2006) Cannabidiol, a Cannabis sativa constituent, as an antipsychotic drug. Braz J Med Biol Res 39:421–429.

Zuardi AW, Morais SL, Guimaraes FS, Mechoulam R (1995) Antipsychotic effect of cannabidiol. J Clin Psychiatry 56:485–486.

Chapter 23
The Cannabinoid Controversy: Cannabinoid Agonists and Antagonists as Potential Novel Therapies for Mood Disorders

Eleni T. Tzavara and Jeffrey M. Witkin

Abstract The last twenty years have been characterized by a huge progress in the field of cannabinoids. Metabotropic (CB_1 and CB_2) and ionotropic ($TRPV_1$) receptors for cannabinoids were discovered and their endogenous ligands (endocannabinoids) were isolated. Cannabinoid research has evolved from studying the effects of exogenous cannabinoid substances to unraveling the functional role of the endocannabinoid system. Potent and selective cannabinoid agonists and antagonists, as well as endocannabinoid inhibitors, have been and are synthesized and characterized for their therapeutic potential. Since cannabis preparations have historically been abused for their psychotropic and mood-altering properties, many research efforts focused on endocannabinoids, regulation of affect, and mood disorders. Interesting but also apparently discordant results and hypothesis have thus emanated. In this chapter we will first review the link between cannabinoids and affective disorders as evidenced by clinical studies on cannabis users or abusers, as well as by genetic, postmortem, and biomarker studies in relevant populations. We will critically discuss the current neurobiological hypotheses of affective disorders and the functional role of the endocannabinoid system in the regulation and dysregulation of neuronal networks mediating emotional responses. We will finally examine the potential value of endocannabinoid targets in the search for novel and improved medications, in particular preclinical behavioral results with CB_1 receptor antagonists, and with indirect cannabinoid agonists/endocannabinoid catabolism inhibitors, as well as recent findings from clinical studies with Rimonabant.

Introduction

In the middle of the nineteenth century, French and English romantic poets were using cannabis to experience "Artificial Paradises." In the middle of the twentieth century, marijuana use inevitably accompanied the stereotyped blissful but unproductive picture of the hippie culture (see Chap. 1). Today, many mainstream professional and lay people alike smoke the occasional joint as a social drug. Thus, cannabis preparations have been used over centuries and cultures for recreational purposes. From these collective experiences along with clinical and preclinical

A. Kofalvi (ed.), *Cannabinoids and the Brain.*

studies, we know that cannabinoid consumption can change perception and expression of emotions. The discovery of an endogenous cannabinoid system that operates largely in brain circuits associated with "pleasure," thought, and regulation of emotions offered scientific basis to this knowledge. The historical link of marijuana and subjective state prompted research on cannabinoids and the neurobiology of mood, and on the possibility of discovering compounds that target endocannabinoid neurotransmission (mainly through CB_1 receptors) for the treatment of mood and anxiety disorders. Ironically, there is probably no other field in cannabinoid research that has generated such a vivid debate as this one. Effects of cannabinoids on a number psychomotor domains including mood can be very different, ranging from a pleasant intoxication to intense anxiety, dysphoria, and negative mood. In line, pharmacological studies in animal models of mood regulation and mood disorders have also led to apparently opposite and contradicting findings. Work from different laboratories suggests endocannabinoid agonism as a novel target for depression (Gobbi et al., 2006; Hill and Gorzalka, 2005a,b; Naidu et al., 2007). On the other hand, we (Tzavara et al., 2001, Tzavara et al., 2003a, Witkin et al., 2005a,b) and others (Griebel et al., 2005; Shearman et al., 2003) proposed that cannabinoid antagonists or inverse agonists may have an antidepressant action. What are the effects of cannabinoids on psychomotor function, motivation, and affect? What is the functional role of endogenous cannabinoids in mood regulation and mood disorders? Are there endocannabinoid-related drugable targets with therapeutic relevance for affective pathologies? The present chapter reviews the state of the art knowledge, focusing on the involvement of the endocannabinoid system in mood disorders and the potential value that cannabinoid agonists and antagonists/inverse agonists might have in the treatment of these disorders. Previous reviews in this and related areas have been published (Ashton et al., 2005; Hill and Gorzalka, 2005a; Viveros et al., 2005; Watjak, 2005; Witkin et al., 2005a; Felder et al., 2006; Pacher et al., 2006; Piomelli et al., 2006; Vinod and Hungund, 2006).

Current Classification and Therapeutics of Mood Disorders

Mood disorders cause serious problems for individuals, families, and societies, and also place large economic pressures on health care systems. Although we have a host of medicines that treat mood disorders to some degree or other, their prevalence continues to be staggering with estimates of 9–20% of the population in the Western world being affected and with increased projections for the future. According to the DSM-IV, mood disorders are characterized by (1) depressed mood, (2) greatly diminished interest and pleasure in life events, (3) weight gain or loss, (4) sleep alterations, (5) agitation, (6) fatigue, loss of energy, (7) thoughts of worthlessness or inappropriate guilt, (8) decreased capacity to concentrate, think, and make decisions, and (9) suicidal ideation. Medical classification as a mood disorder requires that at least five of these manifestations exist, are not transient, and are not due to external causes such as bereavement, other medical conditions, or drug use. Age of onset, duration, frequency, and severity of the episodes permit further refinements in classification within the mood disorders. The current diagnostic criteria of the DSM-IV catalog major depression, dysthymic disorder, psychotic depression, cyclothymic disorder, and a variety of other disorders and secondary mood disorders (e.g., seasonal affective disorder and substance-induced mood disorder) (Dubovsky and Buzan, 1999) as well as bipolar disorder (also referred to as manic depression), for which at least one episode of mania or hypomania (elevated, expansive, irritable mood) must have occurred and for which subtypes are also recognized. Another classification according to DSM-IV distinguishes catatonic, melancholic, and atypical depression based on the predominant symptomatology and the intrinsic value of negative thoughts, as discussed below. The rate of affective disorders among first degree relatives of patients suffering from unipolar or bipolar disorder ranges from 10 to 33%, but is only 4.5–6.5% for control subjects. It is now believed that predisposing genes confer inheritable vulnerability; however, the manifestation of the disease is the result of complex interactions between genetic load and environmental risk factors. Among these risk factors, the most prominent are adverse life effects, exposure to persistent, unpredictable and uncontrollable stressors (especially in early life), cultural patterns, and the existence of social network and support. Personality traits (themselves also shaped by genetic–environmental interaction) such as stress-coping strategies and cognitive styles also affect the genesis and course of the affective disorders. Although ancient remedies have been described and are still in use, modern medicinal treatments for depression have been in use only since the introduction of the monoamine oxidase inhibitors and the tricyclic compounds into clinical practice in the 1950s. In addition to changing remarkably the management of these disorders, a great deal of our understanding of the neurobiology of depression derives from analysis of the biochemical mechanism of action of drugs effective in the treatment of depression. The observation that a structurally and pharmacologically diverse group of antidepressant molecules (as well as electroconvulsive therapy (ECT)) all increase the concentration of biogenic amines (norepinephrine (NE), serotonin (5-HT), and/or dopamine (DA) is one of the foundations of the biogenic amine theory of depression (Iversen, 2005). Although marked improvements in the safety and side effect profile of antidepressants have been engineered into modern medicines, there are still a number of critical dimensions along which improvements are needed. One important dimension is efficacy. Older and yet more toxic or controversial antidepressant treatments such as the tricyclic molecules, monoamine oxidase inhibitors, or electroconvulsive therapy have generally shown better efficacy over the safer and more widely prescribed selective monoamine uptake inhibitors (cf., Dubovsky and Buzan, 1999). However, even with the former agents, some patients continue to be treatment resistant. In addition, although antidepressant effects of compounds can be seen rather soon after dosing, the full efficacy of these compounds is generally observed only after several weeks of treatment (Katz et al., 2004), leaving the risk of suicide incompletely managed. Finally, unwanted side effects, including weight gain and sexual dysfunction, continue to be a problem. Because of these limitations in efficacy, rate of onset of therapeutic effect, and side effect profile, there remains a large unmet need for improved medicines for the treatment of mood disorders.

Clinical Findings on Cannabinoids and Mood Disorders

Evidence from Marihuana Users

Much of our understanding of the role of cannabinoid receptors on mood comes from individual reports after cannabis use for recreational or automedication purposes and from clinical studies with marijuana (cf., Paton and Pertwee, 1973). It is generally agreed that the primary psychoactive substance is (–)-*trans*-Δ^9-tetrahydrocannabinol (Δ^9-THC), a direct CB_1 receptor agonist and direct comparisons of the subjective effects of marijuana smoking or eating and Δ^9-THC in humans have confirmed this idea (cf., Isbel et al., 1967; Wachtel et al., 2002). Marijuana is typically smoked, resulting in subjective effects that may include euphoria, as well as depersonalization, altered time sense, but also lethargy, drowsiness, confusion, and changes in mood and anxiety. Marijuana users often vividly paint pleasant but at times also distressing and frightening emotions the drug provokes (for instance, as quoted in Iversen, 2000). Positive and negative drug effects have also been repeatedly measured in clinical studies with drug-experienced or naïve volunteers (Hart et al., 2001; Wachtel et al., 2002; Ilan et al., 2005). It is widely accepted that the subjective and neurobiological consequences of marijuana depend to a great extent on the history of drug use and the dose used, i.e., the exposure to the drug, as well as the context of use and the genetic makeup of the individual. Still, the data reviewed here point to the conclusion that chronic, heavy use of marijuana and cannabis is often associated with dysphoric states, and cognitive and neurobiological effects that are also encountered in major depressive disorders. Neurobiological correlates of the behavioral effects of marijuana and Δ^9-THC have been investigated with the subjective reports of euphoria having been given the most scrutiny. These effects occur during the rising phase of plasma Δ^9-THC. Correlated with behavioral signs and verbal reports of euphoria were increases in EEG alpha power (Lukas et al., 1995) and bilateral increases in cerebral blood flow most prominent in frontal structures (Mathew and Wilson, 1993) as well as in paralimbic brain areas (O'Leary et al., 2002). On the other hand, significantly lower mean hemispheric and frontal blood flow values were reported in long-term heavy cannabis smokers compared to normal controls (e.g., Tunving et al., 1986; Lundqvist et al., 2001). To this point, the amotivational symptoms that are often observed in heavy marijuana users seem to correlate with symptoms of depression (Musty and Kaback, 1995), a condition that is also associated with hypofrontality, which is a decrease in the function and neuronal activation of the frontal cortex (Galynker et al., 1998). Self-reports from some bipolar patients suggest that marijuana smoking might help to alleviate mania and depression symptoms but controlled studies do not exist (Ashton et al., 2005). In contrast, some of the most compelling data in opposition of cannabinoid agonists having antidepressant effects comes from reports in patients suffering from mood disorders. Cannabis has been reported to induce dysphoria in patients with mood disorders (Ablon and Goodwin, 1974) as well as in recreational cannabis users (Tunving, 1985). Review of the current literature shows a modest association between heavy or problematic cannabis use and depression in cohort studies and well-designed cross-sectional studies in the general population (Degenhardt et al., 2003; Konings and Maharajh, 2006). Prenatal cannabis use is also associated with anxiety and depression in children (Goldschmidt et al., 2004) as well as with downregulation in mesolimbic dopamine (DA) D_2 receptors in fetuses, a potential mechanism for emotional dysregulation (Wang et al., 2004). Marijuana has also been shown to promote risk-taking behavior (Lane et al., 2005) and to increase aspects of impulsivity (McDonald et al., 2003) often seen in mood and anxiety disorders. The CB_1 receptor inverse agonist rimonabant has shown clinical efficacy in decreasing obesity and reducing tobacco smoking, two conditions associated with impulsivity as well as with depression and anxiety.

Genetic Studies

Numerous sites of genetic variance of the *CNR1* gene, the gene that encodes the CB_1 receptor, have been characterized in different pathologies and ethnic backgrounds. These include exonic and intonic single nucleotide polymorphism (SNPs), an alternative promoter producing a novel 5′-untranslated region, and AAT triplet repeats ((AAT)n) in the 3′ flanking region which is considered to modify the transcription of the gene. Surprisingly, up to now, there has been no large systematic screening of the *CNR1*/CB_1 locus in depressive disorders. However, there is evidence linking *CNR1* alleles with conditions and disorders comorbid with depression as well as with polysomnographic symptom clusters relative to affect. Thus, an association of the $(AAT)_n$ repeat with schizophrenia in a Spanish (Martinez-Gras et al., 2006) and a Japanese (Ujike et al., 2002) population, was shown. Schizophrenia is characterized by severe affective disruption and shares genetic factors including overlap in confirmed linkages with bipolar disorder (see Chap. 22). Eating disorders and substance abuse are heavily comorbid with unipolar and bipolar depression. An association between the 3813G allele of the 3813A/G SNP and obesity-related phenotypes was found in Western European males (Russo et al., 2007). A preferential transmission of the (AAT) trinucleotide repeat allele of *CNR1* gene was seen in the binging/purging type of anorexia nervosa (Siegfried et al., 2004). *CNR1*/CB_1 variance has been repeatedly associated with substance (cannabis, alcohol, cocaine) abuse (Schmidt et al., 2002; Ballon et al., 2006; Herman et al., 2006). Two recently published articles call attention to a primary link between the

CNR1 gene and emotional responsiveness. In the first one, occurrence of depression in elderly subjects with Parkinson's disease was related to the length of the polymorphic triplet $(AAT)_n$ (Barrero et al., 2005). The second shows an exciting, although preliminary, association between *CNR1* variance and striatal response to faces exhibiting happiness (Chakrabarti et al., 2006).

Neuroanatomical Studies

Compelling evidence for a role of endocannabinoid dynamics in depression and in particular in suicide comes from postmortem studies in suicide victims. CB_1 receptor expression and coupling efficacy (as evidenced by agonist-stimulated $[^{35}S]GTP\gamma$ binding) are increased in the prefrontal cortex of depressed suicide victims (Hungund et al., 2004). Similarly, elevated levels of endocannabinoids and increased CB_1 receptor-mediated G protein signaling were seen in the prefrontal cortex of alcoholic suicide victims (Vinod et al., 2005). The authors acknowledge that it is not known whether these increases reflect a primary causal pathology or a compensatory adaptation (Vinod and Hungund, 2006). They argue, however, that hyperactivity of the endocannabinoid/CB_1 system in the prefrontal cortex might lead to a reduction of overall neurochemical prefrontal activity and hypofrontality. This is because CB_1 receptor activation is coupled negatively to adenylyl cyclase and globally leads to inhibitory effects on neurotransmission. As discussed above, hypofrontality often accompanies administration of high doses of exogenous cannabinoid agonists, exemplified in chronic heavy marijuana users. Whether overactivation of the endocannabinoid tone can lead to diminished cortical activity, is not known. Hypofrontality impacts the ability to monitor changes in reinforcement contingencies, a disruption that may be related to affective inflexibility and mood alterations. Interestingly, expression and activity of cyclic AMP (cAMP) signaling effectors are diminished in postmortem samples from depressed patients prone to suicide. On the other hand, cannabinoid antagonists increase cAMP production and $[^3H]$DA release in human cortical slices (Mato et al., 2002; Steffens et al., 2004). Increases in both cAMP signaling and DA are predictive of antidepressant activity.

Neurobiological Preclinical Evidence Linking the Endocannabinoid Systems to Mood Disorders

The presynaptic localization of CB_1 receptors to presynaptic terminals and axons of glutamatergic, dopaminergic, and cholinergic primary projection neurons, GABAergic interneurons, and in postsynaptic processes in the basal ganglia, extended amygdala, cerebral cortex, and hippocampus indicates that endocannabinoids regulate the activity of major neural networks involved in the regulation and dysregulation of mood and anxiety (Fig. 1).

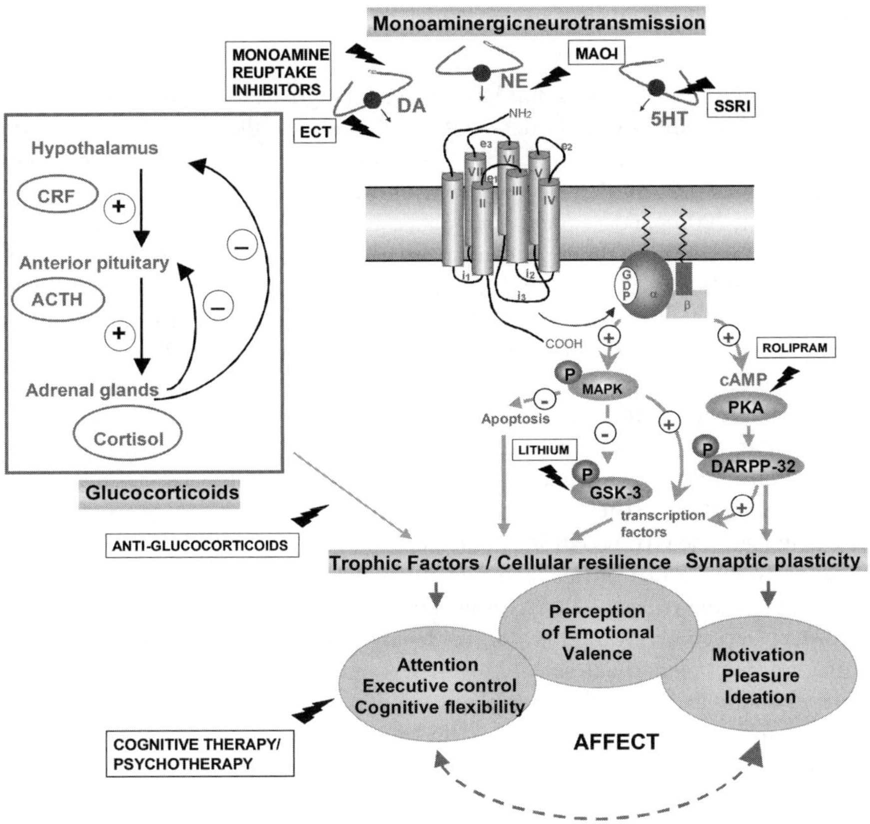

Fig. 1 Predominant neurobiological hypotheses of depression and current targets for conventional and putative antidepressants. Dysregulated monoaminergic neurotransmission, intracellular phosphorylation–dephosphorylation equilibrium, impaired neurotrophin production, and neurogenesis lead to deficits in cellular resilience and synaptic plasticity. In some forms of depression, a dysregulation of the HPA axis leads to hypercortisolemia that contributes to the depressive phenotype. Consequent dysregulation of systems regulating motivation/reward, emotional valence and memory, attention, executive function, and cognitive flexibility contribute to the pathophysiology of the disease. Conventional antidepressants and electroconvulsive therapy impact central neurotransmission of monoamines by increasing their synaptic availability. Other compounds target second messenger cascades such as the phosphodiesterase type 4 inhibitor rolipram with demonstrated efficacy in clinical investigations. Rolipram increases postsynaptic cell responsiveness by shifting the steady-state of the cAMP-PKA cascade. GSK-3 inhibitors have been proposed as mood regulators. Antiglucocorticoids could have a therapeutic value in depressive states associated with HPA hyperactivity. A dysregulation of the HPA axis leads to hypercortisolemia that contributes to the depressive phenotype. Ligands that target the endocannabinoid treatment are predicted to act on the same neurochemical substrates as the above-mentioned antidepressants (see text). *5HT* serotonin; *AC* adenylyl cyclase; *cAMP* adenosine 3′,5′-cyclic phosphate; *DA* dopamine; *DARPP-32* dopamine and cyclic AMP regulated phosphoprotein of 32 kilodaltons; *ECT* electroconvulsive treatment; *GSK-3* glycogen synthase kinase 3; *MAPK* mitogen-activated protein kinase, *MAO-I* monoamine oxidase inhibitor; *NE* norepinephrine; *PDE* phosphodiesterase; *PKA* protein kinase A; PI_3K phosphatidylinositol-3 kinase; *PLA2* phospholipase A2; *SERT* serotonin uptake inhibitor; *SNRI* selective serotonin/norepinephrine inhibitor; *SSRI* selective serotonin uptake inhibitor

Monoaminergic Neurotransmission

Monoaminergic neurotransmission has received the greatest attention in neurobiological studies of affective disorders, essentially because the first effective antidepressant drugs were shown to act by regulating synaptic levels of monoamines. This hypothesis was further endorsed by metabolomic and imaging studies in patients (Nutt, 2006), and by preclinical data establishing a crucial role for DA, NE, and serotonin 5-HT in energy/interest, reinforecement/pleasure, impulse (cf., Nemeroff, 2002). It has been argued that by increasing synaptic availability of monoamines, antidepressants restore the neurochemical milieu of the brain with an environment more conducive to normal affective tone and adaptability (cf., Duman, 2004). In all brain regions, CB_1 receptor agonists presynaptically inhibit the release of most known neurotransmitters (cf., Schlicker and Kathmann, 2001). Endocannabinoids also inhibit presynaptic activity and neurotransmitter release acting as retrograde homeostatic signals (cf., Diana and Marty, 2004). Therefore, a CB_1 receptor antagonist/inverse agonist is predicted to enhance neurotransmitter release. Indeed, the CB_1 receptor, rimonabant increases NE and 5-HT efflux in the medial prefrontal cortex of rats, and selectively the efflux of DA and acetylcholine (ACh) in the same region, whereas it does not have an effect on DA and ACh in a subcortical dopaminergic region, the nucleus accumbens (Tzavara et al., 2003a). Similar findings of cortical selectivity of DA action are observed with classical antidepressants and selective serotonin reuptake inhibitors (SSRIs) that also increase 5-HT and NE efflux in the brain to varying degrees (Tanda et al., 1994; Bymaster et al., 2002). These effects of established and experimental antidepressant medicines are considered as a neurochemical landmark for clinically effective antidepressant activity, notwithstanding the fact that their primary mechanism of action may be different. Another CB_1 receptor antagonist, SLV-319, also increased cortical DA, and the stimulatory action of rimonabant on cortical monoamine efflux was prevented in CB_1 KO mice (Witkin et al., 2005b). Cannabinoid agonists administered at low doses also increased cortical DA (Chen et al., 1990) and ACh (Acquas et al., 2000). On the other hand, prolonged agonist administration, mimicking heavy cannabis consumption in humans, resulted in a reduction of DA metabolism in the prefrontal cortex of rats (Verrico et al., 2003). As we will see dose-dependent actions of cannabionoid agonists are recurrent to a number of neurochemical and behavioral readouts.

Synaptic Plasticity and Neuronal Resilience: Effects on Intacellular Signaling Cascades

Despite rapid changes in monoamine content, the full-blown clinical effects of antidepressants occur only after chronic administration, suggesting that downstream effectors are ultimately responsible for therapeutic effects. These downstream cascades regulate synaptic plasticity and cellular resilience, defined as diverse processes by which the brain perceives, adapts, and responds to a variety of internal and external stimuli by altered activity, synaptic remodeling, long-term potentiation, and even neurogenesis (cf., Manji and Duman, 2001). Intracellular phosphorylation/dephosphorylation balance and underlying kinase/phosphatase signaling pathways appear as important mediators of the action of known and potential antidepressants (e.g., Svenningsson et al., 2002). Glycogen synthase kinase (GSK3) and protein kinase C (PKC) inhibitors have been proposed as putative mood regulators (cf., Payne et al., 2004) and activation of protein kinase A (PKA) and its downstream cascade is associated with antidepressant activity (cf., Duman, 2004). Rolipram, an inhibitor of phosphodiesterase 4 (enzyme that catabolizes cAMP) has demonstrated antidepressant activity in a number of experimental models and in humans. CB_1 receptor agonists are negatively coupled to adenylyl cyclase: CB_1 receptor activation via the $G_{i/o\alpha}$ protein subunit, reduces cAMP production and inhibits PKA, but increases mitogen-activated protein kinases (MAPK) phosphorylation via beta-gamma G-protein subunits in vitro (Bouaboula et al., 1995) and in vivo (Wade et al., 2004; Derkinderen et al., 2003). Rimonabant activates PKA in naïve and Δ^9-THC-dependent animals, though with different patterns (Rubino et al., 2000; Tzavara et al., 2000; Mato et al., 2002). While cAMP/PKA activation by CB_1 receptor antagonists is consistent with antidepressant-like effects, the effect of rimonabant on kinase/phosphatase cascades needs to be investigated.

Synaptic Plasticity and Neuronal Resilience: Effects on Trophic Factors and on Neurogenesis

A major prevailing hypothesis of the etiology of mood disorders or the triggering of mood disorder relapse postulates that neurotrophic factors, in particular brain-derived neurotrophic factor (BDNF), is a primary regulator of mood through its control of neurogenesis (Santarelli et al., 2003; Duman, 2004). One of the most frequently cited structural changes in major depressive disorders is decreased hippocampal volume (Videbach and Ravnkilde, 2004; Campbell and MacQueen, 2006). Stress, known to increase neurotoxic glucocorticoids and reduce the generation of BDNF is postulated to increase the probability of mood disorders. On the other hand, antidepressants have been shown to increase BDNF levels and neurogenesis (Alt et al., 2006). It has been known for some time that cannabinoids may play a neuroprotective role. In fact, BDNF mRNA and protein expression is increased after chronic administration with Δ^9-THC in rat nucleus accumbens, ventral tegmental area, paraventricular nucleus, and medial prefrontal cortex, but not in hippocampus (Butovsky et al., 2005). The lack of induction in hippocampus sets a limitation in the potential antidepressant profile of CB_1 agonists. It is not consistent with studies showing hippocampal BDNF induction with antidepressant agents (cf., Duman, 2004), the fact that BDNF intrahippocampally induces antidepressant-like effects in rats (Shirayama et al., 2002), or findings that neurogenesis in the hippocampal subventricular zone is necessary for antidepressant-like effects (Santarelli et al., 2003). Likewise, the induction of BDNF in the nucleus accumbens is an effect that may oppose antidepressant-like effects (cf., Eisch et al., 2003). Nonetheless, demonstrations of neuroprotective-like effects in hippocampal neurons through endocannabinoid induction have been seen. For example, the excitotoxin kainate induces rapid increases in anandamide levels in mouse hippocampal pyramidal neurons in vitro and engendered cell protective mechanisms. This protective effect was absent in conditional mutant mice that lack CB_1 receptors in principal forebrain neurons but not in adjacent inhibitory cells (Marsicano et al., 2003). More direct evidence supporting a role for hippocampal neurogenesis in cannabinoid regulation of mood comes from a study reporting that hippocampal X-irradiation, which disrupts neurogenesis, prevented the antidepressant-like effects of low doses of HU-210 in rats (Jiang et al., 2005). However, in another study, AM404 failed to enhance neurogenesis per se, although it blocked stress-induced decreases in hippocampal neurogenesis; surprisingly, the CB_1 antagonist AM251 induced robust cell proliferation (Hill et al., 2006a). A similar effect was shown for rimonabant; interestingly, this effect persisted in CB_1 KO mice but not in $TRPV_1$ receptor KO mice, suggesting $TRPV_1$ receptors might be implicated in some of the effects of rimonabant (Jin et al., 2004).

Regulation of the Hypothalamus–Pituitary Axis and Effects of Glucocorticoids

Stress and stress-associated stimuli can engender effects behaviorally and neurochemically that might be related to mood disorders, anhedonia, learned helplessness, and decreased neurogenesis to name a few (cf., Rasmusson et al., 2002; Duman, 2004). Presentation or anticipation of a stressor releases corticotropin-releasing hormone (CRH) from the hypothalamus; CRH acts on the anterior pituitary which starts to secrete adrenocorticotropic hormone (ACHT); ACTH triggers release of adrenal glucocorticoids. In prolonged uncontrollable stress, the hypothalamic–pituitary–adrenal (HPA) axis homeostasis is disrupted, resulting in abnormally elevated glucocorticoid secretion. Increased glucocorticoid secretion in turn inhibits activity in the hippocampus and prefrontal cortex (Gold et al., 2002). Since the HPA axis itself is under corticohippocampal inhibitory control (Sapolsky et al., 1991), this will cause further increase in glucocorticoids. Activity of the amygdala, a region highly involved in the processing of aversive emotions, is also under tonic inhibitory control of cortical circuits; cortical deficiency will cause functional activation of the amygdala (Gold et al., 2002) which in turn will aggravate hypofrontality. Thus, recurrent stress results in unbalanced emotional loops that become more and more inflexible, operating in a feed-forward pattern (Gold and Chrousos, 2002) to sustain depression symptoms. Increased salience of aversive memories due to the activation of the amygdala on one hand, and poor executive control, attention, cognitive inhibition, and cognitive flexibility due to ablated cortical activity on the other could provide one explanatory mechanism for the debilitating, persistent, compulsive rumination of negative thoughts and the increase in negative biasing that parasitize every other activity in depressed individuals. Increases in glucocorticoids, abnormal HPA activation, and increased amygdala activity occur only in a subgroup of depressed patients, who according to DSM-IV are characterized as melancholic (Gold and Chrousos, 2002). Melancholic depression is defined by (1) profound anhedonia; (2) loss of interest in previously rewarding or pleasant activities; (3) depressed mood with suicidal thoughts; (4) depressed mood accentuated in the morning; (5) insomnia or early morning wakening; (6) agitation; (7) anorexia, loss of weight; and (8) excessive guilt. This condition is to be considered in contrast to atypical depression characterized by (1) mood responsive to pleasant incidents; (2) hyperphagia, weight gain; (3) leaden paralysis; and (4) rejection sensitivity. Restraint-stress-induced corticosterone increases were dampended by cannabinoid direct or indirect agonists (Patel et al., 2004), showing clearly that the endocannabinoid system has a large involvement in the control of the physiological reactions to stress and of HPA activity. Endocannabinoids appear to be involved in stress-induced analgesia (Hohmann et al., 2005; Vaughan, 2006), and chronic stress was recently shown to downregulate CB_1 receptors and levels of endocannabinoid 2-arachidonylglycerol within the hippocampus but not in the limbic forebrain (Hill et al., 2006b). On the contrary, disruption of the CB_1 receptor gene was shown to result in HPA axis hyperactivity (Barna et al., 2004), hypophagia (Cota et al., 2003), and reduced neurogenesis (Jin et al., 2004). These findings lead to propose a role for endocannabinoid activation in the therapeutics of melancholic depression (Hill and Gorzalka, 2005a). Nevertheless, the causal link of these changes has not been demonstrated. Also, anxiety-related responses and affective adaptations in response to chronic stress described in CB_1 receptor KO animals are extremely dependent on the experimental context (Haller et al., 2004) and strain used. It should also be emphasized that CB_1 receptor antagonists/inverse agonists reduce corticosterone concentration and corticotrophin release factor (CRF) mRNA expression in control animals (Gonzalez et al., 2004), suggesting a mechanism for the antidepressant effects of rimonabant, similar to that exhibited by CRF antagonists. However, a new perspective, supporting a role for endocannabinoid activation in the therapeutics of melancholic depression, comes from the fact that agonist activation of CB_1 receptors decreases the aversive memory of shock exposure as evidenced by increases in the extinction of fear-related behaviors (Chhatwal et al., 2005) and that the extinction of aversive memories is enhanced in FAAH KO mice (Azad et al., 2004). In contrast, CB_1 receptor blockade has the opposite effect increasing storage of aversive but not positive emotional memories (Marsicano et al., 2002; Holter et al., 2005).

Behavioral Preclinical Evidence Linking the Endocannabinoid Systems to Mood Disorders

Preclinical Data Used to Predict Antidepressant Efficacy

One source of preclinical data implicating the endocannabinoid system in mood disorders comes primarily from data in animal models where effects of compounds are used to predict antidepressant effects in humans. Cryan and colleagues (2002) and McArthur and Borsini (2006) have provided reviews of such models and an equally comprehensive discussion of such models is also available (O'Neill and Moore, 2003). Like most, if not all, of the models used in psychiatry, the depression models are predictive models. That is, although sometimes based upon hypothetical, neural, and/or behavioral predispositions or etiological factors in depression, the models are not functional models of the human disease state. Nonetheless, the different models described below provide predictive power for the potential of a novel chemical entity to produce antidepressant effects in humans. It is also important to note from the outset, that mood disorders, like all psychiatric disorders, are not homogenous. Individuals and patients at different times can have wide-ranging symptoms that encompass not only those of mood but of cognition, anxiety, appetite, sleep, drug, and other dependencies, and the general state of behavioral engagement. The use of behavioral and neurological tests specific to these symptoms could be valuable in supporting the discovery of novel agents for depression. Two of the most commonly employed animal models of antidepressant action include the rodent forced swim test and the mouse tail suspension test (Porsolt et al., 1977; Steru et al., 1985). Although highly predictive of antidepressant efficacy in humans, the predictive validity is based upon the current antidepressants that act through more or less the same mechanism (increasing synaptic monoamine levels). The ability of these models for predicting antidepressant effects of compounds acting through a different mechanism is therefore difficult to estimate. These two methods also detect antidepressant-like effects after acute dosing, whereas full antidepressant effects are observed only after several weeks of dosing. These limitations led to the development of alternative animal models that require repeated dosing before antidepressant-like effects are uncovered. These include the reduction of submissive behavior paradigm (Malatynska et al., 2002), learned helplessness (Maier, 1984), the novelty-suppressed feeding assay that also detects anxiolytics (Bodnoff et al., 1989), the chronic unpredictable stress assay (Willner, 2005), and the chronic social defeat model (Tsankova et al., 2006). Although these assays require subchronic dosing for efficacy to be achieved, their predictive validity has not been better established than the acute assays discussed above.

Cannabinoid Agonists and Mood Disorders

As already noted in the discussion of the endocannabinoid system elsewhere in this book, several molecular targets through agonist effects can be generated. Pharmacologically, the two major classes of cannabinoid agonists are direct-acting and indirect-acting in reference to the CB_1 and CB_2 receptors. Thus, compounds like Δ^9-THC, WIN55212-2, and CP55940 act as direct agonist at CB_1 receptors. Indirect agonists can activate CB_1 and CB_2 receptors and other targets indirectly by increasing the concentration of anandamide and other endogenous molecules at relevant proximities to the target proteins. Compounds that block the uptake of anandamide are one such class of indirect agonists for which compounds like LY2077885 and OMDM-1 have been categorized. Compounds that inhibit FAAH also increase concentrations of anandamide; URB597 is perhaps the best characterized putative selective FAAH inhibitor. Although the indirect agonists have been classified on the basis of their selectivity for anandamide uptake inhibition or FAAH inhibition, the line demarking these actions appears to be hazier than previously reported (Dickason-Chesterfield et al., 2006). If the interpretation of in vitro selectivity is not ambiguous enough, there is almost nothing known about the selectivity of these molecules for their putative targets in vivo. It must also be noted that the independent activity of the transporter (which remains to be cloned) and FAAH also does not appear a viable idea since several pieces of data point to their dynamic interaction (Dickason-Chesterfield et al., 2006; Felder et al., 2006). Discriminative stimulus effects of drugs predict subjective effect profiles in humans (cf., Wiley, 1999). It has been universally clear that the discriminative stimulus effects of Δ^9-THC are qualitatively reproduced by other direct-acting and in a number of cases indirect-acting CB_1 receptor agonists. In rhesus monkeys, CP55940, WIN55212-2, and *R*-methanandamide produced full Δ^9-THC-like discriminative stimulus effects; in contrast, the CB_2 receptor agonist AM1241 and the noncannabinoids cocaine, ketamine, midazolam, and morphine did not (McMahon, 2006). Antagonism studies against these CB_1 agonists with both rimonabant and AM251 revealed competitive antagonism as apparent pA_2 values derived from Schild analysis did not differ from one another (McMahon, 2006). In rats trained to discriminate either Δ^9-THC or *R*-methanandamide from vehicle, both drugs fully substituted for one another and their discriminative stimulus effects were prevented by the CB_1 inverse agonists, rimonabant and AM251, but not by the CB_2 receptor antagonist SR144528 (Jarbe et al., 2005). Anandamide is metabolically unstable and therefore experiments with exogenous administration have utilized stable analogs. In one such study, Δ^9-THC and the stable anandamide analog, O-1812, were used as discriminative stimuli in separate groups of rats. Both compounds substituted for one another and were antagonized by rimonabant. In contrast, a $TRPV_1$ receptor agonist, O-1839, did not substitute (Wiley et al., 2004). In addition to predicting the subjective effects such compounds are likely to have in humans, these assays also have relevance to the prediction of abuse potential as discussed below. There has been some evaluation of effects of cannabinoid agonists in assays predictive of antidepressant efficacy. This work has been carried out to date only in acute models. Direct-acting agonists of CB_1 receptors such as HU-210 and indirect agonists such as AM404 and URB597 have shown efficacy in the forced swim test in rats and the tail suspension test in mice (Table 1). Table 1 also shows that there have also been negative findings in antidepressant tests with this mechanism. Electrophysiological data are also consistent with an antidepressant interpretation of biological activity. URB597

Table 1 Effects of cannabinoid agonists in tests predictive of an impact in mood disorders

Compound	Procedure	Effect	Reference
Positive Impact			
URB597	Modified tail suspension mouse	Decreased immobility	Gobbi et al., 2006
URB597	Modified tail suspension mouse	Decreased immobility	Naidu et al., 2007
None	Modified tail suspension mouse	*FAAH*$^{-/-}$ mice have decreased immobility compared to wild-type	Naidu et al., 2007
URB597	Forced swim rat	Decreased immobility	Gobbi et al., 2005
AM404 HU-210	Forced swim rat	Decreased immobility as with DMI	Hill and Gorzalka, 2005a,b
URB597	Forced swim rat	Decreased immobility	Hill et al., 2007
HU-210	Forced swim rat	Decreased immobility	Jiang et al., 2005
HU-210	Forced swim rat	Hippocampal X-irradiation prevented antidepressant-like effects	Jiang et al., 2005
ACEA	Forced swim mouse	Decreased immobility	Rutkowska and Jachimczuk, 2004
CP55940 AM404 URB597	Restraint rat	Blockade of corticosterone increases	Patel et al., 2004
CP55940 URB597	Restraint rat	Blockade of decreased sucrose intake and preference	Rademacher and Hillard, 2007
URB597	Electrophysiology rat	Increased firing of dorsal raphe 5-HT and locus ceruleus NE neurons	Gobbi et al., 2005
No Impact			
Anandamide	Tail suspension mouse	Increases in immobility	Naidu et al., 2007
URB597	Tail suspension mouse	No effect from 1–10 mg/kg	Naidu et al., 2007
CP47,497	Tetrabenazine-induced ptosis mouse	No blockade up to 10 mg/kg	Weissman et al., 1982

Single dose data are not included in the table. *DMI* desipramine; *5-HT* serotonin; *NE* norepinephrine

induces increased rates of firing of serotonergic and noradrendergic processes from their origin in the dorsal raphe nucleus and the locus ceruleus, respectively. These effects are associated with increased brain anandamide levels, are sensitive to CB_1 receptor blockade, and are not subject to tolerance by subacute dosing (Gobbi et al., 2006). Understanding that these effects might be due to their actions at CB_1 receptors has been achieved in some cases. For example, the antidepressant-like effects of HU-210 and AM404 in the rat forced swim test are attenuated by the CB_1 inverse agonist AM251 (Hill and Gorzalka, 2005b) used at a dose at which it does not affect immobility by itself (Tzavara et al., 2003a). Additional support for the hypothesis that cannabinoid agonism can drive antidepressant-like responses comes from interaction studies with conventional antidepressants. The CB_1 agonists ACEA when given in conjunction with fluoxetine produced greater decreases in immobility in the mouse forced swim test than by doses of either drug alone (Rutkowska and Jachimczuk, 2004). In another report, desipramine treatment for 3 weeks increased densities of CB_1 receptors in rat hippocampus and hypothalamus. These biochemical changes were associated with decreases in forced swim-induced corticosterone secretion and *c-fos* induction in the paraventricular nucleus. Both the corticosterone and *c-Fos* effects were prevented by AM251 pretreatment indicating a CB_1 receptor mediation of these effects (Hill et al., 2006c). A compelling case for the induction of antidepressant-like effects through CB_1 receptor agonism comes from data linking neurogenesis and behavioral effects of the CB_1 receptor agonist HU-210, as noted above. Hippocampal neurogenesis has been linked to antidepressant effects (Santarelli et al., 2003). Jiang and coworkers (2005) demonstrated that HU-210 induced neurogensis and also induced reductions in immobility in the rat forced swim test. X-irradiation of the hippocampus disrupted both the neurogenesis and antidepressant-like effects of HU-210.

Cannabinoid Agonists: Side Effect Liability

Cannabinoid agonists produce a spectrum of behavioral effects that are often considered liabilities. These include sedative-like effects, class-related subjective effects, abuse liability, tolerance, and dependence (Wiley, 1999; Tanda and Goldberg, 2003; Wiley and Martin, 2003; Lichtman and Martin, 2005). Another effect of cannabinoids is their propensity to increase food intake (Wiley et al., 2005), an effect that could be associated with weight gain, already an unwanted side effect in some patients with some conventional antidepressant agents. Cannabis use may also be associated with sexual dysfunction although these data do not cleanly address the causal chain (Johnson et al., 2004). As discussed previously, cannabis is considered an abused substance and Δ^9-THC is also self-administered by primates (cf., Tanda and Goldberg, 2003). Like a host of drugs of abuse, anandamide increases dopamine efflux in rat nucleus accumbens (Murillo-Rodriguez et al., 2007), an effect that is enhanced by URB597 (Solinas et al., 2006). In contrast, cannabinoid agonists are not typical of other drugs of abuse when evaluated for their effects on brain stimulation reinforcement thresholds. Three compounds that increase brain anandamide levels were evaluated for their ability to alter the threshold for reinforcing effects of electrical stimulation of the medial forebrain bundle. Drugs of abuse generally decrease the current required to maintain behavior; in contrast, phenylmethylsulfonyl fluoride and URB597, characterized as FAAH inbibitors, and OMDM-2, characterized as an anandamide transport inhibitor, all increased the reinforcement current thresholds (Vlachou et al., 2006). These effects were CB_1 receptor dependent as demonstrated by their prevention by the CB_1 receptor inverse agonist rimonabant. These data are comparable to those obtained by the same group with the direct-acting CB_1 receptor agonists WIN55212-2 and CP55940.

However, under other procedures which assess abuse potential, direct-acting CB_1 receptor agonists such as Δ^9-THC (Tanda et al., 2000) and WIN55212-2 (Martellotta et al., 1998; Fattore et al., 1999) are self-administered in experimental animals as marijuana and the active consitiuent Δ^9-THC are abused by humans (Chait and Zacney, 1992). Recent findings have suggested that abuse potential of direct- and indirect-acting CB_1 receptor agonists may be different. Under progressive ratio schedules of heroin self-administration of rats, Δ^9-THC and WIN55212-2 but not AM404 or URB597 increased the breakpoint (a measure used to assess reinforcing strength) (Solinas et al., 2005). As noted above, URB597 has been reported to produce antidepressant-like effects (Table 1). Under conditions that engender increases in brain anandamide, URB597 does not substitute for the discriminative stimulus effects of Δ^9-THC in rats or produce conditioned place preference (Gobbi et al., 2006), two indicators of abuse potential. These findings point to the possibility of dissociating antidepressant-like and abuse-related effects of cannabinoid agonists. Given the metabolic liability of anandamide and its often-reported lack of Δ^9-THC-like effects when administered exogenously, the possible lack of abuse liability of compounds like URB597 may be expected. However, if dosed appropriately, anandamide does substitute for Δ^9-THC and its effects are enhanced by URB597 (Solinas et al., 2007). It was surprising that these same authors did not find comparable enhancements with the putative anandamide transport inhibitors AM404 and UCM-707, although the effects of these compounds alone in full dose ranges were not reported nor was there information from the same report on the increases these compounds produce in brain anandamide levels or CB_1 receptor occupancy upon systemic dosing. Profound tolerance develops to the effects of Δ^9-THC and to other cannabionoid agonists. For example, tolerance in mice is engendered to the hypoactivity, hypothermia, antinociception, and cataleptic effects induced by Δ^9-THC, WIN55212-2, and CP55940 (e.g., Fan et al., 1996; Hutcheson et al., 1998). Physical dependence to cannabinoid agonists is also now being appreciated both preclinically and in humans (Lichtman and Martin, 2005). Administration of the cannabinoid inverse agonist rimonabant (i.p. or i.c.v.) induced a profound precipitated withdrawal syndrome in Δ^9-THC-tolerant animals exemplified by alterations in motor sequences but without autonomic signs (Tsou et al., 1995; Hutcheson et al., 1998; Tzavara et al., 2000).

Cannabinoid Antagonists/Inverse Agonists and Mood Disorders

Rimonabant (SR141716A) is the first selective CB_1 receptor antagonist/inverse agonist to be reported (Rinaldi-Carmona et al., 1994) and along with other ligands such as AM251, these agents provide an opportunity to provide additional insight into the potential involvement of endocannabinoid systems in mood. To date, most so-called CB_1 receptor antagonists demonstrate inverse agonist effects in vitro but are all antagonists in the functional sense (Gatley et al., 1998; Felder et al., 1998; Plummer et al., 2005; Stoit et al., 2002). The biological significance of inverse agonism will not be known until a neutral antagonist is fully characterized in vivo. Although a case was made above for the potential for cannabinoid agonists to serve a role as antidepressants, antidepressant potential of CB_1 receptor antagonist/inverse agonists has also been supported by a number of preclinical findings. Rimonabant has EEG-activating effects and produces decreases in REM sleep (Santucci et al., 1996) as seen with conventional antidepressants. Two CB_1 receptor antagonists/inverse agonists have also demonstrated efficacy in a number of models used to predict antidepressant effects in humans (Table 2). Efficacy has been observed in models utilizing both acute and chronic dosing and in several different species. Antidepressant-like effects have been observed in mice, rats, and gerbils, and in models utilizing different dependent measures. However, it is also important to recognize that both rimonabant and AM251, the only inverse agonists well characterized in behavioral experiments, are close structural analogs and therefore, increased confidence in this mechanism would be derived from data on structurally novel molecules. Critical pharmacological experiments have been conducted that support the hypothesis that CB_1 receptor antagonism is responsible for the antidepressant-like activity of these compounds

Table 2 Effects of cannabinoid receptor antagonist/inverse agonists in tests predictive of an impact in mood disorders

Compound	Procedure	Effect	Reference
Positive Impact			
Rimonabant	Forced swim mouse	Decreased immobility	Tzavara et al., 2003a
Rimonabant	Forced swim rat	Decreased immobility as with fluoxetine	Griebel et al., 2005
AM251	Forced swim mouse	Decreased immobility as with DMI	Shearman et al., 2003
AM251	Forced swim mouse	CB_1 receptor deletion prevented antidepressant-like effects	Shearman et al., 2003
AM251	Tail suspension mouse	Decreased immobility as with DMI	Shearman et al., 2003
Rimonabant	Tonic immobility gerbil	Decreased immobility as with fluoxetine	Griebel et al., 2005
Rimonabant	Chronic mild stress mouse	Decreased fur deterioration. Decreased immobility in forced swim test	Griebel et al., 2005
Rimonabant	EEG	Antidepressant-like EEG activation	Santucci et al., 1996
AM281	Sexual behavior male, rough-skinned newts	Blockade of stress and CORT-induced suppression of sexual behavior	Coddington et al., 2007
Rimonabant	Neurochemistry rat	Increased efflux of 5-HT, NE, DA, Ach	Tzavara et al., 2003a

Single dose data are not included in the table. *ACh* acetylcholine; *CORT* corticosterone; *DA* dopamine; *DMI* desipramine; *5-HT* serotonin; *NE* norepinephrine and define the CB_1 receptor as a mechanism for these effects: the CB_1 receptor agonist CP55940 prevented the antidepressant-like effects of AM251 and the antidepressant-like effects of AM251 are absent in CB_1 receptor null mice (Shearman et al., 2003). Other studies with CB_1 receptor KO mice have not provided a fully consistent picture. CB_1 receptor KO mice did not differ significantly from control mice in the control levels of immobility in the forced swim test (Shearman et al., 2003). However, an increase in depressive-like behaviors has also been reported in CB_1 receptor KO mice compared to their wild type controls in studies with the chronic mild stress paradigm (Martin et al., 2002) but opposite findings (i.e., reduced depressive-like behaviors) were seen with chronic rimonabant (Griebel et al., 2005). Some data from CB_1 receptor KO mice have also suggested that CB_1 receptors are important regulators of stress responses in vivo (cf., Barna et al., 2004). CB_1 receptors were downregulated in the hippocampus of rats exposed to repeated stress (Hill et al., 2006b). There have also been limited data suggesting a protective role of CB_1 receptor blockade against stress (Degroot and Nomikos, 2004), but there is also a body of data suggesting the opposite, that an increase in endocannabinoid neurotransmission instead may tonicaly reduce stress reactivity (cf., Tasker, 2004). Again these effects seem to be dose dependent since high doses of exogenous cannabinoid agonists appear to be anxiogenic (Hill et al., 2007). Recent findings have also been disclosed on the interactions of rimonabant with desipramine. Rimonabant did not interfere with the antidepressant-like effects of desipramine in the mouse forced swim test and yet was able to reduce the body weight gain produced by repeated desipramine treatment (Gobshtis et al., 2007). These data have implications for treatment of antidepressant-emergent weight gain (see Chap. 14). In summary, there are several pieces of data implicating CB_1 receptor blockade as a potential novel antidepressant mechanism: the clinical pharmacology of cannabis has commonalities with mood disorders, the CNS localization of endocannabinoid protein targets, CB_1 receptor alterations in brains of suicide victims, CB_1 receptor involvement in related neuropsychiatric disorders, antidepressant-like increases in cortical monoamines produced by CB_1 receptor inverse agonists, antidepressant-like behavioral effects of CB_1 receptor inverse agonists, antidepressant-like EEG effects of CB_1 receptor inverse agonists, procognitive effects of CB_1 receptor inverse agonists that might be of value for mood disorder symptoms, and efficacy of rimonabant against comorbid substance abuse and obesity, disorders often associated with impulsivity, anxiety, and depression.

Cannabinoid Antagonists/Inverse Agonoists: Side-effect Liability

There is little reason to anticipate weight gain, sexual dysfunction, or abuse liability of CB_1 receptor antagonists/inverse agonists. The lack of effect of rimonabant on subcortical DA and ACh release also lends credence to the possibility that overt actions on mood, emotionality, pleasure, and psychomotor regulation in the general population may be negligible. In addition, whereas the agonist WIN55212-2 altered cocaine self-administration by rats, suggesting that its reinforcing effects added with those of cocaine, rimonabant did not alter cocaine self-administration (Fattore et al., 1999). Depressed mood and anxiety were among common adverse effects that affected continuation in a study of effects of rimonabant on weight reduction and cardiovascular risk factors in overweight patients (Van Gaal et al., 2005). It is not clear what factors were responsible for the changes reported in these subjects. However, there were no significant changes in depression or anxiety scores in patients that completed the trial (Van Gaal et al., 2005). Also, in a previous clinical trial with rimonabant in schizophrenic patients (Meltzer et al., 2004), no adverse effects on mood and anxiety were reported. It is also important to note that the subjective effects of a particular treatment do not necessarily have implication for the mood-altering or antidepressant potential of the treatment. Acute dosing with a host of conventional antidepressants such as the tricylic agents or the monoamine uptake blockers does not produce positive mood effects. The NMDA receptor antagonist ketamine induces a dysphoric mood state (Krystal et al., 2006) and yet has been reported to produce antidepressant effects in humans (Berman et al., 2000; Zarate et al., 2006).

Endocannabinoid Agonism and CB_1 Antagonism in Mood Disorders: Contradictory or Complementary Strategies?

A recurrent theme in this review is that with respect to many biochemical (e.g., cortical ACh and DA release) and behavioral (e.g., forced swim) readouts, CB_1 antagonism/inverse agonism and CB_1 agonism (induced by endocannabinoid catabolism inhibition or by low doses of exogenous cannabinoids) seem to have identical and not opposing effects, as one might expect in a classical model of agonism/antagonism. Only, endocannabinoid neurotransmission cannot be considered as a simple, straightforward, one-input, one-output model of neuronal activity. First, endocannabinoids operate as a retrograde feedback system that homeostatically regulates neuronal plasticity. Thus the impact of alterations in endocannabinoid activity depends on the actual steady-state and previous history of the endocannabinoid synaptic module. Second, CB_1 receptors are localized on and presynapticaly inhibit, opposing neuronal populations, GABAergic interneurons on the one hand and projection neurons, including glutamatergic neurons on the other. These express high and low levels of CB_1 receptors, respectively (Marsicano and Lutz, 1999), suggesting that GABAergic circuits might be more sensitive to cannabinoids. In other words, the effects of cannabinoid agonists in experimental models largely depend on the dose used and on the responsiveness of the underlying neuronal circuit. Thus, low levels of cannabinoid agonism would lead in neuronal activation via disinhibition of GABAergic interneurons, while CB_1 antagonism would induce a similar effect by inhibiting direct inhibitory control in projection neurons. Finally, high doses of exogenously applied CB_1 agonists would lead in nonspecific profound dampening of neuronal activity, an undesirable feature for the treatment of affective disorders.

This bimodal regulation has been clearly illustrated with respect to ACh release in the hippocampus, where underlying mechanisms have been demonstrated and decorticated (Tzavara et al., 2003b). A second take-home message is that the diverse biological effects of endocannabinoid catabolism inhibitors and of CB_1 antagonists/inverse agonists might be relevant to the need for differential treatment in different aspects/subtypes of mood disorders. Although it is not easy to understand how seemingly opposite neurochemical modulation could have a common overall therapeutic impact, or if this indeed will be proven by clinical test, it should be emphasized that the mood disorders represent a heterogeneous set of disorders. As stated above, major depressive disorders may present with psychotic symptoms, melancholia, or atypical features. It should be noted that pure melancholic or atypical depression constitute less than 25% of depressive pathologies, and that classical antidepressants fail to differentiate between these two conditions, with the notable exception of ECT efficacious in severe melancholia. However, one might speculate that the different mood syndromes might be best treated by different medicines. From the perspective of melancholic depression, mild endocannabinoid activation might help alleviate some of the unbearable and debilitating tension created by recurrent negative stereotyped thoughts and ideations. Flexibility in affect and cognition relies on an activated prefrontal cortex but also from an optimally fine-tuned stress and amygdalar system. Mild endocannabinoid stimulation might reverse the disruptive unpatterned HPA hyperactivity thus restoring a minimal prefrontal-mediated control on amygdala-dependent bias toward the negative interpretation of emotional information. Endocannabinoid activation might also trigger subcortical circuits to reinstate a sense of pleasure and responsiveness to rewarding stimuli. By alleviating the excessive anxiety imposed by HPA-amygdala overactivation, mild endocannabinoid stimulation might help reinstate a vast palette of thoughts and emotions. On the other hand, CB_1 receptor antagonist-induced increase in monoaminergic neurotransmission might be directly responsible for normalizing a dysregulated subcortical dopaminegic function that contributes to hedonic and motivational allostasis and consequently to affective disorders, impulsivity, and drug dependence. In addition, enhanced dopaminergic neurotransmission in the prefrontal cortex might normalize the persistence of unfulfilling pleasure seeking, fear of rejection as well as overeating, seen in atypical depression. CB_1 antagonist-mediated stimulation of neuronal activity in corticohippocampal circuits, might be a key factor to enhance cognitive ability and clarity; ACh and DA-mediated attentional focus and executive inhibitory control might help to reinstate forceful conscious control of behavior and alleviate the prevalent mental weariness. A third point that we would like to stress, as we examine the antidepressant-like potential of cannabinoid ligands, is the importance of global management of affective disorders, to include comorbid conditions. Mood disorders present with a significant comorbidity with psychiatric and neuroendocrine or metabolic pathologies such as cognitive dysfunction, drug abuse, agitation, impulsivity, obesity, metabolic syndrome, cardiovascular risks, and osteoporosis. These conditions and their relationship to endocannabinoids have been extensively discussed in this book. As discussed, CB_1 antagonists/inverse agonists could be helpful in the management of drug abuse (Cohen et al., 2005; De Vries and

Schoffelmeer, 2005; Fattore et al., 2007). These compounds also effectively decrease weight, obesity, and related cardiometabolic risks (Matias and Di Marzo, 2007; see Chap. 14), and show some promise in the management of osteoporosis (Idris et al., 2005). CB_1 antagonists (Tzavara et al., 2003a) as well as endocannabinoid catabolism inhibitors (Tzavara et al., 2006) reduce hyperlocomotion in pharmacological or genetic rodent models, thought to reflect manic state in humans. We stress again the importance of cognitive function in psychiatric prognosis and management (Fig. 2). Executive deficits (in episodic memory, response to novelty, inhibiting incorrect responses, strategy selection) and cognitive inhibition deficits in depressed patients could reduce their ability to control transient mood changes. We also stress the importance of minimizing side effects in patient compliance. In this respect CB_1 antagonists/inverse agonists that show procognitive potential (see Witkin et al., 2005b) and help reduce weight gain (seen often in depressed patients due to either the condition itself or to antidepressant treatment) might add therapeutic value in depressive pathologies in monotherapy or as adjuvant treatments. Finally we would like to point out the scarcity of literature on $TRPV_1$ receptors, affect, and emotional regulation. However, recent studies implicate $TRPV_1$ receptors in the regulation of

Fig. 2 A procognitive pharmacological profile may be desirable for future antidepressants. Impaired cognition, in particular attention and executive function, are a part of the clinical profile of the mood disorders, probably underpinned by hypofrontality. CB_1 receptor antagonists that increase cortical neurotransmission and arousal could have a role in the treatment of the cognitive dimension of depression

neurochemical functions and of behavioral patterns related to mood disorders. $TRPV_1$ receptors are present in the brain (Toth et al., 2005; see Chap. 10) where they regulate mesocorticolimbic neurotransmission (Marinelli et al., 2005). We have demonstrated that $TRPV_1$ receptors could restore DA-related behavioral deficits, in particular pathological abberant hyperlocomotion (Tzavara et al., 2006). On the other hand, $TRPV_1$ receptor KO mice showed reduced anxiety, loss of conditioned fear, and impairments in hippocampal long-term potentiation (Marsch et al., 2007). We contend that there is a necessity for preclinical and clinical studies on $TRPV_1$ receptor-mediated signaling in depressive disorders.

Concluding Remarks

A broad body of data exists to support the hypothesis that endocannabinoid neurotransmission and neuromodulation are involved in mood disorders. Further, recent data from pharmacological and behavioral studies has suggested the potential of both cannabinoid agonists as well as cannabinoid antagonists/inverse-agonists in the therapeutic management of mood disorders. For both agonists and antagonist/inverse agonists, the data sets are relatively restricted either from the limited compounds that have been explored and/or the limited number of models in which these compounds have been assessed. Data in support of agonist therapy have gained ground in recent years with reports that some indirect activators of cannabinoid receptors may be able to transduce their antidepressant-like effects without inducing the side effects typically observed with direct-acting agonists. The broader base of preclinical support for a CB_1 receptor antagonist/inverse agonist antidepressant is weighted against clinical findings in one of the several clinical reports on rimonabant that a select population of people might be susceptible to mood and anxiety problems with this compound. These data are limited to an imperfect CB_1 receptor antagonist/inverse agonist but suggest the possibility that endocannabinoid tone may be responsible for disparate effects across individuals. Combined with the fact that mood disorders represent a very heterogeneous group of symptoms and comorbidities, a place for both agonist and antagonist/inverse agonist therapies for major depressive disorders would also not be surprising. It is hoped that direct clinical tests of these ideas will be forthcoming in the near future with the promise of novel and improved medicines for the debilitating disorders of mood.

Acknowledgments We are thankful to Marie-Anne El Khoury for critical comments and to the cartoonist ArKas for Fig. 2.

References

Ablon SL, Goodwin FK (1974) High frequency of dysphoric reactions to tetrahydrocannabinol among depressed patients. Am J Psychiatry 13:448–453.

Acquas E, Pisanu A, Marrocu P, Di Chiara G (2000) Cannabinoid CB1 receptor agonists increase rat cortical and hippocampal acetylcholine release *in vivo*. Eur J Pharmacol 401:179–185.

Alt A, Nisenbaum ES, Bleakman D, Witkin JM (2006) A role for AMPA receptors in mood disorders. Biochem Pharmacol 71:1273–1288.

Ashton CH, Moore PB, Gallagher P, Young AH (2005) Cannabinoids in bipolar affective disorder: a review and discussion of their therapeutic potential. J Psychopharmacol 19:293–300.

Azad SC, Monory K, Marsicano G, Cravatt BF, Lutz B, Zieglgansberger W, Rammes G (2004) Circuitry for associative plasticity in the amygdala involves endocannabinoid signaling. J Neurosci 24:9953–9961.

Ballon N, Leroy S, Roy C, Bourdel MC, Charles-Nicolas A, Krebs MO, Poirier MF (2006) $(AAT)_n$ repeat in the cannabinoid receptor gene (CNR1): association with cocaine addiction in an African-Caribbean population. Pharmacogenomics J 6:126–130.

Barna I, Zelena D, Arszovszki AC, Ledent C (2004) The role of endogenous cannabinoids in the hypothalamo-pituitary-adrenal axis regulation: *in vivo* and *in vitro* studies in CB_1 receptor knockout mice. Life Sci 75:2959–2970.

Barrero FJ, Ampuero I, Morales B, Vives F, de Dios Luna Del Castillo J, Hoenicka J, Garcia Yebenes J (2005) Depression in Parkinson's disease is related to a genetic polymorphism of the cannabinoid receptor gene (CNR1). Pharmacogenomics J 5:135–141.

Berman RM, Cappiello A, Anand A, Oren DA, Heninger GR, Charney DS, Krystal JH (2000) Antidepressant effects of ketamine in depressed patients. Biol Psychiatry 47:351–354.

Bodnoff SR, Suranyi-Cadotte B, Quirion R, Meaney MJ (1989) A comparison of the effects of diazepam versus several typical and atypical anti-depressant drugs in an animal model of anxiety. Psychopharmacology (Berl) 97:277–279.

Bouaboula M, Poinot-Chazel C, Bourrie B, Canat X, Calandra B, Rinaldi-Carmona M, Le Fur G, Casellas P (1995) Activation of mitogen-activated protein kinases by stimulation of the central cannabinoid receptor CB_1. Biochem J 312:637–641.

Butovsky E, Juknat A, Goncharov I, Elbaz J, Eilam R, Zangen A, Vogel Z (2005) *In vivo* up-regulation of brain-derived neurotrophic factor in specific brain areas by chronic exposure to Delta-tetrahydrocannabinol. J Neurochem 93:802–811.

Bymaster FP, Zhang W, Carter PA, Shaw J, Chernet E, Phebus L, Wong DT, Perry KW (2002) Fluoxetine, but not other selective serotonin uptake inhibitors, increases norepinephrine and dopamine extracellular levels in prefrontal cortex. Psychopharmacology 160:353–361.

Campbell S, MacQueen G (2006) An update on regional brain volume differences associated with mood disorders. Curr Opin Psychiatry 19:25–33.

Chait LD, Zacny JP (1992) Reinforcing and subjective effects of oral delta 9-THC and smoked marijuana in humans. Psychopharmacology (Berl) 107:255–262.

Chakrabarti B, Kent L, Suckling J, Bullmore E, Baron-Cohen S (2006) Variations in the human cannabinoid receptor (CNR1) gene modulate striatal responses to happy faces. Eur J Neurosci 23:1944–1948.

Chen J, Paredes W, Lowinson JH, Gardner EL (1990) Delta 9-tetrahydrocannabinol enhances presynaptic dopamine efflux in medial prefrontal cortex. Eur J Pharmacol 190:259–262.

Chhatwal JP, Davis M, Maguschak KA, Ressler KJ (2005) Enhancing cannabinoid neurotransmission augments the extinction of conditioned fear. Neuropsychopharmacology 30:516–524.

Coddington E, Lewis C, Rose JD, Moore FL (2007) Endocannabinoids mediate the effects of acute stress and corticosterone on sex behavior. Endocrinology 148:493–500.

Cohen C, Kodas E, Griebel G (2005) CB_1 receptor antagonists for the treatment of nicotine addiction. Pharmacol Biochem Behav 81:387–395.

Cota D, Marsicano G, Tschöp M, Grübler Y, Flachskamm C, Schubert M, Auer D, Yassouridis A, Thöne-Reineke C, Ortmann S, Tomassoni F, Cervino C, Nisoli E, Linthorst AC, Pasquali R, Lutz B, Stalla GK, Pagotto U (2003) The endogenous cannabinoid system affects energy balance via central orexigenic drive and peripheral lipogenesis. J Clin Invest 112:423–431.

Cryan JF, Markou A, Lucki I (2002) Assessing antidepressant activity in rodents: Recent developments and future needs. Trends Pharmacol Sci 23:238–245.

Degenhardt L, Hall W, Lynskey M (2003) Exploring the association between cannabis use and depression. Addiction 98:1493–1504.

Degroot A, Nomikos GG (2004) Genetic deletion and pharmacological blockade of CB_1 receptors modulates anxiety in the shock-probe burying test. Eur J Neurosci 20:1059–1064.

Derkinderen P, Valjent E, Toutant M, Corvol J-C, Enslen H, Ledent C, Trzaskos J, Caboche J, Girault J-A (2003) Regulation of extracellular signal-regulated kinase by cannabinoids in hippocampus. J Neurosci 23:2371–2382.

De Vries TJ, Schoffelmeer AN (2005) Cannabinoid CB_1 receptors control conditioned drug seeking. Trends Pharmacol Sci 26:420–426.

Diana MA, Marty A (2004) Endocannabinoid-mediated shortterm synaptic plasticity: depolarization-induced suppression of inhibition (DSI) and depolarization-induced suppression of excitation (DSE). Br J Pharmacol 142:9–19.

Dickason-Chesterfield AK, Kidd SR, Moore SA, Schauss JM, Liu B, Nomikos GG, Felder CC (2006) Pharmacological characterization of endocannabinoid transport and fatty acid amino-hydrolase inhibitors. Cell Mol Neurobiol 26:407–423.

Dubovsky SL, Buzan R (1999). Mood disorders. In: Hales RE, Yudofsky SC, Talbott JA (eds.), Textbook of Psychiatry, 3rd edn. Washington, DC: American Psychiatric Press, pp. 479–565.

Duman RS (2004) The neurochemistry of depressive disorders: preclinical studies. In: Charney DS, Nestler EJ (eds.), Neurobiology of Mental Illness, 2nd edn. Oxford: Oxford University Press, pp. 421–439.

Eisch AJ, Bolanos CA, de Wit J, Simonak RD, Pudiak CM, Barrot M, Verhaagen J, Nestler EJ (2003) Brain-derived neurotrophic factor in the ventral midbrain-nucleus accumbens pathway: a role in depression. Biol Psychiatry 54:994–1005.

Fan F, Tao Q, Abood M, Martin BR (1996) Cannabinoid receptor down-regulation without alteration of the inhibitory effect of CP55,940 on adenylyl cyclase in the cerebellum of CP55,940-tolerant mice. Brain Res 706:13–20.

Fattore L, Martellotta MC, Cossu G, Mascia MS, Fratta W (1999) CB_1 cannabinoid receptor agonist WIN 55,212-2 decreases intravenous cocaine self-administration in rats. Behav Brain Res 104:141–146.

Fattore L, Spano MS, Deiana S, Melis V, Cossu G, Fadda P, Fratta W (2007) An endocannabinoid mechanism in relapse to drug seeking: a review of animal studies and clinical perspectives. Brain Res Rev 53:1–16.

Felder CC, Joyce KE, Briley EM, Glass M, Mackie KP, Fahey KJ, Cullinan GJ, Hunden DC, Johnson DW, Chaney MO, Koppel GA, Brownstein M (1998) LY320135, a novel cannabinoid CB_1 receptor antagonist, unmasks coupling of the CB1 receptor to stimulation of cAMP accumulation. J Pharmacol Exp Ther 284:291–297.

Felder CC, Dickason-Chesterfield AK, Moore SA (2006) Cannabinoids biology: the search for new therapeutic targets. Mol Interv 6:149–161.

Galynker II, Cai J, Ongseng F, Finestone H, Dutta E, Serseni D (1998) Hypofrontality and negative symptoms in major depressive disorder. J Nucl Med 39:608–612.

Gatley SJ, Lan R, Volkow ND, Pappas N, King P, Wong CT, Gifford AN, Pyatt B, Dewey SL, Makriyannis A (1998) Imaging the brain marijuana receptor: development of a radioligand that binds to cannabinoid CB1 receptors *in vivo*. J Neurochem 70:417–423.

Gobbi G, Bambico FR, Mangieri R, Bortolato M, Campolongo P, Solinas M, Cassano T, Morgese MG, Debonnel G, Duranti A, Tontini A, Tarzia G, Mor M, Trezza V, Goldberg SR, Cuomo V, Piomelli D (2006) Antidepressant-like activity and modulation of brain monoaminergic transmission by blockade of anandamide hydrolysis. Proc Natl Acad Sci USA 102: 18620–18625.

Gobshtis N, Ben-Shabat S, Fride E (2007) Antidepressant-induced undesirable weight gain: prevention with rimonabant without interference with behavioral effectiveness. Eur J Pharmacol 554:155–163.

Gold PW, Chrousos GP (2002) Organization of the stress system and its dysregulation in melancholic and atypical depression: High vs. low CRH/NE states. Mol Psychiatry 7:254–275.

Gold PW, Drevets WC, Charney DS (2002) New Insights into the role of cortisol and the glucocorticoid receptor in severe depression. Biol Psychiatry 52:381–385.

Goldschmidt L, Richardson GA, Cornelius MD, Day NL (2004) Prenatal marijuana and alcohol exposure and academic achievement at age 10. Neurotoxicol Teratol 26:521–532.

Gonzalez S, Fernandez-Ruiz J, Di Marzo V, Hernandez M, Arevalo C, Nicanor C, Cascio MG, Ambrosio E, Ramos JA (2004) Behavioral and molecular changes elicited by acute administration of SR141716 to Delta9-tetrahydrocannabinol-tolerant rats: an experimental model of cannabinoid abstinence. Drug Alcohol Depend 74:159–170.

Griebel G, Stemmelin J, Scatton B (2005) Effects of the cannabinoid CB_1 receptor antagonist rimonabant in models of emotional reactivity in rodents. Biol Psychiatry 57:261–267.

Haller J, Varga B, Ledent C, Barna I, Freund TF (2004) Context-dependent effects of CB_1 cannabinoid gene disruption on anxiety-like and social behaviour in mice. Eur J Neurosci 19:1906–1912.

Hart CL, van Gorp W, Haney M, Foltin RW, Fischman MW (2001) Effects of acute smoked marijuana on complex cognitive performance. Neuropsychopharmacology 25:757–765.

Herman AI, Kranzler HR, Cubells JF, Gelernter J, Covault J (2006) Association study of the CNR1 gene exon 3 alternative promoter region polymorphisms and substance dependence. Am J Med Genet B Neuropsychiatr Genet 141:499–503.

Hill MN, Gorzalka BB (2005a) Is there a role for the endocannabinoid system in the etiology and treatment of melancholic depression? Behav Pharmacol 6:333–352.

Hill MN, Gorzalka BB (2005b) Pharmacological enhancement of cannabinoid CB_1 receptor activity elicits an antidepressant-like response in the rat forced swim test. Eur Neuropsychopharmacol 15:593–599.

Hill MN, Kambo JS, Sun JC, Gorzalka BB, Galea LA (2006a) Endocannabinoids modulate stress-induced suppression of hippocampal cell proliferation and activation of defensive behaviours. Eur J Neurosci 24:1845–1849.

Hill MN, Patel S, Carrier EJ, Rademacher DJ, Ormerod BK, Hillard CJ, Gorzalka BB (2006b) Downregulation of endocannabinoid signaling in the hippocampus following chronic unpredictable stress. Neuropsychopharmacology 31:471–472.

Hill MN, Ho WS, Sinopoli KJ, Viau V, Hillard CJ, Gorzalka BB (2006c) Involvement of the endocannabinoid system in the ability of long-term tricyclic antidepressant treatment to suppress stress-induced activation of the hypothalamic-pituitary-adrenal axis. Neuropsychopharmacology 31:2591–2599.

Hill MN, Karacabeyli ES, Gorzalka BB (2007) Estrogen recruits the endocannabinoid system to modulate emotionality. Psychoneuroendocrinology 32:350–357.

Hohmann AG, Suplita RL, Bolton NM, Neely MH, Fegley D, Mangieri R, Krey JF, Walker JM, Holmes PV, Crystal JD, Duranti A, Tontini A, Mor M, Tarzia G, Piomelli D (2005) An endocannabinoid mechanism for stress-induced analgesia. Nature 435:1108–1112.

Holter SM, Kallnik M, Wurst W, Marsicano G, Lutz B, Wotjak CT (2005) Cannabinoid CB_1 receptor is dispensable for memory extinction in an appetitively-motivated learning task. Eur J Pharmacol 510:69–74.

Hungund BL, Vinod KY, Kassir SA, Basavarajappa BS, Yalamanchili R, Cooper TB, Mann JJ, Arango V (2004). Upregulation of CB_1 receptors and agonist-stimulated [^{35}S]GTPgammaS binding in the prefrontal cortex of depressed suicide victims. Mol Psychiatry 9:184–190.

Hutcheson DM, Tzavara ET, Smadja C, Valjent E, Roques BP, Hanoune J, Maldonado R (1998) Behavioural and biochemical evidence for signs of abstinence in mice chronically treated with delta-9-tetrahydrocannabinol. Br J Pharmacol 125:1567–1577.

Idris AI, van't Hof RJ, Greig IR, Ridge SA, Baker D, Ross RA, Ralston SH (2005) Regulation of bone mass, bone loss and osteoclast activity by cannabinoid receptors. Nat Med 11:774–779.

Ilan AB, Gevins A, Coleman M, ElSohly MA, de Wit H (2005) Neurophysiological and subjective profile of marijuana with varying concentrations of cannabinoids. Behav Pharmacol 16:487–496.

Isbell H, Gorodetzky CW, Jasinski DR, Claussen U,Von Spulak F, Korte F (1967) Effects of (−)Δ,9-trans-tetrahydrocannabinol in man. Psychopharmacologia 11:184–188.

Iversen L (2000) The effects of cannabis in the central nervous system. In: Iversen L (ed.), The Science of Marijuana. Oxford: Oxford University Press, pp. 78–119.

Iversen L (2005) The monoamine hypothesis of depression. In: Licinio J, Wong ML (eds.), Biology of Depression. Weinheim: Wiley-VCH, pp. 71–86.
Jarbe TU, Liu Q, Makriyannis A (2005) Antagonism of discriminative stimulus effects of delta^9-THC and (R)-methanandamide in rats. Psychopharmacology (Berl) 184:36–45.
Jiang W, Zhang Y, Xiao L, Van Cleemput J, Ji SP, Bai G, Zhang X (2005) Cannabinoids promote embryonic and adult hippocampus neurogenesis and produce anxiolytic- and antidepressant-like effects. J Clin Invest 115:3104–3116.
Jin K, Xie L, Kim SH, Parmentier-Batteur S, Sun Y, Mao XO, Childs J, Greenberg DA (2004) Defective adult neurogenesis in CB_1 cannabinoid receptor knockout mice. Mol Pharmacol 66:204–208.
Johnson SD, Phelps DL, Cottler LB (2004) The association of sexual dysfunction and substance use among a community epidemiological sample. Arch Sex Behav 33:55–63.
Katz MM, Tekell JL, Bowden CL, Brannan S, Houston JP, Berman N, Frazer A (2004) Onset and early behavioral effects of pharmacologically different antidepressants and placebo in depression. Neuropsychopharmacology 29:566–579.
Konings M, Maharajh HD (2006) Cannabis use and mood disorders: patterns of clinical presentations among adolescents in a developing country. Int J Adolesc Med Health 18:221–233.
Krystal JH, Madonick S, Perry E, Gueorguieva R, Brush L, Wray Y, Belger A, D'Souza DC (2006) Potentiation of low dose ketamine effects by naltrexone: potential implications for the pharmacotherapy of alcoholism. Neuropsychopharmacology 31:1793–1800.
Lane SD, Cherek DR, Tcheremissine OV, Lieving LM, Pietras CJ (2005) Acute marijuana effects on human risk taking. Neuropsychopharmacology 30:800–809.
Lichtman AH, Martin BR (2005) Cannabinoid tolerance and dependence. Handb Exp Pharmacol 168:691–717.
Lukas SE, Mendelson JH, Benedikt R (1995) Electroencephalographic correlates of marijuana-induced euphoria. Drug Alcohol Depend 37:131–140.
Lundqvist T, Jonsson S, Warkentin S (2001) Frontal lobe dysfunction in long-term cannabis users. Neurotoxicol Teratol 23:437–443.
Maier SF (1984) Learned helplessness and animal models of depression. Prog Neuropsychopharmacol Biol Psychiatry 8:435–446.
Malatynska E, Goldenberg R, Shuck L, Haque A, Zamecki P, Crites G, Schindler N, Knapp RJ (2002) Reduction of submissive behavior in rats: a test for antidepressant drug activity. Pharmacology 64:8–17.
Manji HK, Duman RS (2001) Impairments of neuroplasticity and cellular resilience in severe mood disorders: implications for the development of novel therapeutics. Psychopharmacol Bull 35:5–49.
Marinelli S, Pascucci T, Bernardi G, Puglisi-Allegra S, Mercuri NB (2005) Activation of $TRPV_1$ in the VTA excites dopaminergic neurons and increases chemical- and noxious-induced dopamine release in the nucleus accumbens. Neuropsychopharmacology 30:864–870.
Marsch R, Foeller E, Rammes G, Bunck M, Kossl M, Holsboer F, Zieglgansberger W, Landgraf R, Lutz B, Wotjak CT (2007) Reduced anxiety, conditioned fear, and hippocampal long-term potentiation in transient receptor potential vanilloid type 1 receptor-deficient mice. J Neurosci 27:832–839.
Marsicano G, Lutz (1999) Expression of the cannabinoid receptor CB_1 in distinct neuronal subpopulations in the adult mouse forebrain. Eur J Neurosci 11:4213–4325.
Marsicano G, Wotjak CT, Azad SC, Bisogno T, Rammes G, Cascio MG, Hermann H, Tang J, Hofmann C, Zieglgansberger W, Di Marzo V, Lutz B (2002) The endogenous cannabinoid system controls extinction of aversive memories. Nature 418:530–534.
Marsicano G, Goodenough S, Monory K, Hermann H, Eder M, Cannich A, Azad SC, Cascio MG, Gutierrez SO, van der Stelt M, Lopez-Rodriguez ML, Casanova E, Schutz G, Zieglgansberger W, Di Marzo V, Behl C, Lutz B (2003) Cannabinoid receptors and on-demand defense against excitotoxicity. Science 302:84–88.
Martellotta MC, Cossu G, Fattore L, Gessa GL, Fratta W (1998) Self-administration of the cannabinoid receptor agonist WIN 55,212-2 in drug-naive mice. Neuroscience 85:327–330.

Martin M, Ledent C, Parmentier M, Maldonado R, Valverde O (2002) Involvement of CB_1 cannabinoid receptors in emotional behaviour. Psychopharmacology (Berl) 159:379–387.

Martinez-Gras I, Hoenicka J, Ponce G, Rodriguez-Jimenez R, Jimenez-Arriero MA, Perez-Hernandez E, Ampuero I, Ramos-Atance JA, Palomo T, Rubio G (2006) (AAT)n repeat in the cannabinoid receptor gene, CNR1: association with schizophrenia in a Spanish population. Eur Arch Psychiatry Clin Neurosci 256:437–441.

Mathew RJ, Wilson WH (1993) Acute changes in cerebral blood flow after smoking marijuana. Life Sci 52:757–767.

Matias I, Di Marzo V (2007) Endocannabinoids and the control of energy balance. Trends Endocrinol Metab 18:27–37.

Mato S, Pazos A, Valdizan EM (2002) Cannabinoid receptor antagonism and inverse agonism in response to SR141716A on cAMP production in human and rat brain. Eur J Pharmacol 443:43–46.

McArthur R, Borsini F (2006) Animal models of depression in drug discovery: a historical perspective. Pharmacol Biochem Behav 84:436–452.

McDonald J, Schleifer L, Richards JB, de Wit H (2003) Effects of THC on behavioral measures of impulsivity in humans. Neuropsychopharmacology 28:1356–1365.

McMahon LR (2006) Characterization of cannabinoid agonists and apparent pA2 analysis of cannabinoid antagonists in rhesus monkeys discriminating Delta9-tetrahydrocannabinol. J Pharmacol Exp Ther 319:1211–1218.

Meltzer HY, Arvanitis L, Bauer D, Rein W, Meta-Trial Study Group (2004) Placebo-controlled evaluation of four novel compounds for the treatment of schizophrenia and schizoaffective disorder. Am J Psychiatry 161:975–984.

Murillo-Rodriguez E, Vazquez E, Millan-Aldaco D, Palomero-Rivero M, Drucker-Colin R (2007) Effects of the fatty acid amide hydrolase inhibitor URB597 on the sleep-wake cycle, c-Fos expression and dopamine levels of the rat. Eur J Pharmacol 562:82–91.

Musty RE, Kaback L (1995) Relationships between motivation and depression in chronic marijuana users. Life Sci 56:2151–2158.

Naidu PS, Varvel SA, Ahn K, Cravatt BF, Martin BR, Lichtman AH (2007) Evaluation of fatty acid amide hydrolase inhibition in murine models of emotionality. Psychopharmacology (Berl) 192:61–70.

Nemeroff CB (2002) Recent advances in the neurobiology of depression. Psychopharmacol Bull 36:6–23.

Nutt DJ (2006) The role of dopamine and norepinephrine in depression and antidepressant treatment. J Clin Psychiatry 67:3–8.

O'Leary DS, Block RI, Koeppel JA, Flaum M, Schultz SK, Andreasen NC, Ponto LB, Watkins GL, Hurtig RR, Hichwa RD (2002) Effects of smoking marijuana on brain perfusion and cognition. Neuropsychopharmacology 26:802–816.

O'Neill MF, Moore NA (2003) Animal models of depression: Are there any? Hum Psychopharmacol Clin Exp 18:239–254.

Pacher P, Batkai S, Kunos G (2006) The endocannabinoid system as an emerging target of pharmacotherapy. Pharmacol Rev 58:389–462.

Patel S, Roelke CT, Rademacher DJ, Cullinan WE, Hillard CJ (2004) Endocannabinoid signaling negatively modulates stress-induced activation of the hypothalamic-pituitary-adrenal axis. Endocrinology 45:5431–5438.

Paton WDM, Pertwee RG (1973) Actions of cannabis in man. Marijuana 287–333.

Payne JL, Quiroz JA, Gould TD, Zarate Jr CA, Manji HK (2004) The cellular neurobiology of bipolar disorder. In: Charney DS, Nestler EJ (eds.), Neurobiology of Mental Illness, 2nd edn. Oxford: Oxford University Press, pp. 397–420.

Piomelli D, Tarzia G, Duranti A, Tontini A, Mor M, Compton TR, Dasse O, Monaghan EP, Parrott JA, Putman D (2006) Pharmacological profile of the selective FAAH inhibitor KDS-4103 (URB597). CNS Drug Rev 12:21–38.

Plummer CW, Finke PE, Mills SG, Wang J, Tong X, Doss GA, Fong TM, Lao JZ, Schaeffer MT, Chen J, Shen CP, Stribling DS, Shearman LP, Strack AM, Van der Ploeg LH (2005) Synthesis and activity of 4,5-diarylimidazoles as human CB_1 receptor inverse agonists. Bioorg Med Chem Lett 15:1441–1446.

Porsolt RD, Le Pichon M, Jalfre M (1977) Depression: a new animal model sensitive to antidepressant treatments. Nature 266:730–732.

Rademacher DJ, Hillard CJ (2007) Interactions between endocannabinoids and stress-induced decreased sensitivity to natural reward. Prog Neuropsychopharmacol Biol Psychiatry 31:633–641.

Rasmusson AM, Shi L, Duman R (2002) Downregulation of BDNF mRNA in the hippocampal dentate gyrus after re-exposure to cues previously associated with footshock. Neuropsychopharmacology 27:133–142.

Rinaldi-Carmona M, Barth F, Heaulme M, Shire D, Calandra B, Congy C, Martinez S, Maruani J, Neliat G, Caput D, et al. (1994) SR141716A, a potent and selective antagonist of the brain cannabinoid receptor. FEBS Lett 350:240–244.

Rubino T, Vigano D, Zagato E, Sala M, Parolaro D (2000) *In vivo* characterization of the specific cannabinoid receptor antagonist, SR141716A: behavioral and cellular responses after acute and chronic treatments. Synapse 35:8–14.

Russo P, Strazzullo P, Cappuccio FP, Tregouet DA, Lauria F, Loguercio M, Barba G, Versiero M, Siani A (2007) Genetic variations at the endocannabinoid type 1 receptor gene (CNR1) are associated with obesity phenotypes in men. J Clin Endocrinol Metab 92: 2382–2386.

Rutkowska M, Jachimczuk O (2004) Antidepressant-like properties of ACEA (arachidonyl-2-chloroethylamide), the selective agonist of CB_1 receptors. Acta Pol Pharm 61:165–167.

Santarelli L, Saxe M, Gross C, Surget A, Battaglia F, Dulawa S, Weisstaub N, Lee J, Duman R, Arancio O, Belzung C, Hen R (2003) Requirement of hippocampal neurogenesis for the behavioral effects of antidepressants. Science 301:805–809.

Santucci, V Storme JJ, Soubrie P, Le Fur G (1996) Arousal-enhancing properties of the CB_1 cannabinoid receptor antagonist SR 141716A in rats as assessed by electroencephalographic spectral and sleep-waking cycle analysis. Life Sci 58:PL103–PL110.

Sapolsky RM, Zola-Morgan S, Squire LR (1991) Inhibition of glucocorticoid secretion by the hippocampal formation in the primate. J Neurosci 11:3695–3704.

Schlicker E, Kathmann M (2001) Modulation of transmitter release via presynaptic cannabinoid receptors. Trends Pharmacol Sci 22:565–572.

Schmidt LG, Samochowiec J, Finckh U, Fiszer-Piosik E, Horodnicki J, Wendel B, Rommelspacher H, Hoehe MR (2002) Association of a CB_1 cannabinoid receptor gene (CNR1) polymorphism with severe alcohol dependence. Drug Alcohol Depend 65:221–224.

Shearman LP, Rosko KM, Fleischer R, Wang J, Xu S, Tong XS, Rocha BA (2003) Antidepressant-like and anorectic effects of the cannabinoid CB_1 receptor inverse agonist AM251 in mice. Behav Pharmacol 14:573–582.

Shirayama Y, Chen AC, Nakagawa S, Russell DS, Duman RS (2002) Brain-derived neurotrophic factor produces antidepressant effects in behavioral models of depression. J Neurosci 22:3251–3261.

Siegfried Z, Kanyas K, Latzer Y, Karni O, Bloch M, Lerer B, Berry EM (2004) Association study of cannabinoid receptor gene (CNR1) alleles and anorexia nervosa: differences between restricting and binging/purging subtypes. Am J Med Genet 125B:126–130.

Solinas M, Panlilio LV, Tanda G, Makriyannis A, Matthews SA, Goldberg SR (2005) Cannabinoid agonists but not inhibitors of endogenous cannabinoid transport or metabolism enhance the reinforcing efficacy of heroin in rats. Neuropsychopharmacology 30:2046–2057.

Solinas M, Justinova Z, Goldberg SR, Tanda G (2006) Anandamide administration alone and after inhibition of fatty acid amide hydrolase (FAAH) increases dopamine levels in the nucleus accumbens shell in rats. J Neurochem 98:408–419.

Solinas M, Tanda G, Justinova Z, Wertheim CE, Yasar S, Piomelli D, Vadivel SK, Makriyannis A, Goldberg SR (2007) The endogenous cannabinoid anandamide produces Δ^9-tetrahydrocannabinol-like discriminative and neurochemical effects that are enhanced by inhibition of fatty acid amide hydrolase but not by inhibition of anandamide transport. J Pharmacol Exp Ther 321:370–380.

Steffens M, Engler C, Zentner J, Feuerstein TJ (2004) Cannabinoid CB_1 receptor-mediated modulation of evoked dopamine release and of adenylyl cyclase activity in the human neocortex. Br J Pharmacol 141:1193–1203.

Steru L, Chermat R, Thierry B, Simon P (1985) The tail suspension test: a new method for screening antidepressants in mice. Psychopharmacology (Berl) 85:367–370.

Stoit AR, Lange JH, Hartog AP, Ronken E, Tipker K, Stuivenberg HH, Dijksman JA, Wals HC, Kruse CG (2002) Design, synthesis and biological activity of rigid cannabinoid CB_1 receptor antagonists. Chem Pharm Bull 50:1109–1113.

Svenningsson, P, Tzavara ET, Witkin JM, Fienberg AA, Nomikos GG, Greengard P (2002) Involvement of striatal and extrastriatal DARPP-32 in biochemical and behavioral effects of fluoxetine (Prozac). Proc Natl Acad Sci 99:3182–3187.

Tanda G, Goldberg SR (2003) Cannabinoids: reward, dependence, and underlying neurochemical mechanisms – a review of recent preclinical data. Psychopharmacology (Berl) 169:115–134.

Tanda G, Carboni E, Frau R, Di Chiara G (1994) Increase of extracellular dopamine in the prefrontal cortex: a trait of drugs with antidepressant potential? Psychopharmacology 115:285–288.

Tanda G, Munzar P, Goldberg SR (2000) Self-administration behavior is maintained by the psychoactive ingredient of marijuana in squirrel monkeys. Nat Neurosci 3:1073–1074.

Tasker J (2004) Endogenous cannabinoids take the edge off neuroendocrine responses to stress. Endocrinology 145:5429–5430.

Toth A, Boczan J, Kedei N, Lizanecz E, Bagi Z, Papp Z, Edes I, Csiba L, Blumberg PM (2005) Expression and distribution of vanilloid receptor 1 ($TRPV_1$) in the adult rat brain. Brain Res Mol Brain Res 135:162–168.

Tsankova NM, Berton O, Renthal W, Kumar A, Neve RL, Nestler EJ (2006) Sustained hippocampal chromatin regulation in a mouse model of depression and antidepressant action. Nat Neurosci 9:519–525.

Tsou K, Patrick SL, Walker JM (1995). Physical withdrawal in rats tolerant to D-9 tetrahydrocannabinol precipitated by a cannabinoid receptor antagonist. Eur J Pharmacol 280: R13–R15.

Tunving K (1985) Psychiatric effects of cannabis use. Acta Psychiatr Scand 72:209–217.

Tunving K, Thulin SO, Risberg J, Warkentin S (1986) Regional cerebral blood flow in long-term heavy cannabis use. Psychiatry Res 17:15–21.

Tzavara ET, Valjent E, Firmo C, Mas M, Beslot F, Defer N, Roques BP, Hanoune J, Maldonado R (2000) Cannabinoid withdrawal is dependent upon PKA activation in the cerebellum. Eur J Neurosci 12:1038–1046.

Tzavara ET, Perry KW, Rodriguez DE, Bymaster FP, Nomikos GG (2001) The cannabinoid CB_1 receptor antagonist SR141716A increases norepinephrine outflow in the rat anterior hypothalamus. Eur J Pharmacol 426:R3–R4.

Tzavara ET, Davis RJ, Perry KW, Li X, Salhoff C, Bymaster FP, Witkin JM, Nomikos GG (2003a) The CB_1 receptor antagonist SR141716A selectively increases monoaminergic neurotransmission in the medial prefrontal cortex: implications for therapeutic actions. Br J Pharmacol 138:544–553.

Tzavara ET, Wade M, Nomikos GG (2003b) Biphasic effects of cannabinoids on acetylcholine release in the hippocampus: site and mechanism of action. J Neurosci 23:9374–9384.

Tzavara ET, Li DL, Moutsimilli L, Bisogno T, Di Marzo V, Phebus LA, Nomikos GG, Giros B (2006) Endocannabinoids activate transient receptor potential vanilloid 1 receptors to reduce hyperdopaminergia-related hyperactivity: therapeutic implications. Biol Psychiatry 59:508–515.

Ujike H, Takaki M, Nakata K, Tanaka Y, Takeda T, Kodama M, Fujiwara Y, Sakai A, Kuroda S (2002) CNR1, central cannabinoid receptor gene, associated with susceptibility to hebephrenic schizophrenia. Mol Psychiatry 7:515–518.

Van Gaal LF, Rissanen AM, Scheen AJ, Ziegler O, Rossner S, RIO-Europe Study Group (2005) Effects of the cannabinoid-1 receptor blocker rimonabant on weight reduction and cardiovascular risk factors in overweight patients: 1-year experience from the RIO-Europe study. Lancet 365:1389–1397.

Vaughan CW (2006) Stressed-out endogenous cannabinoids relieve pain. Trends Pharmacol Sci 27:69–71.

Verrico CD, Jentsch JD, Roth RH (2003) Persistent and anatomically selective reduction in prefrontal cortical dopamine metabolism after repeated, intermittent cannabinoid administration to rats. Synapse 49:61–66.

Videbech P, Ravnkilde B (2004) Hippocampal volume and depression: a meta-analysis of MRI studies. Am J Psychiatry 161:1957–1966.

Vinod KY, Hungund BL (2006) Role of the endocannabinoid system in depression and suicide. Trends Pharmacol Sci 27:539–545.

Vinod KY, Arango V, Xie S, Kassir SA, Mann JJ, Cooper T, Hungund BL (2005) Elevated levels of endocannabinoids and CB_1 receptor-mediated G-protein signaling in the prefrontal cortex of alcoholic suicide victims. Biol Psychiatry 57:480–486.

Viveros MP, Marco EM, File SE (2005) Endocannabinoid system and stress and anxiety responses. Pharmacol Biochem Behav 81:331–342.

Vlachou S, Nomikos GG, Panagis G (2006) Effects of endocannabinoid neurotransmission modulators on brain stimulation reward. Psychopharmacology (Berl) 188:293–305.

Wachtel SR, ElSohly MA, Ross SA, Ambre J, De Wit H (2002) Comparison of the subjective effects of Δ9-tetrahydrocannabinol and marijuana in humans. Psychopharmacology 161:331–339.

Wade MR, Tzavara ET, Nomikos GG (2004) Cannabinoids reduce cAMP levels in the striatum of freely moving rats: an *in vivo* microdialysis study. Brain Res 1005:117–123.

Wang X, Dow-Edwards D, Anderson V, Minkoff H, Hurd YL (2004) In utero marijuana exposure associated with abnormal amygdala dopamine D2 gene expression in the human fetus. Biol Psychiatry 56:909–915.

Watjak CT (2005) Role of endogenous cannabinoids in cognition and emotionality. Mini Rev Med Chem 5:659–670.

Weissman A, Milne GM, Melvin LS Jr (1982) Cannabimimetic activity from CP-47, 497, a derivative of 3-phenylcyclohexanol. J Pharmacol Exp Ther 223:516–23.

Wiley JL (1999) Cannabis: discrimination of "internal bliss"? Pharmacol Biochem Behav 64:257–260.

Wiley JL, Martin BR (2003) Cannabinoid pharmacological properties common to other centrally acting drugs. Eur J Pharmacol 471:185–193.

Wiley JL, LaVecchia KL, Karp NE, Kulasegram S, Mahadevan A, Razdan RK, Martin BR (2004) A comparison of the discriminative stimulus effects of delta^9-tetrahydrocannabinol and O-1812, a potent and metabolically stable anandamide analog, in rats. Exp Clin Psychopharmacol 12:173–179.

Wiley JL, Burston JJ, Leggett DC, Alekseeva OO, Razdan RK, Mahadevan A, Martin BR (2005) CB_1 cannabinoid receptor-mediated modulation of food intake in mice. Br J Pharmacol 145:293–300.

Willner P (2005) Chronic mild stress (CMS) revisited: consistency and behavioural-neurobiological concordance in the effects of CMS. Neuropsychobiology 52:90–110.

Witkin JM, Tzavara ET, Nomikos GG (2005a) A role for cannabinoid CB_1 receptors in mood and anxiety disorders. Behav Pharmacol 16:315–331.

Witkin JM, Tzavara ET, Davis RJ, Li X, Nomikos GG (2005b) A therapeutic role for cannabinoid CB1 receptor antagonists in major depressive disorders. Trends Pharmacol Sci 26:609–617.

Zarate Jr CA, Singh JB, Carlson PJ, Brutsche NE, Ameli R, Luckenbaugh DA, Charney DS, Manji HK (2006) A randomized trial of an N-methyl-D-aspartate antagonist in treatment-resistant major depression. Arch Gen Psychiatry 63:856–864.

Chapter 24
Role of Cannabinoid Receptors in Anxiety Disorders

Aldemar Degroot

Abstract Cannabinoid agents modulate anxiety, although their effects vary and depend on regional endogenous tone, basal anxiety levels, environmental context, species differences, type of anxiety, prior exposure, and dose. Cannabinoid receptors are densely located in brain areas that are involved in the regulation of emotional states and induce neurochemical responses that are congruent with anxiolyt ic/anxiogenic effects. The effects on emotion mediated by cannabinoid compounds are believed to be due to a regulation of activity at the cannabinoid CB_1 receptors although there is some limited evidence implicating the cannabinoid CB_2 and a putative novel cannabinoid receptor (GPR55?) in some of the observed emotional responses. Effects on emotion are likely the result of a net effect of the summated neurochemical responses. Compounds that indirectly regulate activity at the cannabinoid receptors more consistently reduce anxiety both in preclinical and clinical models. Consequently, these compounds may be the focus of future pharmaceutical development of anxiolytic compounds.

Introduction

Controversy: Anxiolytic or Anxiogenic?

Numerous studies have demonstrated the effects of cannabinoid agents on the modulation of anxiety (Viveros et al., 2005; Witkin et al., 2005; Hill and Gorzalka, 2006). Typically, increased transmission at CB_1 receptors decreases anxiety, whereas inhibition at CB_1 receptors induces an anxiogenic effect (e.g., Biscaia et al., 2003; Uriguen et al., 2004). However, both human and animal data indicate that cannabinoid agents have inconsistent effects on emotional behavior (see Table 1). Consequently, compounds that facilitate transmission at cannabinoid CB_1 receptors have both anxiolytic and anxiogenic effects in preclinical models of anxiety (e.g., Pamplona et al., 2006). Similarly, smoked marijuana can either increase or decrease anxiety in man. Oral administration of Δ^9-tetrahydrocannabinol (Δ^9-THC) in the form of Dronabinol™ induces similar contradictory findings on anxiety modulation.

A. Kofalvi (ed.), *Cannabinoids and the Brain.*

Table 1 Effects of direct CB_1 receptor activation and inactivation on preclinical models of anxiety

Cannabinoid receptor activation	Animal model	References	Cannabinoid receptor inactivation	Animal model	References
↔anxiety	Fear-potentiated startle	Chhatwal et al., 2005	↔anxiety	T-maze model of anxiety	Takahashi et al., 2005
↔anxiety	Elevated plus-maze	Hill and Gorzalka, 2006	↔anxiety	Elevated plus-maze	Rodgers et al., 2003
↓anxiety	Fear conditioning	Pamplona et al., 2006	↔anxiety	Light–dark box	Rutkowska et al., 2006
↓anxiety	Elevated plus-maze	Biscaia et al., 2003; Haller et al., 2004; Hill and Gorzalka, 2004; Marco et al., 2004; Moreira et al., 2006; Patel and Hillard, 2006	↓anxiety	Elevated plus-maze	Haller et al., 2002; Griebel et al., 2005
↓anxiety	Novelty-suppressed feeding	Jiang et al., 2005	↓anxiety	Vogel conflict test	Griebel et al., 2005
↓anxiety	Open field test	Biscaia et al., 2003	↓anxiety	Mouse defence test battery	Griebel et al., 2005
↓anxiety	Restraint-induced corti-costerone release	Patel et al., 2004; Patel and Hillard, 2006	↓anxiety	Shock-probe burying test	Degroot and Nomikos, 2004
↑anxiety	Light–dark box	Rutkowska et al., 2006	↑anxiety	Elevated plus-maze	Navarro et al., 1997; Haller et al., 2002, 2004; Uriguen et al., 2004; Rodgers et al., 2005
↑anxiety	Elevated plus-maze	Hill and Gorzalka, 2004; Marin et al., 2003	↑anxiety	Light–dark box	Martin et al., 2002; Uriguen et al., 2004

↑anxiety	Defensive withdrawal test	Rodriguez de Fonseca et al., 1996	↑anxiety	Fear-potentiated startle	Chhatwal et al., 2005
↑anxiety	Open field test	Schneider et al., 2005; Hill and Gorzalka, 2006; Pamplona et al., 2006	↑anxiety	Social interaction	Uriguen et al., 2004
↑anxiety	Social interaction test	Genn et al., 2004	↑anxiety	Novelty stress	Haller et al., 2004

Up and down arrows indicate increases and decreases in anxiety, respectively, whereas no change is indicated by ↔

Generally, cannabis-induced anxiety occurs more frequently in drug-naïve subjects and in novel/stressful environments. In addition, whereas large quantities of smoked marijuana induce an anxiogenic effect, smaller quantities tend to decrease anxiety. Relative "large" or "small" quantities of smoked marijuana are subject to individual differences, which likely result from inherent variations in cannabinoid receptor density in key brain structures (Weiser and Noy, 2005). Contradictory effects on anxiety have also been found following decreased transmission at CB_1 receptors (inverse agonists, CB_1 knockout (KO) mice; e.g., Haller et al., 2002; see Table 1). Clinical evidence is limited to the CB_1 inverse agonist rimonabant, which induces an anxiogenic effect in 6–9% of patients that use the drug (http://www.nyrdtc.nhs.uk/docs/nde/NDE_78_Rimonabant.pdf). There is no evidence to date of a rimonabant-induced anxiolytic effect in man. A more subtle stimulation of the cannabinoid receptors may be required for a more consistent anxiolytic effect. In support of this notion, fatty acid amide hydrolase (FAAH; see Chap. 3) inhibitors, which indirectly increase the levels of the endocannabinoid anandamide, have consistently produced anxiolytic effects. Cannabidiol (CBD) which can act as a FAAH inhibitor (Rakhshan et al., 2000) has also repeatedly reduced anxiety in both preclinical and clinical models (see Table 2). Thus, whereas an inconsistent regulation of anxiety limits the clinical usefulness of compounds that directly target activity at the CB_1 receptors, both preclinical and clinical data suggest that an indirect modulation of cannabinoid receptors through increased endocannabinoid tone could represent a novel therapeutic approach for the treatment of clinical anxiety.

Table 2 Effects of indirect cannabinoid receptor activation on preclinical and clinical models of anxiety

	Preclinical	
Effect on anxiety (compound)	Model	References
↓anxiety (CBD)	Avoidance learning	Musty, 1984
↓anxiety (CBD)	Punished response task	Musty, 1984
↓anxiety (CBD)	Taste aversion	Musty et al., 1984
↓anxiety (CBD and CBD derivatives)	Elevated plus-maze	Guimaraes et al., 1990, 1994; Onaivi et al., 1990
↓anxiety (CBD)	Vogel conflict test	Moreira et al., 2006
↓anxiety (CBD)	Contextual fear paradigm	Resstel et al., 2006
↓anxiety (AM404, URB597)	Elevated plus-maze	Patel and Hillard, 2006
↓anxiety (AM404, URB597)	Restraint-induced corticosterone release	Patel et al., 2004
	Clinical	
↓anxiety (CBD)	Effect of CBD on THC-induced anxiety	Zuardi et al., 1982
↓anxiety (CBD)	Simulated public speaking test	Zuardi et al., 1993a
↓anxiety (CBD)	Subjective anxiety	Crippa et al., 2004

Up and down arrows indicate increases and decreases in anxiety, respectively, whereas no change is indicated by ↔

Why the Discrepancy?

There are several plausible possibilities that could explain the often contradictory findings obtained with cannabinoid compounds in anxiety models. For instance, contradictory findings may be due to the level of regional endogenous tone, basal anxiety levels, environmental context, species differences, the type of anxiety that is being examined, prior use, or dose levels. Endocannabinoid tone can be modulated by state- and disease-dependent neurochemical and/or neurophysiological events (Howlett, 2005; Pertwee, 2005). In addition, endocannabinoids can be affected by different neurotransmitter systems (Jung et al., 2005; Kreitzer and Malenka, 2005; Maejima et al., 2005). Endocannabinoid tone, in turn, may alter basal anxiety levels (Bari et al., 2006). Specifically, endocannabinoid production in the amygdala during periods of stress may affect emotion by regulating transmission from the amygdala (Patel et al., 2005). A change in endogenous tone and subsequent emotional state alters the assessment of the environmental stressor and modulates the environmental context. The emotional consequences of cannabinoid agents, in turn, may be dependent upon this environmental context (Patel et al., 2005). Specifically, cannabinoid receptor facilitation is more likely to induce an anxiogenic effect in novel or stressful environmental situations. This notion is supported by clinical evidence in that it has been previously demonstrated that anxiogenic effects mediated by Δ^9-THC can be further exacerbated by oral surgery or cognitive tests accompanied by experimenter harassment (Gregg et al., 1976; Naliboff et al., 1976). Similarly, in animal models, cannabinoid-induced anxiogenic effects are more likely to occur following exposure to novel or stressful environments (Ng et al., 1973; MacLean and Littleton, 1977; Haller et al., 2004). The effect of dosing on anxiety is particularly relevant for cannabinoid agonists. Specifically, cannabinoid agonists, but not antagonists, induce a dose-dependent bidirectional modulation of anxiety that involves different brain regions and perhaps different neurotransmitter systems. These agonists affect neurochemistry and behavior in a bimodal fashion. Consequently, low doses of cannabinoid agonists induce anxiolytic effects in preclinical models, whereas higher doses result in an anxiogenic effect (Manzanares et al., 1999; Giuliani et al., 2000; Berrendero and Maldonado, 2002; Marin et al., 2003). This bimodal effect may be due to distinct CB_1 receptors with different neuroanatomical localizations that have differential sensitivity to cannabinoid agents (Viveros et al., 2005). For instance, a low cannabinoid agonist dose increases hippocampal ACh efflux indirectly through the medial septum, whereas a high agonist dose decreases hippocampal ACh efflux directly through the hippocampus (Tzavara et al., 2003). This differential regulation, in turn, may result in a bimodal regulation of anxiety. Haller et al. (2006) demonstrated the importance of species differences in the effect of cannabinoid agents on emotional behavior. Specifically, enhanced transmission at the cannabinoid receptor has effects on anxiety that are diametrically opposite in rats and mice. Haller et al. (2006) postulated that these conflicting findings resulted from species differences in their relative responsiveness of

GABA and glutamate to cannabinoids. A differential regulation of GABA and glutamate transmission can result in differences in anxiety since both GABA and glutamate have been widely implicated in emotional behavior (Millan, 2003). Inconsistent effects on anxiety following CB_1 receptor blockade have been observed in studies that used conflict tasks to measure anxiety. For instance, in the elevated plus-maze, to elicit anxiolytic behavior, animals have to go against their natural inclination to remain in enclosed, dark places. More consistent results have been obtained with an animal model of anxiety that encompasses multiple aspects of anxiety behavior such as the shock-probe burying t est (Degroot and Nomikos, 2004).

Cannabinoid Receptor Facilitation and Anxiety

A facilitation of transmission at cannabinoid receptors can occur through specific agonists or through compounds that counteract the reuptake mechanism of endocannabinoids (i.e., FAAH inhibitors). As mentioned, direct facilitation of cannabinoid receptors yield variable effects, whereas an indirect activation of these same receptors consistently reduces anxiety (compare Tables 1 and 2). The fact that enhanced transmission at cannabinoid receptors induces anxiolytic effects in animal models of anxiety that use both painful and nonpainful stressors (see Table 1) suggests that the anxiolytic effect induced by a facilitation of transmission at cannabinoid receptors is not due to the analgesic properties of cannabinoid receptor agonism.

The Brain's Own Marijuana and Anxiety

Endocannabinoid tone can be altered by an anxiety provoking stimulus. Typically anxiety raises endocannabinoid tone, and endocannabinoids, in turn, decrease anxiety (coping mechanism). This was demonstrated in an elegant paper by Marsicano et al. (2002). They showed that anandamide levels in the amygdala increase when the animal is conditioned to expect a foot shock after hearing a tone. Endocannabinoids are crucial for the extinction of aversive memories and this process is likely mediated through the amygdala (Azad et al., 2004).

Relation to Neurochemistry/Neuroanatomy

The cannabinoid receptors are prominent in anxiety-related brain regions such as the hippocampus, amygdala, prefrontal cortex, and the nucleus accumbens. A modulation of ACh, GABA, glutamate, and monoamine levels in these brain structures is likely associated with the regulation of anxiety by cannabinoid agonists.

For instance, an anxiolytic effect induced by CB_1 receptor stimulation can be blocked with a 5-HT_{1A} antagonist (Braida et al., 2007). In addition, CB_1 receptor stimulation modulates hippocampal ACh efflux, which has been shown to be associated with an effect on emotional behavior (Degroot and Nomikos, 2005). Differential receptor density or the type of anxiety may result in cannabinoid-induced activation of specific brain regions upon exposure to an anxiety provoking situation. Specifically, Rubino et al. (2007) demonstrated that Δ^9-THC treatment significantly decreased *c-Fos* amounts in the prefrontal cortex and amygdala of rats exposed to the elevated plus-maze without affecting the other cerebral areas investigated. This effect was mediated through the CB_1 receptors since it was reversed by the CB_1 inverse agonist AM251.

CB_1 Receptor Blockade and Anxiety

Table 1 demonstrates that similar to cannabinoid receptor agonism, cannabinoid receptor antagonism yields variable effects on emotional behavior in preclinical models. Clinical data is limited, but indicates that cannabinoid receptor antagonism increases anxiety in a small percentage of the population. It is possible that CB_1 receptor antagonism decreases anxiety in certain situation by acting as a coping mechanism through an interaction with the cholinergic system. CB_1 receptor blockade increases hippocampal ACh efflux (Degroot et al., 2006) and it has been previously proposed that increased hippocampal ACh efflux may allow for an enhanced coping strategy in a fearful situation, thus reducing anxiety (Degroot and Treit, 2002).

Relation to Neurochemistry/Neuroanatomy

An inactivation of activity at CB_1 receptors may regulate anxiety through an increase in neocortex monoamine or hippocampal ACh levels (Degroot et al., 2006; Tzavara et al., 2003). Specifically, pharmacological blockade of CB_1 receptors increases both neocortical and hippocampal ACh efflux, whereas genetic deletion does not affect basal, but increases stress-induced hippocampal ACh efflux. Pharmacological blockade also increases monoamine levels in the prefrontal cortex and increases those of 5-HT, but not norepinephrine or dopamine, in the nucleus accumbens (Degroot and Nomikos, 2007). The nature of the emotional response may be the result of a summation of these neurochemical effects. The septo-hippocampal cholinergic system may be particularly involved in the ultimate emotional response and may serve as the nodal point for the cannabinoid-induced neurochemical events.

CB_2 and CB_3 Receptors?

Since the recent suggestions about the neuronal presence of CB_2 receptors in the CNS under physiological conditions (see Chap. 10), these receptors have been implicated in anxiety, depression, and drug addiction (Onaivi, 2006). In addition, there is evidence that a novel, yet to be identified receptor dubbed GPR55 (Chap. 10) is involved in anxiety modulation.

CB_2 Receptors

To date, few studies have examined the effect of CB_2 receptor modulation on behavioral responses. Onaivi (2006) demonstrated that direct intracerebroventricular microinjection of CB_2 antisense oligonucleotide into the mouse brain reduced anxiety in the elevated plus-maze test. Future studies will need to further explore the potential significance of CB_2 receptors in anxiety modulation.

CB_3 Receptors?

There is evidence that a novel, yet to be identified cannabinoid receptor dubbed GPR55 (Petitet et al., 2006) may be involved in anxiety modulation by cannabinoid compounds. Evidence comes from an array of behavioral studies where CB_1 antagonists can have anxiolytic effects in mice void of CB_1 receptors. For instance, the CB_1 inverse agonist SR141716A decreased anxiety in the elevated plus-maze in both wild-type and CB_1 KO mice (Haller et al., 2002).

Cannabidiol and Anxiety

Both preclinical and clinical studies suggest that the phytocannabinoid cannabidiol (CBD; see Chap. 9) is effective in the treatment of anxiety disorders (see Table 2). An early study by Musty et al. (1984) indicated that a prior administration of CBD effectively reduced anxiety in an avoidance learning task, a punished response task, and a taste aversion model. In addition, Musty found that CBD reduced stress-induced ulcers in mice. Various experiments displayed that CBD effectively reduced anxiety in the elevated plus-maze in rats (Guimarães et al., 1990, 1994; Onaivi et al., 1990). Specifically, Guimarães et al. (1990) indicated that an i.p. administration of CBD reduced anxiety in the elevated plus-maze in a biphasic manner. Consequently, intermediate, but not low or high, doses of CBD significantly increased open arm entries. In addition, Guimarães et al. (1994) tested the dimethylheptyl homolog of CBD (HU-219) in the elevated plus-maze. HU-219 was found to be even more potent in reducing anxiety than CBD. Lastly, CBD increased the number of licks in the Vogel conflict test in rats (Moreira et al., 2006) and decreased anxiety in a contextual fear model (Resstel et al., 2006). Clinical evidence for the anxiolytic effects of CBD comes from four independent studies dating back as far as 1982. CBD at a dose of 1.0 mg/kg was found to counteract Δ^9-THC-induced anxiety (Zuardi et al., 1982). In a later study by the same group, a single oral administration of 300 mg/kg of CBD reduced anxiety in the public speaking test (Zuardi et al., 1993a). In fact, the effect was found to be comparable to that of both diazepam and ipsapirone. In another study, a single administration of 300 or 600 mg was found to interfere with cortisol secretion (Zuardi et al., 1993b). Lastly, in a more recent study, Crippa et al. (2004) determined that a single oral administration of 400 mg CBD significantly decreased subjective anxiety and increased overall mental sedation as determined through imaging techniques.

Method of Action of CBD in Anxiety Disorders

CBD has very low affinity for both cannabinoid receptors 1 and 2 (Pertwee, 1997). In addition, CBD has been reported to inhibit both the FAAH-mediated degradation of the endogenous cannabinoid ligand anandamide (Watanabe et al., 1996) and the RBL-2H3 cell anandamide transporter activity (Rakhshan et al., 2000). Therefore, some of the pharmacological actions of CBD may be due to enhanced anandamide levels, which indirectly modulate activity at the CB_1 receptor. There are several studies that suggest that decreased FAAH activity and a consequent rise in endocannabinoids reduces anxiety (see Table 2). For instance, the FAAH inhibitor URB597 elicits significant anxiolytic, antidepressant, and analgesic effects that can be prevented by pretreatment with a CB_1 antagonist. These therapeutic effects occurred in the absence of typical cannabinoid-related adverse effects such as catalepsy, hypothermia, and hyperphagia. In addition, the compound did not induce place preference, thus limiting abuse potential (Piomelli et al., 2006). Similarly, Gaetani et al. (2003) showed that pharmacological blockade of FAAH produces anxiolytic-like effects in rats without causing the wide spectrum of behavioral responses typical of direct-acting cannabinoid agonists. Therefore, it appears that CBD may reduce anxiety through an indirect increase in anandamide levels. An increase in CBD-induced anandamide tone does not explain why CBD can counteract Δ^9-THC-induced anxiety. However, CBD partially inhibits the CYP 2C catalyzed hydroxylation of Δ^9-THC to 11-OH-THC (Nadulski et al., 2005). 11-OH-THC has been postulated to induce some of the adverse effects associated with cannabinoid consumption such as an increase in anxiety. Therefore, the fact that CBD limits the production of this metabolite serves as a plausible explanation for the effect of CBD on Δ^9-THC-induced anxiety (Zuardi et al., 1982). Another potential molecular mechanism of action underlying the anxiolytic activity of CBD may be mediated through serotonin (5-HT) receptors. It has been proposed that CBD may act as an agonist at $5\text{-}HT_{1A}$ receptors (Russo et al., 2005). Since a facilitation of neurotransmission at the 5-HT_{1A} receptor has been demonstrated to alleviate anxiety (Kataoka et al., 1991; Meneses and Hong, 1993; Dekeyne et al., 2000), this may be another plausible mechanism through which CBD exerts its anxiolytic effect.

Concluding Remarks

Currently, selective-serotonin-reuptake-inhibitors (SSRIs) constitute the most widely prescribed treatment for anxiety disorders. SSRIs are effective and have fewer side effects than the traditionally used benzodiazepines. However, SSRIs also induce sexual dysfunction and require 4–6 weeks of chronic use before inducing an anxiolytic effect. In fact, early use is often associated with a paradoxical increase in anxiety. Therefore, early treatment with SSRIs is often combined with benzodiazepines to counteract a potential anxiogenic response. Moreover, the use of SSRIs may increase the risk of suicide or self harm in certain patient populations that suffer from both anxiety and depression (Gunnell et al., 2005). Therefore, SSRI use in this population needs to be carefully monitored especially during the early stages of treatment. Unlike SSRIs, a careful modulation of the cannabinoid system may be able to alleviate anxiety disorders immediately and with fewer side effects. The rapid onset of action of cannabinoid agents may also make it an effective alternative treatment to β blockers for performance anxiety. Since β blockers must be used with caution in people suffering from asthma, certain heart complications, or diabetes, cannabinoid-based products may provide a more safe treatment strategy for performance anxiety in this population. Modulation at the cannabinoid receptors may regulate anxiety through various neurochemical processes including an effect on serotonergic, cholinergic, or glutamatergic receptors. The net effect of cannabinoid compounds on emotion may be the result of a summation of the neurochemical effects.

References

Azad SC, Monory K, Marsicano G, Cravatt BF, Lutz B, Zieglgansberger W, Rammes G (2004) Circuitry for associative plasticity in the amygdala involves endocannabinoid signaling. J Neurosci 24:9953–9961.

Bari M, Battista N, Fezza F, Gasperi V, Maccarrone M (2006) New insights into endocannabinoid degradation and its therapeutic potential. Mini Rev Med Chem 6:257–268.

Berrendero F, Maldonado R (2002) Involvement of the opioid system in the anxiolytic-like effects induced by Delta^9-tetrahydrocannabinol. Psychopharmacology 163:111–117.

Biscaia M, Marin S, Fernandez B, Marco EM, Rubio M, Guaza C, Ambrosio E, Viveros MP (2003) Chronic treatment with CP 55,940 during the peri-adolescent period differentially affects the behavioural responses of male and female rats in adulthood. Psychopharmacology 170:301–308.

Braida D, Limonta V, Malabarba L, Zani A, Sala M (2007) 5-HT1A receptors are involved in the anxiolytic effect of Delta^9-tetrahydrocannabinol and AM 404, the anandamide transport inhibitor, in Sprague-Dawley rats. Eur J Pharmacol 555:156–163.

Chhatwal JP, Davis M, Maguschak KA, Ressler KJ (2005) Enhancing cannabinoid neurotransmission augments the extinction of conditioned fear. Neuropsychopharmacology 30:516–524.

Crippa JA, Zuardi AW, Garrido GE, Wichert-Ana L, Guarnieri R, Ferrari L, Azevedo-Marques PM, Hallak JE, McGuire PK, Filho Busatto G (2004) Effects of cannabidiol (CBD) on regional cerebral blood flow. Neuropsychopharmacology 29:417–426.

Degroot A, Nomikos GG (2004) Genetic deletion and pharmacological blockade of CB_1 receptors modulates anxiety in the shock-probe burying test. Eur J Neurosci 20:1059–1064.

Degroot A, Nomikos GG (2005) Fluoxetine disrupts the integration of anxiety and aversive memories. Neuropsychopharmacology 30:391–400.

Degroot A, Nomikos GG (2007) *In vivo* neurochemical effects induced by changes in endocannabinoid neurotransmission. Curr Opin Pharmacol 7:62–68.

Degroot A, Treit D (2002) Dorsal and ventral hippocampal cholinergic systems modulate anxiety in the plus-maze and shock-probe tests. Brain Res 949:60–70.

Degroot A, Köfalvi A, Wade MR, Davis RJ, Rodrigues RJ, Rebola N, Cunha RA, Nomikos GG (2006) CB_1 receptor antagonism increases hippocampal acetylcholine release: site and mechanism of action. Mol Pharmacol 70:1236–1245.

Dekeyne A, Brocco M, Adhumeau A, Gobert A, Millan MJ (2000) The selective serotonin $(5\text{-HT})_{1A}$ receptor ligand, S15535, displays anxiolytic-like effects in the social interaction and Vogel models and suppresses dialysate levels of 5-HT in the dorsal hippocampus of freely-moving rats. A comparison with other anxiolytic agents. Psychopharmacology 152:55–66.

Gaetani S, Cuomo V, Piomelli D (2003) Anandamide hydrolysis: a new target for anti-anxiety drugs? Trends Mol Med 9:474–478.

Genn RF, Tucci S, Marco EM, Viveros MP, File SE (2004) Unconditioned and conditioned anxiogenic effects of the cannabinoid receptor agonist CP 55,940 in the social interaction test. Pharmacol Biochem Behav 77:567–573.

Giuliani D, Ferrari F, Ottani A (2000) The cannabinoid agonist HU 210 modifies rat behavioural responses to novelty and stress. Pharmacol Res 41:45–51.

Gregg JM, Small EW, Moore R, Raft D, Toomey TC (1976) Emotional response to intravenous delta9tetrahydrocannabinol during oral surgery. J Oral Surg 34:301–313.

Griebel G, Stemmelin J, Scatton B (2005) Effects of the cannabinoid CB_1 receptor antagonist rimonabant in models of emotional reactivity in rodents. Biol Psychiatry 57:261–267.

Guimaraes FS, Chiaretti TM, Graeff FG, Zuardi AW (1990) Antianxiety effect of cannabidiol in the elevated plus-maze. Psychopharmacology 100:558–559.

Guimaraes FS, de Aguiar JC, Mechoulam R, Breuer A (1994) Anxiolytic effect of cannabidiol derivatives in the elevated plus-maze. Gen Pharmacol 25:161–164.

Gunnell D, Saperia J, Ashby D (2005) Selective serotonin reuptake inhibitors (SSRIs) and suicide in adults: meta-analysis of drug company data from placebo controlled, randomised controlled trials submitted to the MHRA's safety review. BMJ 330:385.

Haller J, Bakos N, Szirmay M, Ledent C, Freund TF (2002) The effects of genetic and pharmacological blockade of the CB1 cannabinoid receptor on anxiety. Eur J Neurosci 16:1395–1398.

Haller J, Varga B, Ledent C, Freund TF (2004) CB_1 cannabinoid receptors mediate anxiolytic effects: convergent genetic and pharmacological evidence with CB_1-specific agents. Behav Pharmacol 15:299–304.

Haller J, Soproni K, Varga B, Nemeth B, Mikics E, Freund TF, Matyas F, Hajos N (2006) Correlated species differences in the effects of cannabinoid ligands on anxiety and on GABAergic/glutamatergic synaptic transmission. Sixteenth Annual Symposium on the Cannabinoids, ICRS, Tihany, Hungary.

Hill MN, Gorzalka BB (2004) Enhancement of anxiety-like responsiveness to the cannabinoid CB_1 receptor agonist HU-210 following chronic stress. Eur J Pharmacol 499:291–295.

Hill MN, Gorzalka BB (2006) Increased sensitivity to restraint stress and novelty-induced emotionality following long-term, high dose cannabinoid exposure. Psychoneuroendocrinology 31:526–536.

Howlett AC (2005) Cannabinoid receptor signaling. Handb Exp Pharmacol 168:53–79.

Jiang W, Zhang Y, Xiao L, Van Cleemput J, Ji SP, Bai G, Zhang X (2005) Cannabinoids promote embryonic and adult hippocampus neurogenesis and produce anxiolytic- and antidepressant-like effects. J Clin Invest 115:3104–3116.

Jung KM, Mangieri R, Stapleton C, Kim J, Fegley D, Wallace M, Mackie K, Piomelli D (2005) Stimulation of endocannabinoid formation in brain slice cultures through activation of group I metabotropic glutamate receptors. Mol Pharmacol 68:1196–1202.

Kataoka Y, Shibata K, Miyazaki A, Inoue Y, Tominaga K, Koizumi S, Ueki S, Niwa M (1991) Involvement of the dorsal hippocampus in mediation of the antianxiety action of tandospirone, a 5-hydroxytryptamine$_{1A}$ agonistic anxiolytic. Neuropharmacology 30:475–480.

Kreitzer AC, Malenka RC (2005) Dopamine modulation of state-dependent endocannabinoid release and long-term depression in the striatum. J Neurosci 25:10537–10545.

MacLean KI, Littleton JM (1977) Environmental stress as a factor in the response of rat brain catecholamine metabolism to delta^8-tetrahydrocannabinol. Eur J Pharmacol 41:171–182.

Maejima T, Oka S, Hashimotodani Y, Ohno-Shosaku T, Aiba A, Wu D, Waku K, Sugiura T, Kano M (2005) Synaptically driven endocannabinoid release requires Ca^{2+}-assisted metabotropic glutamate receptor subtype 1 to phospholipase Cbeta4 signaling cascade in the cerebellum. J Neurosci 25:6826–6835.

Manzanares J, Corchero J, Fuentes JA (1999) Opioid and cannabinoid receptor-mediated regulation of the increase in adrenocorticotropin hormone and corticosterone plasma concentrations induced by central administration of delta^9-tetrahydrocannabinol in rats. Brain Res 839:173–179.

Marco EM, Perez-Alvarez L, Borcel E, Rubio M, Guaza C, Ambrosio E, File SE, Viveros MP (2004) Involvement of 5-HT$_{1A}$ receptors in behavioural effects of the cannabinoid receptor agonist CP 55,940 in male rats. Behav Pharmacol 15:21–27.

Marin S, Marco E, Biscaia M, Fernandez B, Rubio M, Guaza C, Schmidhammer H, Viveros MP (2003) Involvement of the kappa-opioid receptor in the anxiogenic-like effect of CP 55,940 in male rats. Pharmacol Biochem Behav 74:649–656.

Marsicano G, Wotjak CT, Azad SC, Bisogno T, Rammes G, Cascio MG, Hermann H, Tang J, Hofmann C, Zieglgansberger W, Di Marzo V, Lutz B (2002) The endogenous cannabinoid system controls extinction of aversive memories. Nature 418:530–534.

Martin M, Ledent C, Parmentier M, Maldonado R, Valverde O (2002) Involvement of CB1 cannabinoid receptors in emotional behaviour. Psychopharmacology (Berl) 159:379–387.

Meneses A, Hong E (1993) Modification of the anxiolytic effects of 5-HT$_{1A}$ agonists by shock intensity. Pharmacol Biochem Behav 46:569–573.

Millan MJ (2003) The neurobiology and control of anxious states. Prog Neurobiol 70:83–244.

Moreira FA, Aguiar DC, Guimaraes FS (2006) Anxiolytic-like effect of cannabidiol in the rat Vogel conflict test. Prog Neuropsychopharmacol Biol Psychiatry 30:1466–1471.

Musty R (1984) Possible anxiolytic effects of cannabidiol. In: Agurell S, Dewey WL, Willette RE (eds.), The Cannabinoids: Chemical, Pharmacological and Therapeutic Aspects. New York: Academic Press, pp. 795–813.

Musty R, Conti LH, Mechoulam R (1984) Anxiolytic properties of cannabidiol. In: Harvey DJ (ed.), Marihuana 84: Proceedings of the Oxford Symposium on Cannabis. Oxford: IRL Press, pp. 713–719.

Nadulski T, Sporkert F, Schnelle M, Stadelmann AM, Roser P, Schefter T, Pragst F (2005) Simultaneous and sensitive analysis of THC, 11-OH-THC, THC-COOH, CBD, and CBN by GC-MS in plasma after oral application of small doses of THC and cannabis extract. J Anal Toxicol 29:782–789.

Naliboff BD, Rickles WH, Cohen MJ, Naimark RS (1976) Interactions of marijuana and induced stress: forearm blood flow, heart rate, and skin conductance. Psychophysiology 13:517–522.

Navarro M, Hernandez E, Munoz RM, del Arco I, Villanua MA, Carrera MR, Rodriguez de Fonseca F (1997) Acute administration of the CB$_1$ cannabinoid receptor antagonist SR 141716A induces anxiety-like responses in the rat. Neuroreport 8:491–496.

Ng LK, Lamprecht F, Williams RB, Kopin IJ (1973) Delta^9-tetrahydrocannabinol and ethanol: differential effects on sympathetic activity in differing environmental setting. Science 180:1368–1369.

Onaivi ES (2006) Neuropsychobiological evidence for the functional presence and expression of cannabinoid CB_2 receptors in the brain. Neuropsychobiology 54:231–246.
Onaivi ES, Green MR, Martin BR (1990) Pharmacological characterization of cannabinoids in the elevated plus maze. J Pharmacol Exp Ther 253:1002–1009.
Pamplona FA, Prediger RD, Pandolfo P, Takahashi RN (2006) The cannabinoid receptor agonist WIN 55,212-2 facilitates the extinction of contextual fear memory and spatial memory in rats. Psychopharmacology 188:641–649.
Patel S, Hillard CJ (2006) Pharmacological evaluation of cannabinoid receptor ligands in a mouse model of anxiety: further evidence for an anxiolytic role for endogenous cannabinoid signaling. J Pharmacol Exp Ther 318:304–311.
Patel S, Roelke CT, Rademacher DJ, Cullinan WE, Hillard CJ (2004) Endocannabinoid signaling negatively modulates stress-induced activation of the hypothalamic-pituitary-adrenal axis. Endocrinology 145:5431–5438.
Patel S, Cravatt BF, Hillard CJ (2005) Synergistic interactions between cannabinoids and environmental stress in the activation of the central amygdala. Neuropsychopharmacology 30:497–507.
Pertwee RG (1997) Pharmacological of cannabinoid CB_1 and CB_2 receptors. Pharmacol Ther 74:129–180.
Pertwee RG (2005) Pharmacological actions of cannabinoids. Handb Exp Pharmacol 168:1–51.
Petitet F, Donlan M, Michel A (2006) GPR55 as a new cannabinoid receptor: still a long way to prove it. Chem Biol Drug Des 67:252–253.
Piomelli D, Tarzia G, Duranti A, Tontini A, Mor M, Compton TR, Dasse O, Monaghan EP, Parrott JA, Putman D (2006) Pharmacological profile of the selective FAAH inhibitor KDS-4103 (URB597). CNS Drug Rev 12:21–38.
Rakhshan F, Day TA, Blakely RD, Barker EL (2000) Carrier-mediated uptake of the endogenous cannabinoid anandamide in RBL-2H3 cells. J Pharmacol Exp Ther 292:960–967.
Resstel LB, Joca SR, Moreira FA, Correa FM, Guimaraes FS (2006) Effects of cannabidiol and diazepam on behavioral and cardiovascular responses induced by contextual conditioned fear in rats. Behav Brain Res 172:294–298.
Rimonabant. New Drug Evaluation. Regional Drug & Therapeutic Centre (November 2006): No 78. Available at http://www.nyrdtc.nhs.uk/docs/nde/NDE_78_Rimonabant.pdf.
Rodgers RJ, Haller J, Halasz J, Mikics E (2003) 'One-trial sensitization' to the anxiolytic-like effects of cannabinoid receptor antagonist SR141716A in the mouse elevated plus-maze. Eur J Neurosci 17:1279–1286.
Rodgers RJ, Evans PM, Murphy A (2005) Anxiogenic profile of AM-251, a selective cannabinoid CB_1 receptor antagonist, in plus-maze-naive and plus-maze-experienced mice. Behav Pharmacol 16:405–413.
Rodriguez de Fonseca F, Rubio P, Menzaghi F, Merlo-Pich E, Rivier J, Koob GF, Navarro M (1996) Corticotropin-releasing factor (CRF) antagonist [D-Phe12,Nle21,38,C alpha MeLeu37]CRF attenuates the acute actions of the highly potent cannabinoid receptor agonist HU-210 on defensive-withdrawal behavior in rats. J Pharmacol Exp Ther 276:56–64.
Rubino T, Sala M, Vigano D, Braida D, Castiglioni C, Limonta V, Guidali C, Realini N, Parolaro D (2007) Cellular mechanisms underlying the anxiolytic effect of low doses of peripheral $delta^9$-tetrahydrocannabinol in rats. Neuropsychopharmacology 32:2036–2045.
Russo EB, Burnett A, Hall B, Parker KK (2005) Agonistic properties of cannabidiol at $5\text{-}HT_{1a}$ receptors. Neurochem Res 30:1037–1043.
Rutkowska M, Jamontt J, Gliniak H (2006) Effects of cannabinoids on the anxiety-like response in mice. Pharmacol Rep 58:200–206.
Schneider M, Drews E, Koch M (2005) Behavioral effects in adult rats of chronic prepubertal treatment with the cannabinoid receptor agonist WIN 55,212-2. Behav Pharmacol 16:447–454.
Takahashi RN, Pamplona FA, Fernandes MS (2005) The cannabinoid antagonist SR141716A facilitates memory acquisition and consolidation in the mouse elevated T-maze. Neurosci Lett 380:270–275.

Tzavara ET, Wade M, Nomikos GG (2003) Biphasic effects of cannabinoids on acetylcholine release in the hippocampus: site and mechanism of action. J Neurosci 23:9374–9384.

Uriguen L, Perez-Rial S, Ledent C, Palomo T, Manzanares J (2004) Impaired action of anxiolytic drugs in mice deficient in cannabinoid CB_1 receptors. Neuropharmacology 46:966–973.

Viveros MP, Marco EM, File SE (2005) Endocannabinoid system and stress and anxiety responses. Pharmacol Biochem Behav 81:331–342.

Watanabe K, Kayano Y, Matsunaga T, Yamamoto I, Yoshimura H (1996) Inhibition of anandamide amidase activity in mouse brain microsomes by cannabinoids. Biol Pharm Bull 19:1109–1111.

Weiser M, Noy S (2005) Interpreting the association between cannabis use and increased risk for schizophrenia. Dialogues Clin Neurosci 7:81–85.

Witkin JM, Tzavara ET, Nomikos GG (2005) A role for cannabinoid CB_1 receptors in mood and anxiety disorders. Behav Pharmacol 16:315–331.

Zuardi AW, Shirakawa I, Finkelfarb E, Karniol IG (1982) Action of cannabidiol on the anxiety and other effects produced by delta 9-THC in normal subjects. Psychopharmacology 76:245–250.

Zuardi AW, Cosme RA, Graeff FG, Guimareas FS (1993a) Effects of ipsapirone and cannabidiol on human experimental anxiety. J Psychopharmacol 7:82–88.

Zuardi AW, Guimaraes FS, Moreira AC (1993b) Effect of cannabidiol on plasma prolactin, growth hormone and cortisol in human volunteers. Braz J Med Biol Res 26:213–217.

Index

C

Printed in the United States of America